Johnson and Kiokemeister's
Calculus with Analytic Geometry

Richard E. Johnson
University of New Hampshire

Fred L. Kiokemeister

Elliot S. Wolk
University of Connecticut

ALLYN AND BACON, INC.
Boston • London • Sydney • Toronto

Johnson and Kiokemeister's

Calculus with Analytic Geometry

Sixth Edition

470 Atlantic Avenue, Boston, Massachusetts 02210

Library of Congress Cataloging in Publication Data

Johnson, Richard E 1913–
Johnson and Kiokemeister's Calculus with analytic geometry.

Includes index.
1. Calculus. 2. Geometry, Analytic. I. Kiokemeister, Fred Ludwig, joint author. II. Wolk, E. S., joint author. III. Title.
QA303.J58 1978 515'.15 77-22301

ISBN 0-205-05917-1

Contents

Preface

This sixth edition of CALCULUS AND ANALYTIC GEOMETRY preserves the basic features of the previous editions. While maintaining the same logical and critical spirit that has characterized all previous versions of this text, the authors have introduced additional material to provide more intuitive background for the presentation of difficult topics. This interplay between intuition and rigor is obviously essential to a full understanding of the calculus. It has been the authors' purpose to achieve a reasonable balance between these qualities.

As in the fifth edition, derivatives and integrals make an early appearance. Introductory material on algebra and analytic geometry is presented in a preliminary chapter, Chapter 0. This chapter will be a review for many students. After a short chapter on functions, limits are presented in Chapter 2. This chapter is assembled in such a way that the limit theorems can be passed over quickly if the teacher so desires. Chapters 3 and 4 are concerned with derivatives and the usual applications to extrema and motion of a particle. Although integrals are defined in Chapter 5 in terms of upper and lower sums, the fundamental theorem is soon given so that integrals can be evaluated as antiderivatives. By the end of Chapter 6, the calculus of algebraic functions has been introduced.

The next three chapters cover the calculus of transcendental functions. A substantial number of geometrical and physical applications is then presented in Chapter 10. Chapters 11 and 12, on improper integrals, infinite series, and related topics, may be postponed until later without breaking the continuity of the course.

Chapters 13 through 16 focus on two- and three-dimensional vectors, plane and space curves, surfaces, and elementary multidimensional calculus. Chapter 17 is an optional chapter on line integrals, Green's theorem, and change of variable in multiple integrals. The final chapter is on differential equations.

The major changes in the sixth edition follow:

1. A new section on related rates has been added to Chapter 2 in order to provide better coverage of this important topic.
2. The intuitive approach to the exponential and logarithmic functions has been rewritten in a simpler and clearer way.
3. A simpler proof of Taylor's theorem is given (which removes the need for the discussion of Cauchy's formula).
4. Chapter 12 on infinite series has been extensively rewritten. In particular, the material on power series has been reorganized into a clearer and more effective teaching unit. Simpler proofs are given of the theorems on differentiation and integration of power series. These proofs are placed in an appendix to the chapter.
5. The section on tangent planes in Chapter 15 has been rewritten to provide a clearer and more intuitive treatment.
6. A new section on path-independent line integrals has been added to Chapter 17. This will be useful for students encountering physical applications of line integrals in other courses, and will provide more insight into the significance of exact differentials.
7. Two new sections, on complex numbers and Euler's formula, have been added to Chapter 18. These will enable the instructor to give a coherent and self-contained treatment of linear differential equations with constant coefficients.
8. Many changes, too numerous to mention in detail, have been made to improve clarity and readability. Many new illustrative examples have been included. Each type of problem the student will encounter has been illustrated by an example with worked-out details. The exercises have been carefully scrutinized and revised. Inappropriate or excessively difficult problems have been removed and new problems added. All errors in the answers, which occurred in the fifth edition, have been corrected.

Answers to most of the odd-numbered exercises appear at the end of the book. An optional student guide is also available. This useful supplement contains detailed solutions to over 500 of the exercises in the text, as well as over 500 additional problems in the form of self-tests. Answers to the even-numbered problems and sample tests are included in an instructor's supplement.

Many users of previous editions of this book have suggested improvements, for which we are grateful. We shall continue to welcome such suggestions in the future. To all the people who helped so greatly in the formation of this book, we express our deep appreciation. We particularly appreciate the valuable comments of the following reviewers: William J. Bruce (University of Alberta), John Haverhals (Bradley University), Robert Moore (Bucknell University), Jerry F. Reed (Mississippi State University), John W. Schleusner (West Virginia University), James W. Vick (University of Texas at Austin), and W. Thurmon Whitley (Marshall University).

Richard E. Johnson

Elliot S. Wolk

Johnson and Kiokemeister's
Calculus with Analytic Geometry

0 Elements of Analytic Geometry

1 NUMBERS

The number system of elementary calculus is the system of real numbers. Included among the real numbers are the *integers*

$$\ldots, -4, -3, -2, -1, 0, 1, 2, 3, 4, \ldots$$

and the ratios of integers, called *rational numbers*. Thus, each rational number has the form m/n, where m and n are integers and $n \neq 0$. Real numbers which are not rational are called *irrational*. For example, $\sqrt{2}$, $1 + \sqrt[3]{7}$, and π are irrational numbers.

Numbers can be added, subtracted, multiplied, and divided to obtain other numbers. We shall not list the various properties of these operations, since they are familiar to all high school students of algebra.

Every nonzero number is either a *positive number* or a *negative number*. Two different numbers a and b can always be compared: *a is greater than b*, written $a > b$, if $a - b$ is positive, and *a is less than b*, written $a < b$, if $a - b$ is negative. In particular, $a > 0$ means a is a positive number and $a < 0$ means a is a negative number. Relations other than "$>$" and "$<$" are "$\geq$" (greater than or equal to) and "$\leq$" (less than or equal to), defined by:

$$a \geq b \text{ if either } a > b \text{ or } a = b,$$
$$a \leq b \text{ if either } a < b \text{ or } a = b.$$

A meaningful algebraic expression involving relations such as "$>$" and "$\leq$" is called an *inequality*.

The following laws of inequalities are used frequently in the calculus. They are valid for any numbers a, b, c.

(1) If $a > b$ and $b > c$, then $a > c$ (*transitive law*).
(2) $a > b$ if and only if $a + c > b + c$.
(3) If c is positive, $a > b$ if and only if $ac > bc$.
(4) If c is negative, $a > b$ if and only if $ac < bc$.

The laws above are also valid if we replace "$>$" by "$<$" and "$<$" by "$>$".

We write

$$a < b < c \text{ to indicate } a < b \text{ and } b < c,$$
$$a < b \leq c \text{ to indicate } a < b \text{ and } b \leq c,$$

and so on. The set of all numbers strictly between a and b (assuming $a < b$) is denoted by (a,b). Thus, in set notation,

$$(a,b) = \{x \mid a < x < b\}, \qquad \textit{open interval.}$$

We call (a,b) an *open interval* of real numbers. The set of numbers between a and b, including a and b, is called a *closed interval* and is denoted by $[a,b]$; thus,

$$[a,b] = \{x \mid a \leq x \leq b\}, \qquad \textit{closed interval.}$$

There are two *half-open* (or, half-closed) intervals, denoted by $[a,b)$ and $(a,b]$:

$$[a,b) = \{x \mid a \leq x < b\}, \qquad \textit{half-open interval.}$$
$$(a,b] = \{x \mid a < x \leq b\}, \qquad \textit{half-open interval.}$$

The symbol ∞, called *infinity*, is useful in many situations. However, in using this symbol, we must realize that it is not a number in the ordinary sense of the word. For example, we cannot add 3 to ∞ or divide 1 by ∞.

We shall use the notation (a,∞) to designate the set of all numbers greater than a:

$$(a,\infty) = \{x \mid x > a\}.$$

Other infinite intervals which we shall find useful are defined below.

$$(-\infty,a) = \{x \mid x < a\},$$
$$[a,\infty) = \{x \mid x \geq a\},$$
$$(-\infty,a] = \{x \mid x \leq a\}.$$

We shall on occasion use the notation $(-\infty,\infty)$ for the set of all real numbers.

Each set of numbers has a *graph* on a coordinate axis. Some intervals are graphed below.

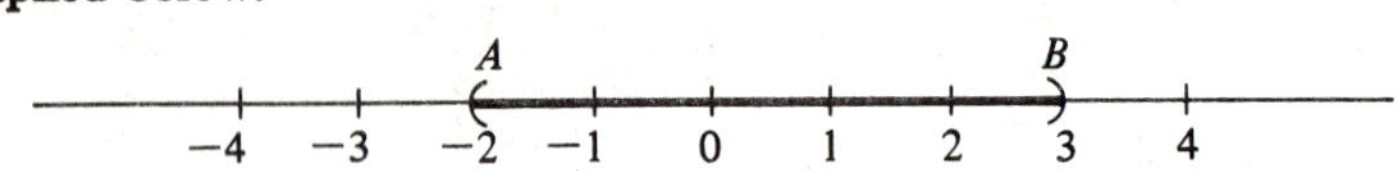

Graph of $(-2,3)$

The graph of $(-2,3)$ consists of all *points* between A and B. The parentheses at A and B indicate that A and B are not part of the graph.

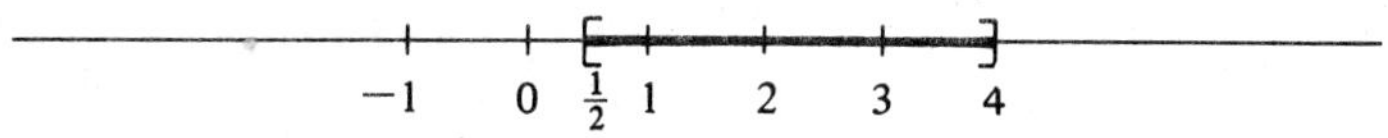

Graph of $[\frac{1}{2}, 4]$

The brackets at $\frac{1}{2}$ and 4 indicate that the points with coordinates $\frac{1}{2}$ and 4 are part of the graph.

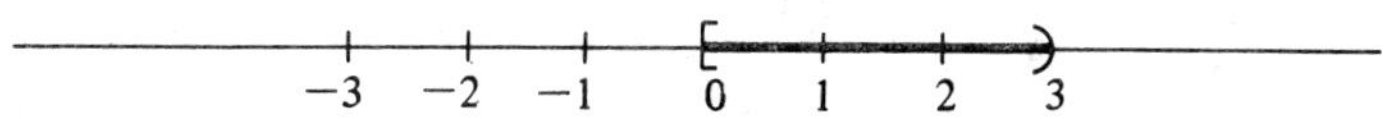

Graph of $[0,3)$

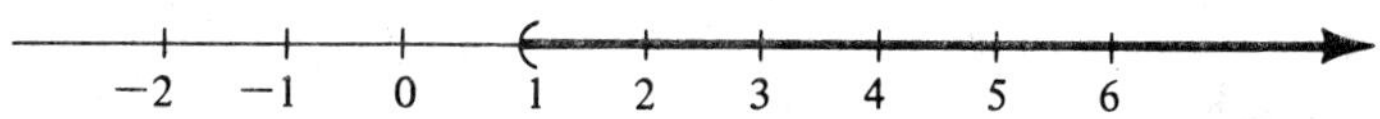

Graph of $(1,\infty)$

Graph of $(-\infty,3]$

We shall have much use for the concept of a *neighborhood* of a number.

Neighborhood of c: Any open interval (a,b) containing c is called a *neighborhood* of c. Graphically:

(a,b) is a neighborhood of c

For example:

$(-1,2)$ is a neighborhood of 0

Deleted neighborhood of c: **Any neighborhood of c, with c removed, is called a** ***deleted neighborhood*** **of c.**

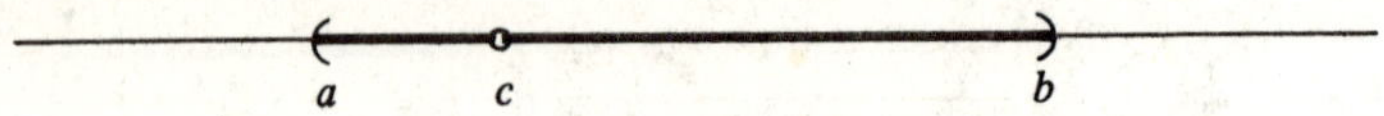

(a,b), with c removed, is a deleted neighborhood of c

Using the union symbol "∪" for sets, if $a < c < b$, then

$$(a,c) \cup (c,b) \text{ is a } \textit{deleted neighborhood} \text{ of } c.$$

For example:

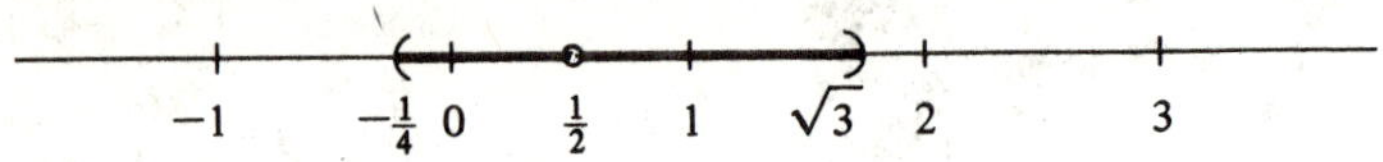

$(-\frac{1}{4}, \sqrt{3})$, with $\frac{1}{2}$ removed, is a deleted neighborhood of $\frac{1}{2}$

The deleted neighborhood above can also be written

$$(-\tfrac{1}{4}, \tfrac{1}{2}) \cup (\tfrac{1}{2}, \sqrt{3}).$$

Symmetric neighborhood of c: **$(c - r, c + r)$ for any $r > 0$.**

Symmetric neighborhood of c

For example:

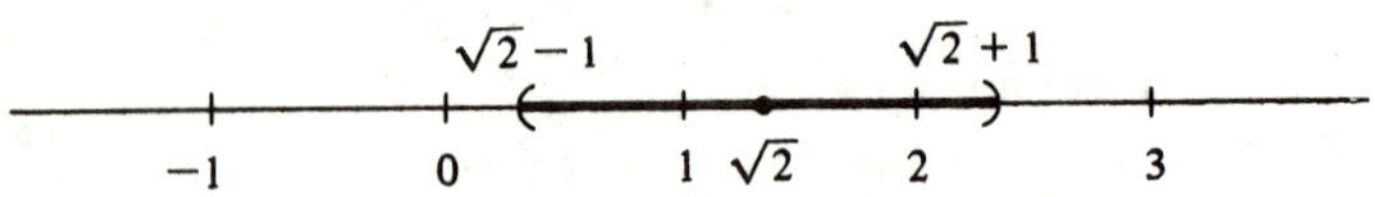

$(\sqrt{2} - 1, \sqrt{2} + 1)$ is a symmetric neighborhood of $\sqrt{2}$

Useful fact: Every neighborhood of c contains a symmetric neighborhood of c.

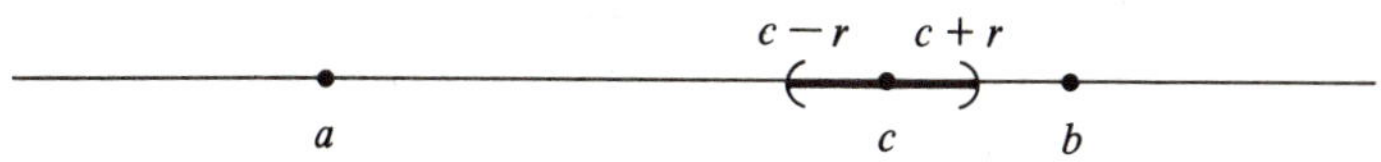

Symmetric neighborhood $(c - r, c + r)$ of c is contained in neighborhood (a,b) of c

For example: (1,5) is a neighborhood of π ($\pi = 3.1416$ approx.). It is not symmetric. However $(\pi - \frac{1}{2}, \pi + \frac{1}{2})$ is a symmetric neighborhood of π contained in (1,5). So also are

$$(\pi - 1, \pi + 1), \qquad (\pi - \tfrac{3}{2}, \pi + \tfrac{3}{2}), \qquad (\pi - .01, \pi + .01),$$

and so on.

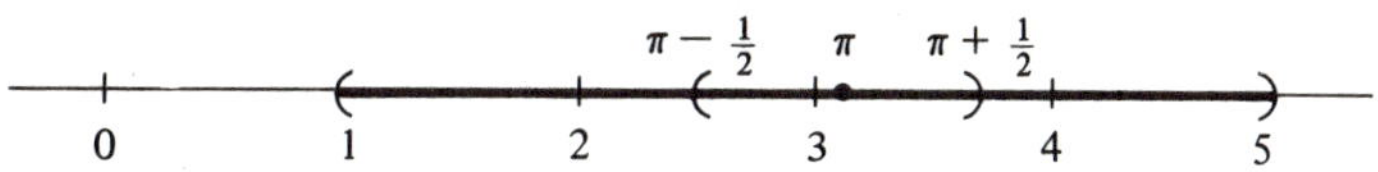

$(\pi - \frac{1}{2}, \pi + \frac{1}{2})$ is a symmetric neighborhood of π contained in (1,5)

Each real number a has an *absolute value* (or *numerical value*) designated by $|a|$ and defined by:

$$|a| = a \qquad \text{if } a \text{ is a nonnegative number.}$$
$$|a| = -a \qquad \text{if } a \text{ is a negative number.}$$

Since $-a$ is a positive number whenever a is a negative number, always $|a|$ is nonnegative. For example,

$$|7| = 7, \qquad 7 \textit{ is nonnegative.}$$
$$|-\tfrac{3}{2}| = -(-\tfrac{3}{2}) = \tfrac{3}{2}, \qquad -\tfrac{3}{2} \textit{ is negative.}$$
$$|1 - \sqrt{3}| = -(1 - \sqrt{3}) = \sqrt{3} - 1, \qquad 1 - \sqrt{3} \textit{ is negative.}$$

Some useful properties of absolute value are listed below. They are valid for all real numbers a and b.

(1) $|a \cdot b| = |a| \cdot |b|$.
(2) $|a + b| \leq |a| + |b|$.
(3) $|a| = |-a|$.

(4) If $b > 0$, $|a| < b$ if and only if $-b < a < b$.

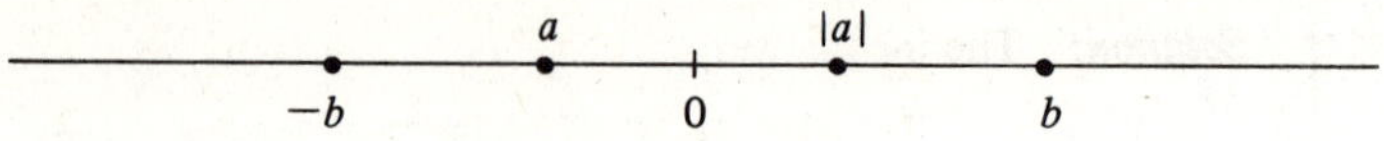

Illustrates (4) above

An inequality containing one variable has as its *solution set* the set of all values of the variable which make the inequality true. For example, the solution set of the inequality

$$|x| < 3$$

is the open interval $(-3,3)$ according to (4) above.

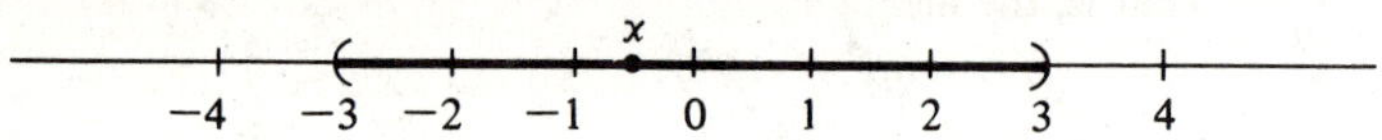

$|x| < 3$ if and only if $-3 < x < 3$

We solve inequalities essentially as we solve equations.

Example 1 Solve the inequality $7x - 5 > 3x + 4$.

Solution:

$$\begin{aligned} 7x - 5 &> 3x + 4, && \textit{given;} \\ -3x + 7x - 5 &> -3x + 3x + 4, && \textit{add } -3x \textit{ to each side;} \\ 4x - 5 &> 4, && \textit{simplify;} \\ 4x - 5 + 5 &> 4 + 5, && \textit{add 5 to each side;} \\ 4x &> 9, && \textit{simplify;} \\ \tfrac{1}{4}(4x) &> \tfrac{1}{4}(9), && \textit{multiply each side by the positive number } \tfrac{1}{4}; \\ x &> \tfrac{9}{4}, && \textit{simplify.} \end{aligned}$$

Thus, the solution set is $(\frac{9}{4},\infty)$.

Example 2 For what set of values of y is $\sqrt{1 - 2y}$ a real number?

Solution: Only nonnegative numbers have real square roots. Thus, $\sqrt{1 - 2y}$ is real if and only if

$$1 - 2y \geq 0.$$

We solve this inequality as follows:

$$\begin{aligned} 2y + 1 - 2y &\geq 2y, && \textit{add } 2y \textit{ to each side;} \\ 1 &\geq 2y, && \textit{simplify;} \\ \tfrac{1}{2} &\geq y, && \textit{multiply each side by } \tfrac{1}{2}. \end{aligned}$$

Thus, the solution set is $\{y \mid y \leq \frac{1}{2}\}$, or

$$(-\infty,\tfrac{1}{2}].$$

Example 3 Solve the inequality

$$|x - 3| > 1.$$

Solution: The inequality

$$|x - 3| \leq 1$$

is equivalent to

$$-1 \leq x - 3 \leq 1$$

or, adding 3 to each member,

$$2 \leq x \leq 4.$$

Thus,

$$|x - 3| \leq 1 \text{ has solution set the interval } [2,4].$$

Isn't every number outside the interval [2,4] a solution of the inequality $|x - 3| > 1$? That is, the inequality

$$|x - 3| > 1$$

has solution set

$$(-\infty,2) \cup (4,\infty).$$

EXERCISES

I

In each of Exercises 1 to 14, solve the given inequality.

1. $3x + 1 < x + 5$

2. $1 - 2x < 5x - 2$

3. $3 - 2x > 0$

4. $\dfrac{3x}{4} - \dfrac{1}{2} < 0$

5. $\dfrac{3x - 5}{2} \leq 0$

6. $.01x - 2.32 \geq 0$

7. $-.1 < x - 5 < .1$

8. $-.01 < x + 3 < .01$

9. $-.03 \leq \dfrac{2x + 3}{5} \leq .03$

10. $-.001 \leq \dfrac{5 - 2x}{4} \leq .001$

11. $x < x + 1$

12. $-\varepsilon < \dfrac{4x - 3}{7} < \varepsilon, \varepsilon > 0$

13. $\dfrac{3}{x} - \dfrac{1}{4} > \dfrac{1}{x} + 1$

14. $\dfrac{3}{6x + 2} \geq -\dfrac{2}{4 - x}$

In each of Exercises 15 to 18, determine the set of all numbers x for which the given square root is a real number.

15. $\sqrt{2x - 8}$

16. $\sqrt{3 + 4x}$

17. $\sqrt{b^2 - 4ax}$ **18.** $\sqrt{16 - 7x}$

In each of Exercises 19 to 22, determine the set of all numbers x for which the given square root is not a real number.

19. $\sqrt{3 - 2x}$ **20.** $\sqrt{x^2 + 5}$

21. $\sqrt{x^2 - 6x}$ **22.** $\sqrt{\dfrac{x + 3}{x - 2}}$

In each of Exercises 23 to 34, find the set of all numbers x satisfying the given conditions.

23. $x + 1 > 0$ and $x - 3 < 0$ **24.** $x - 1 < 0$ and $x + 2 > 0$

25. $\dfrac{1}{x + 3} < 0$ **26.** $\dfrac{1}{2x - 5} > 0$

27. $3x - 2 \geq 0$ and $5x - 1 \leq 0$ **28.** $2x + 1 > 0$ and $x - 1 > 0$

29. $4x + 3 < 0$ or $6x + 7 > -3$ **30.** $-3x < 0$ or $9x > 0$

31. $3x - 7 > 0$ and $4x + 2 < 0$ and $-3x + 5 > 0$

32. $-3x < 0$ and $9x > 0$ **33.** $|x - 3| = x + 7$

34. $|x| = 2x + 1$

If in the following exercises, (a,b) and (c,d) are neighborhoods of m and n, respectively, show that:

35. $(a + c, b + d)$ is a neighborhood of $m + n$.

36. $(a - d, b - c)$ is a neighborhood of $m - n$.

37. (ra,rb) is a neighborhood of rm, if $r > 0$.

38. (ac,bd) is a neighborhood of mn, if $a,c > 0$.

39. $(a/d, b/c)$ is a neighborhood of m/n, if $a,c > 0$.

In each of Exercises 40 to 47, solve the given inequality.

40. $|x - 1| < 3$ **41.** $|x + 2| < 5$

42. $|2x - 1| < .1$ **43.** $|3x - 5| \leq .05$

44. $|4 - 3x| < 5$ **45.** $|1 - 2x| \leq 1$

46. $|1 + 2x| \leq 1$ **47.** $|x + \pi| < 2$

In each of Exercises 48 to 55, solve the given inequality.

48. $|x| > 2$ **49.** $|3x| > 6$

50. $|x - 1| > 3$ **51.** $|x + 2| \geq 5$

52. $|x + 2| < \varepsilon, \varepsilon > 0$ **53.** $|3x + 7| < \delta, \delta > 0$

54. $|x - a| < \delta, \delta > 0$ **55.** $|y - 7| < \dfrac{\delta}{3}, \delta > 0$

II

1. If $a \neq b$, prove that $(a + b)/2$ is strictly between a and b.
2. If $a > 0$ and $b > 0$, and if $a \neq b$, prove that $\sqrt{ab}$ (called the geometric average or mean of a and b) is strictly between a and b.
3. If $a \geq 0$ and $b \geq 0$, prove that $a^n \geq b^n$ if and only if $a \geq b$ (n a positive integer).
4. If $a \geq 0$ and $b \geq 0$, prove that $(a + b)/2 \geq \sqrt{ab}$ with equality if and only if $a = b$.
5. If $a > 0$ and $b > 0$, prove that $a > b$ if and only if $1/a < 1/b$.

2 PLANE ANALYTIC GEOMETRY

In analytic geometry, we identify geometric objects such as lines and circles by equations. Then we are able to solve geometric problems algebraically.

Basic to this study is a *rectangular coordinate system* in the plane (Figure 0.1). This system consists of a pair of perpendicular coordinate axes (the x and y axes of the figure) meeting at the point, called the *origin*, having both coordinates zero. Each point has an ordered *pair of coordinates*, as illustrated in the figure. Also, each ordered pair of numbers identifies a point, called the *graph* of the ordered pair. The first number in a pair is called the *x coordinate*, or *abscissa*, of the point; the second the *y coordinate*, or *ordinate*.

If a point P in the plane has coordinates (a,b), as in Figure 0.2, we

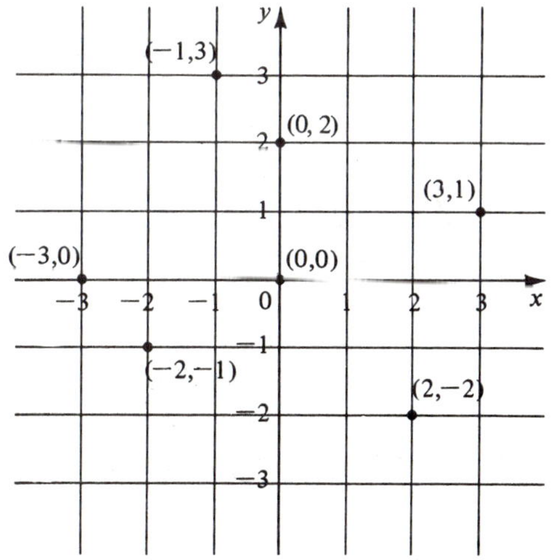

FIGURE 0.1

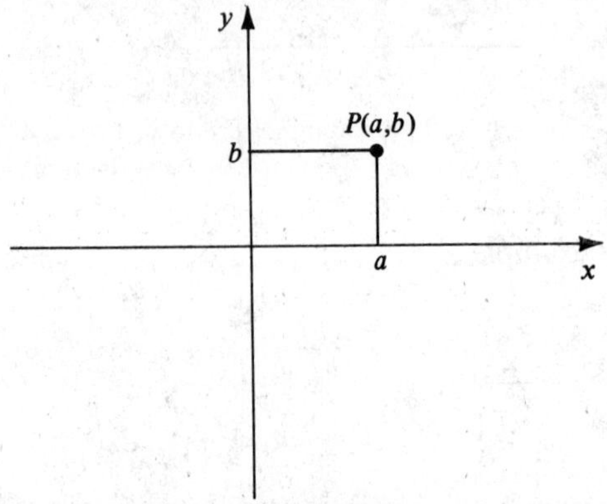

FIGURE 0.2

signify this fact by writing

$$P(a,b).$$

Given points A and B on a coordinate line (Figure 0.3), the *distance*

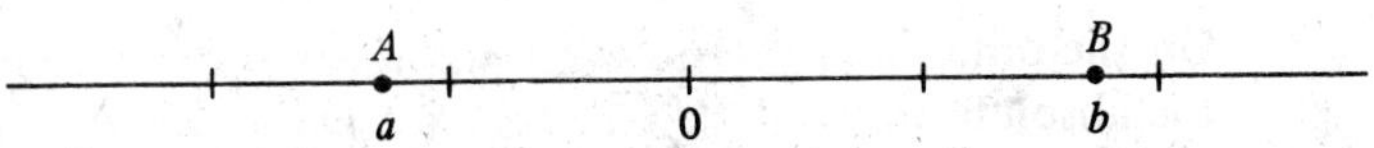

FIGURE 0.3

between A and B, denoted by $|AB|$, is given by

$$|AB| = |b - a|.$$

For example:

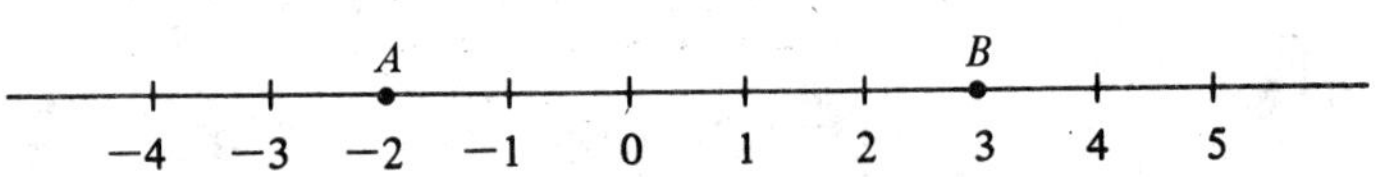

$$|AB| = |3 - (-2)| = 5$$

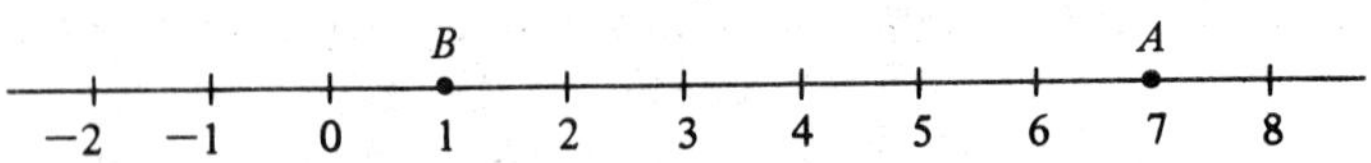

$$|AB| = |1 - 7| = 6$$

The *directed distance* between points A and B on a coordinate line (Figure 0.4), denoted by $\overline{AB}$, is given by

$$\overline{AB} = b - a.$$

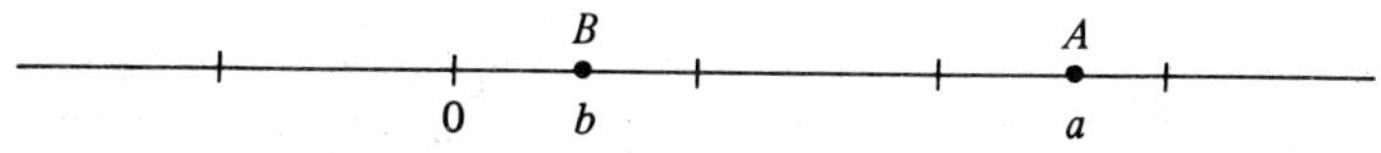

FIGURE 0.4

For example:

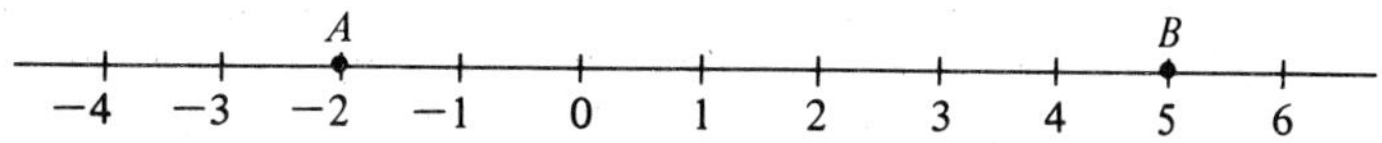

$\overline{AB} = 5 - (-2) = 7$

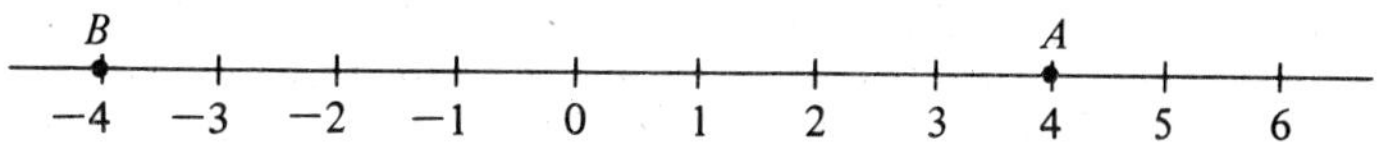

$\overline{AB} = (-4) - 4 = -8$

The distance between two distinct points A and B is always positive,

$$|AB| > 0.$$

On the other hand, the directed distance can be positive or negative. Always, the absolute value of the directed distance is the actual distance:

$$|\overline{AB}| = |AB|.$$

We can compute the distance between two points in a coordinate plane by the following formula. If $P(x_1,y_1)$ and $Q(x_2,y_2)$ are the given points, then

Distance Formula: $$|PQ| = \sqrt{(x_2 - x_1)^2 + (y_2 - y_1)^2}.$$

This follows from the pythagorean theorem, as seen in Figure 0.5.

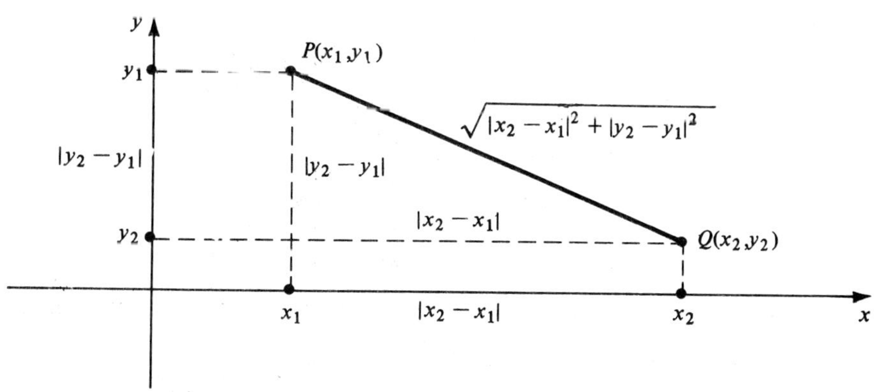

FIGURE 0.5

Example 1 Show that the triangle with vertices $A(-3,1)$, $B(5,4)$, and $C(0,-7)$ is a right triangle, and find its area (see Figure 0.6).

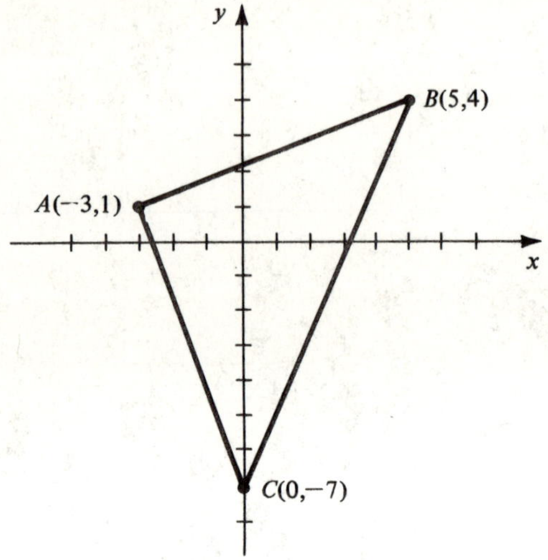

FIGURE 0.6

Solution: We have by the distance formula that

$$|AB| = \sqrt{[5-(-3)]^2 + (4-1)^2} = \sqrt{73},$$
$$|AC| = \sqrt{[0-(-3)]^2 + (-7-1)^2} = \sqrt{73},$$
$$|BC| = \sqrt{(0-5)^2 + (-7-4)^2} = \sqrt{146}.$$

Since

$$|AB|^2 + |AC|^2 = 73 + 73 = 146 = |BC|^2,$$

the triangle has a right angle at vertex A. It happens that this triangle also is isosceles, since $|AB| = |AC|$.

The area of the right triangle ABC is given by

$$\tfrac{1}{2}|AB| \cdot |AC| = \tfrac{1}{2}\sqrt{73} \cdot \sqrt{73} = \tfrac{73}{2}.$$

A set of ordered pairs has a *graph* consisting of the set of graphs of the individual ordered pairs.

Example 2 Graph the set $\{(x,y) \mid 1 \le x \le 3,\ 0 \le y \le 4\}$.

Solution: The graph is the rectangular region of the plane shaded in Figure 0.7.

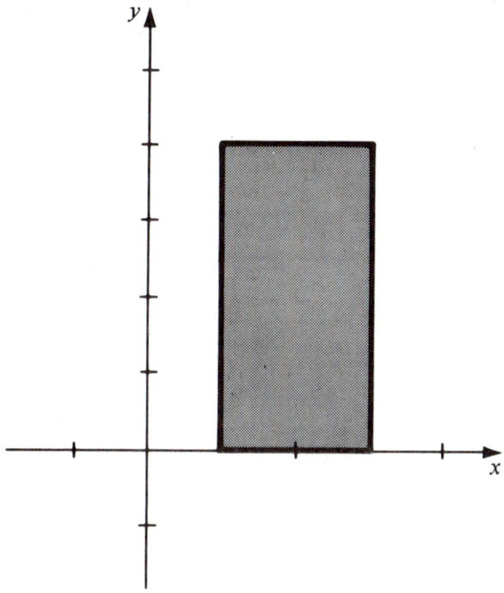

FIGURE 0.7

Given an equation or inequality in two variables, say x and y, a *solution* consists of an ordered pair of numbers (a,b) such that the equation or inequality is true when we let $x = a$ and $y = b$.

For example, (2,4) is a solution of the equation

$$y = x^2,$$

since $4 = 2^2$ is a true equation. Also, $(-\sqrt{7},7)$ is a solution since $7 = (-\sqrt{7})^2$. On the other hand, (3,5) is not a solution since the equation $5 = 3^2$ is false.

The set S of all solutions of an equation or inequality in two variables is called the *solution set* of the equation or inequality. In turn, the graph of S in a coordinate plane is called the *graph of the equation or inequality.*

Given a set S of points in a coordinate plane, any equation (or inequality) whose graph is S is called an *equation* (or inequality) of S.

For example, we can find an equation of the circle with center $C(h,k)$ and radius r (Figure 0.8) as follows. A variable point $P(x,y)$ in the plane is on the circle if and only if

$$|CP| = r,$$

or, equivalently,

$$|CP|^2 = r^2.$$

Using the distance formula to find $|CP|$, we obtain the equation

$$(x - h)^2 + (y - k)^2 = r^2, \qquad \textit{equation of circle, center } C(h,k), \textit{ radius } r.$$

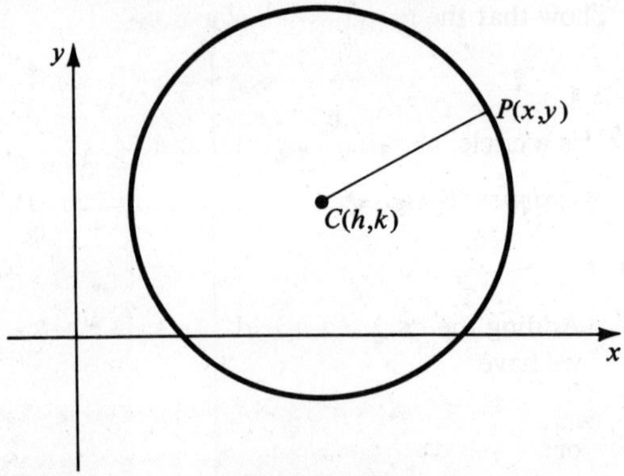

FIGURE 0.8

Since the graph of this equation is the circle in Figure 0.8, this is an equation of the circle with radius r and center $C(h,k)$.

Example 3 Find an equation of the circle with center $C(4,-3)$ and radius 6.

Solution: In this example $h = 4$, $k = -3$, $r = 6$. Hence,

$$(x - 4)^2 + [y - (-3)]^2 = 6^2,$$
$$(x - 4)^2 + (y + 3)^2 = 36$$

is an equation of the circle. After noting that $(x - 4)^2 = x^2 - 8x + 16$ and $(y + 3)^2 = y^2 + 6y + 9$, we may also write the following equation of the circle:

$$x^2 + y^2 - 8x + 6y - 11 = 0.$$

The circle is shown in Figure 0.9.

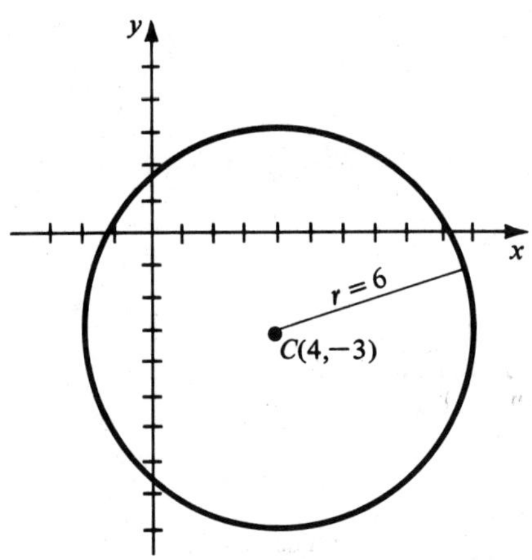

FIGURE 0.9

Example 4 Show that the graph of the equation

$$x^2 + y^2 + 4x - 2y - 1 = 0$$

is a circle, and find its center and radius.

Solution: The given equation is equivalent to the equation

$$x^2 + y^2 + 4x - 2y = 1.$$

Adding the numbers 4 and 1 to each side of this equation to complete the squares, we have

$$x^2 + 4x + 4 + y^2 - 2y + 1 = 1 + 4 + 1,$$

or

$$(x + 2)^2 + (y - 1)^2 = (\sqrt{6})^2.$$

We recognize this as an equation of the circle with center $C(-2,1)$ and radius $r = \sqrt{6}$ (Figure 0.10).

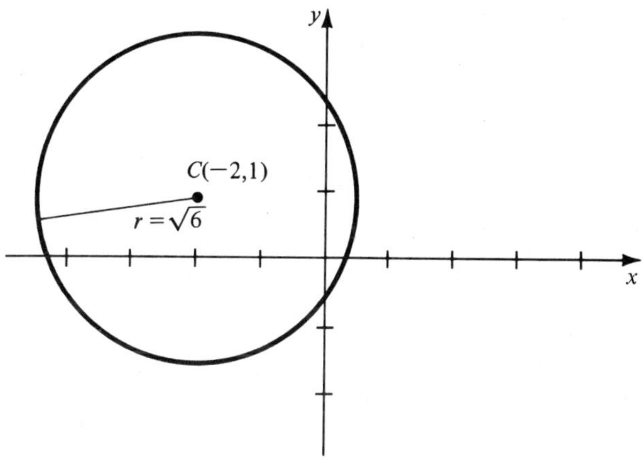

FIGURE 0.10

EXERCISES

I

Graph each of the following sets of ordered pairs.

1. $\{(x,y) \mid 0 < x < 4, 0 < y < 2\}$
2. $\{(x,y) \mid -1 \leq x \leq 1, -1 \leq y \leq 1\}$
3. $\{(x,y) \mid x = 3, -2 \leq y \leq 2\}$
4. $\{(x,y) \mid 1 \leq x \leq 5, y = 0\}$
5. $\{(x,y) \mid x = 2, y \text{ any real number}\}$

6. $\{(x,y) \mid x \text{ any real number}, y = -1\}$
7. $\{(x,y) \mid x \text{ any real number}, |3y + 2| > 1\}$
8. $\{(x,y) \mid (x - 3)^2 + (y + 2)^2 \leq 0\}$
9. $\{(x,y) \mid |x - 3| < 2, |-y + 1| \leq 5\}$
10. $\{(x,y) \mid |x| \leq 3, |y| \leq 1\}$

In each of Exercises 11 to 16, show that ABC is a right triangle. Find the equation of each circumscribed circle, using the fact that the center of the circle circumscribed about ABC is the midpoint of the hypotenuse. Sketch.

11. $A(1,0)$, $B(5,3)$, $C(4,-4)$
12. $A(1,0)$, $B(4,2)$, $C(-3,6)$
13. $A(4,4)$, $B(1,5)$, $C(-2,-4)$
14. $A(3,-2)$, $B(4,3)$, $C(-6,5)$
15. $A(0,1)$, $B(2,3)$, $C(0,5)$
16. $A(1,-2)$, $B(0,0)$, $C(-2,-1)$

In each of Exercises 17 to 22, find an equation of the circle and sketch it.

17. Center $(-1,-3)$, radius $r = 3$.
18. Center $(3,0)$, radius $r = 5$.
19. Center $(2,4)$, passing through the origin.
20. Center $(0,0)$, passing through the point $(3,7)$.
21. Radius $r = 4$, tangent to both coordinate axes, lying in the second quadrant.
22. Radius $r = 1$, tangent to both coordinate axes, lying in the fourth quadrant.

Use set notation to describe:

23. The exterior of the circle with radius $\sqrt{7}$ and center $(-1,2)$.
24. The interior of the circle with radius 3 and center $(4,2)$.

In each of Exercises 25 to 34, find the center and radius of the circle with given equation. Sketch.

25. $x^2 + y^2 - 16 = 0$
26. $2x^2 + 2y^2 + 4y + 1 = 0$
27. $x^2 + y^2 - 2x - 4y + 1 = 0$
28. $x^2 + y^2 + 6x + 8y = 0$
29. $x^2 + y^2 + 6x - 10y + 25 = 0$
30. $9x^2 + 9y^2 - 6x + 12y + 4 = 0$
31. $4x^2 + 4y^2 + 16x + 15 = 0$
32. $4x^2 + 4y^2 + 4x - 12y + 7 = 0$
33. Show that the point $P(x,y)$ is equidistant from the point $(0,4)$ and the x axis if and only if $x^2 - 8y + 16 = 0$.
34. Find the center and radius of the circle with equation

$$x^2 + y^2 + ax + by + c = 0.$$

Under what conditions on a, b, and c will the equation have no graph? Is the graph ever just a point?

II

1. Find the points of intersection of the two circles

$$x^2 + y^2 + 8y = 64, \qquad x^2 + y^2 - 6x = 16.$$

Also, sketch the figure.

2. Determine an equation of a circle tangent to both coordinate axes and passing through the given point (h,k).

3 LINES

A straight line L (simply called a line) in a coordinate plane has *inclination* α as shown in Figure 0.11. Thus, α is the angle drawn from the x axis to L in a

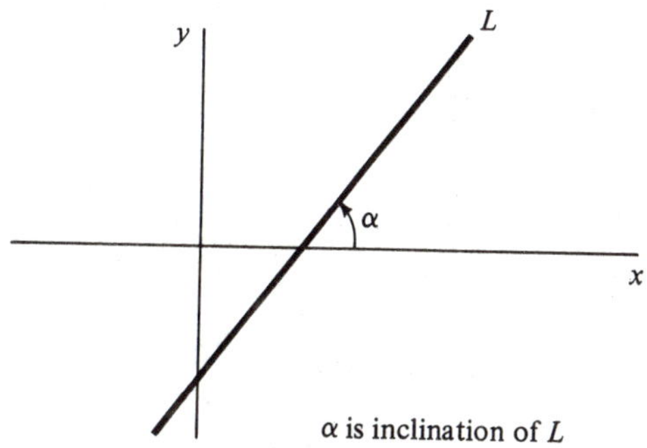

α is inclination of L

FIGURE 0.11

counterclockwise direction. Always,

$$0° \leq \alpha < 180°.$$

If L is perpendicular to the x axis, $\alpha = 90°$; if L is parallel to the x axis, we take $\alpha = 0°$.

A line L with inclination α has *slope m* defined by

$$m = \tan \alpha, \qquad \textit{slope of L.}$$

If $\alpha = 0°$, $m = \tan 0° = 0$. Since $\tan 90°$ is undefined, a line perpendicular to the x axis has no slope.

Knowing two points $P(x_1,y_1)$ and $Q(x_2,y_2)$ on a line L, the slope m of L is easily found by the formula

$$m = \frac{y_2 - y_1}{x_2 - x_1}, \qquad \text{if } x_1 \neq x_2.$$

Why this is so is apparent from the Figures 0.12 and 0.13. Figure 0.12 shows the case when $\alpha < 90°$ and $m > 0$; Figure 0.13 when $\alpha > 90°$ and $m < 0$.

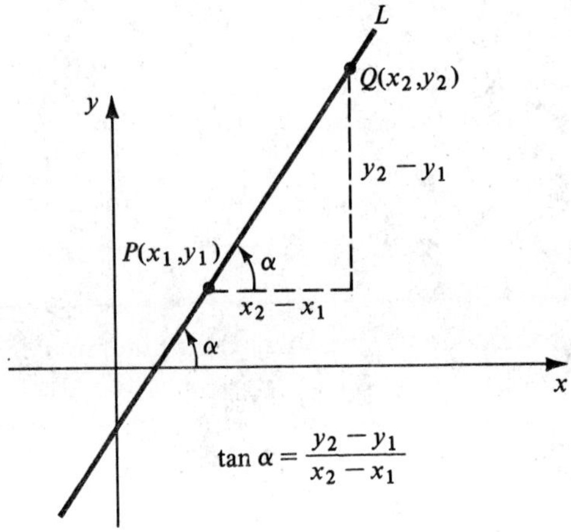

FIGURE 0.12

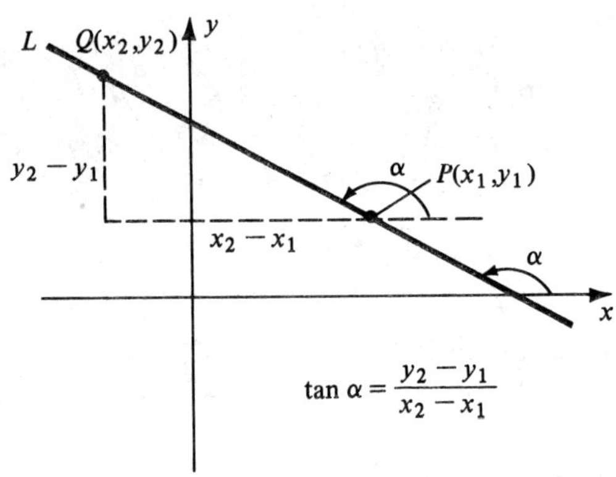

FIGURE 0.13

Example 1 Find the slope and inclination of lines K and L in Figure 0.14.

Solution: By the formula, letting $P(2,1)$ and $Q(5,3)$, line K has slope

$$m_K = \frac{3-1}{5-2} = \frac{2}{3}.$$

Letting $P(2,1)$ and $Q(-1,3)$, line L has slope

$$m_L = \frac{3-1}{-1-2} = -\frac{2}{3}.$$

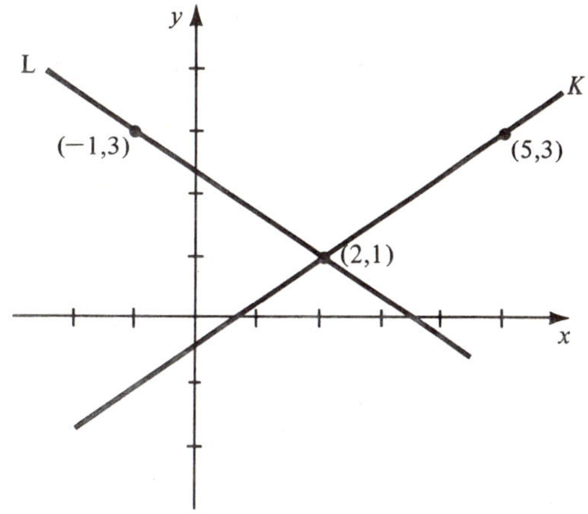

FIGURE 0.14

If α_K is the inclination of K and α_L of L, then

$$\tan \alpha_K = \tfrac{2}{3}, \qquad \tan \alpha_L = -\tfrac{2}{3}.$$

From Table 4 of the appendix,

$$\alpha_K \doteq 34°, \qquad \alpha_L \doteq 180° - 34° \doteq 146°.$$

Here, "$\doteq$" means "is approximately equal to."

Clearly, two lines are parallel if and only if they have the same slope. What is the relationship between the slopes of two perpendicular lines? An inspection of Figure 0.15 gives us the answer. If line L passes through the origin and $P(a,b)$, then the line K through the origin which is perpendicular

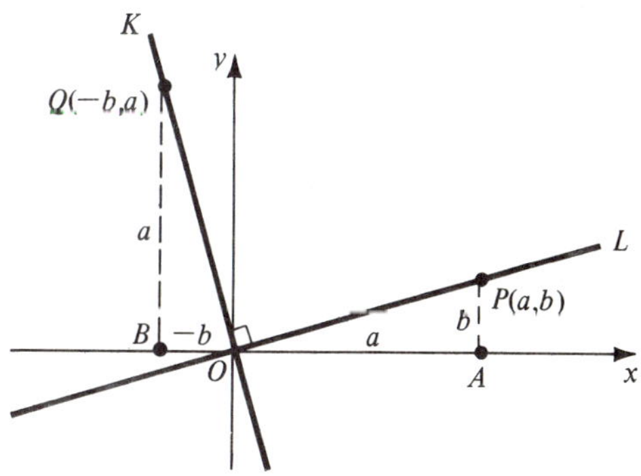

FIGURE 0.15

to L is the line K of the figure. L and K are perpendicular because triangles OAP and QBO are congruent. The slopes of L and K are:

$$m_L = \frac{b}{a}, \qquad m_K = \frac{a}{-b}.$$

The numbers b/a and $-a/b$ are negative reciprocals of each other. The remarks above indicate the validity of the following statement.

Two lines are perpendicular if and only if their slopes are negative reciprocals of each other.

This statement isn't true if one line is parallel to the y axis. Then the other line is perpendicular to it if and only if its slope is zero.

Example 2 Show that the triangle with vertices $A(-1,-1)$, $B(-9,6)$, and $C(-2,14)$ is a right triangle.

Solution: The slopes of the sides of triangle ABC (Figure 0.16) are as follows:

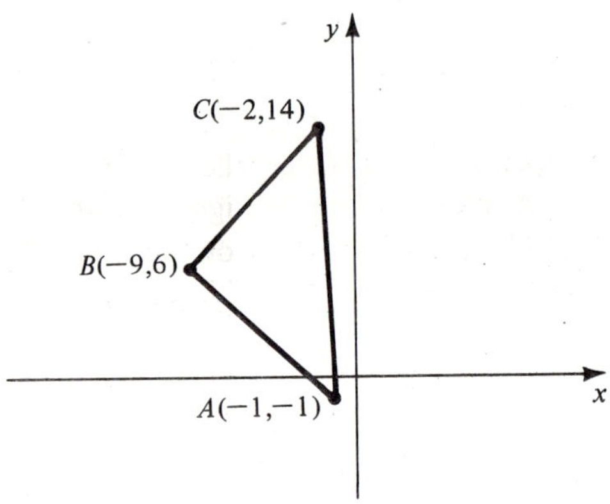

FIGURE 0.16

$$\text{Slope of } AB = \frac{6-(-1)}{-9-(-1)} = -\frac{7}{8},$$

$$\text{Slope of } AC = \frac{14-(-1)}{-2-(-1)} = -15,$$

$$\text{Slope of } BC = \frac{14-6}{-2-(-9)} = \frac{8}{7}.$$

Since $-\frac{7}{8}$ and $\frac{8}{7}$ are negative reciprocals, sides AB and BC are perpendicular. Hence, ABC is a right triangle with right angle at B.

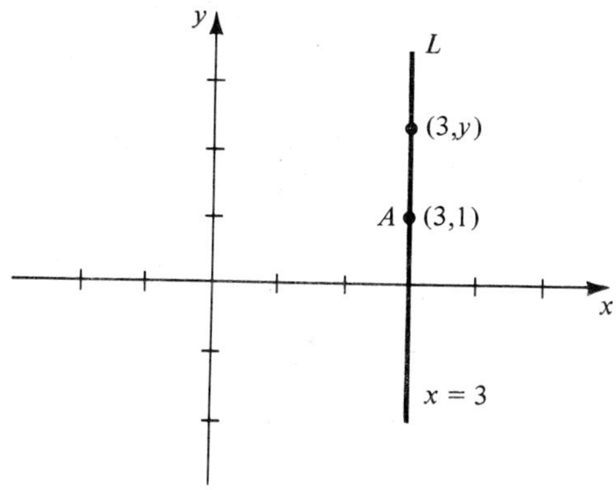

FIGURE 0.17

The vertical line L of Figure 0.17 passes through the point $A(3,1)$. Isn't a point $P(x,y)$ on L if and only if $x = 3$? Thus,

$$x = 3$$

is an equation of L. Similarly,

$$x = a$$

is an *equation* of the line parallel to the y axis and passing through the point (a,b).

A line such as L in Figure 0.18 also has an equation. A variable point $P(x,y)$ (different from Q) is on L if and only if the slope of the line joining P and Q equals m, the slope of L:

$$\frac{y-b}{x-a} = m.$$

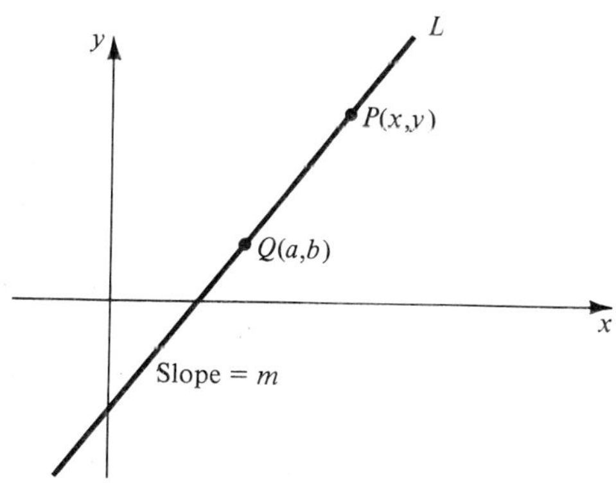

FIGURE 0.18

On multiplying both sides of the equation above by $x - a$, we get

$$y - b = m(x - a), \qquad \textit{point-slope equation of a line,}$$

as an equation of L. This is an *equation of the line passing through the point $Q(a,b)$ and having slope m.*

Example 3 Find an equation of the line on the points $(-1,4)$ and $(5,6)$ (Figure 0.19).

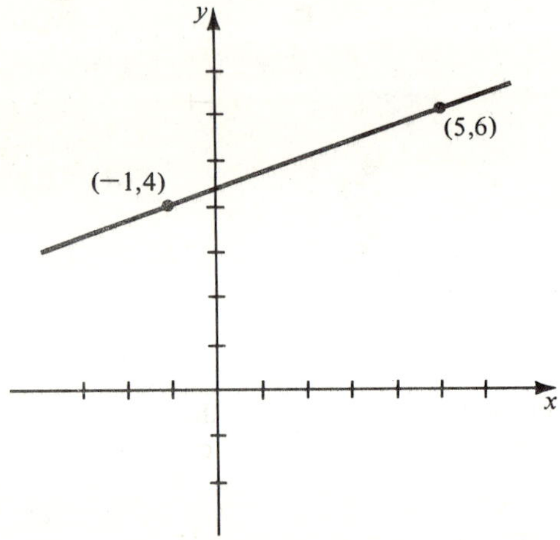

FIGURE 0.19

Solution: The slope of the line is given by

$$m = \frac{6 - 4}{5 + 1} = \frac{1}{3}.$$

Taking $(-1,4)$ as the given point (a,b) and $m = \frac{1}{3}$, we have that an equation of the line is

$$y - 4 = \tfrac{1}{3}(x + 1).$$

This may be simplified to the form

(1) $$x - 3y + 13 = 0.$$

Had we chosen $(5,6)$ for the point (a,b), we would have obtained

$$y - 6 = \tfrac{1}{3}(x - 5)$$

as an equation of the line. However, it may be verified that this reduces to equation (1) above.

Example 4 Find an equation of the line on the point $(3,-2)$ and perpendicular to the line of the previous example (Figure 0.20).

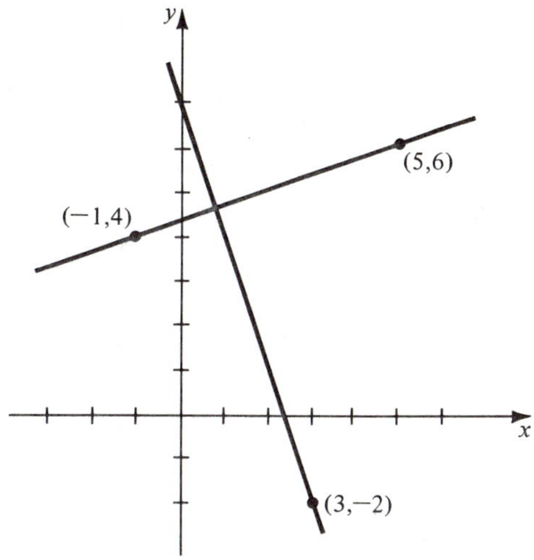

FIGURE 0.20

Solution: Since the slope of the line of Example 3 is $\frac{1}{3}$, the slope of the line in question here is the negative reciprocal of $\frac{1}{3}$, that is, -3. Hence its equation is

$$y + 2 = -3(x - 3)$$

or

$$3x + y - 7 = 0.$$

If a line L is as shown in Figure 0.21, then the numbers a and b are called the *intercepts* of L as labeled. The line L has equation $y - b = m(x - 0)$, or

$$y = mx + b, \qquad \textit{slope-intercept equation of a line.}$$

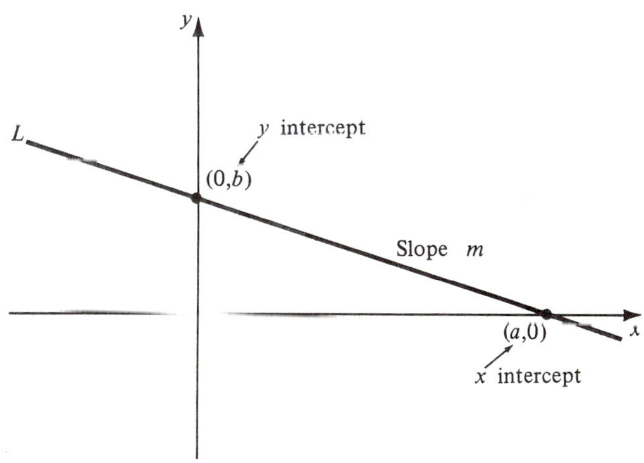

FIGURE 0.21

For example, $y = -3x + 7$ is an equation of a line with slope -3, y intercept 7. It is graphed in Figure 0.20.

An equation of the form

$$Ax + By + C = 0,$$

where A, B, and C are given numbers with not both A and B equal to zero, is called an equation of the first degree in x and y. From the results above, every line has an equation of the first degree. It is true, conversely, that the graph of every first-degree equation is a straight line. For this reason, first-degree equations are often called *linear equations*.

Example 5 Sketch the graph of the linear equation

$$3x + 4y - 6 = 0.$$

Solution: The given equation may be solved for y as follows:

$$4y = -3x + 6,$$
$$y = -\tfrac{3}{4}x + \tfrac{3}{2}.$$

We recognize this equation as the equation of the line with slope $-\frac{3}{4}$ and y intercept $\frac{3}{2}$. Its graph is sketched in Figure 0.22. The x intercept of the line is 2, as may be seen by letting $y = 0$ in the equation and solving the resulting equation for x.

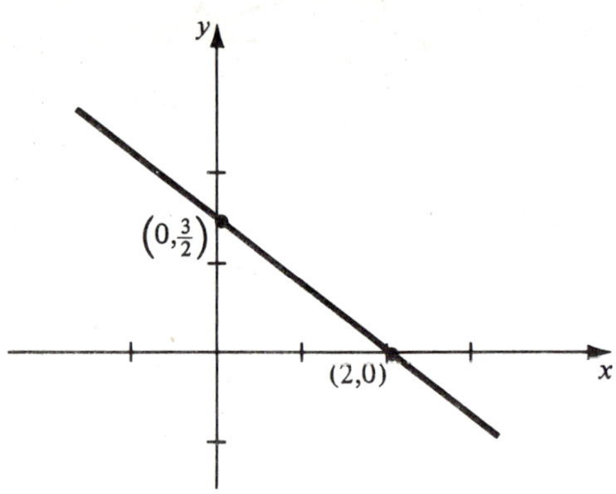

FIGURE 0.22

Two nonparallel lines have a point of intersection. This point can be found by solving simultaneously the equations of the lines, since a point lies on two graphs if and only if its coordinates satisfy the equations of both graphs.

Example 6 Find the point of intersection of the lines with equations

$$6x - 3y - 10 = 0 \quad \text{and} \quad 2x + 6y - 1 = 0.$$

Solution: The coordinates of the point of intersection of these two lines must satisfy both equations; that is, they must be the simultaneous solution of these equations. The given equations have slope-intercept form

$$y = 2x - \tfrac{10}{3} \quad \text{and} \quad y = -\tfrac{1}{3}x + \tfrac{1}{6}.$$

Their simultaneous solution may be found by eliminating y as follows:

$$\begin{aligned} 2x - \tfrac{10}{3} &= -\tfrac{1}{3}x + \tfrac{1}{6}, \\ 12x - 20 &= -2x + 1, \\ 14x &= 21, \\ x &= \tfrac{3}{2}. \end{aligned}$$

Then $$y = 2 \cdot \tfrac{3}{2} - \tfrac{10}{3} = -\tfrac{1}{3},$$

so that $(\frac{3}{2}, -\frac{1}{3})$ is the required point of intersection (see Figure 0.23).

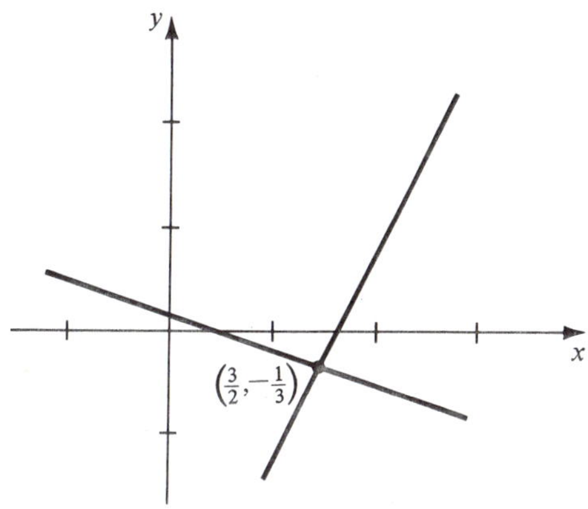

FIGURE 0.23

The points of intersection of any two graphs can be found similarly by solving simultaneously the equations of the graphs. We illustrate this procedure by finding the points of intersection of a circle and a straight line.

Example 7 Find the points of intersection of the circle with equation

$$x^2 - 8x + y^2 + 11 = 0$$

and the line with equation

$$x + y - 5 = 0.$$

Solution: The slope-intercept form of the equation of the line is

$$y = -x + 5.$$

Substituting this value of y in the equation of the circle, we obtain

$$x^2 - 8x + (-x + 5)^2 + 11 = 0,$$

which reduces to

$$x^2 - 9x + 18 = 0.$$

This quadratic equation may be factored to yield

$$(x - 3)(x - 6) = 0.$$

Thus $x = 3$ and $x = 6$ are the abscissas of the points of intersection of the line and the circle. The ordinate of each point may be found by substituting each x value in the second of the given equations and solving for y. Thus, if $x = 3$, $y = -3 + 5 = 2$; if $x = 6$, $y = -6 + 5 = -1$. The points of intersection are (3,2) and (6,−1), as shown in Figure 0.24.

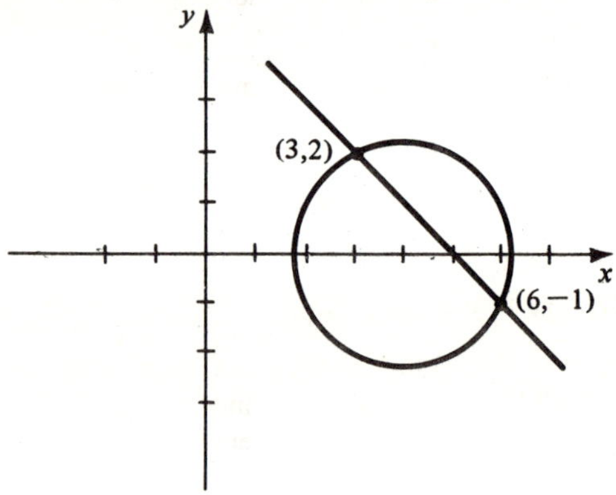

FIGURE 0.24

EXERCISES

I

1. Show that (0,3), (−2,12), (−4,11), and (2,4) are vertices of a rectangle.
2. Determine x so that the line on the points (2,−3) and (x,3) has slope 2.

In each of Exercises 3 to 6, show by means of slopes that the three points are collinear.

3. (1,−1), (−2,5), (3,−5)
4. (2,0), (4,1), (−6,−4)
5. (−1,1), (2,3), (−4,−1)
6. (−6,3), (4,−1), (3,$-\frac{3}{5}$)
7. How could one do Exercises 3 to 6 without using slopes?
8. Show that the point (6,3) is on the perpendicular bisector of the line segment with endpoints (3,2) and (7,6).

In each of Exercises 9 to 20, find an equation of the line satisfying the given conditions.

9. On $(5,1)$ and $(-1,-1)$.
10. On $(3,3)$ and $(7,6)$.
11. With slope 3 and y intercept -3.
12. With slope $-\frac{1}{2}$ and passing through the point $(-3,-5)$.
13. With slope 4 and passing through the point $(1,6)$.
14. With x intercept 3 and y intercept 2.
15. With x intercept $-\frac{1}{3}$ and slope 6.
16. On $(-1,2)$ and $(-1,-3)$.
17. On $(-4,1)$ and $(3,1)$.
18. With x intercept a and y intercept b.
19. With x intercept a and slope m.
20. With inclination 135° and y intercept -2.

In each of Exercises 21 to 28, find the slope and intercepts, and sketch the line with given equation.

21. $2x - y + 3 = 0$
22. $3x - 2y + 2 = 0$
23. $x + 2y + 6 = 0$
24. $5x + y - 2 = 0$
25. $3x + 5 = 0$
26. $4y - 1 = 0$
27. $5x + y + 15 = 0$
28. $2x + 4y - 1 = 0$

In each of Exercises 29 to 34, find the equations of the lines on the given point that are, respectively, parallel and perpendicular to the given line. Sketch.

29. $(3,3)$, $2x + y - 1 = 0$
30. $(-1,4)$, $3x - y + 5 = 0$
31. $(2,-1)$, $3x - 2y - 8 = 0$
32. $(5,1)$, $2x - 4y - 5 = 0$
33. $(0,0)$, $4x + 7y - 1 = 0$
34. $(-1,-1)$, $3x + 7 = 0$

In each of Exercises 35 to 38, find the point of intersection of the lines with given equations.

35. $3x - y + 4 = 0$
    $x - 2y + 18 = 0$
36. $2x + y - 3 = 0$
    $x - 3y - 12 = 0$
37. $3x + 2y - 7 = 0$
    $2x + 3y + 2 = 0$
38. $5x + y - 2 = 0$
    $3x - 2y + 7 = 0$

In each of Exercises 39 to 42, find the points of intersection of the line and circle with given equations.

39. $4x - 3y - 10 = 0$
    $x^2 + y^2 - 2x + 4y - 20 = 0$
40. $x - y + 1 = 0$
    $x^2 + y^2 + 6x - 10y + 9 = 0$
41. $3x - 4y + 10 = 0$
    $x^2 + y^2 - 10x = 0$
42. $x - y - 5 = 0$
    $x^2 + y^2 - 4y - 1 = 0$

43. Show that $Ax + By = Ax_1 + By_1$ is an equation of the line on the point (x_1,y_1) and parallel to the line $Ax + By + C = 0$.

44. Show that $Bx - Ay = Bx_1 - Ay_1$ is an equation of the line on the point (x_1,y_1) and perpendicular to the line $Ax + By + C = 0$.

II

1. Show that the angle of intersection of two lines of slope m_1 and m_2, respectively, is given by

$$\tan \theta = \frac{m_2 - m_1}{1 + m_2 m_1}.$$

2. Find the tangents of the angles of a triangle whose vertices are

$$(0,0),\ (4,10),\ (-8,8).$$

4 PARABOLAS

If a right circular cone is cut by a plane, the curve of intersection is called a *conic section*. There are essentially three types of curves obtainable in this way: parabolas, ellipses, and hyperbolas. We shall study parabolas in this section, postponing until a later chapter the study of ellipses and hyperbolas.

If we think of the cone as lying on a table (like an ice cream cone lying on its side), then a parabola is obtained when we cut the cone with a plane parallel to the table top. Another way of defining a parabola is as follows:

> *Parabola:* Set of all points in a plane equidistant from a fixed point (the *focus*) and a fixed line (the *directrix*).

We may find an equation of a parabola by introducing a coordinate system in its plane. Although it might seem natural to choose the directrix as one axis, a better choice is a line parallel to the directrix and halfway between the focus and the directrix. Then the other axis is chosen through the focus, as in Figure 0.25.

A point $P(x,y)$ is on the given parabola if and only if

$$|PF| = |PA|;$$

that is, if and only if

$$\sqrt{(x - p)^2 + y^2} = \sqrt{(x + p)^2 + (y - y)^2},$$

or, equivalently,

$$(x - p)^2 + y^2 = (x + p)^2.$$

This equation reduces to

$$y^2 = 4px, \qquad \textit{equation of parabola in Figure 0.25.}$$

The *vertex* and *axis* of a parabola are shown in Figure 0.25. For each point $P(x,y)$ on the parabola, the point $P'(x,-y)$ is also on the parabola. We say that the parabola is *symmetric to its axis*; i.e., for every point $P(x,y)$ on

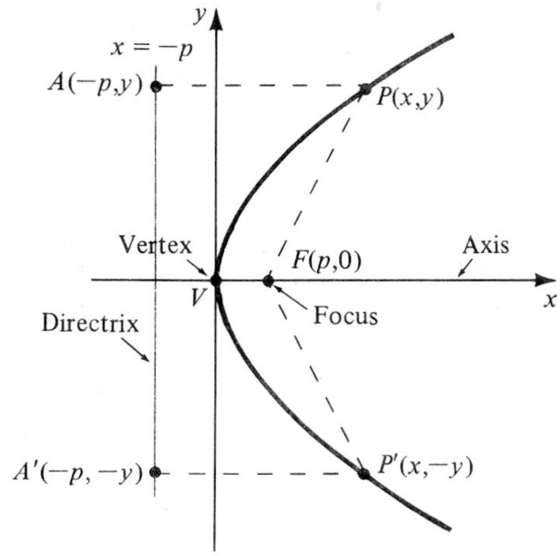

FIGURE 0.25

the parabola, the reflection of this point in the x axis, namely $P'(x, -y)$, is also on the parabola. Thus, if we imagine folding the plane along the x axis until the upper half-plane coincides with the lower half-plane, the graph of the parabola above the x axis will coincide with the part below the x axis.

The number p in Figure 0.25 is positive. If p is negative, then the parabola lies to the left of the y axis, as sketched in Figure 0.26.

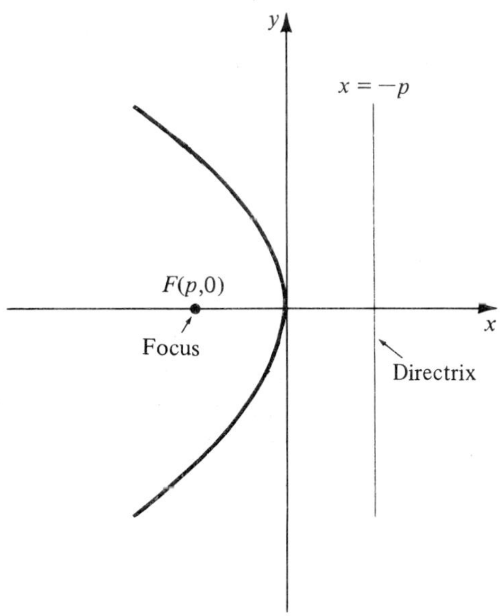

FIGURE 0.26

If we start with the point $F(0,p)$ as the focus and $y = -p$ as the directrix (Figure 0.27), we obtain

$$x^2 = 4py, \qquad \textit{equation of parabola in Figure 0.27.}$$

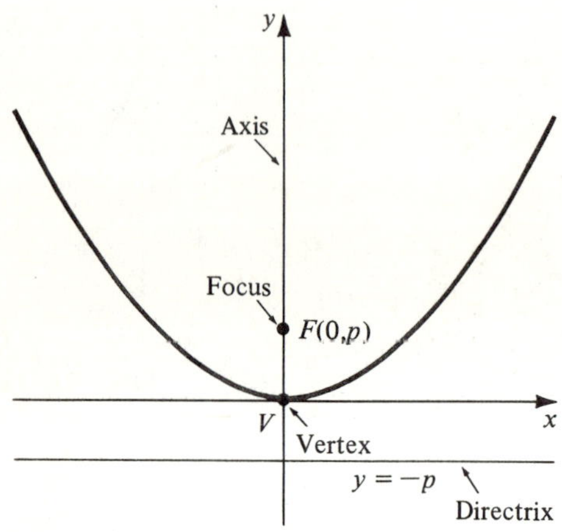

FIGURE 0.27

This is evident since we have simply interchanged the roles of x and y in the two cases. The parabola is now symmetric to the y axis.

Again, $p > 0$ in Figure 0.27. If $p < 0$, the parabola opens downward as in Figure 0.28.

Example 1 Find an equation of the parabola with focus $F(-2,0)$ and directrix $x = 2$.

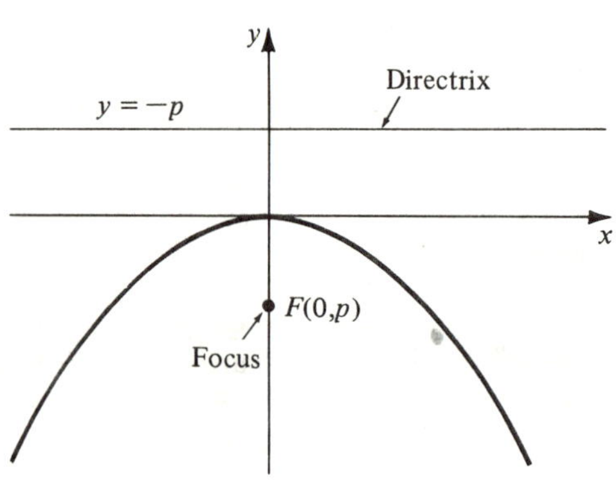

FIGURE 0.28

Solution: This case is shown in Figure 0.26. Its equation is $y^2 = 4px$ with $p = -2$, that is,

$$y^2 = -8x.$$

In sketching this parabola (Figure 0.29), it is convenient to draw in the chord that is on the focus and perpendicular to the axis. This chord is called the ***latus rectum*** of the parabola. In Figure 0.29 AB is the latus rectum. Note that $|AB| = 8$

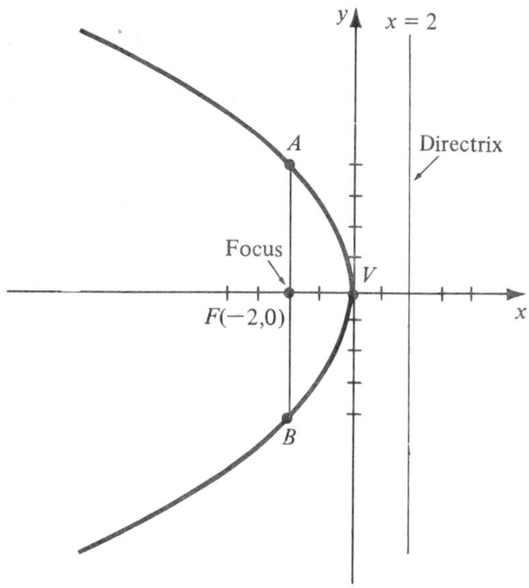

FIGURE 0.29

Example 2 Discuss and sketch the graph of the equation

$$x^2 = 6y.$$

Solution: This equation has the form 2.13 with $4p = 6$, or

$$p = \tfrac{3}{2}.$$

Therefore the graph is a parabola with the y axis as its axis, focus $(0,\frac{3}{2})$, and directrix $y = -\frac{3}{2}$. The parabola is sketched in Figure 0.30. The segment AB, of length 6, is its latus rectum.

It is easy to derive an equation for a parabola even though the coordinate axes are not placed as in Figure 0.25 or Figure 0.27, as long as the directrix is parallel to a coordinate axis. For example, let the directrix D be parallel to the y axis, as in Figure 0.31. If we designate the coordinates of the vertex V by (h,k), then the focus is $F(h + p, k)$ and the directrix has equation $x = h - p$, where the number p has the same meaning as before. Again, a point $P(x,y)$

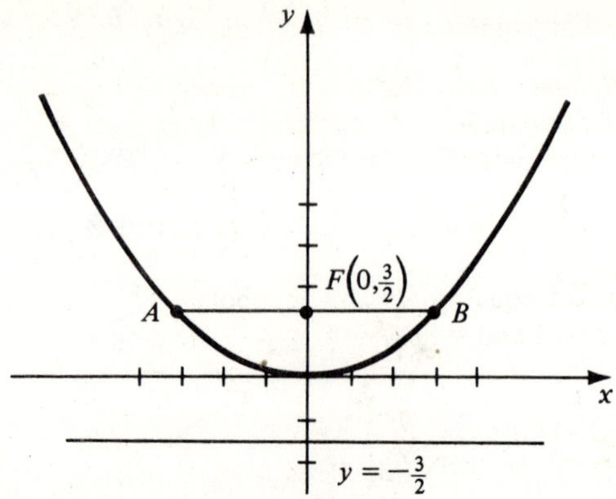

FIGURE 0.30

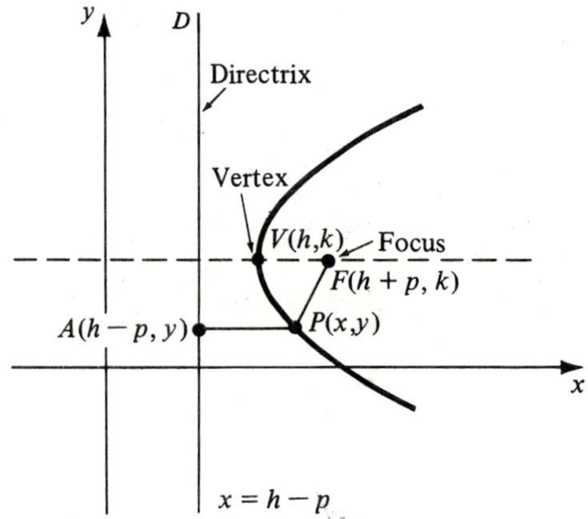

FIGURE 0.31

is on the parabola if and only if $|FP| = |PA|$; that is,

$$\sqrt{(x - h - p)^2 + (y - k)^2} = \sqrt{(x - h + p)^2}.$$

We easily reduce this equation to the form

$$(y - k)^2 = 4p(x - h), \qquad \textit{equation of parabola in Figure 0.31.}$$

By analogy,

$$(x - h)^2 = 4p(y - k)$$

is an equation of the parabola with vertex $V(h,k)$, focus $F(h, k + p)$, and directrix $y = k - p$ parallel to the x axis.

Example 3 Find an equation of the parabola with directrix $y = 1$ and focus $F(3, -5)$.

Solution: An equation will have form $(x - h)^2 = 4p(y - k)$, since the directrix, D, is parallel to the x axis. The vertex V, being midway between D and F, has coordinates $(3, -2)$. Thus $-5 = -2 + p$ and $p = -3$. Hence

$$(x - 3)^2 = -12(y + 2)$$

is an equation of the parabola. An equivalent equation, obtained by squaring $x - 3$ and simplifying, is

$$x^2 - 6x + 12y + 33 = 0.$$

Sketch it!

Example 4 Describe the graph of the equation $y^2 + 4y - 6x + 22 = 0$.

Solution: By completing the square on the y terms, the given equation may be shown to be equivalent to one of form $(y - k)^2 = 4p(x - h)$.

$$\begin{aligned} y^2 + 4y &= 6x - 22, \\ y^2 + 4y + 4 &= 6x - 18, \\ (y + 2)^2 &= 6(x - 3). \end{aligned}$$

Hence, $k = -2$, $h = 3$, and $4p = 6$, or $p = \frac{3}{2}$. Thus the graph of the given equation is a parabola with vertex $V(3, -2)$, focus $F(\frac{9}{2}, -2)$, and directrix $x = \frac{3}{2}$. Sketch it!

EXERCISES

I

Discuss and sketch the graph of each of the following equations.

1. $y^2 = 12x$
2. $x^2 = -4y$
3. $x^2 - y$
4. $4y^2 - x = 0$
5. $y^2 - 6y - 2x - 11 = 0$
6. $x^2 + 2x + 3y - 8 = 0$
7. $x^2 - 4y + 7 = 0$
8. $y^2 - 14y + 8x + 49 = 0$
9. $x^2 - 4x - 8y + 4 = 0$
10. $y^2 + 6y - 4x + 9 = 0$
11. $x^2 + 6y - 12 = 0$
12. $y^2 + 2y + 3x - 5 = 0$

In each of Exercises 13 to 20, find an equation of the parabola having:

13. Focus (4,0), directrix $x = -4$.
14. Focus (0,2), vertex (0,0).
15. Focus $(-3, -3)$, directrix $y = 3$.
16. Focus (0,0), directrix $x = 2$.
17. Directrix $x = 0$, vertex (2,2).
18. Directrix $y = -4$, vertex $(2, -6)$.
19. Focus (2,5), vertex (2,1).
20. Focus $(-2,2)$, vertex (0,2).

II

1. Sketch the graph of $|x - y^2| = 1$.
2. In the parabola $y^2 = 4ax$, an equilateral triangle is inscribed with one vertex at the origin. Find the length of a side.

CHAPTER REVIEW

Solve the following inequalities. Give the answer in interval notation.

1. $3x - 7 < 20x + 5$
2. $|7x + 4| < \varepsilon, \varepsilon > 0$
3. $-10x + 5 \geq 4x + 2$
4. $\left|\dfrac{x - a}{3}\right| < \delta, a > 0, \delta > 0$
5. $\dfrac{1}{x - 2} < 5$
6. $|-6x + \frac{4}{3}| < \dfrac{\varepsilon}{2}, \varepsilon > 0$
7. $|-x + 7| < 3$
8. $x^2 < 9$
9. $\dfrac{1}{|2x - 3|} < 1$
10. $\dfrac{1}{|x - 1|} > 2$

Solve for x.

11. $|x| = x + 4$
12. $|x|^2 - x - 12 = 0$
13. $|9 - x^2| = 7$
14. $|4x - 6| = 2$
15. $|-3x + 6| = -3$
16. $|x - 3| = |3x + 5|$

Graph the following sets of ordered pairs.

17. $\{(x,y) \mid (x - 3)^2 + (y + 2)^2 < 36\}$
18. $\{(x,y) \mid (x - 3)^2 + (y + 2)^2 \geq 36\}$
19. $\{(x,y) \mid x \geq 3 \text{ and } y \leq 1\}$
20. $\{(x,y) \mid x \geq 3 \text{ or } y \leq 1\}$
21. $\{(x,y) \mid 3x^2 - 6x + 3y^2 + 9y = 27\}$
22. $\{(x,y) \mid y > 2x, x^2 + y^2 \leq 1\}$
23. $\{(x,y) \mid y^2 \leq x^2\}$
24. $\{(x,y) \mid 9 - x^2 < y < x^2 + 25\}$
25. $\{(x,y) \mid 2x + y^2 - 7 < 4y\}$
26. $\{(x,y) \mid 3x + y^2 < 2\}$
27. For what values of x is $\sqrt{-3x - 4}$ not a real number?
28. Show that the point $\left(-\dfrac{\sqrt{3}}{2}, \dfrac{1}{2}\right)$ is on the unit circle having its center at the origin.

Find an equation of the line:

29. Through (1,5), having inclination 120°.
30. Having x intercept 3 and parallel to the line $3x - y = 11$.

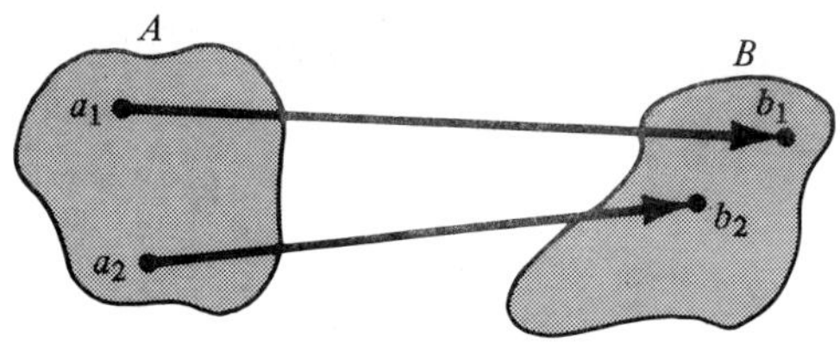

FIGURE 1.1

Functions are often designated by letters such as f, F, g, G, and so on. If f is a function with domain A and range B, then for each number x in set A the corresponding number in set B is designated by

$$f(x) \qquad \text{(read "}f\text{ of }x\text{").*}$$

If we let f denote the function above defined by the equation

$$y = 3x - 5$$

then y is $f(x)$:

$$f(x) = 3x - 5$$

for every real number x. Thus,

$$f(5) = 3 \cdot 5 - 5 = 10,$$
$$f(-2) = 3 \cdot (-2) - 5 = -11,$$

and so on. The domain and range of f is the set of all real numbers.

Example 1 The usual formula

$$S = 6x^2$$

for the area S of the surface of a cube in terms of the length x of each edge defines an area function. If we designate this function by F, then

$$F(x) = 6x^2$$

for each positive number x. For example,

$$F(1) = 6, \qquad F(2) = 6 \cdot 2^2 = 24, \qquad F(10) = 6 \cdot 10^2 = 600.$$

The domain (and the range) of F is the set $(0, \infty)$ of all positive real numbers.

Example 2 The equation

$$s = 16t^2$$

gives the number of feet s an object falls in t seconds from a point of rest if the only force acting on the object is the earth's gravity. This equation defines a

* This functional notation is credited to the Swiss mathematician Leonhard Euler (1707–1783), probably the most prolific mathematician of all time. The form of many of our present-day mathematics textbooks is due to Euler.

function f, the function that maps each number t into the number of feet the object falls in t seconds,

$$f(t) = 16t^2.$$

Thus

$$f(1) = 16, \qquad f(2) = 16 \cdot 2^2 = 64, \qquad f(\tfrac{9}{2}) = 16 \cdot (\tfrac{9}{2})^2 = 324.$$

The natural domain (and range) of f is the set $[0,\infty)$ of all nonnegative real numbers.

Example 3 Let function G be defined by the equation

$$G(u) = \sqrt{u - 3}.$$

When a function is defined by an equation and no mention is made of the domain, it is understood to be the set of all real numbers for which the equation is defined. In this example we must have $u - 3 \geq 0$ and therefore $u \geq 3$. Thus the domain of G is the infinite interval $[3,\infty)$. We note that

$$G(3) = 0, \qquad G(4) = \sqrt{4 - 3} = 1, \qquad G(8) = \sqrt{8 - 3} = \sqrt{5}.$$

Although the letter u was used in the definition of G, we could just as well have defined G by the equation

$$G(x) = \sqrt{x - 3},$$

or by the equation

$$G(z) = \sqrt{z - 3}.$$

A function need not be defined by an explicit equation, as those in the above examples were. One of the celebrated functions of mathematics is the counting function p: for each positive integer n,

$$p(n) = \text{the number of positive prime integers} \leq n.$$

Thus $p(1) = 0$, $p(2) = 1$, $p(3) = 2$, and $p(18) = 7$, the prime integers ≤ 18 being 2, 3, 5, 7, 11, 13, and 17. There is no known formula giving $p(n)$ for any positive integer n.

Example 4 To send a parcel by first-class mail* costs 10 cents per ounce or fraction thereof. Any parcel not exceeding 12 ounces may be sent by first-class mail. If we let $F(w)$ be the cost in cents of sending a parcel of weight w ounces, then

$$\begin{aligned} F(w) &= 10 \quad \text{if } 0 < w \leq 1, \\ F(w) &= 20 \quad \text{if } 1 < w \leq 2, \\ &\cdots\cdots\cdots\cdots \\ F(w) &= 120 \quad \text{if } 11 < w \leq 12. \end{aligned}$$

The postage function F has the interval $(0,12]$ as its domain and $\{10n \mid n \text{ an integer}, 1 \leq n \leq 12\}$ as its range.

We may also describe a function in a more formal way as a certain kind of set of ordered pairs. For this purpose we introduce the following definition.

* As of 1 July 1975.

1.1 Definition

Let A and B be any sets. By the *cartesian product* of A and B, denoted by $A \times B$, we mean the set of all ordered pairs (x,y) such that x is an element of A and y is an element of B. That is,

$$A \times B = \{(x,y) \mid x \text{ is in } A \text{ and } y \text{ is in } B\}.$$

For example, if $A = \{1,2\}$ and $B = \{a,b,c\}$, then $A \times B$ consists of the ordered pairs $(1,a)$, $(1,b)$, $(1,c)$, $(2,a)$, $(2,b)$, and $(2,c)$.

1.2 Definition

Let A and B be any sets. A subset F of $A \times B$ is called a *function* if and only if no two distinct ordered pairs in F have the same first member. That is, F is a function if and only if, whenever (x,y) and (x,z) are both in F, then $y = z$.

If the ordered pair (x,y) is in F, we may write $y = F(x)$. Thus we see that this definition simply makes more precise our intuitive notion of a *mapping*: i.e., the element y is associated or paired with the element x if and only if the ordered pair (x,y) is in F. Also, an essential feature of a mapping is that a *unique* image is associated with every member of the domain of the mapping, and this idea is also formalized in Definition 1.2.

Example 5 Let $A = \{1,2,3\}$, $B = \{a,b,c\}$. Then the set of ordered pairs $\{(1,b), (2,c), (3,b)\}$ is a function whose domain is A and whose range is $\{b,c\}$. However, the set of ordered pairs $\{(1,a), (1,b), (2,c), (3,b)\}$ is not a function. (Why not?)

2 TYPES OF FUNCTIONS

If the range of a function contains only one number, say c, then f is called a *constant function*. Thus $f(x) = c$ for every x in the domain of f. We shall often represent this function by its value c, and speak of the constant function c.

If $a_0, a_1, \ldots, a_n$ are given numbers and n is a given nonnegative integer, then an expression of the form

$$a_0x^n + a_1x^{n-1} + \cdots + a_{n-1}x + a_n$$

is called a *polynomial in x*. A function f defined by a polynomial

$$f(x) = a_0x^n + a_1x^{n-1} + \cdots + a_{n-1}x + a_n,$$

is called a *polynomial function*. The domain of f is taken to be the set of all real numbers.

Special polynomial functions are the *linear function* f defined by

$$f(x) = a_0x + a_1,$$

the *quadratic function* g defined by

$$g(x) = a_0x^2 + a_1x + a_2, \qquad a_0 \neq 0,$$

and the *cubic function* h defined by

$$h(x) = a_0x^3 + a_1x^2 + a_2x + a_3, \qquad a_0 \neq 0.$$

A function defined by a quotient of two polynomials in x is called a *rational function.* For example, the function f defined by

$$f(x) = \frac{x^2 - 3x + 1}{x + 2}$$

is a rational function having as its domain the set of all real numbers except -2.

A function defined in terms of polynomials and roots of polynomials is called an *algebraic function.* For example, the function f defined by

$$f(x) = \frac{x - 1}{x\sqrt{x^2 + 1}}$$

is an algebraic function. Its domain is the set of all nonzero real numbers.

Examples of nonalgebraic functions are the six trigonometric functions sine, cosine, tangent, cotangent, secant, and cosecant; the logarithmic function f defined by

$$f(x) = \log_a x;$$

the exponential function g defined by

$$g(x) = a^x;$$

and combinations of algebraic functions and the functions above, such as

$$h(x) = x \sin x + x^2 - 2^x.$$

These are examples of *transcendental functions.* They are discussed in detail in Chapters 7 and 8.

EXERCISES

I

In Exercises 1 to 3, if $f(x) = x^2 - 3x + 1$, find:

1. $f(0), f(-1), f(-\sqrt{3})$

2. $f(-2 + h), f(x + h)$

3. $\dfrac{f(x) - f(a)}{x - a}, x \neq a; \dfrac{f(x + h) - f(x)}{h}, h \neq 0$

In Exercises 4 and 5, if $g(x) = (x - 1)/(x + 1)$, $x \neq -1$, find:

4. $g(1), g(-\sqrt{2}), g(a^3)$

5. $g(-1 + 2h), g(x + y), g\left(\dfrac{1}{x} - 1\right), g\left(\dfrac{x + 1}{x - 1}\right)$

6. If $f(x) = 2x^2 - 5x$, find an equation of the line passing through the following pairs of points:

   a. $(2,f(2))$ and $(\frac{5}{2},f(\frac{5}{2}))$

   b. $(a,f(a))$ and $(a + h, f(a + h))$

7. If $f(x) = x^2 - 3x + 1$, for what numbers x is $f(x) = f(2x)$? Is $f(x) = f(ax)$? Is $2f(x) = f(2x)$?

8. If $F(x) = \sqrt{x}$, $x > 0$, show that

$$\frac{F(x + h) - F(x)}{h} = \frac{1}{\sqrt{x + h} + \sqrt{x}}, \qquad h \neq 0.$$

Find the domain and range of each function (y is a function of x):

9. $\{(x,y) \mid y = x^2 - 2\}$

10. $\{(x,y) \mid y = 1/\sqrt{9 - x^2}\}$

11. $\{(x,y) \mid x = \sqrt{y - 2}\}$

12. $\{(x,y) \mid y = \sqrt{x^2 - 16}\}$

13. $\left\{(x,y) \mid y = \sqrt{\dfrac{x + 3}{x - 4}}\right\}$

14. $\left\{(x,y) \mid y = \dfrac{x + 2}{x - 7}\right\}$

II

1. If $F(x) = x^4 + 4x^3 + 6x^2 + 4x$, find:

   a. $F(3x - 1)$

   b. $\dfrac{F(x) - F(-1)}{x + 1}$, $x \neq -1$

2. If $F(x) = \sqrt[3]{x}$, show that

$$\frac{F(x + h) - F(x)}{h} = (\sqrt[3]{(x + h)^2} + \sqrt[3]{x^2 + xh} + \sqrt[3]{x^2})^{-1}, \; h \neq 0.$$

3 GRAPHS OF FUNCTIONS

The *graph of a function f* in a coordinate plane is the graph of the set

$$\{(x,f(x)) \mid x \text{ in domain of } f\}.$$

Or, equivalently, it is the graph of the equation

$$y = f(x).$$

That is, the point $P(a,b)$ is in the graph of f if and only if $b = f(a)$.

Some examples of graphs of functions are given below.

Example 1 Determine the graph of the constant function f defined by $f(x) = 3$ for every real number x; of the linear function g defined by $g(x) = 3x - 2$ for every real number x.

Solution: The graph of f is the graph of the equation $y = 3$; the graph of g is the graph of the equation $y = 3x - 2$. Both are sketched in Figure 1.2.

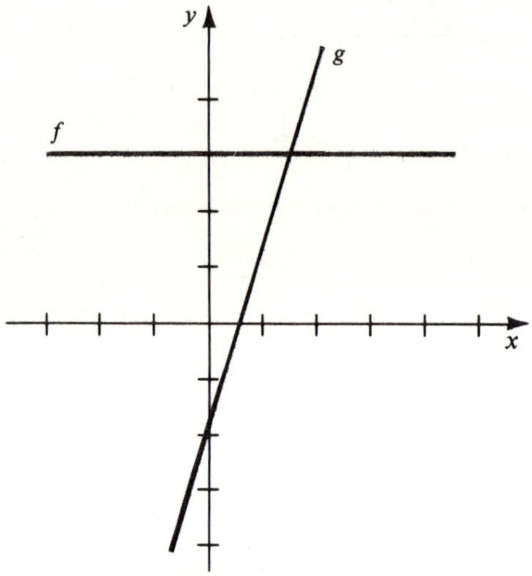

FIGURE 1.2

Example 2 Discuss the graphs of the functions f and g defined by

$$f(x) = \sqrt{4 - x^2}, \qquad g(x) = -\sqrt{4 - x^2},$$

each function having the closed interval $[-2,2]$ as its domain.

Solution: The graph of the equation $y = \sqrt{4 - x^2}$ lies on or above the x axis since $\sqrt{4 - x^2} \geq 0$, while the graph of the equation $y = -\sqrt{4 - x^2}$ lies on or below the x axis since $-\sqrt{4 - x^2} \leq 0$. Thus the graph of f is the upper half of the circle with equation

$$y^2 = 4 - x^2 \qquad \text{or} \qquad x^2 + y^2 = 4;$$

the graph of g is the lower half of the same circle. These graphs are shown in Figure 1.3.

Example 3 Determine the graph of the postage function given in Section 1, Example 4, page 38.

Solution: For this function F we know that

$$F(x) = 10n \qquad \text{if } n - 1 < x \leq n,$$

with $(0,12]$ the domain of F. Thus F has a constant value $10n$ over each half-open interval $(n - 1, n]$. The graph is made up of 12 "steps," as partially shown in Figure 1.4. For this reason a function such as F is frequently called a *step function*.

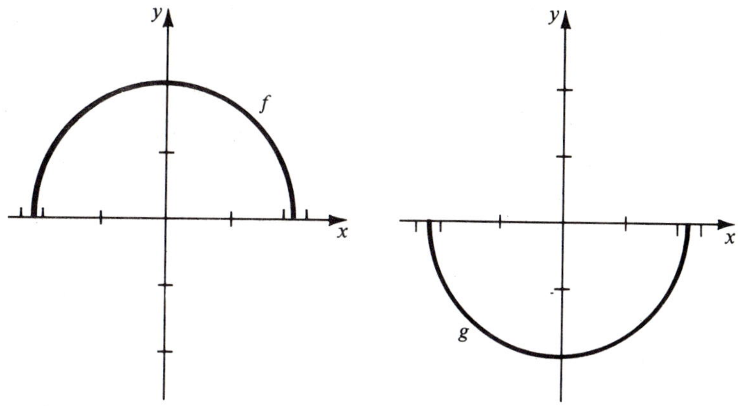

FIGURE 1.3

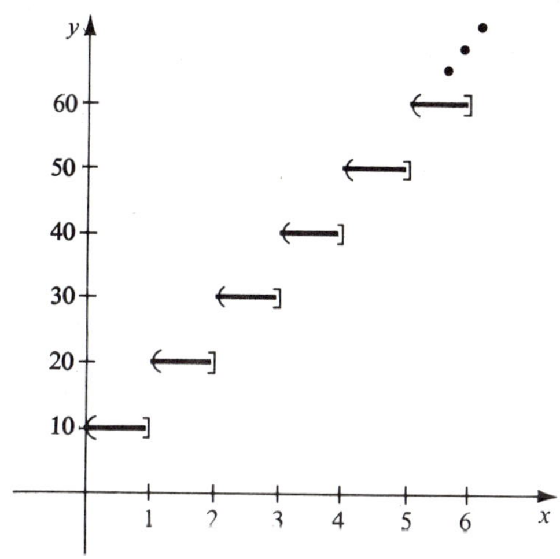

FIGURE 1.4

Example 4 Discuss the graph of the function G defined by

$$G(x) = \frac{1}{x}.$$

Solution: The domain of G is the set of all nonzero numbers, and the graph of G is the graph of the equation

$$y = \frac{1}{x},$$

or

$$xy = 1.$$

Since x and y are reciprocals of each other, y is small when x is large and y is large when x is small. Also, x and y are either both positive or both negative. Some specific points on the graph of G are:

$$(.1,10),\ (.2,5),\ (.5,2),\ (1,1),\ (2,.5),\ (5,.2),\ (10,.1).$$

Then $(-.1,-10)$, $(-.2,-5)$, and so on are also on the graph. From this information the graph is sketched in Figure 1.5.

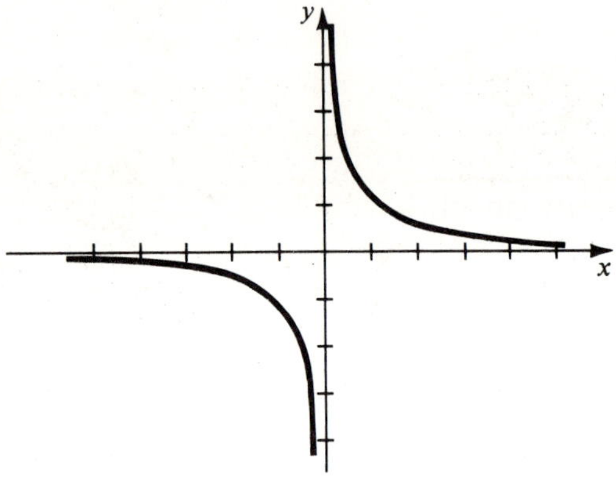

FIGURE 1.5

EXERCISES

I

In each of Exercises 1 to 16, sketch the graph of the function defined by:

1. $f(x) = 2$
2. $F(x) = 1 + x$
3. $g(x) = 3 - 5x$
4. $G(t) = 1 + 2t$
5. $F(r) = \dfrac{2}{r},\ r \neq 0$
6. $f(x) = \sqrt{1 - x^2}$
7. $F(x) = -\sqrt{16 - x^2}$
8. $g(t) = \dfrac{1}{t - 2},\ t \neq 2$
9. $f(x) = \begin{cases} x \text{ if } x \geq 0 \\ -1 \text{ if } x < 0 \end{cases}$
10. $F(x) = \begin{cases} 1 - x \text{ if } x \geq 1 \\ x - 1 \text{ if } x < 1 \end{cases}$
11. $G(y) = y^2/2$
12. $g(x) = \sqrt{x},\ x \geq 0$
13. $r(t) = -|t - 3|$
14. $f(s) = \dfrac{s^2 - 9}{s + 3}$

15. $g(t) = \begin{cases} 3 & \text{if } t < 2 \\ t + 2 & \text{if } 2 \leq t < 4 \\ 0 & \text{if } t \geq 4 \end{cases}$

16. $f(x) = \begin{cases} -1 & \text{if } x < 0 \\ 0 & \text{if } x = 0 \\ 1 & \text{if } x > 0 \end{cases}$

II

In each of Exercises 1 to 4, sketch the graph of the function.

1. $G(x) = \dfrac{3x + 1}{x}, x \neq 0$

2. $B(r) = r + \dfrac{1}{r}, r \neq 0$

3. $B(r) = r - \dfrac{1}{r}, r \neq 0$

4. $H(x) = \dfrac{(x - 1)^2}{x^2}, x \neq 0$

4 COMBINATIONS OF FUNCTIONS

If f and g are functions, with respective domains A and B, then we can define the *sum* $f + g$, the *difference* $f - g$, the *product* fg, and the *quotient* f/g of these two functions in an obvious way:

$$(f + g)(x) = f(x) + g(x), \qquad \text{domain } (f + g) = A \cap B,$$
$$(f - g)(x) = f(x) - g(x), \qquad \text{domain } (f - g) = A \cap B,$$
$$(fg)(x) = f(x)g(x), \qquad \text{domain } fg = A \cap B,$$
$$\frac{f}{g}(x) = \frac{f(x)}{g(x)}, \qquad \text{domain } \frac{f}{g} = \{x \mid x \text{ in } A \cap B, g(x) \neq 0\}.$$

Example 1 Let

$$f(x) = x^2 + 1, \qquad \text{domain } f = (-\infty,\infty),$$
$$g(x) = \sqrt{x}, \qquad \text{domain } g = [0,\infty).$$

Then

$$(f + g)(x) = x^2 + 1 + \sqrt{x}, \qquad \text{domain } (f + g) = [0,\infty),$$
$$(f - g)(x) = x^2 + 1 - \sqrt{x}, \qquad \text{domain } (f - g) = [0,\infty),$$
$$(fg)(x) = (x^2 + 1)\sqrt{x}, \qquad \text{domain } fg = [0,\infty),$$
$$\frac{f}{g}(x) = \frac{x^2 + 1}{\sqrt{x}}, \qquad \text{domain } \frac{f}{g} = (0,\infty).$$

Example 2 Let

$$f(x) = \sqrt{1 - x}, \qquad \text{domain } f = (-\infty,1],$$
$$g(x) = \sqrt{x + 2}, \qquad \text{domain } g = [-2,\infty).$$

Then

$$(f + g)(x) = \sqrt{1 - x} + \sqrt{x + 2}, \qquad \text{domain } (f + g) = [-2,1],$$
$$(fg)(x) = \sqrt{(1 - x)(x + 2)}, \qquad \text{domain } fg = [-2,1],$$
$$\frac{f}{g}(x) = \sqrt{\frac{1 - x}{x + 2}}, \qquad \text{domain } \frac{f}{g} = (-2,1].$$

Customarily, we write f^2 for $f \cdot f$, f^3 for $f \cdot f \cdot f$, and so on. For example:

$$\text{If } f(x) = x - 1, \text{ then } f^3(x) = (x - 1)^3.$$

The function F defined by

$$F(x) = (2x - 1)^2$$

is a combination of a square function and a linear function. If we let f be the square function

$$f(x) = x^2,$$

and g be the linear function

$$g(x) = 2x - 1;$$

then $f(g(x)) = (2x - 1)^2$. That is,

$$F(x) = f(g(x)).$$

This function F above is an example of a composite of two functions. For any two functions f and g, the *composite* of f and g is designated by

$$f \circ g$$

and is defined by

$$(f \circ g)(x) = f(g(x)).$$

The domain of $f \circ g$ is the set of all real numbers x in the domain of g for which $g(x)$ is in the domain of f. Both f and g in the example above have $(-\infty,\infty)$ as domain, so their composite F also has $(-\infty,\infty)$ as its domain.

Example 3 Let

$$f(x) - \sqrt{x}, \qquad g(x) = 4 - x^2.$$

Then

$$(f \circ g)(x) = f(g(x)) = f(4 - x^2) = \sqrt{4 - x^2}.$$

What is the domain of $f \circ g$? It is the set of all numbers x for which $4 - x^2$ has a square root; that is, all x for which $4 - x^2 \geq 0$. Thus,

$$x^2 \leq 4 \qquad \text{and} \qquad -2 \leq x \leq 2.$$

The domain of $f \circ g$ is the interval $[-2,2]$.

The function $g \circ f$ is given by

$$(g \circ f)(x) = g(f(x)) = g(\sqrt{x}) = 4 - (\sqrt{x})^2 = 4 - x.$$

The domain of $g \circ f$ is the set of all numbers x for which $\sqrt{x}$ is in the domain of g. Thus, the domain of $g \circ f$ is $[0,\infty)$. Observe that $f \circ g$ and $g \circ f$ are different functions!

Example 4 Let

$$f(x) = \frac{1-x}{1+x}, \qquad g(x) = \frac{1}{x}.$$

Then

$$(f \circ g)(x) = f(g(x)) = f\left(\frac{1}{x}\right) = \frac{1 - 1/x}{1 + 1/x} = \frac{x-1}{x+1}, \qquad x \neq 0, x \neq -1.$$

Note that

$$(g \circ g)(x) = g(g(x)) = g\left(\frac{1}{x}\right) = \frac{1}{1/x} = x, \qquad x \neq 0.$$

Thus, the function $g \circ g$ makes every nonzero number x correspond to itself!

EXERCISES

1. If $f(x) = x^2 - 1$ and $g(x) = 3x + 1$, give the formulas for the following functions.

   a. $f + g$ b. $f - g$ c. fg
   d. f/g e. $f \circ g$ f. $g \circ f$

2. Give the formulas and domains for the functions in Exercise 1 if $f(x) = \sqrt{x-1}$, $x \geq 1$, $g(x) = x^2 + 1$.
3. Give the formulas and domains for the functions in Exercise 1 if $f(x) = 1/(x+1)$, $x \neq -1$, $g(x) = x/(x-1)$, $x \neq 1$.
4. Give the formulas and domains for the functions in Exercise 1 if $f(x) = 1/x^2$, $x \neq 0$, $g(x) = \sqrt{x}$, $x \geq 0$.
5. If $f(x) = x^2$, find a function g such that $f(g(x)) = x$. Is $g(f(x)) = x$ also?
6. If $f(x) = \sqrt[3]{x}$, find a function g such that $f(g(x)) = x$. Is $g(f(x)) = x$ also?
7. If $f(x) = |x|$, find $f(f(x))$.
8. If $f(x) = \sqrt{1 - x^2}$ and $g(x) = \sin x$, what is $(f \circ g)(x)$? Give its domain and range.
9. If $f(x) = 2x^2 + x$ and $g(x) = 2 - x$, find $f \circ g - g \circ f$.
10. If $f(x) = \dfrac{x}{2-x}$, $x \neq 2$, find $f \circ (f \circ f)$ and $(f \circ f) \circ f$.
11. Prove that $f \circ (g \circ h) = (f \circ g) \circ h$ for any three functions f, g, and h. Verify this for $f(x) = x^2 + 2x$, $g(x) = 2x - 1$, $h(x) = 1 + x$.
12. If $f(x) = x^2 + 1$, $g(x) = \sqrt{x}$, $h(x) = 1 - x$, find $(f \circ g) \circ h$ and $f \circ (g \circ h)$. What is the domain of each?

In each of Exercises 13 to 16, find $F \circ F$ and $F \circ (F \circ F)$ for the function F defined.

13. $F(x) = x^{-1}$ 14. $F(x) = x^n$

15. $F(x) = a + bx$

16. $F(x) = 2/(1 + x)$

17. Find a function f such that $f(x + t) = f(x)f(t)$ for all x and t. Is f unique?
18. If $f(x) = ax^2 + b$, $g(x) = cx + d$, and $f \circ g = g \circ f$, what can be said about a, b, c, d?
19. Are there any nonconstant functions f and g such that $f \circ g = fg$?
20. If f and g are functions such that $f(g(x)) = g(x)$ for all x, does it follow that $f(x) = x$ for all x? Explain and give examples.
21. Given $f(x) = \sqrt{x} + 3$, $g(x) = x^2 + 9$, find $(f \circ g)(4)$.

5 INVERSE FUNCTIONS

Many functions have the following property.

1.3 Definition

A function f is called a 1–1 (one-to-one) function if and only if $f(x_1) \neq f(x_2)$ for any two different elements x_1 and x_2 of the domain of f.

Example 1 If

$$f(x) = 3x - 7$$

then $f(3) = 2$ is different from $f(5) = 8$; $f(-4) = -19$ is different from $f(-3) = -16$; and so on. In fact, if x_1 and x_2 are any unequal numbers then $f(x_1) \neq f(x_2)$. For if $x_1 \neq x_2$ then $3x_1 \neq 3x_2$ and $3x_1 - 7 \neq 3x_2 - 7$. Thus, f is a 1–1 function.

Example 2 The function

$$f(x) = \sqrt{4 - x^2}, \qquad \text{domain } f = [-2,2],$$

is not a 1–1 function. Thus, 1 and -1 are two unequal numbers in the domain of f such that

$$\begin{aligned} f(-1) &= \sqrt{4 - (-1)^2} = \sqrt{3}, \\ f(1) &= \sqrt{4 - 1^2} = \sqrt{3}, \end{aligned} \qquad f(-1) = f(1).$$

For every number x in $[-2,2]$,

$$f(x) = f(-x) = \sqrt{4 - x^2}.$$

If $y = f(x)$ as in Example 1,

$$y = 3x - 7,$$

then we can solve this equation for x in terms of y,

$$x = \tfrac{1}{3}(y + 7).$$

This defines another function g,

$$g(y) = \tfrac{1}{3}(y + 7).$$

The function g is called the *inverse* of function f.

As in this example, a function must be a 1–1 function in order to have an inverse. In terms of ordered pairs, the inverse function is defined as follows.

1.4 Definition

If $\{(x,y) \mid y = f(x)\}$ is a 1–1 function, then its inverse, designated by f^{-1}, is $\{(y,x) \mid y = f(x)\}$.

In other words, if f is a 1–1 function and

$$y = f(x),$$

then when we "solve" the equation above for x in terms of y, we get

$$x = f^{-1}(y).$$

The domain of f^{-1} is the range of f.

An important property of a 1–1 function and its inverse is as follows:

1.5
$$\begin{aligned} f^{-1}(f(x)) &= x \qquad \text{for all } x \text{ in domain } f; \\ f(f^{-1}(y)) &= y \qquad \text{for all } y \text{ in domain } f^{-1}. \end{aligned}$$

Example 3

The function

$$f(x) = \sqrt{x - 1}, \qquad \text{domain } f = [1,\infty),$$

is 1–1. Find its inverse.

Solution: Let $y = f(x)$, that is,

$$y = \sqrt{x - 1}, \qquad x \geq 1.$$

We get the inverse function by solving this equation for x in terms of y. Thus,

$$y^2 = x - 1, \qquad y \geq 0$$

and

$$x = 1 + y^2, \qquad y \geq 0.$$

Hence, $x = f^{-1}(y)$,

$$f^{-1}(y) = 1 + y^2, \qquad \text{domain } f^{-1} = [0,\infty).$$

Example 4

The function

$$f(x) = 9 - x^2,$$

domain $f = [0,\infty)$, is 1–1. Its graph is half a parabola as shown.

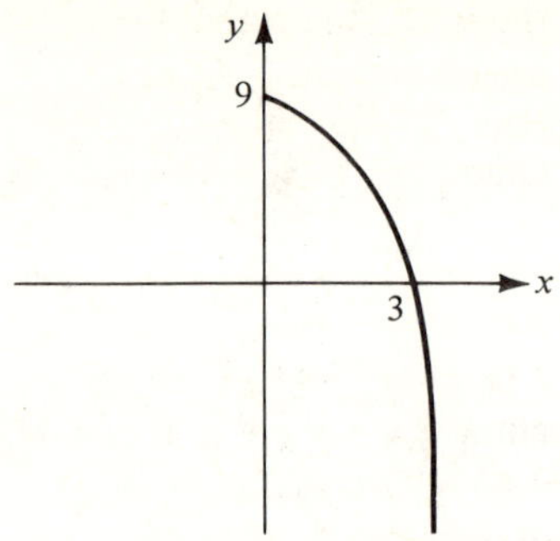

FIGURE 1.6

Let	$y = 9 - x^2,$	$x \geq 0.$
Then	$x^2 = 9 - y,$	$y \leq 9$
	$x = \sqrt{9 - y}.$	

Thus, the inverse of function f is given by

$$f^{-1}(y) = \sqrt{9 - y}, \qquad \text{domain } f^{-1} = (-\infty, 9].$$

EXERCISES

I

In each of Exercises 1–12, find the inverse of the function and give its domain. [Each unspecified domain is understood to be $(-\infty,\infty)$.]

1. $f(x) = 2x - 1$
2. $g(x) = 3 + 6x$
3. $g(x) = \frac{1}{2}x + 5$
4. $f(x) = x^2$, domain $f = (-\infty,0]$
5. $f(x) = x^2$, domain $f = [1,4]$
6. $h(x) = \sqrt[3]{x}$
7. $h(x) = \sqrt[3]{x + 8}$
8. $f(x) = 5x + 3$, domain $f = (-3,5)$
9. $g(x) = (x - 1)^3$
10. $F(x) = \sqrt{4 - x^2}$, domain $F = [0,2]$
11. $f(x) = \sqrt{1 + x^2}$, domain $f = [0,\infty)$
12. $g(x) = \sqrt[3]{1 - x^2}$, domain $g = [0,\infty)$
13. Consider the function $f(x) = \sqrt{x^2 - 9}$.
    a. Find the domain of f.
    b. Is f 1–1? Justify your answer.

c. If your answer to part b is "no," restrict the domain of f so that f will be 1–1 over the restricted domain.
d. Find f^{-1} over this restricted domain.
e. Graph f and f^{-1} on the same coordinate axes.

II

1. Let f be a function with domain A and range B, and let g be a function with domain B and range C. If both f and g are 1–1, prove that the function $g \circ f$ is 1–1 and that $(g \circ f)^{-1} = f^{-1} \circ g^{-1}$.
2. Recall that f^n denotes the function $f \circ f \circ \cdots \circ f$, where f occurs n times. Suppose that f is a function with domain A and range contained in A, and that $f^n = i$ (the identity function). Prove that f is 1–1 and that the range of f is all of A.

CHAPTER REVIEW

Give the domain of each function defined below.

1. $f(x) = 16 - x^2$
2. $g(t) = 4t - 7$
3. $F(x) = \dfrac{1}{x^2 - 1}$
4. $K(r) = -\dfrac{4}{r^2}$
5. $G(h) = \dfrac{1}{\sqrt{6 - h}}$
6. $f(x) = \dfrac{\sqrt{x - 2}}{\sqrt{x^2 - 1}}$
7. $g(y) = \dfrac{y - 1}{(y + 2)(y + 3)}$
8. $F(x) = 7$
9. $F(t) = \dfrac{1}{|t + 3|}$
10. $f(s) = \dfrac{\sqrt{s + 3}}{s + 1}$

Graph each function.

11. $f(x) = 3x^2 + 6x - 2$
12. $g(t) = 3^t$
13. $H(x) = 4x - 3$
14. $F(x) = |x - 2|$
15. $G(r) = |r^2 - 4|$
16. $f(x) = \begin{cases} x^2 \text{ if } x \le 0 \\ \sqrt{x} \text{ if } x > 0 \end{cases}$
17. $g(y) = \dfrac{1}{|y - 3|}$
18. $H(x) = \begin{cases} -x^2 \text{ if } x \ge 0 \\ x^2 \text{ if } x < 0 \end{cases}$
19. A function F is said to be *even* if $F(x) = F(-x)$ for every x in the domain of F. A function F is *odd* if $F(-x) = -F(x)$ for every x in the domain of F. For each of the functions defined below, state whether it is even, odd, or neither.
   a. $f(x) = |x|$
   b. $g(x) = \sin x$

c. $h(x) = \sin x^2$

d. $F(x) = 10^{2x} - 10^{-2x}$

e. $G(x) = x - x^3$

f. $H(x) = x + x^2$

g. $g(x) = \sin x + \cos x$

h. $f(x) = |x| + 1 + 2^x + 2^{-x}$

20. What kind of a function is each of the following? Give examples.

a. The product of an odd function by an odd function.

b. The product of an odd function by an even function.

c. The product of an even function by an even function.

Find the inverse of each function.

21. $f(x) = 3x - 5$

22. $h(t) = \dfrac{1}{t}$

23. $h(x) = (x + 3)^3$

24. $g(r) = \sqrt{16 - r^2}, 0 \le r \le 4$

25. $F(t) = \dfrac{4}{1 + t}$

26. $h(x) = \begin{cases} x & \text{if } x < 0 \\ x^2 & \text{if } 0 \le x \le 4 \\ 8\sqrt{x} & \text{if } x > 4 \end{cases}$

27. Given $f(x) = \frac{2}{3}\sqrt{x^2 + 9}$:

a. Show that f is not 1–1.

b. Restrict the domain of f so that f is 1–1.

c. Find f^{-1}.

d. Graph f and f^{-1} on the same coordinate axes.

28. If

$$f(x) = \begin{cases} 0 & \text{if } x < 0 \\ x^3 & \text{if } 0 \le x \le 2 \\ 2 & \text{if } x > 2 \end{cases} \quad \text{and} \quad g(x) = \begin{cases} -1 & \text{if } x < 0 \\ 3x & \text{if } 0 \le x \le 2 \\ 4 & \text{if } x > 2 \end{cases}$$

find $(f \circ g)(x)$, $(f \circ f)(x)$, $(g \circ g)(x)$.

2

Limits

In this chapter we lay a foundation for the calculus. This foundation will consist of definitions and basic theorems. The new concepts of limit and continuity play a fundamental role throughout the calculus.

Early in our mathematical training we were faced with problems whose solutions involved the use of limits, although we were probably unaware of it at the time. To select just one example: According to a formula of geometry, the area of a circle of radius r is πr^2, where π is a number approximately equal to 3.1416. How is such a formula derived? The usual way is to inscribe regular polygons in the circle, find the areas of these polygons, and then determine the "limiting value" of these areas as the numbers of sides of the polygons increase without bound. Thus, even such a seemingly simple formula as that for the area of a circle depends on the concept of limit for its derivation.

1 INTRODUCTION TO LIMITS

The perimeter p of a square is a function of the length x of a side:

$$p(x) - 4x.$$

When $x = 5$, the perimeter is 20,

$$p(5) = 20.$$

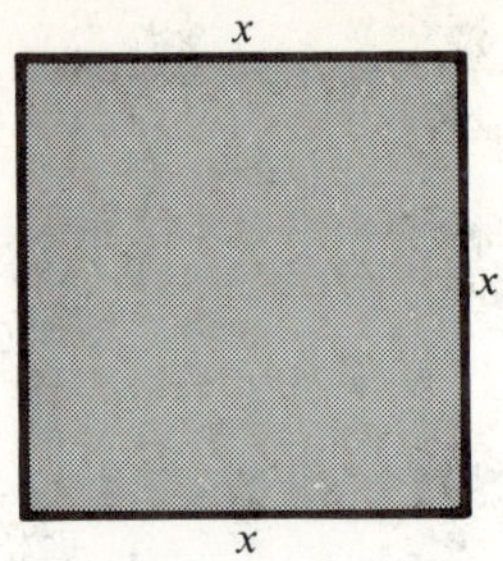

FIGURE 2.1

Suppose we make a small error in constructing the square, so that the length x of a side is not exactly 5. Is the perimeter still close to 20? We might make a table of values of $p(x)$ for x close to 5:

x	4	4.5	4.9	4.99	5.01	5.1	5.5	6
$p(x)$	16	18	19.6	19.96	20.04	20.4	22	24

It certainly seems to be true that $p(x)$ is close to 20 when x is close to 5! Therefore, we shall say that the limit of $p(x)$ as x approaches 5 is 20, and write

$$\operatorname*{limit}_{x \to 5} p(x) = 20.$$

Intuitively, for any function f we write

$$\operatorname*{limit}_{x \to a} f(x) = b$$

if *$f(x)$ is a number close to b when x is close to* (but unequal to) a. A mathematical definition is given in the next section.

Some examples using the intuitive definition are given below.

Example 1 Show that $\operatorname*{limit}_{x \to 3} \dfrac{1/x - \frac{1}{3}}{x - 3} = -\frac{1}{9}.$

Solution: If we define the function g by

$$g(x) = \frac{1/x - \frac{1}{3}}{x - 3}, \qquad x \neq 3,$$

then we must show that $g(x)$ is close to $-\frac{1}{9}$ when x is close to 3. By algebra,

$$g(x) = \frac{1/x - \frac{1}{3}}{x - 3} = \frac{(3 - x)/3x}{x - 3} = \frac{3 - x}{3x(x - 3)} = \frac{-(x - 3)}{3x(x - 3)}.$$

Since $x \neq 3$, we can cancel the nonzero number $x - 3$ from numerator and denominator, obtaining

$$g(x) = -\frac{1}{3x}.$$

Clearly, $3x$ is close to 9 when x is close to 3, so that $-1/3x$ is close to $-\frac{1}{9}$ when x is close to 3. Hence,

$$\underset{x \to 3}{\text{limit}}\, g(x) = -\tfrac{1}{9}$$

as desired.

Example 2 Find $\underset{y \to 1}{\text{limit}}\, \dfrac{\sqrt{y} - 1}{y - 1}$.

Solution: If the function f is defined by

$$f(y) = \frac{\sqrt{y} - 1}{y - 1}, \qquad y \geq 0,\, y \neq 1,$$

then we are asked to find

$$\underset{y \to 1}{\text{limit}}\, f(y).$$

So we are looking for a number b such that $f(y)$ is close to b when y is close to 1. We cannot guess the answer by letting $y = 1$ in $f(y)$, since 1 is not in domain f. By algebra,

$$f(y) = \frac{\sqrt{y} - 1}{y - 1} = \frac{(\sqrt{y} - 1)(\sqrt{y} + 1)}{(y - 1)(\sqrt{y} + 1)} = \frac{y - 1}{(y - 1)(\sqrt{y} + 1)}.$$

Cancelling $y - 1$ (which is nonzero),

$$f(y) = \frac{1}{\sqrt{y} + 1}.$$

Now it is clear that $f(y)$ is close to $\frac{1}{2}$ when y is close to (but unequal to) 1. Hence,

$$\underset{y \to 1}{\text{limit}}\, f(y) - \tfrac{1}{2}.$$

Example 3 Find $\underset{h \to 0}{\text{limit}}\, \dfrac{(2 + h)^2 - 4}{h}$.

Solution: If the function G is defined by

$$G(h) = \frac{(2 + h)^2 - 4}{h}, \qquad h \neq 0,$$

then we wish to find

$$\underset{h \to 0}{\text{limit}}\, G(h).$$

Again, we cannot find this limit by letting $h = 0$, since $G(0)$ is undefined. Thus we proceed as before by simplifying $G(h)$:

$$G(h) = \frac{(4 + 4h + h^2) - 4}{h} = \frac{4h + h^2}{h} = 4 + h,$$

since $h \neq 0$. Then it is clear that $G(h)$ is close to 4 if h is close (but unequal) to 0; that is,

$$\operatorname*{limit}_{h \to 0} \frac{(2 + h)^2 - 4}{h} = 4.$$

The next example illustrates the use of limits in a geometrical problem.

Example 4 Let $P(3,4)$ and $Q(x, \sqrt{25 - x^2})$ be two distinct points on the semicircle (Figure 2.2) with equation

$$y = \sqrt{25 - x^2},$$

and let $M(x)$ designate the slope of the secant line through P and Q. Find

$$\operatorname*{limit}_{x \to 3} M(x).$$

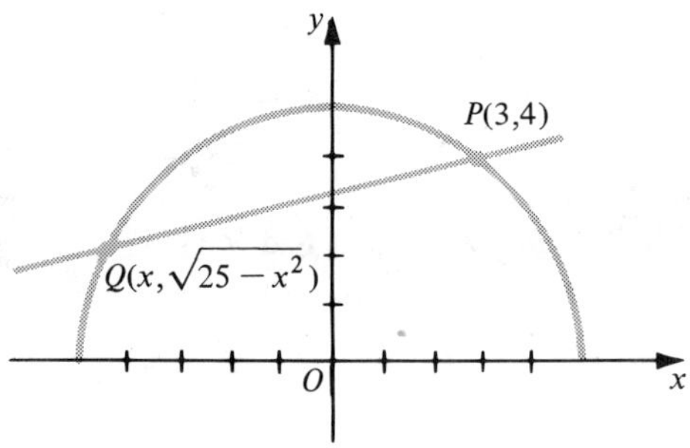

FIGURE 2.2

Solution: We cannot let $x = 3$ in $M(x)$, because the points P and Q coincide if $x = 3$ and the secant line is not well defined. Clearly,

$$M(x) = \frac{\sqrt{25 - x^2} - 4}{x - 3}, \qquad x \neq 3.$$

We can rationalize the numerator of $M(x)$ as follows:

$$M(x) = \frac{(\sqrt{25 - x^2} - 4)(\sqrt{25 - x^2} + 4)}{(x - 3)(\sqrt{25 - x^2} + 4)}$$

$$= \frac{(25 - x^2) - 16}{(x - 3)(\sqrt{25 - x^2} + 4)} = \frac{9 - x^2}{(x - 3)(\sqrt{25 - x^2} + 4)}$$

$$= -\frac{x + 3}{\sqrt{25 - x^2} + 4}, \qquad x \neq 3.$$

If x is close to 3, $\sqrt{25 - x^2}$ is close to 4 and $M(x)$ is close to $-\frac{6}{8}$; that is,

$$\operatorname*{limit}_{x \to 3} M(x) = -\tfrac{3}{4}.$$

The slope of the radius OP is $\frac{4}{3}$, and therefore the slope of the tangent line to the circle at P is $-\frac{3}{4}$, which is the same as the limit of $M(x)$ as x approaches 3. Thus the secant line of the circle approaches the position of the tangent line as Q approaches P.

Example 4 suggests the possibility of defining the tangent line to a curve other than a circle as the limiting position of the secant line. Tangent lines are considered from this standpoint in Chapter 3.

The meaning of the word *close*, which is used so frequently above, was purposely left to the interpretation of the reader. Actually, it is the uncertainty of the meaning of this word that impels us to give a more mathematical definition of limit in the next section.

EXERCISES

I

In each of Exercises 1 to 12, find the limit.

1. $\operatorname*{limit}_{x \to 3} (x^3 - 5x^2 + 2x - 1)$

2. $\operatorname*{limit}_{x \to 2} \dfrac{x^2 - 4}{x - 2}$

3. $\operatorname*{limit}_{x \to -3} \dfrac{x^2 - 9}{x + 3}$

4. $\operatorname*{limit}_{x \to 3} \dfrac{4/x - \frac{4}{3}}{x - 3}$

5. $\operatorname*{limit}_{x \to -2} \dfrac{1/(x + 1) + 1}{x + 2}$

6. $\operatorname*{limit}_{x \to 3} \dfrac{1/(x + 2) - \frac{1}{5}}{x - 3}$

7. $\operatorname*{limit}_{z \to 1} \dfrac{z - 1}{z^2 - 1}$

8. $\operatorname*{limit}_{h \to 0} \dfrac{(a + h)^2 - a^2}{h}$

9. $\operatorname*{limit}_{x \to 2} \dfrac{x^3 - 8}{x - 2}$

10. $\operatorname*{limit}_{x \to 4} \dfrac{x/(x - 3) - 4}{x - 4}$

11. $\operatorname*{limit}_{z \to 0} \dfrac{z^2}{z}$

12. $\operatorname*{limit}_{t \to -4} \dfrac{t + 4}{t^2 + 3t - 4}$

13. If $f(x) = x^2 - 2x + 3$, find

a. $\operatorname*{limit}_{x \to 1} \dfrac{f(x) - f(1)}{x - 1}$

b. $\operatorname*{limit}_{h \to 0} \dfrac{f(1 + h) - f(1)}{h}$

14. If $g(x) = \sqrt{25 - x^2}$, find

$$\lim_{x \to 4} \frac{g(x) - g(4)}{x - 4},$$

and give a geometrical interpretation of this limit.

15. If $G(x) = -\sqrt{16 - x^2}$, find

$$\lim_{x \to 1} \frac{G(x) - G(1)}{x - 1},$$

and give a geometrical interpretation of this limit.

16. If $f(x) = x^3 - 8$, find

$$\lim_{h \to 0} \frac{f(x + h) - f(x)}{h}.$$

For Exercises 17 to 24, follow Example 2 of this section.

17. $\lim_{x \to 4} \frac{\sqrt{x} - 2}{x - 4}$

18. $\lim_{x \to 1} \frac{\sqrt{x + 3} - 2}{x - 1}$

19. $\lim_{h \to 0} \frac{\sqrt{2 + h} - \sqrt{2}}{h}$

20. $\lim_{h \to 0} \frac{\sqrt{a + h} - \sqrt{a}}{h}, a > 0$

21. $\lim_{h \to 0} \frac{\sqrt{7 + h} - \sqrt{7}}{h}$

22. $\lim_{x \to -2} \frac{\sqrt{4 - x} - \sqrt{6}}{x + 2}$

23. $\lim_{z \to 25} \frac{\sqrt{z} - 5}{z - 25}$

24. $\lim_{w \to 0} \frac{3 - \sqrt{9 - w}}{w}$

II

In each of Exercises 1 to 6, find the limit if it exists.

1. $\lim_{x \to 1} \frac{\sqrt{x} - 2}{x - 1}$

2. $\lim_{y \to 2} \frac{\frac{1}{4} - 1/\sqrt{y}}{y - 2}$

3. $\lim_{m \to 0} \frac{\sqrt{4 + m + m^2} - 2}{m}$

4. $\lim_{r \to 1} \frac{1 - r^2}{2 - \sqrt{r^2 + 3}}$

5. $\lim_{t \to 1} \frac{t^3 - 1}{t - 1}$

6. $\lim_{t \to 2} \frac{t^4 - 16}{t - 2}$

2 DEFINITION OF LIMIT

We shall now give a precise definition of the limit concept discussed intuitively in Section 1.

2.1 Definition

The limit of a function f at a equals b, and we write

$$\lim_{x \to a} f(x) = b,$$

if for every neighborhood N of b there exists a deleted neighborhood D of a contained in domain f such that:

For every x in D, $f(x)$ is in N.

Note:

1. a need not be in domain f.
2. Some deleted neighborhood of a is contained in domain f.

We shall say that

$$\lim_{x \to a} f(x) \textit{ exists}$$

if there exists a number b such that

$$\lim_{x \to a} f(x) = b.$$

If a limit exists, it is easily shown to be unique.

According to 2.1,

$$\lim_{x \to a} f(x) = b$$

if for every neighborhood

$$N = (b_1,b_2) \text{ of } b, \textit{ no matter how small,}$$

there exists a deleted neighborhood

$$D = (a_1,a) \cup (a,a_2) \text{ of } a$$

such that for every number x:

$$\text{If } a_1 < x < a_2,\ x \neq a, \text{ then } b_1 < f(x) < b_2.$$

Note:

1. There is no unique choice for D. Once we have found one D, any other deleted neighborhood of a contained in D will work equally well.
2. We can always choose D as a *symmetric* neighborhood of a. For if $D = (a_1,a) \cup (a,a_2)$ as above and we choose a positive number δ such that $\delta \leq a - a_1$ and $\delta \leq a_2 - a$, then $(a - \delta, a) \cup (a, a + \delta)$ is symmetric and contained in D (Figure 2.3).

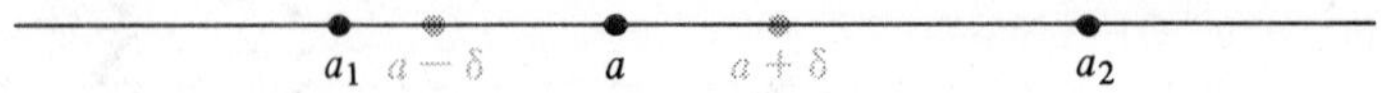

FIGURE 2.3

3. If for a given N you have found D above, then for any neighborhood N' of b containing N, the same neighborhood D of a will work.

From (3), you then can easily prove the following:

4. To show that

$$\operatorname*{limit}_{x \to a} f(x) = b,$$

you need only show that for every *symmetric neighborhood* N of b there exists a deleted neighborhood D of a such that the above conclusion holds.

The meaning of the limit of f at a can be explained geometrically by looking at the graph of f in the vicinity of the point (a,b). Suppose the graph of f is as shown in Figure 2.4. The hollow dot at (a,b) indicates that this point

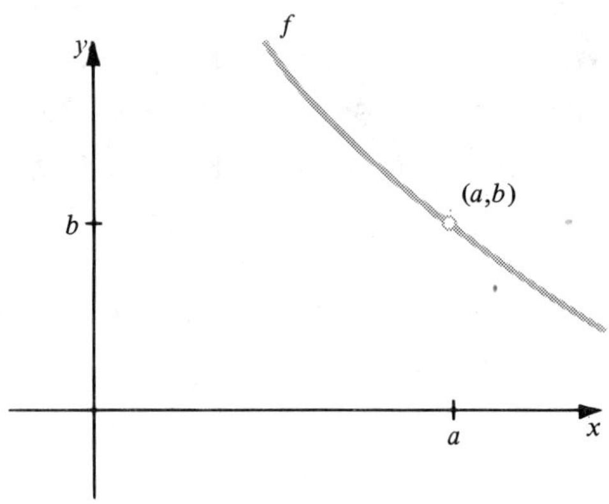

FIGURE 2.4

need not be on the graph of f. Then

$$\operatorname*{limit}_{x \to a} f(x) = b$$

if for any lines $y = b_1$ and $y = b_2$ below and above the line $y = b$, there exist lines $x = a_1$ and $x = a_2$ to the left and right of line $x = a$ such that (Figure 2.5):

The graph of f between $x = a_1$ and $x = a_2$ lies between the line $y = b_1$ and $y = b_2$.

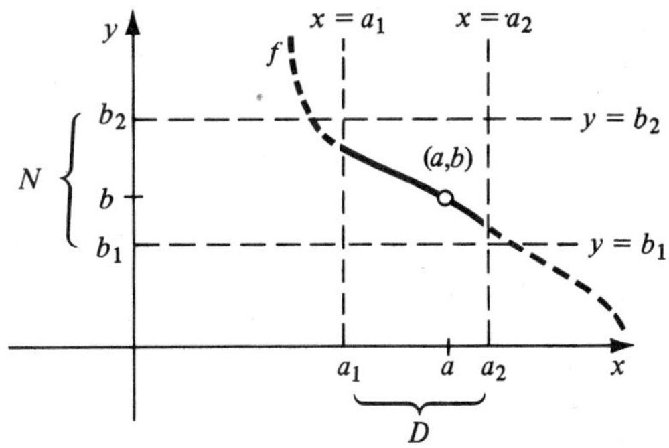

FIGURE 2.5

Example 1 Prove that $\lim\limits_{x \to 5} 4x - 3 = 17$.

Solution: Each symmetric neighborhood of 17 has the form $N = (17 - \varepsilon, 17 + \varepsilon)$ for some positive number ε (epsilon). For every such neighborhood $(17 - \varepsilon, 17 + \varepsilon)$, we must find a deleted neighborhood $D = (a_1,5) \cup (5,a_2)$ of 5 such that for every x in D, $f(x) = 4x - 3$ is in N, that is,

(1) $$17 - \varepsilon < 4x - 3 < 17 + \varepsilon.$$

Adding 3 to each member of (1), we get an equivalent inequality

(2) $$20 - \varepsilon < 4x < 20 + \varepsilon.$$

Now multiplying each member of (2) by $\frac{1}{4}$, we get another equivalent inequality

(3) $$5 - \frac{\varepsilon}{4} < x < 5 + \frac{\varepsilon}{4}.$$

Thus, every number x which is a solution of (3) is also a solution of (2) and of (1). Hence, for every x:

$$\text{If } 5 - \frac{\varepsilon}{4} < x < 5 + \frac{\varepsilon}{4}, \text{ then } 17 - \varepsilon < 4x - 3 < 17 + \varepsilon.$$

Taking

$$D = \left(5 - \frac{\varepsilon}{4}, 5\right) \cup \left(5, 5 + \frac{\varepsilon}{4}\right),$$

we have proved that for every x in D, $f(x)$ is in N. This proves

$$\lim_{x \to 5} 4x - 3 = 17.$$

We notice that the limit, 17, is simply the value of $4x - 3$ when $x = 5$.

In a similar manner, we can prove for any linear function $f(x) = mx + b$,

2.2 Theorem

$$\lim_{x \to a} (mx + b) = ma + b.$$

Special cases of 2.2 are:

2.3 Theorem

$$\lim_{x \to a} b = b, \qquad \textit{case } m = 0.$$

2.4 Theorem

$$\lim_{x \to a} x = a, \qquad \textit{case } m = 1, b = 0.$$

In finding $\lim_{x \to a} f(x)$, we need only look at values of f in some deleted neighborhood of a. To be more precise, the following result holds.

2.5 Theorem

If f and g are functions and a is a number such that $f(x) = g(x)$ at every number x in some deleted neighborhood K of a, and if $\lim_{x \to a} f(x)$ exists, then so does $\lim_{x \to a} g(x)$ and

$$\lim_{x \to a} f(x) = \lim_{x \to a} g(x).$$

Proof: If $\lim_{x \to a} f(x) = b$, then for every neighborhood N of b there exists a deleted neighborhood D of a such that $f(x)$ is in N for every x in D. Hence $D \cap K$ is also a deleted neighborhood of a such that $f(x)$ is in N for every x in $D \cap K$. Since $f(x) = g(x)$ for every x in $D \cap K$, $g(x)$ is in N for every x in $D \cap K$. Therefore $\lim_{x \to a} g(x) = b$ also.

Example 2 Prove that $\lim_{x \to 3} x^2 = 9$.

Solution: The graph of the function

$$f(x) = x^2$$

is a parabola, sketched in Figure 2.6. The graph rises as we move to the right. For every neighborhood $N = (b_1,b_2)$ of 9, with $b_1 > 0$, we can find a deleted neighborhood $D = (a_1,3) \cup (3,a_2)$ of 3 as shown in Figure 2.6. That is, we select a_1 and a_2 so that the points (a_1,b_1) and (a_2,b_2) are on the graph of f. Thus, $b_1 = a_1^2$ and $b_2 = a_2^2$, that is,

$$a_1 = \sqrt{b_1}, \qquad a_2 = \sqrt{b_2}.$$

Clearly, for every number x:

$$\text{If } a_1 < x < a_2,\ x \neq 3, \text{ then } b_1 < x^2 < b_2.$$

This proves

$$\lim_{x \to 3} x^2 = 9.$$

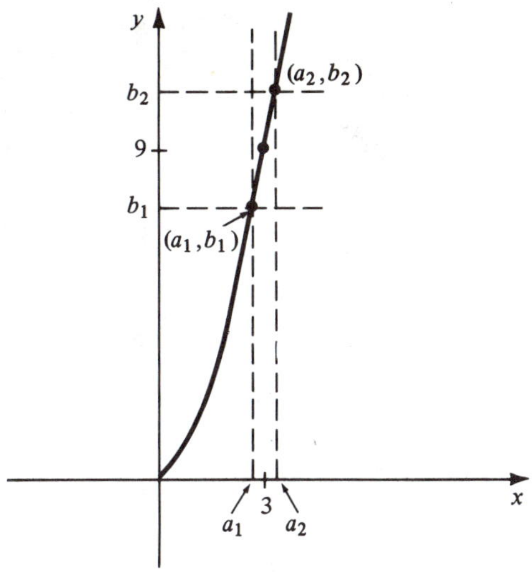

FIGURE 2.6

Similarly, we can prove that for every number a,

$$\operatorname*{limit}_{x \to a} x^2 = a^2.$$

Example 3 Prove that $\operatorname*{limit}_{x \to 4} \sqrt{x} = 2$.

Solution: The graph of the function

$$f(x) = \sqrt{x}, \qquad \text{domain } f = [0,\infty),$$

is half a parabola, as sketched in Figure 2.7. For every neighborhood $N = (b_1,b_2)$ of 2, we can find a deleted neighborhood $D = (a_1,4) \cup (4,a_2)$ of 4 as shown in Figure

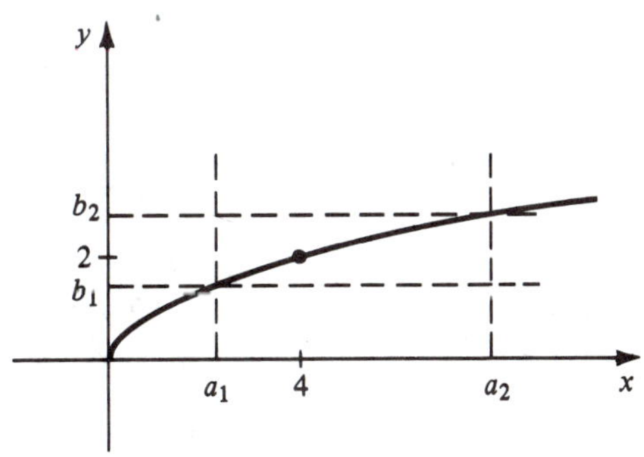

FIGURE 2.7

2.7. Thus, $b_1 = \sqrt{a_1}$ and $b_2 = \sqrt{a_2}$ or, solved for the a's,

$$a_1 = b_1^2, \qquad a_2 = b_2^2.$$

Clearly, for every number x:

$$\text{If } a_1 < x < a_2,\ x \neq 4, \text{ then } b_1 < \sqrt{x} < b_2.$$

This proves

$$\underset{x \to 4}{\text{limit}} \sqrt{x} = 2.$$

Similarly, we can prove that for every number $a > 0$,

$$\underset{x \to a}{\text{limit}} \sqrt{x} = \sqrt{a}.$$

Example 4 Prove that $\underset{x \to 5}{\text{limit}} \dfrac{1}{x} = \dfrac{1}{5}$.

Solution: For every positive number $\varepsilon < \frac{1}{5}$, the inequality

$$\tfrac{1}{5} - \varepsilon < \frac{1}{x} < \tfrac{1}{5} + \varepsilon$$

is equivalent to

$$\frac{1 - 5\varepsilon}{5} < \frac{1}{x} < \frac{1 + 5\varepsilon}{5}$$

or, inverting each member,

$$\frac{5}{1 - 5\varepsilon} > x > \frac{5}{1 + 5\varepsilon}.$$

Thus, for every x in

$$D = \left(\frac{5}{1 + 5\varepsilon}, 5\right) \cup \left(5, \frac{5}{1 - 5\varepsilon}\right)$$

$f(x)$ is in

$$N = (\tfrac{1}{5} - \varepsilon, \tfrac{1}{5} + \varepsilon).$$

This proves

$$\underset{x \to 5}{\text{limit}} \frac{1}{x} = \frac{1}{5}.$$

Similarly, we can prove that for every nonzero number a,

$$\underset{x \to a}{\text{limit}} \frac{1}{x} = \frac{1}{a}.$$

2.6 Theorem

If $\underset{x \to a}{\text{limit}} f(x) = b$ *and* $b > 0$, *then there exists a deleted neighborhood* D *of* a *such that* $f(x) > 0$ *for every* x *in* D. *Similarly, if* $b < 0$ *there exists a* D *such that* $f(x) < 0$ *for every* x *in* D.

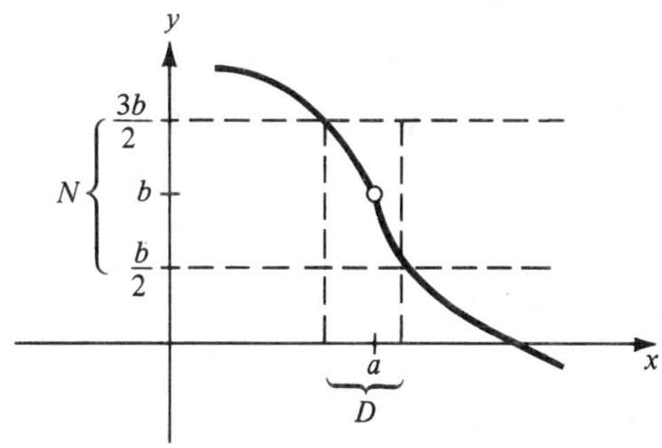

FIGURE 2.8

Proof: The symmetric neighborhood $N = (b/2, 3b/2)$ of b contains only positive numbers if $b > 0$. Since $\lim_{x \to a} f(x) = b$, there exists a deleted neighborhood D of a such that $f(x)$ is in N whenever x is in D. Therefore, $f(x) > 0$ for every x in D. See Figure 2.8.

An example in which the limit does not exist is given below.

Example 5 Let the function f be defined as follows:

$$f\left(\frac{1}{n}\right) = 1 \qquad \text{for every nonzero integer } n,$$

$$f(x) = x \qquad \text{for every other real number } x.$$

Prove that $\lim_{x \to 0} f(x)$ does not exist.

Solution: The graph of f is the line $y = x$, with the exception that when $x = 1/n$, n a nonzero integer, $y = 1$. It is sketched in Figure 2.9. Since $f(0) = 0$ and $f(x)$

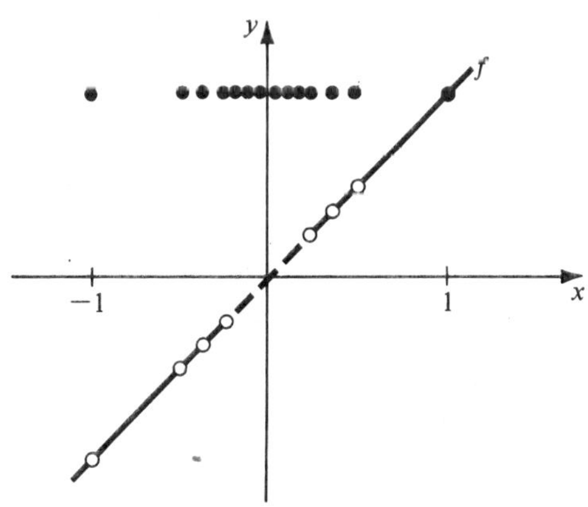

FIGURE 2.9

is close to 0 when x is close to 0, $x \neq 1/n$, we might suspect that

$$\lim_{x\to 0} f(x) = 0.$$

However, this is not correct. For if we take the neighborhood $N = (-\frac{1}{2},\frac{1}{2})$ of 0, there is no deleted neighborhood $D = (-\delta,0) \cup (0,\delta)$ of 0 such that $f(x)$ is in N for every x in D. Thus, we can always find a positive integer n such that $1/n < \delta$; and for this number $f(1/n)$ is not in N since $f(1/n) = 1$. In a similar way, we can show that $\lim_{x\to 0} f(x)$ is not equal to any other number b. Therefore, this limit does not exist.

EXERCISES

I

Using only the definition of a limit, prove that each of the following statements is correct.

1. $\lim_{x\to 2} (2x + 1) = 5$
2. $\lim_{x\to -1} (3x - 4) = -7$
3. $\lim_{x\to 0} (7x + 3) = 3$
4. $\lim_{x\to -1} (4x - 1) = -5$
5. $\lim_{x\to 2} (x^2 + 1) = 5$
6. $\lim_{x\to 0} x^2 = 0$
7. $\lim_{x\to 3} \sqrt{x + 1} = 2$
8. $\lim_{x\to -1} \dfrac{1}{x} = -1$
9. $\lim_{x\to 3} \dfrac{3}{x} = 1$
10. $\lim_{x\to -3} (x^2 + x) = 6$
11. $\lim_{x\to 4} \dfrac{8}{x} = 2$
12. $\lim_{x\to 3} (x^2 - 4) = 5$
13. $\lim_{x\to -1} \left(\dfrac{3}{x} + 5\right) = 2$
14. $\lim_{x\to 5} \sqrt{4 + x} = 3$
15. $\lim_{w\to -4} (w^2 - w) = 20$
16. $\lim_{x\to -5} \dfrac{1}{x + 4} = -1$
17. $\lim_{x\to 5} \dfrac{2x}{x + 4} = \dfrac{10}{9}$
18. $\lim_{x\to 27} \sqrt[3]{x} = 3$

In Exercises 19 to 24, prove that the limits do not exist.

19. $\lim_{x\to -1} \dfrac{x + 2}{x + 1}$
20. $\lim_{x\to -2} \dfrac{|x + 2|}{x + 2}$

21. $\lim\limits_{y \to 0} \dfrac{|y| - y}{y}$

22. $\lim\limits_{y \to 1} \dfrac{|y - 1| - y + 1}{|y - 1| + y - 1}$

23. $\lim\limits_{x \to 0} \dfrac{|x|}{x}$

24. $\lim\limits_{x \to 3} \dfrac{2}{x - 3}$

II

In each of Exercises 1 to 8, prove that the limit exists.

1. $\lim\limits_{x \to 0} \dfrac{x^2 + 4}{x - 2}$

2. $\lim\limits_{x \to 2} \dfrac{x^3 + 1}{x - 1}$

3. $\lim\limits_{x \to a} \sqrt[3]{x}$

4. $\lim\limits_{x \to a} x^3$

5. $\lim\limits_{x \to 1} \dfrac{1}{x^2}$

6. $\lim\limits_{x \to 9} \dfrac{1}{\sqrt{x}}$

7. $\lim\limits_{x \to -1} \dfrac{(x + 1)^2}{x^3 + 1}$

8. $\lim\limits_{x \to 0} \dfrac{|x|^3 + x^3}{x}$

In Exercises 9 to 10, prove that the limits do not exist.

9. $\lim\limits_{x \to 5} x - [x]$ ($[x]$ denotes the greatest integer $\leq x$).

10. $\lim\limits_{x \to 0} \dfrac{|x| - x}{|x|^3 - x^3}$.

11. Prove that if $\lim\limits_{x \to a} f(x)$ exists then it is unique.

12. Prove that if $\lim\limits_{x \to a} f(x) = b$ then $\lim\limits_{x \to a} cf(x) = cb$ for every positive constant c.

3 CONTINUITY

We saw in Examples 1 to 4 of Section 2 that not only did $\lim\limits_{x \to a} f(x)$ exist in each case but also $\lim\limits_{x \to a} f(x) = f(a)$. This situation arises often enough to be given a special name, as stated below.

2.7 Definition

The function f is *continuous* at the number a if (1) a is in the domain of f, (2) $\lim\limits_{x \to a} f(x)$ exists, and (3)

$$\lim_{x \to a} f(x) = f(a).$$

Going back to the definition of limit, we find that f is continuous at a if a is in the domain of f and if for every neighborhood N of $f(a)$ there exists a neighborhood D of a such that $f(x)$ is in N for every x in D. Note that we need no longer delete a from D.

The function f might be continuous at every number in some set A, in which case we say f is *continuous in A*. For example, the function f defined by

$$f(x) = \frac{1}{x}$$

is continuous in the set of all nonzero numbers. The function f is not continuous at 0, since $f(0)$ is undefined.

An example of a function g for which $\lim_{x \to 1} g(x)$ exists and $g(1)$ exists but g is not continuous at 1 is defined as follows:

$$g(x) = \begin{cases} x \text{ if } x \neq 1. \\ 3 \text{ if } x = 1. \end{cases}$$

Its graph is shown in Figure 2.10. Since $g(x) = x$ if $x \neq 1$,

$$\lim_{x \to 1} g(x) = \lim_{x \to 1} x = 1$$

by 2.5. However, $g(1) = 3$. Therefore, g is not continuous at 1.

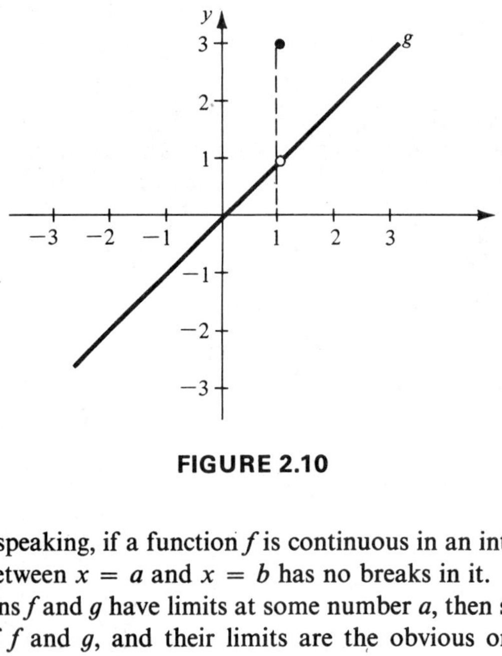

FIGURE 2.10

Roughly speaking, if a function f is continuous in an interval $[a,b]$, then the graph of f between $x = a$ and $x = b$ has no breaks in it.

If functions f and g have limits at some number a, then so do the various combinations of f and g, and their limits are the obvious ones listed below. Thus if

$$\lim_{x \to a} f(x) = b \quad \text{and} \quad \lim_{x \to a} g(x) = c,$$

then

2.8 $$\lim_{x \to a} (f + g)(x) = b + c, \qquad \lim_{x \to a} (f - g)(x) = b - c,$$

2.9 $$\lim_{x\to a} (fg)(x) = bc,$$

2.10 $$\lim_{x\to a} \left(\frac{f}{g}\right)(x) = \frac{b}{c}, \qquad \text{provided } c \neq 0.$$

We shall postpone the proofs of 2.8 to 2.10 until Section 6, but shall use them in the meantime whenever the occasion arises.

An immediate application of 2.8 to 2.10 is to polynomial functions and rational functions. In the first place, if $f(x) = x$, then $\lim_{x\to a} f(x) = a$ by 2.4. Hence, by 2.9,

$$\lim_{x\to a} x^2 = \lim_{x\to a} f^2(x) = a^2.$$

Assuming that

$$\lim_{x\to a} x^k = a^k$$

for some positive integer k, we have, by 2.9,

$$\lim_{x\to a} x^{k+1} = \lim_{x\to a} x^k \cdot x = a^k \cdot a = a^{k+1}.$$

Hence, by mathematical induction,

2.11 $$\lim_{x\to a} x^n = a^n \qquad \text{for every positive integer } n.$$

If f is any polynomial function so that

$$f(x) = c_0x^n + c_1x^{n-1} + \cdots + c_{n-1}x + c_n,$$

then repeated applications of 2.8 give

$$\lim_{x\to a} f(x) = \lim_{x\to a} c_0x^n + \lim_{x\to a} c_1x^{n-1} + \cdots + \lim_{x\to a} c_{n-1}x + \lim_{x\to a} c_n.$$

Since $\lim_{x\to a} c_0 = c_0$ by 2.3, evidently $\lim_{x\to a} c_0x^n = c_0a^n$ by 2.9 and 2.11. A similar argument for each term in the expression above yields

$$\lim_{x\to a} f(x) = c_0a^n + c_1a^{n-1} + \cdots + c_{n-1}a + c_n.$$

Since the right side of the equation above is simply $f(a)$, this proves that a *polynomial function is continuous at every number*.

A rational function is a quotient of two polynomial functions. If f and g are rational functions and a is a number such that $g(a) \neq 0$, then, by 2.10,

$$\lim_{x\to a} \frac{f}{g}(x) = \frac{f(a)}{g(a)}.$$

Since $g(a) \neq 0$ for every a in the domain of f/g, we have proved the following theorem.

2.12 Theorem

A rational function is continuous in its domain.

Since every polynomial function is also a rational function, this theorem also contains the statement above that a polynomial function is continuous.

Example 1 Find $\lim_{x \to 3} \left(x^2 - 3x + \frac{2}{x}\right)$.

Solution: Since

$$f(x) = x^2 - 3x + \frac{2}{x} = \frac{x^3 - 3x^2 + 2}{x},$$

f is a rational function. Evidently, 3 is in the domain of f [that is, $f(3)$ makes sense!]. Hence,

$$\lim_{x \to 3} f(x) = f(3) = 3^2 - 3 \cdot 3 + \tfrac{2}{3} = \tfrac{2}{3}.$$

Example 2 Find $\lim_{x \to 2} \frac{x^3 - 8}{x - 2}$.

Solution: We cannot apply 2.12 directly to the rational function F defined by

$$F(x) = \frac{x^3 - 8}{x - 2},$$

since 2 is not in the domain of F. Note, however, that

$$\frac{x^3 - 8}{x - 2} = x^2 + 2x + 4, \qquad \text{if } x \neq 2.$$

Therefore, if f is the polynomial function

$$f(x) = x^2 + 2x + 4,$$

we have

$$F(x) = f(x) \qquad \text{for every } x \neq 2$$

Hence, by 2.5 and 2.12,

$$\lim_{x \to 2} \frac{x^3 - 8}{x - 2} = \lim_{x \to 2} (x^2 + 2x + 4) = 2^2 + 2 \cdot 2 + 4 = 12.$$

EXERCISES

I

Prove that each limit exists by citing appropriate limit theorems and find the limit.

1. $\lim_{x \to 3} (x^2 - 7x + 5)$
2. $\lim_{t \to 1} (t^3 - 3t^2 + 2t + 6)$

3. $\lim_{y \to 5} (y^2 - 5y + 3)$

4. $\lim_{x \to 10} (x^5 - 10x^4 + x^2 - x)$

5. $\lim_{x \to 3} \dfrac{x^2 - 7}{x^2 + 7}$

6. $\lim_{y \to -2} \dfrac{y^3 + 6}{y^2 - y}$

7. $\lim_{t \to -4} \dfrac{t + 1}{t - 2}$

8. $\lim_{x \to 0} \dfrac{x^3 - 3x}{x^2 + 4}$

9. $\lim_{x \to 6} \left(x^2 - \dfrac{3}{x^2}\right)$

10. $\lim_{y \to x} \dfrac{y - x}{y + x}$

11. $\lim_{t \to a} \dfrac{t^2 - at + a^2}{t^2 + 1}$

12. $\lim_{x \to a} (2x + a)^3$

13. $\lim_{x \to 3} \dfrac{x^2 - 9}{x - 3}$

14. $\lim_{x \to -5} \dfrac{x^3 + 125}{x + 5}$

15. $\lim_{t \to 1} \dfrac{t^2 + 3t - 4}{t^2 + 5t + 4}$

16. $\lim_{y \to -2} \dfrac{y^4 - 16}{y + 2}$

17. Given $f(x) = \begin{cases} x^2 & \text{if } x \neq 2, \\ 5 & \text{if } x = 2, \end{cases}$ is f continuous at $x = 2$? Graph f.

18. Given $g(x) = \dfrac{3}{x - 2}$, is g continuous at $x = 3$? At $x = 2$?

19. Prove that the function $f(x) = \dfrac{x + 4}{2x + 6}$ is continuous at $x = 0$.

20. Prove that the function $G(t) = t^3 - 4/t^2$ is continuous at $t = -2$.

21. Prove that the function $F(x) = x^2 + 2$ is continuous at every number in its domain.

II

1. If $F(x) \leq G(x) \leq H(x)$ for every x in some deleted neighborhood of a, and if $\lim_{x \to a} F(x) = \lim_{x \to a} H(x) = L$, prove that $\lim_{x \to a} G(x) = L$. (This result is often called the *squeeze theorem.*)

2. Use the squeeze theorem to prove that

$$\lim_{x \to 0} x \sin x = 0.$$

(*Hint:* $|\sin x| \leq 1$ for every number x.)

3. Is the function

$$F(x) = \frac{|x|}{x}$$

continuous at the number 0? Prove your answer.

4. Let f and g be two functions with a common domain. Prove or disprove each statement below.

a. If the product function fg is continuous at $x = a$, then both f and g are continuous at $x = a$.
b. If both f and g are discontinuous at $x = a$, then the function $f + g$ is discontinuous at $x = a$.

4 ONE-SIDED LIMITS

The function f defined by

$$f(x) = \sqrt[n]{x}$$

has as its domain $[0,\infty)$ if the positive integer n is even, and the set $(-\infty,\infty)$ if n is odd. Therefore

$$\operatorname*{limit}_{x \to 0} f(x)$$

does not exist if n is even, because no deleted neighborhood of 0 is contained in the domain of f (see 2.1). However, for every positive number ε,

$$f(x) \text{ is in } (0,\varepsilon) \text{ for every } x \text{ in } (0,\varepsilon^n),$$

since $0 < \sqrt[n]{x} < \varepsilon$ if and only if $0 < x < \varepsilon^n$. Thus $f(x)$ is "close" to 0 when x is "close" to 0. This example illustrates part of the following definition.

2.13 Definition

The *right-hand limit* of a function f at a equals b, and we write

$$\operatorname*{limit}_{x \to a^+} f(x) = b$$

if for every neighborhood N of b there exists an open interval $(a, a + \delta)$ contained in the domain of f such that $f(x)$ is in N for every x in $(a, a + \delta)$.

The definition of the *left-hand limit*

$$\operatorname*{limit}_{x \to a^-} f(x) = b$$

is the same, except that the open interval $(a, a + \delta)$ is replaced by $(a - \delta, a)$.

By our remarks above,

$$\operatorname*{limit}_{x \to 0^+} \sqrt{x} = 0,$$

since for every neighborhood $(-\varepsilon,\varepsilon)$ of 0 there exists an open interval of the form $(0,\delta)$, namely $(0,\varepsilon^2)$, such that $\sqrt{x}$ is in N for every x in $(0,\varepsilon^2)$.

The limit theorems of Section 2 hold for one-sided limits as well as for ordinary limits, with the obvious changes when necessary. For example, 2.5 becomes: If $f(x) = g(x)$ at every x in some open interval (a,c) and if $\operatorname*{limit}_{x \to a^+} f(x)$ exists, then so does $\operatorname*{limit}_{x \to a^+} g(x)$ and

$$\operatorname*{limit}_{x \to a^+} f(x) = \operatorname*{limit}_{x \to a^+} g(x).$$

An illustration of the use of one-sided limits is afforded by the *greatest-integer function*. The greatest integer in a number x, designated by $[x]$, is defined to be the greatest integer $\leq x$; that is,

$$[x] = n \text{ if } n \leq x < n + 1, \qquad n \text{ an integer.}$$

For example,

$$[7.3] = 7, \qquad [4] = 4, \qquad [\pi] = 3, \qquad [-\tfrac{4}{3}] = -2.$$

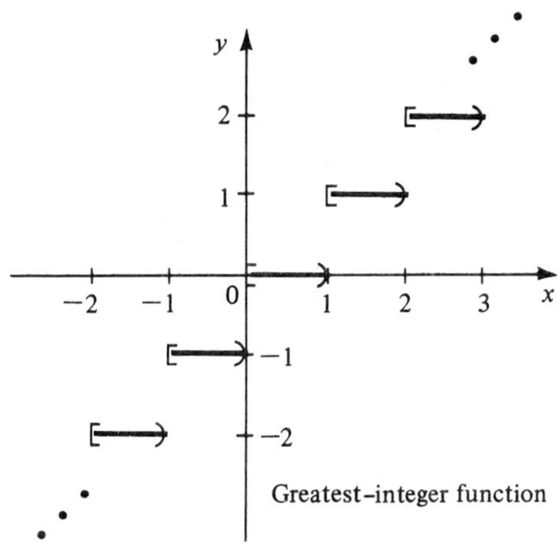

FIGURE 2.11

The graph of the greatest-integer function (Figure 2.11) is similar to that of the postage function (page 43). It consists of an infinite number of "steps," each step having a left-hand endpoint but not a right-hand endpoint.

Wc observe that

$$[x] = 2 \text{ for every } x \text{ in the open interval } (2,3).$$

Hence, by 2.5,

$$\operatorname*{limit}_{x \to 2^+} [x] = 2.$$

Similarly,

$$[x] = 1 \text{ for every } x \text{ in the open interval } (1,2).$$

Hence, by 2.5,

$$\operatorname*{limit}_{x \to 2^-} [x] = 1.$$

Thus, the one-sided limits at 2 are different. More generally,

$$\operatorname*{limit}_{x \to n^-} [x] = n - 1, \quad \operatorname*{limit}_{x \to n^+} [x] = n \qquad \text{for every integer } n.$$

One-sided limits are related to ordinary limits in the manner stated below.

2.14 Theorem

For a given function f, $\operatorname*{limit}_{x \to a} f(x) = b$ *if and only if*

$$\operatorname*{limit}_{x \to a^+} f(x) = \operatorname*{limit}_{x \to a^-} f(x) = b.$$

Proof: If $\operatorname*{limit}_{x \to a} f(x) = b$, then for every neighborhood N of b there exists a deleted neighborhood $D = (a - \delta, a) \cup (a, a + \delta)$ of a such that $f(x)$ is in N for each x in D. Hence $f(x)$ is in N for every x in $(a, a + \delta)$ and also for every x in $(a - \delta, a)$. This proves that $\operatorname*{limit}_{x \to a^+} f(x) = b$ and also $\operatorname*{limit}_{x \to a^-} f(x) = b$.

Conversely, if $\operatorname*{limit}_{x \to a^+} f(x) = b$ and $\operatorname*{limit}_{x \to a^-} f(x) = b$, then for every neighborhood N of b there exists an open interval $(a, a + \delta)$ such that $f(x)$ is in N for every x in $(a, a + \delta)$, and also an open interval $(a - \delta, a)$ such that $f(x)$ is in N for every x in $(a - \delta, a)$. Therefore, for every neighborhood N of b there exists a deleted neighborhood $D = (a - \delta, a) \cup (a, a + \delta)$ of a such that $f(x)$ is in N for every x in D. This proves that $\operatorname*{limit}_{x \to a} f(x) = b$.

As an application of this theorem, we see that $\operatorname*{limit}_{x \to n} [x]$ does not exist if n is an integer, since $\operatorname*{limit}_{x \to n^-} [x] = n - 1$ whereas $\operatorname*{limit}_{x \to n^+} [x] = n$. Thus, the greatest-integer function is discontinuous (i.e., not continuous) at each integer n. It is continuous at every other number.

We shall follow the usual practice of saying that a function f is continuous in a closed interval $[a,b]$ if it is continuous in the open interval (a,b) and

$$\operatorname*{limit}_{x \to a^+} f(x) = f(a), \qquad \operatorname*{limit}_{x \to b^-} f(x) = f(b).$$

For example, the function

$$g(x) = \sqrt{4 - x^2}, \qquad \text{domain } g = [-2,2],$$

is continuous in the closed interval $[-2,2]$. Its graph is a semicircle of radius 2 (Figure 1.3).

Similarly, the function

$$h(x) = \sqrt{x}, \qquad \text{domain } h = [0,\infty),$$

is continuous in the infinite interval $[0,\infty)$. Its graph is a parabola (Figure 2.7).

Example

Does $\lim_{x \to 3} \frac{|x-3|}{x-3}$ exist?

Solution: You recall that

$$|x-3| = \begin{cases} x-3 & \text{if } x-3 \geq 0, \quad \text{that is, } x \geq 3. \\ -(x-3) & \text{if } x-3 \leq 0, \quad \text{that is, } x \leq 3. \end{cases}$$

Thus,

$$\frac{|x-3|}{x-3} = \begin{cases} 1 & \text{if } x > 3 \\ -1 & \text{if } x < 3 \end{cases}$$

and

$$\lim_{x \to 3^+} \frac{|x-3|}{x-3} = \lim_{x \to 3^+} 1 = 1$$

$$\lim_{x \to 3^-} \frac{|x-3|}{x-3} = \lim_{x \to 3^-} -1 = -1.$$

Since

$$\lim_{x \to 3^+} \frac{|x-3|}{x-3} \neq \lim_{x \to 3^-} \frac{|x-3|}{x-3}$$

the limit

$$\lim_{x \to 3} \frac{|x-3|}{x-3}$$

does not exist according to 2.14.

EXERCISES

I

Find each of the following limits, if it exists.

1. $\lim_{x \to 1^+} (x^2 - 3x)$

2. $\lim_{x \to 4^-} \frac{x^2 - 16}{x - 4}$

3. $\lim_{x \to 0^+} \sqrt{3x}$

4. $\lim_{x \to 1^-} \sqrt{1 - x}$

5. $\lim_{x \to 0^-} \frac{x}{|x|}$

6. $\lim_{x \to 7^+} \frac{|x - 7|}{x - 7}$

7. $\lim_{x \to 3^-} 3 + [x]$

8. $\lim_{x \to 1^+} [x + 3]$

9. $\lim_{x \to -3^+} (1 + \sqrt{x + 3})$

10. $\lim_{x \to 2^+} \sqrt{(x - 2)(x + 3)}$

11. $\lim_{x \to 4^-} (\sqrt{4 - x} + [x - 1])$

12. $\lim_{x \to 1^+} \frac{2x|x - 1|}{x - 1}$

Find the indicated limits for each function below.

13. $f(x) = \begin{cases} 3 \text{ if } x < 2 \\ x^2 \text{ if } x \geq 2 \end{cases}$ a. $\lim_{x \to 2^+} f(x)$ b. $\lim_{x \to 2^-} f(x)$ c. $\lim_{x \to 2} f(x)$

14. $g(t) = \begin{cases} 4 + t^3 \text{ if } t \leq 1 \\ 5 \quad \text{if } 1 < 1 \end{cases}$ a. $\lim_{t \to 1^-} g(t)$ b. $\lim_{t \to 1^+} g(t)$ c. $\lim_{t \to 1} g(t)$

15. $h(x) = \begin{cases} 4x + 1 \text{ if } x < 2 \\ 6 - x \text{ if } x \geq 2 \end{cases}$ a. $\lim_{x \to 2^-} h(x)$ b. $\lim_{x \to 2^+} h(x)$ c. $\lim_{x \to 2} h(x)$

16. $g(x) = \begin{cases} 9 - x^2 \text{ if } x \leq -2 \\ 3x + 11 \text{ if } x > -2 \end{cases}$ a. $\lim_{x \to -2^-} g(x)$ b. $\lim_{x \to -2^+} g(x)$ c. $\lim_{x \to -2} g(x)$

17. Let $f(x) = \begin{cases} x^2 + 4 \text{ if } x \leq 2 \\ x + 2 \text{ if } x > 2 \end{cases}$ and $g(x) = \begin{cases} x^2 \text{ if } x \leq 2 \\ 8 \text{ if } x > 2 \end{cases}$

  a. Show that $\lim_{x \to 2} f(x)$ and $\lim_{x \to 2} g(x)$ do not exist.
  b. Show that $\lim_{x \to 2} f(x)g(x)$ does exist by using one-sided limits.

Discuss the continuity of each function at the given number.

18. $f(x) = \begin{cases} 1 + x \text{ if } x \leq -4 \\ x^2 \quad \text{if } x > -4 \end{cases}$ $x = -4.$

19. $g(t) = \begin{cases} 3 + t \text{ if } t \leq 2 \\ t^2 + 1 \text{ if } t > 2 \end{cases}$ $t = 2.$

20. $F(x) = \begin{cases} x^2 \text{ if } x \text{ is an integer} \\ 0 \text{ if } x \text{ is not an integer} \end{cases}$ $x = 3, x = \frac{1}{2}$

II

Find each limit (if it exists):

1. $\lim_{x \to 4^+} [[x]]$

2. $\lim_{x \to 2^+} \dfrac{x - [x]}{x - 2}$

3. $\lim_{x \to 2^-} \dfrac{x - [x]}{x - 2}$

4. $\lim_{x \to 1^+} \dfrac{[x^2] - [x]^2}{x^2 - 1}$

5 INFINITE LIMITS

It is convenient to say that the limit of a function f at some number a is infinite if $f(x)$ is "large" when x is "close" to a. A more precise definition is given below.

2.15 Definition

The right-hand limit of the function f at a is infinite, and we write

$$\underset{x \to a^+}{\text{limit}}\, f(x) = \infty$$

if for every positive number k, no matter how large, there exists an open interval $(a, a + \delta)$ such that

$$f(x) \text{ is in } (k,\infty) \text{ for every } x \text{ in } (a, a + \delta).$$

If for every positive number k there exists an open interval $(a - \delta, a)$ such that

$$f(x) \text{ is in } (k,\infty) \text{ for every } x \text{ in } (a - \delta, a),$$

then we write

$$\underset{x \to a^-}{\text{limit}}\, f(x) = \infty.$$

For example,

$$\underset{x \to 0^+}{\text{limit}}\, \frac{1}{x} = \infty,$$

since for every positive number k, $1/x > k$ if and only if $0 < x < 1/k$. Thus,

$$\frac{1}{x} \text{ is in } (k,\infty) \text{ for every } x \text{ in the interval } \left(0, \frac{1}{k}\right).$$

Geometrically, this means that the graph of f "goes to infinity" as x approaches 0 from the right (Figure 2.12).

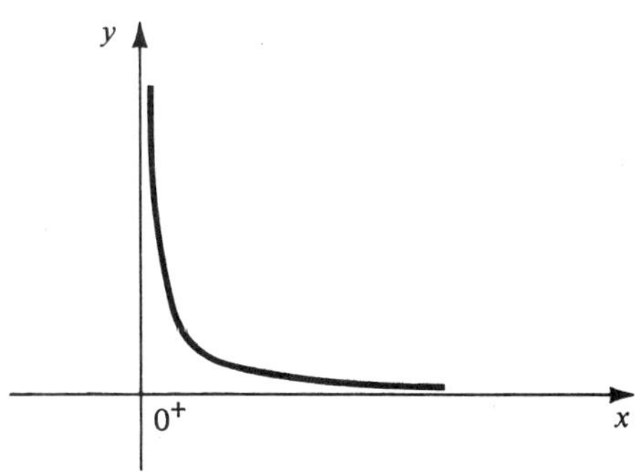

FIGURE 2.12

Negative infinite limits are defined in the obvious way. Thus,

$$\operatorname*{limit}_{x\to a^+} f(x) = -\infty$$

if for every negative number k there exists an open interval $(a, a + \delta)$, such that

$$f(x) \text{ is in } (-\infty,k) \text{ for every } x \text{ in } (a, a + \delta).$$

We leave the definition of

$$\operatorname*{limit}_{x\to a^-} f(x) = -\infty$$

to the reader.

For example,

$$\operatorname*{limit}_{x\to 0^-} \frac{1}{x} = -\infty,$$

since for every negative number k, $1/x < k$ if and only if $1/k < x < 0$. Thus,

$$\frac{1}{x} \text{ is in } (-\infty,k) \text{ for every } x \text{ in the interval } \left(\frac{1}{k}, 0\right).$$

Most of the infinite limits that we shall encounter are of the following type.

2.16
$$\operatorname*{limit}_{x\to a^+} \frac{1}{(x-a)^p} = \infty, \qquad \text{if } p > 0.$$

Proof: We must show that for every positive integer k, no matter how large, there exists an open interval $(a, a + \delta)$ such that

$$\frac{1}{(x-a)^p} \text{ is in } (k,\infty) \text{ for every } x \text{ in } (a, a + \delta).$$

If $x - a > 0$, then the inequality $1/(x - a)^p > k$ is equivalent to

$$0 < (x-a)^p < \frac{1}{k},$$

which in turn is equivalent to

$$0 < x - a < \left(\frac{1}{k}\right)^{1/p}.$$

The solution set of this last inequality is evidently $(a, a + \delta)$, where $\delta = (1/k)^{1/p}$. Hence $1/(x - a)^p$ is in (k,∞) if x is in the open interval $(a, a + \delta)$. This proves 2.16.

If $p = m/n$, where m is an even integer and n is odd, then $(x - a)^p$ is defined for $x < a$ as well as $x > a$. Then 2.16 can be shown to hold for x approaching a from the left:

$$\lim_{x \to a^-} \frac{1}{(x - a)^p} = \infty \qquad \text{if } p = \frac{m}{n} > 0,\ m \text{ even integer, } n \text{ odd.}$$

In a similar vein,

$$\lim_{x \to a^-} \frac{1}{(x - a)^p} = -\infty \qquad \text{if } p = \frac{m}{n} > 0,\ m, n \text{ odd integers.}$$

For example,

$$\lim_{x \to 4^+} \frac{1}{(x - 4)^2} = \lim_{x \to 4^-} \frac{1}{(x - 4)^2} = \infty,$$

$$\lim_{x \to -2^+} \frac{1}{(x + 2)^{1/3}} = \infty, \qquad \lim_{x \to -2^-} \frac{1}{(x + 2)^{1/3}} = -\infty.$$

The graph of the function f defined by

$$f(x) = \frac{1}{(x - a)^p}, \qquad p > 0,$$

is, according to 2.16, unbounded to the right of the line $x = a$. That is, for any line $y = k$ (no matter how large k is), there is some point on the graph of f above this line. The graph near $x = a$ is roughly as indicated in Figure 2.13.

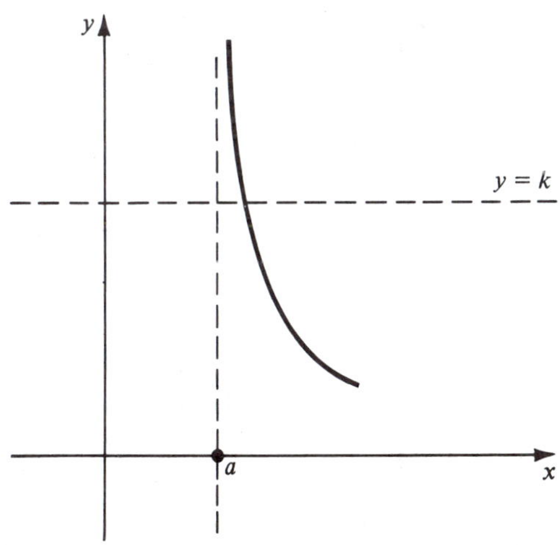

FIGURE 2.13

Evidently, the graph is "approaching" the line $x = a$ as it gets farther and farther from the x axis. A line such as $x = a$ is called an *asymptote* of the graph of f.

2.17 Definition

A line $x = a$ is called a *vertical asymptote* of the graph of a function f if one of the limits

$$\operatorname*{limit}_{x\to a^+} f(x), \qquad \operatorname*{limit}_{x\to a^-} f(x)$$

is either ∞ or $-\infty$.

Clearly f is discontinuous at a if $x = a$ is an asymptote.

Modified limit theorems hold for infinite limits. We state without proof two such theorems that will be useful to us in the sequel.

If $\operatorname*{limit}_{x\to a^+} f(x) = \infty$ or $-\infty$, and $\operatorname*{limit}_{x\to a^+} g(x) = b$, then

2.18 $$\operatorname*{limit}_{x\to a^+} (f + g)(x) = \operatorname*{limit}_{x\to a^+} f(x)$$

and

2.19 $$\operatorname*{limit}_{x\to a^+} (fg)(x) = \begin{cases} \operatorname*{limit}_{x\to a^+} f(x) \text{ if } b > 0. \\ -\operatorname*{limit}_{x\to a^+} f(x) \text{ if } b < 0. \end{cases}$$

Similar theorems hold for left-hand limits.

Example 1 Find $\operatorname*{limit}_{x\to 1^+} \dfrac{1}{x^2 - x}$ and $\operatorname*{limit}_{x\to 1^-} \dfrac{1}{x^2 - x}$.

Solution: We first express

$$\frac{1}{x^2 - x} = \frac{1}{x} \cdot \frac{1}{x - 1}.$$

Then, by 2.16 and 2.19, with $f(x) = 1/(x - 1)$ and $g(x) = 1/x$,

$$\operatorname*{limit}_{x\to 1^+} \frac{1}{x^2 - x} = \operatorname*{limit}_{x\to 1^+} \frac{1}{x - 1} = \infty,$$

$$\operatorname*{limit}_{x\to 1^-} \frac{1}{x^2 - x} = \operatorname*{limit}_{x\to 1^-} \frac{1}{x - 1} = -\infty.$$

For the function f defined by

$$f(x) = 3 - \frac{1}{x},$$

it is clear that $f(x)$ is "close" to 3 when x is "large." This type of a limiting value of a function is defined as follows.

2.20 Definition

If for every neighborhood N of b there exists a positive number k such that $f(x)$ is in N for every x in (k,∞), then we say

$$\operatorname*{limit}_{x\to\infty} f(x) = b.$$

Similarly, we say for the function f that

$$\operatorname*{limit}_{x\to-\infty} f(x) = b$$

if for every neighborhood N of b there exists a negative number k such that $f(x)$ is in N for every x in $(-\infty,k)$.

A useful instance of an infinite limit of type 2.20 is:

2.21
$$\operatorname*{limit}_{x\to\infty} \frac{1}{x^p} = 0, \qquad \text{if } p > 0.$$

Proof: Since we are interested only in the behavior of $1/x^p$ when x is large, we might as well let $x > 0$. Then for each neighborhood $(-\varepsilon,\varepsilon)$ of 0, we have $-\varepsilon < 1/x^p < \varepsilon$ if and only if

$$0 < \frac{1}{x^p} < \varepsilon.$$

This inequality is equivalent to

$$x^p > \frac{1}{\varepsilon},$$

or

$$x > \left(\frac{1}{\varepsilon}\right)^{1/p}$$

Hence, letting $k = (1/\varepsilon)^{1/p}$, $1/x^p$ is in $(-\varepsilon,\varepsilon)$ for every $x > k$.

The ordinary limit theorems hold for limits as x approaches ∞ or $-\infty$. Their proofs are but slight modifications of those for finite limits.

It is convenient to write

$$\operatorname*{limit}_{x\to\infty} f(x) = \infty$$

if $f(x)$ is large when x is large, or more precisely, if for every number k (no matter how large) there exists a number n such that $f(x) > k$ for every $x > n$. The many variations of this limit involving $-\infty$ are defined analogously. Clearly,

$$\operatorname*{limit}_{x\to\infty} x^p = \infty, \qquad \text{if } p > 0.$$

Example 2 Find $\lim_{x\to\infty} \dfrac{2 - x + 3x^2}{1 + x^2}$.

Solution: In order to use 2.21, we divide numerator and denominator of the given expression by x^2, and then proceed as indicated below:

$$\lim_{x\to\infty} \frac{2 - x + 3x^2}{1 + x^2} = \lim_{x\to\infty} \frac{2/x^2 - 1/x + 3}{1/x^2 + 1}$$

$$= \frac{2 \lim_{x\to\infty} 1/x^2 - \lim_{x\to\infty} 1/x + \lim_{x\to\infty} 3}{\lim_{x\to\infty} 1/x^2 + \lim_{x\to\infty} 1}$$

$$= \tfrac{3}{1} = 3.$$

The existence of a limit involving ∞ for a function f tells us something about the graph of f far away from the origin. If

$$\lim_{x\to\infty} f(x) = b,$$

then $f(x)$ is close to b when x is large, and the graph of f gets close to the line $y = b$ as x gets large. Such a line $y = b$ is an asymptote of the graph.

2.22 Definition

A line $y = b$ is called a *horizontal asymptote* of the graph of a function f if either

$$\lim_{x\to\infty} f(x) = b \qquad \text{or} \qquad \lim_{x\to-\infty} f(x) = b.$$

Example 3 Discuss and sketch the graph of the equation

$$y = \frac{4x^2}{x^2 + 1}.$$

Solution: The graph is symmetric to the y axis, since x occurs to even powers only.

Since

$$0 \le \frac{x^2}{x^2 + 1} < 1,$$

$0 \le y < 4$ for every number x. Thus the graph lies between the lines $y = 0$ and $y = 4$.

Since

$$\lim_{x\to\pm\infty} \frac{4x^2}{x^2 + 1} = \lim_{x\to\pm\infty} \frac{4}{1 + 1/x^2} = 4,$$

the line $y = 4$ is an asymptote of the graph of the given equation [i.e., of the graph of the function f defined by $f(x) = 4x^2/(x^2 + 1)$]. When a few points are plotted, the graph may be drawn as in Figure 2.14.

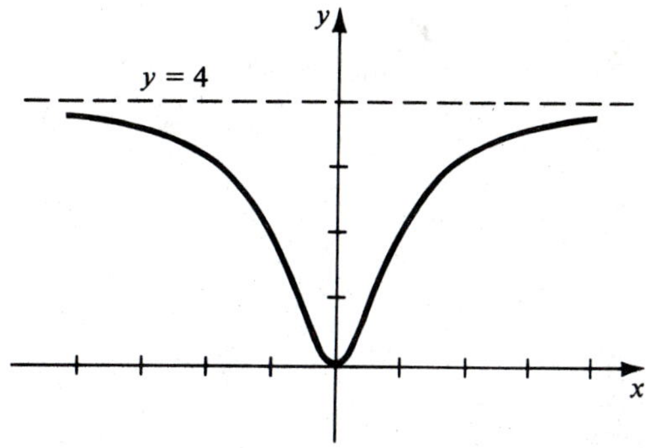

FIGURE 2.14

EXERCISES

I

Find each limit.

1. $\lim\limits_{x\to 3^+} \dfrac{2}{(x-3)^2}$

2. $\lim\limits_{x\to 3^-} \dfrac{2}{(x-3)^2}$

3. $\lim\limits_{x\to 2^-} \dfrac{1}{x^2-2x}$

4. $\lim\limits_{x\to 2^+} \dfrac{1}{x^2-2x}$

5. $\lim\limits_{x\to -3^+} \dfrac{1}{x^2+x-6}$

6. $\lim\limits_{x\to 0^-} \dfrac{2}{x^2+7x}$

7. $\lim\limits_{x\to \infty} \dfrac{1}{\sqrt{x}}$

8. $\lim\limits_{x\to -\infty} \dfrac{3}{\sqrt[3]{x}}$

9. $\lim\limits_{x\to \infty} \dfrac{7x^2-2x+5}{x^2+3}$

10. $\lim\limits_{x\to \infty} \dfrac{x^3+1}{3x^3-x}$

11. $\lim\limits_{x\to \infty} \dfrac{4x^4+3x^2-1}{x^3+x}$

12. $\lim\limits_{x\to \infty} \dfrac{-6x^3+x^2-1}{2x^3+x+5}$

13. $\lim\limits_{x\to \infty} (\sqrt{x^2+9}-x)$

14. $\lim\limits_{x\to 4^-} \dfrac{5}{(x-4)^3}$

15. $\lim\limits_{x\to 0^+} \dfrac{1}{x^{2/3}}$

16. $\lim\limits_{x\to 3^+} \dfrac{x^2+x+1}{x^2-x-6}$

Find the vertical and horizontal asymptotes of the graph of the given function F and give a rough sketch of the graph.

17. $F(x) = \dfrac{1}{x-2}$

18. $F(x) = \dfrac{1}{(x-2)^2}$

19. $F(x) = \dfrac{2x}{x+2}$

20. $F(x) = x + \dfrac{1}{x}$

21. $F(x) = \dfrac{2x^2}{(x+2)^2}$

22. $F(x) = \dfrac{2}{x^2 - 9}$

Using only the definitions of $\lim_{x\to a^+} f(x) = \infty$, $\lim_{x\to\infty} f(x) = b$, and so on, prove the following limits.

23. $\lim_{x\to 3^+} \dfrac{1}{x-3} = \infty$

24. $\lim_{x\to\infty} \dfrac{x+3}{x-2} = 1$

25. $\lim_{x\to 4^-} \dfrac{1}{x^2-16} = -\infty$

26. $\lim_{x\to 1^+} \dfrac{3x}{x-1} = \infty$

II

Sketch the graph of each of the following functions, indicating the vertical and horizontal asymptotes.

1. $F(x) = \dfrac{x^2-1}{x^2-4}$

2. $F(x) = \dfrac{x^2-1}{x^2+4}$

3. $F(x) = \dfrac{x^2+1}{x^2-4}$

4. $F(x) = \dfrac{x^2+1}{x^2+4}$

6 THE LIMIT THEOREMS

This section will be devoted to proofs of the tundamental limit theorems of the calculus.

First, let us prove 2.8, which we now restate.

2.8 Theorem

If $\lim_{x\to a} f(x) = b$ *and* $\lim_{x\to a} g(x) = c$, *then*

$$\lim_{x\to a} (f+g)(x) = b + c, \qquad \lim_{x\to a} (f-g)(x) = b - c.$$

Proof: Let us prove that $\lim_{x\to a} (f+g)(x) = b + c$, leaving the analogous proof for the function $f - g$ to the reader. We must show that for every neighborhood $(b + c - \varepsilon, b + c + \varepsilon)$ of $b + c$ there exists a deleted neighborhood D of a such that $f(x) + g(x)$ is in $(b + c - \varepsilon, b + c + \varepsilon)$ for every x in D. Since $\lim_{x\to a} f(x) = b$, there is corresponding to the neighborhood $(b - \varepsilon/2, b + \varepsilon/2)$ of b some deleted neighborhood D_1 of a such that $f(x)$ is in $(b - \varepsilon/2, b + \varepsilon/2)$ for every x in D_1, and since

$\lim_{x \to a} g(x) = c$, $g(x)$ is in $(c - \varepsilon/2, c + \varepsilon/2)$ for every x in some deleted neighborhood D_2 of a. Then $D = D_1 \cap D_2$ is a deleted neighborhood of a such that

$$b - \frac{\varepsilon}{2} < f(x) < b + \frac{\varepsilon}{2} \quad \text{and} \quad c - \frac{\varepsilon}{2} < g(x) < c + \frac{\varepsilon}{2}$$

for every x in D. Adding like members of these inequalities, we have

$$b + c - \varepsilon < f(x) + g(x) < b + c + \varepsilon$$

for every x in D. Thus we have shown that $f(x) + g(x)$ is in $(b + c - \varepsilon, b + c + \varepsilon)$ for every x in D.

By mathematical induction, 2.8 can be extended to a sum of any finite number of functions. We state such a result as follows:

2.23 Theorem

If $\lim_{x \to a} f_1(x) = b_1$, $\lim_{x \to a} f_2(x) = b_2, \ldots, \lim_{x \to a} f_n(x) = b_n$, *then*

$$\lim_{x \to a} (f_1 + f_2 + \cdots + f_n)(x) = b_1 + b_2 + \cdots + b_n.$$

Before proving the corresponding product limit theorem (2.9), we state and prove the special case when one function is a constant.

2.24 Theorem

If $\lim_{x \to a} f(x) = b$, *then* $\lim_{x \to a} kf(x) = kb$.

Proof: If $k = 0$, then the theorem follows from 2.3. So let us assume $k > 0$. The proof for $k < 0$ is similar and hence omitted. We must show that for every neighborhood $(kb - \varepsilon, kb + \varepsilon)$ of kb there exists a deleted neighborhood D of a such that $kf(x)$ is in $(kb - \varepsilon, kb + \varepsilon)$ for every x in D. Since $\lim_{x \to a} f(x) = b$, there is corresponding to the neighborhood $(b - \varepsilon/k, b + \varepsilon/k)$ of b some deleted neighborhood D of a such that $f(x)$ is in $(b - \varepsilon/k, b + \varepsilon/k)$ for every x in D. Thus

$$b - \frac{\varepsilon}{k} < f(x) < b + \frac{\varepsilon}{k} \qquad \text{for every } x \text{ in } D,$$

and on multiplying each member of the above inequality by k, we have

$$kb - \varepsilon < kf(x) < kb + \varepsilon \qquad \text{for every } x \text{ in } D.$$

Hence $kf(x)$ is in $(kb - \varepsilon, kb + \varepsilon)$ for every x in D.

Another special case of the product limit theorem is as follows:

2.25 Theorem

If $\lim_{x\to a} f(x) = 0$ *and* $\lim_{x\to a} g(x) = 0$, *then* $\lim_{x\to a} (fg)(x) = 0$.

Proof: We must show that for every neighborhood $(-\varepsilon,\varepsilon)$ of 0 there exists a deleted neighborhood D of a such that $f(x)g(x)$ is in $(-\varepsilon,\varepsilon)$ for every x in D. Since $\lim_{x\to a} f(x) = \lim_{x\to a} g(x) = 0$, there are corresponding to the neighborhood $(-\sqrt{\varepsilon}, \sqrt{\varepsilon})$ of 0 deleted neighborhoods D_1 and D_2 of a such that $f(x)$ is in $(-\sqrt{\varepsilon}, \sqrt{\varepsilon})$ for every x in D_1 and $g(x)$ is in $(-\sqrt{\varepsilon}, \sqrt{\varepsilon})$ for every x in D_2. Therefore, if we let $D = D_1 \cap D_2$,

$$-\sqrt{\varepsilon} < f(x) < \sqrt{\varepsilon} \text{ and } -\sqrt{\varepsilon} < g(x) < \sqrt{\varepsilon} \qquad \text{for every } x \text{ in } D.$$

Hence $|f(x)| < \sqrt{\varepsilon}$, $|g(x)| < \sqrt{\varepsilon}$, and $|f(x) \cdot g(x)| < \varepsilon$, or

$$-\varepsilon < f(x) \cdot g(x) < \varepsilon \qquad \text{for every } x \text{ in } D.$$

We can now prove 2.9 algebraically.

2.9 Theorem

If $\lim_{x\to a} f(x) = b$ *and* $\lim_{x\to a} g(x) = c$, *then* $\lim_{x\to a} (fg)(x) = bc$.

Proof: Since $\lim_{x\to a} [f(x) - b] = 0$ and $\lim_{x\to a} [g(x) - c] = 0$ by 2.8, we have, by 2.25,

$$\lim_{x\to a} [f(x) - b][g(x) - c] = 0,$$

or

$$\text{(1)} \qquad \lim_{x\to a} [f(x)g(x) - bg(x) - cf(x) + bc] = 0.$$

We know that

$$\lim_{x\to a} [bg(x) + cf(x) - bc]$$

exists, since the limit of each of the three functions $bg(x)$, $cf(x)$, bc exists. Hence,

$$\lim_{x\to a} \{[f(x)g(x) - bg(x) - cf(x) + bc] + [bg(x) + cf(x) - bc]\}$$

exists. But this sum is simply $f(x)g(x)$! Thus, $\lim_{x\to a} f(x)g(x)$ exists and from (1) above

$$\lim_{x\to a} f(x)g(x) - \lim_{x\to a} bg(x) - \lim_{x\to a} cf(x) + \lim_{x\to a} bc = 0.$$

Thus by 2.24,

$$\lim_{x \to a} f(x)g(x) - bc - cb + bc = 0,$$

and 2.9 follows.

The product limit theorem can be extended to any finite number of functions by mathematical induction.

2.26 Theorem

If $\lim_{x \to a} f_1(x) = b_1, \lim_{x \to a} f_2(x) = b_2, \ldots, \lim_{x \to a} f_n(x) = b_n$, *then*

$$\lim_{x \to a} (f_1 \cdot f_2 \cdot \cdots \cdot f_n)(x) = b_1 \cdot b_2 \cdot \cdots \cdot b_n.$$

A special case of 2.26 is that in which $f_1 = f_2 = \cdots = f_n = f$.

2.27 Theorem

If $\lim_{x \to a} f(x) = b$, *then* $\lim_{x \to a} f^n(x) = b^n$ *for every positive integer n.*

The composite of functions f and g has the following limit property.

2.28 Theorem

If $\lim_{x \to a} g(x) = b$ *and* f *is continuous at* b, *then* $\lim_{x \to a} f(g(x)) = f(b)$.

Proof: Since f is continuous at b, for every neighborhood N of $f(b)$ there exists a neighborhood C of b such that $f(y)$ is in N for every y in C. In turn, since $\lim_{x \to a} g(x) = b$, there exists a deleted neighborhood D of a such that $g(x)$ is in C for every x in D. Combining these remarks, we see that for every neighborhood N of $f(b)$ there exists a deleted neighborhood D of a such that $g(x)$ is in C, and therefore $f(g(x))$ is in N, for every x in D.

A function such as F, where

$$F(x) = \frac{1}{x^2 + 3},$$

is a composite of the polynomial function

$$g(x) = x^2 + 3$$

and the reciprocal function

$$f(x) = \frac{1}{x};$$

that is,

$$f(g(x)) = \frac{1}{g(x)} = \frac{1}{x^2 + 3}.$$

Thus $F = f \circ g$. Therefore, since $\lim_{x\to a} g(x) = g(a)$ and f is continuous at $g(a)$ (by Example 4, page 63),

$$\lim_{x\to a} F(x) = \frac{1}{a^2 + 3},$$

by 2.28.

In the same way, we can prove that if $\lim_{x\to a} g(x) = c \neq 0$ then

$$\lim_{x\to a} \frac{1}{g(x)} = \frac{1}{c}.$$

If, also,

$$\lim_{x\to a} f(x) = b,$$

then

$$\begin{aligned} \lim_{x\to a} \frac{f(x)}{g(x)} &= \lim_{x\to a} f(x) \cdot \frac{1}{g(x)} \\ &= \lim_{x\to a} f(x) \cdot \lim_{x\to a} \frac{1}{g(x)} \qquad \textit{by 2.9} \\ &= b \cdot \frac{1}{c}, \quad \text{or } \frac{b}{c}. \end{aligned}$$

This proves 2.10, which we now restate.

2.10 Theorem

If $\lim_{x\to a} f(x) = b$ *and* $\lim_{x\to a} g(x) = c$, *then*

$$\lim_{x\to a} \frac{f(x)}{g(x)} = \frac{b}{c}, \qquad \textit{provided } c \neq 0.$$

Such limits as

$$\lim_{x\to 2} \sqrt{x^2 + x + 1}$$

occur quite often. We easily compute them by using 2.28. Thus if $f(x) = \sqrt{x}$ and $g(x) = x^2 + x + 1$,

$$f(g(x)) = \sqrt{x^2 + x + 1}.$$

Hence by 2.28 and Example 3, page 62,

$$\lim_{x\to 2} \sqrt{x^2 + x + 1} = \sqrt{2^2 + 2 + 1} = \sqrt{7}.$$

By the same argument, we may prove the following result.

2.29 Theorem

If $\lim_{x\to a} f(x) = b$, $b \neq 0$, *then* $\lim_{x\to a} \sqrt[n]{f(x)} = \sqrt[n]{b}$, *provided, of course,* $\sqrt[n]{b}$ *exists.*

EXERCISES

I

By referring to the appropriate limit theorems, prove that each of the following limits exists.

1. $\lim\limits_{x\to a} (x^5 + 3x^2 + 2x + 1)$.

2. $\lim\limits_{x\to 2} \dfrac{x+2}{\sqrt{x^3 - 4x + 1}}$.

3. $\lim\limits_{x\to 1+} \sqrt{\dfrac{x^2-1}{x^2+1}}$.

4. $\lim\limits_{t\to 1} \dfrac{\sqrt{t}}{t+1} - \dfrac{t^2-t}{t^4-1}$.

5. $\lim\limits_{x\to 1} \dfrac{\sqrt{x}(x+2)}{x+1}$.

6. $\lim\limits_{y\to -3} \sqrt[3]{(y+2)^5}$.

7. $\lim\limits_{k\to 0} \dfrac{k^{15}-1}{k^{15}+1}$.

8. $\lim\limits_{s\to 2} \sqrt{s+1}\cdot\sqrt[3]{s-1}$.

9. $\lim\limits_{h\to 0} \dfrac{\sqrt{h^3+1}-1}{h^2}$.

10. $\lim\limits_{x\to 0} \dfrac{\sqrt{4+x}-2}{x}$.

11. $\lim\limits_{z\to -1} \dfrac{z^{-2}-1}{z+1}$.

12. $\lim\limits_{y\to b^2} \dfrac{\dfrac{y}{b^2}-\dfrac{b^4}{y^2}}{y-b^2}$.

13. $\lim\limits_{x\to a} \sqrt{x^2+2\,ax}$.

14. $\lim\limits_{y\to b} (y^2-1)(y^2+2b^2)$.

15. $\lim\limits_{s\to 2} \dfrac{s^2+2s+3}{s+1}$.

16. $\lim\limits_{t\to -3} (3-2t)^{5/2}$.

17. $\lim\limits_{x\to a} \dfrac{1/x - 1/a}{x-a}$.

18. $\lim\limits_{h\to 0} \dfrac{\sqrt[3]{x+h}-\sqrt[3]{x}}{h}$.

II

In each of Exercises 1-6, prove that the limit exists and find it.

1. $\lim\limits_{r\to 1} \dfrac{1-r^2}{2-\sqrt{r^2+3^2}}$.

2. $\lim\limits_{t\to 2} \dfrac{\sqrt{1+\sqrt{2+t}}-\sqrt{3}}{t-2}$.

3. $\lim\limits_{m\to 0} \dfrac{\sqrt{4+m+m^2}-2}{\sqrt{4+m-m^2}+2}$.

4. $\lim\limits_{x\to 0} \dfrac{x}{\sqrt{9-x+x^3}-3}$.

CHAPTER REVIEW

I

Prove that each of the limits exists by referring to the appropriate limit theorems and find the limit.

1. $\lim\limits_{x\to 2} (3x^2 - 4x + 1)$

2. $\lim\limits_{x\to 1} (7x^5 - 2x^4 + x - 5)$

3. $\lim\limits_{t\to 5} \dfrac{t+1}{t-1}$

4. $\lim\limits_{y\to -2} \dfrac{2-y}{y^2}$

5. $\lim\limits_{x\to 4} \sqrt{x^2-8}$

6. $\lim\limits_{x\to -3} \sqrt{3-x}$

7. $\lim\limits_{y\to 3} \dfrac{y^2-2y-3}{y^2+y-12}$

8. $\lim\limits_{x\to 2} \dfrac{x^2+2x-8}{x^2-x-2}$

9. $\lim\limits_{x\to 0} |x|$

10. $\lim\limits_{t\to 1} \dfrac{1-\sqrt{t}}{t-1}$

Find each of the following limits.

11. $\lim\limits_{x\to \infty} \dfrac{2}{x\sqrt{x}}$

12. $\lim\limits_{y\to \infty} \dfrac{y+2y^3}{y^3+1}$

13. $\lim\limits_{t\to 1^-} \dfrac{1-t}{|t-1|}$

14. $\lim\limits_{x\to -2^+} \dfrac{x^2-4}{\sqrt{x+2}}$

15. $\lim\limits_{x\to \infty} \dfrac{3x^2-5x}{2x^2+7}$

16. $\lim\limits_{t\to -\infty} \dfrac{t+6}{t-2}$

17. $\lim\limits_{x\to 3^-} \dfrac{x}{3-x}$

18. $\lim\limits_{x\to -5^+} \dfrac{x+2}{x^2+3x-10}$

Discuss the continuity of each function at the given values.

19. $f(x) = x^2 + 5,\ x = 3$

20. $g(t) = \sqrt{t+6},\ t = 4$

21. $f(t) = \begin{cases} \dfrac{1}{t} & \text{if } t < 0 \\ -|t| & \text{if } t \geq 0 \end{cases}$ $\quad t = 0$

22. $G(x) = \begin{cases} 4 - x & \text{if } x \leq 3 \\ \dfrac{x^2}{9} & \text{if } x > 3 \end{cases}$ $\quad x = 3$

23. $F(t) = \begin{cases} 7 & \text{if } t \leq 4 \\ 5 & \text{if } 4 < t \leq 6 \\ 8 & \text{if } t > 6 \end{cases}$ $\quad t = 4, t = 5, t = 6$

24. $f(r) = \dfrac{|r + 4|}{r}, r = -3$

Give a rough sketch of the graph of each function, showing the vertical and horizontal asymptotes.

25. $f(x) = 1 + \dfrac{2}{x}$

26. $g(x) = 3 - \dfrac{1}{x^2}$

27. $F(x) = \dfrac{3x}{x + 3}$

28. $G(x) = \dfrac{x + 1}{x - 1}$

29. $f(x) = \dfrac{x^2}{x^2 - 4}$

30. $g(x) = \dfrac{2}{x^2 - 1}$

Using only limit definitions, prove:

31. $\underset{x \to 3}{\text{limit}}\ (4x + 1) = 13$

32. $\underset{x \to 2}{\text{limit}}\ (x^2 + x) = 6$

33. $\underset{x \to (1/3)^+}{\text{limit}}\ \dfrac{4}{x} - 2 = 10$

34. $\underset{x \to 3^-}{\text{limit}}\ \dfrac{1}{x - 3} = -\infty$

35. $\underset{x \to \infty}{\text{limit}}\ \dfrac{4x^2 + x - 1}{5x^2} = \dfrac{4}{5}$

36. $\underset{x \to 4}{\text{limit}}\ \left(\dfrac{1}{x} - 1\right) = -\dfrac{3}{4}$

37. Find $\underset{h \to 0}{\text{limit}}\ \dfrac{\sqrt{4 + h} - 2}{h}$.

38. Find $\underset{h \to 0}{\text{limit}}\ \dfrac{1}{h}\left(\dfrac{1}{3 + h} - \dfrac{1}{3}\right)$.

39. Let $f(x) = \begin{cases} -x & \text{if } x \leq 1 \\ 3 + x & \text{if } x > 1 \end{cases}$ and $g(x) = \begin{cases} x^2 & \text{if } x \leq 1 \\ 2 - x & \text{if } x > 1 \end{cases}$. Find:

a. $\underset{x \to 1^-}{\text{limit}}\ f(x)$

b. $\underset{x \to 1^+}{\text{limit}}\ f(x)$

c. $\underset{x \to 1}{\text{limit}}\ f(x)$

d. $\underset{x \to 1^-}{\text{limit}}\ g(x)$

e. $\underset{x \to 1^+}{\text{limit}}\ g(x)$

f. $\underset{x \to 1}{\text{limit}}\ g(x)$

g. $\underset{x \to 1^+}{\text{limit}}\ f(g(x))$

h. $\underset{x \to 1^-}{\text{limit}}\ f(g(x))$

40. Give an example of a function f such that:

a. $\lim_{x\to 1^+} f(x)$ does not exist but $f(1)$ exists.

b. $\lim_{x\to 2} f(x)$ exists but $f(2)$ does not exist.

c. $\lim_{x\to 2} f(x)$ exists and $f(2)$ exists, but f is not continuous at 2.

II

Find each limit.

1. $\lim_{t\to\infty}\left(\frac{1}{\sqrt{t^2+1}} - \frac{1}{\sqrt{t^2-1}}\right)$

2. $\lim_{x\to\infty}(\sqrt{2x^2+1} - \sqrt{x^2-x})$

3. $\lim_{x\to 1^-}(1 - x + [x] - [1 - x])$

4. $\lim_{x\to 0^+}(1 - x + [x] - [1 - x])$

5. Show that if $\lim_{x\to a} g(x) = b$, with $g(x) \neq b$ for every x in some deleted neighborhood of a, and if $\lim_{x\to b} f(x) = c$, then $\lim_{x\to a} f(g(x)) = c$. (This theorem is sometimes useful if f is discontinuous at b. Compare this theorem with 2.28.)

3

Derivatives

Our first application of the limit concept will be to the problem of determining the instantaneous rate of change of a function. Geometrically, this problem is equivalent to that of finding a tangent line to the graph of the function. Both of these problems are solved by finding the derivative of the given function, as will be described in this chapter.

1 DEFINITIONS

Assume that a chemical reaction is taking place and that $f(t)$ designates the amount of a substance present t units of time after the reaction starts. Then the change in the amount of the substance from time t_1 to time t_2 is $f(t_2) - f(t_1)$, and the average rate of change of the amount per unit of time during the time interval $[t_1,t_2]$ is

$$\frac{f(t_2) - f(t_1)}{t_2 - t_1}.$$

It is natural to call

$$\operatorname*{limit}_{t_1 \to t_2} \frac{f(t_2) - f(t_1)}{t_2 - t_1}$$

the instantaneous rate of change of the amount of the substance at time t_2. This number is the derivative of the function f at t_2, according to the following definition.

3.1 Definition

The *derivative* of a function f is the function f' defined by

$$f'(a) = \operatorname*{limit}_{x \to a} \frac{f(x) - f(a)}{x - a}$$

The domain of f' is the set consisting of every number a at which the above limit exists.

Example 1 If $f(x) = x^2$, find the derivative f' of f.

Solution: By 3.1,

$$f'(a) = \operatorname*{limit}_{x \to a} \frac{x^2 - a^2}{x - a}.$$

Hence,

$$f'(a) = \operatorname*{limit}_{x \to a} (x + a) = 2a.$$

Since the above limit exists at every number a, the domain of f' is the set of all real numbers.

Example 2 If $g(x) = x^3 - 3x$, find $g'(2)$.

Solution: By 3.1,

$$g'(2) = \operatorname*{limit}_{x \to 2} \frac{(x^3 - 3x) - (2^3 - 3 \cdot 2)}{x - 2}.$$

By algebra,

$$\frac{(x^3 - 3x) - (2^3 - 3 \cdot 2)}{x - 2} = \frac{(x^3 - 2^3) - 3(x - 2)}{x - 2}$$

$$= x^2 + 2x + 4 - 3 \qquad [x^3 - 8 = (x - 2)(x^2 + 2x + 4).]$$

Therefore,

$$g'(2) = \operatorname*{limit}_{x \to 2} (x^2 + 2x + 1)$$

$$= 2^2 + 2 \cdot 2 + 1$$

$$= 9.$$

If in 3.1 we let $x - a = h$, then $x = a + h$ and

$$\frac{f(x) - f(a)}{x - a} = \frac{f(a + h) - f(a)}{h}.$$

To let x approach a is the same as letting h approach 0. Thus, it is intuitively clear that also

$$f'(a) = \operatorname*{limit}_{h \to 0} \frac{f(a + h) - f(a)}{h}.$$

In terms of a variable x,

3.1′ $$f'(x) = \underset{h\to 0}{\text{limit}} \frac{f(x+h) - f(x)}{h}$$

Example 3 If $f(x) = 1/x$, find $f'(x)$.

Solution: By 3.1′,

$$f'(x) = \underset{h\to 0}{\text{limit}} \frac{1/(x+h) - 1/x}{h}$$

$$= \underset{h\to 0}{\text{limit}} \frac{x - (x+h)}{hx(x+h)}$$

$$= \underset{h\to 0}{\text{limit}} \frac{-1}{x(x+h)}$$

$$= -\frac{1}{x^2}.$$

Another common notation for the derivative of f is Df. Then

$$D_x f(x) = f'(x)$$

denotes the derivative of f evaluated at the number x. If $f'(x)$ exists, the function f is said to be *differentiable* at the number x.

Example 4 Find $D_x\sqrt{x}$.

Solution: By 3.1′,

$$D_x\sqrt{x} = \underset{h\to 0}{\text{limit}} \frac{\sqrt{x+h} - \sqrt{x}}{h}.$$

If we "rationalize" the numerator, we get

$$D_x\sqrt{x} = \underset{h\to 0}{\text{limit}} \frac{(\sqrt{x+h} - \sqrt{x})(\sqrt{x+h} + \sqrt{x})}{h(\sqrt{x+h} + \sqrt{x})}$$

$$= \underset{h\to 0}{\text{limit}} \frac{(x+h) - x}{h(\sqrt{x+h} + \sqrt{x})}$$

$$= \underset{h\to 0}{\text{limit}} \frac{1}{\sqrt{x+h} + \sqrt{x}} = \frac{1}{2\sqrt{x}}.$$

The domain of $f(x) = \sqrt{x}$ is $[0,\infty)$, whereas the domain of its derivative f' is $(0,\infty)$, since $1/(2\sqrt{x})$ is undefined when $x = 0$.

EXERCISES

In each of Exercises 1 to 14, find $F'(x)$ for the given function F.

1. $F(x) = 3x + 1$
2. $F(x) = 1 - 2x^2$
3. $F(x) = 2x - x^2$
4. $F(x) = x + 5x^2$
5. $F(x) = 2x^2 - 3$
6. $F(x) = x^3$
7. $F(x) = (3x + 1)^3$
8. $F(x) = x^{-2}$
9. $F(x) = \sqrt{5x - 6}$
10. $F(x) = 1/\sqrt{x}$
11. $F(x) = \dfrac{1}{x - 2}$
12. $F(x) = \dfrac{x^2 + 1}{x}$
13. $F(x) = x^{3/2}$
14. $F(x) = \dfrac{1}{1 - 2x^2}$

Using Definition 3.1, find $f'(a)$ at the given value of a.

15. $f(x) = 2x^2 + x,\ a = 1$
16. $f(x) = x^{-3},\ a = -2$
17. $f(x) = \sqrt{x - 4},\ a = 13$
18. $f(x) = \sqrt{x^2 - 4},\ a = 5$

Using Definition 3.1′, find $f'(x)$ at the given value of x.

19. $f(x) = (x^2 + x)^2,\ x = 2$
20. $f(x) = \dfrac{3}{x^4} + \dfrac{1}{x},\ x = 2$
21. $f(x) = -2x^3,\ x = 0$
22. $f(x) = \dfrac{1}{\sqrt{1 - 3x}},\ x = -8$
23. Is the function $g(x) = \sqrt{x - 2}$ differentiable at $x = 3$? At $x = 1$?
24. Is the function $F(x) = 3/\sqrt{x} + 3$ differentiable at $x = 4$? At $x = 0$?

2 TANGENT LINES

One of the many applications of the derivative is to the problem of finding the tangent line to the graph of a function at some point on the graph.

It is possible to define the tangent line to a circle at a point P on the circle either as the line perpendicular to the radius at P or as the line intersecting the circle in only one point P. Neither of these definitions of a tangent line carries over to a general curve. In Figure 3.1, for example, what we would like to consider to be the tangent line T at P intersects the curve in another point M.

The tangent line T at P to a curve such as that shown in Figure 3.1 should be the line on P that is nearest to the curve in the neighborhood of P. We interpret this to mean that each secant line S on P and some other point Q of the curve should be close to T when Q is close to P. The limiting position of the secant line S as Q approaches P should be the tangent line T.

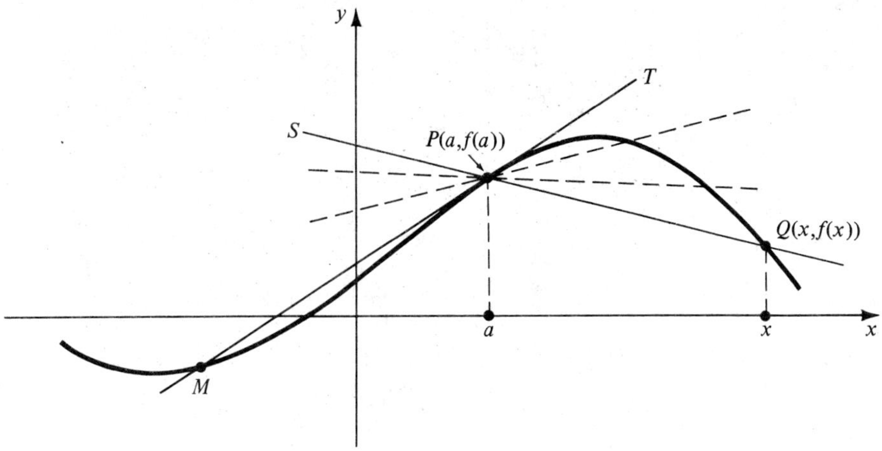

FIGURE 3.1

These rather vague statements as to the nature of the tangent line to the graph of f at a point $P(a,f(a))$ can be made precise by the use of slopes as follows.

3.2 Definition

The tangent line to the graph of a function f at a point $(a,f(a))$ has slope

$$\operatorname*{limit}_{x\to a} m(x),$$

if this limit exists, where for every $x \neq a$ in domain f, $m(x)$ is the slope of the secant line joining $(a,f(a))$ and $(x,f(x))$. (See Figure 3.1.)

Evidently,

$$m(x) = \frac{f(x) - f(a)}{x - a}.$$

Thereforc,

$$\operatorname*{limit}_{x\to a} m(x) = \operatorname*{limit}_{x\to a} \frac{f(x) - f(a)}{x - a} = f'(a),$$

if a is in domain f'. In this case, the *tangent line to the graph of f at the point* $(a,f(a))$ has slope $f'(a)$ and equation

3.3
$$y - f(a) = f'(a)(x - a).$$

Example 1 Find an equation of the tangent line T to the graph of $f(x) = x^2$ at the point $(2,4)$.

Solution: By Example 1, page 91, $f'(a) = 2a$. Therefore $f'(2) = 4$ and, by 3.3,

$$y - 4 = 4(x - 2) \qquad \text{or} \qquad y = 4x - 4$$

is an equation of T. The tangent line T is sketched in Figure 3.2.

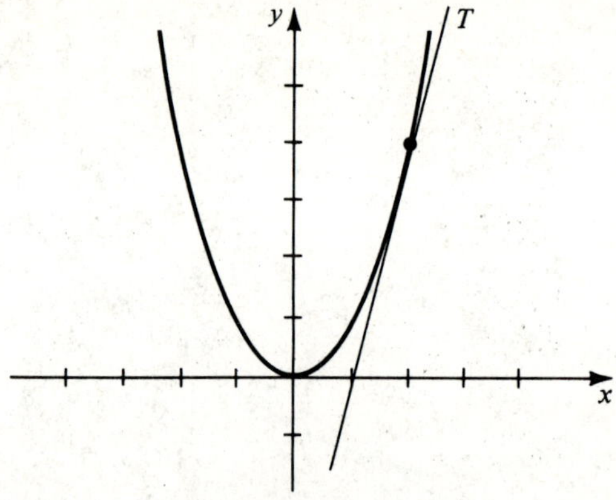

FIGURE 3.2

The *normal line* N to the graph of a function f at the point $P(a,f(a))$ is defined to be the line through P perpendicular to the tangent line. It follows that if $f'(a) \neq 0$ the slope of N is $-1/f'(a)$ and

3.4 $$y - f(a) = -\frac{1}{f'(a)}(x - a)$$

is an equation of N. If $f'(a) = 0$, then N is the vertical line $x = a$.

Example 2 Find equations of the tangent line T and normal line N to the graph of the function f defined by $f(x) = 2x - x^2$ at the point $(3,-3)$.

Solution: We first find $f'(3)$, the slope of T:

$$f'(3) = \operatorname*{limit}_{h\to 0}\frac{f(3+h) - f(3)}{h}$$

$$= \operatorname*{limit}_{h\to 0}\frac{[2(3+h) - (3+h)^2] - [2\cdot 3 - 3^2]}{h}$$

$$= \operatorname*{limit}_{h\to 0}(-4 - h) = -4.$$

Thus, $$y + 3 = -4(x - 3) \quad \text{or} \quad y = -4x + 9$$

is an equation of T and

$$y + 3 = \tfrac{1}{4}(x - 3) \quad \text{or} \quad 4y = x - 15$$

is an equation of N. The graph of f is a parabola. T and N are shown in Figure 3.3.

If, in Definition 3.2, $\operatorname*{limit}_{x\to a} m(x)$ does not exist in the ordinary sense but

$$\operatorname*{limit}_{x\to a} |m(x)| = \infty,$$

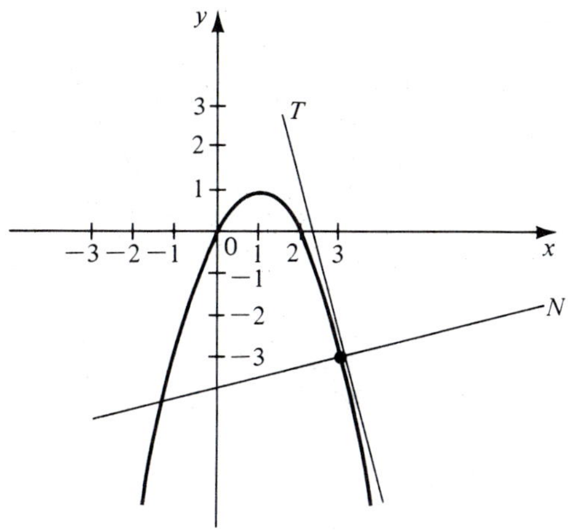

FIGURE 3.3

then the secant line S (Figure 3.1) is getting steeper and steeper as x approaches a. In this case, the tangent line T to the graph of f at the point $(a,f(a))$ is apparently vertical. Thus, we *define* T to be the vertical line

$$x = a$$

and, hence, the normal line N to be the horizontal line

$$y = f(a)$$

in case $\lim_{x \to a} |m(x)| = \infty$.

Example 3 Find equations of the tangent line T and the normal line N to the graph of the function $F(x) = \sqrt[3]{x}$ at the origin.

Solution: We have, by 2.16,

$$\lim_{x \to 0} m(x) = \lim_{x \to 0} \left| \frac{F(x) - F(0)}{x - 0} \right| = \lim_{x \to 0} \left| \frac{\sqrt[3]{x}}{x} \right| = \lim_{x \to 0} \frac{1}{x^{2/3}} = \infty.$$

Therefore, T is the vertical line $x = 0$ and N is the horizontal line $y = 0$. In other words, T is the y axis and N is the x axis. The graph of F is shown in Figure 3.4.

Example 4 Find equations of the tangent line T and the normal line N to the graph of the function $g(x) = \sqrt{1 + x}$ at the point $P(0,1)$.

Solution: By 3.1,

$$g'(0) = \lim_{x \to 0} \frac{\sqrt{1 + x} - \sqrt{1 + 0}}{x - 0} = \lim_{x \to 0} \frac{\sqrt{1 + x} - 1}{x}.$$

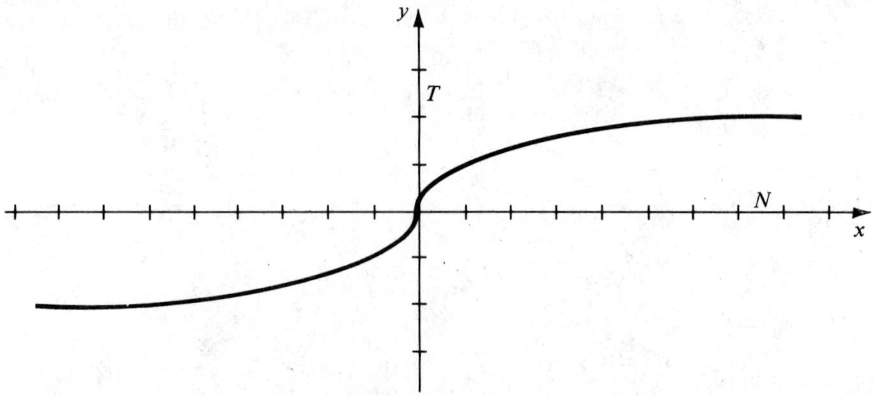

FIGURE 3.4

Since $(\sqrt{1+x}-1)(\sqrt{1+x}+1)=x$, we have

$$g'(0) = \lim_{x\to 0} \frac{1}{\sqrt{1+x}+1} = \frac{1}{2}.$$

Therefore, by 3.3 and 3.4,

$$y - 1 = \tfrac{1}{2}(x - 0) \quad \text{or} \quad x - 2y + 2 = 0$$

is an equation of T, and

$$y - 1 = -2(x - 0) \quad \text{or} \quad 2x + y - 1 = 0$$

is an equation of N.

EXERCISES

I

Find equations of the tangent line and normal line to the graph of the function at the given point.

1. $f(x) = (x-1)^2$, point (2,1)
2. $g(x) = 3x + 2x^2$, point $(-1,-1)$
3. $F(x) = \sqrt{x}$, point (4,2)
4. $G(x) = 2 - x^2$, point (0,2)
5. $g(x) = x^3 - 3x$, point (0,0)
6. $f(x) = \dfrac{1}{x}$, point $(2,\frac{1}{2})$
7. $F(x) = \sqrt{x-1}$, point (2,1)
8. $G(x) = \sqrt[4]{x}$, point (0,0)
9. $f(x) = \sqrt{1-4x}$, point $(-2,3)$
10. $F(x) = x^3 - 1$, point (1,0)

11. Find an equation of the normal line to the graph of $y = \sqrt{x + 4}$ which is parallel to the line $3x + 6y = 7$.
12. Find an equation of the tangent line to the curve $y = x^2 + x$ which is perpendicular to the line $x - y = 4$.
13. Find an equation of the normal line to the graph of $y = \sqrt[3]{x + 4}$ which is perpendicular to the line $y - \frac{1}{3}x = 5$.
14. Find the points on the curve $y = x^2 + 2x$ at which the normal line has slope $-\frac{1}{4}$.

II

1. Show that the graphs of the equations

$$y = 3x^2, \qquad y = 2x^3 + 1$$

are tangent at the point (1,3), that is, that they have a common tangent line at this point.
2. Show that there are exactly two tangent lines to the graph of $y = (x + 1)^3$ which pass through the origin, and find their equations.
3. Show that $y_1 y = 2p(x + x_1)$ is an equation of the tangent line to the parabola $y^2 = 4px$ at the point (x_1, y_1) on the parabola.

3 CONTINUITY OF A DIFFERENTIABLE FUNCTION

The condition that a function have a derivative is stronger than that of being continuous, as we shall now prove.

3.5 Theorem

If the function f is differentiable at a, then f is continuous at a.

Proof: By assumption and by 3.1, a is in the domain of f and

$$\operatorname*{limit}_{x \to a} \frac{f(x) - f(a)}{x - a} = f'(a).$$

It is possible to write $f(x) - f(a)$ as a product in the following way:

$$f(x) - f(a) = \frac{f(x) - f(a)}{x - a} \cdot (x - a), \qquad x \neq a.$$

Hence, by the product limit theorem,

$$\operatorname*{limit}_{x \to a} [f(x) - f(a)] = \operatorname*{limit}_{x \to a} \frac{f(x) - f(a)}{x - a} \cdot \operatorname*{limit}_{x \to a} (x - a)$$

$$= f'(a) \cdot 0 = 0.$$

Thus

$$\operatorname*{limit}_{x \to a} f(x) = f(a)$$

and the theorem is proved.

The converse of this theorem is not true. That is, there are functions continuous but not differentiable at some number. The example below illustrates this fact.

Example If f is the absolute value function,

$$f(x) = |x|,$$

show that f is continuous but not differentiable at the number 0.

Solution: The graph of f, consisting of two half-lines meeting at the origin, is shown in Figure 3.5. By definition,

$$f(x) = x \quad \text{if } x \geq 0;$$

$$f(x) = -x \text{ if } x \leq 0.$$

Hence

$$\lim_{x \to 0^+} f(x) = \lim_{x \to 0^+} x = 0,$$

$$\lim_{x \to 0^-} f(x) = \lim_{x \to 0^-} (-x) = 0,$$

and therefore, by 2.14,

$$\lim_{x \to 0} f(x) = 0.$$

Since $f(0) = 0$, this proves that f is continuous at 0.

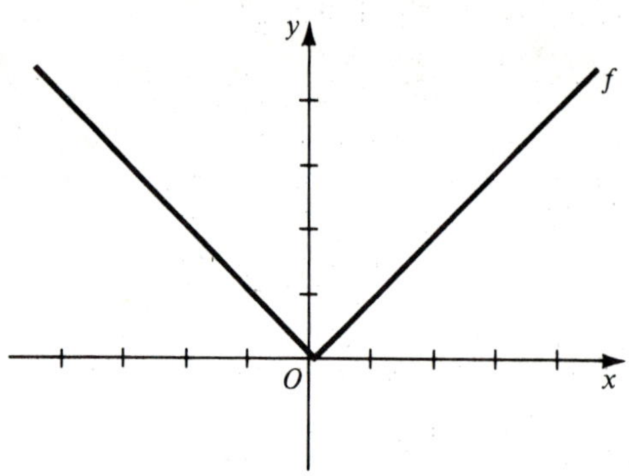

FIGURE 3.5

In order to prove that f is not differentiable at 0, we must show that

$$\lim_{x \to 0} \frac{f(x) - f(0)}{x - 0} = \lim_{x \to 0} \frac{|x|}{x}$$

does not exist. We will do this by showing that the one-sided limits at 0 are different. Thus

$$\lim_{x \to 0^+} \frac{|x|}{x} = \lim_{x \to 0^+} \frac{x}{x} = \lim_{x \to 0^+} 1 = 1,$$

whereas $$\lim_{x\to 0^-} \frac{|x|}{x} = \lim_{x\to 0^-} \frac{-x}{x} = \lim_{x\to 0^-} (-1) = -1.$$

Hence the derivative of f at 0 does not exist.

It is geometrically clear that the graph of f (Figure 3.5) has no tangent line at the origin O. For each secant line on O and a point P to the right of O has slope 1, whereas each secant line on O and a point P to the left of O has slope -1. Obviously, these slopes are not approaching some fixed number as P approaches O.

4 DIFFERENTIATION FORMULAS

If f is a constant function, say

$$f(x) = k$$

for every x in some interval A, then the derivative of f at a, $Df(a)$, is given by

3.6 $$Df(a) = \lim_{x\to a} \frac{k - k}{x - a} = \lim_{x\to a} 0 = 0.$$

Thus Df is the zero function with domain A, indicated as follows:

$$D_x k = 0 \qquad \text{for every constant function } k.$$

The linear function g defined by

$$g(x) = x$$

has derivative Dg given by

$$Dg(a) = \lim_{x\to a} \frac{x - a}{x - a} = \lim_{x\to a} 1 = 1.$$

We will often indicate this fact by writing

$$D_x x = 1.$$

Let us call a function f defined by

$$f(x) = x^n, \qquad n \text{ a given real number},$$

a *power function*. When $n = 1$, f is the linear function g discussed above. Other examples of power functions are the functions f, G, and F defined below:

$$f(x) = x^3, \qquad \text{domain } f = (-\infty,\infty),$$

$$G(x) = x^{-2} = 1/x^2, \qquad \text{domain } g = (-\infty,0) \cup (0,\infty),$$

$$F(x) = x^{3/2} = \sqrt{x^3}, \qquad \text{domain } F = [0,\infty).$$

Before attempting to differentiate a power function, we mention the identity

3.7 $$u^n - v^n = (u - v)(u^{n-1} + u^{n-2}v + \cdots + uv^{n-2} + v^{n-1}),$$

which holds for every integer $n > 1$. Identity 3.7 may be verified by multiplying out the right side of the equation and observing that all terms in the identity cancel with the exception of the two on the left side. We note that the second factor on the right side of 3.7 is a sum of n terms.

The special cases of 3.7 for $n = 2$, 3, and 4 are listed below.

$$u^2 - v^2 = (u - v)(u + v)$$
$$u^3 - v^3 = (u - v)(u^2 + uv + v^2)$$
$$u^4 - v^4 = (u - v)(u^3 + u^2v + uv^2 + v^3).$$

3.8 Theorem

$$D_x x^n = nx^{n-1}, \qquad n \text{ a positive integer.}$$

Proof: By 3.1′,

$$D_x x^n = \lim_{h\to 0} \frac{(x + h)^n - x^n}{h}$$

$$= \lim_{h\to 0} \frac{(x + h - x)[(x + h)^{n-1} + (x + h)^{n-2}x + \cdots + (x + h)x^{n-2} + x^{n-1}]}{h}$$

$$= \lim_{h\to 0} [(x + h)^{n-1} + (x + h)^{n-2}x + \cdots + (x + h)x^{n-2} + x^{n-1}]$$

$$= \underbrace{x^{n-1} + x^{n-2}\cdot x + \cdots + x\cdot x^{n-2} + x^{n-1}}_{n \text{ terms}}$$

$$= nx^{n-1}.$$

For example,

$$D_x x^2 = 2x, \qquad D_x x^7 = 7x^6, \qquad D_x x^{29} = 29x^{28}.$$

The derivative of a constant k times a function can be computed as follows.

3.9 Theorem

$$D_x kf(x) = kD_x f(x).$$

Proof:

$$D_x kf(x) = \lim_{h\to 0} \frac{kf(x + h) - kf(x)}{h}$$

$$= k \lim_{h\to 0} \frac{f(x + h) - f(x)}{h}$$

$$= kD_x f(x).$$

For example,

$$D_x(7x^6) = 7D_x x^6 = 7 \cdot 6x^5 = 42x^5$$

$$D_x(-8x^3) = (-8)D_x x^3 = (-8) \cdot 3x^2 = -24x^2.$$

It is easy to find the derivative of the sum of two functions if you know the derivative of each function:

3.10 Theorem

$$D_x[f(x) + g(x)] = D_x f(x) + D_x g(x).$$

Proof: We have

$$\begin{aligned} D_x[f(x) + g(x)] &= \lim_{h \to 0} \frac{[f(x+h) + g(x+h)] - [f(x) + g(x)]}{h} \\ &= \lim_{h \to 0} \left[\frac{f(x+h) - f(x)}{h} + \frac{g(x+h) - g(x)}{h}\right] \\ &= \lim_{h \to 0} \frac{f(x+h) - f(x)}{h} + \lim_{h \to 0} \frac{g(x+h) - g(x)}{h} \\ &= D_x f(x) + D_x g(x). \end{aligned}$$

Of course,

$$\text{domain } D(f + g) = (\text{domain } Df) \cap (\text{domain } Dg).$$

Formulas 3.9 and 3.10 can be extended by mathematical induction to the following form.

3.11 Theorem

For any numbers $k_1, k_2, \ldots, k_n$ and any functions $f_1, f_2, \ldots, f_n$,

$$\begin{aligned} &D_x[k_1 f_1(x) + k_2 f_2(x) + \cdots + k_n f_n(x)] \\ &\qquad = k_1 D_x f_1(x) + k_2 D_x f_2(x) + \cdots + k_n D_x f_n(x). \end{aligned}$$

For example,

$$\begin{aligned} D_x(4x^2 - 8x + 3) &= 4D_x x^2 - 8D_x x + D_x 3 \\ &= 4 \cdot 2x - 8 \cdot 1 + 0 = 8x - 8. \end{aligned}$$

$$\begin{aligned} D_x(x^6 - x^4 + \sqrt{3}\, x^3 + x - 7) &= D_x x^6 - D_x x^4 + \sqrt{3}\, D_x x^3 + D_x x - D_x 7 \\ &= 6x^5 - 4x^3 + 3\sqrt{3}\, x^2 + 1. \end{aligned}$$

As these examples show, we are now able to find with ease the derivative f' of any polynomial function f. Note that the degree of f' is exactly one less than the degree of f.

5 PRODUCT AND QUOTIENT DIFFERENTIATION FORMULAS

The derivative of the product of two functions is not the product of the derivatives of the two functions, as one might suspect from the sum formula (3.10). The correct formula is derived below.

If f and g are functions, then

$$D(fg)(a) = \operatorname*{limit}_{x \to a} \frac{f(x)g(x) - f(a)g(a)}{x - a}$$

at every number a at which the above limit exists. The evaluation of this limit is not evident. We would like to separate the functions f and g, if possible, as in the proof of 3.10. The following device of adding and subtracting a term allows us to make such a separation.

$$\frac{f(x)g(x) - f(a)g(a)}{x - a} = \frac{f(x)g(x) - f(x)g(a) + f(x)g(a) - f(a)g(a)}{x - a}$$

$$= f(x)\frac{g(x) - g(a)}{x - a} + \frac{f(x) - f(a)}{x - a}g(a).$$

Hence, by the limit theorems,

$$D(fg)(a) = \operatorname*{limit}_{x \to a} f(x) \operatorname*{limit}_{x \to a} \frac{g(x) - g(a)}{x - a} + g(a) \operatorname*{limit}_{x \to a} \frac{f(x) - f(a)}{x - a}.$$

If $f'(a)$ and $g'(a)$ exist, then, by 3.5, f is continuous at a, and we have

$$D(fg)(a) = f(a)g'(a) + g(a)f'(a).$$

The result above, called the *product differentiation formula*, may be stated as follows.

3.12 Theorem

$$D_x[f(x)g(x)] = f(x)D_xg(x) + g(x)D_xf(x).$$

As was the case with the derivative of a sum of two functions,

$$\text{domain } D(fg) = \text{domain } Df \cap \text{domain } Dg.$$

Example 1 Find $D_x(x^3 - x)(x^3 + x)$.

Solution: We are asked to find $D_x[f(x)g(x)]$, where

$$f(x) = x^3 - x, \qquad g(x) = x^3 + x.$$

By 3.12,

$$\begin{aligned} D_x[(x^3 - x)(x^3 + x)] &= (x^3 - x)D_x(x^3 + x) + (x^3 + x)D_x(x^3 - x) \\ &= (x^3 - x)(3x^2 + 1) + (x^3 + x)(3x^2 - 1) \\ &= 6x^5 - 2x. \end{aligned}$$

Perhaps an easier way to find the derivative is to first multiply

$$(x^3 - x)(x^3 + x) = x^6 - x^2$$

and then take the derivative,

$$D_x(x^6 - x^2) = 6x^5 - 2x.$$

Example 2 Find $D_x[\sqrt{x}(x^2 - 3)]$.

Solution: By 3.12,

$$\begin{aligned} D_x[\sqrt{x}(x^2 - 3)] &= \sqrt{x}\, D_x(x^2 - 3) + (x^2 - 3)D_x\sqrt{x} \\ &= \sqrt{x}(2x) + (x^2 - 3)\frac{1}{2\sqrt{x}} \qquad \text{(using Example 4, page 92)} \\ &= \frac{5x^2 - 3}{\sqrt{x}}. \end{aligned}$$

There is no easy way, as in the example above, to do this problem.

There is a *quotient differentiation formula* as well as a product formula. It is stated below.

3.13 Theorem

$$D_x \frac{f(x)}{g(x)} = \frac{g(x)D_x f(x) - f(x)D_x g(x)}{[g(x)]^2}.$$

Proof: The formula follows easily from the equation

$$\frac{f(x)}{g(x)} - \frac{f(a)}{g(a)} = \frac{1}{g(x)g(a)}\{g(a)[f(x) - f(a)] - f(a)[g(x) - g(a)]\}.$$

You carry out the steps!

The domain of $D(f/g)$ is the intersection of the domains of f/g, Df, and Dg.

Example 3 Find $D_x \dfrac{x^2}{x^2 - 4}$.

Solution: We are asked to find

$$D_x \frac{f(x)}{g(x)}, \qquad \text{where } f(x) = x^2,\ g(x) = x^2 - 4.$$

By 3.13,

$$D_x \frac{x^2}{x^2 - 4} = \frac{(x^2 - 4)D_x x^2 - x^2 D_x(x^2 - 4)}{(x^2 - 4)^2}$$

$$= \frac{(x^2 - 4) \cdot 2x - x^2 \cdot 2x}{(x^2 - 4)^2}$$

$$= \frac{-8x}{(x^2 - 4)^2}.$$

Example 4 Find $D_x \frac{1}{x^3}$.

Solution: By 3.13,

$$D_x \frac{1}{x^3} = \frac{x^3 D_x 1 - 1 D_x x^3}{(x^3)^2}$$

$$= \frac{x^3 \cdot 0 - 1 \cdot 3x^2}{x^6}$$

$$= -\frac{3}{x^4}.$$

Using negative exponents, we have shown

$$D_x x^{-3} = -3x^{-4}.$$

In precisely the same way, we prove

$$D_x x^{-n} = -nx^{-n-1} \qquad \text{for every positive integer } n.$$

This result together with 3.8 proves the following formula.

3.14 Theorem

$$D_x x^r = rx^{r-1} \qquad \textit{for every integer } r.$$

For example,

$$D_x x^4 = 4x^3, \qquad D_x x^0 = 0x^{-1} = 0, \qquad D_x x^{-4} = -4x^{-5}.$$

EXERCISES

I

Differentiate each function.

1. $f(x) = x^4 - 4x^2 + 2$
2. $g(x) = \frac{1}{5}x^5 - 4x^3 + 7x$
3. $g(x) = 5x^6 - 3x^4$
4. $f(t) = t^4 - 4t^3 + 2$

5. $F(t) = t^2(3t^2 - 7t + 5)$

6. $G(x) = x^7(x^7 + 1)$

7. $f(y) = \dfrac{y - 1}{y + 1}$

8. $g(t) = \dfrac{t^2}{t + 1}$

9. $g(x) = x^2 - \dfrac{1}{x^2}$

10. $F(y) = 3y + \dfrac{2}{y^3}$

11. $G(t) = \dfrac{3t^2}{t + 1}$

12. $f(x) = \left(x - \dfrac{1}{x}\right)^2$

13. $f(x) = \left(1 + \dfrac{2}{x}\right)\left(2 + \dfrac{1}{x}\right)$

14. $g(t) = t^4\left(1 - \dfrac{3}{t + 1}\right)$

15. $F(y) = \dfrac{\sqrt{y}}{y + 1}$

16. $f(y) = \dfrac{y^2 - 1}{\sqrt{y}}$

17. $f(x) = (x + 1)(2x^2 + 1)(3x^3 + 1)$

18. $F(x) = \dfrac{\sqrt{x}}{1 + \sqrt{x}}$

19. $G(t) = 3 - \dfrac{4}{t} + \dfrac{7}{t^2}$

20. $g(x) = \dfrac{x^4 - 16}{x^4 + 16}$

21. $s(t) = \dfrac{4t^3}{t^2 + 1}$

22. $s(t) = -16t^2 + 4t - 2$

23. $H(w) = (w + 1)(3w^2 + 2)$

24. $T(z) = \dfrac{z^3 + 2}{1 - \sqrt{z}}$

Find equations of the tangent line and normal line to the graph of:

25. $f(x) - 4x^2 - x^{-4}$ at $x = 1$

26. $g(x) = 3x^3 + 3$ at $x = -3$

27. $h(x) = \dfrac{3x + 1}{\sqrt{x}}$ at $x = 4$

28. $G(y) = 1 - \dfrac{2}{y} + \dfrac{3}{y^2}$ at $y = -2$

II

Find the derivative of each of the following functions and state the domain of each derivative.

1. $f(x) = |x^2 - 4|$

2. $G(x) = |x^3 - 1|$

3. $F(x) = [x]$

4. $G(x) = \begin{cases} x^2, & x \geq 0 \\ -x^2, & x < 0 \end{cases}$

6 THE CHAIN RULE

If f and g are functions, then the *composite function* $f \circ g$ was defined in Chapter 1 by

$$(f \circ g)(x) = f(g(x)).$$

For example, if

$$f(x) = x^{20}, \qquad g(x) = x^2 + 1$$

then

$$\begin{aligned}(f \circ g)(x) &= f(g(x)) \\ &= f(x^2 + 1) \\ &= (x^2 + 1)^{20}.\end{aligned}$$

Knowing the derivatives of functions f and g, we can find the derivative of $f \circ g$ by the following rule.

3.15 The Chain Rule

$$D_x f(g(x)) = f'(g(x))g'(x).$$

Proof: Let $F = f \circ g$, the composite of f and g. Then, $F(x) = f(g(x))$, and

$$F'(a) = \operatorname*{limit}_{x \to a} \frac{f(g(x)) - f(g(a))}{x - a}.$$

If $g(x) \neq g(a)$ for every x in some deleted neighborhood of a, then

$$\frac{f(g(x)) - f(g(a))}{x - a} = \frac{f(g(x)) - f(g(a))}{g(x) - g(a)} \cdot \frac{g(x) - g(a)}{x - a}$$

and

(1) $$F'(a) = \operatorname*{limit}_{x \to a} \frac{f(g(x)) - f(g(a))}{g(x) - g(a)} \cdot \operatorname*{limit}_{x \to a} \frac{g(x) - g(a)}{x - a}.$$

Since

$$\operatorname*{limit}_{z \to g(a)} \frac{f(z) - f(g(a))}{z - g(a)} = f'(g(a))$$

it follows that as x approaches a, $z = g(x)$ approaches $g(a)$. Hence, from (1),

$$F'(a) = f'(g(a))g'(a).$$

If we replace a by x, we get the desired result.

It should be pointed out that the above proof is based on the assumption that there is *some* deleted neighborhood of a throughout which the function g never takes on the value $g(a)$. Such an hypothesis holds for most functions ordinarily considered in elementary calculus. A general proof, without using this assumption, can also be given, but it is not very instructive and hence is omitted.

Example 1 Find $D_x(x^2 + 1)^{20}$.

Solution: We saw above that

$$(x^2 + 1)^{20} = f(g(x)), \qquad \text{where } f(x) = x^{20},\ g(x) = x^2 + 1.$$

By the chain rule,

$$D_x(x^2 + 1)^{20} = f'(g(x))g'(x).$$

Since

$$f'(x) = 20x^{19}, \qquad f'(g(x)) = 20(g(x))^{19} = 20(x^2 + 1)^{19},$$

and

$$g'(x) = 2x,$$

we have

$$D_x(x^2 + 1)^{20} = 20(x^2 + 1)^{19} \cdot 2x = 40x(x^2 + 1)^{19}.$$

Example 2 Find $D_x \dfrac{1}{(x^3 - 4x^2 + 1)^2}$.

Solution: In our mind, we see

$$\frac{1}{(x^3 - 4x^2 + 1)^2} = (x^3 - 4x^2 + 1)^{-2} = f(g(x)),$$

where

$$f(x) = x^{-2}, \qquad g(x) = x^3 - 4x^2 + 1.$$

Since $f'(x) = -2x^{-3}$, we have

$$f'(g(x)) = -2(x^3 - 4x^2 + 1)^{-3}, \qquad g'(x) = 3x^2 - 8x.$$

Therefore, by the chain rule,.

$$D_x \frac{1}{(x^3 - 4x^2 + 1)^2} = -2(x^3 - 4x^2 + 1)^{-3}(3x^2 - 8x) = \frac{-6x^2 + 16x}{(x^3 - 4x^2 + 1)^3}.$$

The two examples above are powers of functions: In Example 1, we have $[g(x)]^{20}$, where $g(x) = x^2 + 1$; in Example 2, we have $[g(x)]^{-2}$, where $g(x) = x^3 - 4x^2 + 1$.

Whenever function F is a power of another function f, say, $F = f^n$, then the derivative of F can be found as follows.

3.16 Theorem

$$D_x f^n(x) = nf^{n-1}(x)D_x f(x) \qquad \textit{for every integer } n.$$

Proof: If $F = f^n$, so that

$$F(x) = [f(x)]^n,$$

then

$$F(x) = h(f(x)), \qquad \text{where } h(x) = x^n.$$

Hence, $h'(x) = nx^{n-1}$ and by the chain rule,

$$\begin{aligned} D_x F(x) &= h'(f(x))f'(x) \\ &= n(f(x))^{n-1}D_x f(x). \end{aligned}$$

Example 3 Find $D_x(3 - 2x + x^2)^4$.

Solution: We have

$$D_x(3 - 2x + x^2)^4 = 4(3 - 2x + x^2)^3 D_x(3 - 2x + x^2)$$
$$= 4(3 - 2x + x^2)^3(-2 + 2x)$$
$$= 8(x - 1)(3 - 2x + x^2)^3.$$

7 ALGEBRAIC FUNCTIONS

The nth root function

$$f(x) = \sqrt[n]{x}, \qquad n \text{ an integer, } n > 1$$

has derivative

$$f'(a) = \operatorname*{limit}_{x \to a} \frac{\sqrt[n]{x} - \sqrt[n]{a}}{x - a}.$$

We note that the quotient $(\sqrt[n]{x} - \sqrt[n]{a})/(x - a)$ has the form $(u - v)/(u^n - v^n)$. Hence, by 3.7,

$$\frac{u - v}{u^n - v^n} = \frac{1}{\underbrace{u^{n-1} + u^{n-2}v + \cdots + uv^{n-2} + v^{n-1}}_{n \text{ terms}}}.$$

Thus

$$f'(a) = \operatorname*{limit}_{x \to a} \frac{1}{\sqrt[n]{x^{n-1}} + \sqrt[n]{x^{n-2}}\sqrt[n]{a} + \cdots + \sqrt[n]{x}\sqrt[n]{a^{n-2}} + \sqrt[n]{a^{n-1}}}.$$

Since $\operatorname*{limit}_{x \to a} \sqrt[n]{x^k} = \sqrt[n]{a^k}$, each term in the denominator above has the same limit $\sqrt[n]{a^{n-1}}$. Therefore,

$$f'(a) = \frac{1}{n\sqrt[n]{a^{n-1}}} = \frac{1}{n} a^{1/n - 1}.$$

Hence, the formula

$$D_x x^r = r x^{r-1}$$

holds for r the reciprocal of a positive integer as well as for r an integer. Actually, a stronger result holds.

3.17 Theorem

$$D_x x^r = r x^{r-1}, \qquad r \text{ a rational number.}$$

Proof: If $r = m/n$, m, n integers with $n > 0$, then

$$x^r = (x^{1/n})^m$$

and by the chain rule,

$$\begin{aligned} D_x x^r &= m(x^{1/n})^{m-1} D_x x^{1/n} \\ &= m x^{(m-1)/n} \frac{1}{n} x^{1/n-1} \\ &= \frac{m}{n} x^{m/n-1} \\ &= r x^{r-1}. \end{aligned}$$

For example, if f is the power function

$$f(x) = \sqrt[3]{x^2} = x^{2/3},$$

then

$$f'(x) = \tfrac{2}{3}x^{-1/3}$$

or

$$f'(x) = \frac{2}{3\sqrt[3]{x}}.$$

The domain of f is $(-\infty,\infty)$ and of f' is $(-\infty,0) \cup (0,\infty)$.

Using 3.17 and the chain rule, we obtain the following general form of 3.16.

3.18 Theorem

$$D_x f^r(x) = r f^{r-1}(x) D_x f(x), \qquad r \textit{ a rational number.}$$

Example 1 Find $D_x\sqrt{x^2+1}$.

Solution:

$$\begin{aligned} D_x\sqrt{x^2+1} &= D_x(x^2+1)^{1/2} \\ &= \tfrac{1}{2}(x^2+1)^{-1/2} D_x(x^2+1) \\ &= \frac{x}{\sqrt{x^2+1}}. \end{aligned}$$

Example 2 Find equations of the tangent line T and the normal line N to the graph of the function F defined by

$$F(x) = \frac{x}{\sqrt[3]{2x-1}}$$

at the point $P(1,1)$.

Solution: We first find Df as follows:

$$\begin{aligned} D_xF(x) &= \frac{\sqrt[3]{2x-1}\cdot D_x x - x\cdot D_x(2x-1)^{1/3}}{(\sqrt[3]{2x-1})^2} \\ &= \frac{\sqrt[3]{2x-1}\cdot 1 - x\cdot\frac{1}{3}\cdot(2x-1)^{-2/3}\cdot 2}{\sqrt[3]{(2x-1)^2}}. \end{aligned}$$

Hence $f'(1) = (1 - \frac{2}{3})/1 = \frac{1}{3}$. Therefore T has equation

$$y - 1 = \tfrac{1}{3}(x - 1) \quad \text{or} \quad x - 3y + 2 = 0,$$

and N has equation

$$y - 1 = -3(x - 1) \quad \text{or} \quad 3x + y - 4 = 0.$$

EXERCISES

I

In each of Exercises 1 to 16, differentiate the function.

1. $f(x) = (1 + 2x^3)^4$
2. $f(x) = (1 + \sqrt{x})^3$
3. $f(x) = \sqrt{x^3 + x}$
4. $f(x) = (2x^2 + 1)^{-2}$
5. $f(x) = (4x^2 - x)^{-3/4}$
6. $f(x) = (2 + x^2 + x^3)^{-1/2}$
7. $f(x) = x\sqrt{1 - x^2}$
8. $h(z) = \sqrt{\dfrac{1 - z}{1 + z}}$
9. $F(t) = (2t + t^2)^{3/2}$
10. $G(x) = \dfrac{\sqrt{x^2 + 1}}{x + 1}$
11. $S(t) = \sqrt[3]{3t + 1}$
12. $F(y) = \left(y - \dfrac{1}{y}\right)^{3/2}$
13. $g(x) = \sqrt{1 + (1/x^2)}$
14. $f(x) = (1 + \sqrt{x})\sqrt[3]{x^2 + x + 1}$
15. $G(z) = \sqrt{2z} + \sqrt{z/2}$
16. $H(x) = \dfrac{(x^3 + 1)^2\sqrt{1 + x^2}}{1 + \sqrt{x}}$

In each of Exercises 17 to 22, find equations of the tangent line and the normal line to the graph of the equation at the indicated point.

17. $y = \sqrt{x + 1}$, $(3,2)$
18. $y = \dfrac{x}{\sqrt{7 - 3x}}$, $(-3, -\frac{3}{4})$
19. $y = x\sqrt{x^2 + 1}$, $(0,0)$
20. $y = \dfrac{x}{\sqrt{5x + 1}}$, $(5, \frac{5}{6})$
21. $y = \sqrt{2 + \sqrt{x}}$, $(4,2)$
22. $y = \sqrt[n]{x^n + 1}$, $(0,1)$

II

In each of Exercises 1 to 4, find the derivative of the given function.

1. $f(x) = \sqrt{x + \sqrt{x + \sqrt{x}}}$
2. $f(x) = \sqrt{x + (x^2 + 1)^3}$
3. $f(x) = \sqrt[3]{x + |x|}$
4. $f(x) = [(x^2 + (2x + 1)^7 + 1)^3 + 1]^{1/3}$

8 IMPLICIT DIFFERENTIATION

Most of the functions we have discussed so far have been *explicitly* defined by an algebraic equation. For example, the equation

$$y = x^3 + 1$$

defines a function f where $f(x) = x^3 + 1$. The graph of the function f is simply the graph of the given equation.

Not all functions are defined in such an explicit way. For example, an equation in x and y such as

$$x^3 - x = y^3 - y^2 + 24$$

is not easily solved for y in terms of x (or, for that matter, x in terms of y). However, there might exist a function f such that the equation

$$x^3 - x = f^3(x) - f^2(x) + 24$$

is true for every x in the domain of f. Such a function is said to be defined *implicitly* by the given equation.

The derivative of a function defined implicitly by an equation in x and y can often be found without explicitly solving the equation for y in terms of x. The process of finding a derivative in this case is called *implicit differentiation*. We illustrate this process in the following examples.

Example 1 On the assumption that there is a differentiable function f defined implicitly by the equation

$$x^3 - x = f^3(x) - f^2(x) + 24,$$

find its derivative.

Solution: If we let functions F and G be defined by

$$F(x) = x^3 - x, \qquad G(x) = f^3(x) - f^2(x) + 24,$$

with domain of F = domain of G = domain of f, then $F = G$ by what is given. Hence $DF = DG$; that is,

$$D_x(x^3 - x) = D_x[f^3(x) - f^2(x) + 24]$$

and

$$3x^2 - 1 = 3f^2(x)D_xf(x) - 2f(x)D_xf(x).$$

Solving this equation for $D_xf(x)$, we get

$$D_xf(x) = \frac{3x^2 - 1}{3f^2(x) - 2f(x)}.$$

If, for example, we are given that (3,1) is a point on the graph of f, then

$$f'(3) = \frac{3 \cdot 3^2 - 1}{3 \cdot 1^2 - 2 \cdot 1} = 26.$$

Hence $\quad y - 1 = 26(x - 3) \quad$ or $\quad y = 26x - 77$

is an equation of the tangent line to the graph of f at (3,1). This gives us an indication of the nature of the graph of f near (3,1); it must be a steep curve in the neighborhood of this point.

Example 2 Two differentiable functions are defined by the equation

$$x^2 + y^2 = 16$$

of a circle, namely those defined by the equations

$$y = \sqrt{16 - x^2} \quad \text{and} \quad y = -\sqrt{16 - x^2}.$$

Find the derivative of each function.

Solution: If f designates either of these two functions, then

$$x^2 + f^2(x) = 16$$

for every x in $[-4,4]$, the domain of f. By implicit differentiation,

$$D_x[x^2 + f^2(x)] = D_x16,$$

$$2x + 2f(x)D_xf(x) = 0,$$

$$D_xf(x) = -\frac{x}{f(x)}.$$

Note that -4 and 4 must be excluded from the domain of Df, since $f(-4) = f(4) = 0$. By the above formula for Df, we have

$$D_x\sqrt{16 - x^2} = -\frac{x}{\sqrt{16 - x^2}} \quad \text{and} \quad D_x(-\sqrt{16 - x^2}) = \frac{x}{\sqrt{16 - x^2}}.$$

Example 2 illustrates the fact that implicit differentiation gives the derivative of every differentiable function defined by the given equation.

We can also use implicit differentiation to find derivatives of inverse functions.

3.19 Theorem

If f and g are inverse functions which are differentiable, then

$$g'(x) = \frac{1}{f'(g(x))}, \quad \textit{whenever } f'(g(x)) \neq 0.$$

Proof: Since f and g are inverses of each other,

$$f(g(x)) = x \quad \text{for every } x \text{ in domain } g.$$

By implicit differentiation,

$$D_xf(g(x)) = D_xx.$$

Hence, by the chain rule,

$$f'(g(x))g'(x) = 1$$

and

$$g'(x) = \frac{1}{f'(g(x))}.$$

This result shows that, in a certain sense, the derivatives of a function and its inverse are *reciprocals*.

Example 3 The functions

$$f(x) = x^3, \qquad g(x) = \sqrt[3]{x}$$

are 1–1 functions and inverses of each other:

$$f(g(x)) = (\sqrt[3]{x})^3 = x.$$

Since $f'(x) = 3x^2$, $f'(g(x)) = 3(\sqrt[3]{x})^2$ and

$$g'(x) = \frac{1}{3x^{2/3}} = \frac{1}{3}x^{-2/3}, \qquad x \neq 0,$$

by 3.19. This agrees with 3.17.

9 HIGHER DERIVATIVES

We call the derivative f' of a function f the *first derivative* of f. The function f' might itself have a derivative, which is called the *second derivative* of f; and so on.

f	first derivative of f,
$f'' = (f')'$	second derivative of f,
$f''' = (f'')'$	third derivative of f,
$f^{[4]} = (f''')'$	fourth derivative of f,

and so on. In the D notation,

$$f' = Df,$$
$$f'' = D^2f,$$
$$f''' = D^3f,$$
$$f^{[4]} = D^4f,$$

and so on.

Example 1

$$D_x x^4 = 4x^3$$

$$D_x^2 x^4 = D_x(4x^3) = 12x^2$$

$$D_x^3 x^4 = D_x(12x^2) = 24x$$

$$D_x^4 x^4 = D_x(24x) = 24 = 4!$$

$$D_x^5 x^4 = D_x 24 = 0.$$

Similarly, we could show that

$$D_x^n x^n = n!, \qquad n \text{ a positive integer.}$$

Example 2

If $f(x) = \sqrt{9 + x^2}$, find $f''(x)$.

Solution: We have $f(x) = (9 + x^2)^{1/2}$ and

$$f'(x) = \tfrac{1}{2}(9 + x^2)^{-1/2} \cdot 2x$$

$$= \frac{x}{\sqrt{9 + x^2}}$$

by 3.18. Using the quotient differentiation formula,

$$f''(x) = \frac{\sqrt{9 + x^2} \cdot D_x x - x \cdot D_x\sqrt{9 + x^2}}{(\sqrt{9 + x^2})^2}$$

$$= \frac{\sqrt{9 + x^2} - x^2/\sqrt{9 + x^2}}{9 + x^2}$$

$$= \frac{9}{(9 + x^2)^{3/2}}.$$

We could also use implicit differentiation, starting with $y = \sqrt{9 + x^2}$ and therefore

$$y^2 = 9 + x^2.$$

Then (considering y as a function of x)

$$D_x y^2 = D_x(9 + x^2)$$

(1) $$2yD_x y = 2x$$

(2) $$D_x y = \frac{x}{y}.$$

From (1)

$$D_x(yD_x y) = D_x x.$$

Using the product differentiation formula on the left side,

$$y \cdot D_x(D_x y) + D_x y \cdot D_x y = 1$$

$$y D_x^2 y + (D_x y)^2 = 1,$$

$$D_x^2 y = \frac{1 - (D_x y)^2}{y}$$

$$= \frac{y^2 - x^2}{y^3} = \frac{9}{y^3}, \qquad \textit{using (2).}$$

EXERCISES

In each of Exercises 1 to 4, use implicit differentiation to find the derivative at the indicated point of each differentiable function f [letting $y = f(x)$] defined by the equation.

1. $x^3 + y^3 = 9$, (2,1)
2. $x^2 + xy + y^3 = 3$, (1,1)
3. $x + xy^2 + x^2y - y = 7$, (2,1)
4. $\sqrt{x} + x\sqrt{y} + y = 10$, (4,4)

Find $f'(x)$ by implicit differentiation, where $y = f(x)$ satisfies each of the following equations.

5. $(x + y)^2 + (x - y)^2 = 4y^3$
6. $(x + y)^3 - x^2 = y^2$
7. $x^2(2x - y)^{1/2} = x - 3y$
8. $\sqrt{xy} - 3x = \sqrt{y}$
9. $\dfrac{1}{y} + \dfrac{1}{x} = 2$
10. $\sqrt{x} - \sqrt{y} = 5$

In each of Exercises 11 to 18, find the indicated derivative of the given function.

11. $f''(x), f(x) = (3x^2 + 1)^4$
12. $f'''(x), f(x) = \sqrt{x^2 + 4}$
13. $f'''(x), f(x) = (1 + x)^{3/2}$
14. $f^{[4]}(x), f(x) = \dfrac{1}{x - 2}$
15. $D_t^2 s(t), s(t) = 3t^3 - \dfrac{1}{t} + 1$
16. $D_w^2 h(w), h(w) = \sqrt{w^3 - w}$
17. $D_v^3(v\sqrt{9 - v})$
18. $D_x^n x^{1/2}$, n a positive integer

In each of Exercises 19 to 22, use implicit differentiation to find the first and second derivative at the indicated point of each differentiable function f [letting $y = f(x)$] defined by the equation.

19. $\sqrt{x} + \sqrt{y} = 3$; (1,4)
20. $x^2 + y^2 = 25$, (4,3)

21. $x^3 + 3xy + y^2 = 5, (1,1)$

22. $\dfrac{y^2}{x + y} = 1 - x^2, (1,0)$

In each of Exercises 23 to 26, find f' and f'' for each differentiable function f [letting $y = f(x)$] defined by the equation.

23. $x = \dfrac{y^5 + y + 1}{y^2 + y + 1}$

24. $y + \sqrt{xy} = x^2$

25. $x = \dfrac{1 - \sqrt{y}}{1 + \sqrt{y}}$

26. $x^2y^2 + xy = 2$

Find an equation of the tangent line, at the indicated point, to the graph of each of the following equations.

27. $xy = 4, (-2,-2)$

28. $x + x^2y^2 - y = 1, (1,1)$

29. $x^2 + xy + y^2 = 3, (-1,-1)$

30. $x^3 + y^3 = 6xy, (3,3)$

31. If $f(x) = \begin{cases} -x^3 \text{ if } x < 0 \\ x^3 \text{ if } x > 0 \end{cases}$, find $f'(x)$ and $f''(x)$. What are the domains of f' and f''?

32. If $f(x) = \begin{cases} \dfrac{x^3}{|x|} \text{ if } x \neq 0 \\ 0 \text{ if } x = 0 \end{cases}$, find $D_xf(x)$ and $D_x^2f(x)$. What are the domains of Df and D^2f?

33. If $f(x) = |x^2 - 4|$ find $f'(x)$ and $f''(x)$. What are the domains of f' and f''?

10 DIFFERENTIALS

Present-day symbolism in the calculus has evolved from the work of many mathematicians over the past three centuries. The notation f', f'', and f''' for derivatives of a function f, for example, is credited to the great eighteenth-century French mathematician Lagrange. However, the greatest inventor of symbolism for the calculus was Leibnitz, one of the founders of the subject. His contributions to this facet of the calculus date back as early as 1675. Let us devote some time to a description of his notation.

Given a function and the equation

$$y = f(x),$$

Let Δx (read delta x) denote a variable, and Δy be defined by

$$\Delta y = f(x + \Delta x) - f(x).$$

We think of Δy as the *change* in y caused by a *change* of Δx in x. Clearly, Δy depends on f, x, and Δx.

Example 1 If $f(x) = x^2 - 1/x$, then

$$\Delta y = \left[(x + \Delta x)^2 - \frac{1}{x + \Delta x}\right] - \left[x^2 - \frac{1}{x}\right]$$

$$= x^2 + 2x\,\Delta x + \Delta x^2 - x^2 - \left[\frac{1}{x + \Delta x} - \frac{1}{x}\right].$$

Thus, $$\Delta y = 2x\,\Delta x + \Delta x^2 + \frac{\Delta x}{x(x + \Delta x)}.$$

The derivative of f was denoted by dy/dx by Leibnitz:

$$\frac{dy}{dx} = f'(x) = \operatorname*{limit}_{\Delta x \to 0} \frac{\Delta y}{\Delta x}.$$

We note that Δx plays the role of h in 3.1′. We think of dy/dx as "the derivative of y with respect to x."

For the function f of Example 1,

$$\frac{dy}{dx} = \operatorname*{limit}_{\Delta x \to 0} \frac{\Delta y}{\Delta x}$$

$$= \operatorname*{limit}_{\Delta x \to 0} \left[2x + \Delta x + \frac{1}{x(x + \Delta x)}\right]$$

$$= 2x + \frac{1}{x^2}.$$

Of course, this is simply $f'(x)$.

If the notation dy/dx is used for the first derivative of a function, then the second, third, and higher derivatives of the function are denoted as follows:

$$f'(x) = \frac{dy}{dx}, \quad f''(x) = \frac{d^2y}{dx^2}, \quad f'''(x) = \frac{d^3y}{dx^3}, \quad \ldots, \quad f^{[n]}(x) = \frac{d^ny}{dx^n}.$$

In the notation of Leibnitz, the sum and product formulas can be written as follows:

$$\frac{d}{dx}(u + v) = \frac{du}{dx} + \frac{dv}{dx}, \qquad \frac{d}{dx}(uv) = u\frac{dv}{dx} + v\frac{du}{dx},$$

where $u = f(x)$ and $v = g(x)$ for some functions f and g. The chain rule has a particularly simple form in this notation: If

$$y = f(z) \text{ and } z = g(x) \qquad [\text{so that } y = f(g(x))],$$

then $$\frac{dy}{dx} = \frac{dy}{dz}\frac{dz}{dx} \qquad \left[\frac{dy}{dz} = f'(g(x)), \frac{dz}{dx} = g'(x)\right].$$

Example 2 If $y = z^4 - 3z$ and $z = x^5 + 1$, then

$$\frac{dy}{dz} = 4z^3 - 3, \qquad \frac{dz}{dx} = 5x^4$$

and

$$\frac{dy}{dx} = \frac{dy}{dz}\frac{dz}{dx} = (4z^3 - 3) \cdot 5x^4.$$

We could express dy/dx in terms of x alone by replacing z by $x^5 + 1$.

The other founder of the calculus, Newton, used a completely different notation. His derivatives (which he called *fluxions*) were designated by $\dot{y}$ and $\ddot{y}$, in place of dy/dx and d^2y/dx^2. This notation is still used today, primarily to show derivatives with respect to time [i.e., if $y = f(t)$, where t designates time, then $\dot{y} = f'(t)$].

Leibnitz also introduced variables dx and dy which he called *differentials*. If

$$y = f(x),$$

then he defined

$$dy = f'(x)\, dx.$$

In his notation, we have

$$dy = \left(\frac{dy}{dx}\right) dx.$$

Thus, dy depends on f, x, and dx for its value.

Example 3 If $f(x) = \sqrt{2x + 3}$, then

$$\begin{aligned} f'(x) &= D_x(2x + 3)^{1/2} \\ &= \tfrac{1}{2}(2x + 3)^{-1/2} \cdot 2 \\ &= \frac{1}{\sqrt{2x + 3}} \end{aligned}$$

and

$$dy = \frac{1}{\sqrt{2x + 3}}\, dx.$$

If, for example, $x = 11$ and $dx = .01$, then

$$dy = \tfrac{1}{5} \cdot (.01) = .002.$$

Since

$$f'(x) = \lim_{\Delta x \to 0} \frac{\Delta y}{\Delta x},$$

we have

$$\lim_{\Delta x \to 0} \left[\frac{\Delta y}{\Delta x} - f'(x)\right] = 0$$

and $$\lim_{\Delta x \to 0} \frac{\Delta y - f'(x)\,\Delta x}{\Delta x} = 0.$$

Hence, the numerator of this fraction must be small in comparison to Δx when Δx is close to 0. That is,

$$\Delta y \doteq f'(x)\,\Delta x \qquad (\doteq \text{ means "is approximately equal to"}).$$

If we let

$$dx = \Delta x,$$

then $f'(x)\,dx$ is dy by definition, and we have

$$\Delta y \doteq dy \qquad \text{when } dy \text{ is close to } 0.$$

Example 4 If $f(x) = x^2 - 1/x$ as in Example 1, then

$$dy = f'(x)\,dx = \left(2x + \frac{1}{x^2}\right) dx$$

and (letting $dx = \Delta x$),

$$\Delta y = 2x\,dx + dx^2 + \frac{dx}{x(x + dx)}.$$

If we let $x = 2$, then

$$dy = 4.25\,dx, \qquad \Delta y = \left(4 + dx + \frac{1}{x(x + dx)}\right) dx.$$

For various values of dx, we compute the following approximate table of values.

dx	1	.5	.1	.01
dy	4.25	2.13	.425	.0425
Δy	5.17	2.35	.434	.0426

We note that as Δx gets close to 0, the difference between dy and Δy is much smaller than dx.

The relationship between dy and Δy is indicated geometrically in Figure 3.6. Thus Δy is the actual difference in ordinates between points P and Q, whereas dy is the rise (or fall) in the tangent line at P when x changes from x to $x + dx$.

The fact that the differential dy is an approximation of Δy when dx is small may be used to approximate errors, as is indicated in the following example.

Example 5 A box in the form of a cube has an edge of length $x = 4$ in. with a possible error of .05 in. What is the possible error in the volume V of the box?

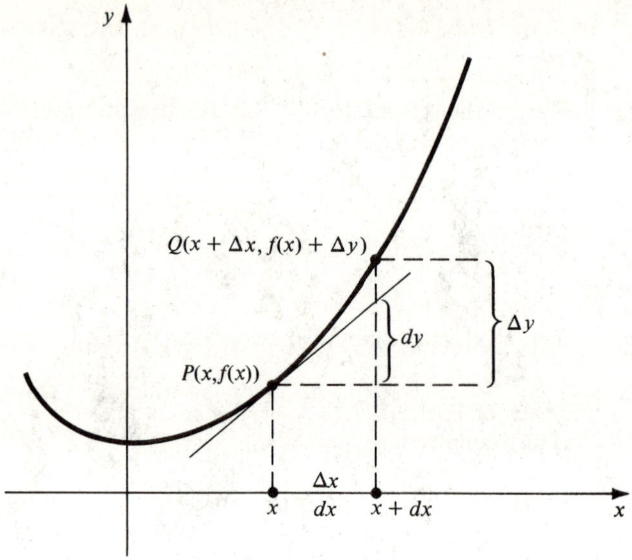

FIGURE 3.6

Solution: If the length of the edge is changed from x to $x + dx$, then the change in the volume ΔV is given by

$$\Delta V = (x + dx)^3 - x^3.$$

By our results above, $\Delta V \doteq dV$, where $V = x^3$, $V' = 3x^2$, and

$$dV = 3x^2\, dx.$$

If we let $x = 4$ and $dx = \pm.05$, then

$$dV = 3 \cdot 16 \cdot (\pm.05) = \pm 2.4.$$

Hence the possible error in the volume is approximately ± 2.4 in.^{3}

The superiority of Lagrange's notation over that of Leibnitz is that it indicates the basic fact that the derivative of a function f is another function f'. The notation of Leibnitz for the derivative dy/dx focuses our attention on the definition of the derivative of a function at a particular value of x. It makes us tend to lose sight of the fact that the derivative is again a function.

Example 6 Use differentials to approximate $\sqrt[3]{1001}$.

Solution: If $y = \sqrt[3]{x}$, then if x changes from 1000 to 1001, y changes by $\Delta y = \sqrt[3]{1001} - \sqrt[3]{1000}$. We approximate Δy by dy,

$$dy = \left(\frac{dy}{dx}\right) dx \text{ evaluated at } x = 1000, \qquad dx = 1.$$

Now $dy/dx = \frac{1}{3}x^{-2/3} = \frac{1}{3\sqrt[3]{x^2}}$ and when $x = 1000$, $dx = 1$, we have

$$dy = \frac{1}{3\sqrt[3]{(1000)^2}} \cdot 1 = \tfrac{1}{300}.$$

Thus $$\sqrt[3]{1001} - \sqrt[3]{1000} \doteq \tfrac{1}{300}$$

and $$\sqrt[3]{1001} \doteq 10 + \tfrac{1}{300} \doteq 10.003.$$

EXERCISES

In each of Exercises 1 to 4, express $\Delta y = f(x + \Delta x) - f(x)$ in terms of x and Δx (as in Example 1) and compare it with the value of dy.

1. $f(x) = x^3 + 2x^2$

2. $f(x) = x + \dfrac{1}{x^2}$

3. $f(x) = \dfrac{x}{x^2 + 1}$

4. $f(x) = \sqrt{2x + 1}$

For the given values of x and Δx find Δy, dy, and $\Delta y - dy$:

5. $y = 3x^2 + x - 2$; $x = 1$, $\Delta x = .01$

6. $y = \dfrac{2}{x^3}$; $x = 2$, $\Delta x = .002$

7. $y = x^3 - 1$; $x = 1$, $\Delta x = -.5$

8. $y = x^2 - x$; $x = -1$, $\Delta x = -.0001$

9. $y = \sqrt[3]{x}$; $x = 64$, $\Delta x = 1$

10. $y = \sqrt{7x + 1}$; $x = 5$, $\Delta x = .1$

Use differentials to approximate each of the following:

11. $\sqrt{9.05}$

12. $(8.1)^{2/3}$

13. $1/\sqrt{26}$

14. $37^{3/2}$

15. $\sqrt[5]{33}$

16. $1/\sqrt[3]{123}$

17. A circle has a radius of 6.0 in. If the possible error in the measurement of this radius is at most $\pm .3$ in., what is the possible error in the area of the circle?

18. A sphere has a radius of 15.0 in. If the possible error in the radius is $\pm .4$ in., what is the possible error in the surface area of the sphere?

19. A piece of wire 20.0 in. long is bent into the shape of a rectangle. If the length of one side is 4.0 in. with a possible error of $\pm .2$ in., what is the possible error in the area of the rectangle?

20. A box with vertical sides and a square base is made so that the height is exactly twice the width of the base. If the width of the base is 8.5 in. with a possible error of $\pm .3$ in., what is the possible error in the volume of the box?

21. A balloon has a diameter of 10 ft. Approximately how much does the volume increase if the diameter increases by 1 in.?

CHAPTER REVIEW

I

Use the definition of the derivative to find:

1. $g'(2)$, where $g(x) = \dfrac{x+3}{2x-5}$
2. $f'(3)$, where $f(x) = 7 - 4x^2$
3. $G'(-4)$, where $G(x) = 3 - \sqrt{5+x}$
4. $F'(x)$, where $F(x) = (3x-2)^2 - 3$

Find the derivative of the given function.

5. $f(x) = x^5 - 5x + 1$
6. $f(x) = x^4 + 3x^2 - 2x + 5$
7. $g(t) = t^3(t^3 - 1)$
8. $g(t) = t^4(1 - 2t + t^3)$
9. $F(x) = \dfrac{1}{2x+3}$
10. $F(x) = \dfrac{x}{2x+3}$
11. $G(y) = \dfrac{1}{y} - \dfrac{2}{y^3}$
12. $G(y) = 3 - \dfrac{4}{y^2}$
13. $g(x) = (2x - x^2)^5$
14. $g(x) = (3 - x^3)^6$
15. $f(t) = \sqrt{1-t^2}$
16. $f(t) = \dfrac{1}{\sqrt{1+t^2}}$
17. $F(y) = y\sqrt{y+1}$
18. $F(y) = y\sqrt{1-y}$
19. $s(t) = (t^3 + \sqrt{t^4+9})^{1/3}$
20. $s(x) = \left(\dfrac{2x-1}{8x+3}\right)^{4/3}$

Find equations of the tangent line and normal line to the graph of the function at the given point.

21. $f(x) = (x^2-1)^2$, (1,0)
22. $g(x) = \sqrt{2x+1}$, (4,3)
23. $g(x) = \dfrac{\sqrt{x}}{x+1}$, (4,.4)
24. $f(x) = (3x-1)^9$, $(1,2^9)$
25. Find an equation of the tangent line to the graph of $f(x) = \frac{1}{3}x^2 - 3x$ which is parallel to the line $6x - y = 4$.
26. Find an equation of the normal line to the graph of $y = 3x^2 + 4$ which is parallel to the line $2x + y = 1$.

Use implicit differentiation to find $f'(x)$ [letting $y = f(x)$] if f is defined by the equation.

27. $x^3 - 3xy + y^2 = 4$
28. $x^2 - 3y^3 = 4$
29. $\sqrt{x+1} - \sqrt{y+1} = 2$
30. $x + xy^2 + x^2y + y = 3$
31. $(3x^2 + y)^2 + (x+y)^{1/2} = x$
32. $\dfrac{1}{x+y} = y + x$

33. At what points is the slope of the tangent line to the curve $y = x^3$ equal to 12?
34. Determine the slope of the tangent line to the circle $(x - 1)^2 + (y + 3)^2 = 17$ at the point (2,1).
35. At what points of the curve $y = x^3 + x - 2$ is the tangent line parallel to the line $y = 4x - 1$?

Find Δy, dy, and $\Delta y - dy$ for:

36. $f(x) = x^3 + 6x - 2$; $x = 3$, $\Delta x = .01$
37. $F(x) = \sqrt{x^2 + 5}$; $x = 2$, $\Delta x = .1$
38. $xy - x = \dfrac{6}{y}$; $x = 3$, $y = 2$, $\Delta x = .03$
39. $y + \dfrac{1}{x} = \dfrac{10}{x + y}$; $x = \frac{1}{2}$, $y = 2$, $\Delta x = -.2$

Find an equation of the tangent line to the graph of:

40. $xy - \sqrt{x} + \sqrt{y} = 5$, at (1,4)
41. $(2x + y)^{1/4} = \sqrt[3]{x} + y + 1$, at $(1, -1)$
42. If $F(x) = kx^2$, show that the tangent line to the graph of F at the point $(c, F(c))$ is parallel to the secant line on the points $(a, F(a))$ and $(b, F(b))$ if and only if $c = (a + b)/2$.
43. Find f', f'', and f''' for each differentiable function f [letting $y = f(x)$] defined by $x = y^5 + y + 1$.
44. Discuss the continuity and differentiability of the function $f(x) = |x|^3$ at $x = 0$.
45. Approximate: (a) $\sqrt[3]{996}$ and (b) $\sqrt{99}$.
46. If $y = F(x)$, $x = G(z)$, and $z = H(t)$, find an expression for dy/dt, assuming differentiability of all functions involved.

II

1. How many tangent lines can be drawn to a given parabola from a given point? Prove your answer.
2. Determine equations of the common tangent line and common normal line to:

$$x^2 + y^2 = 2ax \qquad \text{and} \qquad y^2 = 2ax.$$

3. Prove the focusing property of parabolas, i.e., all rays which come in parallel to the axis and are reflected such that the angle of incidence equals the angle of reflection pass through the focus. It will be a subsequent problem to show this is the only smooth curve having this property. Give some applications of this property.
4. Let L be the tangent line at an arbitrary point A of a parabola. If F is the focus of the parabola and B is the point of intersection of L and the directrix, prove that the lines AF and BF are perpendicular.

5. Let L be the tangent line at a point A other than the vertex of a parabola, K the tangent line at the vertex, and F the focus of the parabola. If B is the point of intersection of K and L, prove that the lines AB and BF are perpendicular.

6. One application of tangent lines is Newton's method for approximating the real roots of an arbitrary equation $F(x) = 0$ (F, however, must be differentiable). The method consists of first making an initial approximation x_0 (sufficiently close) to the root to be approximated. For a "better" approximation, a tangent line is drawn to the graph of $y = F(x)$ at the point $(x_0, F(x_0))$. The x intercept

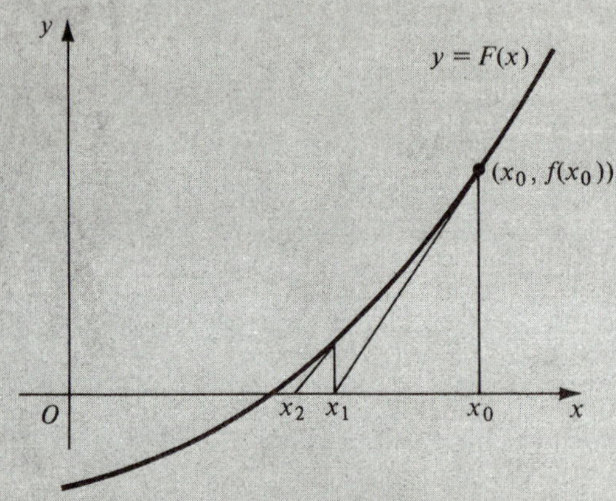

of this tangent line is the "better" approximation x_1. Starting at x_1, this process is repeated again and is continued until the desired degree of accuracy is obtained. If the initial approximation is too coarse (not sufficiently close), the successive approximations could progressively get worse each time. One should avoid starting at a point which is close to a local maximum or minimum value of the function.

Show that two successive approximations are related by (difference equation)

$$x_{n+1} = x_n - \frac{F(x_n)}{F'(x_n)}, \qquad n = 0, 1, 2, 3, \ldots.$$

7. Use Exercise 6 to approximate to $\sqrt[3]{1001}$, correct to four decimal places. How is this method related to that in Section 10 of this chapter?

8. Using Newton's method, approximate to four decimal places the root of the equation $x^3 - 3x^2 + 3 = 0$ that lies between 2 and 3.

Approximate to four decimal places the real root of the equation:

9. $x^3 + 2x + 1 = 0$

10. $x^3 - 3x = 3$

4

Applications of the Derivative

We have already discussed one application of the derivative, namely, the problem of finding tangent lines to the graph of a function.

One of the most useful and interesting applications of the derivative is to aid in the determination of the maximum and minimum values of a function. Many practical problems seeking the "best" way to do something can be formulated as problems to find maximum or minimum values of a function. Much of this chapter is devoted to the study of maxima and minima.

When a derivative is thought of as the instantaneous rate of change of a function, many physical applications of the derivative present themselves. The most obvious application of the derivative of this type is that of finding the velocity and acceleration of a moving object. This is discussed toward the end of the present chapter.

1 EXTREMA OF A FUNCTION

The graph of the function

$$f(x) = x^2 + 1$$

is sketched over the interval $[-1,2]$ in Figure 4.1. It is evident from the figure that the maximum value of f over $[-1,2]$ is $f(2) = 5$ and the minimum value is $f(0) = 1$ according to the following definition.

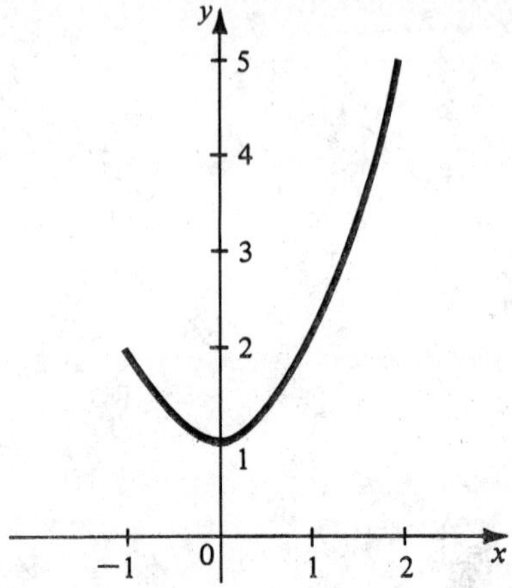

FIGURE 4.1

4.1 Definition

Let f be a function and A be a set of numbers contained in the domain of f. A number k is called the *maximum value of f over A* if

$$f(x) \leq k \text{ for every } x \text{ in } A \qquad \text{and} \qquad f(c) = k \text{ for some } c \text{ in } A.$$

The *minimum value of f over A* is defined similarly. If k is either a maximum or minimum value of f over A, then k is called an *extremum* of f over A.

Example 1 The function $g(x) = 1/x$ is sketched in Figure 4.2. The domain of g is $D = (-\infty,0) \cup (0,\infty)$. Over D, g has neither a maximum value nor a minimum value. Over the interval $(0,1]$, g doesn't have a maximum value but it does have a minimum value

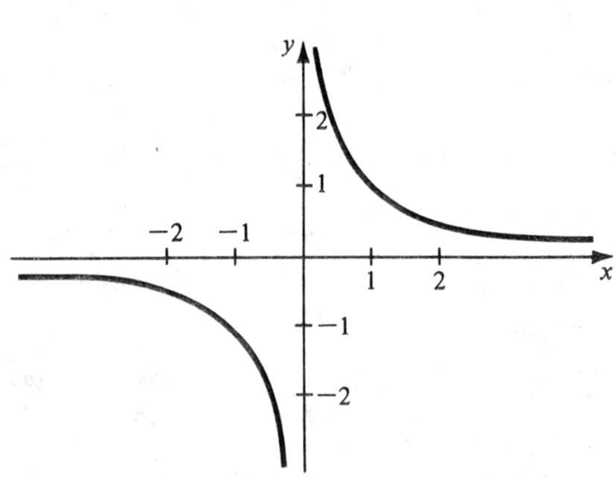

FIGURE 4.2

$g(1) = 1$. Over the interval $[-2,-1]$, g has a maximum value $g(-2) = -\frac{1}{2}$ and a minimum value $g(-1) = -1$.

The following basic theorem is given here without proof. Its proof is to be found in Chapter 5, Section 10, after a more detailed analysis of the real number system has been carried out.

4.2 Theorem

If a function f is continuous in a closed interval I, then f has both a maximum value and a minimum value over I.

For example, the function $f(x) = x^2 + 1$ of Figure 4.1 is continuous everywhere, and hence it is continuous in the closed interval $[-1,2]$. We saw that f has a maximum value of 5 and a minimum value of 1 over $[-1,2]$.

The function $g(x) = 1/x$ of Figure 4.2 is also continuous at every number in its domain D. Therefore, it is continuous over every closed interval contained in D. For example, g is continuous in $[-2,-1]$, and has maximum value of $-\frac{1}{2}$ and a minimum value of -1 over $[-2,-1]$. Note that 4.2 does not apply to a nonclosed interval such as $(0,1]$; the function g does not have a maximum value over $(0,1]$.

There is a basic relationship, stated below, between an extremum of a function f and the derivative of f.

4.3 Theorem

If the derivative f' of f exists at a number c and if $f'(c) \neq 0$, then $f(c)$ is not an extremum of f in any neighborhood of c.

Proof: If $f'(c) > 0$, then

$$\operatorname*{limit}_{x \to c} \frac{f(x) - f(c)}{x - c} > 0.$$

Hence, by 2.6, there exists a neighborhood N of c such that

$$\frac{f(x) - f(c)}{x - c} > 0 \text{ for every } x \text{ in } N, \qquad x \neq c.$$

This means that $f(x) - f(c)$ and $x - c$ have the same sign in N; that is, for every x in N,

$$(1) \qquad f(x) < f(c) \text{ if } x < c, \qquad f(x) > f(c) \text{ if } x > c$$

as illustrated in Figure 4.3. Thus, $f(c)$ is neither a maximum nor a minimum value of f in N. For any other neighborhood N' of c, every x in the neighborhood $N \cap N'$ satisfies (1). Hence, $f(c)$ is not an extremum of f in $N \cap N'$ or in N'. A similar argument holds if $f'(c) < 0$.

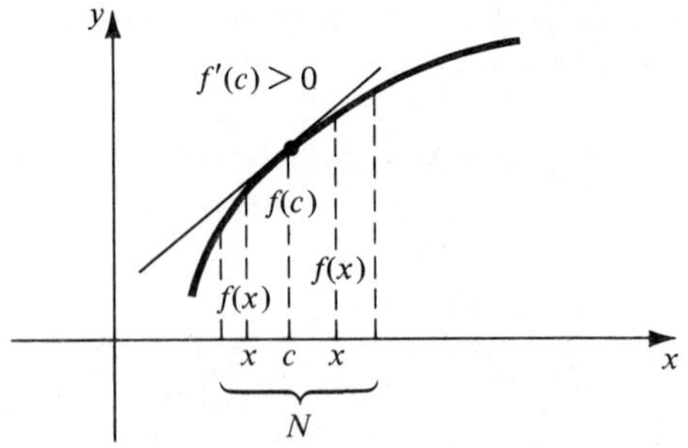

FIGURE 4.3

Another way of stating 4.3 is as follows (this theorem is called the *contrapositive* of 4.3).

4.4 Theorem

If $f(c)$ is an extremum of a function f in some neighborhood of c, then either $f'(c)$ does not exist or $f'(c) = 0$.

Try drawing a curve over some interval I so that the curve has an extremum at some point *inside* the interval. See if you don't always have a horizontal tangent line or no tangent line at the extremum! Some examples are shown in Figure 4.4.

For convenience, we make the following definition.

4.5 Definition

A number c in the domain of a function f is called a *critical number* of f if either $f'(c)$ does not exist or $f'(c) = 0$.

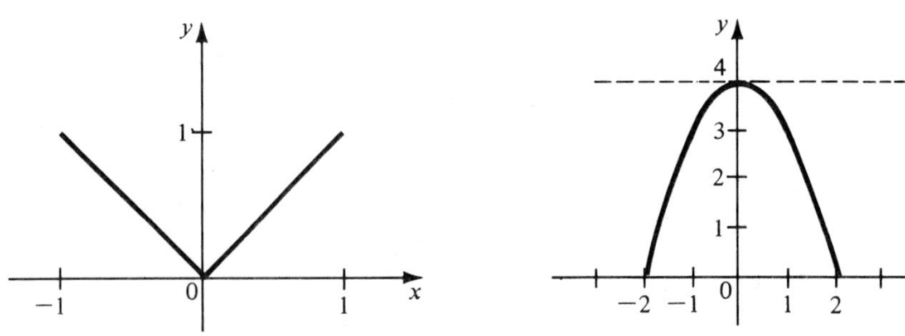

$f(x) = |x|, f'(0)$ does not exist

$f(x) = 4 - x^2, f'(0) = 0$

FIGURE 4.4

By 4.4, if $f(c)$ is an extremum of a continuous function f in a closed interval $[a,b]$ and if $a < c < b$, then necessarily either $f'(c) = 0$ or $f'(c)$ does not exist.

It is not true, however, that if c is a critical number of f then $f(c)$ is an extremum of f in some neighborhood of c. Consider the following example.

Example 2 If $f(x) = x^3$, then

$$f'(x) = 3x^2, \qquad f'(0) = 0.$$

Thus, 0 is a critical number of f. Actually, it is the only critical number. We note that

$$f(0) = 0; \qquad f(x) < f(0) \text{ if } x < 0; \qquad f(x) > f(0) \text{ if } x > 0.$$

Therefore, $f(0)$ is not an extremum of f over any neighborhood N of 0 (see Figure 4.5).

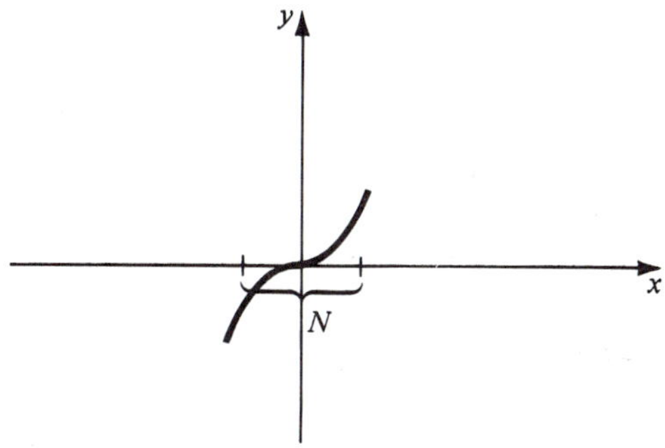

FIGURE 4.5

A useful test for finding critical numbers follows.

4.6 Theorem

If a function f is continuous in an interval I and if $f'(a) > 0, f'(b) < 0$ for some a, b in I, then f has a critical number in I.

To prove 4.6 assume that $a < b$ and then select c in $[a,b]$ so that $f(c)$ is the maximum value of f in $[a,b]$. Do you see why c cannot equal a or b? Then c is a critical number by 4.4.

Example 3 Show that the function f defined by

$$f(x) = x^4 - 2x^3 + 3x - 1$$

has a critical number in the interval $(-1,2)$.

Solution: We have

$$f'(x) = 4x^3 - 6x^2 + 3$$

and $f'(-1) = -7$, $f'(2) = 11$. Since f clearly is continuous in $[-1,2]$, f has, by 4.6, a critical number c in $(-1,2)$. Evidently, $f'(c) = 0$. Thus c is a root of the equation $4x^3 - 6x^2 + 3 = 0$.

4.7 Rolle's Theorem*

If f is a continuous function in the closed interval $[a,b]$ and if $f(a) = f(b)$, then f has at least one critical number in the open interval (a,b).

Proof: If $f(x) = f(a)$ for every x in $[a,b]$, then f is a constant function and $f'(x) = 0$. Hence every x in (a,b) is a critical number. If $f(x) \neq f(a)$ for some x in (a,b), then either the maximum value [if $f(x) > f(a)$] or the minimum value [if $f(x) < f(a)$] of f occurs at a number c in (a,b). The number c is, by 4.4, a critical number of f.

Examples illustrating Rolle's theorem are shown in Figure 4.4.

(1) $f(x) = |x|$ is continuous in $[-1,1]$, $f(1) = f(-1) = 1$, 0 is a critical number in $(-1,1)$.
(2) $f(x) = 4 - x^2$ is continuous in $[-2,2]$, $f(-2) = f(2) = 0$, 0 is a critical number in $(-2,2)$.

A result similar to Rolle's theorem holds for any function f such that f is continuous in a closed interval $[a,b]$ and f' exists in the open interval (a,b). Geometrically, it states that some tangent T is parallel to the secant line S on the points $(a,f(a))$ and $(b,f(b))$ (Figure 4.6). Line T has slope $f'(c)$ and S has

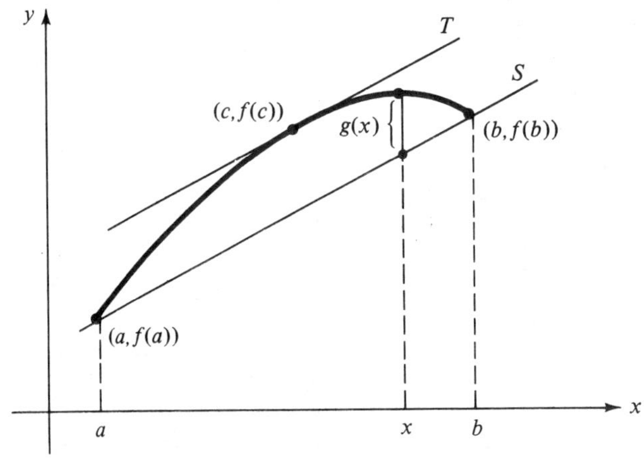

FIGURE 4.6

* Michel Rolle (1652–1719) was a French mathematician principally known for his book, *Traite d'algebre*, published in 1690.

slope $[f(b) - f(a)]/(b - a)$; so it states

$$\frac{f(b) - f(a)}{b - a} = f'(c) \qquad \text{for some } c \text{ in } (a,b).$$

This result is stated below.

4.8 Mean Value Theorem

If f is a continuous function in a closed interval $[a,b]$ and if f' contains the open interval (a,b) in its domain, then there exists a number c in (a,b) such that

$$f(b) - f(a) = (b - a)f'(c).$$

Proof: Let S be the secant line of the graph of f passing through the points $(a,f(a))$ and $(b,f(b))$, as shown in Figure 4.6. Evidently,

$$y = m(x - a) + f(a), \qquad \text{where } m = \frac{f(b) - f(a)}{b - a},$$

is an equation of S. For each number x in $[a,b]$ let $g(x)$ be the directed distance from line S to the graph of f, measured parallel to the y axis (Figure 4.6). Clearly,

$$g(x) = f(x) - [m(x - a) + f(a)]$$

for every x in $[a,b]$, and $g(a) = g(b) = 0$. Since g is continuous in $[a,b]$ and g' exists in (a,b), we may apply Rolle's theorem to the function g to conclude that $g'(c) = 0$ for some c in (a,b). We have

$$g'(x) = f'(x) - m \qquad \text{for every } x \text{ in } (a,b),$$

and therefore

$$g'(c) = f'(c) - m = 0$$

for some c in (a,b). Thus

$$f'(c) = m = \frac{f(b) - f(a)}{b - a}$$

and

$$f(b) - f(a) = (b - a)f'(c).$$

Example 4 If $f(x) = 1 + 4/x$, then f is continuous in $[1,4]$. Clearly,

$$f'(x) = -\frac{4}{x^2}$$

and f' exists in $(1,4)$. Hence, by the mean value theorem,

$$f(4) - f(1) = (4 - 1)f'(c)$$

for some c in (1,4). We compute $f(4) = 2, f(1) = 5$. Therefore,

$$2 - 5 = 3\left(-\frac{4}{c^2}\right),$$

$$c^2 = 4,$$

$$c = 2.$$

Thus,
$$f(4) - f(1) = (4 - 1)f'(2).$$

EXERCISES

In each of Exercises 1 to 10, find the critical numbers of the given function in the indicated interval.

1. $F(x) = 3 - 5x - x^2$, $[-3,0]$

2. $G(z) = 2z^3 + 3z - 1$, $[-3,3]$

3. $f(x) = x^3 + x^2 + x - 4$, $[-1,1]$

4. $g(x) = x^4 - x^3$, $[0,1]$

5. $f(r) = r^2 + \dfrac{4}{r^2}$, $[-2,2]$

6. $g\ x) = \dfrac{x - 1}{x^2 + 3}$, $[-4,2]$

7. $f(x) = \begin{cases} \sqrt{4 - x^2}, & |x| \le 2, \\ 0, & |x| > 2, \end{cases}$ in the interval $[-3,3]$

8. $g(x) = |x^2 - 9|$, $[-4,4]$

9. $s(t) = \begin{cases} -t \text{ if } t \le 0, \\ 4t \text{ if } 0 < t \le 1, \\ 4t^2 \text{ if } t > 1, \end{cases}$ $[-1,2]$

10. $f(x) = \begin{cases} 2x + 6 \text{ if } -3 \le x \le 1, \\ 8x \quad \text{if } 11 < x \le 4, \end{cases}$ $[-3,4]$

Show that the function f has a critical number in the given interval.

11. $f(x) = x^5 + 50x^2 + x$, $(-2,1)$

12. $f(x) = 3x^4 - 20x^3 + 7x$, $(2,10)$

In each of Exercises 13 to 16, verify Rolle's theorem by finding a critical number of the given function in the indicated interval (a,b).

13. $f(x) = 2x + \dfrac{1}{x}$; $a = \frac{1}{4}$, $b = 2$

14. $f(x) = x^3 - x - 1$; $a = -1$, $b = 1$

15. $f(x) = \sqrt{|x - 1|}$; $a = -1$, $b = 3$

16. $f(x) = |1 - x^2|$; $a = 0$, $b = \sqrt{2}$

In each of Exercises 17 to 19, verify Rolle's theorem by finding the values of x for which $F(x)$ and $F'(x)$ vanish.

17. $F(x) = 3x - x^3$

18. $F(x) = x^3 - ax^2$

19. $F(x) = x^n(x - a)$, n any positive integer

20. If $(y + 4)^3 = x^2$, then $y = 0$ when $x = 8$ or $x = -8$. Does Rolle's theorem justify the conclusion that $dy/dx = 0$ for some number x in the interval $(-8,8)$? Check by sketching the graph of the given equation.

In each of Exercises 21 to 26, verify the mean value theorem for the given function in the interval $[a,b]$ by finding a point in (a,b) where the secant and tangent lines are parallel.

21. $f(x) = x^2$; $a = -2$, $b = 3$

22. $g(x) = \dfrac{1}{x - 1}$; $a = 2$, $b = 4$

23. $F(x) = \sqrt{x + 1}$; $a = 0$, $b = 3$

24. $G(x) = 1 - 2x^3$; $a = -1$, $b = 1$

25. $h(x) = x + \dfrac{1}{x}$; $a = 1$, $b = 3$

26. $f(x) = \sqrt{x^2 + x - 2}$; $a = 1$, $b = 4$

Verify that the hypothesis of the mean value theorem is satisfied by the given function over the given interval. Find a value of c as predicted by the conclusion of the mean value theorem.

27. $f(x) = \sqrt{x + 5}$, $[-1,4]$

28. $F(x) = x^3 + x^2 - 1$, $[-2,-1]$

29. $g(x) = x^2 + 2x - 5$, $[1,5]$

30. $s(t) = t + 2 + \dfrac{1}{t - 3}$, $[-1,2]$

In each of Exercises 31 to 34, either verify that the mean value theorem holds for the given function or give a reason why it does not.

31. $g(x) = \dfrac{x - 1}{x}$; $a = 1$, $b = 3$

32. $f(x) = |x|$; $a = -1$, $b = 2$

33. $F(x) = [x]$; $a = -\frac{1}{2}$, $b = \frac{3}{2}$

34. $G(x) = \begin{cases} 1 + x^2, & x \geq 0; \\ 1 - x^2, & x < 0; \end{cases}$ $a = -1$, $b = 1$

35. Let $f(x) = x^2$ for $x \leq 1$, $f(x) = x$ for $x > 1$. Does the mean value theorem hold for f in the interval $[\frac{2}{3},2]$? Why or why not?

36. In Exercise 35 suppose that the definition of $f(x)$ is changed to $f(x) = 2x$ for $x > 1$. Answer the same questions.

37. If function F is constant in an open interval I, then $F'(x) = 0$ for every x in I. Prove, conversely, that if $F'(x) = 0$ for every x in I then F is constant in I.

2 MONOTONIC FUNCTIONS

The derivative of a function often indicates whether a function is increasing or decreasing, as we shall see below.

4.9 Definition

Let f be a function and I be an interval contained in the domain of f. Then:

(i) f is *increasing* in I if $f(x_1) \leq f(x_2)$, whenever $x_1 \leq x_2$,
(ii) f is *strictly increasing* in I if $f(x_1) < f(x_2)$, whenever $x_1 < x_2$,

for all x_1,x_2 in I. *Decreasing* and *strictly decreasing* functions are defined similarly.

Some examples of increasing and decreasing functions are shown in Figure 4.7.

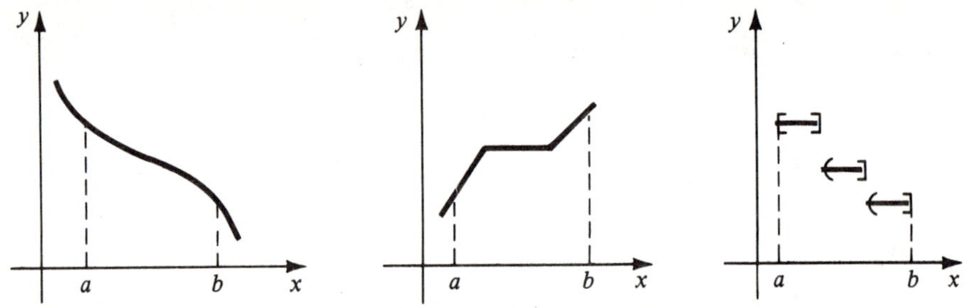

Strictly decreasing in $[a,b]$ Increasing in $[a,b]$ Decreasing $[a,b]$

FIGURE 4.7

4.10 Definition

A function f is *monotonic* in an interval I if f is either increasing in I or decreasing in I. Similarly, f is *strictly monotonic* in an interval I if f is either strictly increasing in I or strictly decreasing in I.

A test to tell if a function is strictly monotonic is as follows.

4.11 Theorem

Let I be an interval in which function f is continuous.

(1) *If $f'(x) > 0$ for every x in I, x not an endpoint of I, then f is strictly increasing in I.*
(2) *If $f'(x) < 0$ for every x in I, x not an endpoint of I, then f is strictly decreasing in I.*

Proof: (1) If x_1,x_2 are in I and $x_1 < x_2$, then, by the mean value theorem,

$$f(x_2) - f(x_1) = (x_2 - x_1)f'(c)$$

for some c in (x_1,x_2). Since $x_2 - x_1 > 0$ and $f'(c) > 0$ by assumption, evidently $f(x_2) - f(x_1) > 0$ and $f(x_1) < f(x_2)$. Hence, f is strictly increasing in I.

The proof of (2) is similar and hence is omitted.

4.12 Theorem

If a function f is continuous in an interval I and has no critical numbers in I except possibly at an endpoint of I, then f is strictly monotonic in I.

Proof: By 4.6, either $f'(x) > 0$ for every x in I, or $f'(x) < 0$ for every x in I, x not an endpoint. The theorem now follows from 4.11.

Once we have found the critical numbers of a function f, we can easily find the intervals in which f is monotonic, as shown below.

Example 1 Find the intervals in which the function g defined by

$$g(x) = x^3 - 3x + 1$$

is strictly monotonic.

Solution: The function g has derivative $g'(x) = 3x^2 - 3$, or

$$g'(x) = 3(x - 1)(x + 1).$$

Therefore, -1 and 1 are the only critical numbers of g. Since g is continuous everywhere, we conclude g is strictly monotonic in each of the intervals

$$(-\infty,-1], \qquad [-1,1], \qquad [1,\infty)$$

by 4.12. To tell whether g is increasing or decreasing, we need only check the value of g at two points in each interval. From the table of values

x	-2	-1	1	2
$g(x)$	-1	3	-1	3

g is strictly increasing in $(-\infty,-1]$ since $g(-2) < g(-1)$.
g is strictly decreasing in $[-1,1]$ since $g(-1) > g(1)$.
g is strictly increasing in $[1,\infty)$ since $g(1) < g(2)$.

The graph of g is sketched in Figure 4.8.

Example 2 Find the intervals in which the function f defined by

$$f(x) = x + \frac{4}{x^2}$$

is strictly monotonic.

Solution: The function f is undefined at $x = 0$, but continuous elsewhere. Evidently $x = 0$ is a vertical asymptote of the graph. Since

$$f'(x) = 1 - \frac{8}{x^3},$$

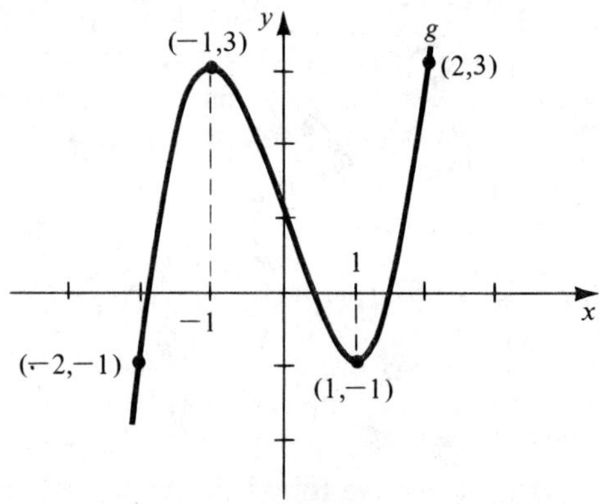

FIGURE 4.8

$f'(x) = 0$ if, and only if, $x = 2$. Thus 2 is the only critical number. Therefore, by 4.12, f is strictly monotonic in each of the intervals

$$(-\infty,0), \qquad (0,2], \qquad [2,\infty).$$

By 4.11,

f is strictly increasing in $(-\infty,0)$ since $f'(-1) > 0$.
f is strictly decreasing in $(0,2]$ since $f'(1) < 0$.
f is strictly increasing in $[2,\infty)$ since $f'(3) > 0$.

The graph of f is sketched in Figure 4.9.

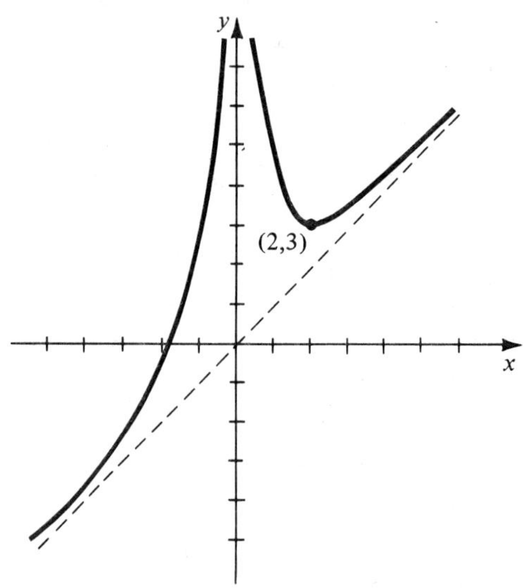

FIGURE 4.9

EXERCISES

In each of Exercises 1 to 14, find the intervals in which the given function is strictly increasing and those in which it is strictly decreasing. From this information sketch the graph of each function.

1. $f(x) = x^2 - 2x - 8$

2. $G(x) = 4 - 4x - x^2$

3. $f(x) = x^3 - 6x^2 + 9x + 2$

4. $f(x) = x^3 + 3x^2 + 1$

5. $f(x) = x^3 - x^4$

6. $f(x) = \dfrac{x}{x^2 + 1}$

7. $f(x) = \dfrac{x^2 + 1}{x^2 - 1}$

8. $F(x) = (x - 2)^3(x + 1)^2$

9. $f(x) = \sqrt{x} + \sqrt{x + 1}$

10. $F(x) = (x - 1)^{1/3} + \frac{1}{2}(x + 1)^{2/3}$

11. $G(x) = x + \dfrac{1}{x}$

12. $g(x) = x + \dfrac{5}{2x + 3}$

13. $g x) = \begin{cases} (x + 3)^2 - 4 \text{ if } x \le -1 \\ 3x + 3 \quad \text{if } x > -1 \end{cases}$

14. $F(t) = \begin{cases} 2t - 8 \text{ if } t \le 2 \\ -t^2 \quad \text{if } t > 2 \end{cases}$

15. If a function f is increasing in $[a,b]$ and also in $[b,c]$, is it increasing in $[a,c]$? Prove your answer.

16. If functions f and g are increasing in $[a,b]$, are $f + g$ and fg also increasing in $[a,b]$? Prove your answer.

17. If functions f and g are increasing in $[a,b]$ and $f \circ g$ exists in $[a,b]$, is $f \circ g$ increasing in $[a,b]$? Prove your answer.

3 RELATIVE EXTREMA OF A FUNCTION

The graph of a function may have many "relative" maximum and minimum points. In Figure 4.8, for example, $P(-1,3)$ is a "relative" maximum point and $Q(1,-1)$ is a "relative" minimum point of the graph of g. Note that P is not the highest point on the graph of g, but it is the highest point of the graph in a neighborhood of P. Similarly, Q is not the lowest point on the graph, but it is the lowest point in a neighborhood of Q. With this example in mind, let us make the following definition.

4.13 Definition

The number $f(c)$ is called a *relative extremum* of a function f if there exists a neighborhood N of c contained in the domain of f such that $f(c)$ is an extremum of f in N and $f(x) \neq f(c)$ if $x \neq c$, x in N.

If $f(c)$ is a relative extremum of a function f, then by 4.4 either $f'(c) = 0$ or $f'(c)$ does not exist; that is, c is a critical number of f. Whether or not $f(c)$ is a relative extremum of f in case c is a critical number may be decided by use of 4.11 and 4.12 as follows.

(1) Find (if possible) a neighborhood (a,b) of c such that c is the only critical number of f in (a,b) and f is continuous in $[a,b]$. By 4.12, f is strictly monotonic in $[a,c]$ and in $[c,b]$.

(2) $f(c)$ is a *relative maximum* value of f if $f(a) < f(c), f(c) > f(b)$.
$f(c)$ is a relative minimum value of f if $f(a) > f(c), f(c) < f(b)$.
$f(c)$ is not an extremum otherwise.

In place of (2), we can also apply (3):

(3) *First derivative test for extrema.*
$f(c)$ is a *relative maximum* value of f, if

$$f'(d) > 0, f'(e) < 0 \text{ for some } d \text{ in } [a,c),\ e \text{ in } (c,b].$$

$f(c)$ is a relative minimum value of f, if

$$f'(d) < 0, f'(e) > 0 \text{ for some } d \text{ in } [a,c),\ e \text{ in } (c,b].$$

$f(c)$ is not an extremum otherwise.

Example 1

The function

$$f(x) = x^3$$

has derivative

$$f'(x) = 3x^2.$$

Clearly, 0 is the only critical number. Select any neighborhood of 0, say $(-1,1)$. Using test (2) above, $a = -1$, $c = 0$, $b = 1$:

$$f(-1) = -1, f(0) = 0, f(1) = 1: \qquad \text{so } f(0) \text{ is not an extremum.}$$

Using the first derivative test,

$$f'(-1) > 0, f'(1) > 0: \qquad \text{so } f(0) \text{ is not an extremum.}$$

Example 2

Find the relative extrema of the function f defined by

$$f(x) = x^3 + 3x^2 - 1.$$

Solution: The functions f and f' have $(-\infty,\infty)$ as their domain. Clearly,

$$f'(x) = 3x(x + 2)$$

and $\{-2,0\}$ is the set of critical numbers of f. There is a unique critical number in each of the intervals $[-3,-1]$ and $[-1,1]$. Using (2) above,

$$f(-3) = -1, f(-2) = 3, f(-1) = 1:$$

$$\text{so } f(-2) = 3 \text{ is a relative maximum value of } f.$$

$$f(-1) = 1, f(0) = -1, f(1) = 3:$$

$$\text{so } f(0) = -1 \text{ is a relative minimum value of } f.$$

The graph is sketched in Figure 4.10.

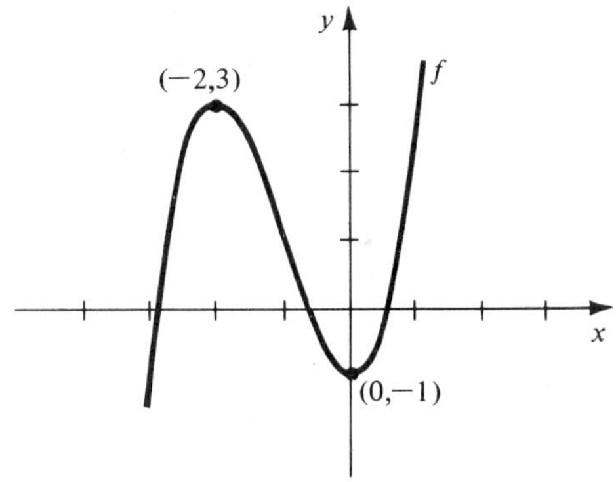

FIGURE 4.10

Example 3 Find the relative extrema of the function g defined by

$$g(x) = x^3 + \frac{3}{x}.$$

Solution: We have

$$g'(x) = 3x^2 - \frac{3}{x^2} = \frac{3(x^4 - 1)}{x^2}.$$

Evidently, the domain of both g and g' is the set of nonzero real numbers. Thus the set of critical numbers of g is $\{x \mid g'(x) = 0\}$ or $\{x \mid x^4 - 1 = 0\}$. That is, $\{-1,1\}$ is the set of critical numbers of g. The function g is continuous in its domain; so it is continuous in $[-2,-\frac{1}{2}]$ and $[\frac{1}{2},2]$. Each of these intervals contains one critical number. Using (3), the first derivative test,

$g'(-2) > 0, g'(-\frac{1}{2}) < 0$:

hence $g(-1) = -4$ is a relative maximum value of g.

$g'(\frac{1}{2}) < 0, g'(2) > 0$:

hence $g(1) = 4$ is a relative minimum value of g.

Note that the relative minimum value of g is greater than the relative maximum value! The graph of g in Figure 4.11 shows how this can happen. Of course, the y axis is an asymptote.

Example 4 Find the relative extrema of the function F defined by

$$F(x) = \sqrt[3]{x}(x - 7)^2.$$

Solution: The function F is continuous in the interval $(-\infty,\infty)$. Its derivative is given by

$$F'(x) = \tfrac{1}{3}x^{-2/3}(x - 7)^2 + 2x^{1/3}(x - 7) = \frac{7(x - 7)(x - 1)}{3x^{2/3}}.$$

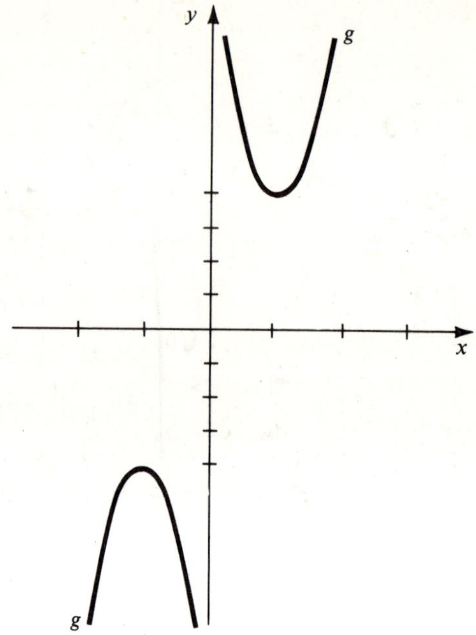

FIGURE 4.11

Thus, the domain of F' is the set of nonzero numbers. The critical numbers of F are 0, at which $F'(x)$ is undefined, and 1 and 7, at which $F'(x) = 0$. Each of the intervals

$$[-1,\tfrac{1}{8}], \qquad [\tfrac{1}{8},6], \qquad [6,8]$$

contains one critical number. Using (2) above,

$$F(-1) < 0,\ F(0) = 0,\ F(\tfrac{1}{8}) > 0:$$

hence, $F(0)$ is not an extremum.

$$F(\tfrac{1}{8}) < \tfrac{49}{2},\ F(1) = 36,\ F(6) = \sqrt[3]{6} < 2:$$

hence, $F(1) = 36$ is a relative maximum value of F.

$$F(6) > 0,\ F(7) = 0,\ F(8) > 0:$$

hence, $F(7) = 0$ is a relative minimum value of F.

The graph of F is sketched in Figure 4.12. Evidently, F has a vertical tangent line at the origin.

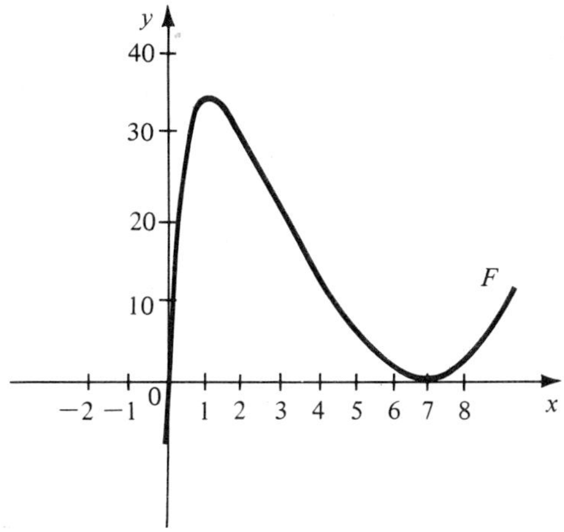

FIGURE 4.12

EXERCISES

I

In each of Exercises 1 to 20, find the extrema of the function and sketch the graph.

1. $f(x) = x^2 - 4x + 4$

2. $g(x) = 4 - x^2$

3. $F(x) = 3 - 5x - x^2$

4. $G(x) = 3x^2 - 4x + 7$

5. $g(x) = 2x^3 + 3x - 1$

6. $f(x) = x^3 - 6x^2 + 9x - 2$

7. $G(x) = x^3 + x^2 + x - 4$

8. $F(x) = x^4 - x^3$

9. $g(x) = (x + 2)^2(x - 3)^3$

10. $h(x) = (x + 2)^2(x - 3)^2$

11. $f(x) = (x^2 - 4)^2$

12. $g(x) = x^2(x^2 - 1)^2$

13. $F(x) = 3x^5 - 25x^3 + 60x$

14. $G(x) = x^4 - 4x$

15. $f(x) = x^3 + 3x^2 - 9x + 1$

16. $g(x) = x^3 + 3x^2 + 3x - 2$

17. $h(x) = x^3 + 3x$

18. $F(x) = x^2 + 2$

19. $F(x) = \begin{cases} 2x - 3 & \text{if } x \le 2 \\ x^2 - 3 & \text{if } x > 2 \end{cases}$

20. $f(x) = \begin{cases} \sqrt{15 + x^2} & \text{if } x \le -1 \\ 3 + x^2 & \text{if } x > -1 \end{cases}$

In each of Exercises 21 to 32, find the extrema of the function and sketch the graph.

21. $g(x) = x^4 - 2x^2$

22. $G(x) = 2x^3 - 3x^2$

23. $f(x) = x^2(x - 3)^3$

24. $g(x) = (2x^2 - 3x - 2)^2$

25. $g(x) = x^{2/3}(x - 5)$

26. $h(x) = x^{1/3}(x - 8)$

27. $F(x) = x\sqrt[3]{3x - 4}$

28. $f(x) = \sqrt[3]{x - 1} + \sqrt{x + 1}$

29. $f(x) = \sqrt[3]{x^3 - 9x}$

30. $F(x) = \sqrt[3]{x^2 - 2x}$

31. $G(x) = x^2 + \dfrac{1}{x^2}$

32. $g(x) = \dfrac{x - 1}{x^2 + 3}$

II

1. If p and q are integers and $f(x) = (x - 1)^p(x + 1)^q$, $p \geq 2$, $q \geq 2$, show that f has the three critical numbers, -1, $(q - p)/(q + p)$, 1. Find the extrema of f for the following cases:

- **a.** p and q are both even.
- **b.** p is even and q is odd.
- **c.** p is odd and q is even.
- **d.** p and q are both odd.

2. Show that the function G defined by $G(x) = (ax + b)/(cx + d)$ has no extrema regardless of the values of a, b, c, and d. Sketch the graph of G.

4 CONCAVITY

Just as the sign of the first derivative of a function f tells us about the rising and falling of the graph of f, so the sign of the second derivative of f tells us about the concavity of the graph of f. The concavity of a graph is defined below.

4.14 Definition

The graph of a function f is said to be *concave upward* at the point $(c,f(c))$ if $f'(c)$ exists and if there exists a deleted neighborhood D of c such that the graph of f in D is above the tangent line at $(c,f(c))$. *Downward concavity* is defined analogously.

For example, the graph of f is concave upward at $(c,f(c))$ in Figure 4.13 and is concave downward at $(c,f(c))$ in Figure 4.14. The equation of the tangent line T to the graph of f at point $(c,f(c))$, assuming $f'(c)$ exists, is given by

$$y = f(c) + f'(c)(x - c).$$

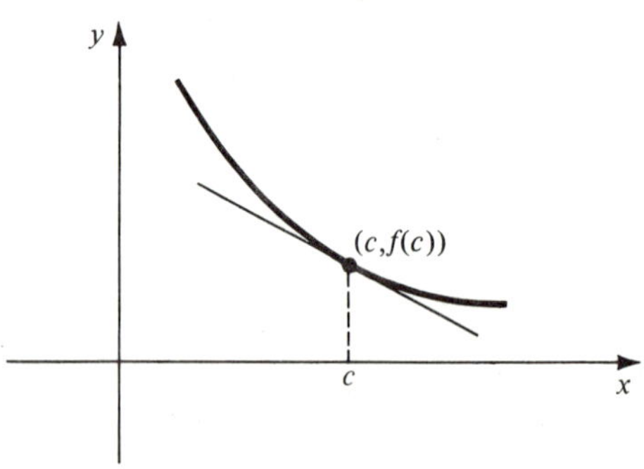

FIGURE 4.13

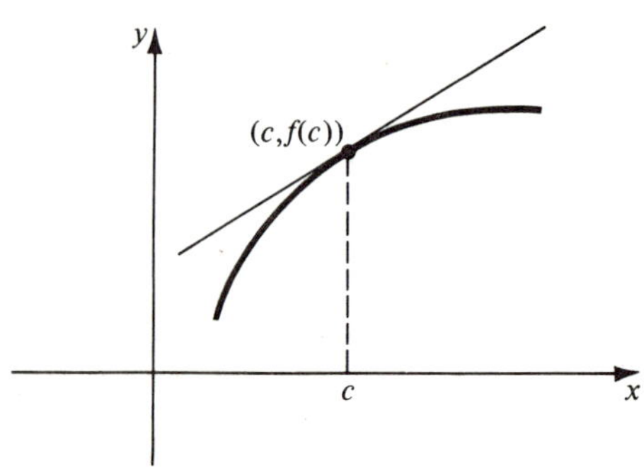

FIGURE 4.14

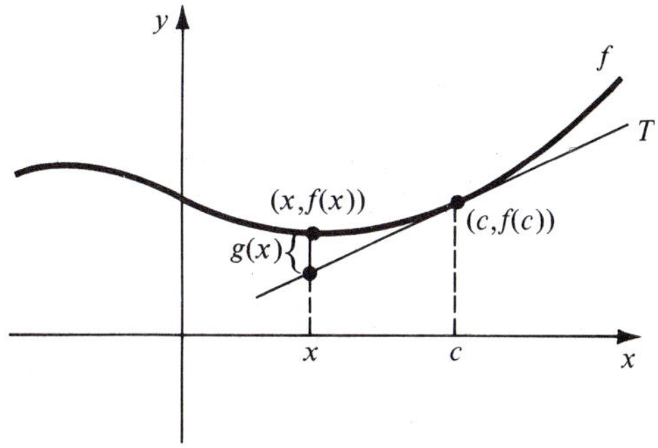

FIGURE 4.15

Hence the directed distance $g(x)$ from T to the graph of f at x, measured parallel to the y axis, is given by (Figure 4.15)

4.15
$$g(x) = f(x) - [f(c) + f'(c)(x - c)].$$

Since $g(x)$ is a directed distance, the point $(x,f(x))$ is above T if $g(x) > 0$ and below T if $g(x) < 0$. Clearly, $g(c) = 0$.

The concavity of a graph is easily stated in terms of the function g of 4.15. Thus the graph of f at $(c,f(c))$ is concave upward if $g(x) > 0$ for every x in some deleted neighborhood of c, and concave downward if $g(x) < 0$ for every x in some deleted neighborhood of c.

Test for Concavity 4.16

If f is a function and c is a number such that the derivative f' is defined in some neighborhood of c, then:

(1) *The graph of f is concave upward at $(c,f(c))$ if $f''(c) > 0$.*
(2) *The graph of f is concave downward at $(c,f(c))$ if $f''(c) < 0$.*

Proof: If $f''(c) > 0$, then there exists a neighborhood N of c such that $f'(x) < f'(c)$ if $x < c$, $f'(x) > f'(c)$ if $x > c$, for every x in N. Using the mean value theorem in 4.15, we have

$$\begin{aligned} g(x) &= [f(x) - f(c)] - f'(c)(x - c) \\ &= f'(d)(x - c) - f'(c)(x - c) \\ &= [f'(d) - f'(c)](x - c) \end{aligned}$$

for some number d between x and c. If x is in N and $x < c$, then $x < d < c$, $f'(d) < f'(c)$, and $g(x) > 0$. If x is in N and $x > c$, then $c < d < x$, $f'(c) < f'(d)$, and once again $g(x) > 0$. Hence $g(x) > 0$ for every x in N, $x \neq c$, and the graph of f is concave upward at $(c,f(c))$.

The proof of (2) is similar and is therefore omitted.

A useful corollary of 4.16 is given in the following theorem.

4.17 Second Derivative Test for Extrema

If f is a function and c a critical number of f such that the derivative f' is defined in some neighborhood of c, then:

(1) *$f(c)$ is a relative maximum value of f if $f''(c) < 0$.*
(2) *$f(c)$ is a relative minimum value of f if $f''(c) > 0$.*

These results follow easily from 4.16, since the tangent line is now horizontal and the graph of f is below the tangent line in (1) and above the tangent line in (2) for some neighborhood of c.

If c is a critical number of f for which either $f''(c) = 0$ or $f''(c)$ does not exist, then the second derivative test cannot be applied.

A special name is given below to those points on the graph of a function at which the tangent line cuts the graph in the sense indicated in Figure 4.16.

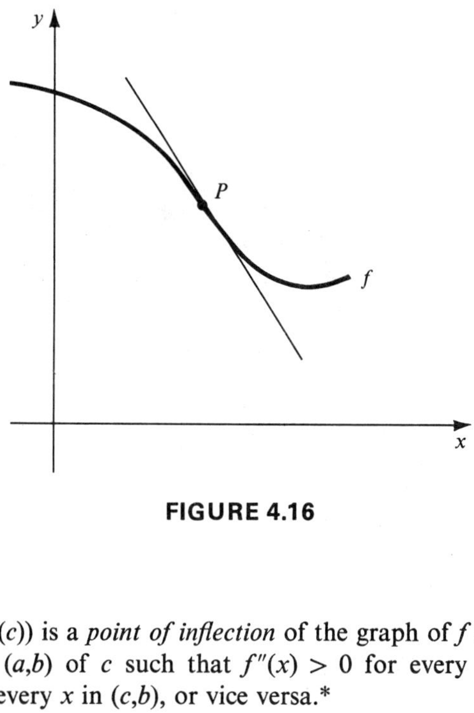

FIGURE 4.16

4.18 Definition

The point $(c,f(c))$ is a *point of inflection* of the graph of f if there exists a neighborhood (a,b) of c such that $f''(x) > 0$ for every x in (a,c) and $f''(x) < 0$ for every x in (c,b), or vice versa.*

The first property we notice for points of inflection is as follows.

---

* An interesting discussion of possible definitions of points of inflection may be found in an article by A. M. Bruckner, *The American Mathematical Monthly*, **69** (1962), 787.

4.19 Theorem

If $(c, f(c))$ is a point of inflection of the graph of f and if $f''(c)$ exists, then necessarily $f''(c) = 0$.

Proof: Let (a,b) be a neighborhood of c having the property stated in 4.18. If $g = f'$, then the function g has a critical number in (a,b) by 4.6. Since $g'(x) \neq 0$ for every x other than c in (a,b), the critical number of g must be c. Since $g'(c) = f''(c)$ exists, necessarily $g'(c) = f''(c) = 0$.

Example 1 The function

$$f(x) = x^4$$

has derivatives

$$f'(x) = 4x^3, \qquad f''(x) = 12x^2.$$

Clearly $f''(0) = 0$ and $f''(x) > 0$ for every $x \neq 0$. Thus, the graph of f is concave upward everywhere, and $(0,0)$ is not a point of inflection.

This example shows that having the second derivative zero is not enough to ensure a point of inflection. We need something more, such as the following theorem. Its proof is omitted, since it is a consequence of the theory in Section 2.

4.20 Theorem

If f is a function and c is a number such that $f''(c) = 0$, then $(c, f(c))$ is a point of inflection of the graph of f provided there exists a neighborhood (a,b) of c such that (1) $f''(x)$ exists and is nonzero for every x in $[a,b]$, $x \neq c$, and (2) $f''(a)$ and $f''(b)$ differ in sign.

Example 2 If $f(x) = x^3 - 3x^2$, find the relative extrema of f and the points of inflection of the graph of f.

Solution: We have

$$f'(x) = 3x^2 - 6x = 3x(x - 2), \qquad f''(x) = 6x - 6 = 6(x - 1).$$

Thus, $\{0,2\}$ is the set of critical numbers of f; and 1 is the only number for which $f''(x) = 0$. We conclude:

$f(0) = 0$ is a relative maximum value of f because $f''(0) < 0$.
$f(2) = -4$ is a relative minimum value of f because $f''(2) > 0$.
$(1, -2)$ is a point of inflection of the graph of f by 4.20.

The graph of f is sketched in Figure 4.17.

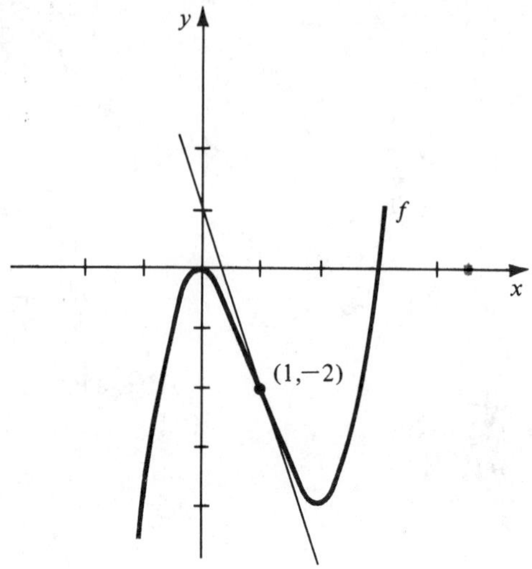

FIGURE 4.17

Example 3 If $f(x) = x^4 - 4x^3 + 10$, find the relative extrema of the function f and the points of inflection of its graph.

Solution: Since

$$f'(x) = 4x^3 - 12x^2 = 4x^2(x - 3),$$

$\{0,3\}$ is the set of critical numbers of f. The second derivative of f is given by

$$f''(x) = 12x^2 - 24x = 12x(x - 2).$$

Thus the points of inflection can occur only at $x = 0$ and $x = 2$.

Since $f''(3) > 0$, the number $f(3) = -17$ is a relative minimum value of f. On the other hand, $f''(0) = 0$ and the second derivative test does not apply to the critical number 0. However,

$$f''(x) > 0, \text{ if } x < 0, \qquad f''(x) < 0, \text{ if } 0 < x < 2,$$

and therefore $(0,10)$ is an inflection point of the graph of f. Thus $f(0)$ is not an extremum of f. Also,

$$f''(x) < 0, \text{ if } 0 < x < 2, \qquad f''(x) > 0, \text{ if } x > 2,$$

and we conclude that $(2,-6)$ is a point of inflection of the graph. The graph of f is sketched in Figure 4.18.

Some functions are defined on closed or half-closed intervals. Then, their graphs have one or two *endpoints* which will usually be extrema.

For example, the function

$$f(x) = \sqrt{4 - x^2}$$

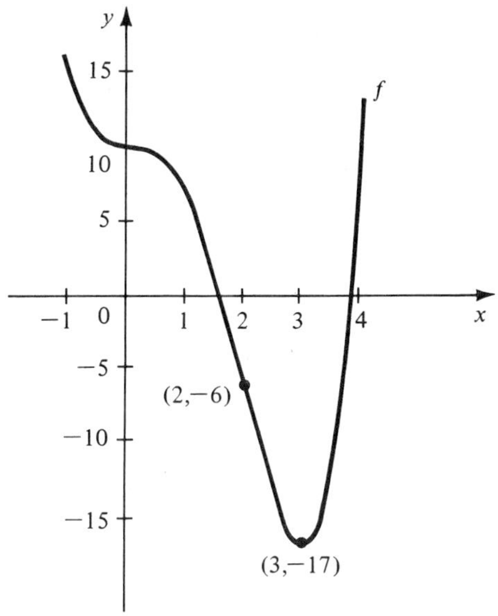

FIGURE 4.18

has domain $[-2,2]$. Points $(-2,0)$ and $(2,0)$ are endpoints of the graph. They are minimum points of the graph, which is a semicircle of radius 2 on and above the x axis.

Example 4 Find the extrema of the function f defined by

$$f(x) = 3x - (x - 1)^{3/2}.$$

Solution: Since $(x - 1)^{3/2}$ is a real number only if $x \geq 1$, the domain of f is the set $[1,\infty)$. Thus, $(1,3)$ is an endpoint of the graph of f.

We have

$$f'(x) = 3 - \tfrac{3}{2}(x - 1)^{1/2}, \qquad f''(x) = -\tfrac{3}{4}(x - 1)^{-1/2}.$$

Clearly, $f'(x) = \frac{3}{2}[2 - \sqrt{x - 1}] = 0$ only if $x = 5$. Since $f''(5) < 0, f(5) = 7$ is a maximum value of f. Thus, the function f must be strictly increasing in $[1,5]$ and hence, $f(1) = 3$ must be a minimum value of f. The graph of f is sketched in Figure 4.19 with the aid of the accompanying table of values.

x	1	2	5	10	17
$f(x)$	3	5	7	3	−13

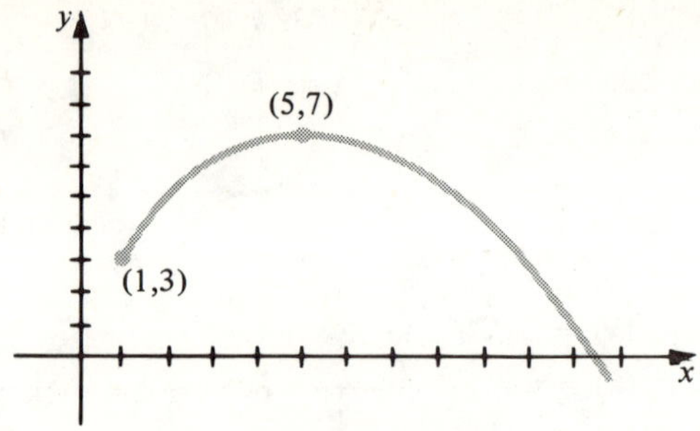

FIGURE 4.19

EXERCISES

In each of Exercises 1 to 10, find the extrema of the function, using the second derivative test.

1. $f(x) = 5x^2 - 2x + 1$

2. $g(x) = 3 + x - x^2$

3. $g(x) = x^3 + 3x^2 - 9x + 10$

4. $F(x) = x^3 + 4x^2 - 3x - 9$

5. $F(x) = \dfrac{x^2}{4} + \dfrac{4}{x}$

6. $f(x) = x^3 + \dfrac{3}{x}$

7. $f(x) = \dfrac{1}{\sqrt{x}} + \dfrac{\sqrt{x}}{9}$

8. $g(x) = \dfrac{x^3}{x^2 + 1}$

9. $h(x) = \dfrac{x^2}{\sqrt{x^2 + 4}}$

10. $F(x) = \dfrac{x}{x^2 + 1}$

In each of Exercises 11 to 16, find the intervals where the graph of the given function is concave upward and those where it is concave downward. Find the points of inflection and sketch the graph.

11. $f(x) = x^4 + x^3 - 3x^2 + 2x - 1$

12. $f(x) = \dfrac{1}{\sqrt{1 + x^2}}$

13. $f(x) = 2\sqrt{x^2 + 1} - x^2$

14. $f(x) = \dfrac{x}{a^2x^2 + 1}$

15. $f(x) = \dfrac{x^2 - 1}{x}$

16. $F(x) = \begin{cases} -x^2, \text{ if } x < 0 \\ x^2, \text{ if } x \geq 0 \end{cases}$

In each of Exercises 17 to 22, find the extrema of the function, using the second derivative test. Also find the points of inflection and sketch the graph.

17. $G(x) = 3 - 2x - x^2$

18. $F(x) = x^2 - 4x + 5$

19. $F(x) = x^3 - x^2 - x + 2$

20. $F(x) = x^3 - 5x^2 - 8x + 20$

21. $f(x) = 3x^4 - 4x^3 - 12x^2 + 3$

22. $g(x) = x^5 - 5x + 2$

In each of Exercises 23 to 28, find the maximum and minimum values of the function on the specified interval.

23. $f(x) = 3x - (x - 1)^{3/2}$ on $[1,17]$

24. $f(x) = 2x^3 + 3x^2 - 12x - 1$ on $[-3,3]$

25. $f(x) = x^{1/3}(x - 3)^{2/3}$ on $[-1,4]$

26. $f(x) = \sqrt{x}(x - 5)^{1/3}$ on $[0,6]$

27. $f(x) = x^{2/3} + 4$ on $[-3,3]$

28. $f(x) = (x - 6)^{2/3}$ on $[-5,1]$

In each of Exercises 29 to 44, find the extrema of the function, using whatever test is more convenient.

29. $g(x) = x^3 - 15x$

30. $F(x) = x^{7/5}$

31. $G(x) = x^{5/3}(x - 1)$

32. $g(x) = \dfrac{ax}{x^2 + a^2}, a \neq 0$

33. $F(x) = x\sqrt{x + 3}$

34. $G(x) = 6\sqrt[3]{x} + x^2$

35. $f(x) = x^{2/3}(x - 2)^2$

36. $f(x) = 5x^{2/5} + 2x$

37. $g(x) = \dfrac{x + a}{\sqrt{x^2 + 1}}, a \neq 0$

38. $g(x) = \dfrac{x + 2}{x^2 + 2x + 4}$

39. $f(x) = |x^3 + 1|$

40. $f(x) = x + |x^2 - 1|$

41. $F(x) = x^2 + |x|$

42. $g(x) = \dfrac{|x|}{x^2 + 1}$

43. $g(x) = \dfrac{x^2 + 3}{x^2 - 9}$

44. $F(x) = \dfrac{x}{x^2 - 4} + |x^3 - 8|$

45. Find the points of inflection of the graph of

$$y = \frac{1}{x^2 + 3}.$$

Sketch this graph, showing the tangent lines at the points of inflection.

46. Find the points of inflection and extrema of the graph of

$$y = x^{1/3}(x - 4).$$

Sketch this graph, showing the tangent lines at the points of inflection.

47. Find the point of inflection of the graph of

$$y = x^2 - \frac{1}{6x^3}.$$

Find the equation of the tangent line to the graph at this point.

48. Let $f(x) = 2x^3 - 3(a + b)x^2 + 6abx$. Find the extrema of f if:

a. $a < b$ **b.** $a = b$

49. Determine a and b so that 1 is a critical number of the function f defined by

$$f(x) = x^3 + ax^2 + bx, \qquad f(1) = -3.$$

Is $(1, -3)$ a maximum or minimum point on the graph of f?

5 APPLICATIONS OF THE THEORY OF EXTREMA

It is frequently possible to solve a problem which asks for the largest area or the least volume or the lowest cost by recognizing that the solution of the problem is a maximum or minimum value of some function. We illustrate the procedure with the following examples.

Example 1 A rectangular field is to be adjacent to a river and is to have fencing on three sides, the side on the river requiring no fencing. If 100 rods of fencing is available, find the dimensions of the field with largest area.

Solution: Let the two sides of the field which are perpendicular to the river each have length x rods. Then the side parallel to the river has length $100 - 2x$ rods, since the sum of the lengths of the three sides is 100 rods (Figure 4.20). Clearly, $0 < x < 50$.

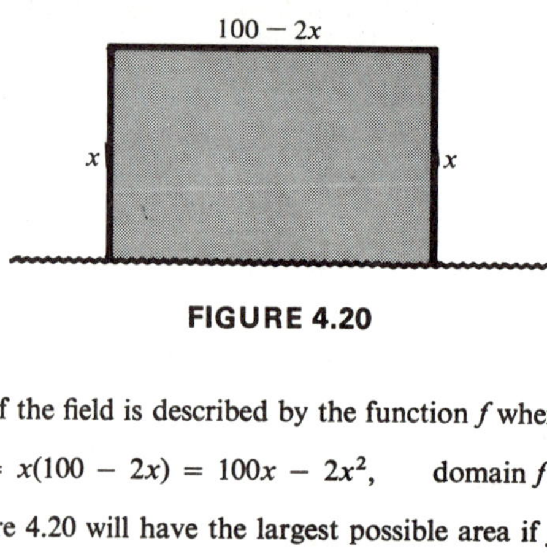

FIGURE 4.20

The area of the field is described by the function f where

$$f(x) = x(100 - 2x) = 100x - 2x^2, \qquad \text{domain } f = (0,50).$$

The field of Figure 4.20 will have the largest possible area if $f(x)$ is the maximum value of f.

Now

$$f'(x) = 100 - 4x,$$

and $f'(x) = 0$ only if $x = 25$. Thus 25 is the critical number of f. Since

$$f''(x) = -4,$$

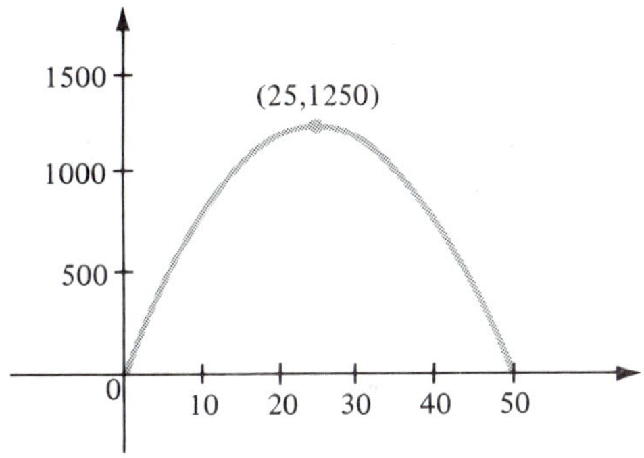

FIGURE 4.21

$f''(25) = -4$ and $f(25)$ is a maximum value of f. The dimensions of this field of largest area are $x = 25$ rods and $100 - 2x = 50$ rods; its area is 1250 square rods.

The graph of the function f is sketched in Figure 4.21. The ordinate $f(x)$ of each point $(x, f(x))$ on the graph is the area of the field with given width x.

We might have started differently, letting z be the length of the side of the field parallel to the river. Then each of the other two sides would have length $(100 - z)/2 = 50 - z/2$, and the area of the field would be given by

$$g(z) = z\left(50 - \frac{z}{2}\right) = 50z - \tfrac{1}{2}z^2.$$

Since

$$g'(z) = 50 - z,$$

we would discover that $g(50)$ is a maximum value of g, and since

$$g(50) = 1250,$$

the same field would be determined. Thus there are at least two functions which give us a solution of the problem.

In some problems, it may be that one function is easier to recognize or easier to handle than another. However, in the example above, the choice would not be important.

Example 2 Show that a tin can of specified volume K will be made of the least amount of metal if its height equals the diameter of its base.

Solution: We are assuming, of course, that the tin can is in the form of a right circular cylinder and has both top and bottom. The least amount of metal will be used when the surface area of the cylinder is a minimum. Let r be the radius of the base and h the height of the cylinder (Figure 4.22). Then the area of the base is πr^2, as is the area of the top. The lateral area is $2\pi rh$, the product of the circumference and the height of the cylinder. Thus the total area A is given by

$$A = 2\pi rh + 2\pi r^2.$$

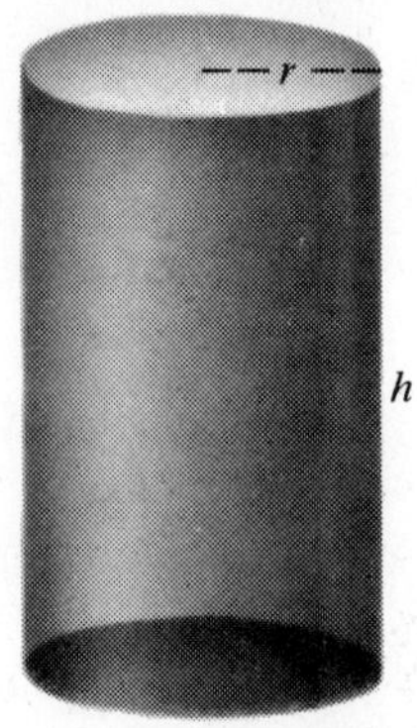

FIGURE 4.22

The volume of the cylinder has been specified as a (positive) number K, and hence

$$K = \pi r^2 h.$$

We can solve this equation for h,

$$h = \frac{K}{\pi r^2},$$

and then express A in terms of r (by replacing h by $K/\pi r^2$) as follows:

$$A = \frac{2K}{r} + 2\pi r^2.$$

We can now say that the area of the cylinder is described by the function f where

$$f(r) = \frac{2K}{r} + 2\pi r^2, \qquad \text{domain } f = (0,\infty).$$

We seek a minimum value of the function f. Since

$$f'(r) = -\frac{2K}{r^2} + 4\pi r$$

for every positive number r, the critical numbers of f are the positive solutions of the equation

$$-\frac{2K}{r^2} + 4\pi r = 0,$$

or

$$\frac{2}{r^2}(-K + 2\pi r^3) = 0.$$

Thus the only critical number of f is

$$r = \sqrt[3]{\frac{K}{2\pi}}.$$

Since

$$f''(r) = \frac{4K}{r^3} + 4\pi,$$

$f''(r) > 0$ for every positive number r. Hence

$$f\left(\sqrt[3]{\frac{K}{2\pi}}\right)$$

is a minimum value of f.

We have shown above that $h = K/\pi r^2$, and therefore

$$\frac{h}{r} = \frac{K}{\pi r^3}.$$

At $r = \sqrt[3]{K/2\pi}$,

$$\frac{h}{r} = \frac{K}{\pi(K/2\pi)} = 2,$$

and

$$h = 2r.$$

This proves that the height and diameter of the tin can of least area are equal.

We might have solved the equation $K = \pi r^2 h$ for r, getting

$$r = \sqrt{\frac{K}{\pi h}},$$

and then used the function g given by

$$g(h) = 2\sqrt{K\pi h} + \frac{2K}{h}$$

to describe the area of the cylinder. However, the function f is somewhat easier to handle than g.

Example 3 Find a triangle of maximum area inscribed in a circle of radius r.

Solution: If triangle ABC is inscribed in a circle and if $AC \neq BC$, then there exists a triangle $AC'B$ inscribed in the circle with $AC' = BC'$ and with greater area than the given triangle, as can be seen from Figure 4.23. Thus it is clear geometrically that a triangle of maximum area inscribed in a circle is equilateral.

We have shown above that if there is a triangle of maximum area inscribed in a circle then it must be equilateral. A more difficult problem, which we will not discuss, is to show that an inscribed triangle of maximum area actually exists.

We see from Example 3 that not all problems in maxima and minima require calculus for their solution. As another example, it is obvious that if

$$f(x) = \frac{1}{1 + x^2}$$

then $f(0) = 1$ is the only maximum value of f.

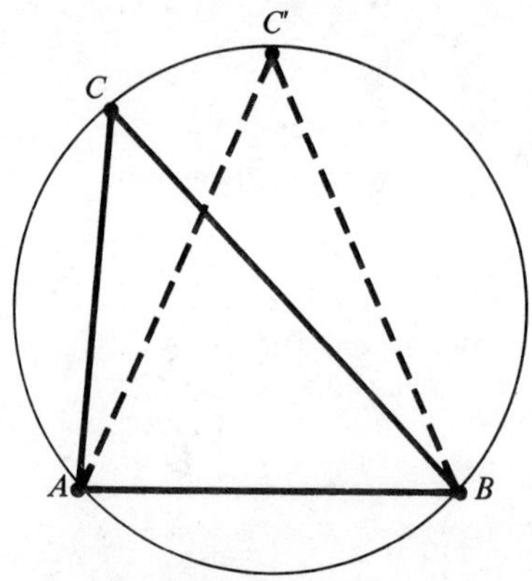

FIGURE 4.23

Example 4 Find the point of the graph of the equation

$$y = x^2$$

that is nearest the point $A(3,0)$.

Solution: For every real number x, the point $P(x,x^2)$ is on the graph of the given equation (Figure 4.24), and

$$|PA|^2 = (x - 3)^2 + (x^2)^2$$
$$= x^4 + x^2 - 6x + 9.$$

If we let

$$g(x) = \sqrt{x^4 + x^2 - 6x + 9},$$

then $g(x)$ is the distance between P and A, and we wish to find a minimum value of g.

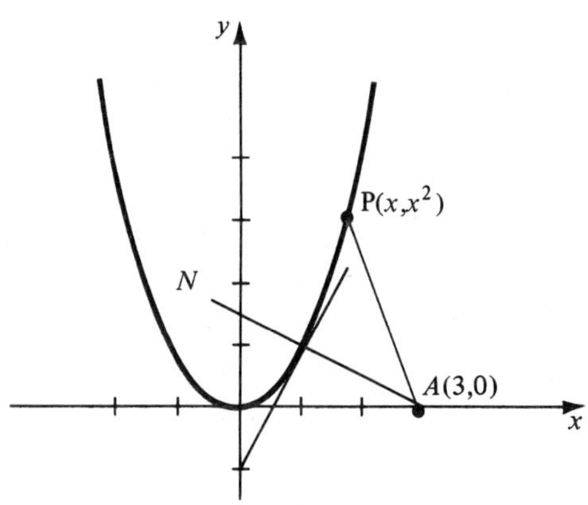

FIGURE 4.24

Now

$$g'(x) = \tfrac{1}{2}(4x^3 + 2x - 6)(x^4 + x^2 - 6x + 9)^{-1/2},$$

and $g'(x) = 0$ only if

$$4x^3 + 2x - 6 = 0.$$

It is clear by inspection that $x = 1$ is a solution of this equation. By dividing the polynomial $4x^3 + 2x - 6$ by $x - 1$, we see that

$$4x^3 + 2x - 6 = (x - 1)(4x^2 + 4x + 6),$$

and, since the equation

$$4x^2 + 4x + 6 = 0$$

has no real solution, that 1 is the only critical number of g.

Clearly, $g'(0) < 0$ and $g'(2) > 0$, and therefore $g(1) = \sqrt{5}$ is a minimum value of g. The point on the graph nearest A is (1,1).

Since $D_x x^2 = 2x$, the slope of the tangent line T to the graph of $y = x^2$ at the point $P(1,1)$ is 2. The slope of the line AP is $-\frac{1}{2}$, and therefore A lies on the normal line N of the graph at $P(1,1)$ (Figure 4.24).

Example 5 An apple orchard now has 30 trees per acre, and the average yield is 400 apples per tree. For each additional tree planted per acre, the average yield per tree is reduced by approximately 10 apples. How many trees per acre will give the largest crop of apples?

Solution: If x is the number of new trees planted per acre, then there are $30 + x$ trees per acre having an average yield of $400 - 10x$ apples per tree. Hence the total yield y of apples per acre is given by

$$\begin{aligned} y &= (30 + x)(400 - 10x) \\ &= 12{,}000 + 100x - 10x^2. \end{aligned}$$

Since

$$\frac{dy}{dx} = 100 - 20x,$$

$dy/dx = 0$ if $x = 5$. Hence an addition of 5 trees per acre will give the largest crop; then there will be 35 trees per acre, each yielding an average of 350 apples.

EXERCISES

I

1. A man has 600 yd of fencing which he is going to use to enclose a rectangular field and then subdivide the field into two plots with a fence parallel to a side. Of all the possible fields that can be so fenced, what are the dimensions of the one of maximum area?
2. Generalize Exercise 1 by dividing the field into n plots.

3. An open box is to be made by cutting out squares from the corners of a rectangular piece of cardboard and then turning up the sides. If the piece of cardboard is 12 in. by 24 in., what are the dimensions of the box of largest volume made in this way?

4. A rectangular box with square base and open top is to be made from 12 ft^2 of cardboard. What is the maximum possible volume of such a box?

5. A cylindrical cup (open top) is to hold a half-pint. How should it be made so as to use the least amount of material?

6. A wire 24 in. long is cut in two, and then one part is bent into the shape of a circle and the other into the shape of a square. How should it be cut if the sum of the areas of the circle and the square is to be a minimum? A maximum?

7. A rectangle of perimeter p is rotated about one of its sides so as to form a cylinder. Of all such possible rectangles, which generates a cylinder of maximum volume?

8. A rectangle has two of its vertices on the x axis and the other two above the x axis and on the graph of the parabola $y = 16 - x^2$. Of all such possible rectangles, what are the dimensions of the one of maximum area?

9. Show that (2,2) is the point on the graph of the equation $y = x^3 - 3x$ that is nearest the point (11,1).

10. Find the point on the graph of the equation $y^2 = 4x$ which is nearest to the point (2,1).

11. A ladder is to reach over a fence 8 ft high to a wall 1 ft behind the fence. What is the length of the shortest ladder that can be used?

12. Find the rectangle of maximum area that can be inscribed in a semicircle of radius r.

13. Show that the greatest area of any rectangle inscribed in a triangle is one-half that of the triangle.

14. A real estate office handles 80 apartment units. When the rent of each unit is \$60 per month, all units are occupied. However, for each \$2 increase in rent, one of the units becomes vacant. Each occupied unit requires an average of \$6 per month for service and repairs. What rent should be charged to realize the most profit?

15. A man in a motorboat 4 miles from the nearest point P on the shore wishes to go to a point Q 10 miles from P along the straight shoreline. The motorboat can travel 18 mph and a car, which can pick up the man at any point between P and Q, can travel 30 mph. At what point should the man land so as to reach Q in the least amount of time?

16. Three sides of a trapezoid have the same length a. Of all such possible trapezoids, show that the one of maximum area has its fourth side of length $2a$.

17. A Boston lodge has asked the railroad company to run a special train to New York for its members. The railroad company agrees to run the train if at least 200 people will go. The fare is to be \$8 per person if 200 go, and will decrease by 1¢ for everybody for each person over 200 that goes (thus, if 250 people go, the fare will be \$7.50). What number of passengers will give the railroad maximum revenues?

18. Find the rectangle of maximum area that can be inscribed in the ellipse

$$\frac{x^2}{a^2} + \frac{y^2}{b^2} = 1.$$

19. Find a point on the graph of $xy = 1$ that is closest to the origin.

20. Find the point on the graph of

$$\frac{x^2}{9} + \frac{y^2}{16} = 1$$

that is the greatest distance from (3,0).

II

1. Find the dimensions of the right circular cylinder of maximum volume inscribed in a sphere of radius r.
2. Find the dimensions of the right circular cone of maximum volume inscribed in a sphere of radius r.
3. Of all the right circular cylinders that can be inscribed in a given right circular cone, show that the one of greatest volume has altitude one-third that of the cone.
4. Find the greatest value of $x^m y^n$ where x and y are positive and $x + y = a$, $m, n > 0$.
5. A line is drawn through the fixed point (h,k) to meet the axes OX and OY in points R and S. Find the minimum value of

   a. RS **b.** $OR + OS$ **c.** $OR \cdot OS$

6 VELOCITY AND ACCELERATION

The derivative is closely related to the rate of change of a function as defined below. Before discussing rates of change in general, let us consider the special case of the motion of a point on a straight line.

We shall assume for the present that L is a horizontal coordinate line with unit point to the right of the origin, as in Figure 4.25. At the time t the expression $s(t)$ will designate the coordinate of the point P which is in motion on L. The function s so determined by the position of P on L is called the *position function* of the moving point P.

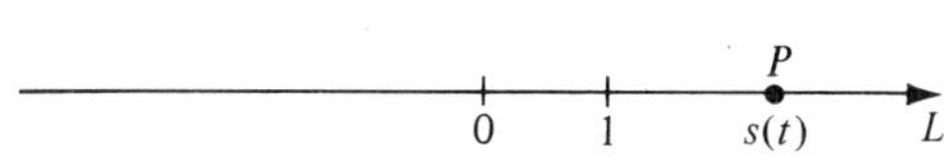

FIGURE 4.25

Example 1 Let the position function s be defined by

$$s(t) = t^2 - 2t.$$

The position of the moving point P on a coordinate line L at various times is shown in the following table of values:

t	0	1	2	3	4
$s(t)$	0	-1	0	3	8

The position of the moving point P on L is indicated in Figure 4.26.

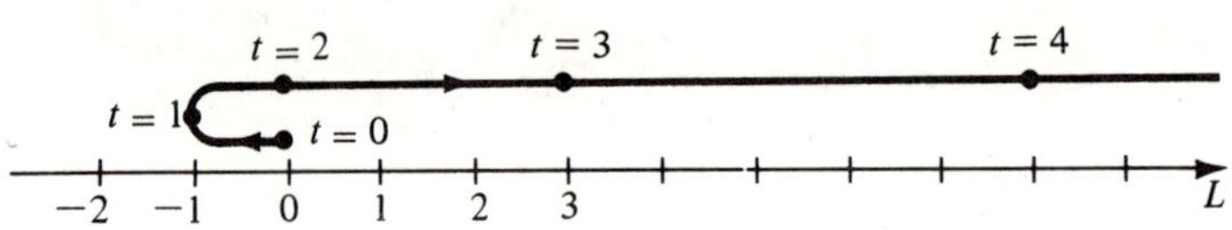

FIGURE 4.26

In Example 1, the point P moves from 0 to 8 on L between $t = 2$ and $t = 4$. Thus, P moves 8 units of distance in 2 units of time. The ratio

$$\tfrac{8}{2} = 4$$

is the average velocity of the point P from time $t = 2$ to $t = 4$ as defined below.

4.21 Definition

If s is the position function of a moving point P on a line, then the *average velocity* of P from time t to time $t + h$ is

$$\frac{s(t + h) - s(t)}{h}$$

Knowledge of the average velocity of a moving point P gives us little if any information about the motion of the point at a particular instant. We cannot conclude, for example, that a jet airplane did not break the sound barrier if its average velocity on a certain flight was 500 miles per hour, nor can we conclude that it did break the sound barrier. The average velocity does not describe the "momentary" or "instantaneous" character of the motion.

This need for knowledge of the instantaneous character of the motion of a point leads us to the following definition.

4.22 Definition

If s is the position function of a moving point P, the *velocity of P at the time t* is designated by $v(t)$ and is defined by

$$v(t) = \lim_{h \to 0} \frac{s(t + h) - s(t)}{h}.$$

We do not append to 4.22 the condition "if this limit exists," since we are working in the realm of physical motion and it is to be assumed that any point P in motion has a velocity at any time t.

What we have done above with the position function can be done with any function f. Thus, $f(b) - f(a)$ is the change in f from $x = a$ to $x = b$ and $f(b) - f(a)$ divided by $b - a$ is the change in f per unit change in x.

4.23 Definition

If a function f is defined in an interval $[a,b]$, then

$$\frac{f(b) - f(a)}{b - a}$$

is the *average rate of change* of f in $[a,b]$. For each x in $[a,b]$,

$$\lim_{h \to 0} \frac{f(x + h) - f(x)}{h}$$

is the *instantaneous rate of change* of f at x, if this limit exists.

We recognize the limit above to be the derivative of f at x. Thus,

$$f'(x)$$

may be considered to be the instantaneous rate of change of f at x.

To return to the position function s, we recognize the limit in 4.22 to be the derivative of s, so that

$$v(t) = s'(t).$$

Thus s is increasing at t if $v(t) > 0$ and decreasing at t if $v(t) < 0$. To say that s is increasing is to say that the point P is moving to the right on L; and to say that s is decreasing is to say that P is moving to the left on L. When $v(t) = 0$, the point P is said to be (momentarily) *at rest*.

Thus we have:

(1) *If $v(t) > 0$, then P is moving to the right.*
(2) *If $v(t) < 0$, then P is moving to the left.*
(3) *If $v(t) = 0$, then P is at rest.*

We are assuming that $v(t)$ exists for every time t, and hence $s'(t)$ always exists. Therefore the only critical numbers of s are those for which $s'(t) = 0$, and we may make the following observation: *A point in motion on a straight line cannot change its direction without coming to rest.*

If s is the position function of a jet plane which leaves at time t_1 and arrives at its destination at time t_2, and if its average velocity is 500 miles per hour, then

$$\frac{s(t_2) - s(t_1)}{t_2 - t_1} = 500$$

By the mean value theorem 4.8, the left side of the equation above is equal to $s'(t) = v(t)$ for some time t between t_1 and t_2. Thus the velocity of the plane is exactly 500 miles per hour at some instant t in its flight.

In Example 1 above,

$$s(t) = t^2 - 2t, \qquad v(t) = s'(t) = 2t - 2.$$

Since $v(0) = -2$, the point P is moving to the left at time $t = 0$. P is at rest at $t = 1$, because $v(1) = 0$. In fact, P is at rest at no other time, since $v(t) \neq 0$ if $t \neq 1$. Clearly, $v(t) > 0$ and P is moving to the right if $t > 1$. This is as indicated in Figure 4.26.

4.24 Definition

The *speed* of a moving point P at time t is given by

$$|v(t)|.$$

Thus, the speed simply tells how fast P is moving—it does not tell the direction of motion.

Average acceleration is defined to be the average rate of change in velocity relative to time. Thus, if v is the velocity function of the moving point P, the *average acceleration* of P from time t to time $t + h$ is

$$\frac{v(t + h) - v(t)}{h}.$$

The acceleration of P at a given time is defined as follows.

4.25 Definition

If v is the velocity function of a moving point P, the *acceleration of P at the time t* is designated by $a(t)$ and defined by

$$a(t) = \operatorname*{limit}_{h \to 0} \frac{v(t + h) - v(t)}{h}.$$

It follows that

$$a(t) = v'(t) = s''(t).$$

In Example 1 above,

$$s(t) = t^2 - 2t, \qquad v(t) = 2t - 2, \qquad a(t) = 2.$$

If distance is measured in feet and time in seconds, the units of velocity are feet per second (ft/sec) and those of acceleration are feet per second per second (ft/sec^2). Thus, in Example 1 the acceleration is a constant of 2 ft/sec^2.

The *second law of newtonian mechanics* states that the force acting on an object is the product of the mass of the object and its acceleration. If m is the mass of the object P moving along a straight line with position function s, then the force $F(t)$ acting on P at time t is given by the equation

$$F(t) = m \cdot a(t),$$

or

$$F(t) = m \cdot s''(t).$$

If at some time t, $a(t) = 0$, then also $F(t) = 0$ and no force is acting on P. In this event P can be said to be coasting.

If the point P is at rest at time t, so that $v(t) = 0$, and if $a(t) < 0$, then $s(t)$ is a relative maximum value of s. That is, P approached the position $s(t)$ from the left, stopped, and then moved to the left. A similar statement holds if $a(t) > 0$. We summarize the motion as follows:

If $v(t) = 0$ and $a(t) \neq 0$, then the moving point P reverses its direction of motion at the position $s(t)$.

Example 2 Discuss the motion of the point P if its position at time t on a coordinate line L is given by

$$s(t) = t^3 - 6t^2 + 20, \qquad \text{domain } s = [-2,6].$$

Solution: Since $v(t) = s'(t)$ and $a(t) = v'(t)$,

$$v(t) = 3t^2 - 12t, \qquad a(t) = 6t - 12.$$

The point P will be at rest when

$$v(t) = 3t(t - 4) = 0,$$

that is, when $t = 0$ and $t = 4$. Thus, 0 and 4 are the critical times of s. Since $a(0) < 0$ and $a(4) > 0$, point P changes its direction of motion at each critical time; and $s(0) = 20$ is a maximum value of s and $s(4) = -12$ is a minimum value of s.

An indication of the motion of P can be obtained from the accompanying table of values and Figure 4.27.

t	$s(t)$	$v(t)$	$a(t)$
-2	-12	36	-24
-1	13	15	-18
0	20	0	-12
2	4	-12	0
4	-12	0	12
6	20	36	24

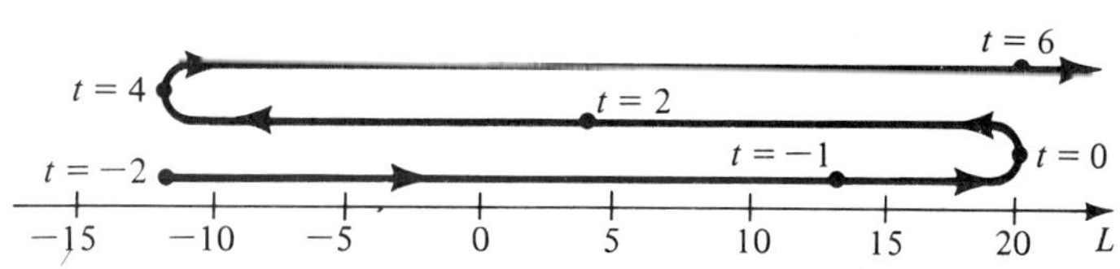

FIGURE 4.27

EXERCISES

In each of Exercises 1 to 10 the position function of a point moving on a straight line is given. Discuss the motion of the point.

1. $s(t) = -16t^2 + 80t,\ 0 \le t \le 5$
2. $s(t) = -16t^2 + 32t + 20,\ 0 \le t \le 4$
3. $s(t) = t^2 - 8t + 4,\ -2 \le t \le 6$
4. $s(t) = 12 + 6t + t^2,\ -6 \le t \le 0$
5. $s(t) = t^3 - 3t$
6. $s(t) = t^3 - 3t^2 - 24t$
7. $s(t) = 2 + t - t^2 - t^3$
8. $s(t) = t^4 - 4t^3$
9. $s(t) = t^2 + \dfrac{16}{t},\ t \ge 1$
10. $s(t) = 4t + \dfrac{9}{t},\ t \ge 1$

7 ANTIDERIVATIVES

Corresponding to each function f is its derived function f'. For example, if f is defined by

$$f(x) = x^3 - 6x + 12,$$

then f' is defined by

$$f'(x) = 3x^2 - 6.$$

We now ask the following question. Given a function g, does there exist a function f such that g is the derivative of f? If, for example, g is defined by

$$g(x) = 3x^2 - 6,$$

then g is the derivative of f defined by

$$f(x) = x^3 - 6x + 12.$$

The answer to this question is yes if g is a continuous function. The proof of this fact will come in Chapter 5 after the concept of the integral has been introduced. Here we shall be concerned only with formal properties of the antiderivative, defined as follows:

4.26 Definition

The function f is called an *antiderivative* of the function g if $f' = g$.

For example, if

$$g(x) = 6x^2,$$

then the function f_1 defined by

$$f_1(x) = 2x^3$$

is an antiderivative of g since

$$f_1'(x) = g(x).$$

It is equally true that the function f_2 defined by

$$f_2(x) = 2x^3 + 7$$

is an antiderivative of g, since

$$f_2'(x) = g(x).$$

It is clear that if f_1 is an antiderivative of g, and if $f_2(x) = f_1(x) + c$ (c a real number), then

$$f_2'(x) = f_1'(x) = g(x)$$

and f_2 also is an antiderivative of g. The converse of this statement is also true, as we shall now prove.

4.27 Theorem

If for the functions f_1 and f_2 there exists an interval I, such that

$$f_1'(x) = f_2'(x) \qquad \textit{for every } x \textit{ in } I,$$

then there exists a number c, such that

$$f_1(x) = f_2(x) + c \qquad \textit{for every } x \textit{ in } I.$$

Proof: Let the function F be the difference between f_1 and f_2,

$$F(x) = f_1(x) - f_2(x).$$

By hypothesis,

$$F'(x) = 0 \qquad \text{for every } x \text{ in } I.$$

Let a be some number in I. By the mean value theorem, for every x in I there exists a number z between a and x such that

$$F(x) - F(a) = (x - a)F'(z) = 0.$$

Thus $F(x) = F(a)$ for every x in I, and

$$f_1(x) = f_2(x) + c \qquad \text{for every } x \text{ in } I,$$

where $c = F(a)$.

If

$$f'(x) = 6x^2,$$

as in the example above, then

$$f(x) = 2x^3 + c$$

for some number c, according to 4.27.

By previous differentiation formulas,

$$D_x\left(\frac{a}{m+1}\,x^{m+1}\right) = ax^m, \qquad m \text{ rational number, } m \neq -1.$$

This leads to the following result.

4.28 Theorem

The function

$$g(x) = ax^m, \qquad m \neq -1,$$

has antiderivative

$$f(x) = \frac{a}{m+1} x^{m+1} + c.$$

For example,

$$g(x) = 5x^9 \text{ has antiderivative } f(x) = \tfrac{1}{2}x^{10} + c.$$

4.29 Theorem

If functions g_1 and g_2 have antiderivatives f_1 and f_2, respectively, then $f_1 + f_2$ is an antiderivative of $g_1 + g_2$.

Proof: We need only show that the derivative of $f_1 + f_2$ is $g_1 + g_2$:

$$D_x[f_1(x) + f_2(x)] = D_xf_1(x) + D_xf_2(x) = g_1(x) + g_2(x).$$

More generally, *an antiderivative of a sum of functions is the sum of antiderivatives of the functions.*

Example 1 Find $f(x)$ if

$$f'(x) = 4x^3 - 2x^2 + 5x + 3.$$

Solution: Since

$$D_x(x^4 - \tfrac{2}{3}x^3 + \tfrac{5}{2}x^2 + 3x) = 4x^3 - 2x^2 + 5x + 3,$$

$$f(x) = x^4 - \tfrac{2}{3}x^3 + \tfrac{5}{2}x^2 + 3x + c$$

for some number c.

If f is a function such that

$$f'(x) = 4x - 3 \qquad \text{and} \qquad f(1) = 3,$$

then

$$f(x) = 2x^2 - 3x + c$$

for some number c. Since

$$f(1) = 2 - 3 + c = 3,$$

$c = 4$. Thus,

$$f(x) = 2x^2 - 3x + 4$$

is the unique function satisfying the given conditions.

4.30 Definition

For functions f and g, an equation of the form

(1) $$f'(x) = g(x)$$

is called a *differential equation.* A value of f is called a *boundary condition* of f. A function f satisfying (1) is called a *solution* of (1).

For example, the equation above,

$$f'(x) = 4x - 3$$

is a differential equation, and

$$f(1) = 3$$

is a boundary condition. Then,

$$f(x) = 2x^2 - 3x + 4$$

is the solution of the differential equation.

Example 2 Solve the differential equation

$$f'(x) = x^2 + 5$$

with boundary condition

$$f(0) = -1.$$

Solution: Clearly,

$$f(x) = \tfrac{1}{3}x^3 + 5x + c$$

for some number c. Since

$$f(0) = 0 + 0 + c = -1,$$

$c = -1$ and

$$f(x) = \tfrac{1}{3}x^3 + 5x - 1$$

is the unique solution.

If s, v, and a are the respective position, velocity, and acceleration functions of some point P in motion on a coordinate line L, then s is an antiderivative of v and v is an antiderivative of a, since

$$s'(t) = v(t), \qquad v'(t) = a(t).$$

Hence, given the velocity or acceleration function and some boundary conditions, called *initial conditions* if given for $t = 0$, it is possible to determine the position function. This is illustrated in the following example.

Example 3 Find $s(t)$ if it is known that

$$a(t) = 6t - 2,$$

and

$$v(0) = 3, \qquad s(0) = -1.$$

Solution: Since $a(t) - v'(t)$,

$$v'(t) = 6t - 2,$$

and

$$v(t) = 3t^2 - 2t + c_1,$$

for some number c_1. However,

$$v(0) = 0 - 0 + c_1 = 3,$$

and $c_1 = 3$. Thus

$$v(t) = 3t^2 - 2t + 3.$$

Again, $v(t) = s'(t) = 3t^2 - 2t + 3$, and

$$s(t) = t^3 - t^2 + 3t + c_2$$

for some number c_2. Since

$$s(0) = c_2 = -1,$$

$c_2 = -1$ and

$$s(t) = t^3 - t^2 + 3t - 1.$$

An object P is pulled toward the earth by a *force of gravity*. The *acceleration of gravity* due to this force is designated by g. The number g varies with the distance of P from the center of the earth, but is essentially a constant over a small range of distances. An approximate value of g is

$$g = 32 \text{ ft/sec}^2$$

if the object P is near sea level.

Example 4 A ball is thrown directly upward from a point 24 ft above the ground with an initial velocity of 40 ft/sec. Assuming no air resistance, how high will the ball rise and when will it return to the ground?

Solution: In Figure 4.28 the vertical line L indicates the path of the ball and the horizontal line represents the ground. The units on L are feet. If s is the position function of the ball,

$$s(0) = 24,$$

by the assumption that the ball starts 24 ft above the ground. Also,

$$v(0) = 40,$$

since the initial velocity is 40 ft/sec.

Since L is directed upward whereas the force of gravity pulls the ball toward the earth, the velocity will decrease, and therefore

$$a(t) = -32.$$

[$a(t) = 32$ if L is directed downward.] Since v is an antiderivative of a,

$$v(t) = -32t + c_1.$$

However, $v(0) = 40$, and therefore $c_1 = 40$ and

$$v(t) = -32t + 40.$$

Also, s is an antiderivative of v, so that

$$s(t) = -16t^2 + 40t + c_2.$$

Again, $s(0) = 24$ and $c_2 = 24$. Thus

$$s(t) = -16t^2 + 40t + 24.$$

The ball reaches its maximum height when $v(t) = 0$, that is, when

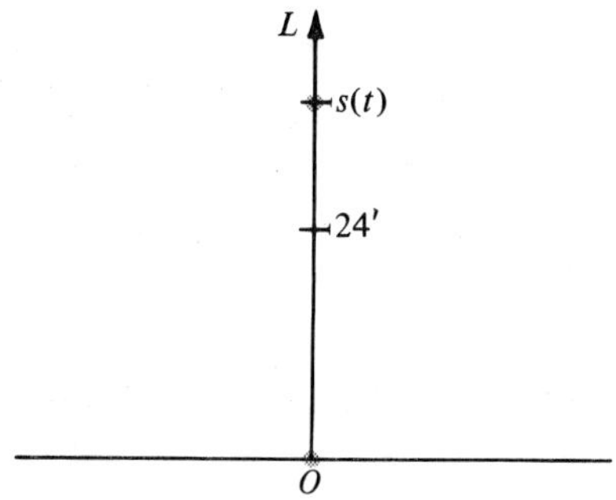

FIGURE 4.28

$$-32t + 40 = 0$$

or

$$t = \tfrac{5}{4} \text{ sec.}$$

The actual maximum height is given by

$$s(\tfrac{5}{4}) = -16(\tfrac{5}{4})^2 + 40(\tfrac{5}{4}) + 24 = 49 \text{ ft.}$$

The ball reaches the ground when $s(t) = 0$, that is, when

$$-16t^2 + 40t + 24 = 0.$$

Since

$$-16t^2 + 40t + 24 = -8(2t + 1)(t - 3),$$

the ball reaches the ground when $t = 3$, that is, after 3 sec. The number $t = -\frac{1}{2}$ is not in the domain of s, since the ball was thrown at the time $t = 0$.

EXERCISES

In each of Exercises 1 to 10, find an antiderivative of the given function.

1. $f(x) = 1 - 4x + 9x^2$

2. $g(x) = 5x^4 - 6x^2 + 3x$

3. $g(x) = 4 - 3x^2 + x^4$

4. $f(x) = \sqrt[3]{x} + x^3$

5. $F(x) = x\sqrt{x} + \sqrt{x} - 5$

6. $h(x) = 3(x - 2)^4$

7. $g(x) = \dfrac{1}{2x^2}$

8. $H(x) = x - x^{-2/3}$

9. $h(x) = \dfrac{(x - 1)^2}{x^4}$

10. $f(x) = \dfrac{x^2 - 5x + 7}{x^5}$

In each of Exercises 11 to 16, find the solution of the given differential equation that satisfies the given boundary conditions.

11. $F'(x) = 4x^{1/3} - x^{1/2}, F(0) = 0$ 12. $G'(x) = \sqrt{2} - 2x + x^2, G(\sqrt{2}) = 1$

13. $s''(t) = 8, s'(0) = 7, s(-1) = -3$

14. $f''(x) = \dfrac{1}{\sqrt{x}} - 10, f'(1) = 3, f(4) = 0$

15. $G'(x) = 3 + x^2 - x^3, G(2) = 3$ 16. $h'(x) = \sqrt{x} - x + x^2, h(1) = -1$

17. A ball is thrown directly upward from the ground with an initial velocity of 56 ft/sec. Assuming no air resistance, how high will the ball rise and when will it return to the ground?

18. An object slides down an inclined plane with an acceleration of 16 ft/sec^2. If the object is given an initial velocity of 4 ft/sec from the top of the inclined plane, find the position function of the object. If the plane is 60 ft long, when does the object reach the end of the plane?

19. A ball is thrown directly downward from a point 144 ft above the ground with enough initial velocity so that it reaches the ground in 2 sec. Ignoring air resistance, find the initial velocity.

20. A car is coasting along a level road at an initial speed of 30 ft/sec. If the car is retarded by friction at the rate of 2 ft/sec^2 (that is, an acceleration of -2 ft/sec^2), in how many seconds will the car stop? How far will it have coasted?

21. A ball rolls down an inclined plane 200 ft long with an acceleration of 8 ft/sec^2. Find the position function of the ball if it is given no initial velocity. How long does it take the ball to reach the end of the plane? What initial velocity must it be given to reach the end of the plane in 4 sec?

22. Starting from rest, with what constant acceleration must a car proceed to go 200 ft in 4 sec?

23. A ball is thrown directly up from the ground and reaches the ground 5 sec later. With what initial velocity was it thrown?

8 RELATED RATES

Many applied problems are concerned with the rate of change of a function with respect to time, for a function other than the position of a moving point on a line. Consider the following examples.

Example 1 Water is running into the conical tank shown in Figure 4.29 at a constant rate of 2 cubic feet per minute (ft^3/min). How fast is the water rising in the tank at any instant?

Solution: At any time t minutes after the water starts running, let $h(t)$ be the depth of the water and $r(t)$ the radius of the surface of the water in the tank (Figure 4.28). Clearly, $r(t) = h(t)/2$, and the volume $V(t)$ of the water in the tank is $2t$, $V(t) = 2t$. By geometry,

$$V = \tfrac{1}{3}\pi r^2 h;$$

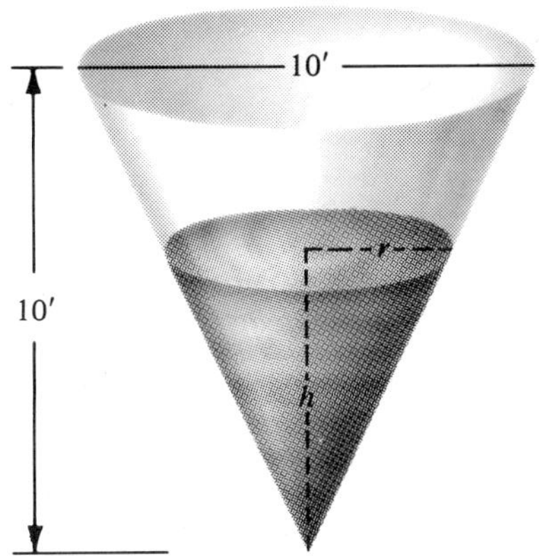

FIGURE 4.29

thus
$$2t = \tfrac{1}{3}\pi\left(\frac{h}{2}\right)^2 h$$

and
$$24t = \pi h^3 \qquad \text{or} \qquad h^3 = \frac{24}{\pi}t.$$

Hence
$$h(t) = \left(2\sqrt[3]{\frac{3}{\pi}}\right)t^{1/3}.$$

The rate of rise of the water is the instantaneous rate of change of h, which is h', by our discussion above. Thus

$$h'(t) = \frac{2}{3}\sqrt[3]{\frac{3}{\pi}}\,t^{-2/3} \text{ ft/min}$$

is the rate at which the water is rising in the tank t minutes after the water starts running. Since $\sqrt[3]{3/\pi} = .98$ approx.,

$$h'(t) \doteq \frac{.65}{\sqrt[3]{t^2}}.$$

Thus, at $t = 8$, $\sqrt[3]{t^2} = 4$ and

$$h'(t) \doteq .16 \text{ ft/min}.$$

Example 2 Highways 29 and 90 meet at right angles at Sioux Falls, S.D. Car A traveling south on route 29 at 60 mph passes through the intersection 3 minutes before car B traveling east on route 90 at 45 mph. How fast are the cars separating 2 minutes after car B passes through the intersection?

Solution: At any time t hours after B passes through the intersection, let x be the distance, in miles, B has traveled from the intersection, and y the distance, in miles,

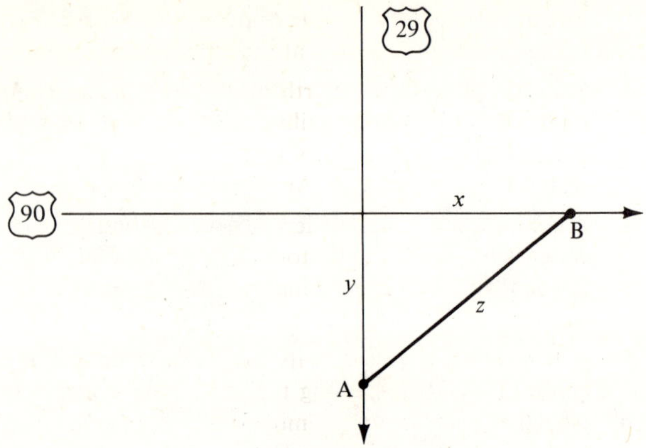

FIGURE 4.30

A has traveled (Fig. 4.30). If z is the distance, in miles, between A and B, then we are asked to find dz/dt, the rate of change of z with respect to time t, at $t = \frac{1}{30}$ hr (2 minutes).

Car A travels 3 miles south of the intersection before B reaches it. Then t hours later,

$$y = 3 + 60t, \qquad x = 45t.$$

Hence,

$$z^2 = x^2 + y^2 = (45t)^2 + (3 + 60t)^2$$

and

$$(1) \qquad 2z\frac{dz}{dt} = 2(45t) \cdot 45 + 2(3 + 60t) \cdot 60.$$

When $t = \frac{1}{30}$, $x = 1.5$, $y = 5$, and

$$z = \sqrt{x^2 + y^2} = \sqrt{27.25} \doteq 5.22 \text{ miles.}$$

Thus, letting $t = \frac{1}{30}$ in (1),

$$\frac{dz}{dt} = (\tfrac{135}{2} + 300) \div \sqrt{27.25}$$

$$\doteq 70.4.$$

The cars are separating at the approximate rate of 70.4 mph.

EXERCISES

1. Gas is being pumped into a balloon so that its volume is constantly increasing at the rate of 4 in.^{3}/sec. Find the rate of increase of the radius of the balloon at any time t seconds after the inflation begins. Approximate this rate at $t = 8$.
2. A ladder 25 ft long is leaning against a wall, with the bottom of the ladder 7 ft from the base of the wall. If the lower end is pulled away from the wall at the

rate of 1 ft/sec, find the rate of descent of the upper end along the wall. Approximate this rate of descent at the end of 8 sec.

3. Ship A is 60 miles due north of ship B at 10 A.M. Ship A is sailing due east at 20 mph, while ship B is sailing due north at 16 mph. Find the distance $d(t)$ between the ships t hours after 10 A.M. Also find the rate of change of the distance between them. At what time are the ships closest together?

4. A conical tank, full of water, is 12 ft high and 20 ft in diameter at the top. If the water is let out at the bottom at the rate of 4 ft^3/min, find the rate of change of the depth of the water t minutes after the water starts running out. Approximate this rate at $t = 10$ min.

5. A trough 9 ft long has as its cross section an isosceles right triangle with hypotenuse of length 2 ft along the top of the trough. If water is pouring into the trough at the rate of 2 ft^3/min, find the depth $h(t)$ of the water t minutes after the water is turned on. Also find the rate at which the depth is increasing when $t = 2$.

6. A right circular cylinder originally has radius 2 in. and height 10 in. The dimensions of the cylinder change in such a way that its volume increases at the rate of 5 $in.^3$/min while its height decreases at the rate of $\frac{1}{2}$ in./min. How fast is its radius changing after 1 min?

7. A rhombus has sides 10 in. long. Two of its opposite vertices are pulled apart at a rate of 2 in. per second. How fast is the area changing when the vertices are 16 in. apart?

8. Triangle ABC is initially a right triangle with $|AB| = 15$ in., $|AC| = 20$ in., and $|BC| = 25$ in. The point A is pulled away perpendicularly from BC (so that the foot of the altitude from BC remains unchanged) at the rate of 3 in./sec. How fast is the perimeter of the triangle changing after 5 sec?

9. A conical tank 10 ft high and 6 ft across at the top is being filled with water at a constant rate. When it is filled halfway to the top, the depth of the water is increasing at 1 ft/min. How long does it take to fill the tank?

10. A spherical drop of water loses moisture by evaporation at a rate proportional to its surface area. Show that the radius decreases at a constant rate.

11. An airplane, cruising at an altitude of 8 miles, flies directly over a radar station on the ground. Compute the speed of the plane, knowing that when the plane is 10 miles from the station (in a direct line) the distance between the plane and the station is increasing at the rate of 330 mph.

12. A balloon is rising vertically at the rate of 8 ft/sec. When the balloon is up 400 ft, a car passes directly under it at 60 mph. (88 ft/sec). How fast are the car and balloon separating 6 seconds later?

13. Three miles above ground, a plane is flying due north at 750 mph. If it flies directly over a car traveling due west at 80 mph, how fast is the distance between the car and the plane changing 1 min later?

14. How fast is the end of a man's shadow moving if he is 6 feet tall and is approaching a 10-ft street light at 5 mph?

15. A man on a pier is pulling in his boat using a rope tied to the boat at water level. His rate of pull is 60 ft/min and he is gripping the rope 20 ft above water level. How fast is the boat approaching the pier when 30 ft of rope remains to be pulled?

CHAPTER REVIEW

In each of Exercises 1 to 8, sketch the graph of the polynomial function. Find extremal points and points of inflection.

1. $f(x) = x^2 - 4$
2. $g(x) = (x^2 - 4)^2$
3. $f(x) = x(x^2 - 1)$
4. $g(x) = x^2(x^2 - 1)^2$
5. $F(x) = 3x^5 - 40x^3 - 135x$
6. $G(x) = x^4 - 4x$
7. $g(x) = x^3 - x^2 - x + 5$
8. $f(x) = 15x^5 - x^3 + 4$

Find the extrema and points of inflection of each of the following functions. Sketch each graph.

9. $G(x) = \dfrac{x}{x^2 + 3}$
10. $F(x) = x^{1/3}(x - 3)$
11. $h(x) = \dfrac{x - 3}{x + 4}$
12. $f(x) = \dfrac{1}{x} - \dfrac{1}{x + 2}$
13. $F(x) = x^3 - x^2 + 2|x| - 1$
14. $G(x) = |16 - x^2|$
15. $g(x) = x\sqrt{9 - x^2}$
16. $f(x) = (x + 4)^{4/3} + 2(x + 4)^{2/3}$
17. Let $g(x) = x^{1/2} + mx$. Find the extrema of g when $m > 0$, $m = 0$, $m < 0$.
18. Let $f(x) = x^4 - 8x^2 + 4$. Sketch graphs of f, f', and f'' on the same coordinate axes and interpret the relationships among these graphs in terms of the theorems of this chapter.
19. Given $f(x) = x^3 - 2x^2 + x - 1$, find a value of c predicted by the mean value theorem on the interval $[-1,1]$.
20. Let $f(x) = x^{2/3}$ on $[-1,1]$. Does f satisfy the conditions for Rolle's theorem? Explain.
21. Let $f(x) = (x - a)^n g(x)$, where $g'(a)$ exists, $g(a) \neq 0$, and n is a positive integer. Under what conditions on n will the graph of f cross the x axis at $(a,0)$? Under what conditions is the x axis the tangent line to the graph of f at $(a,0)$?
22. Find all the normal lines to the graph of the equation $y = x^2$ which pass through the point $(0,2)$.
23. Let $f(x) = x^2|x|$. Sketch the graph of f. Is $(0,0)$ a point of inflection of this graph? Show that $f'''(0)$ does not exist.
24. Analyze the graph of the function g defined by $g(x) = x|x - 1|$.
25. The total area of a page in a book is to be 96 in.^{2} The margins at the top and bottom of the page are each to have width $1\frac{1}{2}$ in. and those at the sides are to have width 1 in. For what dimensions of the page is the printed area the greatest, and what is this maximum area?
26. A truck has a minimum speed of 10 mph in high gear. When traveling x mph in high gear, the truck burns diesel oil at the rate of

$$\frac{1}{300}\left(\frac{900}{x} + x\right) \text{ gal/mile.}$$

The truck cannot be driven over 50 mph. If diesel oil costs 20¢ a gallon, find:

a. The steady speed that will minimize the cost of fuel for a 500-mile trip.

b. The steady speed that will minimize the total cost of the trip if the driver is paid \$1.50 an hour.

27. A boat leaves an island at 11 A.M. and travels due east at 25 knots. The next day at 11 A.M. another boat leaves the island from the same place and travels northeast at 20 knots. At what rate are the two boats separating when the second boat has gone 110 nautical miles?

28. A ball is dropped from a tower 1000 ft high. Assuming no air friction:

a. How long will it take the ball to hit the ground?

b. At what speed will the ball hit the ground?

29. A rock is thrown vertically upward from the top of a cliff 100 ft above the ground with an initial velocity of 60 ft/sec. When will the rock reach its highest point and what is its greatest height (assuming no air resistance)? What is the velocity of the rock as it passes the top of the cliff on its descent? How long will it take the rock to hit the ground?

30. Determine the minimum area of a triangle formed by the x axis, the y axis, and a line through the point (h,k).

31. Two points A and B lie on the same side of the line l. If P be a point on l, determine its position such that $AP + PB$ is a minimum. Show that for this point AP and BP make equal angles with l.

In Exercises 32 to 35, find the maxima and minima of the given functions in the stated intervals.

32. $y = \sqrt{25 - x^2},\ -3 \le x \le 4$

33. $y = \dfrac{1 - x + x^2}{1 + x - x^2},\ 0 \le x \le 1$

34. $y = \dfrac{x - 2}{x + 2},\ 0 \le x \le 4$

35. $y = \dfrac{a^2}{x} + \dfrac{b^2}{1 - x},\ 0 < x < 1$

36. Find $f(x)$ if:

a. $f'(x) = 3x^2 + x - 4, f(1) = 2$

b. $f''(x) = x^{1/3} - x^{1/2}, f(0) = 0, f'(1) = 1$

Find an antiderivative of each of the following functions.

37. $f(x) = 4x^4 + x^3 - x^2 - x + 3$

38. $g(x) = (x - 2)^3 + \sqrt{x}$

39. $h(x) = 3x^{1/3} - 2x^{4/7}$

40. $s(t) = -16t^2 + t - 1$

41. $g(y) = (y - 3)^{-1/4}$

42. $F(x) = \dfrac{x^2 - 3x + 2}{x^4}$

Discuss the motion of a point on a line, given each of the following position function s.

43. $s(t) = 2 - t + 2t^2 - 2t^4$

44. $s(t) = t^3 + 4t,\ 0 \le t \le 4$

45. $s(t) = 3t + \dfrac{6}{t},\ t \ge 2$

46. $s(t) = -10t^2 + 15t + 5,\ 0 \le t \le 6$

5

Integrals

We interrupt our study of the derivative to describe the other basic concept of the calculus, namely the integral. Before defining the integral, we must state another property of the real number system that will enter into this definition.

1 COMPLETENESS OF THE REAL NUMBER SYSTEM

The open interval (1,4) has the property

$$x < 4 \qquad \text{for every } x \text{ in } (1,4).$$

It is also true that

$$x < 7 \qquad \text{for every } x \text{ in } (1,4),$$

$$x < 486 \qquad \text{for every } x \text{ in } (1,4),$$

and so on. Each of the numbers 4, 7, and 486 is an upper bound of the set (1,4) as defined below.

5.1 Definition

A number c is called an *upper bound* of a set A of numbers if

$$x \leq c \qquad \text{for every } x \text{ in } A.$$

In the other direction, a number d is called a *lower bound* of a set B of numbers if

$$d \leq x \qquad \text{for every } x \text{ in } B.$$

For example, 1, 0, and -17 are lower bounds of the set (1,4).

The upper bound 4 of the open interval (1,4) is special in that it is the least one of all the upper bounds of (1,4). And the lower bound 1 of (1,4) is the greatest one of all the lower bounds of (1,4). This illustrates the following definition.

5.2 Definition

A number c is called the *least upper bound* (abbreviated l.u.b.) of a set A of numbers if c is an upper bound of A and no other upper bound of A is less than c. Similarly, a number d is called the *greatest lower bound* (abbreviated g.l.b.) if d is a lower bound of B and no other lower bound of B is greater than d.

Example 1 The infinite interval $(-\infty,8]$ has 8 as its l.u.b. Thus, $x \leq 8$ for every x in $(-\infty,8]$. Furthermore, if a is any number less than 8, then a is not an upper bound of $(-\infty,8]$. For the number

$$b = \frac{a + .8}{2}$$

is larger than a but less than 8. That is, b is in $(-\infty,8]$ and $b > a$. The set $(-\infty,8]$ has no lower bound ($-\infty$ is *not* a number!) and hence no g.l.b.

The l.u.b. of a set, if it exists, is unique; and the same for the g.l.b. If a set A has a l.u.b. c, then c might or might not belong to A. For example,

$$8 \text{ is the l.u.b. of } (-\infty,8] \text{ and also of } (-\infty,8).$$

The number 8 is in $(-\infty,8]$, but not in $(-\infty,8)$.

Example 2 Consider the set

$$A = \{1, 1.4, 1.41, 1.414, 1.4142, 1.41421, \ldots\}$$

of successively closer rational approximations of $\sqrt{2}$. Since $x < \sqrt{2}$ for every x in A, $\sqrt{2}$ is an upper bound of A. Actually, from the way set A is chosen,

$$\sqrt{2} = \text{l.u.b. } A.$$

We are now ready to state a property of the real number system which will be used frequently in this chapter.

5.3 The Completeness Property

Every non-empty set S of real numbers that has a lower bound has a greatest lower bound, and every non-empty set S that has an upper bound has a least upper bound.

This is not a theorem to be proved. Rather, we take it to be one of the basic defining properties of the real number system.

The least upper bound of a set S is somewhat like the limit of a function.

If

$$c = \text{l.u.b. } S$$

then either c is in S or for every positive number ε, no matter how small, the interval $(c - \varepsilon, c)$ contains some element of S. If this statement were false for some ε, then $c - \varepsilon$ would be an upper bound of set S, contrary to the fact that c is the *least* upper bound. Analogous statements hold for lower bounds.

EXERCISES

I

Find the l.u.b. and g.l.b., if they exist, for each of the following sets.

1. $[0,3)$

2. $(-2,\infty)$

3. $[3,5) \cup (7,\sqrt{53}]$

4. $(-\infty,1]$

5. $\{5\}$

6. $\{-3,4,17,105\}$

7. $\{x \mid x \text{ a rational number}, x^2 < 3\}$

8. $\{x \mid x \text{ an irrational number}, x^2 > 5\}$

9. $\left\{x \mid x = \frac{(-1)^n}{n}, n = 1, 2, 3, \ldots\right\}$

10. $\left\{x \mid x = 4 - \frac{1}{n}, n = 1, 2, 3, \ldots\right\}$

11. $\{7,14,21,28, \ldots\}$

12. $\{1,-1,\frac{1}{2},-\frac{1}{2},\frac{1}{3},-\frac{1}{3}, \ldots\}$

13. $\{1,-2,3,-4,5,-6, \ldots\}$

14. $\{\frac{3}{2},\frac{5}{3},\frac{7}{4},\frac{9}{5},\frac{11}{6}, \ldots\}$

II

1. If S is a set of real numbers and b is a number such that (1) b is an upper bound of S, and (2) for every upper bound b' of S the relation $b \leq b'$ holds, then b is a least upper bound (l.u.b.) of S. Prove that b is unique. (*Hint:* Assume that b and c are least upper bounds of S. Show that $b \leq c$ and $c \leq b$ both hold.) State a similar theorem for greatest lower bound (g.l.b.).

2. It is shown in the text that a is the g.l.b. and b is the l.u.b. of the open interval (a,b). Prove that a is the g.l.b. and b is the l.u.b. of the closed interval $[a,b]$ also.

3. Give examples of sets which contain their l.u.b. and sets which do not. Similarly, give examples for g.l.b.

4. If a set S contains a greatest element b, prove that b is the l.u.b. of S. State and prove a similar result for g.l.b.

5. Prove that every set S consisting of a finite number $a_1, a_2, \ldots, a_n$ of real numbers contains its l.u.b. and g.l.b.

6. If $c > 0$ and S is the set of all multiples $c, 2c, 3c, \ldots, nc, \ldots$ of c, prove that S has no upper bound, and hence no l.u.b.

7. Use the preceding exercise to prove the archimedean principle: Let c and d be any numbers with $c > 0$. Then there exists some positive integer n such that $nc > d$. Illustrate this principle for the following cases:

a. $c = 2, d = 131$

b. $c = 10^{-6}, d = 10^6$.

8. Let S be the set consisting of the rational numbers $\frac{1}{2}, \frac{2}{3}, \frac{3}{4}, \frac{4}{5}, \ldots, n/(n+1), \ldots$. Prove that 1 is the l.u.b. of S. [*Hint:* If $0 < c < 1$, use Exercise 7 to show that there exists a positive integer m such that $c < m/(m+1)$.]

2 AREA

Associated with each rectangle A having length l and width w is the number lw called the *area* of A. Every triangle is "half" a rectangle (Figure 5.1) and hence is defined to have area $bh/2$ where b is its base and h its altitude. Since any polygon can be subdivided into triangles (Figure 5.2), the area of a polygon can be defined to be the sum of the areas of the triangles into which it is subdivided. It can be proved that the area so obtained is the same no matter how the polygon is divided up into triangles.

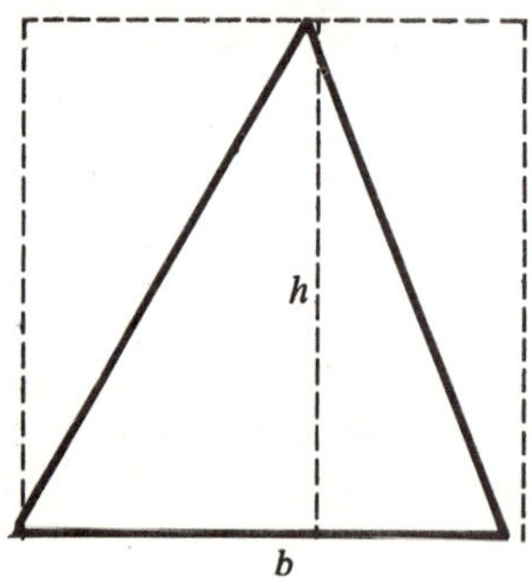

FIGURE 5.1

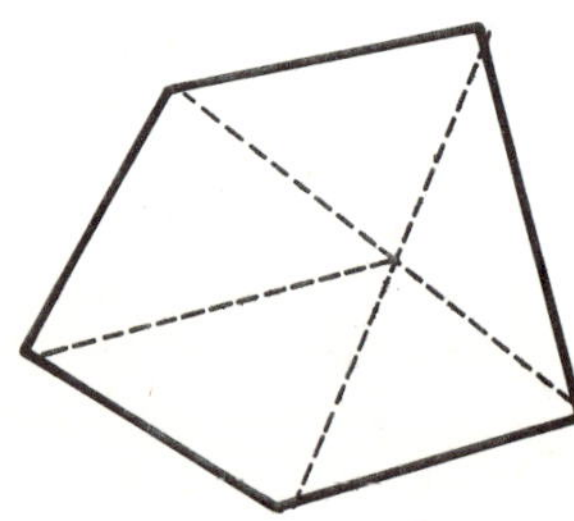

FIGURE 5.2

If a region R of the plane has a curved boundary, then it is not clear if the region has area. However, if the region is placed on a piece of graph paper (Figure 5.3), then we can underapproximate the area A of R by counting the number of squares of the graph paper completely contained in R and overapproximate its area by counting the number of squares containing some part of R. The actual area of R should lie between these two numbers. By using finer and finer graph paper, we should be able to get a closer and closer approximation of the area of R.

To make the problem of finding the area of a region with a curved boundary mathematically tractable, we shall assume that the curved part of the boundary is the graph of a continuous function. Thus, let f be a function that is continuous and nonnegative (that is, $f(x) \geq 0$) in an interval $[a,b]$. We shall call the region R (Figure 5.4) in the coordinate plane bounded by the graph of f, the lines $x = a$ and $x = b$, and the x axis, *the region under the graph of f from a to b*. We wish to associate with R a number A to be called the *area* of R. The area A should be consistent with the area of a polygon. That is, A should

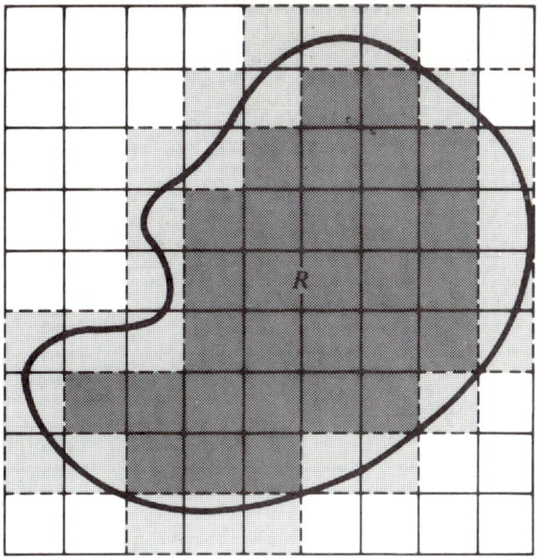

FIGURE 5.3

be greater than or equal to the area of each polygon contained in R, and less than or equal to the area of each polygon containing R. We shall see in due course that there is one and only one number A having this property.

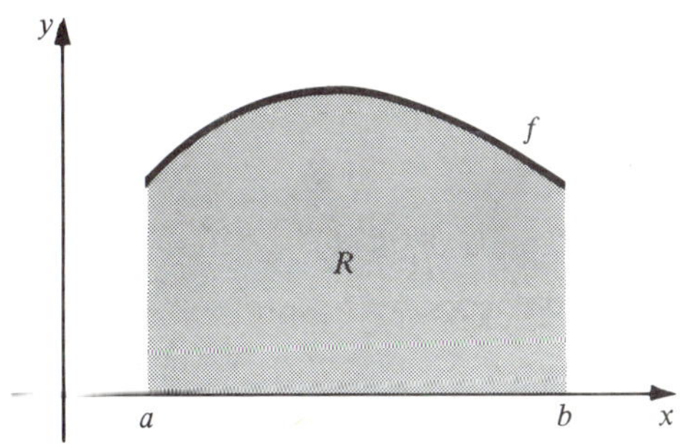

FIGURE 5.4

Before discussing the area in detail, we need to introduce the concept of a partition of an interval.

5.4 Definition

A *partition* of a closed interval $[a,b]$ is a division of $[a,b]$ into subintervals. If $x_0, x_1, \ldots, x_{n-1}, x_n$ are numbers in $[a,b]$ such that

$$a = x_0 < x_1 < \cdots < x_{n-1} < x_n = b,$$

then these $n + 1$ numbers partition $[a,b]$ into n subintervals

$$[x_0,x_1], [x_1,x_2], \ldots, [x_{n-1},x_n].$$

The notation

$$P = \{x_0, x_1, \ldots, x_n\}$$

is used to denote this set of n subintervals of $[a,b]$. Partition P is called *regular* if every subinterval has the same length, i.e.,

$$x_1 - x_0 = x_2 - x_1 = \cdots = x_n - x_{n-1} = \frac{b - a}{n}.$$

Example 1 The partition $P = \{1,3,4,\frac{9}{2},6\}$ of $[1,6]$ consists of the 4 subintervals

$$[1,3], [3,4], [4,\tfrac{9}{2}], [\tfrac{9}{2},6].$$

It is sketched in Figure 5.5.

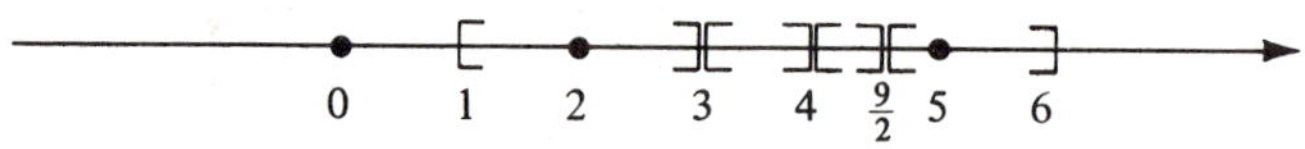

FIGURE 5.5

Example 2 The partition $P = \{-1,2,5,8,11,14\}$ of $[-1,14]$ consists of the 5 subintervals

$$[-1,2], [2,5], [5,8], [8,11], [11,14].$$

It is regular; each subinterval has length 3.

Returning to the problem of finding the area of region R (Figure 5.4), let

$$P = \{x_0, x_1, \ldots, x_n\}$$

be any partition of $[a,b]$. Since f is a continuous function in $[a,b]$, we can select (by 4.2) numbers $u_1, u_2, \ldots, u_n$ in the n subintervals of P such that

$f(u_1)$ is the minimum value of f in $[x_0,x_1]$,

$f(u_2)$ is the minimum value of f in $[x_1,x_2]$,

....

$f(u_n)$ is the minimum value of f in $[x_{n-1},x_n]$.

Then we can construct n rectangles whose bases are the subintervals of P and whose respective heights are $f(u_1), f(u_2), \ldots, f(u_n)$. The case in which P consists of 4 subintervals ($n = 4$) is illustrated in Figure 5.6.

In general, the first rectangle has area

$$f(u_1)(x_1 - x_0),$$

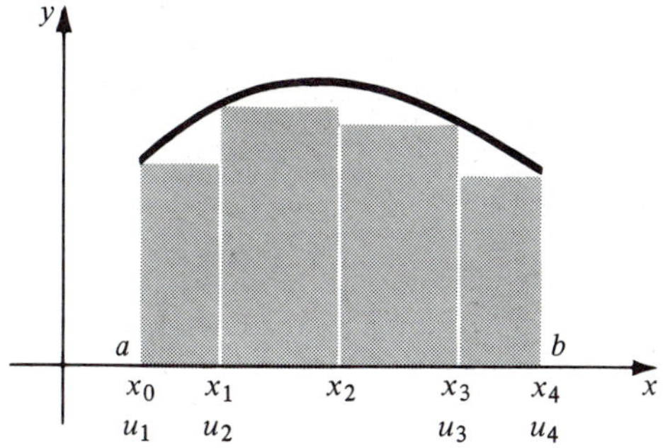

FIGURE 5.6

the product of its height and the length of its base; the second has area

$$f(u_2)(x_2 - x_1),$$

and so on. These n rectangles taken together form a polygon, called a *rectangular polygon*, inscribed in R (it is shown in color in Figure 5.6). Its area, denoted by $I(P)$, is the sum of the areas of the n rectangles,

$$I(P) = f(u_1)(x_1 - x_0) + f(u_2)(x_2 - x_1) + \cdots + f(u_n)(x_n - x_{n-1}).$$

Similarly, we can select numbers $v_1, v_2, \ldots, v_n$ in the n subintervals of P such that

$f(v_1)$ is the maximum value of f in $[x_0,x_1]$,

$f(v_2)$ is the maximum value of f in $[x_1,x_2]$,

. . . .

$f(v_n)$ is the maximum value of f in $[x_{n-1},x_n]$.

Then we can construct n rectangles whose bases are the subintervals of P and whose respective heights are $f(v_1), f(v_2), \ldots, f(v_n)$. The case $n = 4$ is shown in Figure 5.7.

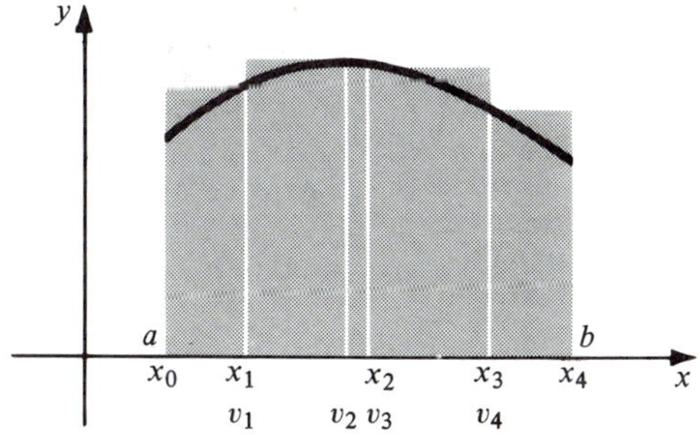

FIGURE 5.7

The rectangular polygon made up of these n rectangles is circumscribed about region R. Its area, denoted by $C(P)$, is given by

$$C(P) = f(v_1)(x_1 - x_0) + f(v_2)(x_2 - x_1) + \cdots + f(v_n)(x_n - x_{n-1}).$$

We can find the areas $I(P)$ and $C(P)$ associated with region R (Figure 5.4) for every partition P of $[a,b]$. Since each inscribed rectangular polygon is contained in R and in turn each circumscribed polygon contains R, each inscribed rectangular polygon is contained in every circumscribed one. Hence,

5.5 $$I(P_1) \leq C(P_2) \qquad \text{for all partitions } P_1, P_2 \text{ of } [a,b].$$

Consider now the set of all areas of inscribed polygons,

$$L = \{I(P) \mid P \text{ a partition of } [a,b]\},$$

and the set of all areas of circumscribed polygons,

$$U = \{C(P) \mid P \text{ a partition of } [a,b]\}.$$

By 5.5, each number in set L is less than or equal to every number in set U. In other words, U is a set of upper bounds of L and L is a set of lower bounds of U.

Since set L has each $C(P)$ as an upper bound, L has a l.u.b. by the completeness property 5.3. Similarly, U has a g.l.b. Let

$$A_l = \text{l.u.b. } L, \qquad A_u = \text{g.l.b. } U.$$

By its choice, $A_l \leq C(P)$ for every $C(P)$ in U. Hence,

$$A_l \leq A_u.$$

The area A of region R (Figure 5.4), if it exists, should be somewhere between A_l and A_u. That is, A should be an upper bound of L and a lower bound of U. However, we shall prove in Chapter 6 that $A_l = A_u$! Thus, we can *define* the area A of region R to be

$$A = A_l = A_u.$$

The question of how to compute the area A of region R will be answered later. In the meantime, we compute some $I(P)$ and $C(P)$ discussed above.

Example 3 If $f(x) = 4 - x^2$, let R be the region under the graph of f from -2 to 2 (Figure 5.8). Find $I(P)$ and $C(P)$ in case:

(a) $P = \{-2, -\frac{1}{2}, 1, 2\}$.

(b) P is the regular partition of $[-2,2]$ into 8 subintervals.

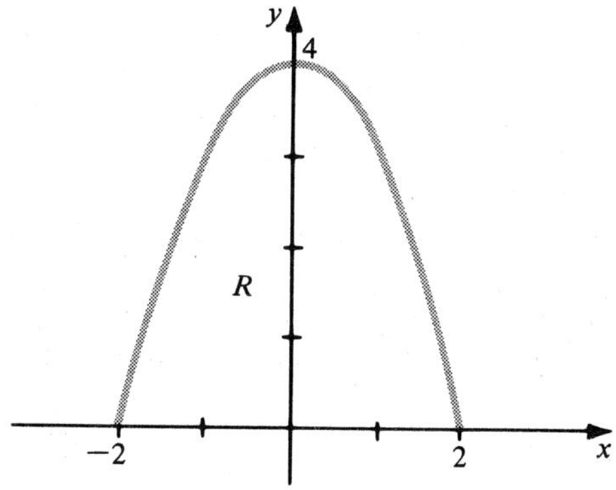

$f(x) = 4 - x^2$

FIGURE 5.8

Solution: (a) The minimum value of f in $[-2,-\frac{1}{2}]$ is $f(-2) = 0$; in $[-\frac{1}{2},1]$ is $f(1) = 3$; in $[1,2]$ is $f(2) = 0$. Hence

$$I(P) = 0\cdot(-\tfrac{1}{2} + 2) + 3\cdot(1 + \tfrac{1}{2}) + 0\cdot(2 - 1) = \tfrac{9}{2}.$$

The maximum value of f in $[-2,-\frac{1}{2}]$ is $f(-\frac{1}{2}) = \frac{15}{4}$; in $[-\frac{1}{2},1]$ is $f(0) = 4$; in $[1,2]$ is $f(1) = 3$. Hence

$$C(P) = \tfrac{15}{4}\cdot(-\tfrac{1}{2} + 2) + 4\cdot(1 + \tfrac{1}{2}) + 3(2 - 1) = \tfrac{117}{8}.$$

Evidently $I(P)$ and $C(P)$ differ considerably.

(b) In this case

$$P = \{-2,-\tfrac{3}{2},-1,-\tfrac{1}{2},0,\tfrac{1}{2},1,\tfrac{3}{2},2\},$$

and

$$\begin{aligned} I(P) &= \tfrac{1}{2}f(-2) + \tfrac{1}{2}f(-\tfrac{3}{2}) + \tfrac{1}{2}f(-1) + \tfrac{1}{2}f(-\tfrac{1}{2}) + \tfrac{1}{2}f(\tfrac{1}{2}) \\ &\quad + \tfrac{1}{2}f(1) + \tfrac{1}{2}f(\tfrac{3}{2}) + \tfrac{1}{2}f(2) \\ &= \tfrac{17}{2}, \end{aligned}$$

$$\begin{aligned} C(P) &= \tfrac{1}{2}f(-\tfrac{3}{2}) + \tfrac{1}{2}f(-1) + \tfrac{1}{2}f(-\tfrac{1}{2}) + \tfrac{1}{2}f(0) + \tfrac{1}{2}f(0) \\ &\quad + \tfrac{1}{2}f(\tfrac{1}{2}) + \tfrac{1}{2}f(1) + \tfrac{1}{2}f(\tfrac{3}{2}) \\ &= \tfrac{25}{2}. \end{aligned}$$

Now $I(P)$ and $C(P)$ are closer together. If we would select a regular partition P of $[-2,2]$ with many more subintervals, we would obtain $I(P)$ and $C(P)$ which were still closer to each other.

Example 4 Let R be the trapezoidal region under the graph of $f(x) = x$ from 1 to 4 (Figure 5.9). Find $I(P)$ and $C(P)$ for any partition $P = \{x_0, x_1, \ldots, x_n\}$ of $[1,4]$.

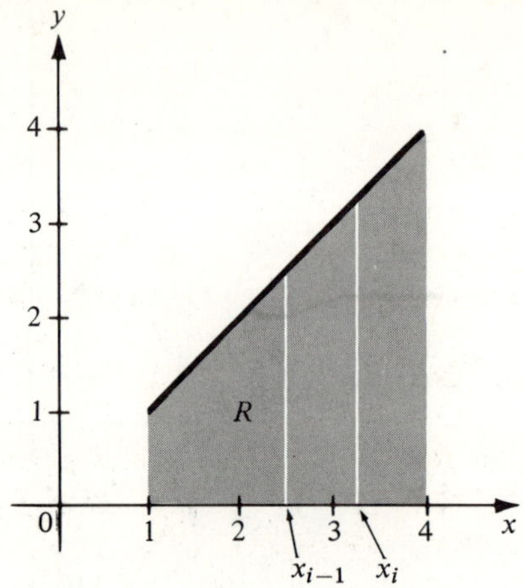

FIGURE 5.9

Solution: In each subinterval $[x_{i-1},x_i]$ of P, the minimum value of f occurs at x_{i-1} and the maximum value at x_i. Also $f(x_{i-1}) = x_{i-1}$ and $f(x_i) = x_i$. Thus,

$$I(P) = x_0(x_1 - x_0) + x_1(x_2 - x_1) + x_2(x_3 - x_2) + \cdots + x_{n-1}(x_n - x_{n-1}),$$

$$C(P) = x_1(x_1 - x_0) + x_2(x_2 - x_1) + x_3(x_3 - x_2) + \cdots + x_n(x_n - x_{n-1}).$$

The arithmetic average of two numbers is between them. Hence, each term of the sum below

$$A(P) = \frac{x_1 + x_0}{2}(x_1 - x_0) + \frac{x_2 + x_1}{2}(x_2 - x_1) + \cdots + \frac{x_n + x_{n-1}}{2}(x_n - x_{n-1})$$

is between the corresponding terms of $I(P)$ and $C(P)$, and

$$I(P) < A(P) < C(P).$$

Now

$$A(P) = \frac{x_1^2 - x_0^2}{2} + \frac{x_2^2 - x_1^2}{2} + \frac{x_3^2 - x_2^2}{2} + \cdots + \frac{x_{n-1}^2 - x_{n-2}^2}{2} + \frac{x_n^2 - x_{n-1}^2}{2}.$$

Do you see that all but two terms of $A(P)$ cancel out? Thus,

$$A(P) = \frac{-x_0^2 + x_n^2}{2} = \frac{-1^2 + 4^2}{2} = \frac{15}{2}$$

and $$I(P) < \tfrac{15}{2} < C(P).$$

We easily check that $\frac{15}{2}$ is the area of the trapezoid according to the usual formula. From our remarks above,

$$\tfrac{15}{2} = \text{l.u.b. } L = \text{g.l.b. } U.$$

EXERCISES

I

1. For $f(x) = x^2$, sketch the region under the graph of f from 2 to 4.
   a. Compute $I(P)$ and $C(P)$ associated with the partition $\{2,\frac{5}{2},3,\frac{7}{2},4\}$ of $[2,4]$.
   b. Compute $I(P)$ and $C(P)$ for the regular partition of $[2,4]$ into 8 subintervals.
2. For $f(x) = x^3$, sketch the region under the graph of f from 1 to 2.
   a. Compute $I(P)$ and $C(P)$ for the partition $\{1,2\}$ of $[1,2]$.
   b. Compute $I(P)$ and $C(P)$ for the partition $\{1,\frac{3}{2},2\}$ of $[1,2]$.
   c. Compute $I(P)$ and $C(P)$ for the regular partition of $[1,2]$ into 4 subintervals.
3. For $f(x) = 1 - x^3$, sketch the region bounded by the graph of f and the coordinate axes.
   a. Compute $I(P)$ and $C(P)$ for the partition $\{0,\frac{1}{2},1\}$ of $[0,1]$.
   b. Compute $I(P)$ and $C(P)$ for the regular partition of $[0,1]$ into 8 subintervals.
4. For $f(x) = 1/x$, sketch the region under the graph of f from $\frac{1}{2}$ to 2. (Use a large unit segment.)
   a. Compute $I(P)$ and $C(P)$ for the regular partition of $[\frac{1}{2},2]$ into 6 subintervals.
   b. Same as (a) except 12 subintervals.
5. For $f(x) = 1/x^2$, sketch the region under the graph of f from 1 to 4 (use a large unit segment).
   a. Compute $I(P)$ and $C(P)$ for the regular partition of $[1,4]$ into 3 subintervals.
   b. Same as (a) except 6 subintervals.
6. If R is the region under the graph of $f(x) = b$ from 0 to a, then R is a rectangle. Show that $I(P)$ and $C(P)$ equal the area of R for every partition P of $[0,a]$.

Compute $I(P)$ and $C(P)$, using a regular partition P:

7. $f(x) = [x]$, interval $[-1,2]$, 6 subintervals
8. $f(x) = x^2 - 1$, interval $[0,1]$, 8 subintervals
9. $f(x) = 1 - x^3$, interval $[-3,-1]$, 4 subintervals
10. $f(x) = \begin{cases} x \text{ if } 0 \le x \le 1, \\ -x \text{ if } 1 < x \le 2, \end{cases}$ interval $[0,2]$, 4 subintervals

II

1. Let R be the region under the graph of $f(x) = x$ from a to b $(0 \le a < b)$. Show that for any partition P of $[a,b]$,

$$I(P) < \frac{b^2 - a^2}{2} < C(P).$$

2. Let R be the region under the graph of $f(x) = x^2$ from a to b $(0 \le a < b)$. Show that for any partition P of $[a,b]$,

$$I(P) < \frac{b^3 - a^3}{3} < C(P).$$

3. Let R be the region under the graph of $f(x) = x^3$ from a to b ($0 \le a < b$). Show that for any partition P of $[a,b]$,

$$I(P) < \frac{b^4 - a^4}{4} < C(P).$$

4. Let R be the region under the graph of $f(x) = 1/x^2$ from a to b, $0 < a < b$, and let $P = \{x_0, x_1, \ldots, x_{n-1}, x_n\}$ be a partition of $[a,b]$.
   a. Write down the sums $I(P)$ and $C(P)$ associated with this partition.
   b. Show that for each integer i between 0 and $n - 1$,

$$\frac{1}{x_{i+1}^2} < \frac{1}{x_i x_{i+1}} < \frac{1}{x_i^2}$$

   and hence that

$$\frac{x_{i+1} - x_i}{x_{i+1}^2} < \frac{1}{x_i} - \frac{1}{x_{i+1}} < \frac{x_{i+1} - x_i}{x_i^2}.$$

   Then show that

$$I(P) < \frac{1}{a} - \frac{1}{b} < C(P).$$

3 THE SIGMA NOTATION

A sum of n terms such as $a_1 + a_2 + \cdots + a_n$ is designated by

$$\sum_{i=1}^{n} a_i$$

in the handy *sigma notation*; that is,

$$\sum_{i=1}^{n} a_i = a_1 + a_2 + \cdots + a_n.$$

Some examples of the use of the sigma notation are given below.

$$\sum_{i=1}^{n} i = 1 + 2 + \cdots + n.$$

$$\sum_{j=2}^{m} j^2 = 2^2 + 3^2 + \cdots + m^2.$$

$$\sum_{i=1}^{n} c = c + c + \cdots + c = nc.$$

$$\sum_{k=3}^{6} k(k-2) = 3 \cdot 1 + 4 \cdot 2 + 5 \cdot 3 + 6 \cdot 4 = 50.$$

$$\sum_{i=0}^{5} \frac{i-1}{i+1} = \frac{-1}{1} + \frac{0}{2} + \frac{1}{3} + \frac{2}{4} + \frac{3}{5} + \frac{4}{6} = 1.1.$$

Listed below are a few useful summation formulas for future reference. These may be proved by mathematical induction.

5.6
$$\sum_{i=1}^{n} i = \frac{n(n+1)}{2}.$$

5.7
$$\sum_{i=1}^{n} i^2 = \frac{n(n+1)(2n+1)}{6}.$$

5.8
$$\sum_{i=1}^{n} i^3 = \left[\frac{n(n+1)}{2}\right]^2.$$

It follows directly from the associative, commutative, and distributive laws that:

5.9
$$\sum_{i=1}^{n} (a_i + b_i) = \sum_{i=1}^{n} a_i + \sum_{i=1}^{n} b_i$$

5.10
$$\sum_{i=1}^{n} ka_i = k \sum_{i=1}^{n} a_i.$$

We may use 5.6 to 5.10 to work problems of the following type.

Example 1 Find $\sum_{i=1}^{20} i(2i - 3)$.

Solution: We have

$$\begin{aligned}
\sum_{i=1}^{20} i(2i-3) &= \sum_{i=1}^{20} (2i^2 - 3i) \\
&= \sum_{i=1}^{20} 2i^2 + \sum_{i=1}^{20} (-3)i && \textit{(by 5.9)} \\
&= 2\sum_{i=1}^{20} i^2 - 3\sum_{i=1}^{20} i && \textit{(by 5.10)} \\
&= 2\,\frac{(20)(21)(41)}{6} - 3\,\frac{(20)(21)}{2} && \textit{(by 5.6, 5.7)} \\
&= 20 \cdot 287 - 10 \cdot 63 = 5110.
\end{aligned}$$

Example 2 Find $\sum_{i=1}^{n} (3 - i)(2 + i)$.

Solution: We have

$$\sum_{i=1}^{n}(3 - i)(2 + i) = \sum_{i=1}^{n}(6 + i - i^2)$$

$$= \sum_{i=1}^{n} 6 + \sum_{i=1}^{n} i - \sum_{i=1}^{n} i^2$$

$$= 6n + \frac{n(n + 1)}{2} - \frac{n(n + 1)(2n + 1)}{6}$$

$$= \tfrac{1}{3}n(19 - n^2).$$

EXERCISES

Using 5.6 to 5.10, evaluate the following sums.

1. $\sum_{i=1}^{n}(2i - 1)$

2. $\sum_{i=7}^{12}(i^2 - i)$

3. $\sum_{i=4}^{10}(i - 1)$

4. $\sum_{i=1}^{n}\frac{i^2}{n^3}$

5. $\sum_{i=1}^{n}\left(1 + \frac{i}{n}\right)\cdot\frac{1}{n}$

6. $\sum_{i=1}^{n}\left(a + \frac{(b - a)i}{n}\right)\left(\frac{b - a}{n}\right)$

7. $\sum_{j=2}^{5}(-1)^j\left(\frac{2}{j + 1}\right)$

8. $\sum_{t=1}^{n}(st^2 + r)$

9. $\sum_{i=0}^{6}(i^2 + 6i - t)$

10. $\sum_{i=1}^{n}(3^i - 3^{i-1})$

11. $\sum_{j=2}^{8}(2j - 1)^2$

12. $\sum_{j=1}^{n}(j - 1)^3\left(\frac{b - a}{n}\right)$

13. $\sum_{i=1}^{n}(ai + b)^2$

14. $\sum_{i=1}^{n}(4i^2 - 4i + 1)$

15. $\sum_{i=1}^{n} i\ i + 1)$

16. $\sum_{i=1}^{n} i(i + 1)(i + 2)$

17. Show that $\sum_{i=1}^{n}[F(i) - F(i - 1)] = F(n) - F(0)$.

18. Using Exercise 17 with $F(i) = i^5$ and 5.6 to 5.10, evaluate $\sum_{i=1}^{n} i^4$.

19. Prove that $\sum_{i=1}^{n} ar^i = \frac{a(1 - r^{n+1})}{1 - r}$ for $r \neq 1$.

20. Let $\bar{x} = \frac{1}{n}\sum_{i=1}^{n} x_i$. Prove that $\sum_{i=1}^{n} (x_i - \bar{x})^2 = \sum_{i=1}^{n} x_i^2 - \frac{1}{n}\left(\sum_{i=1}^{n} x_i\right)^2$.

4 INTEGRALS

What was done in Section 2 for nonnegative continuous functions can be done for any continuous function. We will not try to interpret our results geometrically at this time.

Let function f be continuous in a closed interval $[a,b]$. For each partition

$$P = \{x_0, x_1, x_2, \ldots, x_{n-1}, x_n\}$$

of $[a,b]$, we can form the *lower sum*

$$S(P) = f(u_1)(x_1 - x_0) + f(u_2)(x_2 - x_1) + \cdots + f(u_n)(x_n - x_{n-1}),$$

and the *upper sum*

$$T(P) = f(v_1)(x_1 - x_0) + f(v_2)(x_2 - x_1) + \cdots + f(v_n)(x_n - x_{n-1})$$

where $f(u_1)$ is the minimum value and $f(v_1)$ the maximum value of f in $[x_0,x_1]$, $f(u_2)$ is the minimum value and $f(v_2)$ the maximum value of f in $[x_1,x_2]$, and so on. Since

$$f(u_i) \leq f(v_i) \qquad \text{for each subscript } i = 1, 2, \ldots, n,$$

evidently,

5.11 $$S(P) \leq T(P).$$

Of course, $S(P)$ is what we called $I(P)$ and $T(P)$ what we called $C(P)$ in Section 2.

It is convenient to use the *delta notation* in describing the lengths of the subintervals of P:

$$\Delta x_1 = x_1 - x_0, \Delta x_2 = x_2 - x_1, \ldots, \Delta x_n = x_n - x_{n-1}.$$

Clearly

$$\sum_{i=1}^{n} \Delta x_i = b - a.$$

Then the *lower sum* $S(P)$ has the compact form

5.12 $$S(P) = \sum_{i=1}^{n} f(u_i)\, \Delta x_i$$

and the *upper sum* $T(P)$ the form

5.13 $$T(P) = \sum_{i=1}^{n} f(v_i)\, \Delta x_i.$$

Example 1 The function $f(x) = 2x - x^2$ is continuous in $[-1,3]$. Find $S(P)$ and $T(P)$ for the partition $P = \{-1, \frac{1}{2}, \frac{3}{2}, 2, 3\}$.

Solution: From

$$f'(x) = 2 - 2x, \qquad f''(x) = -2,$$

we see that $f(1) = 1$ is a maximum value of f. The curve is sketched in Figure 5.10. From the figure, we see that the minimum values of f in the four subintervals $[-1,\frac{1}{2}]$, $[\frac{1}{2},\frac{3}{2}]$, $[\frac{3}{2},2]$, $[2,3]$ occur at

$$u_1 = -1,\ u_2 = \tfrac{1}{2}\ (\text{or } \tfrac{3}{2}),\ u_3 = 2,\ u_4 = 3.$$

Then $$f(u_1) = -3,\ f(u_2) = \tfrac{3}{4},\ f(u_3) = 0,\ f(u_4) = -3.$$

Thus, $$S(P) = (-3)\cdot\tfrac{3}{2} + \tfrac{3}{4}\cdot 1 + 0\cdot\tfrac{1}{2} + (-3)\cdot 1 = -\tfrac{27}{4}.$$

The maximum values occur at

$$v_1 = \tfrac{1}{2},\ v_2 = 1,\ v_3 = \tfrac{3}{2},\ v_4 = 2$$

and $$f(v_1) = \tfrac{3}{4},\ f(v_2) = 1,\ f(v_3) = \tfrac{3}{4},\ f(v_4) = 0.$$

Thus, $$T(P) = \tfrac{3}{4}\cdot\tfrac{3}{2} + 1\cdot 1 + \tfrac{3}{4}\cdot\tfrac{1}{2} + 0\cdot 1 = \tfrac{5}{2}.$$

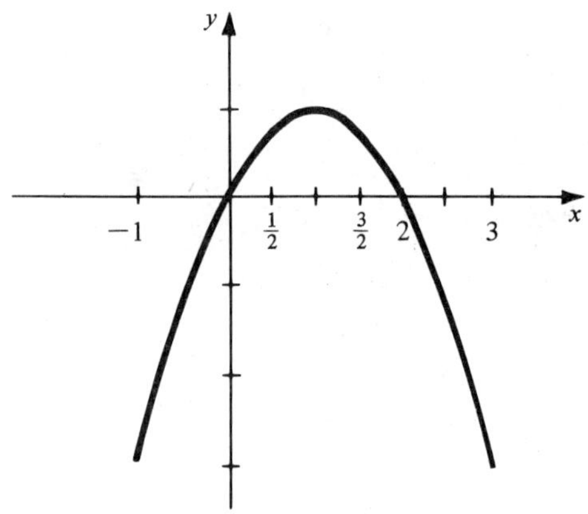

FIGURE 5.10

5.14 Definition

Let P be a partition of $[a,b]$. If one or more of the subintervals of P are further subdivided, the resulting partition P' of $[a,b]$ is called a *refinement* of P.

Example 2 If $P = \{1,3,4,6\}$, then $P' = \{1,\frac{3}{2},2,\frac{7}{3},3,4,5,6\}$ is a refinement of P. Thus, P consists of the 3 subintervals

$$[1,3],\ [3,4],\ [4,6]$$

and P' of the 7 subintervals

$$[1,\tfrac{3}{2}],\ [\tfrac{3}{2},2],\ [2,\tfrac{7}{3}],\ [\tfrac{7}{3},3],\ [3,4],\ [4,5],\ [5,6].$$

The first 4 are subdivisions of [1,3] and the last 2 of [4,6].

One way to get a refinement is to start with two partitions P_1 and P_2 of $[a,b]$. Then the intersections of the subintervals of P_1 and P_2 form a refinement of both P_1 and P_2. For example, if

$$P_1 = \{1,2,3,5,6\}, \qquad P_2 = \{1,2,4,4.5,6\}$$

then the union P of sets P_1 and P_2,

$$P = \{1,2,3,4,4.5,5,6\}$$

is a refinement of both P_1 and P_2.

5.15 Theorem

Let function f be continuous in $[a,b]$, P be a partition of $[a,b]$, and $S(P)$, $T(P)$ be as defined in 5.11 and 5.12. For every refinement P' of P,

$$S(P') \geq S(P),$$

$$T(P') \leq T(P).$$

Proof: We need only consider a partition P' having one more subinterval than P. Every other refinement of P can be arrived at by a succession of such simple partitions. For simplicity, assume that the first subinterval $[x_0,x_1]$ of $P = \{x_0, x_1, \ldots, x_n\}$ is subdivided into two subintervals $[x_0,x]$ and $[x,x_1]$ in P'. Then, $P' = \{x_0, x, x_1, \ldots, x_n\}$ and

$$S(P') = f(u)(x - x_0) + f(u')(x_1 - x) + \sum_{i=2}^{n} f(u_i)\,\Delta x_i,$$

where $f(u)$ is the minimum value of f in $[x_0,x]$ and $f(u')$ in $[x,x_1]$. Evidently, $f(u_1) \leq f(u)$, since $f(u_1)$ is the minimum value of f over a larger interval $[x_0,x_1]$ than is $f(u)$. Similarly, $f(u_1) \leq f(u')$. Hence,

$$\begin{aligned} f(u_1)\,\Delta x_1 &= f(u_1)[(x - x_0) + (x_1 - x)] \\ &\leq f(u)(x - x_0) + f(u')(x_1 - x) \end{aligned}$$

and

$$\begin{aligned} S(P) &= f(u_1)\,\Delta x_1 + \sum_{i=2}^{n} f(u_i)\,\Delta x_i \\ &\leq f(u)(x - x_0) + f(u')(x_1 - x) + \sum_{i=2}^{n} f(u_i)\,\Delta x_i = S(P'). \end{aligned}$$

Thus, $S(P') \geq S(P)$. The proof that $T(P') \leq T(P)$ is similar, and hence is omitted.

The following theorem was proved geometrically in Section 2.

5.16 Theorem

If f is a continuous function in $[a,b]$, then for all partitions P_1 and P_2 of $[a,b]$

$$S(P_1) \leq T(P_2).$$

Proof: Let P be a refinement of P_1 and P_2. By 5.11,

$$S(P) \leq T(P).$$

Hence, by 5.15,

$$S(P_1) \leq S(P) \quad \text{and} \quad T(P) \leq T(P_2).$$

Putting these inequalities together,

$$S(P_1) \leq S(P) \leq T(P) \leq T(P_2).$$

Thus, $S(P_1) \leq T(P_2)$.

By 5.16, every lower sum of f over $[a,b]$ is a lower bound for the set of all upper sums; and every upper sum is an upper bound for the set of all lower sums.

5.17 Theorem

If f is a continuous function in a closed interval $[a,b]$ and if L is the set of all lower sums and U of upper sums of f over $[a,b]$, then

$$\text{l.u.b. } L = \text{g.l.b. } U.$$

Proof: Since L has upper bounds, namely the elements of U, the l.u.b. L exists by the completeness property, and similarly for g.l.b. U. Also,

$$\text{l.u.b. } L \leq \text{g.l.b. } U$$

because every element of U is an upper bound of L.

We shall prove the theorem in case f is monotonic in $[a,b]$. The general proof is in an appendix (page 225). To this end, assume f is increasing in $[a,b]$. For each positive integer n, let $P_n = \{x_0, x_1, \ldots, x_n\}$ be the regular partition of $[a,b]$ into n subintervals. Thus,

$$\Delta x_1 = \Delta x_2 = \cdots = \Delta x_n = \frac{b-a}{n}.$$

In the ith subinterval $[x_{i-1},x_i]$ of P_n, $f(x_{i-1})$ is the minimum value of f and $f(x_i)$ the maximum value. Hence,

$$S(P_n) = \sum_{i=1}^{n} f(x_{i-1})\,\Delta x_i = \left(\sum_{i=1}^{n} f(x_{i-1})\right)\frac{b-a}{n}$$

and

$$T(P_n) = \sum_{i=1}^{n} f(x_i)\,\Delta x_i = \left(\sum_{i=1}^{n} f(x_i)\right)\frac{b-a}{n}.$$

When we subtract $S(P_n)$ from $T(P_n)$, most of the terms cancel; in fact,

$$T(P_n) - S(P_n) = [f(x_n) - f(x_0)]\frac{b-a}{n}$$

$$= \frac{(b-a)[f(b) - f(a)]}{n}.$$

For each positive number ε, we can find a positive integer n such that for the increasing function f

$$0 \leq \frac{(b-a)[f(b)-f(a)]}{n} < \varepsilon.$$

Thus we need only select n so that

$$n > \frac{(b-a)[f(b)-f(a)]}{\varepsilon}.$$

We conclude that *for each* $\varepsilon > 0$ *there exists an integer* n *such that*

$$0 \leq T(P_n) - S(P_n) < \varepsilon.$$

Since

$$S(P_n) \leq \text{l.u.b. } L \leq \text{g.l.b. } U \leq T(P_n)$$

we also have for each $\varepsilon > 0$

$$0 \leq \text{g.l.b. } U - \text{l.u.b. } L < \varepsilon.$$

There is only one number greater than or equal to 0 and at the same time less than every positive number ε, namely 0. Hence,

$$\text{g.l.b. } U - \text{l.u.b. } L = 0$$

as desired. The proof is similar if f is decreasing in $[a,b]$.

5.18 Definition

Let f be a continuous function in $[a,b]$ and L be the set of lower sums of f over $[a,b]$. Then l.u.b. L is called the *integral of f from a to b* and is denoted by $\int_a^b f(x)\,dx$,

$$\int_a^b f(x)\,dx = \text{l.u.b. } L.$$

Of course, by 5.17, the integral of f from a to b is also g.l.b. U.

Example 3 The function $f(x) = 4 - x$ is continuous and decreasing in $[1,3]$. Find the integral of f from 1 to 3.

Solution: The regular partition P_n of $[1,3]$ into n subintervals has the form

$$P_n = \{1, 1 + \Delta x, 1 + 2\,\Delta x, \ldots, 1 + (n-1)\,\Delta x, 3\},$$

where $\Delta x = 2/n$. The minimum value of f in the ith subinterval

$$[1 + (i - 1)\,\Delta x, 1 + i\,\Delta x]$$

is $f(1 + i\,\Delta x) = 4 - (1 + i\,\Delta x) = 3 - i\,\Delta x$, $i = 1, 2, \ldots, n$. Hence,

$$\begin{aligned} S(P_n) &= \sum_{i=1}^{n} (3 - i\,\Delta x)\,\Delta x \\ &= \sum_{i=1}^{n} 3\,\Delta x - \sum_{i=1}^{n} i\,(\Delta x)^2 \\ &= (3\,\Delta x)n - \frac{n(n+1)}{2}(\Delta x)^2. \end{aligned}$$

Using the fact that $\Delta x = 2/n$, this becomes

$$S(P_n) = 6 - \left(2 + \frac{2}{n}\right) = 4 - \frac{2}{n}.$$

By the proof of 5.17 [interchanging $f(b)$ and $f(a)$ because f is now decreasing], we find that

$$T(P_n) - S(P_n) = \frac{(b - a)[f(a) - f(b)]}{n} = \frac{2 \cdot 2}{n}.$$

Hence,

$$T(P_n) = 4 + \frac{2}{n}.$$

Since the integral is between $S(P_n)$ and $T(P_n)$,

$$4 - \frac{2}{n} \le \int_1^3 f \le 4 + \frac{2}{n}$$

for every positive integer n. Therefore

$$\int_1^3 f(x)\,dx = 4.$$

Geometrically, we expect $\int_1^3 f(x)\,dx$ to be the *area* of the region R of Figure 5.11. This region is a trapezoid whose area A is commonly given by the formula

$$A = \tfrac{1}{2}(b_1 + b_2)h,$$

where b_1 and b_2 are the lengths of the parallel sides and h is the distance between them. For region R,

$$A = \tfrac{1}{2}(3 + 1) \cdot 2 = 4,$$

which agrees with $\int_1^3 f(x)\,dx$.

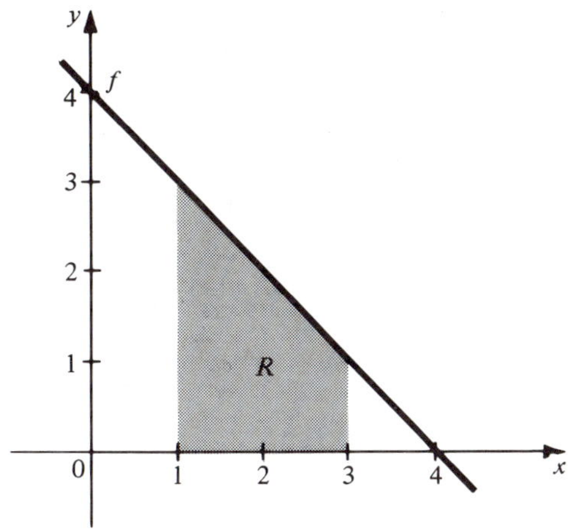

FIGURE 5.11

EXERCISES

I

Find $S(P)$ and $T(P)$ for each of the following functions and partitions P. (You may approximate your answers to two decimal places.)

1. $f(x) = 1 - x^2$, $P = \{0,\frac{1}{2},1,\frac{3}{2},2\}$
2. $f(x) = 1 - x^2$, $P = \{0,\frac{1}{4},\frac{3}{4},1,\frac{7}{4},2\}$
3. $f(x) = 2x^2$, $P = \{-1,-\frac{1}{2},0,\frac{1}{2},1\}$
4. $f(x) = 2x^2$, $P = \{-1,-\frac{5}{8},\frac{1}{8},\frac{1}{2},\frac{3}{4},1\}$
5. $f(x) = x^3$, $P = \{-2,-\frac{5}{3},-\frac{4}{3},-1,-\frac{2}{3},-\frac{1}{3},0\}$
6. $f(x) = x^3$, $P = \{-2,-\frac{3}{2},-\frac{5}{4},-1,-\frac{1}{2},-\frac{1}{4},0\}$
7. $f(x) = 1/x$, $P = \{-4,-3,-2,-1\}$
8. $f(x) = 1/x$, $P = \{-4,-\frac{7}{2},-\frac{13}{4},-3,-\frac{5}{2},-\frac{3}{2},-1\}$
9. $f(x) = 1/x^2$, $P = \{1,\frac{3}{2},2,\frac{5}{2},3,\frac{7}{2},4\}$
10. $f(x) = 1/x^2$, $P = \{1,\frac{5}{4},\frac{7}{4},\frac{9}{4},3,\frac{13}{4},4\}$

Use the method of Example 3 to find the integral of each of the following functions.

11. $f(x) = 2 + x$ from 1 to 5
12. $f(x) = 1 + 3x$ from 0 to 2
13. $f(x) = 4 - 2x$ from -2 to 2
14. $f(x) = x^2$ from 0 to 2
15. $f(x) = 1 - x^2$ from -1 to 0

II

In Exercises 1 to 4, consider each of the functions on the interval $[a,b]$. Use the corresponding results of Exercises 1 to 4, Part II, Section 2, to obtain each of the indicated integrals.

1. If $f(x) = x$, then for every partition P of $[a,b]$,

$$S(P) < \frac{b^2 - a^2}{2} < T(P),$$

and so $\int_a^b f(x)\,dx = \frac{b^2 - a^2}{2}$

2. If $f(x) = x^2$, then for every partition P of $[a,b]$,

$$S(P) < \frac{b^3 - a^3}{3} < T(P),$$

and so $\int_a^b f(x)\,dx = \frac{b^3 - a^3}{3}.$

3. If $f(x) = x^3$, then for every partition P of $[a,b]$,

$$S(P) < \frac{b^4 - a^4}{4} < T(P),$$

and so $\int_a^b f(x)\,dx = \frac{b^4 - a^4}{4}.$

4. If $f(x) = 1/x^2$, then for every partition P of $[a,b]$,

$$S(P) < \frac{1}{a} - \frac{1}{b} < T(P),$$

and so $\int_a^b f(x)\,dx = \frac{1}{a} - \frac{1}{b}.$

5 FUNDAMENTAL THEOREM OF THE CALCULUS

The symbol $\int$ for the integral was introduced by Leibnitz in 1675. This elongated S stood for a "sum" in his notation. Thus $\int_a^b f(x)\,dx$ indicated a sum of elements of the form $f(x)\,dx$, just as the lower sum is a sum of elements of the form $m_i\,\Delta x_i$ and similarly for the upper sum. If $f(x) \geq 0$, then we can think of $f(x)\,dx$ as the product of the height $f(x)$ and width dx of one of the rectangles inscribed in the graph of f.

We realize now that only part of the notation $\int_a^b f(x)\,dx$ is needed to indicate the integral of f from a to b, namely $\int_a^b f$. However, because there are many useful features of the notation of Leibnitz, we shall use it throughout this book.

The letter x occurring in $\int_a^b f(x)\,dx$ is of no special significance. Thus

$$\int_a^b f(x)\,dx = \int_a^b f(t)\,dt = \int_a^b f(u)\,du, \text{ etc.}$$

So far, $\int_a^b f(x)\,dx$ has been defined only if $a < b$. For convenience, let us define

$$\int_a^b f(x)\,dx = 0 \qquad \text{if } a = b,$$

and

$$\int_a^b f(x)\,dx = -\int_b^a f(x)\,dx \qquad \text{if } a > b.$$

Areas are additive in the sense that if a polygon is made up of nonoverlapping smaller polygons, then the area of the larger polygon is the sum of the areas of the smaller polygons. The integral is also additive in the following sense.

5.19 Theorem

If the function f is continuous in an interval containing the numbers a, b, and c, then

$$\int_a^b f(x)\,dx + \int_b^c f(x)\,dx = \int_a^c f(x)\,dx.$$

Proof: See the appendix, page 226, for a proof.

The concepts of the derivative and the integral were well known before the time of Newton and Leibnitz. However, the great English physicist and mathematician Isaac Newton (1642–1727) and the great German mathematician and philosopher Gottfried Wilhelm von Leibnitz (1646–1716) established the intimate relationship between these two concepts. Because these men gained this insight independently of each other, they are both usually credited with the discovery of the calculus.

Before stating the relationship between the derivative and the integral of a function, now known as the fundamental theorem of the calculus, let us study functions defined by integrals.

If a function f is continuous in a closed interval I and a is some number in I, then the integral of f from a to x exists for every number x in I. This allows us to define an *integral function* F by

$$(1) \qquad F(x) = \int_a^x f(t)\,dt, \qquad \text{domain } F = I.$$

We observe that

$$F(a) = \int_a^a f(t)\,dt = 0.$$

Also, if c and x are numbers in I, then by 5.19

$$\int_a^x f(t)\,dt = \int_a^c f(t)\,dt + \int_c^x f(t)\,dt.$$

Hence,

$$(2) \qquad F(x) = F(c) + \int_c^x f(t)\,dt.$$

5.20 Fundamental Theorem of the Calculus

Let function f be continuous in an interval I, a be a number in I, and F be the integral function

$$F(x) = \int_a^x f(t)\,dt, \qquad \textit{domain } F = I.$$

Then:

(i) *F' exists in I and $F'(c) = f(c)$ for every c in I.*
(ii) *If g is any antiderivative of f in I and b is in I,*

$$\int_a^b f(t)\,dt = g(b) - g(a).$$

Proof: (i) Assume $c < x$ in (2) above. Select u, v in $[c,x]$ such that $f(u)$ is the minimum value and $f(v)$ the maximum value of f over $[c,x]$. If we consider $[c,x]$ as being partitioned into one subinterval, then $f(u)(x - c)$ is a lower sum and $f(v)(x - c)$ is an upper sum of f from c to x, and hence

(3) $$f(u)(x - c) \le \int_c^x f(t)\,dt \le f(v)(x - c).$$

From (2) and (3), we get

(4) $$f(u) \le \frac{F(x) - F(c)}{x - c} \le f(v).$$

When x is close to c, both $f(u)$ and $f(v)$ are close to $f(c)$ by the continuity of f in I. Therefore, it seems reasonable that

(5) $$\operatorname*{limit}_{x \to c^+} \frac{F(x) - F(c)}{x - c} = f(c).$$

A similar argument with $x < c$ convinces us that

(6) $$\operatorname*{limit}_{x \to c^-} \frac{F(x) - F(c)}{x - c} = f(c).$$

Putting (5) and (6) together and recalling the definition of the derivative, we are led to believe that

(7) $$F'(c) = f(c) \qquad \text{for every } c \text{ in } I.$$

That is, F is an antiderivative of f in I.

(ii) Let g be an antiderivative of f in I. Since F is also an antiderivative, by 4.27 there exists a constant k such that

$$g(x) - F(x) = k \qquad \text{for every } x \text{ in } [a,b].$$

Evidently, $F(a) = 0$ and therefore

$$g(a) = k.$$

Thus $g(b) - F(b) = k = g(a)$ and $F(b) = g(b) - g(a)$; that is,

$$F(b) = \int_a^b f(t)\,dt = g(b) - g(a).$$

Example 1 Find $\int_{-3}^{0} x^2\,dx$.

Solution: If $f(x) = x^2$, then f is continuous and has g defined by

$$g(x) = \tfrac{1}{3}x^3$$

as an antiderivative. Hence, by the fundamental theorem of the calculus,

$$\int_{-3}^{0} x^2\,dx = g(0) - g(-3) = 0 - \tfrac{1}{3}(-3)^3 = 9.$$

Example 2 Find $\int_1^3 (3 - 2x + x^2)\,dx$.

Solution: The function f defined by

$$f(x) = 3 - 2x + x^2$$

is continuous and has antiderivative g defined by

$$g(x) = 3x - x^2 + \tfrac{1}{3}x^3.$$

Therefore, by the fundamental theorem of the calculus,

$$\begin{aligned}\int_1^3 (3 - 2x + x^2)\,dx &= g(3) - g(1)\\ &= (9 - 9 + 9) - (3 - 1 + \tfrac{1}{3})\\ &= \tfrac{20}{3}.\end{aligned}$$

Example 3 Find $\int_4^9 t\sqrt{t}\,dt$.

Solution: The function f defined by

$$f(t) = t\sqrt{t} = t^{3/2}$$

is continuous and increasing in $[0,\infty)$ and has antiderivative

$$g(t) = \tfrac{2}{5}t^{5/2} = \tfrac{2}{5}t^2\sqrt{t}$$

in $(0,\infty)$. Therefore

$$\begin{aligned}\int_4^9 t\sqrt{t}\,dt &= g(9) - g(4)\\ &= \tfrac{2}{5}(81\sqrt{9} - 16\sqrt{4})\\ &= \tfrac{422}{5}.\end{aligned}$$

EXERCISES

Evaluate each of the following integrals by finding an antiderivative g of the integrand f.

1. $\int_2^5 2x\,dx$

2. $\int_{-1}^3 7\,dx$

3. $\int_{-2}^0 3x^2\,dx$

4. $\int_{-10}^{10} x^3\,dx$

5. $\int_{-6}^6 (4x^2+1)\,dx$

6. $\int_{-100}^{100} (2x^5-3x^3+2x+1)\,dx$

7. $\int_4^9 2(\sqrt{x}-x)\,dx$

8. $\int_1^5 \frac{1}{\sqrt{5t}}\,dt$

9. $\int_1^4 (x-2)^2\,dx$

10. $\int_2^3 \frac{(x-1)^2}{x^5}\,dx$

11. $\int_{-2}^{-1} \left(2x^3+\frac{1}{x^2}\right)dx$

12. $\int_3^5 \left(\sqrt{x}+\frac{1}{\sqrt{x}}\right)dx$

13. $\int_{-3}^{-1} \left(\frac{1}{x^2}+\frac{1}{x^3}\right)dx$

14. $\int_{-1}^4 \pi\,dx$

15. $\int_1^2 \left(\sqrt[3]{x}+\frac{1}{\sqrt[4]{x}}\right)dx$

16. $\int_2^x (t^2+t)\,dt$

17. $\int_0^x (\sqrt{y}+y)\,dy$

18. $\int_{-1}^x (t-1)^2\,dt$

19. $\int_a^b (x+4)^3\,dx$

20. $\int_1^2 \frac{1}{x^2}\,dx$

21. Show that $D_x\sqrt{2x+1}=\frac{1}{\sqrt{2x+1}}$. Then find $\int_0^4 \frac{1}{\sqrt{2x+1}}\,dx$.

22. Show that $D_x\sqrt{1+2x^2}=\frac{2x}{\sqrt{1+2x^2}}$. Then find $\int_0^2 \frac{x}{\sqrt{1+2x^2}}\,dx$.

23. Show that $D_x(x^2+1)^3=6x(x^2+1)^2$, and then find $\int_{-1}^2 x(x^2+1)^2\,dx$.

24. Show that $D_x(3x-2)^{-2}=-6(3x-2)^{-3}$, and then find $\int_{-2}^0 \frac{1}{(3x-2)^3}\,dx$.

25. Show that $D_x\sqrt{\frac{x+1}{x-1}}=\frac{-1}{(x-1)\sqrt{x^2-1}}$, and then find $\int_2^3 \frac{dx}{(x-1)\sqrt{x^2-1}}$.

In Exercises 26 to 30, each of the integrals defines a function F of the variable x. Find $F'(x)$ in each case. (*Hint:* Use 5.20.)

26. $F(x)=\int_1^x \sqrt{t^4+1}\,dt$

27. $F(x)=\int_x^2 \frac{1}{1+s^3}\,ds$

28. $F(x) = \int_1^{3x^2} \sqrt{t^4 + 1}\, dt$ [*Hint:* Write $u(x) = 3x^2$ and $g(u) = \int_1^u \sqrt{t^4 + 1}\, dt$. Then $F(x) = g(u(x))$.]

29. $F(x) = \int_{x^3}^{3} (1 + t^2)^{1/3}\, dt$

30. $F(x) = \int_{x^2}^{x^3} (1 + t^2)^{1/3}\, dt$ [*Hint:* Write $\int_{x^2}^{x^3} f(t)\, dt = \int_{x^2}^{0} f(t)\, dt + \int_0^{x^3} f(t)\, dt$.]

6 INTEGRATION FORMULAS

We know antiderivatives of many continuous functions, and therefore we can actually compute integrals of such functions. Some examples will be given in this section. But before giving examples, let us introduce the convenient notation

$$g(x)\Big|_a^b = g(b) - g(a).$$

Thus, if g is an antiderivative of f,

(1) $$\int_a^b f(x)\, dx = g(x)\Big|_a^b$$

according to the fundamental theorem of the calculus.

An integral of the form $\int_a^b f(x)\, dx$ is commonly called a *definite integral.* An indefinite integral is defined as follows.

5.21 Definition

An *indefinite integral* of a function f, denoted by

$$\int f(x)\, dx,$$

is any antiderivative of f.

Thus, if g is an antiderivative of f, we have

$$\int f(x)\, dx = g(x) + C$$

for some number C by 4.27. By the definition of the indefinite integral,

$$D_x \int f(x)\, dx = f(x),$$

$$\int D_x f(x)\, dx = f(x) + C.$$

In terms of indefinite integrals, 4.28 has the form

5.22 $$\int x^r\, dx = \frac{1}{r+1} x^{r+1} + C, \qquad r \neq -1, r \text{ rational}.$$

We must exclude the case $r = -1$ in 5.22 since the right side of this equation involves the impossible operation of division by zero when $r = -1$. However, this does not mean that $\int x^{-1}\,dx$ does not exist. Actually, this integral does exist in every interval I not containing 0. We shall discuss its value in Chapter 7.

If in 5.22 r is a negative rational number other than -1, then x^r is undefined if $x = 0$. Since the definite integral of f from a to b is defined only if the closed interval $[a,b]$ is contained in the domain of f, we must take a and b both to be positive or both negative in this case.

Example 1 Find $\int_2^4 x^3\,dx$.

Solution: By 5.22,

$$\int_2^4 x^3\,dx = \left.\frac{x^4}{4}\right|_2^4 = \tfrac{1}{4}(4^4 - 2^4) = 60.$$

Example 2 Find $\int_1^4 \sqrt{u}\,du$.

Solution: Since $\sqrt{u} = u^{1/2}$, we have, by 5.22,

$$\int_1^4 u^{1/2}\,du = \left.\frac{u^{3/2}}{\frac{3}{2}}\right|_1^4 = \tfrac{2}{3}(4^{3/2} - 1^{3/2}) = \tfrac{14}{3}.$$

Example 3 Find $\int \frac{1}{x^3}\,dx$.

Solution: Since $1/x^3 = x^{-3}$, we have, by 5.22,

$$\int x^{-3}\,dx = \frac{x^{-2}}{-2} + C = -\frac{1}{2x^2} + C.$$

If a function f is continuous, then so is cf for any number c, and

5.23 $$\int cf(x)\,dx = c\int f(x)\,dx.$$

By this equation we mean that c times an antiderivative of f is an antiderivative of cf. It is true because $D(cg) = cDg$ for any differentiable function g. Another useful integration formula is as follows:

5.24 $$\int [f(x) + g(x)]\,dx = \int f(x)\,dx + \int g(x)\,dx.$$

Proof: If $\int f(x)\,dx = F(x)$ and $\int g(x)\,dx = G(x)$, then

$$D_x[F(x) + G(x)] = D_xF(x) + D_xG(x) = f(x) + g(x)$$

and hence

$$\int [f(x) + g(x)]\,dx = F(x) + G(x) = \int f(x)\,dx + \int g(x)\,dx.$$

Using 5.22 to 5.24, we can easily integrate any polynomial function, as illustrated in the following examples.

Example 4 Find $\int_{-1}^{1} (x^2 + 2x - 3)\,dx$.

Solution:

$$\int_{-1}^{1} (x^2 + 2x - 3)\,dx = (\tfrac{1}{3}x^3 + x^2 - 3x)\Big|_{-1}^{1}$$
$$= (\tfrac{1}{3} + 1 - 3) - (-\tfrac{1}{3} + 1 + 3) = -\tfrac{16}{3}.$$

Example 5 Find $\int (15u^4 - 12u^3 + 3u^2 + 6u + 1)\,du$.

Solution:

$$\int (15u^4 - 12u^3 + 3u^2 + 6u + 1)\,du = (3u^5 - 3u^4 + u^3 + 3u^2 + u) + C.$$

Example 6 Find $\int \frac{x^2 + 2}{x^2}\,dx$.

Solution: We note that

$$\frac{x^2 + 2}{x^2} = \frac{x^2}{x^2} + \frac{2}{x^2} = 1 + 2x^{-2}.$$

Therefore

$$\int \frac{x^2 + 2}{x^2}\,dx = \int (1 + 2x^{-2})\,dx = (x - 2x^{-1}) + C = \left(x - \frac{2}{x}\right) + C.$$

A somewhat different example is as follows.

Example 7 Find $\int_0^3 |x - 1|\,dx$.

Solution: The function

$$f(x) = |x - 1|$$

is continuous, so the integral exists. You know that

$$|x - 1| = \begin{cases} x - 1 & \text{if } x \geq 1, \\ -(x - 1) & \text{if } x \leq 1. \end{cases}$$

This suggests breaking up the integral into two parts:

$$\begin{aligned}\int_0^3 |x-1|\,dx &= \int_0^1 |x-1|\,dx + \int_1^3 |x-1|\,dx \qquad \textit{(by 5.19)}\\ &= \int_0^1 -(x-1)\,dx + \int_1^3 (x-1)\,dx\\ &= -\left(\frac{x^2}{2}-x\right)\Bigg|_0^1 + \left(\frac{x^2}{2}-x\right)\Bigg|_1^3\\ &= [-(\tfrac{1}{2}-1)+0] + [(\tfrac{9}{2}-3)-(\tfrac{1}{2}-1)]\\ &= \tfrac{5}{2}.\end{aligned}$$

EXERCISES

I

In each of Exercises 1 to 18, evaluate the integral.

1. $\int_0^2 (x^3 - 1)\,dx$

2. $\int_{-1}^1 (x + 1)\,dx$

3. $\int_1^4 (u^2 - 2u + 3)\,du$

4. $\int_0^5 (-3u^2 + 2u + 5)\,du$

5. $\int_{-5}^{-1} (t^2 + 1)\,dt$

6. $\int_0^1 (\sqrt{x} + 1)^2\,dx$

7. $\int_4^1 x(\sqrt{x} - 3)\,dx$

8. $\int_{-1}^{-8} z(\sqrt[3]{z} - 2z)\,dz$

9. $\int_4^9 \left(\sqrt{x} - \frac{1}{\sqrt{x}}\right) dx$

10. $\int_1^4 \frac{x^2 - 3x + 4}{\sqrt{x}}\,dx$

11. $\int (t^{2/3} - 2t^{3/4})\,dt$

12. $\int \left(\frac{1}{z^3} + \frac{2}{z^2}\right) dz$.

13. $\int \left(x^2 - \frac{2}{x^2} + x\sqrt{x}\right) dx$

14. $\int \frac{\sqrt{x} - 1}{x^3}\,dx$

15. $\int_2^5 a\,dx$

16. $\int_1^4 (2x^{-2} + x)\,dx$

17. $\int (tx^2 + r)\,dx$

18. $\int (gt^{-2} + v_0 t + a)\,dt$

II

Evaluate each of the following integrals.

1. $\int_{-1}^1 |x|\,dx$

2. $\int_0^3 |x^3 - 2|\,dx$

3. $\int_{-2}^3 |x^3 - x|\,dx$

4. $\int_{-4}^4 \sqrt{1 + 2|x|}\,dx$

7 CHANGE OF VARIABLE

Integrals that cannot be evaluated directly by known formulas may sometimes be evaluated after a "change of variable." A formula for this transformation follows directly from the chain rule.

By the chain rule,

$$D_x F(g(x)) = F'(g(x))g'(x).$$

Hence,

$$\int F'(g(x))g'(x)\,dx = F(g(x)) + C.$$

Letting $f = F'$, this equation becomes

$$\int f(g(x))g'(x)\,dx = F(g(x)) + C,$$

where F is an antiderivative of f. A convenient form of this equation is obtained by letting $u = g(x)$ and $du = g'(x)\,dx$:

5.25 **Change of Variable Integration Formula**

$$\int f(g(x))g'(x)\,dx = \int f(u)\,du\bigg|_{u=g(x)}.$$

The vertical bar followed by $u = g(x)$ indicates u is to be replaced by $g(x)$ after the integration is performed.

For definite integrals 5.25 has the form

5.26
$$\int_a^b f(g(x))g'(x)\,dx = \int_{g(a)}^{g(b)} f(u)\,du,$$

which is easily verified.

We may remember 5.25 and 5.26 in the following way. We "change variables" in the left-hand integral by letting $u = g(x)$ and then by formally letting du be the derivative of g times dx, $du = g'(x)\,dx$. If a and b are thought of as x limits of integration, then the u limits are $g(a)$ and $g(b)$.

Some examples of the usefulness of the change of variable formula are given below.

Example 1 Find $\int 2x(x^2 + 1)^3\,dx$.

Solution: Let $u = x^2 + 1$, so that $du = 2x\,dx$. Then, by 5.25,

$$\int 2x(x^2 + 1)^3\,dx = \int u^3\,du\bigg|_{u=x^2+1} = \tfrac{1}{4}u^4 + C\bigg|_{u=x^2+1} = \tfrac{1}{4}(x^2 + 1)^4 + C.$$

Example 2 Find $\int_1^3 x\sqrt{x^2 - 1}\,dx$.

Solution: We change variables by letting

$$u = x^2 - 1, \qquad du = 2x\,dx.$$

If we supply a factor of 2 in the integrand and multiply the integral by $\frac{1}{2}$, we have

$$\int_1^3 x\sqrt{x^2 - 1}\,dx = \frac{1}{2}\int_1^3 \sqrt{x^2 - 1}(2x\,dx).$$

Since $u = 0$ when $x = 1$, and $u = 8$ when $x = 3$, we have, by 5.26,

$$\int_1^3 x\sqrt{x^2 - 1}\,dx = \frac{1}{2}\int_0^8 u^{1/2}\,du = \frac{1}{2}\left(\frac{2}{3}u^{3/2}\right)\Bigg|_0^8 = \frac{1}{3}(8^{3/2} - 0^{3/2}) = \frac{16\sqrt{2}}{3}.$$

A slight change in the integral of Example 2 makes it nonintegrable by the present methods. Thus we cannot evaluate

$$\int_1^3 \sqrt{x^2 - 1}\,dx$$

by use of 5.26. If we proceed as before and let $u = x^2 - 1$, $du = 2x\,dx$, then $u^{1/2}\,du = \sqrt{x^2 - 1}\,(2x\,dx)$. We can supply a factor of 2 in the integrand, but we cannot supply an x. This integral will be evaluated later by other methods.

Example 3 Find $\displaystyle\int \frac{u^2}{(u^3 + 1)^2}\,du.$

Solution: Since the variable is already u, let us change the variable to y by letting

$$y = u^3 + 1, \qquad dy = 3u^2\,du.$$

Then

$$\int \frac{u^2}{(u^3 + 1)^2}\,du = \frac{1}{3}\int \frac{1}{(u^3 + 1)^2}(3u^2\,du) = \frac{1}{3}\int \frac{1}{y^2}\,dy\Bigg|_{y=u^3+1}$$

$$= \frac{1}{3}\int y^{-2}\,dy\Bigg|_{y=u^3+1} = \frac{1}{3}\frac{y^{-1}}{-1} + C\Bigg|_{y=u^3+1}$$

$$= -\frac{1}{3(u^3 + 1)} + C.$$

Example 4 Find $\displaystyle\int \frac{x}{\sqrt{4 - x}}\,dx.$

Solution: If we let

$$u = 4 - x$$

then

$$du = (-1)\,dx \qquad \text{or} \qquad dx = -du.$$

What do we do with the x in the numerator of the integrand? It is simply $4 - u$!

$$x = 4 - u.$$

Thus, $$\int \frac{x}{\sqrt{4-x}}\,dx = \int \frac{4-u}{\sqrt{u}}(-du)\bigg|_{u=4-x}$$
$$= -\int (4u^{-1/2} - u^{1/2})\,du\bigg|_{u=4-x}$$
$$= -(8u^{1/2} - \tfrac{2}{3}u^{3/2})\bigg|_{u=4-x}$$
$$= -8\sqrt{4-x} + \tfrac{2}{3}(4-x)\sqrt{4-x} + C.$$

EXERCISES

I

In each of Exercises 1 to 28, evaluate the integral.

1. $\int_0^3 \sqrt{x+1}\,dx$

2. $\int 2\sqrt{2x+1}\,dx$

3. $\int_{-1}^1 \frac{2x}{(4+x^2)^2}\,dx$

4. $\int_{-1}^2 9x^2(1+3x^3)^2\,dx$

5. $\int_0^{-5} \sqrt{1-3u}\,du$

6. $\int \frac{1}{\sqrt{4u+1}}\,du$

7. $\int_0^1 y\sqrt{1-y^2}\,dy$

8. $\int_{-1}^{-2} 3z\sqrt[3]{z^2+1}\,dz$

9. $\int_{-2}^2 \frac{x}{\sqrt{1+8x^2}}\,dx$

10. $\int_{-3}^{-1} \frac{1}{(4x-1)^2}\,dx$

11. $\int \frac{x^3}{\sqrt{x^4+2}}\,dx$

12. $\int \frac{x}{(x^2+1)^{3/2}}\,dx$

13. $\int \frac{(2+\sqrt{x})^3}{\sqrt{x}}\,dx$

14. $\int \frac{x}{(x^2-1)^3}\,dx$

15. $\int (2x-1)\sqrt{1+(2x-1)^2}\,dx$

16. $\int \frac{1}{(5x-2)^3}\,dx$

17. $\int \frac{1}{\sqrt{x}(1-\sqrt{x})^2}\,dx$

18. $\int_1^3 \frac{x}{\sqrt{9+x^2}}\,dx$

19. $\int \frac{y^6}{\sqrt[3]{10+y^7}}\,dy$

20. $\int_{-3}^{-2} \sqrt{x^4+x^2}\,dx$

21. $\int_0^b (b^{1/2}-x^{1/2})^2\,dx$

22. $\int \frac{s^3+1}{\sqrt{s^4+4s}}\,ds$

23. $\int_{-3}^{-1} \frac{1}{(2-3t)^3}\,dt$

24. $\int_{-2}^{-1} (2x+3)^{100}\,dx$

25. $\int_{-1}^{1} u(1-u^2)^5\,du$

26. $\int_{1}^{2} \frac{1}{x^2}\sqrt{1-\frac{1}{x}}\,dx$

27. $\int \frac{\sqrt{1+\sqrt{x}}}{\sqrt{x}}\,dx$

28. $\int_{0}^{1} \sqrt{u}\sqrt{1+u\sqrt{u}}\,du$

II

Evaluate the following integrals.

1. $\int \frac{x}{\sqrt{1+x}}\,dx$

2. $\int_{0}^{3} \frac{x^3}{\sqrt{1+x}}\,dx$

3. $\int \frac{\sqrt{2x^2+3}}{x^4}\,dx$

4. $\int_{0}^{1} \frac{1}{\sqrt{1+\sqrt{x}}}\,dx$

5. $\int \frac{x}{x+\sqrt{x^2-1}}\,dx$

6. If $I = \int_{-1}^{1} dy$, then clearly $I = 2$. However, if we first make the change of variable $y = x^{5/2}$ in the integral, we get $I = \frac{5}{2}\int_{1}^{1} x^{3/2}\,dx = 0$. Explain.

8 SEQUENCES

The functions known as sequences are of particular interest in mathematics.

5.27 Definition

A sequence is a function having as its domain the set of positive integers.

In order to define any function, we must somehow describe the correspondence that associates with each number in the domain of the function a number in its range. This may be done by means of a table, an equation, or a combination of a table and an equation.

The customary way to define a sequence a is to list in order the values of a at the successive positive integers, as indicated below.

$$a(1),\ a(2),\ a(3),\ \ldots .$$

The dots are used to suggest that the "table" is an infinite one. A further convention is the use of subscripts rather than the usual functional notation. Thus the sequence above will be written

$$a_1,\ a_2,\ a_3,\ \ldots .$$

The numbers a_1, a_2, a_3, and so on are called the *elements* of the sequence a, a_k being the kth element.

In essence, a sequence is just an unending succession of numbers described by some rule of formation. The following are examples of sequences:

(1) $\frac{2}{1}, \frac{2}{2}, \frac{2}{3}, \frac{2}{4}, \ldots;$ $\quad a_n = \dfrac{2}{n}.$

(2) $1^2, 2^2, 3^2, 4^2, \ldots;$ $\quad a_n = n^2.$

(3) $-1, -2, -3, -4, \ldots;$ $\quad a_n = -n.$

(4) $2, 4, 6, 8, \ldots;$ $\quad a_n = 2n.$

(5) $1, \sqrt{2}, \sqrt{3}, \sqrt{4}, \ldots;$ $\quad a_n = \sqrt{n}.$

(6) $2, 2, 2, 2, \ldots;$ $\quad a_n = 2.$

(7) $-2, 2, -2, 2, \ldots;$ $\quad a_n = (-1)^n 2.$

(8) $\frac{1}{2}, \frac{2}{3}, \frac{3}{4}, \frac{4}{5}, \ldots;$ $\quad a_n = \dfrac{n}{n+1}.$

(9) $-\frac{1}{2}, \frac{1}{4}, -\frac{1}{8}, \frac{1}{16}, \ldots;$ $\quad a_n = (-\frac{1}{2})^n.$

(10) $2, \frac{5}{2}, \frac{8}{3}, \frac{11}{4}, \ldots;$ $\quad a_n = 3 - \dfrac{1}{n}.$

In each of these examples we have displayed the nth element in order to give the general rule of formation of the sequence. For example, in (1), knowing that $a_n = 2/n$, we find

$$a_1 = \tfrac{2}{1},\ a_2 = \tfrac{2}{2},\ a_3 = \tfrac{2}{3},\ a_4 = \tfrac{2}{4},\ a_5 = \tfrac{2}{5},$$

and so on. Thus the 32nd element of the sequence is $a_{32} = \frac{2}{32} = \frac{1}{16}$. In most of the examples above we could guess the rule of formation used just by looking at the first few elements of the sequence. However, given that a sequence starts out as follows,

$$5, 7, 7, 3, \ldots,$$

we would not be able to continue it without some clue as to how the first four elements were chosen. The rule we used in forming this sequence was to let a_n be the nth digit in the decimal expansion of $\sqrt{3}/3$,

$$\frac{\sqrt{3}}{3} = .577350\cdots.$$

It is important to distinguish between a sequence and the *set of elements* of the sequence. For example, the set of elements of sequence (5) above is the set of square roots of all the positive integers. An entirely different sequence is the following:

$$1, \sqrt{2}, 1, \sqrt{3}, 1, \sqrt{4}, \ldots; \qquad a_n = \begin{cases} 1, & n \text{ odd} \\ \sqrt{\dfrac{n}{2} + 1}, & n \text{ even.} \end{cases}$$

Although these two sequences are different, they consist of the same set of elements. In other words, the ranges of the two sequences are the same. The set of elements in sequence (7) above consists of two elements, -2 and 2.

The nth element $2/n$ of sequence (1) above is close to zero if n is large. For this reason, we shall say that the limit of sequence (1) is zero.

The limit of sequence (6) is 2, since the nth element is equal to 2 (and hence close to 2) when n is large.

Since the nth element of sequence (10) above is $3 - 1/n$, evidently the nth element is close to 3 when n is large. Thus the limit of this sequence is 3, according to the following definition of the limit of a sequence.

5.28 Definition

The sequence

$$a_1, a_2, \ldots, a_n, \ldots$$

has b as its limit, and we write

$$\operatorname*{limit}_{n \to \infty} a_n = b,$$

if for every neighborhood N of b there exists a positive number k such that a_n is in N for every integer n in (k,∞).

If a sequence has a limit, we say its limit exists. It should be noted that the definition of the limit of a sequence is almost word for word the same as the definition of

$$\operatorname*{limit}_{x \to \infty} f(x)$$

given in 2.20. Hence, we can expect the methods used in Chapter 2 for infinite limits of functions to carry over to limits of sequences.

Example 1 Show that sequence (10) above has limit 3, that is, that

$$\operatorname*{limit}_{n \to \infty} \left(3 - \frac{1}{n}\right) = 3.$$

Solution: We must show that for every neighborhood N of 3 there exists a positive number k such that

$$3 - \frac{1}{n} \text{ is in } N \text{ for every integer } n \text{ in } (k,\infty).$$

Let $N = (3 - \varepsilon, 3 + \varepsilon)$. Then

$$3 - \varepsilon < 3 - \frac{1}{n} < 3 + \varepsilon \text{ if and only if } -\varepsilon < -\frac{1}{n} < \varepsilon.$$

In turn, the inequality $-\varepsilon < -1/n$ is equivalent to $\varepsilon > 1/n$ and also $n > 1/\varepsilon$. Thus, we need only select $k = 1/\varepsilon$ in order that $3 - 1/n$ be in N for every n in (k,∞).

Of the ten examples of sequences listed at the beginning of this section, it is intuitively clear that (2), (3), (4), (5), and (7) do not have limits. Each of the other sequences has a limit.

The sequence

$$\frac{d}{1^p}, \frac{d}{2^p}, \frac{d}{3^p}, \ldots, \frac{d}{n^p}, \ldots$$

has the limit 0 for any real number d and any positive rational number p, that is,

5.29 $$\lim_{n\to\infty} \frac{d}{n^p} = 0 \qquad \text{if } p > 0.$$

We shall not prove 5.29, since its proof is almost identical with that of 2.21. As a consequence of 5.29 each of the following sequences has zero as its limit:

$$1, \frac{1}{2}, \frac{1}{3}, \frac{1}{4}, \ldots, \frac{1}{n}, \ldots,$$

$$-2, \frac{-2}{\sqrt{2}}, \frac{-2}{\sqrt{3}}, \frac{-2}{\sqrt{4}}, \ldots, \frac{-2}{\sqrt{n}}, \ldots,$$

$$3, \frac{3}{2^2}, \frac{3}{3^2}, \frac{3}{4^2}, \ldots, \frac{3}{n^2}, \ldots.$$

Limits of sequences have many of the properties of limits of functions given in Chapter 2. Before stating these properties, we observe that from the sequences

$$a_1, a_2, \ldots, a_n, \ldots,$$

$$b_1, b_2, \ldots, b_n, \ldots,$$

we can form many new sequences; for example,

$$a_1 + b_1, a_2 + b_2, \ldots, a_n + b_n, \ldots,$$

$$a_1b_1, a_2b_2, \ldots, a_nb_n, \ldots,$$

$$\frac{a_1}{b_1}, \frac{a_2}{b_2}, \ldots, \frac{a_n}{b_n}, \ldots \qquad \text{if each } b_n \neq 0,$$

$$|a_1|, |a_2|, \ldots, |a_n|, \ldots,$$

and so on.

The constant sequence $c, c, \ldots, c, \ldots$ has c as its limit; that is,

5.30 $$\operatorname*{limit}_{n\to\infty} c = c.$$

If

$$\operatorname*{limit}_{n\to\infty} a_n = a, \qquad \operatorname*{limit}_{n\to\infty} b_n = b,$$

then

5.31 $$\operatorname*{limit}_{n\to\infty} (a_n + b_n) = a + b,$$

5.32 $$\operatorname*{limit}_{n\to\infty} a_nb_n = ab,$$

5.33 $$\operatorname*{limit}_{n\to\infty} \frac{a_n}{b_n} = \frac{a}{b} \qquad \text{if all } b_n \neq 0 \text{ and } b \neq 0.$$

Since the proofs of these limit theorems are similar to the corresponding ones for functions, they will be omitted.

Some useful limit theorems of a slightly different nature are as follows:

5.34 Theorem

If $a_n \le c_n \le b_n$ *for each integer* n *and if* $\operatorname*{limit}_{n\to\infty} a_n = \operatorname*{limit}_{n\to\infty} b_n = a$, *then* $\operatorname*{limit}_{n\to\infty} c_n = a$.

Proof: Since $\operatorname*{limit}_{n\to\infty} a_n = a$, for every neighborhood N of a there exists a positive number k_1 such that a_n is in N for every n in (k_1,∞). Similarly, $\operatorname*{limit}_{n\to\infty} b_n = a$, and hence there exists a positive number k_2 such that b_n is in N for every n in (k_2,∞). If we let k be the larger of k_1 and k_2, then both a_n and b_n are in N for every n in (k,∞). Therefore, c_n is in N for every n in (k,∞) and we conclude $\operatorname*{limit}_{n\to\infty} c_n = a$.

5.35 Theorem

If $\operatorname*{limit}_{n\to\infty} a_n = a$, *then* $\operatorname*{limit}_{n\to\infty} |a_n| = |a|$.

5.36 Theorem

If $\lim_{n\to\infty} |a_n| = 0$, *then* $\lim_{n\to\infty} a_n = 0$.

The proofs of 5.35 and 5.36 are left for the reader to supply.

A sequence $a_1, a_2, \ldots, a_n, \ldots$ is said to be *increasing* if $a_1 \leq a_2 \leq a_3 \leq \cdots \leq a_n \leq \cdots$ and *decreasing* if $a_1 \geq a_2 \geq a_3 \geq \cdots \geq a_n \geq \cdots$. As for other functions, a sequence is called *monotonic* if it is either increasing or decreasing. A number M is called an *upper bound* of a sequence $a_1, a_2, \ldots, a_n, \ldots$ if $a_n \leq M$ for every n, and similarly for a lower bound. We call a sequence *bounded* if it has both an upper bound and a lower bound.

5.37 Theorem

Every bounded monotonic sequence has a limit.

Proof: Assume that $a_1, a_2, \ldots, a_n, \ldots$ is a bounded increasing sequence. Hence, the set $S = \{a_n \mid n = 1, 2, \ldots\}$ has a l.u.b., which we denote by a. For every number $\varepsilon > 0$, $a - \varepsilon$ is not an upper bound of S and, therefore, $a_N > a - \varepsilon$ for some integer N. Since the sequence is increasing, $a_n \geq a_N$ for every $n > N$. Thus, for every $\varepsilon > 0$ there exists an integer N such that

$$0 \leq a - a_n < \varepsilon \qquad \text{for every } n > N.$$

This proves that

$$\lim_{n\to\infty} a_n = a.$$

A similar proof holds if the sequence is decreasing.

For example, we can use 5.37 to prove 5.29. If $d > 0$ and $p > 0$, the sequence

$$\frac{d}{1^p}, \frac{d}{2^p}, \ldots, \frac{d}{n^p}, \ldots$$

is decreasing, with g.l.b. 0. Therefore, it has a limit by 5.37, and the limit is its g.l.b.

With thc aid of the above limit theorems, it is possible to evaluate directly the limits of many sequences, as shown in the following examples.

Example 2 Find $\lim_{n\to\infty} \dfrac{n}{n+1}$.

Solution: Since

$$\frac{n}{n+1} = \frac{1}{1 + 1/n},$$

the given sequence is the quotient of two sequences having nth terms 1 and $1 + 1/n$, respectively. Thus, by 5.33,

$$\lim_{n\to\infty} \frac{n}{n+1} = \lim_{n\to\infty} \frac{1}{1 + 1/n} = \frac{\lim_{n\to\infty} 1}{\lim_{n\to\infty} (1 + 1/n)}.$$

However, by 5.29, 5.30, and 5.31,

$$\lim_{n\to\infty}\left(1+\frac{1}{n}\right) = \lim_{n\to\infty} 1 + \lim_{n\to\infty}\frac{1}{n} = 1.$$

We conclude that

$$\lim_{n\to\infty}\frac{n}{n+1} = 1.$$

Example 3 Find $\lim\limits_{n\to\infty}\dfrac{1-2n+3n^2}{5n^2}$.

Solution: We note that

$$\frac{1-2n+3n^2}{5n^2} = \frac{\frac{1}{5}}{n^2} - \frac{\frac{2}{5}}{n} + \frac{3}{5}.$$

Hence

$$\lim_{n\to\infty}\frac{1-2n+3n^2}{5n^2} = \lim_{n\to\infty}\frac{\frac{1}{5}}{n^2} - \lim_{n\to\infty}\frac{\frac{2}{5}}{n} + \lim_{n\to\infty}\tfrac{3}{5} = 0 - 0 + \tfrac{3}{5} = \tfrac{3}{5}.$$

Example 4 Find $\lim\limits_{n\to\infty}\dfrac{5n}{n^2+1}$.

Solution: We again change $5n/(n^2+1)$ into a form containing powers of $1/n$:

$$\frac{5n}{n^2+1} = \frac{5/n}{1+1/n^2}.$$

Hence

$$\lim_{n\to\infty}\frac{5n}{n^2+1} = \frac{\lim\limits_{n\to\infty}(5/n)}{\lim\limits_{n\to\infty}1 + \lim\limits_{n\to\infty}(1/n^2)} = \frac{0}{1+0} = 0.$$

EXERCISES

I

In each of Exercises 1 to 22 the nth element a_n of a sequence is given. Write down the first five numbers of the sequence, and then find $\lim\limits_{n\to\infty} a_n$, if it exists. (You may use any of the limit theorems.)

1. $a_n = \dfrac{3}{n}$

2. $a_n = 1 - \dfrac{2}{n}$

3. $a_n = \dfrac{2n}{n+3}$

4. $a_n = \dfrac{1-2n}{1+n}$

5. $a_n = \dfrac{(-1)^n}{n^2}$

6. $a_n = \dfrac{(-1)^{n+1}(n+1)}{2n}$

7. $a_n = \dfrac{n^2 - 2}{n^2 + 2}$

8. $a_n = \dfrac{n^3 - 1}{n^3 + 1}$

9. $a_n = \dfrac{2n}{n + 1} - \dfrac{n + 1}{2n}$

10. $a_n = \dfrac{n^2}{2n + 1} - \dfrac{n^2}{2n - 1}$

11. $a_n = \dfrac{n^3}{n^2 + 2} - \dfrac{n^3}{n^2 - 2}$

12. $a_n = \dfrac{n^2 + 5n - 2}{2n^2}$

13. $a_n = \dfrac{1}{\sqrt{n^2 + 1}}$

14. $a_n = n - \dfrac{1}{n}$

15. $a_n = \dfrac{1}{3^n}$

16. $a_n = \dfrac{n^2}{3n + 2}$

17. $a_n = \sqrt{n + 2} - \sqrt{n + 1}$

18. $a_n = \dfrac{n}{\sqrt{2n^2 - 1}}$

19. $a_n = \dfrac{3^n + (-1)^n}{3^{n+1} + (-1)^{n+1}}$

20. $a_n = n(\frac{1}{4})^n$

21. $a_n = \begin{cases} 2, & \text{if } n \text{ is odd} \\ \dfrac{3}{n + 2}, & \text{if } n \text{ is even} \end{cases}$

22. $a_n = n^{(-1)^n}$

Prove each of the following limits.

23. $\displaystyle \operatorname*{limit}_{n \to \infty} \frac{n}{2n + 1} = \frac{1}{2}$

24. $\displaystyle \operatorname*{limit}_{n \to \infty} \left(1 + \frac{1}{n}\right) = 1$

25. $\displaystyle \operatorname*{limit}_{n \to \infty} \left[2 + \frac{(-1)^n}{n}\right] = 2$

26. $\displaystyle \operatorname*{limit}_{n \to \infty} \frac{4n + 1}{5n - 4} = \frac{4}{5}$

27. If $0 < k < 1$, prove that $\displaystyle \operatorname*{limit}_{n \to \infty} k^n = 0$.

28. For $|r| < 1$, let $a_n = \displaystyle\sum_{i=1}^{n} ar^i$. Find $\displaystyle \operatorname*{limit}_{n \to \infty} a_n$. (See Exercise 19, Section 3.)

II

Find the limits of the following sequences.

1. $a_n = \dfrac{10{,}000^n}{n!}$

2. $a_n = \dfrac{n!}{n^n}$

3. $a_n = \sqrt[n]{2}$

4. $a_n = \sqrt{\dfrac{n}{n + 1}} - \sqrt{\dfrac{n}{n - 1}}$

5. $a_n = \dfrac{n}{\sqrt{n} - 1} - \dfrac{n}{\sqrt{n} + 1}$

9 RIEMANN SUMS

We recall that for a function f continuous in a closed interval $[a,b]$ and a partition

$$P = \{x_0, x_1, \ldots, x_n\}$$

of $[a,b]$, the upper sum $T(P)$ and lower sum $S(P)$ of f over $[a,b]$ are defined by

$$S(P) = \sum_{i=1}^{n} f(u_i)\, \Delta x_i, \qquad T(P) = \sum_{i=1}^{n} f(v_i)\, \Delta x_i,$$

where $f(u_i)$ is the minimum value and $f(v_i)$ the maximum value of f in $[x_{i-1},x_i]$ for $i = 1, 2, \ldots, n$. Closely related to these sums are the following sums.

5.38 Definition

Let f be a continuous function in a closed interval $[a,b]$. For every partition $P = \{x_0, x_1, \ldots, x_n\}$ of $[a,b]$ and every set of numbers $z_1, z_2, \ldots, z_n$, such that

$$x_0 \leq z_1 \leq x_1 \leq z_2 \leq x_2 \leq \cdots \leq x_{n-1} \leq z_n \leq x_n$$

the sum

$$R(P) = \sum_{i=1}^{n} f(z_i)\, \Delta x_i$$

is called a *Riemann* sum* of f over $[a,b]$.

If $R(P)$ is a Riemann sum as above, then clearly

$$f(u_i) \leq f(z_i) \leq f(v_i) \qquad \text{in each subinterval } [x_{i-1},x_i].$$

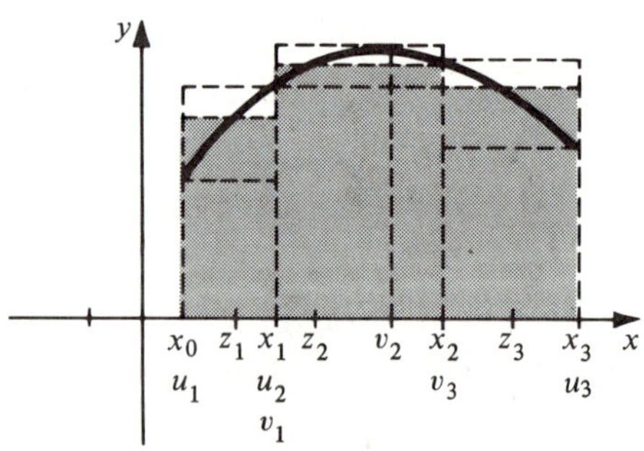

FIGURE 5.12

* Georg Friedrich Riemann was a famous German mathematician of the nineteenth century. In one of his early papers on the foundations of analysis (1850) is given the first rigorous definition of an integral as a limit of a sum.

Hence,

$$S(P) \leq R(P) \leq T(P).$$

This is illustrated in Figure 5.12 for a partition having three subintervals.

What we wish to indicate in this section is how an integral may be expressed as a limit of a sequence of Riemann sums. This will aid us in certain applications. First, we make the following definition.

5.39 Definition

The *norm* of a partition $P = \{x_0, x_1, \ldots, x_n\}$ of $[a,b]$ is denoted by $\|P\|$ and defined by

$$\|P\| = \max \{\Delta x_1, \Delta x_2, \ldots, \Delta x_n\}.$$

Example 1 If $P = \{1,4,6,7,11\}$, a partition of $[1,11]$, then

$$\|P\| = \max \{3,2,1,4\} = 4.$$

In words, $\|P\|$ is the maximum length of any subinterval of P.

The main theorem is as follows.

5.40 Theorem

Let function f be continuous in $[a,b]$ and

$$P_1, P_2, \ldots, P_n, \ldots$$

be a sequence of partition of $[a,b]$ for which

$$\operatorname*{limit}_{n \to \infty} \|P_n\| = 0.$$

If

$$R(P_1), R(P_2), \ldots, R(P_n), \ldots$$

is any sequence of Riemann sums of f associated with the given sequence of partitions, then

$$\operatorname*{limit}_{n \to \infty} R(P_n) = \int_a^b f(x)\, dx.$$

The gist of the proof, given in the appendix on page 226, is to show that $T(P_n) - S(P_n)$ is small when n is large. Both $R(P_n)$ and $\int_a^b f(x)\, dx$ are between $S(P_n)$ and $T(P_n)$, and hence $R(P_n)$ must approach $\int_a^b f(x)\, dx$ when n is large.

The conclusion of 5.40 may be expressed in the following form if P_n is the regular partition of $[a,b]$ into n subintervals:

5.41 $$\int_a^b f(x)\, dx = \operatorname*{limit}_{n \to \infty} \sum_{i=1}^{n} f(z_i)\, \Delta x.$$

In this notation $\sum_{i=1}^{n} f(z_i)\,\Delta x$ represents a Riemann sum of f relative to $P_n = \{x_0, x_1, \ldots, x_n\}$, with

$$x_{i-1} \le z_i \le x_i \text{ and } \Delta x = x_i - x_{i-1} = \frac{b-a}{n}, \qquad i = 1, 2, \ldots, n.$$

Let us illustrate with an example how 5.41 might be used to compute an integral.

Example 2 Find $\int_0^2 x^2\,dx$ by use of 5.41.

Solution: The function f defined by

$$f(x) = x^2$$

is continuous in [0,2]. For each positive integer n, the regular partition P_n is given by

$$P_n = \{0, \Delta x, 2\,\Delta x, \ldots, (n-1)\,\Delta x, 2\}$$

where $\Delta x = 2/n$. Hence

$$\int_0^2 x^2\,dx = \operatorname*{limit}_{n\to\infty} \sum_{i=1}^{n} f(z_i)\,\Delta x$$

for any choice of $z_1, z_2, \ldots, z_n$, one in each of the subintervals of P_n. If, for example, we choose $z_1 = \Delta x$, $z_2 = 2\,\Delta x, \ldots, z_n = n\,\Delta x$, then we have

$$\int_0^2 x^2\,dx = \operatorname*{limit}_{n\to\infty} \sum_{i=1}^{n} f(i\,\Delta x)\,\Delta x.$$

The case $n = 4$ is shown in Figure 5.13, in which the Riemann sum is the area of the colored rectangular polygon. We may compute the above sum as follows:

$$\begin{aligned}
\sum_{i=1}^{n} f(i\,\Delta x)\,\Delta x &= \sum_{i=1}^{n} \left(i\frac{2}{n}\right)^2 \frac{2}{n} \\
&= \frac{8}{n^3}\sum_{i=1}^{n} i^2 \\
&= \frac{8}{n^3}\,\frac{n(n+1)(2n+1)}{6} \qquad \textit{(by 5.7)} \\
&= \frac{4}{3}\left(2 + \frac{3}{n} + \frac{1}{n^2}\right).
\end{aligned}$$

Therefore
$$\int_0^2 x^2\,dx = \operatorname*{limit}_{n\to\infty} \frac{4}{3}\left(2 + \frac{3}{n} + \frac{1}{n^2}\right) = \frac{8}{3}.$$

This result naturally checks with that obtained by using the fundamental theorem of the calculus:

$$\int_0^2 x^2\,dx = \left.\frac{x^3}{3}\right|_0^2 = \frac{8}{3}.$$

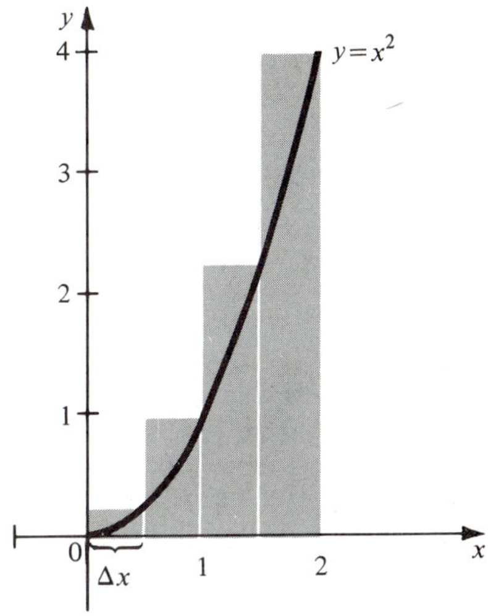

FIGURE 5.13

Example 3 Find $\int_0^1 (2x^2 - 4x)\,dx$ by use of 5.41.

Solution: We have the continuous function

$$f(x) = 2x^2 - 4x$$

in the interval [0,1] and the regular partition

$$P_n = \{0, \Delta x, 2\,\Delta x, \ldots, (n-1)\,\Delta x, 1\}, \qquad \Delta x = \frac{1}{n}.$$

Let us take $z_1, z_2, \ldots, z_n$ to be the midpoints of the subintervals of P_n; so

$$z_1 = \tfrac{1}{2}\,\Delta x,\ z_2 = \tfrac{3}{2}\,\Delta x,\ z_3 = \tfrac{5}{2}\,\Delta x, \ldots, z_n = \frac{2n-1}{2}\,\Delta x.$$

Now,

$$\begin{aligned}\sum_{i=1}^{n} f(z_i)\,\Delta x &= \left[\sum_{i=1}^{n} 2\left(\frac{2i-1}{2}\,\Delta x\right)^2 - 4\left(\frac{2i-1}{2}\,\Delta x\right)\right]\frac{1}{n}\\ &= \frac{1}{n}\left[\frac{2}{n^2}\sum_{i=1}^{n}\left(i^2 - i + \frac{1}{4}\right) - \frac{2}{n}\sum_{i=1}^{n}(2i-1)\right]\\ &= \frac{2}{n^3}\left[\frac{n(n+1)(2n+1)}{6} - \frac{n(n+1)}{2} + \frac{1}{4}n - n[n\ n+1) - n]\right]\\ &= -\frac{4}{3} - \frac{1}{6n^2}.\end{aligned}$$

Hence $$\int_0^1 (2x^2 - 4x)\,dx = \lim_{n\to\infty}\left(-\frac{4}{3} - \frac{1}{6n^2}\right) = -\frac{4}{3}.$$

Check: $$\int_0^1 (2x^2 - 4x)\,dx = \left(\frac{2}{3}x^3 - 2x^2\right)\Big|_0^1 = \frac{2}{3} - 2 = -\frac{4}{3}.$$

EXERCISES

Use 5.41 to evaluate the following definite integrals (as in Examples 2 and 3).

1. $\int_1^2 x\,dx$

2. $\int_0^2 (3x^2 + 2)\,dx$

3. $\int_0^1 x(x + 1)\,dx$

4. $\int_0^1 x^3\,dx$. (*Hint:* Recall Exercise 12, Section 3.)

5. $\int_0^2 x(x^2 + 1)\,dx$

6. $\int_1^3 (x^2 + 2)\,dx$

7. $\int_{-1}^1 (x^3 + x)\,dx$

8. $\int_a^b (4x^2 - 1)\,dx$

9. $\int_1^5 (sx + r)\,dx$

10. $\int_1^3 (-x^2 + 1)\,dx$

CHAPTER REVIEW

I

Find each integral.

1. $\int \frac{x}{(3 - x^2)^2}\,dx$

2. $\int \sqrt{4t + 1}\,dt$

3. $\int \left(x^2 - \frac{4}{x^2}\right)dx$

4. $\int \frac{3x}{\sqrt{4 - x^2}}\,dx$

5. $\int \frac{t^2}{(t^3 + 1)^3}\,dt$

6. $\int x\sqrt{1 + x}\,dx$

7. $\int_1^2 \frac{7 - x^{1/3}}{x^{2/3}}\,dx$

8. $\int_0^1 \frac{y^3}{(y^4 + 5)^4}\,dy$

9. $\int (4x - 3)^6\,dx$

10. $\int \sqrt{6 - (x^2 + 3)^4}(x^2 + 3)^3 x\,dx$

11. $\int_{-1}^3 |2x + 1|\,dx$

12. $\int_{-1}^1 \sqrt{4 + |x|}\,dx$

13. $\int_1^4 \left(\frac{1}{\sqrt{x}} + \frac{1}{\sqrt{2}}\right)dx$

14. $\int_0^1 (1 - x + x^3)^2(3x^2 - 1)\,dx$

15. $\int_0^1 (1 + 5x - x^5)^4(x^2 - 1)(x^2 + 1)\,dx$

16. $\int_1^a x(x^2 - 1)^n\,dx,\ n > 0$

17. $\int_0^1 x(x^2 + a^2)^n\,dx,\ n > 0$

18. $\int_0^a \frac{x}{(x^2 + a^2)^n}\,dx;\ n > 1,\ a \neq 0$

In each of Exercises 19 to 22, find the limit if it exists.

19. $\lim_{t\to\infty} \int_1^t \frac{1}{x^2}\,dx$

20. $\lim_{t\to\infty} \int_1^t \frac{1}{\sqrt{x}}\,dx$

21. $\lim_{t\to 0} \int_t^1 \frac{1}{x^2}\,dx$

22. $\lim_{t\to 0} \int_t^1 \frac{1}{\sqrt{x}}\,dx$

Find the l.u.b. and g.l.b., if they exist, of each of the following sets.

23. $\{x \mid \frac{1}{3} < x < 2\}$

24. $\{y \mid y^2 \leq 3\}$

25. $\left\{y \mid y = \frac{1 + n}{2 + n},\ n = 1, 2, 3, 4, \ldots\right\}$

26. $\{z \mid 16 - z^2 \geq 0\}$

27. $(-\infty,\pi) \cap (3.1,4)$

28. $\left\{x \mid x = (-1)^{n-1}\left(2 + \frac{4}{n}\right),\ n = 1, 2, 3, 4, \ldots\right\}$

Find:

29. $\lim_{n\to\infty} \frac{3n^2 + n}{4n^2 - 1}$

30. $\lim_{n\to\infty} \left(\frac{6}{n^2} - \frac{1}{3}\right)$

31. $\lim_{n\to\infty} \frac{4n + 1}{5n - 3}$

32. $\lim_{n\to\infty} \frac{3 - 2n + n^2}{4 + 3n - 2n^2}$

Find, using 5.41:

33. $\int_1^2 (3x - 7)\,dx$

34. $\int_{-1}^1 (3x - x^2)\,dx$

35. $\int_{-1}^0 (2 + 5x)\,dx$

36. $\int_0^2 (x^2 + 5x)\,dx$

37. Let $f(x) = \begin{cases} 2 - x, & \text{if } x < 2 \\ 0, & \text{if } 2 \leq x < 4. \\ \sqrt{x - 2}, & \text{if } x \geq 4 \end{cases}$ Find $\int_{-1}^5 f(x)\,dx$.

38. Let $f(x) = \begin{cases} x^2, & \text{if } x \leq 0 \\ x, & \text{if } 0 < x \leq 2. \\ \sqrt{x + 2}, & \text{if } x > 2 \end{cases}$ Find $\int_{-3}^3 f(x)\,dx$.

II

1. Write a formula for $D_x\left\{\int_a^{g(x)} f(z)\,dz\right\}$, assuming that $g'(x)$ exists and that f is continuous in a closed interval containing a and $g(x)$.
2. Let $f(x) = 1/x^2$ and $F(x) = -1/x$. Find $F(1) - F(-1)$. Does $\int_{-1}^{1} f(x)\,dx = F(1) - F(-1)$? Explain.
3. Since the integrand of the integral $I = \int_{-1}^{1} \frac{1}{1+x^2}\,dx$ is positive, it follows that $I > 0$. However, if we make the change of variable $x = 1/u$, then

$$I = \int_{-1}^{1} \frac{1}{1+x^2}\,dx = -\int_{-1}^{1} \frac{1}{1+u^2}\,du = -I,$$

whence $I = 0$. Explain.

10 APPENDIX ON THEORY

The theorems of Chapters 4 and 5 whose proofs were not included in the text are proved in this section. A few new theorems are also included. First, we develop the background of the proof of Theorem 4.2.

A set of C of open intervals is called an (open) *covering* of an interval I if every number x in I is also in some open interval of C.

For example, if $C = \{(-1,0), (-\frac{1}{2},1), (\frac{1}{3},2), (\frac{3}{2},3)\}$, then C is a covering of the interval $[0,2]$. Thus each number x, $(0 \le x \le 2)$, is in one or more of the open intervals of C. In particular, 0 is in $(-\frac{1}{2},1)$, $\frac{1}{2}$ is in both $(-\frac{1}{2},1)$ and $(\frac{1}{3},2)$, and $\frac{7}{4}$ is in $(\frac{3}{2},3)$.

As another example, let us associate with each rational number r $(0 \le r \le 1)$ the open interval $(r - .01, r + .01)$. The set C of all such open intervals is a covering of the closed interval $[0,1]$. Thus each real number x in $[0,1]$ is within .01 of a unit of some rational number r, and therefore x is in $(r - .01, r + .01)$. For example, $\sqrt{2} - 1$ is in $(.41,.43)$. Since there are infinitely many rational numbers in $(0,1)$, there are infinitely many open intervals in the covering C of $[0,1]$. However, we can find a finite subset C' of C that is also a covering of $[0,1]$. For example, let

$$C' = \{(-.01,.01), (0,.02), (.01,.03), \ldots, (.98,1), (.99,1.01)\}.$$

Since any two adjacent intervals of C' overlap, it is clear that C' is a covering of $[0,1]$. Evidently, C' contains 100 open intervals.

The example above illustrates the following fundamental property of coverings.

5.42 Theorem

If a set C of open intervals is a covering of a closed interval $[a,b]$, then some finite subset C' of C is also a covering of $[a,b]$.

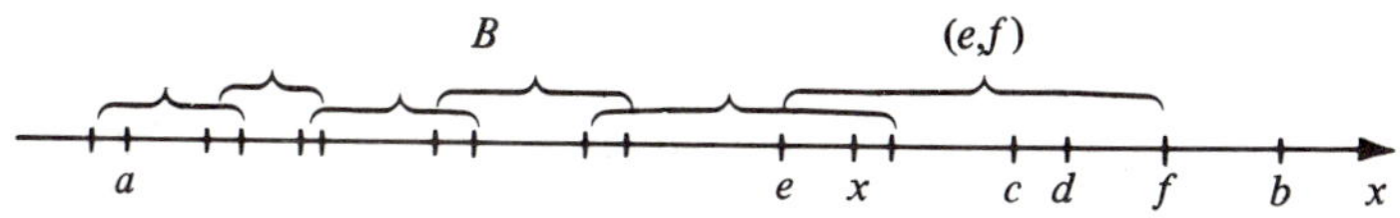

FIGURE 5.14

Proof: Let A be the set consisting of every number x of $[a,b]$ such that the interval $[a,x]$ is covered by some finite subset of C, and let

$$c = \text{l.u.b. } A.$$

Clearly, $c \leq b$ by the definition of A. By assumption, some interval (e,f) of C contains c. Since $c = \text{l.u.b. } A$, necessarily some number x of A is also in (e,f).

Let B be a finite subset of C covering $[a,x]$. Then

$$C' = B \cup \{(e,f)\}$$

is a finite covering of $[a,c]$ (see Figure 5.14). Therefore c is in set A. Actually, $c = b$ and C' is a covering of $[a,b]$. For if $c < b$ and d is in $(c,b) \cap (e,f)$ (see Figure 5.14), then d is in A since $[a,d]$ is also covered by C'. However, this contradicts the fact that $c = \text{l.u.b. } A$. This proves 5.42.

We shall use 5.42 to demonstrate the following boundedness property of a continuous function.

5.43 Theorem

A function that is continuous in a closed interval is bounded in that interval.

Proof: Let function f be continuous in a closed interval $[a,b]$. For each number c in $[a,b]$, there exists a neighborhood N_c of c such that $f(x)$ is close to $f(c)$, say within 1 of $f(c)$, for every x in $N_c \cap [a,b]$. In symbols,

$$f(c) - 1 < f(x) < f(c) + 1 \qquad \text{for every } x \text{ in } N_c \cap [a,b].$$

We note that f is *bounded* in each interval N_c, having lower bound $f(c) - 1$ and upper bound $f(c) + 1$.

The set $C - \{N_c \mid c \text{ in } [a,b]\}$ is an infinite covering of $[a,b]$ by open intervals. Hence, by 5.42, some finite subset C' of C is also a covering of $[a,b]$. Since f is bounded in each of the intervals of C', it is bounded in $[a,b]$. For if $k_1, k_2, \ldots, k_n$ are the lower bounds of f in the intervals of C', then the least of the numbers $k_1, k_2, \ldots, k_n$ is a lower bound of f in $[a,b]$. A similar argument shows that f also has an upper bound.

We should realize that 5.43 is not true for open intervals. For example, the function f defined by $f(x) = 1/x$ is continuous but not bounded in the open interval $(0,1)$.

Knowing that a continuous function f is bounded in a closed interval $[a,b]$, we can prove even more, namely that f assumes a maximum value and a minimum value. We state this result as follows. This is Theorem 4.2 of the text, which we restate as follows.

5.44 Theorem

If a function f is continuous in a closed interval $[a,b]$, then there exist numbers u and v in $[a,b]$ such that $f(u)$ is the minimum value and $f(v)$ is the maximum value of f in $[a,b]$.

Proof: Let $R = \{f(x) \mid x \text{ in } [a,b]\}$ be the range of f in $[a,b]$. By 5.43, R has both an upper bound and a lower bound. Hence R has a l.u.b. and a g.l.b. Let

$$k = \text{l.u.b. } R.$$

If $k = f(v)$ for some v in $[a,b]$, then $f(v)$ is the maximum value of f in $[a,b]$ and the desired conclusion follows.

On the other hand, if $f(x) \neq k$ for every x in $[a,b]$, then $k - f(x) > 0$ for every x in $[a,b]$. The function g defined by

$$g(x) = \frac{1}{k - f(x)}, \qquad \text{domain } g = [a,b],$$

is continuous in $[a,b]$. Hence, by 5.43, g has an upper bound w in $[a,b]$, so that

$$0 < \frac{1}{k - f(x)} \leq w \qquad \text{for every } x \text{ in } [a,b].$$

Solving this inequality for $f(x)$, we have

$$f(x) \leq k - \frac{1}{w} \qquad \text{for every } x \text{ in } [a,b].$$

Thus $k - 1/w$ is an upper bound of R less than l.u.b. R. This contradiction shows that $f(x)$ cannot be different from k for every x in $[a,b]$. Hence, there exists a number v in $[a,b]$ such that $f(v)$ is the maximum value of f in $[a,b]$.

The existence of a number u in $[a,b]$ such that $f(u)$ is the minimum value of f in $[a,b]$ is proved similarly.

Another useful result is as follows.

5.45 Intermediate Value Theorem for Functions

If a function f is continuous in a closed interval $[a,b]$, then f assumes every value between $f(a)$ and $f(b)$.

Proof: We wish to show that for every number d between $f(a)$ and $f(b)$, there exists a number c in $[a,b]$ such that $f(c) = d$. For convenience, let us assume that $f(a) < d < f(b)$. Let S denote the set of all numbers x in $[a,b]$ for which $f(x) < d$. Clearly a is in S, b is not in S, and b is an upper bound of S. Hence, S has a l.u.b. c by the completeness property.

One of three possible situations holds: $f(c) < d, f(c) = d, f(c) > d$. If $f(c) - d < 0$, then $f(x) - d < 0$ in some neighborhood $(c - \delta, c + \delta)$ of c by 2.6. However, then $f(x) < d$ for some $x > c$, contrary to the fact that $c =$ l.u.b. S. If $f(c) - d > 0$, then $f(x) - d > 0$ in some neighborhood $(c - \delta, c + \delta)$ of c by 2.6, contrary to the fact that $f(x) < d$ for some x in *every* neighborhood of c. Therefore, we must have $f(c) = d$.

We recall that a function f is said to be continuous in an interval I if it is continuous at each number c in I. This means that for every c in I and every $\varepsilon > 0$, there exists some $\delta > 0$ (depending on both c and ε) such that:

For every x in $I \cap (c - \delta, c + \delta)$, $f(x)$ is in $(f(c) - \varepsilon, f(c) + \varepsilon)$.

5.46 Definition of Uniform Continuity

A function f is said to be uniformly continuous in an interval I if for every $\varepsilon > 0$, there exists some $\delta > 0$ (depending only on ε) such that:

For every c in I and every x in $I \cap (c - \delta, c + \delta)$,

$$f(x) \text{ is in } (f(c) - \varepsilon, f(c) + \varepsilon).$$

The main result of interest to us on uniform continuity is given below.

5.47 Theorem

A function f that is continuous in a closed interval I is uniformly continuous in I.

Proof: For every $\varepsilon > 0$, we can choose a symmetric neighborhood $(t - r, t + r)$ of each t in I (with r depending on t as well as ε) such that $f(x)$ is in $(f(t) - \varepsilon/2, f(t) + \varepsilon/2)$ for every x in $(t - r, t + r) \cap I$. The set

$$C = \left\{\left(t - \frac{r}{2}, t + \frac{r}{2}\right) \mid t \text{ in } I\right\}$$

of open intervals is a covering of I. By 5.42, some finite subset C' of C is also a covering of I, say

$$C' = \left\{\left(t_1 - \frac{r_1}{2}, t_1 + \frac{r_1}{2}\right), \left(t_2 - \frac{r_2}{2}, t_2 + \frac{r_2}{2}\right), \ldots, \left(t_n - \frac{r_n}{2}, t_n + \frac{r_n}{2}\right)\right\}.$$

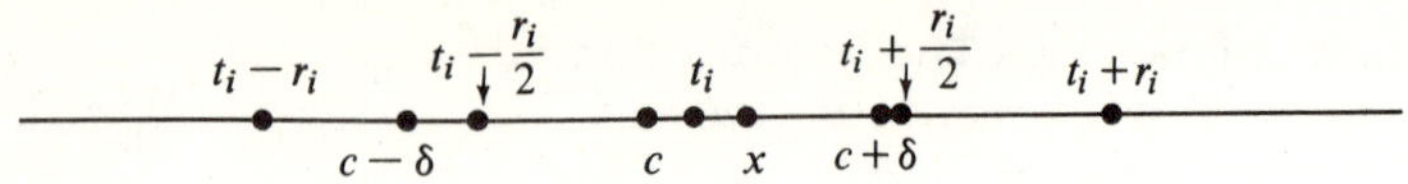

FIGURE 5.15

Now choose δ as the smallest of the numbers $r_1/2, r_2/2, \ldots, r_n/2$. Let us show that this δ satisfies the requirements of the definition of uniform continuity. Each c in I is in some interval $(t_i - r_i/2, t_i + r_i/2)$ of C'. If x is in $(c - \delta, c + \delta)$, then both x and c are in $(t_i - r_i, t_i + r_i)$ since $\delta \leq t_i$ (see Figure 5.15). This means that both $f(x)$ and $f(c)$ lie between $f(t_i) - \varepsilon/2$ and $f(t_i) + \varepsilon/2$. Thus, the difference between $f(x)$ and $f(c)$ is less than ε; that is,

$$f(x) \text{ is in } (f(c) - \varepsilon, f(c) + \varepsilon) \qquad \text{for every } x \text{ in } (c - \delta, c + \delta).$$

This proves that f is uniformly continuous in I.

We are now ready to prove 5.17, which we restate here for convenience.

5.17 Theorem

If f is a continuous function in a closed interval $[a,b]$ and if L is the set of all lower sums and U of upper sums of f over $[a,b]$, then

$$\text{l.u.b. } L = \text{g.l.b. } U.$$

Proof: We know that

$$S(P) \leq \text{l.u.b. } L \leq \text{g.l.b. } U \leq T(P) \qquad \text{for every partition } P \text{ of } [a,b].$$

Hence,
$$0 \leq \text{g.l.b. } U - \text{l.u.b. } L \leq T(P) - S(P).$$

We shall show that for every $\varepsilon > 0$ there exists a partition P of $[a,b]$ such that

$$T(P) - S(P) < \varepsilon.$$

By the uniform continuity of f, for every $\varepsilon > 0$ there exists some $\delta > 0$ such that $f(x)$ is in $(f(c) - \varepsilon/(b - a), f(c) + \varepsilon/(b - a))$ for every c in $[a,b]$ and every x in $(c - \delta, c + \delta) \cap [a,b]$. If $P = \{x_0, x_1, \ldots, x_n\}$ is any partition of $[a,b]$ for which $\|P\| < \delta$, and $f(u_i)$ is the minimum value and $f(v_i)$ the maximum value of f in $[x_{i-1}, x_i]$, then

$$T(P) - S(P) = \sum_{i=1}^{n} [f(v_i) - f(u_i)] \, \Delta x_i$$

Since u_i and v_i are in $[x_{i-1},x_i]$ and $x_i - x_{i-1} < \delta$, we know that $f(v_i)$ is in $(f(u_i) - \varepsilon/(b - a),\ f(u_i) + \varepsilon/(b - a))$. Hence $f(v_i) - f(u_i) < \varepsilon/(b - a)$ for each i, and

$$0 \leq T(P) - S(P) < \sum_{i=1}^{n} \frac{\varepsilon}{b - a} \Delta x_i$$

$$= \frac{\varepsilon}{b - a} \sum_{i=1}^{n} \Delta x_i = \frac{\varepsilon}{b - a}(b - a) = \varepsilon.$$

Combining the results above, we have proved for every $\varepsilon > 0$,

$$0 \leq \text{g.l.b. } U - \text{l.u.b. } L < \varepsilon.$$

This can happen only if g.l.b. U − l.u.b. $L = 0$, that is,

$$\text{l.u.b } L = \text{g.l.b. } U.$$

A somewhat different proof is needed for 5.40, which we restate below.

5.40 Theorem

Let function f be continuous in $[a,b]$ and $P_1, P_2, \ldots, P_n, \ldots$ be a sequence of partitions of $[a,b]$ for which $\lim_{n\to\infty} \|P_n\| = 0$. If $R(P_1), R(P_2), \ldots, R(P_n), \ldots$ is any sequence of Riemann sums associated with the given sequence of partitions, then $\lim_{n\to\infty} R(P_n) = \int_a^b f(x)\,dx$.

Proof: By previous remarks, $S(P_n) \leq R(P_n) \leq T(P_n)$ and $S(P_n) \leq \int_a^b f(x)\,dx \leq T(P_n)$ for every n. If we can show that for every $\varepsilon > 0$ there exists a number N such that $T(P_n) - S(P_n) < \varepsilon$ for every $n > N$, then it will follow that $R(P_n)$ is in the interval

$$\left(\int_a^b f(x)\,dx - \varepsilon,\ \int_a^b f(x)\,dx + \varepsilon\right)$$

for every $n > N$. This implies that $\lim_{n\to\infty} R(P_n) = \int_a^b f(x)\,dx$.

To this end, we know by 5.17 that for every $\varepsilon > 0$ there exists a partition $P = \{x_0, x_1, \ldots, x_k\}$ of $[a,b]$, such that

$$0 \leq T(P) - S(P) < \frac{\varepsilon}{2}.$$

Clearly we can choose P so that $k > 1$. Since f is bounded by 5.43, there exists a positive number M such that

$$-M \leq f(x) \leq M \qquad \text{for every } x \text{ in } [a,b].$$

For each P_n, $k - 1$ or fewer of the subintervals of P_n contain the numbers $x_1, x_2, \ldots, x_{k-1}$ of P and the rest of the subintervals of P_n are contained in subintervals of P. Now each term of $T(P_n) - S(P_n)$ has the form

$[f(v_i) - f(u_i)]\,\Delta z_i$, where $f(u_i)$ is the minimum value and $f(v_i)$ is the maximum value of f in the ith subinterval $[z_{i-1},z_i]$ of P_n. Since $-M \leq f(u_i) \leq f(v_i) \leq M$ and $\Delta z_i \leq \|P_n\|$, each term of $T(P_n) - S(P_n)$ is less than or equal to $2M\|P_n\|$. Hence, the sum of the terms of $T(P_n) - S(P_n)$ which correspond to subintervals containing the numbers $x_1, x_2, \ldots, x_{k-1}$ is at most $2M(k-1)\|P_n\|$.

For each subinterval $[z_{i-1},z_i]$ of P_n which is contained in a subinterval $[x_{j-1},x_j]$ of P, the term $[f(v_i) - f(u_i)]\,\Delta z_i$ of $T(P_n) - S(P_n)$ is less than or equal to the corresponding term $[f(v_j') - f(u_j')]\,\Delta x_j$ of $T(P) - S(P)$ because $f(u_j') \leq f(u_j)$, $f(v_j') \geq f(v_j)$, and $\Delta z_i \leq \Delta x_j$. Hence, the sum of the terms of $T(P_n) - S(P_n)$ which do not involve the numbers $x_1, x_2, \ldots, x_k$ is at most $T(P) - S(P)$.

Combining the two types of subintervals of P_n discussed above, we obtain

$$0 \leq T(P_n) - S(P_n) \leq [T(P) - S(P)] + 2M(k-1)\|P_n\|.$$

Since $\lim_{n\to\infty} \|P_n\| = 0$, for every $\varepsilon > 0$ there exists a number N such that

$$\|P_n\| < \frac{\varepsilon}{4M(k-1)} \quad \text{for every } n > N.$$

Hence, $2M(k-1)\|P_n\| < \varepsilon/2$ for every $n > N$ and

$$0 \leq T(P_n) - S(P_n) < \frac{\varepsilon}{2} + \frac{\varepsilon}{2} = \varepsilon \qquad \text{for every } n > N.$$

Another theorem stated without proof earlier in this chapter is as follows.

5.19 Theorem

If the function f is continuous in an interval containing the numbers a, b, and c, then

$$\int_a^b f(x)\,dx + \int_b^c f(x)\,dx = \int_a^c f(x)\,dx.$$

Proof: We shall assume $a < b < c$. The other possible arrangements of a, b, and c can be handled similarly.

For every $\varepsilon > 0$ there exist partitions P_1 of $[a,b]$ and P_2 of $[b,c]$ such that

$$(1) \qquad 0 \leq T(P_1) - S(P_1) < \frac{\varepsilon}{2}, \qquad 0 \leq T(P_2) - S(P_2) < \frac{\varepsilon}{2}.$$

We also have

$$S(P_1) \leq \int_a^b f(x)\,dx \leq T(P_1), \qquad S(P_2) \leq \int_b^c f(x)\,dx \leq T(P_2)$$

and hence

(2) $S(P_1) + S(P_2) \leq \int_a^b f(x)\,dx + \int_b^c f(x)\,dx \leq T(P_1) + T(P_2).$

Now $S(P_1) + S(P_2)$ is a lower sum and $T(P_1) + T(P_2)$ an upper sum of f over $[a,c]$. Hence,

(3) $$S(P_1) + S(P_2) \leq \int_a^c f(x)\,dx \leq T(P_1) + T(P_2).$$

Since

$$0 \leq [T(P_1) + T(P_2)] - [S(P_1) + S(P_2)] < \varepsilon$$

by (1), we have by (2) and (3) that for all $\varepsilon > 0$,

$$-\varepsilon < \int_a^b f(x)\,dx + \int_b^c f(x)\,dx - \int_a^c f(x)\,dx < \varepsilon.$$

Therefore, $$\int_a^b f(x)\,dx + \int_b^c f(x)\,dx - \int_a^c f(x)\,dx = 0.$$

If a function F is continuous in a closed interval $[a,b]$ and if the domain of F' contains the open interval (a,b), then the mean value theorem 4.8 tells us that there exists a number c in (a,b) such that

$$F(b) - F(a) = (b - a)F'(c).$$

Now if f is any function which is continuous in $[a,b]$, then f has an antiderivative F which has the properties stated above. Since $F(b) - F(a)$ is the integral of f from a to b by 5.20 and $F'(c) = f(c)$, we have established the following integral analog of the mean value theorem.

5.48 Intermediate Value Theorem for Integrals

If function f is continuous in an interval $[a,b]$, then there exists a number c in (a,b) such that

$$\int_a^b f(x)\,dx = (b - a)f(c).$$

6

Applications of the Integral

Like the derivative, the integral has many interesting applications to geometrical and physical problems. Some of these applications are given in the present chapter.

1 AREAS

If f is a continuous, nonnegative-valued function in a closed interval $[a,b]$, then the lines $x = a$, $x = b$, the x axis, and the graph of f bound a region R of the plane (Figure 6.1). As we saw in Section 2 of Chapter 5, each lower sum $S(P)$ of f over $[a,b]$ is the area of a polygon inscribed in R and each upper sum $T(P)$ is the area of a polygon circumscribed about R. Therefore, it is natural to consider the area $A(R)$ of region R, if it exists, to be between $S(P)$ and $T(P)$,

$$S(P) \leq A(R) \leq T(P).$$

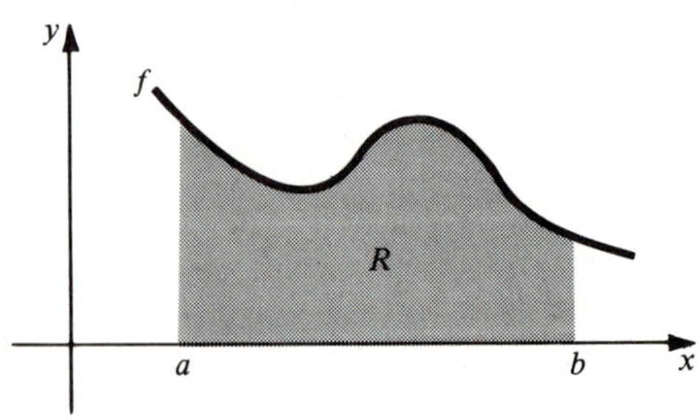

FIGURE 6.1

However, there is precisely one number which is between $S(P)$ and $T(P)$ for every partition P of $[a,b]$, namely $\int_a^b f(x)\,dx$. This fact suggests the following definition for area.

6.1 Definition of Area

If a region R of the plane is bounded by the lines $x = a$, $x = b$, the x axis, and the graph of a continuous, nonnegative-valued function f (Figure 6.1), then the area $A(R)$ of R is defined to be

$$A(R) = \int_a^b f(x)\,dx.$$

We call the region R of Figure 6.1 *the region under the graph of f between a and b.*

Example 1 If function g is defined by

$$g(x) = \sqrt[3]{x},$$

find the area of the region R under the graph of g from 1 to 8.

Solution: The region R, shown in Figure 6.2, has area

$$A(R) = \int_1^8 \sqrt[3]{x}\,dx = \int_1^8 x^{1/3}\,dx = \tfrac{3}{4}x^{4/3}\Big|_1^8$$

$$= \tfrac{3}{4}(8^{4/3} - 1^{4/3}) = \tfrac{45}{4}.$$

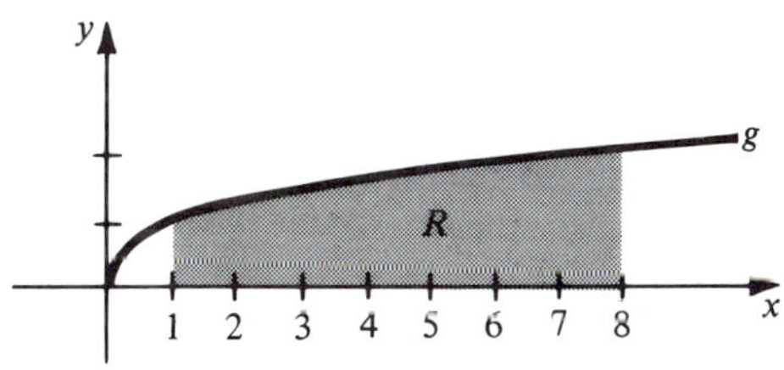

FIGURE 6.2

Is our answer reasonable? Clearly, R can be inscribed in a rectangle of base 7 and height 2, so that we expect $A(R)$ to be less than 14. On the other hand, a trapezoid of height 7 and bases 1 and 2 can be inscribed in R. Thus we expect $A(R)$ to be greater than $\frac{1}{2}[7 \cdot (1 + 2)]$, or $\frac{21}{2}$. Since

$$\tfrac{42}{4} < \tfrac{45}{4} < \tfrac{56}{4},$$

our answer seems reasonable.

If f is a continuous, nonpositive-valued function in an interval $[a,b]$, then $\int_a^b f(x)\,dx < 0$ and the area of the region R bounded by the lines $x = a$, $x = b$, the x axis, and the graph of f is given by

$$A(R) = -\int_a^b f(x)\,dx.$$

This is a special case of the following result.

6.2 Theorem

If f and g are continuous functions in $[a,b]$ and if

$$f(x) \geq g(x) \qquad \textit{for every } x \textit{ in } [a,b],$$

then the area $A(R)$ of the region R between the graphs of f and g from a to b (Figure 6.3) is given by

$$A(R) = \int_a^b [f(x) - g(x)]\,dx.$$

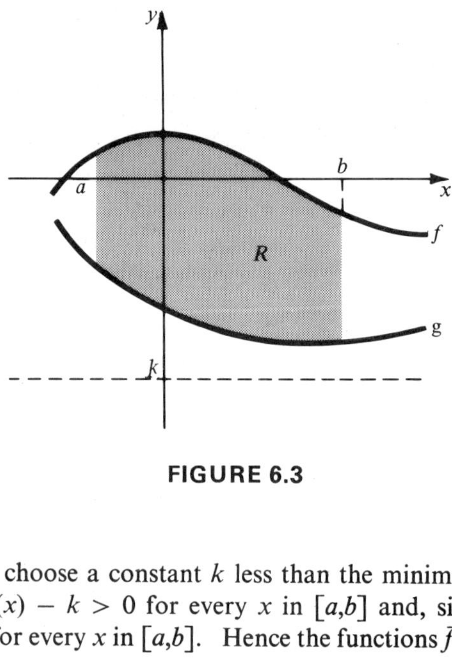

FIGURE 6.3

Proof: Let us choose a constant k less than the minimum value of g in $[a,b]$. Then $g(x) - k > 0$ for every x in $[a,b]$ and, since $f(x) \geq g(x)$, $f(x) - k > 0$ for every x in $[a,b]$. Hence the functions $\bar{f}$ and $\bar{g}$ defined by

$$\bar{f}(x) = f(x) - k, \qquad \bar{g}(x) = g(x) - k$$

are continuous and nonnegative-valued in $[a,b]$. If k is chosen as in Figure 6.3, then the graphs of $\bar{f}$ and $\bar{g}$ are as shown in Figure 6.4.

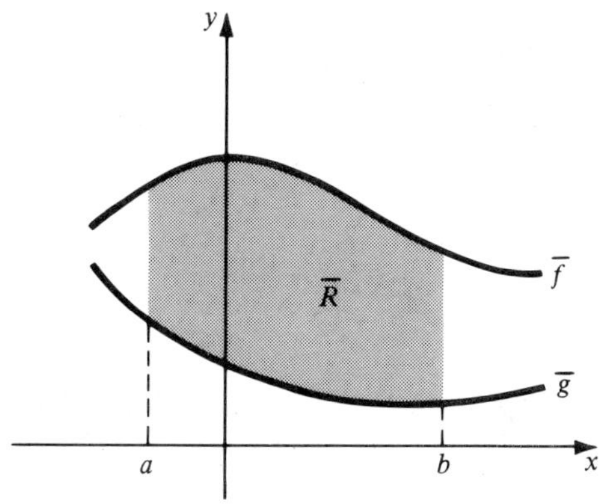

FIGURE 6.4

The area of the region $\bar{R}$ between the graphs of $\bar{f}$ and $\bar{g}$ is the same as that of R, since the regions are congruent. Thus the area of $\bar{R}$, and hence also that of R, is seen to be the difference of the areas of the regions under the graphs of $\bar{f}$ and $\bar{g}$ from a to b,

$$A(R) = \int_a^b \bar{f}(x)\,dx - \int_a^b \bar{g}(x)\,dx.$$

However,

$$\int_a^b \bar{f}(x)\,dx - \int_a^b \bar{g}(x)\,dx = \int_a^b [\bar{f}(x) - \bar{g}(x)]\,dx,$$

$$= \int_a^t \{[f(x) - k] - [g(x) - k]\}\,dx,$$

$$= \int_a^b [f(x) - g(x)]\,dx.$$

Example 2 Find the area of the region over the graph of f between -1 and 2 if

$$f(x) = x^3 - 3x - 3.$$

Solution: The graph of f has a maximum point at $(-1, -1)$ and a minimum point at $(1, -5)$, as shown in Figure 6.5. The area $A(R)$ of the shaded region R is given by

$$A(R) = -\int_{-1}^{2} (x^3 - 3x - 3)\,dx = -\left(\frac{x^4}{4} - 3\frac{x^2}{2} - 3x\right)\Bigg|_{-1}^{2} = \frac{39}{4}.$$

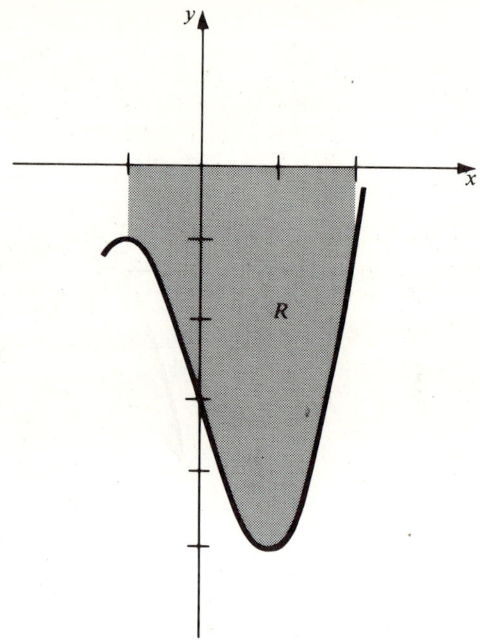

FIGURE 6.5

Example 3 Find the area of the region between the graphs of the equations

$$y = x - 2 \qquad \text{and} \qquad y = 2x - x^2.$$

Solution: The graphs will intersect in those points whose coordinates are the simultaneous solutions of the given equations. Eliminating y between these equations, we have

$$x - 2 = 2x - x^2,$$

or

$$x^2 - x - 2 = (x - 2)(x + 1) = 0.$$

Thus $x = 2$ or $x = -1$, and the common points of the two graphs are (2,0) and $(-1,-3)$.

The graph of $y = x - 2$ is a straight line. The graph of $y = 2x - x^2$ is a parabola that is concave downward and has its vertex at (1,1). The region R whose area is sought is shaded in Figure 6.6. This area is given by

$$A(R) = \int_{-1}^{2} [(2x - x^2) - (x - 2)]\, dx = \int_{-1}^{2} (-x^2 + x + 2)\, dx$$

$$= \left(-\tfrac{1}{3}x^3 + \tfrac{1}{2}x^2 + 2x\right)\Big|_{-1}^{2} = \tfrac{9}{2}.$$

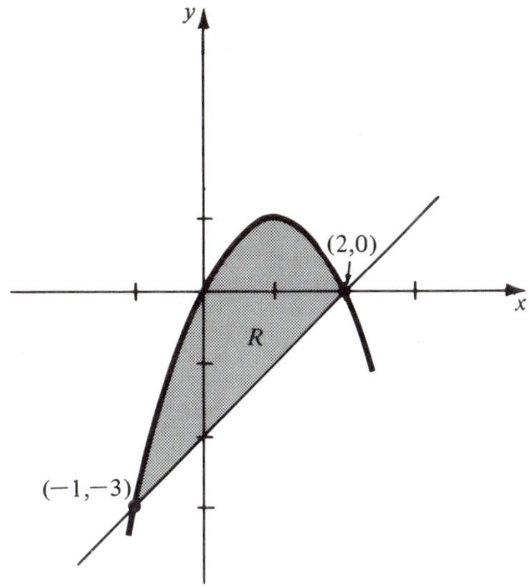

FIGURE 6.6

Example 4 Find the area of the region bounded by one loop of the graph of the equation

$$y^2 = 4x^2 - x^4.$$

Solution: The graph is symmetric to both axes. Since

$$y = \pm x\sqrt{4 - x^2},$$

it is clear that the total graph lies between $x = -2$ and $x = 2$. The point $(\sqrt{2},2)$ is a maximum point on the graph. We can plot a few points and sketch the rest of the curve by symmetry, as indicated in Figure 6.7.

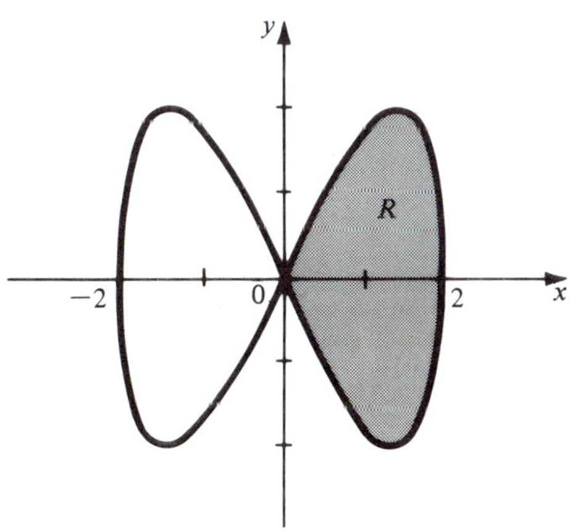

FIGURE 6.7

Let us find the area of the shaded region R bounded by the loop between $x = 0$ and $x = 2$. The equation of the top half of this loop is

$$y = x\sqrt{4 - x^2},$$

whereas that of the lower half is

$$y = -x\sqrt{4 - x^2}.$$

Thus

$$A(R) = \int_0^2 [(x\sqrt{4 - x^2}) - (-x\sqrt{4 - x^2})]\,dx = \int_0^2 2x\sqrt{4 - x^2}\,dx.$$

In order to evaluate this integral, we let

$$u = 4 - x^2, \qquad du = -2x\,dx.$$

Then $u = 4$ when $x = 0$, and $u = 0$ when $x = 2$. Hence

$$A(R) = -\int_4^0 u^{1/2}\,du = -\tfrac{2}{3}u^{3/2}\Big|_4^0 = \tfrac{16}{3}.$$

Example 5 Find the area of a circle C_r of radius r.

Solution: Evidently, the area of C_r is four times the area of region R in Figure 6.8.

$$A(C_r) = 4A(R).$$

Now the curved boundary of R has equation

$$y = \sqrt{r^2 - x^2},$$

so that

$$A(C_r) = 4\int_0^r \sqrt{r^2 - x^2}\,dx.$$

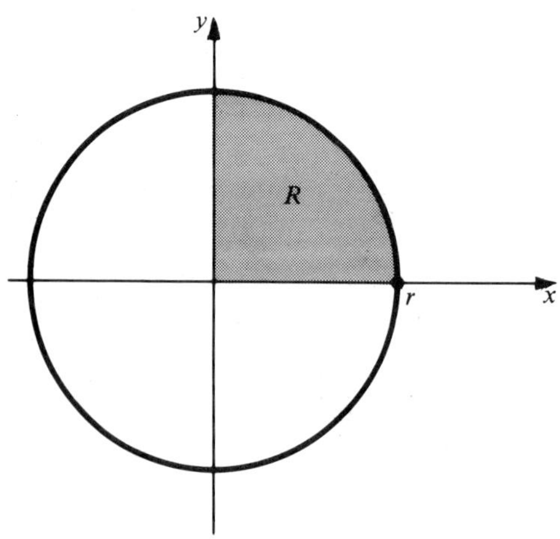

FIGURE 6.8

We cannot evaluate this integral in the sense that we can find an antiderivative. However, we can do the following. We can change variables by letting

$$u = \frac{1}{r}x, \qquad du = \frac{1}{r}dx.$$

Then
$$A(C_r) = 4\int_0^r r\sqrt{1 - \left(\frac{x}{r}\right)^2}\,dx$$
$$= 4r^2 \int_0^r \sqrt{1 - \left(\frac{x}{r}\right)^2}\left(\frac{1}{r}\,dx\right) = 4r^2 \int_0^1 \sqrt{1 - u^2}\,du.$$

After the number π is defined in Chapter 8 (as the circumference of a unit circle), we will show that

$$\pi = 4\int_0^1 \sqrt{1 - u^2}\,du.$$

Hence we obtain the familiar formula

$$A(C_r) = \pi r^2$$

for the area of a circle of radius r.

EXERCISES

I

Find the area of the region bounded by the graphs of the following equations. Sketch each region.

1. $y = x^3, y = 0, x = 1, x = 3$

2. $y = 9 - x^2, y = 0, x = -2, x = 1$

3. $y = x^2, x = y^2$

4. $y^2 = x, 2y = x$

5. $y = \frac{3}{x^2}, y = 4 - x^2$; for $x > 0$

6. $y = x^3 - 2x^2 - 1, y = x^2 + 6x - 9$ (two regions)

7. $y = x^{4/3}, 5x - 3y + 8 = 0$

8. $xy^2 = 1, x - 6y + 5 = 0, y = 0, x = 4$

9. $y = \sqrt{x}, y = -\sqrt{x}, x = 4$

10. $y = \sqrt{x + 4}, y = 0, x = 0$

11. $y = x^2 - 4, y = 4 - x^2$

12. $y = \sqrt[3]{x^2}, y = 0, x = 8$

13. $y = x^2, y = 1$

14. $4y = x^2, x - 4y + 2 = 0$

15. $x^2y = 4, 3x + y - 7 = 0$

16. $y = x(x - 2)^2, y = 0$

17. $y^2 = 4x, x = 1$

18. $y = x^3 - x, y = 0$, in fourth quadrant

19. $y = x + \frac{4}{x^2}, y = 3 + 4x - 2x^2$

20. $y = \frac{4x}{(x^2 + 1)^2}, y = x^2$

II

1. Sketch and find the area of the region bounded by the parabola $y = -x^2 + 4x - 3$ and the tangents to it at the points $(0, -3)$ and $(4, -3)$.
2. Sketch and find the area of the loop of the graph of $y^2 = x(x - a)^2$, $a > 0$.
3. Sketch and find the area of the region bounded by the graph of $y^2 = a^2x^6 - x^8$.

2 VOLUME

Just as integrals can be used to find areas of certain regions in a plane, so they can be used to find volumes of certain regions in space, as we shall show in this section.

We shall call a solid C a *cylinder* if C is bounded by two congruent regions R_1 and R_2 lying in parallel planes and by a lateral surface S composed of line segments that connect corresponding points of the boundaries of R_1 and R_2 and that are perpendicular to the planes of R_1 and R_2 (Figure 6.9). Each of the regions R_1 and R_2 is called a *base* of cylinder C and the distance between the planes of R_1 and R_2 is called the *height* of C.

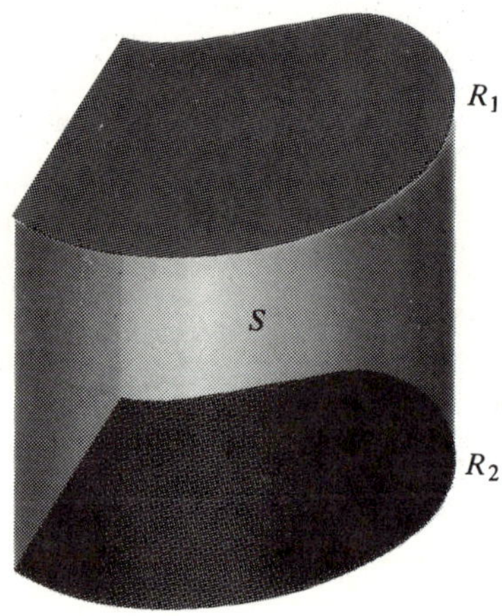

FIGURE 6.9

The common right circular cylinder is, of course, a cylinder having a circle as base. A rectangular parallelepiped is also, according to our general definition, a cylinder having a rectangle as base.

If C is a cylinder with base B and height h, then we define the *volume* $V(C)$ of C to be the area of the base $A(B)$ times the height h; that is,

6.3 $$V(C) = A(B) \cdot h.$$

Formula 6.3 yields the familiar formula

$$V(C) = \pi r^2 h$$

for the volume of a right circular cylinder of radius r and height h. Also, the volume of a rectangular parallelepiped P with edges of length a, b, and c is given by

$$V(P) = abc,$$

as we expected.

If a solid S is a composite of cylinders $C_1, C_2, \ldots, C_n$, then we define the volume $V(S)$ of S to be

$$V(S) = V(C_1) + V(C_2) + \cdots + V(C_n).$$

For example, the solid S of Figure 6.10 is composed of three right circular cylinders and has volume

$$V(S) = \pi r_1^2 h_1 + \pi r_2^2 h_2 + \pi r_3^2 h_3.$$

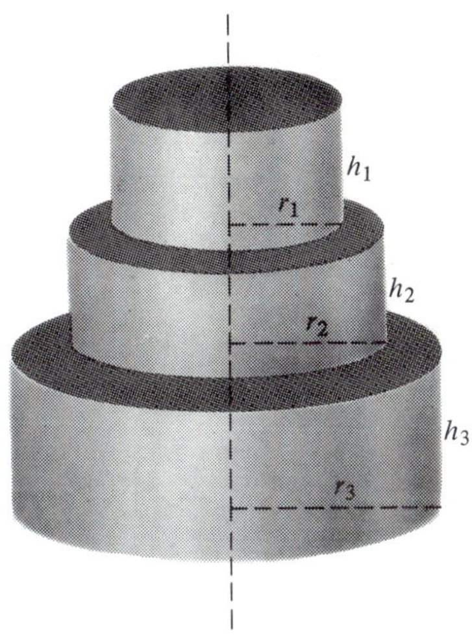

FIGURE 6.10

We turn now to the problem of defining the volume of a solid S which is not composed of cylinders. First, we remark that a plane intersecting S cuts S in a plane region called a *cross section* of S. We shall attempt to define the volume of S only under the added assumption that the areas of all cross sections of S perpendicular to some fixed line are known and change continuously. That is, there exists a coordinate line L such that the solid S lies between the planes drawn perpendicular to L at some numbers a and b, and the cross section of S in the plane perpendicular to L at each number x in $[a,b]$ has a known area $A(x)$ (see Figure 6.11) such that the area function A is continuous in $[a,b]$.

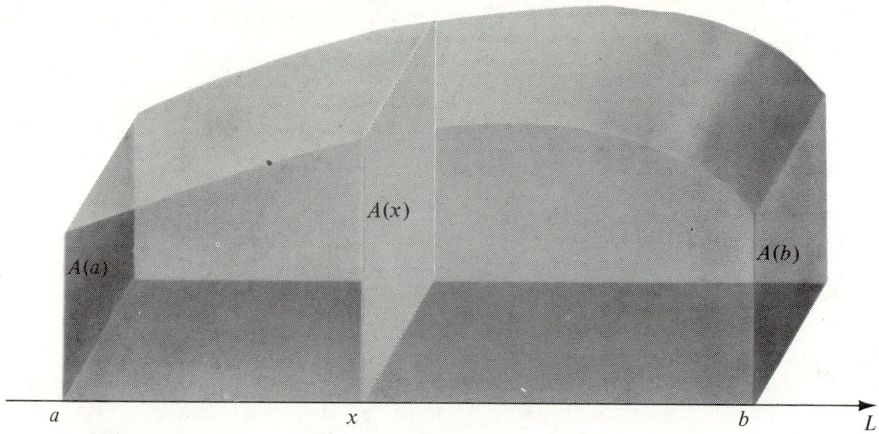

FIGURE 6.11

For each partition $P = \{x_0, x_1, \ldots, x_k\}$ of $[a,b]$, we can approximate the volume $V(S)$ of solid S (Figure 6.11) by selecting numbers $z_1, z_2, \ldots, z_k$ in $[x_0,x_1], [x_1,x_2], \ldots, [x_{k-1},x_k]$, respectively, and then constructing cylinders of heights $\Delta x_1 = x_1 - x_0, \Delta x_2 = x_2 - x_1, \ldots, \Delta x_k = x_k - x_{k-1}$ and respective cross section areas $A(z_1), A(z_2), \ldots, A(z_k)$. Then the Riemann sum

6.4
$$R(P) = \sum_{i=1}^{k} A(z_i)\,\Delta x_i$$

is an approximation of $V(S)$. The solid S of Figure 6.11 is approximated by the three cylinders shown in Figure 6.12.

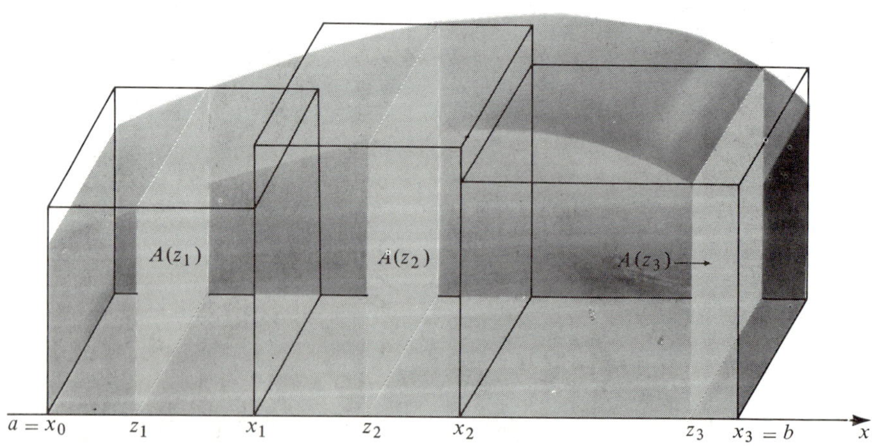

FIGURE 6.12

If we select a sequence of partitions $P_1, P_2, \ldots, P_n, \ldots$ of $[a,b]$ such that

$$\lim_{n\to\infty} \|P_n\| = 0,$$

and if $R(P_1), R(P_2), \ldots, R(P_n), \ldots$ is a corresponding sequence of Riemann sums of the area function A of type 6.4, then we can define the volume $V(S)$ of solid S to be

$$V(S) = \lim_{n\to\infty} R(P_n).$$

By 5.40, the limit above exists and equals $\int_a^b A(x)\,dx$. Thus the volume of S may be defined as follows.

6.5 Definition of Volume

Let S be a bounded solid and L a coordinate line such that S lies between planes drawn perpendicular to L at numbers a and b (Figure 6.11). For each x in $[a,b]$ let $A(x)$ be the area of the cross section of S in the plane drawn perpendicular to L at x. If the function A is continuous in $[a,b]$, then the volume $V(S)$ of S is given by

$$V(S) = \int_a^b A(x)\,dx.$$

Let us use this definition to find the volumes of some easily described solids.

Example 1 Find the volume of a sphere S of radius r.

Solution: We select coordinate axes as shown in Figure 6.13. For each number x in $[-r,r]$ the area $A(x)$ of the circular cross section of S perpendicular to the x

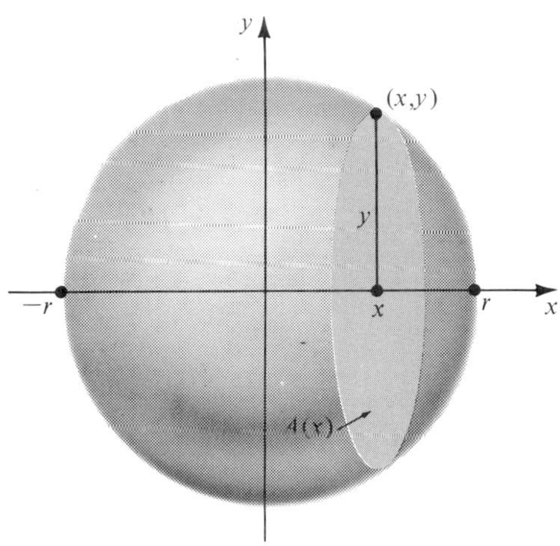

axis is given by $A(x) = \pi y^2$. However, $x^2 + y^2 = r^2$ for every point (x,y) on the circle in the plane of the axes. Thus

$$A(x) = \pi(r^2 - x^2) \qquad \text{for every } x \text{ in } [-r,r].$$

Evidently, A is a continuous function. Hence, by 6.5,

$$\begin{aligned} V(S) &= \int_{-r}^{r} \pi(r^2 - x^2)\,dx \\ &= \pi\left(r^2x - \frac{x^3}{3}\right)\Bigg|_{-r}^{r} = \pi\left(r^3 - \frac{r^3}{3}\right) - \pi\left(-r^3 + \frac{r^3}{3}\right) \\ &= \tfrac{4}{3}\pi r^3. \end{aligned}$$

This is the usual formula for the volume of a sphere.

Example 2 Find the volume of the solid generated by rotating about the x axis the region bounded by the line $x = 4$ and the parabola

$$y^2 = x.$$

Solution: This solid of revolution S, shaped somewhat like a headlight of a car, is shown in Figure 6.14. For each x in $[0,4]$, evidently

$$A(x) = \pi y^2 = \pi x.$$

Hence

$$V(S) = \int_0^4 \pi x\,dx = 8\pi.$$

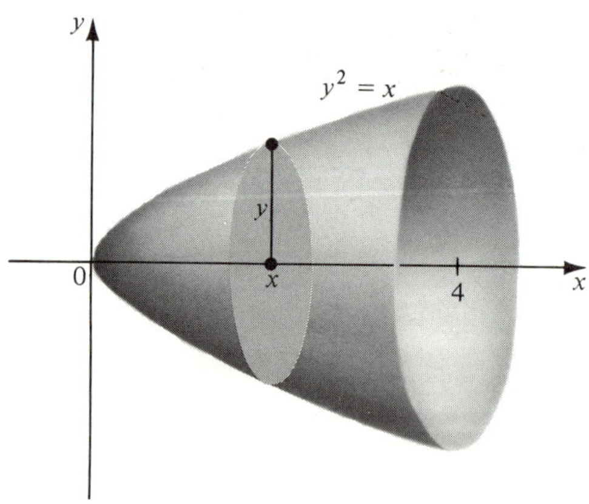

FIGURE 6.14

Example 3 Find the volume of the solid S of intersection of two right circular cylinders of radius r, assuming that their axes meet at right angles.

Solution: Each cross section of S in a plane parallel to both axes is a square. A quarter of this square and an eighth of solid S is shown in Figure 6.15. This

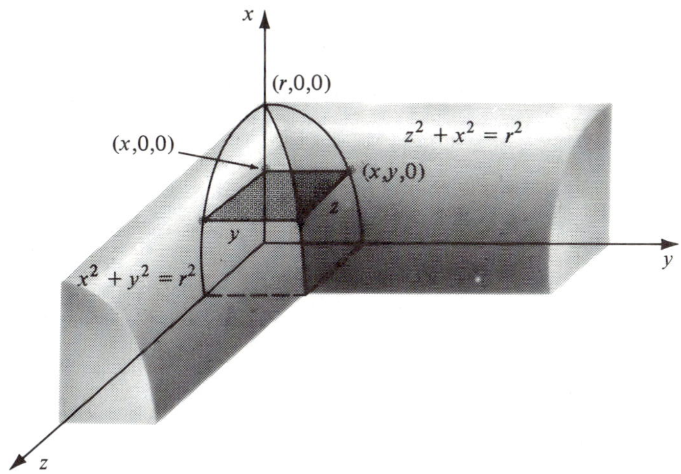

FIGURE 6.15

quarter-square, in the notation of the figure, has area $y^2 = yz = z^2 = r^2 - x^2$. Hence the cross-sectional area of solid S is given by

$$A(x) = 4(r^2 - x^2) \qquad \text{for each } x \text{ in } [-r,r].$$

By 6.5,

$$V(S) = \int_{-r}^{r} 4(r^2 - x^2)\,dx = 4\left(r^2x - \frac{x^3}{3}\right)\Bigg|_{-r}^{r} = \tfrac{16}{3}r^3.$$

EXERCISES

I

In each of Exercises 1 to 8 the graphs of the given equations bound a region of the plane. Find the volume of the solid obtained by rotating the region about the x axis. Sketch the solid.

1. $y = x^2; y = 0, x = 2$

2. $y = 2x^2; y = 0, x = 3$

3. $y = \sqrt{4 + x}; x = 0, y = 0$

4. $y = 4 - x^2; y = 0$

5. $y = \dfrac{1}{x}; x = 1, x = 3, y = 0$

6. $y = \sqrt{9 - x}; x = 0, y - 0$

7. $y = x^2 - x; y = 0$

8. $y = \dfrac{4}{x + 1}; x = -5, x = -2, y = 0$

In Exercises 9 and 10 find the volume of the solid obtained by rotating the indicated region about the y axis.

9. The region bounded by $y = x^2$, the y axis, and the line $y = 4$.
10. The region bounded by $y = x^3$, the x axis, and the line $x = 1$.

Find the volume of the solid of revolution when the given region in the figure is revolved about the stated line.

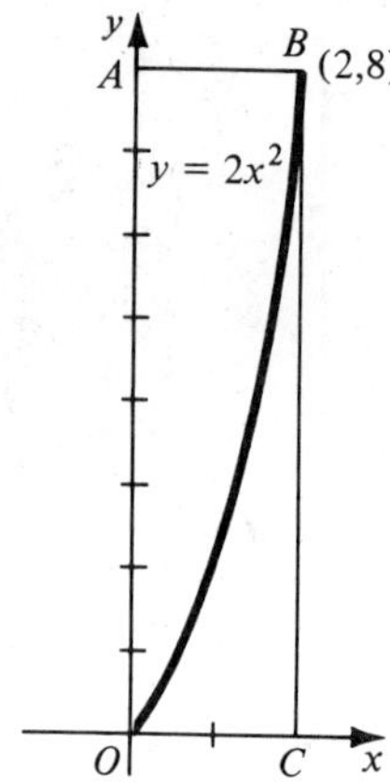

11. Region OBC, about the x axis.
12. Region OBA, about the x axis.
13. Region OBC, about the y axis.
14. Region OBA, about the y axis.
15. Region OBC, about the line BC.
16. Region OBC, about the line AB.
17. Region OAB, about the line BC.
18. Region OAB, about the line AB.
19. Find the volume of a frustum of a cone by the methods of this section. Take the radius of the upper base to be r, of the lower base to be R, and the altitude to be h.
20. Let a sphere of radius r be cut by a plane, thereby forming a segment of the sphere of height h. Prove that the volume of the segment is $\pi h^2(r - h/3)$.
21. Find the volume of a parabolic disc formed by rotating the region bounded by a parabola and its latus rectum about the latus rectum. Let p be the distance between the focus and the vertex of the parabola.
22. The base of a solid is a circle of radius r. All cross sections of the solid perpendicular to a fixed diameter of the base are squares. Find the volume of the solid.

II

1. A hole of radius R is drilled through the center of a sphere of radius r. Find the volume of the solid remaining.
2. The region bounded by a parabola and its latus rectum is rotated about a line through its vertex perpendicular to the axis. Find the volume of the solid generated. Let p be the distance between the focus and the vertex.

3 WORK

If a constant force of F lb is applied to an object in moving it a distance of d feet, then the *work* done on the object has magnitude W defined by

$$W = Fd.$$

If the unit of force is pounds and the unit of distance is feet, then the unit of work is foot-pounds. Other possible units of work are inch-pounds, foot-tons, and the like.

Example 1 An object of weight 110 lb is lifted (at a constant velocity) a distance of 23 ft. Find the amount of work done on the object.

Solution: Since a force of 110 lb is needed to lift the object, the amount of work done on the object in lifting it 23 ft is given by

$$W = 110 \cdot 23 = 2530 \text{ ft-lb}.$$

The calculus comes into play when we wish to define the work done on an object by a variable force. Let us assume that an object A is being moved along a coordinate line L (Figure 6.16), and a force of $F(x)$ units is being applied to A when A is at the point with coordinate x on L. We assume that the object A moves from a to b and that the force function F so defined is continuous in the interval $[a,b]$.

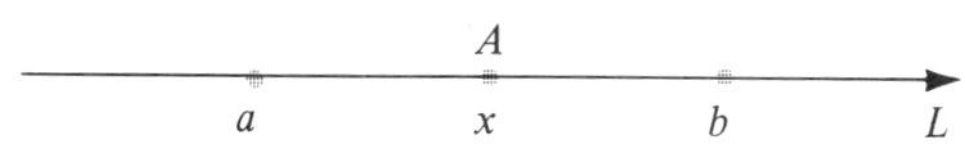

FIGURE 6.16

In order to define the amount of work done on the object A as it moves from a to b, we select a partition

$$P = \{x_0, x_1, \ldots, x_m\}$$

of $[a,b]$. In each subinterval $[x_{i-1}, x_i]$ of P, let $F(u_i)$ be the minimum and $F(v_i)$ be the maximum value of the force F. It is reasonable to assume that the amount W_i of work done on A as it moves from x_{i-1} to x_i is between the minimum value of the force times the distance Δx_i and the maximum value of the force times Δx_i:

(1) $$F(u_i)\,\Delta x_i \leq W_i \leq F(v_i)\,\Delta x_i.$$

Since the total amount W of work done on A as it moves from a to b is the sum of the W_i,

$$W = \sum_{i=1}^{m} W_i,$$

we obtain from (1) that

(2) $$\sum_{i=1}^{m} F(u_i)\,\Delta x_i \leq W \leq \sum_{i=1}^{m} F(v_i)\,\Delta x_i.$$

The inequality (2) holds for every partition P of $[a,b]$. If we take a sequence of partitions $P_1, P_2, \ldots, P_n, \ldots$ with norms of limit zero, then the limit of each sum in (2) as n approaches infinity is equal to

$$\int_a^b F(x)\,dx,$$

according to 5.40. Their common limit must be W. This leads us to the following definition.

The amount W of work done on an object in moving it from a to b along a coordinate line is given by

6.6
$$W = \int_a^b F(x)\,dx,$$

where $F(x)$ is the force applied to the object at position x.

Example 2 Find the amount of work done in stretching a spring from its natural length of 6 in. to double that length if a force of 20 lb is needed to hold the spring at double its natural length.

Solution: The force $F(x)$ required to hold a spring extended (within its elastic limit) x units beyond its natural length is given by

$$F(x) = kx, \qquad k \text{ a constant,}$$

according to *Hooke's law*. We are given that $F(6) = 20$; therefore

$$20 = k \cdot 6 \qquad \text{and} \qquad k = \tfrac{10}{3}.$$

Thus
$$F(x) = \tfrac{10}{3}x$$

for this particular spring.

The amount W of work done in stretching the spring from its natural length ($x = 0$) to double its natural length ($x = 6$) is given by

$$W = \int_0^6 \tfrac{10}{3}x\,dx = \tfrac{5}{3}x^2\Big|_0^6 = 60 \text{ in.-lb.}$$

If we wish to find the amount of work done in stretching this spring from a position already 2 in. extended to a position 4 in. extended, we evaluate the integral

$$\int_2^4 \tfrac{10}{3}x\,dx = 20 \text{ in.-lb,}$$

and so on.

Example 3 Find the amount W of work done in removing all the water at the top from a vertical cylindrical tank 4 ft in diameter and 6 ft high.

Solution: Let us think of the water as being pushed out of the tank by a piston starting out from the bottom of the tank (Figure 6.17). The force $F(x)$ on the piston after it has moved a distance of x feet is the weight of the water remaining in the tank, i.e.,

$$F(x) = \pi \cdot 2^2 \cdot (6 - x) \cdot k,$$

where $k = 62.5$ lb, the weight of a cubic foot of water. Hence

$$W = \int_0^6 4\pi k(6 - x)\,dx = 4\pi k\left(6x - \frac{x^2}{2}\right)\Bigg|_0^6$$

$$= 72\pi k \doteq 14{,}140 \text{ ft-lb}.$$

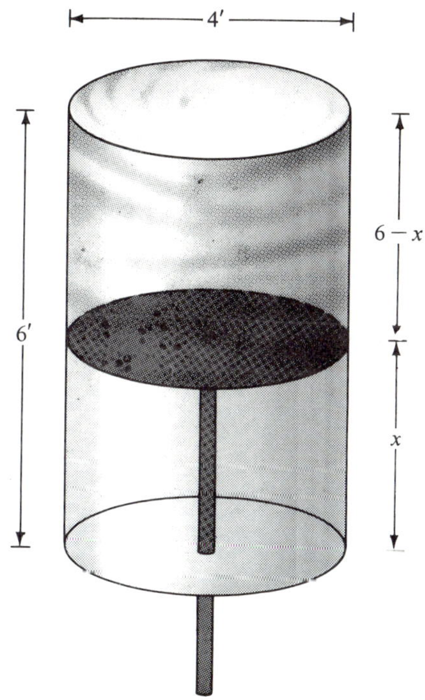

FIGURE 6.17

Example 4 Water is to be pumped out of the conical tank in Figure 6.18 to a point 10 ft above the top of the tank. Find the amount of work required to pump out 4 ft of water from the tank.

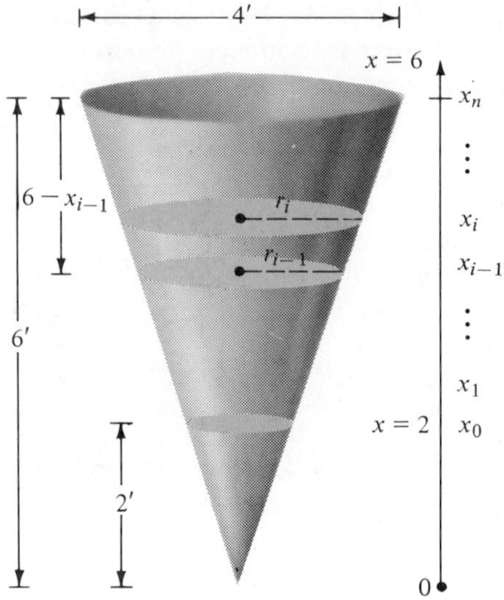

FIGURE 6.18

Solution: Choose a vertical axis with 0 at the bottom of the tank and 6 at the top. Since we only want to pump out 4 ft of water, we want to remove the water between $x = 2$ and $x = 6$. Select a partition $P = \{x_0, x_1, \ldots, x_n\}$ of the interval [2,6]. Suppose the water has been removed down to $x = x_i$. Then we imagine we remove the "slice" of water between $x = x_i$ and $x = x_{i-1}$. The volume of this slice is between $\pi r_{i-1}^2 \, \Delta x_i$ and $\pi r_i^2 \, \Delta x_i$; hence, its weight M_i satisfies the inequality

$$62.5\pi r_{i-1}^2 \, \Delta x_i \leq M_i \leq 62.5\pi r_i^2 \, \Delta x_i.$$

This slice must be lifted to the top of the tank and then 10 ft more. Thus it is lifted a distance $10 + (6 - x_{i-1}) = 16 - x_{i-1}$. Hence the work W_i done in emptying the water from this slice satisfies the inequality

$$62.5\pi r_{i-1}^2 \, \Delta x_i(16 - x_{i-1}) \leq W_i \leq 62.5\pi r_i^2 \, \Delta x_i(16 - x_{i-1}).$$

The total work W done in emptying the tank therefore satisfies the inequality

$$(1) \qquad \sum^{n} 62.5\pi r_{i-1}^2(16 - x_{i-1}) \, \Delta x_i \leq W_i \leq \sum_{i=1}^{n} 62.5\pi r_i^2(16 - x_{i-1}) \, \Delta x_i.$$

By similar triangles,

$$\frac{r_i}{x_i} = \frac{2}{6} \qquad \text{or} \qquad r_i = \frac{x_i}{3}.$$

Then the limit of each summation in (1) taken over a sequence of partitions of [2,6] with norms having limit zero is by 5.40

$$\int_2^6 62.5\pi \left(\frac{x}{3}\right)^2 (16 - x) \, dx.$$

Thus, the integral must be the work done,

$$W = \frac{62.5\pi}{9} \int_2^6 (16x^2 - x^3)\, dx$$

$$= \frac{62.5\pi}{9} \left(\frac{16}{3} x^3 - \frac{1}{4} x^4 \right) \Bigg|_2^6$$

$$= \frac{62.5 \cdot 789.3}{9} \pi \doteq 17{,}220 \text{ ft-lb.}$$

EXERCISES

1. Find the work done in stretching a spring from its natural length of 12 in. to a length of 18 in. if a force of 4 lb is needed to hold the spring extended 1 in.
2. A spring of natural length 5 in. requires a force of 9 oz to hold it at a length of 7 in. Find the work done in stretching the spring from a length of 7 in. to a length of 10 in.
3. A vertical cylindrical tank 6 ft in diameter and 10 ft high is half full of water. Find the amount of work done in pumping all the water out at the top of the tank.
4. A vertical cylindrical tank 6 ft in diameter and 10 ft high is full of water. Find the amount of work done in pumping half the water out at the top of the tank.
5. Any two electrons repel each other with a force inversely proportional to the square of the distance between them. If two electrons are held stationary at the points $(\pm 10,0)$ on the x axis, find the work done in moving a third electron:
   a. From $(8,0)$ to $(-2,0)$ along the x axis
   b. From $(7,0)$ to $(-7,0)$ along the x axis.
6. Water is being pumped into the conical tank of Example 4 (Figure 6.18) at the bottom. Find the amount of work required to fill the tank.
7. Water is being pumped into the cylindrical tank of Example 3 above (Figure 6.17) at the bottom. Find the amount of work required to fill the tank half full.
8. A tank has the shape of a paraboloid of revolution. The radius of the circular top is 4 ft and its depth is 10 ft. If the tank is full of water, find the work required to empty the water from the tank at a point 5 ft above the top of the tank. (*Hint:* If the parabola has its vertex at the origin, its equation is $y = kx^2$ for some determinable number k. Then "slice" the water into disks as in Example 4 above.)
9. Find the work done in pulling a 30 lb bucket filled with 75 lb of sand to the top of a well if the bucket is suspended at the end of a 100 ft chain which weighs 40 lb.

4 ARC LENGTH

We need to assume more than continuity of a function in finding the length of a curve.

6.7 Definition

A function f is called *smooth* in an interval I if f' exists and is continuous in I.

If f is a smooth function in a closed interval $[a,b]$, then it is possible to assign a meaningful length to the graph of f between $(a,f(a))$ and $(b,f(b))$. In order to describe how this is done, let us first simplify our notation. For each number x in the domain of f, let $P(x)$ denote the point $(x,f(x))$ on the graph of f. If $P(c)$ and $P(d)$ are two distinct points on the graph of f, let $\widehat{P(c)P(d)}$ denote the graph, or *arc*, between $P(c)$ and $P(d)$.

An obvious way to approximate the length of arc $\widehat{P(a)P(b)}$ is to inscribe a broken line in the arc and measure its length. Thus, if $p = \{x_0, x_1, \ldots, x_n\}$ is a partition of $[a,b]$, the broken line $P(x_0)P(x_1)\cdots P(x_n)$ made up of n line segments (indicated in Figure 6.19 for $n = 4$) is denoted by I_p and is called an

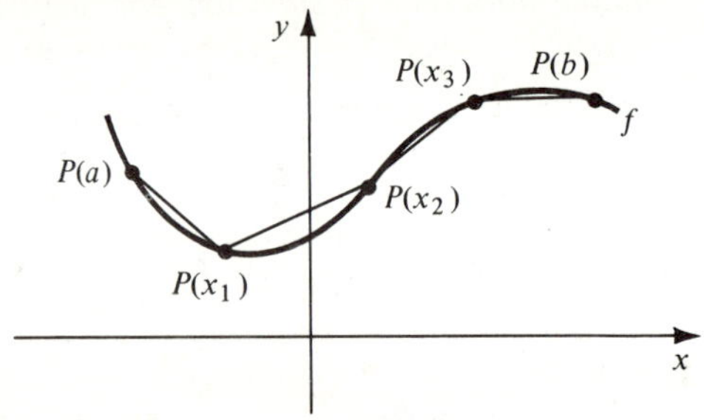

FIGURE 6.19

inscripture of arc $\widehat{P(a)P(b)}$. The length of I_p is denoted by $|I_p|$ and is clearly given by

$$|I_p| = \sum_{i=1}^{n} |P(x_{i-1})P(x_i)|.$$

Using the distance formula, this may be expressed in the form

$$|I_p| = \sum_{i=1}^{n} \sqrt{(x_i - x_{i-1})^2 + [f(x_i) - f(x_{i-1})]^2}. \tag{1}$$

We may put (1) in a more useful form with the aid of the mean value theorem. Thus, for each subinterval $[x_{i-1},x_i]$ of p,

$$f(x_i) - f(x_{i-1}) = (x_i - x_{i-1})f'(z_i)$$

for some number z_i in (x_{i-1},x_i). Letting $x_i - x_{i-1} = \Delta x_i$ as usual, we now obtain

$$|I_p| = \sum_{i=1}^{n} \sqrt{1 + f'^2(z_i)}\, \Delta x_i. \tag{2}$$

It is clear that $|I_p|$ is less than or equal to the number we wish to assign as the length of arc $\widehat{P(a)P(b)}$. Furthermore, we expect $|I_p|$ to be close to the length of this arc if $|p|$ is small. Therefore, if $p_1, p_2, \ldots, p_m, \ldots$ is any sequence of partitions of $[a,b]$ such that limit $\|p_m\| = 0$, we might reasonably expect

$$(3) \qquad \lim_{m \to \infty} |I_{p_m}|$$

to be called the length of arc $\widehat{P(a)P(b)}$, provided that this limit exists.

We can easily evaluate (3) once we observe that $|I_p|$ in (2) is a Riemann sum, not of the function f but of the function

$$\sqrt{1 + f'^2}.$$

This function is continuous because of our assumption that f is smooth. Hence, by 5.40,

$$(4) \qquad \lim_{m \to \infty} |I_{p_m}| = \int_a^b \sqrt{1 + f'^2(x)}\, dx.$$

This leads us to make the following definition.

6.8 Definition of Arc Length

If function f is smooth in the closed interval $[a,b]$, then the length of arc $\widehat{P(a)P(b)}$ of the graph of f is given by

$$L = \int_a^b \sqrt{1 + f'^2(x)}\, dx.$$

A further discussion of arc length is presented in Chapter 13.

Example 1 Find the arc length of the graph of the function

$$f(x) = x^{3/2}$$

over the interval [0,4].

Solution: Since $f'(x) = \frac{3}{2}x^{1/2}$, the arc length is given by

$$L - \int_0^4 \sqrt{1 + \tfrac{9}{4}x}\, dx.$$

We may evaluate this integral by making the change of variable

$$u = 1 + \tfrac{9}{4}x, \qquad du = \tfrac{9}{4}\, dx.$$

This gives

$$L = \tfrac{4}{9}\int_1^{10} \sqrt{u}\, du = \tfrac{4}{9} \cdot \tfrac{2}{3}u^{3/2}\Big|_1^{10} = \tfrac{8}{27}(10\sqrt{10} - 1).$$

Example 2 Find the length of an arc $\widehat{P(a)P(b)}$ of a parabola

$$x^2 = 4py.$$

Solution: We have

$$\frac{dy}{dx} = \frac{x}{2p}.$$

Therefore the arc length of the parabola joining the points $(a, a^2/4p)$ and $(b, b^2/4p)$ (assuming $a < b$) is given by

$$L = \int_a^b \sqrt{1 + \frac{x^2}{4p^2}}\,dx = \frac{1}{2p}\int_a^b \sqrt{4p^2 + x^2}\,dx.$$

We cannot evaluate this integral by means of the formulas available to us at this time. However, in Chapter 9 we shall develop methods for evaluating such integrals.

Example 3 Find the arc length of the graph of $9x^2 = (y^2 + 2)^3$ from the point $(\sqrt{3},1)$ to the point $(2\sqrt{6},2)$.

Solution: Here the graph is more conveniently expressed as the graph of the function

$$x = g(y) = \frac{1}{3}(y^2 + 2)^{3/2}.$$

With the roles of x and y interchanged, Definition 6.8 assumes the form

$$L = \int_c^d \sqrt{1 + g'^2(y)}\,dy,$$

where the limits of integration c and d now refer to y.

For the given function $g(y)$, we have $g'(y) = y(y^2 + 2)^{1/2}$ and hence

$$L = \int_1^2 \sqrt{1 + y^2(y^2 + 2)}\,dy = \int_1^2 \sqrt{1 + 2y^2 + y^4}\,dy$$

$$= \int_1^2 (1 + y^2)\,dy = \frac{10}{3}.$$

EXERCISES

I

Find the arc length of the graph of each of the following equations between the indicated points:

1. $y = 4 - 2\sqrt{x^3}$, $(0,4)$ to $(4,-12)$
2. $y = x^{2/3}$, $(1,1)$ to $(8,4)$
3. $y = \dfrac{x^3}{6} + \dfrac{1}{2x}$, $(1,\frac{2}{3})$ to $(3,\frac{14}{3})$
4. $y = \dfrac{x^6 + 2}{8x^2}$, $x = a$ to $x = b$, where $0 < a < b$
5. Find the length of arc of the semicubical parabola $5y^3 = x^2$ lying inside the circle $x^2 + y^2 = 6$.

6. $y = \frac{2}{3}(x-1)^{3/2}$, $(1,0)$ to $(5,\frac{16}{3})$
7. $y = \frac{1}{3}x^{3/2} - x^{1/2}$, $(1,-\frac{2}{3})$ to $(4,\frac{2}{3})$
8. $y = \frac{2}{3}(x^2+1)^{3/2}$, $(0,\frac{2}{3})$ to $\left(2, \frac{10\sqrt{5}}{3}\right)$
9. $8x = y^4 + 2y^{-2}$, $(\frac{3}{8},1)$ to $(\frac{33}{16},2)$

II

1. Sketch and find the length of the loop of the graph of $3ay^2 = x(x-a)^2$.
2. Sketch and find the length of the hypocycloid $x^{2/3} + y^{2/3} = a^{2/3}$.

5 APPROXIMATIONS BY THE TRAPEZOIDAL RULE

If no antiderivative of f is known, then we cannot evaluate

$$\int_a^b f(x)\,dx$$

by use of the fundamental theorem of the calculus. However, it is possible to approximate the value of this integral as closely as we desire. An obvious approximation method will be given in this section, and a subtler one will be developed in Section 6.

Let f be a continuous function in a closed interval $[a,b]$, and let $P_1, P_2, \ldots, P_n, \ldots$ be any sequence of partitions of $[a,b]$ such that

$$\lim_{n\to\infty} \|P_n\| = 0.$$

If $R(P_1), R(P_2), \ldots, R(P_n), \ldots$ is any sequence of Riemann sums associated with the given sequence of partitions, then, by 5.40,

$$\lim_{n\to\infty} R(P_n) = \int_a^b f(x)\,dx.$$

Hence for every number $\varepsilon > 0$ there exists a number k such that

$$R(P_n) \text{ is in } \left(\int_a^b f(x)\,dx - \varepsilon,\ \int_a^b f(x)\,dx + \varepsilon\right) \quad \text{for every } n \text{ in } (k,\infty).$$

That is, $R(P_n)$ is an approximation within ε of $\int_a^b f(x)\,dx$ for every n in (k,∞).

Such considerations lead to the following approximation formula.

6.9 Trapezoidal Rule

If f is a continuous function in an interval $[a,b]$ and $P_n = \{x_0, x_1, \ldots, x_n\}$ is the regular partition of $[a,b]$ into n subintervals, then

$$\int_a^b f(x)\,dx \doteq \frac{b-a}{2n}\left[f(x_0) + 2f(x_1) + 2f(x_2) + \cdots + 2f(x_{n-1}) + f(x_n)\right].$$

Proof: Two possible Riemann sums of f over $[a,b]$ are

$$R(P_n) = \sum_{i=1}^{n} f(x_{i-1})\,\Delta x, \qquad \bar{R}(P_n) = \sum_{i=1}^{n} f(x_i)\,\Delta x,$$

where $\Delta x = (b-a)/n$. We know that for a large enough integer n, both $R(P_n)$ and $\bar{R}(P_n)$ are close approximations of $\int_a^b f(x)\,dx$. It seems reasonable that the arithmetic average of $R(P_n)$ and $\bar{R}(P_n)$, $[R(P_n) + \bar{R}(P_n)]/2$, might often be a better approximation of $\int_a^b f(x)\,dx$ than either $R(P_n)$ or $\bar{R}(P_n)$. Now,

$$\begin{aligned}\tfrac{1}{2}[R(P_n) + \bar{R}(P_n)] &= \frac{\Delta x}{2}\left[\sum_{i=1}^{n} f(x_{i-1}) + \sum_{i=1}^{n} f(x_i)\right]\\ &= \frac{\Delta x}{2}\left[f(x_0) + \sum_{i=1}^{n-1} 2f(x_i) + f(x_n)\right].\end{aligned}$$

This is the approximation given.

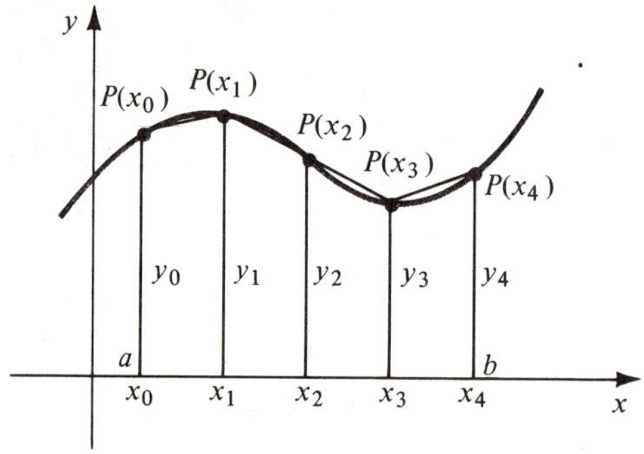

FIGURE 6.20

If the graph of f is above the x axis between $x = a$ and $x = b$, then the right side of 6.9 is the sum of the areas of n trapezoids, each of thickness Δx, as indicated in Figure 6.20. Thus, if $n = 4$ as in the figure, the sum S of the areas of the four trapezoids is given by

$$S = \frac{y_0 + y_1}{2}\,\Delta x + \frac{y_1 + y_2}{2}\,\Delta x + \frac{y_2 + y_3}{2}\,\Delta x + \frac{y_3 + y_4}{2}\,\Delta x,$$

where $y_i = f(x_i)$ and $\Delta x = (b - a)/4$. We may reduce S to the form

$$S = \frac{b - a}{8}(y_0 + 2y_1 + 2y_2 + 2y_3 + y_4),$$

which is the right side of 6.9 if $n = 4$.

Example Use the trapezoidal rule to approximate the area of the region under the graph of the equation

$$y = \sqrt{x^2 + 1}$$

between $x = 0$ and $x = 3$.

Solution: The area in question is given by the integral

$$\int_0^3 \sqrt{x^2 + 1}\, dx,$$

an integral that cannot be evaluated exactly by any of our previous methods.

If we let

$$f(x) = \sqrt{x^2 + 1} \qquad \text{and} \qquad n = 6$$

then the partition P_6 of $[0,3]$ is given by

$$P_6 = \{0,\tfrac{1}{2},1,\tfrac{3}{2},2,\tfrac{5}{2},3\}.$$

The following table of values of f is approximated to two decimal places of accuracy:

x	0	.5	1	1.5	2	2.5	3
$f(x)$	1	1.12	1.41	1.80	2.24	2.69	3.16

If we let $n = 6$ in 6.9, we get

$$\int_0^3 \sqrt{x^2 + 1}\, dx \doteq \tfrac{1}{4}[1 + 2.24 + 2.82 + 3.60 + 4.48 + 5.38 + 3.16],$$

or

$$\int_0^3 \sqrt{x^2 + 1}\, dx \doteq 5.67.$$

It may be shown that the correct result is 5.65, accurate to two decimal places.

EXERCISES

Use the trapezoidal rule to approximate each of the following integrals, taking the suggested value of n:

1. $\int_1^3 \frac{1}{x}\, dx,\ n = 8$

2. $\int_{-4}^{-1} \frac{1}{x}\, dx,\ n = 6$

3. $4\int_0^1 \sqrt{1 - x^2}\, dx,\ n = 4$

4. $\int_0^1 \frac{4}{1 + x^2}\, dx,\ n = 4$

5. $\int_{-1}^{3} \sqrt{4 + x^3}\, dx$, $n = 4$

6. $\int_{-4}^{-1} \frac{x}{3x + 1}\, dx$, $n = 6$

7. $\int_{2}^{4} \frac{x}{1 - x^3}\, dx$, $n = 4$

8. Sketch the graph of

$$x^2 + 16y^2 = 25,$$

and find the length of the arc in the first quadrant between (3,1) and (5,0), using the trapezoidal rule with $n = 4$.

6 APPROXIMATIONS BY SIMPSON'S RULE

Simpson's rule is as easy to apply as the trapezoidal rule and is generally more accurate.

6.10 Simpson's Rule

If f is a continuous function in an interval $[a,b]$, if n is an even integer, and if $P_n = \{x_0, x_1, \ldots, x_n\}$ is the regular partition of $[a,b]$ into n subintervals, then

$$\int_a^b f(x)\, dx \doteq \frac{b - a}{3n}[f(x_0) + 4f(x_1) + 2f(x_2) + 4f(x_3) + \cdots + 2f(x_{n-2}) + 4f(x_{n-1}) + f(x_n)].$$

This rule might well be called the *parabolic rule*, since the basic idea behind the rule is the approximating of the graph of f by arcs of parabolas. Before proving 6.10, let us make a few preliminary remarks on parabolas.

There is a unique curve with equation of the form

$$y = ax^2 + bx + c$$

passing through the three points $(-\Delta x, y_0)$, $(0, y_1)$, and $(\Delta x, y_2)$. That is, the three equations

$$\begin{aligned} y_0 &= a(-\Delta x)^2 + b(-\Delta x) + c, \\ y_1 &= c, \\ y_2 &= a(\Delta x)^2 + b\,\Delta x + c, \end{aligned} \tag{1}$$

have a unique solution for a, b, and c. The area under this curve from $-\Delta x$ to Δx (see Figure 6.21) is given by

$$\int_{-\Delta x}^{\Delta x} (ax^2 + bx + c)\, dx = \frac{ax^3}{3} + \frac{bx^2}{2} + cx \Big|_{-\Delta x}^{\Delta x}$$

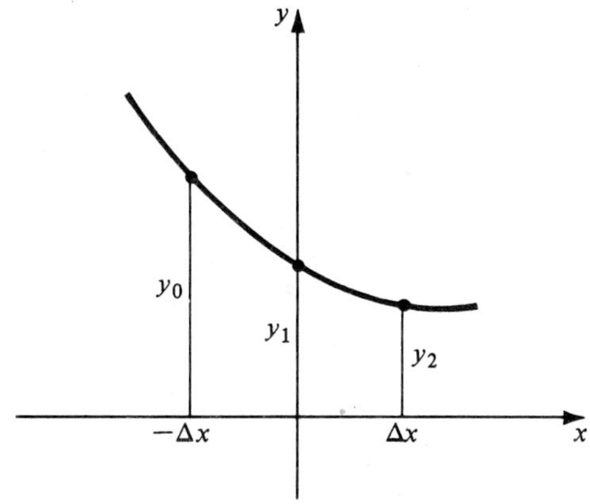

FIGURE 6.21

$$= \frac{\Delta x}{3}\left[2a(\Delta x)^2 + 6c\right].$$

From equations (1) it is easily verified that

$$y_0 + 4y_1 + y_2 = 2a(\Delta x)^2 + 6c.$$

Thus $$\int_{-\Delta x}^{\Delta x} (ax^2 + bx + c)\,dx = \frac{\Delta x}{3}(y_0 + 4y_1 + y_2)$$

gives the area under the parabola of Figure 6.21.

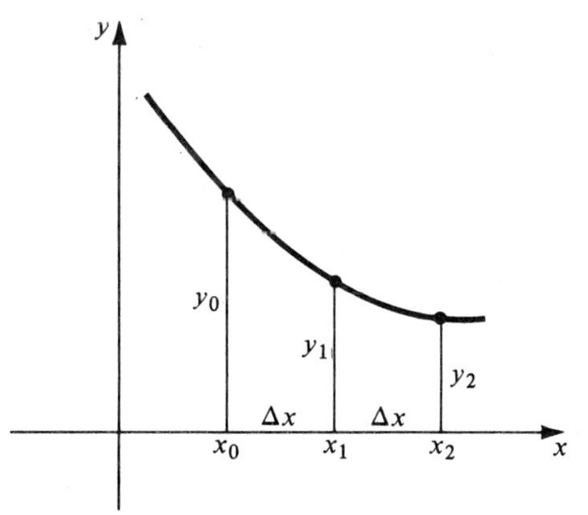

FIGURE 6.22

If a congruent parabola is located elsewhere in the plane, as in Figure 6.22, say a parabola of the form

$$y = ax^2 + bx + c$$

passing through the three points (x_0,y_0), (x_1,y_1), and (x_2,y_2), where $\Delta x = x_2 - x_1 = x_1 - x_0$, then it is clear that the area under the parabola is the same; i.e.,

$$\int_{x_0}^{x_2} (ax^2 + bx + c)\, dx = \frac{\Delta x}{3}(y_0 + 4y_1 + y_2).$$

We are now ready to prove Simpson's rule.

Proof: Let $\Delta x = (b - a)/n$, the norm of P_n. Since n is even, we may write

$$\int_a^b f(x)\, dx = \int_{x_0}^{x_2} f(x)\, dx + \int_{x_2}^{x_4} f(x)\, dx + \cdots + \int_{x_{n-4}}^{x_{n-2}} f(x)\, dx + \int_{x_{n-2}}^{x_n} f(x)\, dx.$$

Each of the integrals

$$\int_{x_i}^{x_{i+2}} f(x)\, dx$$

may be approximated by the area under a parabola through the three points $(x_i,f(x_i))$, $(x_{i+1},f(x_{i+1}))$, and $(x_{i+2},f(x_{i+2}))$. (See Figure 6.23.)

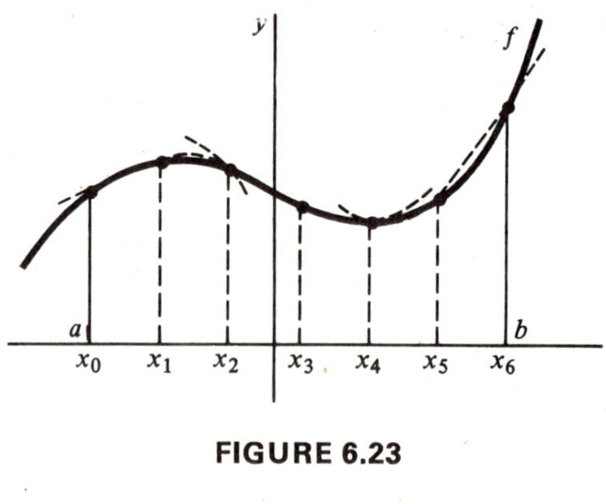

FIGURE 6.23

Thus by our remarks above we have

$$\int_{x_0}^{x_2} f(x)\, dx \doteq \frac{\Delta x}{3}[f(x_0) + 4f(x_1) + f(x_2)],$$

$$\int_{x_2}^{x_4} f(x)\,dx \doteq \frac{\Delta x}{3}[f(x_2) + 4f(x_3) + f(x_4)];$$

. . . .

$$\int_{x_{n-4}}^{x_{n-2}} f(x)\,dx \doteq \frac{\Delta x}{3}[f(x_{n-4}) + 4f(x_{n-3}) + f(x_{n-2})],$$

$$\int_{x_{n-2}}^{x_n} f(x)\,dx \doteq \frac{\Delta x}{3}[f(x_{n-2}) + 4f(x_{n-1}) + f(x_n)].$$

Adding these together, we obtain 6.10.

We shall not give the proof, but it may be shown that the error E incurred by the use of Simpson's rule satisfies the inequality

6.11 $$E \leq \frac{(b-a)^5 K}{180n^4},$$

where the number K is chosen so that $K \geq |f^{[4]}(x)|$ for every x in $[a,b]$.

Example 1 With $n = 4$, approximate $\int_0^1 \sqrt{1 - x^2}\,dx$ by Simpson's rule.

Solution: Clearly, $P_4 = \{0, \frac{1}{4}, \frac{1}{2}, \frac{3}{4}, 1\}$. We make the following table of values, accurate to three decimal places:

x	0	$\frac{1}{4}$	$\frac{1}{2}$	$\frac{3}{4}$	1
$\sqrt{1-x^2}$	1	.968	.866	.661	0

Then, by Simpson's rule,

$$\int_0^1 \sqrt{1-x^2}\,dx \doteq \tfrac{1}{12}[1 + 4(.968) + 2(.866) + 4(.661) + 0]$$
$$\doteq .771.$$

By Example 5, page 236, the given integral equals $\pi/4$. Hence $4(.771) = 3.084$ is an approximation of π. A better approximation could be obtained by letting $n = 8$ or a larger integer.

Example 2 With $n = 6$, approximate $\int_1^4 \frac{1}{x}\,dx$ by Simpson's rule.

Solution: Using the following table of values,

x	1	$\frac{3}{2}$	2	$\frac{5}{2}$	3	$\frac{7}{2}$	4
$1/x$	1	$\frac{2}{3}$	$\frac{1}{2}$	$\frac{2}{5}$	$\frac{1}{3}$	$\frac{2}{7}$	$\frac{1}{4}$

we have, by Simpson's rule,

$$\int_1^4 \frac{1}{x}\,dx \doteq \tfrac{3}{18}\left(1 + 4\cdot\tfrac{2}{3} + 2\cdot\tfrac{1}{2} + 4\cdot\tfrac{2}{5} + 2\cdot\tfrac{1}{3} + 4\cdot\tfrac{2}{7} + \tfrac{1}{4}\right)$$
$$\doteq 1.388.$$

We shall find the antiderivative of this function in Chapter 7. It can be shown that our approximation above is within .002 of the value of the integral.

Example 3 Approximate the arc length of the graph of

$$y = \frac{1}{x}$$

between the points (1,1) and $(5,\frac{1}{5})$.

Solution: Since $dy/dx = -1/x^2$, evidently (6.8)

$$L = \int_1^5 \sqrt{1 + \frac{1}{x^4}}\,dx = \int_1^5 \frac{\sqrt{1 + x^4}}{x^2}\,dx.$$

Using the following table of values, we may approximate L by Simpson's rule.

x	1	2	3	4	5
$\sqrt{1+x^4}/x^2$	1.414	1.031	1.006	1.002	1.001

Thus

$$L \doteq \tfrac{1}{3}[1.414 + 4.124 + 2.012 + 4.008 + 1.001],$$

or $L \doteq 4.186$. This answer seems reasonable, since the chord joining (1,1) and $(5,\frac{1}{5})$ has length approximately equal to 4.08.

EXERCISES

I

In each of Exercises 1 to 6, use Simpson's rule to approximate the integral.

1. Exercise 1, Section 5, $n = 8$

2. Exercise 2, Section 5, $n = 6$

3. $\int_1^{10} \frac{dx}{x}$, $n = 10$

4. $\int_0^1 \sqrt{1 + x^4}\,dx$, $n = 10$

5. Exercise 4, Section 5 with $n = 4$ (the exact answer is π which will be shown subsequently and $\doteq 3.14159265$ to 8 decimal places).

6. Redo Exercise 5, using $n = 10$.

7. Redo Exercise 5, using the trapezoidal rule and $n = 10$.

8. Approximate the area of the region bounded by the hypocycloid $x^{2/3} + y^{2/3} = a^{2/3}$ by using Simpson's rule with $n = 6$ for one quarter of the region. (*Hint:*

First show that the area is proportional to a^2 and then approximate the constant of proportionality.)

9. Approximate the arc length of the graph of the equation $5y = \sqrt{x^5}$ between $x = 0$ and $x = 1$ by using Simpson's rule with $n = 4$.

II

1. Show that the approximation

$$\int_a^b F(x)\,dx \doteq \frac{b-a}{6}\left[F(a) + 4F\left(\frac{a+b}{2}\right) + F(b)\right],$$

which is Simpson's rule for $n = 2$, is exact for $F(x)$ a cubic polynomial.

CHAPTER REVIEW

In each of Exercises 1 to 4 find the area of the region bounded by the graphs of the given equations. Sketch.

1. $y = x^2 - 4x + 1,\ x + y = 5$
2. $y = 3 - 2x - x^2,\ x + y = 1$
3. $y^2(4 - x^2) = x^2,\ x = 1$
4. $y = (x - 1)/\sqrt{x^2 - 2x + 4},\ x = 0,\ y = 0$

In each of Exercises 5 to 10, find the volume of the solid generated by revolving the region with given boundaries about the given line.

5. The region with boundaries $x^2 = 4py$ and $y = b$ (p and b positive) about the x axis.
6. The region in Exercise 5 about the line $y = b$.
7. The region with boundaries $y = \sqrt{x}$, $y = 4$, and the y axis about the line $x = 16$.
8. The region with boundaries $x = y^3$ and $y = x^3$, in the first quadrant, about the line $y = 1$.
9. The region with boundaries $x^2 = 4py$, $y = 0$, and $x = a$ (p and a positive) about the x axis.
10. The region in Exercise 9 about the line $x = a$.
11. Do Exercise 22, Section 2 of this chapter, considering all the cross sections regular pentagons instead of squares.
12. The base of a solid is a segment of a parabola cut off by a chord perpendicular to its axis. The chord has length $2L$ and is at a distance of L from the vertex. Find the volume of the solid if every cross section perpendicular to the axis of the parabola is (a) a semicircle, (b) an equilateral triangle.
13. Find the volume of the solid generated by revolving the region bounded by the hypocycloid $x^{2/3} + y^{2/3} = a^{2/3}$ about the y axis; about the x axis.

14. A parabolic mirror, hollowed out of a cylindrical piece of glass of radius 4 in. and height 2 in., is 1 in. thick at the center. Find the volume of the mirror and the position of the focal point.

15. Find the approximate length of the curve $y^2 = x^5/25$ from $x = 0$ to $x = 2$, using Simpson's rule, with $n = 4$. Also using the trapezoidal rule with $n = 4$.

16. Find the approximate length of the curve $y = x^3/3$ from $(0,0)$ to $(1,\frac{1}{3})$, using Simpson's rule. Also approximate the length using the trapezoidal rule.

17. Find the length of the curve $y = (\frac{4}{9}x^2 + 1)^{3/2}$ between $(0,1)$ and $(2,\frac{125}{27})$.

18. How much work is required to raise a 500 lb safe 200 ft up a side of a building using a nylon rope which weighs $\frac{1}{2}$ lb/ft?

19. How much work is required to pump all the water out of a swimming pool to level 5 ft above the top of the pool, if the pool is 30 ft long, 15 ft wide, and 8 ft deep?

20. Find the length of one loop of the curve $y^2 = x^2(1 - x)$.

7

Exponential and Logarithmic Functions

Although the theory of the derivative and the integral developed in the previous chapters holds for many functions, the examples illustrating the theory have been drawn from the set of algebraic functions.

We shall enlarge our set of functions in this and the next chapter to include the elementary transcendental functions. Thus our new set of functions will include the algebraic, exponential, logarithmic, trigonometric, and inverse trigonometric functions.

1 INTUITIVE APPROACH

The function f defined by

$$f(x) = 2^x, \qquad \text{domain } f = (-\infty,\infty),$$

is called the *exponential function* with base 2. Thus,

$$f(-5) = 2^{-5} = \tfrac{1}{32},$$
$$f(0) = 2^0 = 1,$$
$$f(\tfrac{3}{2}) = 2^{3/2} = 2\sqrt{2},$$

and so on. While it might not be clear what

$$f(\sqrt{3}) = 2^{\sqrt{3}}$$

means, we assume that this is a definite number.

7.1 Definition

For any positive number a, $a \neq 1$, the *exponential function* with base a is defined by

$$f(x) = a^x, \qquad \text{domain} f = (-\infty,\infty).$$

We shall assume that the usual laws of exponents hold. Thus, for all numbers x and y,

$$a^x a^y = a^{x+y},$$

$$\frac{a^x}{a^y} = a^{x-y},$$

$$(a^x)^y = a^{xy}.$$

Does the exponential function

$$f(x) = 2^x$$

have a derivative? By definition, for any number x

$$\begin{aligned} f'(x) &= \operatorname*{limit}_{h\to 0} \frac{f(x+h) - f(x)}{h} \\ &= \operatorname*{limit}_{h\to 0} \frac{2^{x+h} - 2^x}{h} \\ &= \operatorname*{limit}_{h\to 0} \frac{2^x 2^h - 2^x}{h} \\ &= \operatorname*{limit}_{h\to 0} 2^x \left(\frac{2^h - 1}{h}\right). \end{aligned}$$

Since 2^x does not depend on h, we may take 2^x outside the limit:

(1) $$f'(x) = 2^x \operatorname*{limit}_{h\to 0} \frac{2^h - 1}{h}.$$

We conclude that the exponential function $f(x) = 2^x$ does have a derivative provided the limit

(2) $$\operatorname*{limit}_{h\to 0} \frac{2^h - 1}{h}$$

exists. Observe that this limit, if its exists, is simply some number k; that is, the limit does not depend on x.

Using a hand computer having an x^y key, we were able to construct the following table:

h	1	-1	$\frac{1}{2}$	$-\frac{1}{2}$	0.1	-0.1	0.01	-0.01	0.001	-0.001
$\frac{2^h - 1}{h}$	1	$\frac{1}{2}$	0.83	0.59	0.72	0.67	0.696	0.691	0.693	0.693

It certainly appears that the limit in (2) exists and

$$\lim_{h\to 0} \frac{2^h - 1}{h} = k, \qquad \text{where } k \doteq 0.693.$$

Thus, by (1)

$$D_x 2^x = k \cdot 2^x, \qquad \text{where } k \doteq 0.693.$$

Going through the same procedure for the exponential function $f(x) = 3^x$, we find

$$D_x 3^x = k \cdot 3^x, \qquad \text{where } k = \lim_{h\to 0} \frac{3^h - 1}{h} \doteq 1.099.$$

Using any positive number $a \neq 1$ as a base,

7.2 $$D_x a^x = k \cdot a^x$$

where k only depends on a,

7.3 $$k = \lim_{h\to 0} \frac{a^h - 1}{h}.$$

The function

$$f(x) = 2^x$$

is strictly increasing, as indicated by its graph (Figure 7.1). Thus, f is a 1–1 function (see Chapter 1, Section 5). As such, f has an inverse g defined by

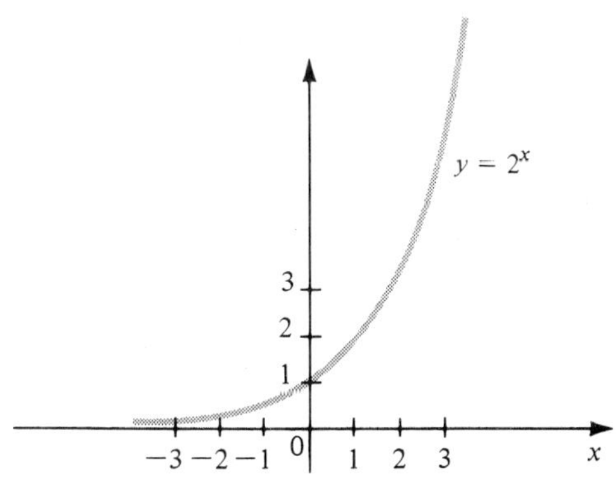

FIGURE 7.1

$$g(x) = y \qquad \text{if and only if } f(y) = x.$$

The exponential function

$$f(x) = a^x$$

is also strictly increasing if $a > 1$ (and strictly decreasing if $0 < a < 1$) and our remarks above hold.

7.4 Definition

The inverse of the exponential function $f(x) = a^x$ is called the *logarithm* to the base a, abbreviated $\log_a$. Thus,

$$\log_a x = y \qquad \text{if and only if } a^y = x,$$

$$\text{domain } \log_a = (0,\infty).$$

For example,

$$\log_2 8 = 3 \qquad \text{because} \qquad 2^3 = 8$$

$$\log_2 \tfrac{1}{4} = -2 \qquad \text{because} \qquad 2^{-2} = \tfrac{1}{4}$$

$$\log_2 \sqrt{2} = \tfrac{1}{2} \qquad \text{because} \qquad 2^{1/2} = \sqrt{2}$$

$$\log_2 1 = 0 \qquad \text{because} \qquad 2^0 = 1.$$

From the definition,

7.5 $$\log_a a^y = y \qquad \text{for all } y,$$

7.6 $$a^{\log_a x} = x \qquad \text{for all } x > 0.$$

For example,

$$\log_2 2^7 = 7,$$

$$2^{\log_2 .3} = .3.$$

What is the derivative of the logarithm function? We can find out by using Theorem 3.19, which relates the derivative of a function and its inverse. It says that if f and g are inverse functions, then

$$g'(x) = 1/f'(g(x)), \qquad \text{if } f'(g(x)) \neq 0.$$

Letting $g(x) = \log_a x$ and $f(x) = a^x$, and knowing that $f'(x) = ka^x$ for some constant k by 7.2, we have

$$g'(x) = \frac{1}{ka^{g(x)}} = \frac{1}{ka^{\log_a x}} = \frac{1}{kx}.$$

Thus, if the logarithm function has a derivative, it is given by

7.7 $$D_x \log_a x = \frac{1}{kx} \qquad \text{for all } x > 0,$$

where k is defined by 7.3.

Equation 7.7 says that the function $\log_a$ is an *antiderivative* of the continuous function

$$F(x) = \frac{1}{kx}, \qquad x > 0.$$

Hence, by the fundamental theorem of the calculus,

$$\int_1^x F(t)\,dt = \log_a t\Big|_1^x = \log_a x - \log_a 1.$$

Since $\log_a 1 = 0$, this yields

7.8 $$\log_a x = \frac{1}{k}\int_1^x \frac{1}{t}\,dt \qquad \begin{array}{l}(k \text{ defined by 7.3}),\\ \text{for all } x > 0.\end{array}$$

Equation 7.8 gives us a clue as to how we might *define* the logarithm function independently of the exponential function. The rational function $F(t) = 1/t$ is continuous in $(0,\infty)$, and therefore, the integral in 7.8 exists for all x in $(0,\infty)$. Hence, we can *define* $\log_a x$ by 7.8, and the resulting function $\log_a$ is differentiable in its domain $(0,\infty)$. Its derivative is given by 7.7. This approach to the logarithm function is used in the next section.

2 THE NATURAL LOGARITHM

You recall that the number k in 7.8 is a function of a:

$$k = \lim_{h\to 0} \frac{a^h - 1}{h}.$$

We saw when $a = 2$, $k \doteq 0.693$; and when $a = 3$, $k \doteq 1.099$. Therefore, it seems reasonable that $k = 1$ for some base a between 2 and 3. We call the function defined by 7.8 with $k = 1$ the natural logarithm.

7.9 Definition of the Natural Logarithm

The natural logarithm function, denoted by ln, is defined by

$$\ln x = \int_1^x \frac{1}{t}\,dt, \qquad \text{domain ln} = (0,\infty).$$

We can derive properties of the function ln by using properties of the integral. Since the integrand $1/t$ is positive,

$$\int_1^x \frac{1}{t}\,dt < 0 \qquad \text{if } 0 < x < 1;$$

$$\int_1^x \frac{1}{t}\,dt = 0 \qquad \text{if } x = 1;$$

$$\int_1^x \frac{1}{t}\,dt > 0 \qquad \text{if } x > 1.$$

Thus,

$$\ln x < 0 \qquad \text{if } 0 < x < 1;$$

$$\ln 1 = 0;$$

$$\ln x > 0 \qquad \text{if } x > 1.$$

Geometrically, if $x > 1$, then $\ln x$ is the area of the region R in Figure 7.2. If $0 < x < 1$, then $-\ln x$ is the area of region R in Figure 7.3.

We can easily approximate $\ln x$ by inscribing rectangles in region R or circumscribing rectangles about region R in Figures 7.2 or 7.3. The case $\ln 2$ is illustrated in the enlarged Figure 7.4. The two inscribed rectangles have total area $\frac{7}{12}$, the two circumscribed $\frac{5}{6}$. Therefore

$$\tfrac{7}{12} < \ln 2 < \tfrac{5}{6}.$$

A more accurate estimate is that $\ln 2 \doteq .69$.

By the fundamental theorem of the calculus (5.20),

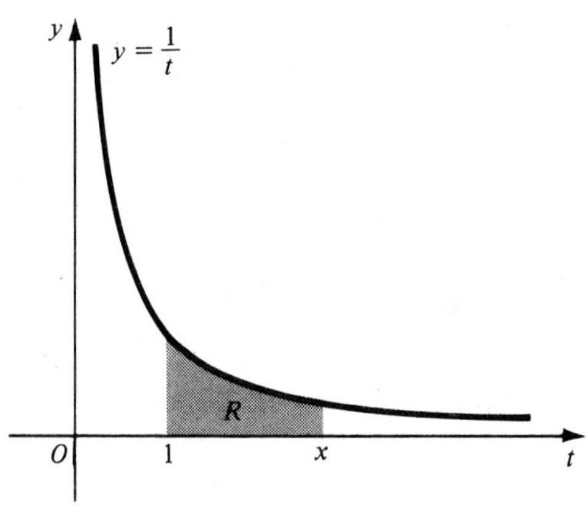

FIGURE 7.2

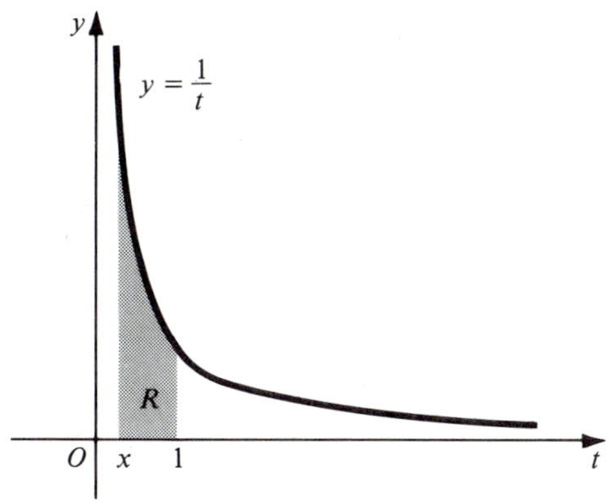

FIGURE 7.3

7.10 $$D_x \ln x = \frac{1}{x} \quad \text{for all } x \text{ in } (0,\infty).$$

Since $1/x > 0$ in $(0,\infty)$, we have by 4.11 that the function ln is *strictly increasing* in its domain $(0,\infty)$. The second derivative of ln is as follows:

7.11 $$D_x^2 \ln x = -\frac{1}{x^2} \quad \text{for all } x \text{ in } (0,\infty).$$

Thus, $D_x^2 \ln x < 0$ and the graph of ln is concave downward at every point $(x, \ln x)$ by 4.16. A rough sketch of the graph of ln is given in Figure 7.5.

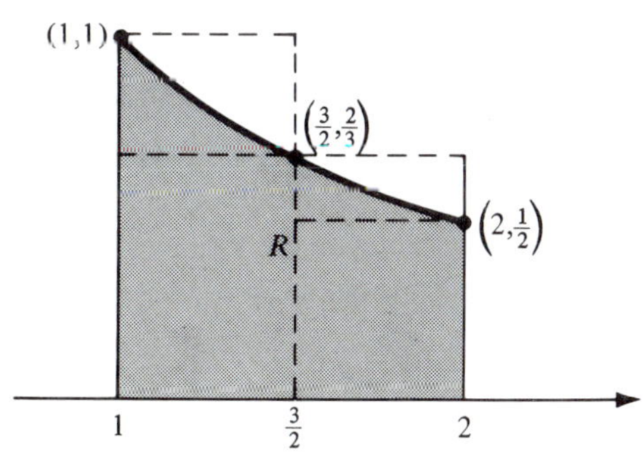

FIGURE 7.4

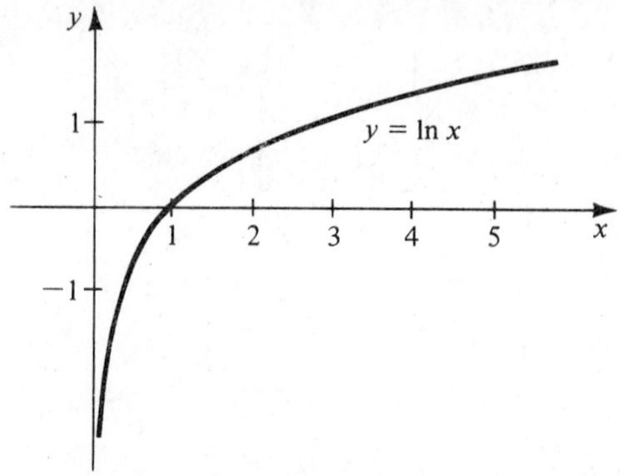

FIGURE 7.5

Some basic properties of ln are as follows: for all $a,b \in (0,\infty)$ and $r \in (-\infty,\infty)$,

7.12 $$\ln(ab) = \ln a + \ln b$$

7.13 $$\ln \frac{a}{b} = \ln a - \ln b$$

7.14 $$\ln a^r = r \ln a.$$

To prove 7.12, observe that if

$$f(x) = \ln(ax)$$

then

$$f'(x) = \frac{1}{ax} D_x(ax) = \frac{1}{ax} \cdot a = \frac{1}{x}.$$

Thus, f and ln have the same derivative. This means $f(x) = \ln x + c$ for some constant c by 4.27; that is

$$\ln (ax) = \ln x + c.$$

Letting $x = 1$, $\ln a = \ln 1 + c = c$. Therefore,

$$\ln (ax) = \ln a + \ln x.$$

This proves 7.12 if we let $x = b$.

The proof of 7.13 is similar, once we observe

$$\ln \frac{1}{b} = -\ln b.$$

For if $c = 1/b$, then $1 = bc$ and $\ln 1 = \ln (bc)$, $0 = \ln b + \ln c$, and $\ln c = -\ln b$.

The proof of 7.14 is more involved. Of course, by 7.12

$$\ln a^2 = \ln a + \ln a = 2 \ln a,$$

$$\ln a^3 = \ln (a^2 \cdot a) = \ln a^2 + \ln a = 3 \ln a,$$

and so on, to give,

$$\ln a^n = n \ln a \qquad \text{for every positive integer } n.$$

If $b = \sqrt[n]{a} = a^{1/n}$, then $a = b^n$ and $\ln a = \ln b^n = n \ln b$. Thus,

$$\ln b = \frac{1}{n} \ln a \quad \text{and} \quad \ln a^{1/n} = \frac{1}{n} \ln a \qquad \text{for every positive integer } n.$$

Now it is easy to see that for every positive rational number m/n, where m and n are positive integers,

$$\ln a^{m/n} = \ln (a^m)^{1/n} = \frac{1}{n} \ln a^m = \frac{1}{n} (m \ln a)$$

$$= \frac{m}{n} \ln a.$$

We omit the remainder of the proof in case r is negative or irrational.

The equations 7.12–7.14 are often called the *laws of logarithms*.

What is the range of ln? By Figure 7.4, $\ln 2 > \frac{1}{2}$ and hence $\ln 4 = \ln 2^2 > 1$. Therefore, by 7.14,

$$\ln 4^r = r \ln 4 > r \qquad \text{for every number } r > 0.$$

Thus, $\ln x$ is large when x is large. Similarly, $\ln 4^{-r} < -r$ for every number $r > 0$; and hence $\ln x$ is a large negative number when x is a small positive number. These remarks indicate what is actually the case:

$$\text{range } \ln = (-\infty, \infty).$$

This fact means that for every number y there exists some x such that $\ln x = y$. The number x is unique since ln is a strictly increasing function. The unique number whose natural logarithm is 1 is designated by e.

7.15 e is the unique number such that $\ln e = 1$.

Since $\ln 2 < 1 < \ln 4$, e is between 2 and 4. A better approximation, to be proved later, is

$$e \doteq 2.71828.$$

Other than roots of rational numbers, e and π are the two most important irrational numbers of mathematics.

The laws of logarithms together with a table of logarithms allowed early astronomers to reduce complicated calculations to a manageable form.

Example 1 Given that $\ln 2 \doteq .69$ and $\ln 7 \doteq 1.95$, find (i) $\ln 28$, (ii) $\ln 3.5$, and (iii) $\ln 98$.

Solution:

(i)
$$\begin{aligned}\ln 28 &= \ln 2^2 \cdot 7\\ &= \ln 2^2 + \ln 7\\ &= 2 \ln 2 + \ln 7\\ &\doteq 2 \cdot (.69) + 1.95 \doteq 3.33.\end{aligned}$$

(ii)
$$\begin{aligned}\ln 3.5 &= \ln \tfrac{7}{2}\\ &= \ln 7 - \ln 2\\ &\doteq 1.95 - .69 \doteq 1.26.\end{aligned}$$

(iii)
$$\begin{aligned}\ln 98 &= \ln 2 \cdot 7^2\\ &= \ln 2 + \ln 7^2\\ &= \ln 2 + 2 \ln 7\\ &\doteq .69 + 2 \cdot (1.95) \doteq 4.59.\end{aligned}$$

Example 2 Solve for x: (i) $\ln (2x + 1) = 2$ and (ii) $\ln (x - 1) < 1$.

Solution: (i) $\ln e^2 = 2 = \ln (2x + 1)$, so $e^2 = 2x + 1$, and $x = (e^2 - 1)/2$.

(ii) $\ln (x - 1) < 1 = \ln e$, so $x - 1 < e$ since ln is a strictly increasing function. Thus,

$$x < 1 + e.$$

EXERCISES

1. Make a table of values for $\ln x$ with x taking on the following values:
$$e, e^2, e^3, e^4, 1/e, 1/e^2, 1/e^3, \sqrt{e}, \sqrt[3]{e}.$$
2. Make a table of values for $\ln x$ with x taking on the following values:
$$10, 4, \tfrac{5}{2}, .5, .2, .1, .25, .01, .001, .0001,$$
given that $\ln 2 \doteq .693$, $\ln 5 \doteq 1.609$. [*Hint:* $\ln 20 = \ln (2^2 \cdot 5) = \ln 2^2 + \ln 5 = 2 \ln 2 + \ln 5 \doteq 2.995$, etc.]
3. Solve each of the following equations:
   - **a.** $\ln x = 0$
   - **b.** $\ln x = 1$
   - **c.** $\ln x = -1$
   - **d.** $\ln x = -3$
   - **e.** $\ln (x - 2) = 3$
   - **f.** $\ln (5 - x^2) = -2$
4. Solve each of the following inequalities:
   - **a.** $\ln x > 2$
   - **b.** $\ln x < -3$
   - **c.** $\ln |x| < -1$
   - **d.** $\ln (x + 2) < 0$

In each of Exercises 5 to 8 sketch the graph of the equation.

5. $y = \ln (-x)$, $x < 0$
6. $y = \ln |x|$, $x \neq 0$
7. $y = \ln (1 + x)$, $x > -1$
8. $y = \ln (1 - x)$, $x < 1$
9. Prove that there exists a unique number u such that $\ln u + u = 0$.
10. Find $D_x^n \ln x$ for any positive integer n.

11. If $x < 0$, express $\int_{-1}^{x} \frac{1}{t}\,dt$ in terms of the ln function.

12. If $x > -1$, express $\int_{0}^{x} \frac{1}{1+t}\,dt$ in terms of the ln function.

3 THE NATURAL EXPONENTIAL FUNCTION

The inverse of the natural logarithm function is the natural exponential function.

7.16 Definition of exp

The *natural exponential function* is denoted by exp and defined by

$$\exp x = e^x, \qquad \text{domain exp} = (-\infty,\infty).$$

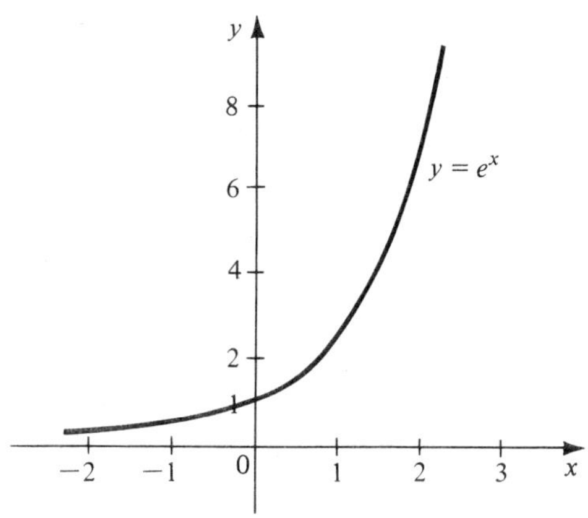

FIGURE 7.6

Using the approximation $e \doteq 2.7$, the graph of exp is sketched in Figure 7.6 from the following table of values.

x	0	1	2	3	−1	−2
$\exp x$	1	2.7	7.3	19.7	.37	.14

The familiar laws of exponents can be derived from the laws of logarithms.

7.17 Laws of Exponents

(i) $a^r a^s = a^{r+s}$ (ii) $\dfrac{a^r}{a^s} = a^{r-s}$

(iii) $a^r b^r = (ab)^r$ (iv) $\dfrac{a^r}{b^r} = \left(\dfrac{a}{b}\right)^r$

(v) $(a^r)^s = a^{rs}$

Proof: (i) It is enough to show that

$$\ln (a^r a^s) = \ln a^{r+s},$$

because ln is a 1–1 function. This is done as follows.

$$\begin{aligned} \ln (a^r a^s) &= \ln a^r + \ln a^s \\ &= r \ln a + s \ln a \\ &= (r + s) \ln a \\ &= \ln a^{(r+s)}. \end{aligned}$$

The others are proved similarly.

Knowing

$$D_x \ln x = \frac{1}{x},$$

we can find the derivative of its inverse function, exp, by 3.19:

$$\begin{aligned} D_x \exp x &= \frac{1}{f'(\exp x)}, \qquad \text{where } f(x) = \ln x, f'(x) = \frac{1}{x} \\ &= \frac{1}{1/(\exp x)} = \exp x. \end{aligned}$$

In exponential notation, we have proved:

7.18 Theorem

$$D_x e^x = e^x.$$

Thus, we have a function, namely exp, which is its own derivative.

Example

Sketch the graph of

$$f(x) = x^2 e^x,$$

showing maximum points, minimum points, and points of inflection.

Solution: We have

$$\begin{aligned} f'(x) &= 2xe^x + x^2 e^x = e^x \cdot x \cdot (2 + x) \\ f''(x) &= (2e^x + 2xe^x) + (2xe^x + x^2 e^x) \\ &= e^x(x^2 + 4x + 2). \end{aligned}$$

The critical values are $x = 0$ and $x = -2$ (e^x is never 0). Since $f''(0) > 0$ and $f''(-2) < 0$, (0,0) is a minimum point and $(-2, 4/e^2)$ is a maximum point. Setting $f''(x) = 0$, we get

$$x^2 + 4x + 2 = 0.$$

$$x = \frac{-4 \pm \sqrt{16 - 8}}{2} = -2 \pm \sqrt{2}.$$

Thus, there are two critical points occurring at $x = -2 + \sqrt{2}$ and $x = -2 - \sqrt{2}$. The graph is sketched in Figure 7.7.

One of the famous limits in mathematics is

$$\operatorname*{limit}_{h \to 0} (1 + h)^{1/h} = e.$$

We are able to prove this easily as follows.

$$\begin{aligned}(1 + h)^{1/h} &= \exp [\ln (1 + h)^{1/h}] \\ &= \exp \left[\frac{1}{h} \ln (1 + h)\right] \\ &= \exp \left[\frac{\ln (1 + h) - \ln 1}{h}\right].\end{aligned}$$

Since

$$\operatorname*{limit}_{h \to 0} \frac{\ln (1 + h) - \ln 1}{h} = \ln' (1),$$

we have, by continuity of the function exp,

$$\operatorname*{limit}_{h \to 0} (1 + h)^{1/h} = \exp [\ln' (1)] = \exp 1 = e.$$

Replacing h by $1/n$, where n is a positive integer, we also have

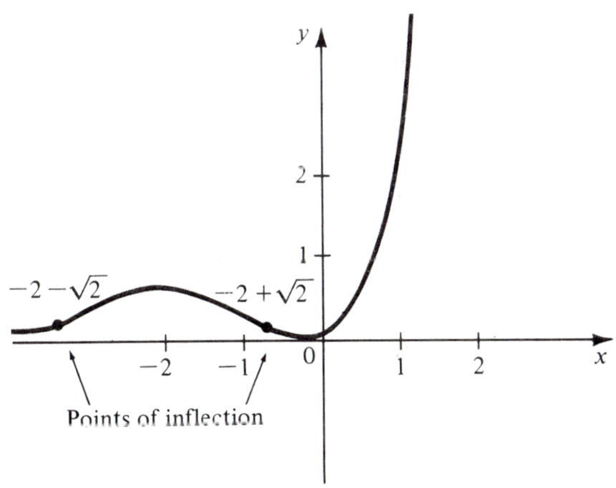

FIGURE 7.7

$$e = \lim_{n\to\infty}\left(1 + \frac{1}{n}\right)^n.$$

In many calculus texts this is taken as the *definition* of the number e.

EXERCISES

1. Prove that the function exp is strictly increasing.

2. Prove $\lim_{h\to 0}\dfrac{e^h - 1}{h} = 1$. (*Hint:* Consider the derivative of e^x at $x = 0$.)

3. Evaluate $\int_0^2 e^{-x}\,dx$.

4. Verify by differentiation that $\int xe^x\,dx = xe^x - e^x + c$.

In each of Exercises 5 to 8, sketch the graph of the given function. Find extrema and points of inflection (if any).

5. $f(x) = xe^x$

6. $g(x) = x^2e^{-x}$

7. $f(x) = \dfrac{e^x}{x}$

8. $f(x) = x^3e^x$

9. Using the mean value theorem, show that

$$e^x \geq 1 + x, \quad e^{-x} \geq 1 - x \qquad \text{for all } x \geq 0.$$

10. If $D_xf(x) = f(x)$ for all x in $(-\infty,\infty)$, is $f(x) = e^x$?

11. Using only the definition of $\lim_{x\to\infty} f(x)$, prove that:

    a. $\lim_{x\to\infty} e^x = \infty$.

    b. $\lim_{x\to-\infty} e^x = 0$.

12. Sketch the graph of the equation

$$y = e^{-x^2},$$

showing extrema and points of inflection.

13. Prove that $\int_a^{ax}\frac{1}{u}\,du = \int_1^x \frac{1}{t}\,dt$. (*Hint:* Make an appropriate change of variable in the first integral.) Use this to give another proof that $\ln ax = \ln a + \ln x$.

14. Approximate ln 3 using Simpson's rule.

15. Verify by differentiating that $\int \ln x\,dx = x\ln x - x + c$.

16. Prove that $\dfrac{x-1}{x} < \ln x < x - 1$, $x \neq 1$.

4 DERIVATIVES AND INTEGRALS

Knowing the derivatives of ln and exp, we can find derivatives of composites of these functions with other differentiable functions by use of the chain rule.

However, in working with a function of the form $f(x) = \ln g(x)$, we must restrict x so that $g(x) > 0$. This suggests that we consider $\ln |g(x)|$ rather than $\ln g(x)$. Actually, $D_x|g(x)|$ exists whenever $D_x g(x)$ exists, as long as $g(x) \neq 0$:

$$D_x|g(x)| = D_x g(x) \quad \text{if } g(x) > 0,$$
$$D_x|g(x)| = D_x[-g(x)] = -D_x g(x) \qquad \text{if } g(x) < 0.$$

From these equations and the chain rule, we have

$$D_x \ln |g(x)| = \frac{1}{|g(x)|} D_x|g(x)| = \frac{D_x g(x)}{g(x)} \qquad \text{if } g(x) \neq 0.$$

Thus we have the formulas

7.19 $$D_x \ln |g(x)| = \frac{D_x g(x)}{g(x)} \qquad \text{if } g(x) \neq 0.$$

7.20 $$\int \frac{g'(x)}{g(x)}\, dx = \ln |g(x)| + C.$$

In using the definite integral analogue of 7.20, we must always remember to integrate over an interval $[a,b]$ in which g is differentiable and $g(x) \neq 0$.

Example 1 Find $D_x \ln |x^2 - 5x + 4|$.

Solution: We have, by 7.19,

$$D_x \ln |x^2 - 5x + 4| = \frac{D_x(x^2 - 5x + 4)}{x^2 - 5x + 4} = \frac{2x - 5}{x^2 - 5x + 4}$$

as long as $x^2 - 5x + 4 \neq 0$. Since

$$x^2 - 5x + 4 = (x - 1)(x - 4)$$

the derivative above is valid for all x except 1 and 4.

Example 2 Find $D_x \ln \sqrt{\dfrac{x^2 + 1}{x^2 - 1}}$.

Solution: Clearly, the domain of the function being differentiated is $(-\infty, -1) \cup (1, \infty)$. Before differentiating, we algebraically simplify the given function as follows:

$$\ln \sqrt{\frac{x^2 + 1}{x^2 - 1}} = \ln \left(\frac{x^2 + 1}{x^2 - 1}\right)^{1/2} = \tfrac{1}{2} \ln \left(\frac{x^2 + 1}{x^2 - 1}\right)$$
$$= \tfrac{1}{2}[\ln (x^2 + 1) - \ln (x^2 - 1)].$$

Therefore $$D_x \ln \sqrt{\frac{x^2+1}{x^2-1}} = \tfrac{1}{2}[D_x \ln (x^2+1) - D_x \ln (x^2-1)]$$

$$= \frac{1}{2}\left[\frac{2x}{x^2+1} - \frac{2x}{x^2-1}\right] = \frac{-2x}{x^4-1}.$$

Example 3 Find $\int \frac{x}{x^2-1}\,dx$.

Solution: If we multiply the integral by $\frac{1}{2}$ and the integrand by 2, then we can use 7.20 directly.

$$\int \frac{x}{x^2-1}\,dx = \frac{1}{2}\int \frac{2x}{x^2-1}\,dx$$

$$= \tfrac{1}{2} \ln |x^2 - 1| + C.$$

This is well defined over any interval not containing -1 or 1.

At times it is convenient to make a change of variables before integrating, as illustrated below.

Example 4 Find $\int_{-3}^{-1} \frac{x^2}{3x-1}\,dx$.

Solution: If we let

$$u = 3x - 1, \qquad du = 3\,dx,$$

then

$$x = \tfrac{1}{3}(u+1)$$

and

$$\int_{-3}^{-1} \frac{x^2}{3x-1}\,dx = \frac{1}{27}\int_{-10}^{-4} \frac{(u+1)^2}{u}\,du$$

$$= \frac{1}{27}\int_{-10}^{-4} \left(u + 2 + \frac{1}{u}\right) du$$

$$= \frac{1}{27}\left(\frac{u^2}{2} + 2u + \ln |u|\right)\Bigg|_{-10}^{-4}$$

$$= -\tfrac{1}{27}(30 + \ln \tfrac{5}{2}).$$

If $f(x) = \exp x$, then $f(g(x)) = \exp g(x)$

$$D_x f(g(x)) = f'(g(x))g'(x) = [\exp g(x)]g'(x).$$

This proves the following formulas.

7.21 $$D_x e^{g(x)} = e^{g(x)} D_x g(x).$$

7.22 $$\int e^{g(x)} g'(x)\,dx = e^{g(x)} + C.$$

Example 5

Find $D_x(e^{3x} + e^{-3x})$.

Solution: By 7.21,

$$\begin{aligned} D_x(e^{3x} + e^{-3x}) &= D_x e^{3x} + D_x e^{-3x} \\ &= e^{3x} \cdot 3 + e^{-3x} \cdot (-3) \\ &= 3(e^{3x} - e^{-3x}). \end{aligned}$$

Example 6

Find $D_x x e^{x^2}$.

Solution: We have

$$\begin{aligned} D_x x e^{x^2} &= (D_x x)(e^{x^2}) + x(D_x e^{x^2}) \\ &= e^{x^2} + x(e^{x^2} D_x x^2) \\ &= e^{x^2} + 2x^2 e^{x^2} \\ &= e^{x^2}(1 + 2x^2). \end{aligned}$$

Example 7

Find $\int x e^{x^2}\, dx$.

Solution: We have by 7.22:

$$\begin{aligned} \int x e^{x^2}\, dx &= \frac{1}{2} \int (2x) e^{x^2}\, dx \\ &= \tfrac{1}{2} e^{x^2} + C. \end{aligned}$$

It is not necessary to remember 7.22 if we use the change of variable formula (5.25 or 5.26).

Example 8

Sketch the graph

$$f(x) = x \exp\left(-\frac{x}{2}\right).$$

Solution: We have

$$\begin{aligned} f'(x) &= \exp\left(-\frac{x}{2}\right) + x\left(-\frac{1}{2}\right)\exp\left(-\frac{x}{2}\right) \\ &= \left(1 - \frac{x}{2}\right)\exp\left(-\frac{x}{2}\right). \end{aligned}$$

Thus, $f'(x) = 0$ if and only if $x = 2$. Now

$$\begin{aligned} f''(x) &= \left(-\frac{1}{2}\right)\exp\left(-\frac{x}{2}\right) + \left(1 - \frac{x}{2}\right)\left(-\frac{1}{2}\right)\exp\left(-\frac{x}{2}\right) \\ &= \left(-1 + \frac{x}{4}\right)\exp\left(-\frac{x}{2}\right) \end{aligned}$$

so that $f''(2) < 0$. Thus, $(2, 2/e)$ is a maximum point of the curve. Clearly a point of inflection occurs at $x = 4$. The graph is sketched in Figure 7.8.

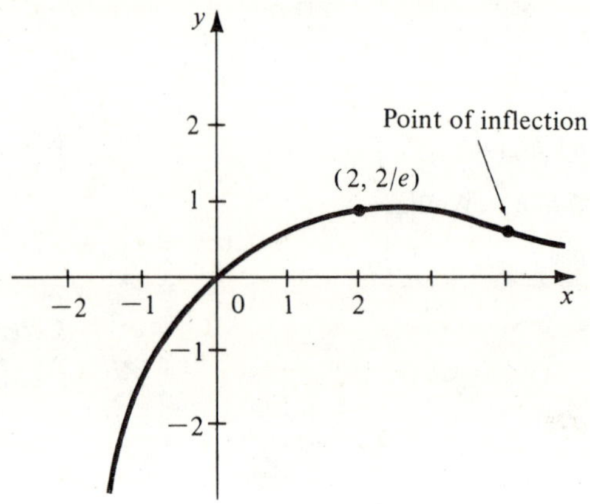

FIGURE 7.8

Example 9 Find $f'(x)$ if

$$f(x) = \frac{(x^2 + 1)^7(x^3 + 2)^5}{x^4 + 1}.$$

Solution: This type of problem is frequently simplified by a technique known as *logarithmic differentiation.* Let $y = f(x)$ and take the logarithm of both sides of the equation. We obtain

$$\ln y = 7 \ln (x^2 + 1) + 5 \ln (x^3 + 2) - \ln (x^4 + 1).$$

Differentiating both sides with respect to x,

$$\frac{1}{y}\frac{dy}{dx} = \frac{14x}{x^2 + 1} + \frac{15x^2}{x^3 + 2} - \frac{4x^3}{x^4 + 1}.$$

Now we replace y by its equivalent expression in terms of x and solve for $\frac{dy}{dx}$:

$$\frac{dy}{dx} = \left[\frac{(x^2 + 1)^7(x^3 + 2)^5}{x^4 + 1}\right]\left(\frac{14x}{x^2 + 1} + \frac{15x^2}{x^3 + 2} - \frac{4x^3}{x^4 + 1}\right).$$

EXERCISES

I

Differentiate each of the following functions.

1. $f(x) = x^2e^{2x}$

2. $g(x) = \sqrt{1 + e^x}$

3. $G(x) = (1 + e^{-3x^2})^3$

4. $F(x) = \dfrac{e^x}{\sqrt{e^{2x} - 1}}$

5. $(x + 1) \ln |x^2 - 1|$

6. $\ln \sqrt{x^4 + 4}$

7. $\dfrac{x^2}{\ln x}$

8. $e^x \ln x$

9. $\ln (e^x + 1)$

10. $\ln \sqrt{e^x + 1}$

11. $\ln (e^{3x} + x^2)$

12. $e^{x^2} \ln (1 + \sqrt{x})$

13. $\exp \dfrac{x}{\sqrt{x - 1}}$

14. $\ln (\ln x)$

15. $f(x) = \ln \left(\dfrac{e^{2x} - 5}{x}\right)$

16. $g(x) = e^{x \ln x}$

In each of Exercises 17 to 20 find all maximum points, minimum points and points of inflection (if any) of the graph of the given function and sketch it.

17. $F(x) = x \ln x$

18. $g(x) = x^2 \ln x$

19. $F(x) = \dfrac{\ln x}{x}$

20. $G(x) = \dfrac{e^{2x} - 1}{e^{2x} + 1}$

In each of Exercises 21 to 34 evaluate the given integral.

21. $\int_0^2 e^{-5x}\, dx$

22. $\int_0^1 x^2 \exp(-x^3)\, dx$

23. $\int \dfrac{e^{\sqrt{x}}}{\sqrt{x}}\, dx$

24. $\int (x + 2) \exp(x^2 + 4x)\, dx$

25. $\int \dfrac{x^3}{x + a}\, dx$

26. $\int_0^3 \dfrac{x}{2x^2 + 3}\, dx$

27. $\int \dfrac{\ln x}{x}\, dx$

28. $\int \dfrac{e^x}{1 - e^x}\, dx$

29. $\int_0^1 \dfrac{x}{x^2 + 4}\, dx$

30. $\int \dfrac{e^x}{e^x + 7}\, dx$

31. $\int_0^2 \dfrac{x^2 + 1}{x + 1}\, dx$

32. $\int \dfrac{x^3 + 1}{x^4 + 4x}\, dx$

33. $\int_0^1 e^{\sqrt{2}}\, dx$

34. $\int e^{3+x}\, dx$

In each of Exercises 35 to 40 determine $f'(x)$ by first taking logs of both sides of the equation and then differentiating implicitly. (This is the logarithmic differentiation technique of Example 9.)

35. $f(x) = e^{\ln^2 x}$

36. $f(x) = x^x$

37. $f(x) = x^{x^2} \cdot e^{x^3}$

38. $f(x) = (\ln x)^x$

39. $f(x) = (x^2 + 1)^{3x}$

40. $f(x) = \dfrac{x^7(x + 1)^5}{(x + 2)(x + 3)^6}$

Find $D_x y$ by implicit differentiation.

41. $\ln (xy^2) - x + y = 2$

42. $\ln \dfrac{y}{x} + x^2 y = 3$

43. $xe^y + \ln y - x^2 = 1$

44. $\ln (e^y + 1) - xy = x$

II

Sketch the graph of each of the following equations and determine y' and y''.

1. $y = \exp (e^x)$

2. $y = \exp (e^{-x})$

3. $y = \exp (-e^x)$

4. $y = \exp (-e^{-x})$

5. $y = \ln (\ln (x + 1))$

In each of Exercises 6 to 9 evaluate the integral.

6. $\displaystyle\int \frac{1}{1 + e^x}\, dx$

7. $\displaystyle\int \frac{1}{x \ln x \ln \ln x}\, dx$

8. $\displaystyle\int \frac{1}{e^x + e^{-x} + 2}\, dx$

9. $\displaystyle\int \exp (x + e^x)\, dx$

10. Find the arc length of the graph of the equation $y = \ln (x^2 - 1)$ between $x = 2$ and $x = 3$. $\left(\textit{Hint:}\ \dfrac{2}{x^2 - 1} = \dfrac{1}{x - 1} - \dfrac{1}{x + 1}.\right)$

5 HYPERBOLIC FUNCTIONS

Some simple combinations of exponential functions have proved useful in physical applications of mathematics. These are the hyperbolic functions defined below.

7.23 Definition

The *hyperbolic sine*, designated by sinh, and the *hyperbolic cosine*, designated by cosh, are defined by

$$\sinh x = \frac{e^x - e^{-x}}{2}, \qquad \text{domain sinh} = (-\infty,\infty),$$

$$\cosh x = \frac{e^x + e^{-x}}{2}, \qquad \text{domain cosh} = (-\infty,\infty).$$

From this definition, evidently sinh is an *odd function* [that is, $\sinh(-x) = -\sinh x$] and cosh is an even function [that is, $\cosh(-x) = \cosh x$]. An easy computation shows that

7.24 $$e^x = \sinh x + \cosh x.$$

In functional notation, exp = sinh + cosh.

By adding ordinates of the graphs of $y = e^x/2$ and $y = -e^{-x}/2$, we can easily sketch the graph of sinh, as shown in Figure 7.9. The graph of cosh is sketched in Figure 7.10 in a similar way.

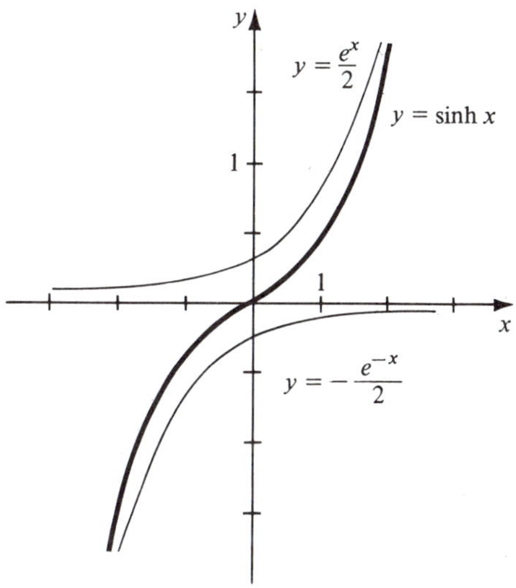

FIGURE 7.9

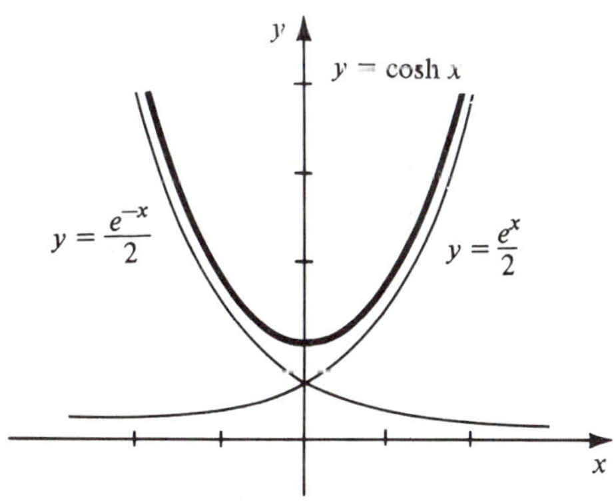

FIGURE 7.10

As is indicated by their names, the hyperbolic functions are related to both the hyperbola (which will be discussed in Chapter 10) and the trigonometric functions. Corresponding to the trigonometric identity

$$\sin^2 x + \cos^2 x = 1$$

is the identity

7.25 $$\cosh^2 x - \sinh^2 x = 1.$$

This identity is easily proved from the definitions of sinh and cosh. Other identities that are similar to trigonometric identities are as follows:

7.26 $$\sinh (x + y) = \sinh x \cosh y + \cosh x \sinh y,$$
$$\cosh (x + y) = \cosh x \cosh y + \sinh x \sinh y.$$

Proofs of these identities will be left as exercises.

Just as the trigonometric functions tangent, cotangent, secant, and cosecant may be defined in terms of sine and cosine, so might we define the corresponding hyperbolic functions:

$$\tanh = \frac{\sinh}{\cosh}, \quad \coth = \frac{\cosh}{\sinh}, \quad \text{sech} = \frac{1}{\cosh}, \quad \text{csch} = \frac{1}{\sinh}.$$

Since $D_x \exp(-x) = -\exp(-x)$, evidently

7.27 $$D_x \sinh x = \cosh x;$$

7.28 $$D_x \cosh x = \sinh x.$$

As we shall see in Chapter 8, these formulas are closely related to the formulas for the derivatives of the sine and the cosine. The integral analogs of 7.27 and 7.28 are as follows:

7.29 $$\int \sinh x\, dx = \cosh x + C.$$

7.30 $$\int \cosh x\, dx = \sinh x + C.$$

Example 1 Find (a) $D_x \tanh x$; (b) $\int \frac{\sinh x}{\cosh^2 x}\, dx$.

Solution: (a) We have

$$D_x \tanh x = D_x \frac{\sinh x}{\cosh x} = \frac{\cosh x D_x \sinh x - \sinh x D_x \cosh x}{\cosh^2 x}$$
$$= \frac{\cosh^2 x - \sinh^2 x}{\cosh^2 x} = \frac{1}{\cosh^2 x} = \text{sech}^2\, x.$$

(b) Letting $u = \cosh x$ and $du = \sinh x\, dx$, we have

$$\int \frac{\sinh x}{\cosh^2 x}\, dx = \int u^{-2}\, du \Big|_{u=\cosh x}$$
$$= -\frac{1}{u} + C \Big|_{u=\cosh x}$$
$$= -\text{sech}\, x + C.$$

Example 2 Find the volume V of the solid generated by rotating about the x axis the region bounded by the graph of $y = \sinh x$, the x axis, and the line $x = 2$ (Figure 7.11).

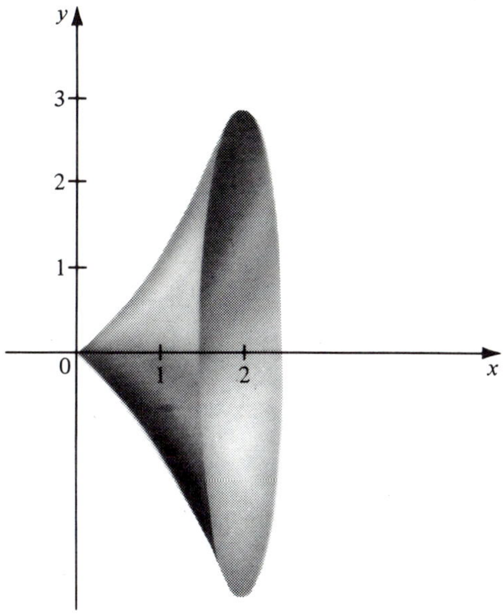

FIGURE 7.11

Solution: Each cross section of the solid is a circle of area πy^2. Thus, by 6.5,

$$V = \pi \int_0^2 \sinh^2 x \, dx.$$

How do we integrate $\sinh^2 x$? Using 7.26 (with $y = x$) and then 7.25, we get

$$\begin{aligned} \cosh 2x &= \cosh^2 x + \sinh^2 x \\ &= (1 + \sinh^2 x) + \sinh^2 x = 1 + 2 \sinh^2 x. \end{aligned}$$

Hence,
$$\sinh^2 x = \frac{\cosh 2x - 1}{2}.$$

Thus,
$$\begin{aligned} \pi \int_0^2 \sinh^2 x \, dx &= \frac{\pi}{2} \int_0^2 (\cosh 2x - 1) \, dx \\ &= \frac{\pi}{4} \int_0^2 \cosh 2x \cdot 2 \, dx - \frac{\pi}{2} \int_0^2 dx \\ &= \frac{\pi}{4} \sinh 2x \Big|_0^2 - \frac{\pi}{2} x \Big|_0^2 \end{aligned}$$

$$= \frac{\pi}{4} \sinh 4 - \pi$$

$$\doteq 5.8\pi.$$

Thus, $V \doteq 18$ cubic units.

EXERCISES

I

Sketch the graph of each of the following equations and find y' and y''.

1. $y = \sinh x$
2. $y = \cosh x$
3. $y = \tanh x$
4. $y = \coth x$
5. $y = \operatorname{csch} x$
6. $y = \operatorname{sech} x$

Establish the following identities.

7. $\cosh^2 x - \sinh^2 x = 1$
8. $\operatorname{sech}^2 x + \tanh^2 x = 1$
9. $\sinh (x - y) = \sinh x \cosh y - \cosh x \sinh y$
10. $\cosh (x - y) = \cosh x \cosh y - \sinh x \sinh y$
11. $(\cosh x + \sinh x)^n = \cosh nx + \sinh nx$ for every integer n

Differentiate:

12. $y = \sinh (e^{2x} + 1)$
13. $y = \cosh (\ln x + 1)$
14. $y = \ln (\tanh x^2)$
15. $y = e^{\tanh (x^2+1)}$
16. $y = \ln (\sin (x + 1))$
17. $y = 3 \sinh 2x - 2 \cosh 3x$

Evaluate the following integrals:

18. $\int \sinh x \cosh x \, dx$
19. $\int \tanh x \operatorname{sech} x \, dx$
20. $\int \coth x \operatorname{csch} x \, dx$
21. $\int \tanh x \operatorname{sech}^2 x \, dx$
22. $\int x \cosh x^2 \, dx$
23. $\int (\cosh x) e^{\sinh x} \, dx$
24. $\int \sinh^3 x \cosh x \, dx$
25. $\int \frac{\cosh \sqrt{x}}{\sqrt{x}} \, dx$
26. $\int \frac{\cosh x}{4 \sinh x + 9} \, dx$
27. $\int_0^1 [\cosh (e^x - e^{-x})](e^x + e^{-x}) \, dx$

28. Find the arc length of the catenary $y = \cosh x$ from $x = 0$ to $x = a$.
29. Find the volume of the solid generated by revolving the region bounded by the catenary $y = a/2 \cosh x/a$, $x = a$, $x = -a$, and $y = 0$ about the x axis.
30. Using Simpson's rule with $n = 4$, approximate the volume of the solid generated by revolving the region bounded by $y = \sinh x$, $y = \sinh 1$, and $x = 0$, about the y axis.

II

Establish the following identities:

1. $\sinh 2x = 2 \sinh x \cosh x$
2. $\cosh 2x = 2 \cosh^2 x - 1 = 1 + 2 \sinh^2 x = \cosh^2 x + \sinh^2 x.$
3. $\tanh 2x = \dfrac{2 \tanh x}{1 + \tanh^2 x}.$
4. $\sinh 3x = 3 \sinh x + 4 \sinh^3 x$
5. $\cosh 3x = 4 \cosh^3 x - 3 \cosh x$

Evaluate the following integrals:

6. $\int \cosh^2 x \, dx$
7. $\int \sinh^3 x \, dx$
8. $\int \cosh^3 x \, dx$
9. $\int \operatorname{sech}^2 x \, dx$

6 CHANGE OF BASE

The exponential function

$$f(x) = a^x, \qquad a > 0, a \neq 1$$

was discussed intuitively along with its inverse, $\log_a$, in Section 1. Using the functions ln and exp, which were developed rigorously in Sections 2 and 3, we can give meaningful definitions of the function f above and $\log_a$.

7.31 Definition

The exponential function f to the base a is given by

$$f(x) = e^{x \ln a}, \qquad \text{domain } f = (-\infty, \infty).$$

Its inverse, $\log_a$, is defined by

$$y = \log_a x \text{ if and only if } x = f(y), \qquad \text{domain } \log_a = (0, \infty).$$

Of course, we denote $f(x)$ by a^x,

$$f(x) = a^x = e^{x \ln a}.$$

Thus,

$$y = \log_a x \text{ if and only if } x = a^y.$$

Hence,

$$\ln x = \ln a^y = y \ln a,$$

and, solving for y,

7.32
$$\log_a x = \frac{\ln x}{\ln a}.$$

We can use 7.31 and 7.32 to differentiate a^x and $\log_a x$. Thus,

$$D_x a^x = D_x e^{x \ln a} = (\ln a)e^{x \ln a} = (\ln a)a^x,$$

that is,

7.33
$$D_x a^x = (\ln a)a^x.$$

Also,

$$D_x \log_a x = D_x \left(\frac{\ln x}{\ln a}\right) = \frac{1}{\ln a} D_x \ln x = \frac{1}{\ln a} \cdot \frac{1}{x},$$

that is,

7.34
$$D_x \log_a x = \frac{1}{x \ln a}.$$

Example 1 Find $D_x \log_{10} |3x + 1|$.

Solution: By 7.34 and 7.19,

$$D_x \log_{10} |3x + 1| = \frac{1}{\ln 10} D_x \ln |3x + 1|$$
$$= \frac{1}{\ln 10} \frac{3}{3x + 1}.$$

The power differentiation formula holds even if the power is irrational according to the following result.

7.35 Theorem $D_x x^r = r x^{r-1}$ *for every real number r.*

Proof: Since

$$x^r = e^{r \ln x},$$

we have
$$\begin{aligned} D_x x^r &= D_x e^{r \ln x} \\ &= e^{r \ln x} D_x(r \ln x) \\ &= e^{r \ln x} \cdot \frac{r}{x} = x^r \cdot \frac{r}{x} = r x^{r-1}. \end{aligned}$$

Example 2 Find $\int 10^x \, dx$.

Solution: We have by 7.31

$$\int 10^x \, dx = \int e^{x \ln 10} \, dx.$$

This may be evaluated by letting

$$u = x \ln 10, \qquad du = \ln 10 \, dx,$$

so that
$$\begin{aligned} \int 10^x \, dx &= \frac{1}{\ln 10} \int e^u \, du \Big|_{u = x \ln 10} \\ &= \frac{1}{\ln 10} e^u + C \Big|_{u = x \ln 10} \\ &= \frac{1}{\ln 10} e^{x \ln 10} + C = \frac{10^x}{\ln 10} + C. \end{aligned}$$

EXERCISES

Differentiate the following.

1. $\log_2 |5 - 2x|$

2. $\log_{10} (x^2 + 2)$

3. $x^2 \log_{10} (3 + 2x^2)$

4. $\dfrac{\log_{10} x}{x}$

5. $(\ln x + 1)^\pi$

6. $\left(\dfrac{x}{x^2 + 1}\right)^e$

7. $(1 + \sqrt{x})^{1+e}$

8. $(1 + e)^{1+\sqrt{x}}$

9. 2^{2x}

10. $\exp (e^x)$

11. $x^{\sqrt{x}}$

12. 3^{4x}

13. $\sqrt{\log_{10} (x^2 + 1)}$

14. $\sinh (4^x + 4)$

15. $\cosh (\log_{10} x)$

16. 10^{x^2}

Find the following integrals.

17. $\int_1^2 2^x \, dx$

18. $\int_{-1}^{1} (1 + e)^x \, dx$

19. $\int_{-3}^{-1} 10^{-x}\,dx$ **20.** $\int x3^{x^2}\,dx$

21. $\int \frac{3^{\ln x}}{x}\,dx$ **22.** $\int (\cosh x)10^{\sinh x}\,dx$

23. $\int_0^1 10^x e^x\,dx$ **24.** $\int_0^1 (x^3+1)^{\pi} x^2\,dx$

25. $\int 7^{3x}\,dx$ **26.** $\int (1+x)^e\,dx$

27. Prove that:

a. $\log_a x = (\log_a b)(\log_b x)$ b. $\log_a x = \frac{\log_b x}{\log_b a}$

7 EXPONENTIAL LAWS OF GROWTH AND DECAY

It sometimes happens that the rate of change of the amount of a given substance at any time is proportional to the amount of the substance present at that time. The first example that comes to mind is the decomposition of a radioactive substance, the rate of decomposition being proportional to the amount of radioactive substance present. The growth of a culture of bacteria obeys this same law under ideal conditions.

In order to give a mathematical analysis of this phenomenon, let $f(t) > 0$ be the amount of a substance present at time t. Then to say that the rate of change of f at time t is proportional to $f(t)$ is to say that

7.36 $$f'(t) = kf(t)$$

for some constant k and every time t in some interval. This differential equation may be solved for f in the following way.

We first put 7.36 in the form

$$\frac{f'(t)}{f(t)} = k,$$

and then integrate as follows:

$$\int_0^t \frac{f'(t)}{f(t)}\,dt = \int_0^t k\,dt,$$

[changing variables by letting $u = f(t)$, $du = f'(t)\,dt$]

$$\int_{f(0)}^{f(t)} \frac{1}{u}\,du = \int_0^t k\,dt,$$

$$\ln u\Big|_{f(0)}^{f(t)} = kt\Big|_0^t,$$

$$\ln f(t) - \ln f(0) = kt,$$

$$\ln \frac{f(t)}{f(0)} = kt,$$

$$\frac{f(t)}{f(0)} = e^{kt},$$

$$f(t) = f(0)e^{kt}.$$

According to this equation, the amount $f(t)$ present at time t equals the initial amount $f(0)$ (at time $t = 0$) times e^{kt}. The constant k depends on the substance in question, and can be found if sufficient data are given. It is permissible to express the equation above in the form

7.37
$$f(t) = f(0)a^{ct}$$

using any base a we wish. Evidently, $c = k \log_a e$.

Example 1 The half-life of radium is approximately 1600 years (i.e., a given amount of radium will be half gone after 1600 years). Starting with 150 mg of pure radium, find the amount left after t years. After how many years is only 30 mg left?

Solution: Using 7.37 with $a = 2$, and the unit of time a year, we have

$$f(t) = 150 \cdot 2^{ct}.$$

Since $f(1600) = 75$,

$$75 = 150 \cdot 2^{1600c},$$
$$\tfrac{1}{2} = 2^{1600c}.$$

Thus $1600c = -1$ and $c = -\frac{1}{1600}$. It is evident now that the base 2 was chosen so as to make the evaluation of c easy. We have proved that the amount $f(t)$ of radium left after t years is given by

$$f(t) = 150 \cdot 2^{-t/1600}.$$

The solution of the equation

$$30 = 150 \cdot 2^{-t/1600}$$

or

$$\tfrac{1}{5} = 2^{-t/1600}$$

will give the number t of years that must elapse before only 30 mg is left. We solve this equation by logarithms, obtaining

$$t = \frac{1600 \log_{10} 5}{\log_{10} 2} \doteq 3715 \text{ years.}$$

The above is an example of an exponential law of decay. The following example is an exponential law of growth.

Example 2 The number of bacteria in a culture was 1000 at a certain instant and 8000 two hours later. Assuming ideal conditions for growth, how many bacteria are there after t hours?

Solution: The number $f(t)$ of bacteria after t hours is given by (using $a = 8$ in 7.37)

$$f(t) = 1000 \cdot 8^{ct}.$$

It is given that $f(2) = 8000$; thus

$$8000 = 1000 \cdot 8^{2c}$$
$$8 = 8^{2c}$$

and $2c = 1$. Hence

$$f(t) = 1000 \cdot 8^{t/2}.$$

After five hours, for example,

$$f(5) = 1000 \cdot 8^{5/2} = 64{,}000\sqrt{8} \doteq 180{,}000.$$

EXERCISES

1. Show that the law of decay for radium may be written in the form

$$A = A_0(\tfrac{1}{2})^{t/1600},$$

where A_0 is the initial amount of radium and A is the amount left after t years. The percentage of radium left at any time t is $100A/A_0$. Find the percentage of radium left after 800 years; after 6400 years. In how many years is 10 percent of the radium decomposed?

2. When bacteria grow under ideal circumstances, their rate of growth is proportional to the number of bacteria present. Express the number of bacteria present in terms of time t. In a certain culture of bacteria the number of bacteria present at a certain instant was 1000 and the number present 10 hours later was 8000. Find the law of growth for this culture, and find the number of bacteria present 15 hours after the first count.

3. a. Under normal conditions the rate of change of population is considered to be proportional to the population at any time. If P_0 is the population at time $t = 0$, express the population P in terms of time t.

   b. A town had a population of 18,000 in 1945 and 25,000 in 1955. Assuming the exponential law of growth, what population was expected in 1965?

4. The radioactive element polonium has a half-life of 140 days. Express the amount A of polonium left after t days in terms of the initial amount A_0 of polonium and t. Approximately what percentage of a given supply of polonium is left after one year?

5. In solution, sugar decomposes into other substances at a rate proportional to the amount x still unchanged. Show that

$$x = x_0e^{kt}$$

where x_0 is the amount unchanged at time $t = 0$. If 30 lb of sugar reduces to 10 lb in 4 hours, when will 95 percent of the sugar be decomposed?

6. Eighty percent of a sample of a radioactive element decays in 10 years. What is its half-life?
7. The population of the U.S. was 100 million in 1920 and 200 million in 1970. Assuming the exponential law of growth, what will be the population in 1990?
8. A town now has a population of 5000. If it grows to 6000 in 1 year, when will the population reach 10,000?
9. A compound A decomposes to form compounds B and C at a rate proportional to the amount of A present, and in such a way that 0.4 of a gram of B is produced for every gram of A that decomposes. If there are initially 50 grams of A and no B or C present, and 40 grams of A are left after 15 minutes, how long will it take to produce 15 grams of B?

A function y satisfies the differential equation $y' = ky$. In each of Exercises 10 to 13, find $y = y(x)$ if the initial condition given is satisfied.

10. $y(0) = 1, y(1) = e^2$
11. $y(0) = 1, y(1) = e^{-2}$
12. $y(0) = 100, y(1) = 200$
13. $y(0) = 100, y(1) = 50$
14. Find $y(5)$ in Exercise 12 above.
15. Find $y(5)$ in Exercise 13 above.

CHAPTER REVIEW

Differentiate (find dy/dx):

1. $y = e^{3x} + 1$
2. $y = \ln (3x + 1)$
3. $y = \cosh (x^2 + 1)$
4. $y = \exp (x^3 + 1)$
5. $y = (x^2)^\pi$
6. $y = \sqrt{\sinh x + 1}$
7. $y = (e^x + x)^{\sqrt{2}}$
8. $y = e^{3x} \ln (2x + 1)$
9. $y = \log_{10} 8^x$
10. $y = \dfrac{\ln x}{\cosh x + 1}$
11. $y = \ln \left(\dfrac{x}{e^{3x}}\right)$
12. $2xe^{-y} = \ln x + e^x$
13. $x = y^{3x+1}$
14. $y = \dfrac{e^x}{e^x - e^{-x}}$
15. $y = x^2 \sinh x^2$
16. $y = \dfrac{\ln x}{e^x}$

In each of Exercises 17 to 24, find dy/dx and sketch the graph.

17. $y = x^2 \ln x$
18. $y = \dfrac{\ln x}{\sqrt{x}}$

19. $y = \ln^2 x$

20. $y = \dfrac{\ln x}{x}$

21. $y = \sqrt{x} \ln x$

22. $y = \dfrac{\ln x}{x^2}$

23. $y = \ln (x + \sqrt{x^2 - 1}),\ x \geq 1$

24. $y = \ln (x + \sqrt{x^2 + 1})$

In each of Exercises 25 to 30, find dy/dx by logarithmic differentiation.

25. $y = \sqrt{\dfrac{1 - x^2}{1 + x^2}}$

26. $y = (x^2 + 1)^{\ln x}$

27. $xe^y y^2 = 1$

28. $y = \sqrt{\dfrac{(a - x)(b - x)}{(a + x)(b + x)}}$

29. $y = (x)^{10^x} - (10)^{x^{10}}$

30. $y = (x)^{x^\pi}$

31. Find the area of the region bounded by the curve $y = a \cosh x/a$, the x axis, and the lines $x = \pm a$.

32. Find the area of the region bounded by the curve $y = a \tanh x/a$, the x axis, and the lines $x = m$ and $x = n$.

33. Find the volume of the solid generated by rotating the region bounded by $y = ae^{bx}$, $x = 0$, $x = c$, $y = 0$ about the x axis.

34. Find the arc length of the graph of the equation $y = x^2/4 - \frac{1}{2} \ln x$ between $x = 1$ and $x = 2$.

In each of Exercises 35 to 48, evaluate the integral.

35. $\displaystyle\int \frac{x}{x + a}\, dx$

36. $\displaystyle\int \frac{x^3 - a^3}{x + a}\, dx$

37. $\displaystyle\int \frac{e^{-3x} - 1}{e^{-3x} + 1}\, dx$

38. $\displaystyle\int_e^{e^3} \frac{\ln x}{x}\, dx$

39. $\displaystyle\int \frac{x + 2}{x + 3}\, dx$

40. $\displaystyle\int \exp (3x + 1)\, dx$

41. $\displaystyle\int \frac{1}{\sqrt{x}} \sinh \sqrt{x}\, dx$

42. $\displaystyle\int \frac{2 + e^{4x}}{e^{2x}}\, dx$

43. $\displaystyle\int \frac{e^{4x}}{e + e^{4x}}\, dx$

44. $\displaystyle\int \sqrt{e^x}\, dx$

45. $\displaystyle\int_1^e \frac{(\ln t)^{1/2}}{t}\, dt$

46. $\displaystyle\int e^3\, dx$

47. $\displaystyle\int \frac{e^x}{2^x}\, dx$

48. $\displaystyle\int_e^{e^2} \frac{1}{x(\ln x)^2}\, dx$

8

Trigonometric and Inverse Trigonometric Functions

We continue in this chapter the study of the elementary transcendental functions started in Chapter 7. The emphasis in the present chapter is on the trigonometric functions.

1 RADIAN MEASURE

It is clear that an arc of a circle has length. Thus an arc smaller than a semicircle can be considered to be the graph of a smooth function and hence, by 6.8, has length. An arc that is a semicircle or larger can be considered to be a union of smaller arcs, and its length to be the sum of the lengths of these smaller arcs.

If $\widehat{A_1B_1}$ and $\widehat{A_2B_2}$ are arcs of concentric circles intercepted by the same central angle (Figure 8.1), then the lengths s_1 and s_2 of $\widehat{A_1B_1}$ and $\widehat{A_2B_2}$ are proportional to the corresponding radii of the circles; that is,

$$\frac{s_1}{s_2} = \frac{r_1}{r_2}.$$

This may be proved in the manner suggested by Figure 8.1. Thus, for each inscripture I_1 of $\widehat{A_1B_1}$ there exists a corresponding inscripture I_2 of $\widehat{A_2B_2}$, and vice versa. By similar triangles,

$$|I_1| = k|I_2|, \qquad \text{where } k = \frac{r_1}{r_2}.$$

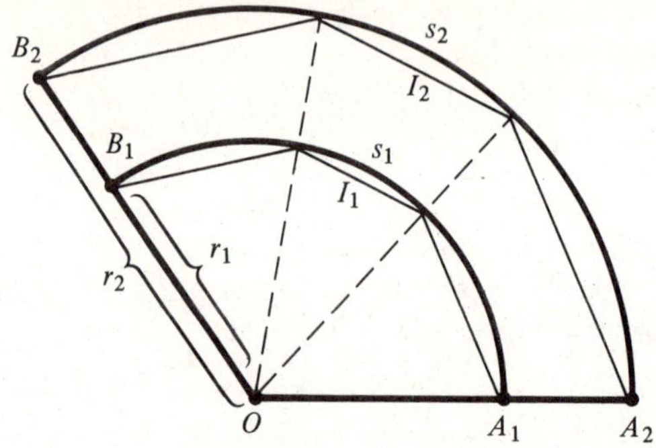

FIGURE 8.1

Taking limits as in Chapter 6, Section 4, we obtain $s_1/s_2 = k = r_1/r_2$.

As is the usual practice, we designate the circumference of a circle of diameter 1 by π. Then if C is the circumference of a circle of radius r, we have $C/\pi = r/\frac{1}{2}$, or

$$C = 2\pi r,$$

in view of our remarks above. This is the usual formula for the circumference of a circle of radius r.

We can think of an angle AOB as being formed by rotating its initial side OA, perhaps more than a complete revolution, to its terminal side OB. Under such a rotation the point A travels on a circle having center O and radius $|OA|$ to the point B. We shall speak of the length of the arc of the circle traversed by the point A as the distance traveled by A.

8.1 Definition

Let AOB be a central angle of a circle of radius 1 having initial side OA and terminal side OB (Figure 8.2). The *radian measure* of angle AOB is

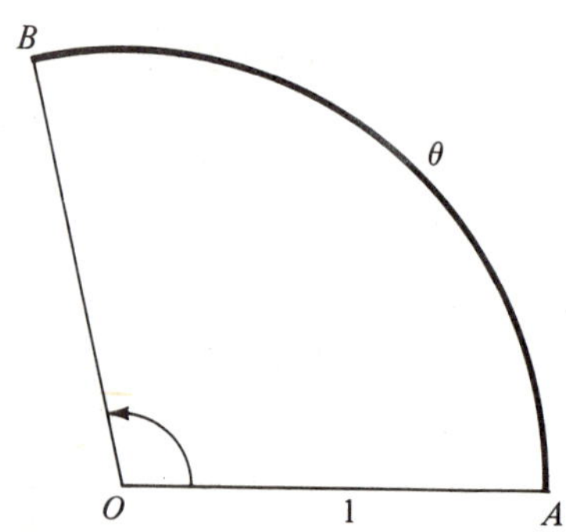

FIGURE 8.2

the distance θ traveled on the unit circle by the point A under the formation of the angle. The number θ is taken to be positive if the rotation is counterclockwise; otherwise, θ is taken to be negative.

Since the circumference of a unit circle is 2π, a semicircular arc has length π. Thus π is the radian measure of a straight angle and $\pi/2$ is the radian measure of a right angle. The relationship between degree and radian measure of an angle is indicated in the following table.

Degree measure	360	180	90	60	45	30
Radian measure	2π	π	$\pi/2$	$\pi/3$	$\pi/4$	$\pi/6$

2 THE SINE AND COSINE FUNCTIONS

For each real number s, $\sin s$ and $\cos s$ are defined as shown in Figure 8.3. Thus, starting from the point $A(1,0)$ on the unit circle with center at the origin O, we traverse the circle in a counterclockwise direction a distance s to the point P if $s \geq 0$; and a distance $|s|$ in a clockwise direction to the point P if $s < 0$. If we think of s as the radian measure of the central angle that intercepts the arc $\widehat{AP}$, we see that $\sin s = \overline{BP}$ and $\cos s = \overline{OB}$ in Figure 8.3. Since $|OP| = 1$, these are the usual definitions of sine and cosine for acute angles. The sine and cosine functions have the set of all real numbers as their domain and the interval $[-1,1]$ as their range.

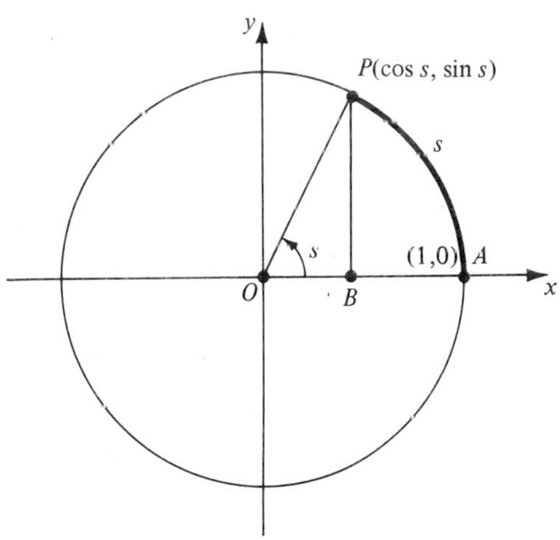

FIGURE 8.3

An important property of the sine and cosine is that they are *periodic functions*, with period 2π. Thus

$$\sin (s + 2\pi) = \sin s, \qquad \cos (s + 2\pi) = \cos s$$

for every real number s, and 2π is the least positive number for which each of these equations is true for every s. Some of the basic trigonometric identities are listed in the appendix (page **B**1).

If the sine has a derivative, it is given by

$$D_x \sin (x) = \operatorname*{limit}_{h\to 0} \frac{\sin (x + h) - \sin x}{h}.$$

By a trigonometric identity, $\sin (x + h) = \sin x \cos h + \cos x \sin h$. Hence,

$$\frac{\sin (x + h) - \sin x}{h} = \frac{\sin x(\cos h - 1) + \cos x \sin h}{h}$$

and

$$D_x \sin (x) = (\sin x) \operatorname*{limit}_{h\to 0} \frac{\cos h - 1}{h} + (\cos x) \operatorname*{limit}_{h\to 0} \frac{\sin h}{h}.$$

Thus, the problem of finding the derivative of the sine reduced to the problem of evaluating the limits

(1) $$\operatorname*{limit}_{h\to 0} \frac{\cos h - 1}{h},$$

(2) $$\operatorname*{limit}_{h\to 0} \frac{\sin h}{h}.$$

In evaluating (1) and (2), we might as well restrict h to be small, say $0 < h < \pi/2$. So select an arc $\widehat{AP}$ of a unit circle of length h and construct the tangent lines PC and AC, and the point D as shown in Figure 8.4. Looking at the figure, we see that

$$|BP| < |AP| < h < |AC| + |CP| < |AD|.$$

Since $|BP| = \sin h$ and $|AD| = \tan h$, we have

$$\sin h < h < \frac{\sin h}{\cos h}.$$

This easily yields the inequality

(3) $$\cos h < \frac{\sin h}{h} < 1.$$

We can approximate the size of $\cos h$ by the half-angle formula:

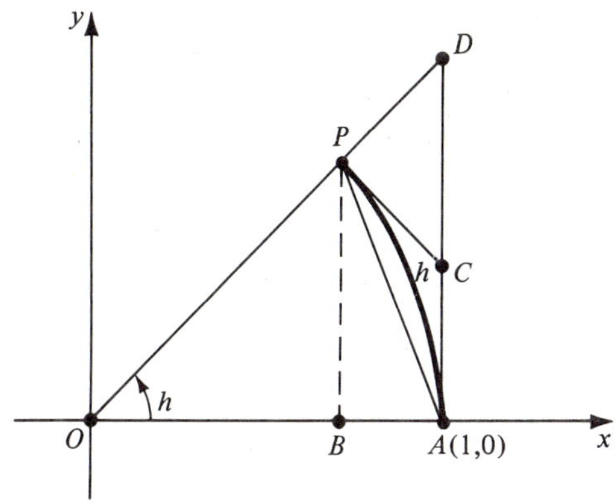

FIGURE 8.4

$$\sin^2 \frac{h}{2} = \frac{(1 - \cos h)}{2}.$$

Since $0 < \sin (h/2) < h/2$ by (3), $\sin^2 (h/2) < (h/2)^2$ and

(4) $$\frac{1 - \cos h}{2} < \frac{h^2}{4}.$$

Solving (4) for $\cos h$,

(5) $$\cos h > 1 - \frac{h^2}{2}.$$

Thus,

(6) $$1 - \frac{h^2}{2} < \frac{\sin h}{h} < 1,$$

(7) $$0 < \frac{1 - \cos h}{h} < \frac{h}{2}.$$

Similar inequalities hold if h is negative. Since $\underset{h \to 0}{\text{limit}} [1 - (h^2/2)] = 1$ and $\underset{h \to 0}{\text{limit}}\, h/2 = 0$, we derive from (6) and (7) that:

8.2 $$\underset{h \to 0}{\text{limit}} \frac{\sin h}{h} = 1.$$

8.3 $$\underset{h \to 0}{\text{limit}} \frac{\cos h - 1}{h} = 0.$$

The two limits above allow us to find:

8.4 $$D_x \sin(x) = \cos x.$$

Using the same procedure for the cosine, we have:

$$D_x \cos(x) = \operatorname*{limit}_{h\to 0} \frac{\cos(x+h) - \cos x}{h}$$

$$= (\cos x) \operatorname*{limit}_{h\to 0} \frac{\cos h - 1}{h} - (\sin x) \operatorname*{limit}_{h\to 0} \frac{\sin h}{h}$$

$$= (\cos x) \cdot 0 - (\sin x) \cdot 1.$$

Thus, using 8.2 and 8.3,

8.5 $$D_x \cos(x) = -\sin x.$$

The integral analogues of 8.5 and 8.4 are

8.6 $$\int \sin x\, dx = -\cos x + C;$$

8.7 $$\int \cos x\, dx = \sin x + C.$$

If in place of $\sin x$ we have $\sin f(x)$, then we can differentiate the resulting function by use of the chain rule:

8.8 $$D_x \sin f(x) = \cos f(x) D_x f(x);$$

8.9 $$D_x \cos f(x) = -\sin f(x) D_x f(x).$$

Example 1 Find (a) $D_x \sin(3x - 1)$; (b) $D_x \cos^2 2x$.

Solution: (a) By 8.8,

$$D_x \sin(3x - 1) = \cos(3x - 1) D_x(3x - 1) = 3\cos(3x - 1).$$

(b) We have

$$\begin{aligned} D_x(\cos 2x)^2 &= 2(\cos 2x)^1 D_x \cos 2x \\ &= 2\cos 2x(-\sin 2x \cdot 2) \\ &= -4 \sin 2x \cos 2x. \end{aligned}$$

Example 2 Find the extrema in one period of the function

$$f(x) = \sin^2 x - \cos x.$$

Solution: The period of f is clearly 2π. We have

$$\begin{aligned} f'(x) &= D_x(\sin x)^2 - D_x \cos x \\ &= 2 \sin x \cos x + \sin x \\ &= 2 \sin x(\cos x + \tfrac{1}{2}). \end{aligned}$$

Hence $f'(x) = 0$ if and only if $\sin x = 0$ or $\cos x = -\frac{1}{2}$. Therefore

$$\left\{0, \pi, \frac{2\pi}{3}, \frac{4\pi}{3}\right\}$$

is the set of critical numbers of f in $[0,2\pi)$.

The second derivative of f is given by

$$\begin{aligned} f''(x) &= 2D_x(\sin x \cos x) + D_x \sin x \\ &= 2[\cos x \cdot \cos x + \sin x \cdot (-\sin x)] + \cos x \\ &= 2(\cos^2 x - \sin^2 x) + \cos x. \end{aligned}$$

We easily verify that $f''(0) = 3$, $f''(\pi) = 1$, $f''(\frac{2}{3}\pi) = -\frac{3}{2}$, $f''(\frac{4}{3}\pi) = -\frac{3}{2}$. Therefore $f(0) = -1$ and $f(\pi) = 1$ are minimum values of f and $f(\frac{2}{3}\pi) = \frac{5}{4}$ and $f(\frac{4}{3}\pi) = \frac{5}{4}$ are maximum values of f. The graph of f in one period is sketched in Figure 8.5 from the following table of values.

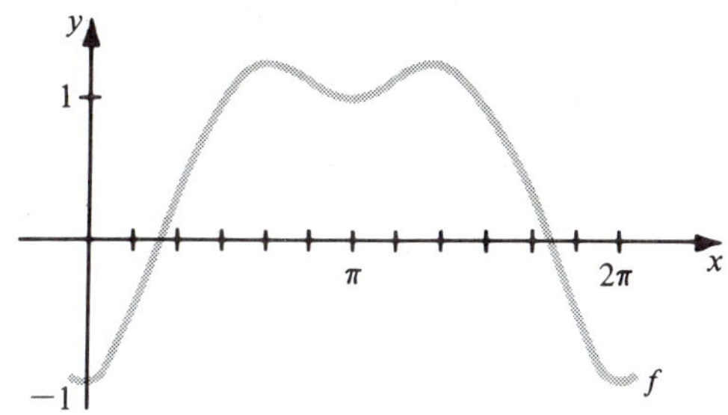

FIGURE 8.5

x	0	$\frac{1}{2}\pi$	$\frac{2}{3}\pi$	π	$\frac{4}{3}\pi$	$\frac{3}{2}\pi$	2π
$f(x)$	-1	1	$\frac{5}{4}$	1	$\frac{5}{4}$	1	-1

Example 3 Find $\int \sin^3 2x \cos 2x\, dx$.

Solution: If we let

$$u = \sin 2x, \qquad du = 2 \cos 2x\, dx,$$

then

$$\begin{aligned} \int \sin^3 2x \cos 2x\, dx &= \tfrac{1}{2} \int u^3\, du \Big|_{u=\sin 2x} \\ &= \tfrac{1}{8} u^4 + C \Big|_{u=\sin 2x} \\ &= \tfrac{1}{8} \sin^4 2x + C. \end{aligned}$$

Example 4 Find the area of the region R bounded by the x axis and one arch of the graph of $y = \sin ax$, a a positive number.

Solution: The function f defined by

$$f(x) = \sin ax$$

has period $2\pi/a$ (let $ax = 2\pi$, the period of sine, and solve for x.) One arch of the graph of f occurs in the interval $[0, \pi/a]$, as shown in Figure 8.6. The area of R is given by

$$A(R) = \int_0^{\pi/a} \sin ax\, dx.$$

We can evaluate this integral directly or by changing variables as follows. Let

$$u = ax, \qquad du = a\,dx.$$

Then
$$\int_0^{\pi/a} \sin ax\,dx = \frac{1}{a}\int_0^{\pi} \sin u\,du$$
$$= \frac{1}{a}(-\cos u)\Big|_0^{\pi} = \frac{2}{a}.$$

Thus $A(R) = 2/a$. For example, if $a = 1$, then we see that the area of the region under an arch of the sine curve is 2.

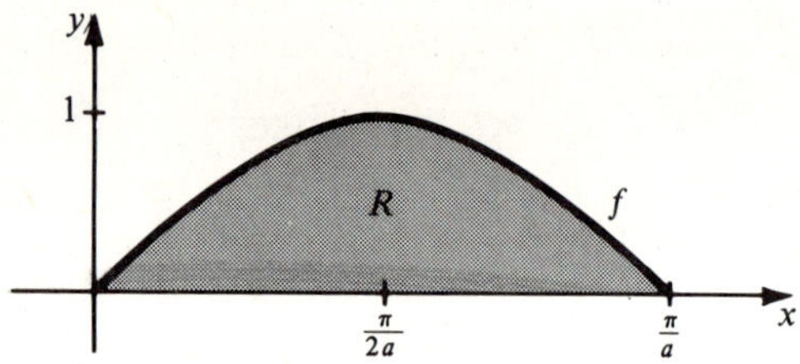

FIGURE 8.6

EXERCISES

Differentiate the following:

1. $\sin 4x$
2. $3 \cos 2x$
3. $\sin^2 3x$
4. $\cos^3 5x$
5. $2 \sin 3x \cos 3x$
6. $\dfrac{\sin x}{1 - \cos x}$
7. $\dfrac{1 + \sin x}{\cos x}$
8. $3 \sin 2x - 4 \sin^2 2x$
9. $\sqrt{1 - \sin x}$
10. $1/\sqrt{\cos^2 x + 2}$
11. $\ln \sin x$
12. $\sinh (\cos x + \sin x)$
13. $\ln (\cos e^x)$
14. $4 \cos x \sin (x^2 + 1)$
15. $\sqrt{\cos^3 x + \pi x}$
16. $\dfrac{\cos x}{4 + 3 \sin x}$

Find $D_x y$:

17. $y \cos x + x \cos y = 4$
18. $\sin (x + y) = y \cos x$
19. $x \sin^2 y + x^3 = y$
20. $y + e^{\cos y} + x = x^2$

Find the integral:

21. $\int_0^{\pi/2} \sin x \, dx$

22. $\int_0^{\pi/2} \cos x \, dx$

23. $\int \sin (2t + 1) \, dt$

24. $\int_0^{\pi/3} \sin 2x \cos^2 2x \, dx$

25. $\int_0^{\pi} \frac{\cos t}{1 + \sin t} \, dt$

26. $\int \frac{\sin x}{(1 + \cos x)^2} \, dx$

27. $\int (\sin 3x - \cos 3x) \, dx$

28. $\int x \sin (x^2 + 1) \, dx$

29. $\int e^x \sin e^x \, dx$

30. $\int (\sin x) e^{\cos x} \, dx$

31. $\int_0^{\pi} (\sin x + x)^4 (\cos x + 1) \, dx$

32. $\int (\sin 4x - \cos 6x) \, dx$

33. $\int e^{\sin x} \cos x \, dx$

34. $\int_1^2 \sin \pi \cos \pi x \, dx$

35. $\int_0^{\pi/2} \sin 2x \, dx$

36. $\int \frac{\cos \sqrt{x}}{\sqrt{x}} \, dx$

37. $\int \sinh (\sin x) \cos x \, dx$

38. $\int \frac{\sin (\ln x)}{x} \, dx$

Find the extrema of the following functions and sketch their graphs in the interval $[0, 2\pi]$.

39. $f(x) = \sin x + \cos x$

40. $F(x) = \sin^2 x + \cos x$

41. $g(x) = 2 \cos x + \cos 2x$

42. $f(x) = 2 \sin x - \sin^2 x$

Find the area of the region bounded by the following and sketch each region:

43. $x = 0, x = \pi/3, y = 0, y = \sin x$

44. The sine curve, the cosine curve, $x = \pi/4$, $x = 5\pi/4$

45. $x = -\pi/4, x = \pi/4, y = 0, y = \cos x$

46. $x = 0, x = \pi/4, y = 0, y = \sin 2x + \cos 2x$

3 THE OTHER TRIGONOMETRIC FUNCTIONS

The other trigonometric functions, tangent, cotangent, secant, and cosecant, are simple combinations of sine and cosine. Thus

$$\tan x = \frac{\sin x}{\cos x}, \qquad \cot x = \frac{\cos x}{\sin x}, \qquad \sec x = \frac{1}{\cos x}, \qquad \csc x = \frac{1}{\sin x}.$$

The domain of tangent and secant is the set of all real numbers except the zeros of cosine, $\{\pm\frac{1}{2}\pi, \pm\frac{3}{2}\pi, \pm\frac{5}{2}\pi, \ldots\}$; that of cotangent and cosecant the set of all real numbers except the zeros of sine, $\{0, \pm\pi, \pm 2\pi, \pm 3\pi, \ldots\}$.

We can find the derivatives of tangent, cotangent, secant, and cosecant in a straightforward manner. Thus

$$D_x \tan x = D_x \frac{\sin x}{\cos x} = \frac{\cos x\, D_x \sin x - \sin x\, D_x \cos x}{\cos^2 x}$$

$$= \frac{\cos^2 x + \sin^2 x}{\cos^2 x} = \frac{1}{\cos^2 x} = \sec^2 x.$$

Also, $$D_x \sec x = D_x (\cos x)^{-1} = -1(\cos x)^{-2} D_x \cos x$$

$$= \frac{\sin x}{\cos^2 x} = \frac{\sin x}{\cos x}\frac{1}{\cos x} = \tan x \cdot \sec x.$$

In this way we derive the following formulas.

8.10 $$D_x \tan x = \sec^2 x.$$

8.11 $$D_x \cot x = -\csc^2 x.$$

8.12 $$D_x \sec x = \sec x \tan x.$$

8.13 $$D_x \csc x = -\csc x \cot x.$$

Example 1 Find $D_x(\sec 2x + \tan 2x)^2$.

Solution: We proceed as follows:

$$\begin{aligned} D_x(\sec 2x + \tan 2x)^2 &= 2(\sec 2x + \tan 2x)D_x(\sec 2x + \tan 2x) \\ &= 2(\sec 2x + \tan 2x)(\sec 2x \tan 2x D_x 2x + \sec^2 2x D_x 2x) \\ &= 2(\sec 2x + \tan 2x)2 \sec 2x(\tan 2x + \sec 2x) \\ &= 4 \sec 2x(\sec 2x + \tan 2x)^2. \end{aligned}$$

Corresponding to the differentiation formulas 8.10 to 8.13 are the following integration formulas:

8.14 $$\int \sec^2 x\, dx = \tan x + C,$$

8.15 $$\int \csc^2 x\, dx = -\cot x + C,$$

8.16 $$\int \sec x \tan x\, dx = \sec x + C,$$

8.17 $$\int \csc x \cot x\, dx = -\csc x + C.$$

Example 2 Find $\int \sec^3 3x \tan 3x\, dx$.

Solution: If we let

$$u = \sec 3x, \qquad du = \sec 3x \tan 3x D_x 3x = 3 \sec 3x \tan 3x\, dx,$$

then

$$\begin{aligned}\int \sec^3 3x \tan 3x\, dx &= \tfrac{1}{3}\int \sec^2 3x(3 \sec 3x \tan 3x\, dx)\\ &= \tfrac{1}{3}\int u^2\, du\Big|_{u=\sec 3x}\\ &= \frac{1}{3}\cdot\frac{u^3}{3} + C\Big|_{u=\sec 3x}\\ &= \tfrac{1}{9}\sec^3 3x + C.\end{aligned}$$

Example 3 Find dy/dx if $\sin(2x - y) = y \cos x$.

Solution: By implicit differentiation,

$$\begin{aligned}D_x \sin(2x - y) &= D_x y \cos x\\ \cos(2x - y) D_x(2x - y) &= (D_x y)\cos x + y D_x \cos x\\ \cos(2x - y)\cdot\left(2 - \frac{dy}{dx}\right) &= \frac{dy}{dx}\cos x - y \sin x\\ 2\cos(2x - y) - \frac{dy}{dx}\cos(2x - y) &= \frac{dy}{dx}\cos x - y\sin x\\ \frac{dy}{dx} &= \frac{2\cos(2x - y) + y \sin x}{\cos(2x - y) + \cos x}.\end{aligned}$$

EXERCISES

I

Differentiate the following:

1. $\tan 4x$
2. $\tan^3 \dfrac{x}{2}$
3. $\dfrac{\sec 3x}{1 + \tan 3x}$
4. $(\csc 3x - \cot 3x)^2$
5. $\sec^2 x \tan^2 x$
6. $\dfrac{\cot 2x - 1}{\csc 2x}$
7. $\ln|\sec x + \tan x|$
8. $e^{2x}\cos 2x$
9. $\sqrt{1 - \tan^2 2x}$
10. $x \sin \dfrac{1}{x}$

11. $e^{\tan x} \sec x$

12. $\ln \left| \cot \frac{x}{4} \right|$

13. $4 \sec \sqrt[3]{t}$

14. $\csc^2 2t - \tan^2 2t$

15. $\ln \dfrac{\tan x}{1 + \sec x}$

16. $\tan (x^2 - 2)$

Find dy/dx and d^2y/dx^2 if:

17. $y = \csc^3 2x$

18. $y = \dfrac{1 - \sin x}{1 + \sin x}$

19. $y = \tan x \sec^2 x$

20. $y = x^2 \tan^2 2x$

Find dy/dx if:

21. $x = \sin y$

22. $x = \tan y$

23. $\cos (x + y) = y \sin x$

24. $\tan (x^2 + y) = 4 + \cot y$

25. $\sec (xy) + y = 1$

26. $\tan (x + y) + \csc (x + y) = y$

Find the extrema of the following functions and sketch their graphs for $0 \leq x \leq 2\pi$:

27. $F(x) = \sec x + \tan x$

28. $f(x) = \tan x - 2x$

29. $g(x) = 2 \sec x - \tan x$

30. $F(x) = \sec x + 2 \cos x$

In each of Exercises 31 to 38 find the integral.

31. $\int_0^{\pi/4} \sec^2 \theta \, d\theta$

32. $\int_{-\pi/4}^{0} \sec x \tan x \, dx$

33. $\int \cot \theta \, d\theta$

34. $\int_1^2 \csc^2 \frac{\pi\theta}{4} \, d\theta$

35. $\int \csc \frac{\pi\theta}{4} \cot \frac{\pi\theta}{4} \, d\theta$

36. $\int \cos^3 x \sin x \, dx$

37. $\int \sec^2 x \tan^2 x \, dx$

38. $\int \csc^2 x \cot x \, dx$

39. Find the area of the region bounded by the graph of $y = \tan x$, the line $x = \pi/4$, and the x axis.

40. Find the area of one of the regions bounded by the graphs of the equations $y = \sec^2 x$ and $y = 2$.

41. Find the area of the region in the first quadrant bounded by the graph of the equation $y = \sin x/(1 + \cos x)$, the x axis and the line $x = \pi/2$.

42. Using Simpson's rule with $n = 6$, approximate the length of one arc of the sine curve.

II

Starting with the identities $\sin(x + y) = \sin x \cos y + \cos x \sin y$ and $\cos(x + y) = \cos x \cos y - \sin x \sin y$, prove each of the following identities:

1. $\tan(x + y) = \dfrac{\tan x + \tan y}{1 - \tan x \tan y}$

2. $2\sin^2 x = 1 - \cos 2x$

3. $2\cos^2 x = 1 + \cos 2x$

4. $\sin 3x = 3\sin x - 4\sin^3 x$

5. $\cos 3x = 4\cos^3 x - 3\cos x$

Using Exercises 1 to 5, evaluate the integrals in Exercises 6 to 11.

6. $\int \sin^2 ax\,dx$

7. $\int x\cos^2 ax^2\,dx$

8. $\int \sin^3 bx\,dx$

9. $\int x^2\cos^3 2x^3\,dx$

10. $\int \dfrac{1 - \cos 2x}{1 + \cos 2x}\,dx$

11. $\int \cos^4 ax\,dx$

12. If the position function s of a moving point on a line satisfies the differential equation $d^2s/dt^2 = -k^2s$, (k a constant), then the point is said to undergo *simple harmonic motion.* Show that $s(t) = A\sin kt + B\cos kt$ satisfies the differential equation for any constants A and B.

13. Using Exercise 12, solve the differential equation $d^2s/dt^2 = -4s$ if s and $v = ds/dt$ have the following values when $t = 0$:

a. $s = 0, v = 10$
b. $s = 5, v = 0$
c. $s = 5, v = 12$

4 INVERSE TRIGONOMETRIC FUNCTIONS

Since the sine function is differentiable, it is everywhere continuous. As we remarked previously, its range is the closed interval $[-1,1]$. From the periodicity of the sine (graphed in Figure 8.7), it is clear that for each number y in $[-1,1]$ there exist infinitely many numbers x such that $\sin x = y$. In fact, for each y in $[-1,1]$ there exists a unique number x in each of the intervals

8.18 $$\ldots, \left[\frac{-3\pi}{2}, \frac{-\pi}{2}\right], \left[\frac{-\pi}{2}, \frac{\pi}{2}\right], \left[\frac{\pi}{2}, \frac{3\pi}{2}\right], \ldots$$

of length π such that $y = \sin x$.

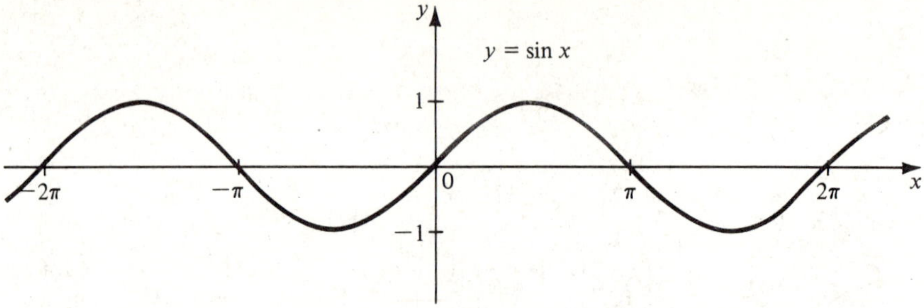

FIGURE 8.7

We now ask if the sine function has an inverse, i.e., if there exists a function f with domain $[-1,1]$ and range the set of all real numbers such that

$$y = f(x) \text{ if and only if } x = \sin y.$$

Since the sine is not a 1–1 function, it is evident that the answer to this question is no. However, if we restrict the sine function to one of the intervals listed in 8.18, then it is strictly monotone and hence has an inverse. It is common practice to restrict the sine to the interval $[-\pi/2, \pi/2]$ in defining its inverse; however, we must realize that any other interval of 8.18 would be equally suitable.

8.19 Definition

The *inverse sine function* is designated by $\sin^{-1}$ and is defined as follows:

$$y = \sin^{-1} x \text{ if and only if } x = \sin y \text{ and } -\frac{\pi}{2} \le y \le \frac{\pi}{2}.$$

By this definition, the function $\sin^{-1}$ has domain $[-1,1]$ and range $[-\pi/2, \pi/2]$. Its graph is sketched in Figure 8.8.

Example 1 Find $\sin^{-1}(-1)$, $\sin^{-1}\frac{1}{2}$, $\sin^{-1} 0$, $\sin^{-1}(-(\sqrt{3}/2))$.

Solution: We have

$$\sin^{-1}(-1) = -\frac{\pi}{2} \text{ because } \sin\left(-\frac{\pi}{2}\right) = -1 \quad\text{and } -\frac{\pi}{2} \le -\frac{\pi}{2} \le \frac{\pi}{2}.$$

$$\sin^{-1}\frac{1}{2} = \frac{\pi}{6} \quad\text{because}\quad \sin\frac{\pi}{6} = \frac{1}{2} \quad\text{and } -\frac{\pi}{2} \le \frac{\pi}{6} \le \frac{\pi}{2}$$

$$\sin^{-1} 0 = 0 \quad\text{because}\quad \sin 0 = 0 \quad\text{and } -\frac{\pi}{2} \le 0 \le \frac{\pi}{2}$$

$$\sin^{-1}\left(-\frac{\sqrt{3}}{2}\right) = -\frac{\pi}{3} \text{ because } \sin\left(-\frac{\pi}{3}\right) = -\frac{\sqrt{3}}{2} \text{ and } -\frac{\pi}{2} \le -\frac{\pi}{3} \le \frac{\pi}{2}.$$

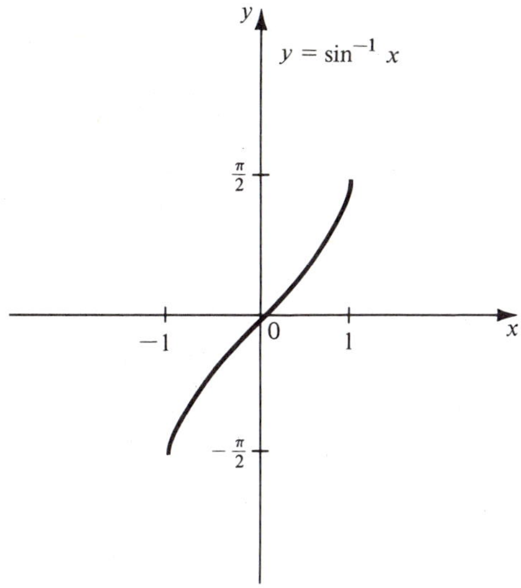

FIGURE 8.8

Because of the possible confusion between

$$\sin^{-1} x, \qquad \text{the inverse sine of } x,$$

and $$(\sin x)^{-1}, \qquad \text{which equals } \frac{1}{\sin x},$$

we shall sometimes call the inverse sine function the *arcsine* function and write

$$\arcsin x \qquad \text{for } \sin^{-1} x.$$

By definition,

8.20
$$\sin(\sin^{-1} x) = \sin(\arcsin x) = x, \qquad x \text{ in } [-1,1].$$
$$\sin^{-1}(\sin y) = \arcsin(\sin y) = y, \qquad y \text{ in } \left[\frac{-\pi}{2}, \frac{\pi}{2}\right]$$

The other inverse trigonometric functions may be defined similarly. Let us carry through the definition of one other such function, namely the inverse tangent.

The tangent function is continuous at every number other than an odd multiple of $\pi/2$. The lines

$$\ldots, \quad x = -\frac{3\pi}{2}, \quad x = -\frac{\pi}{2}, \quad x = \frac{\pi}{2}, \quad x = \frac{3\pi}{2}, \quad \ldots$$

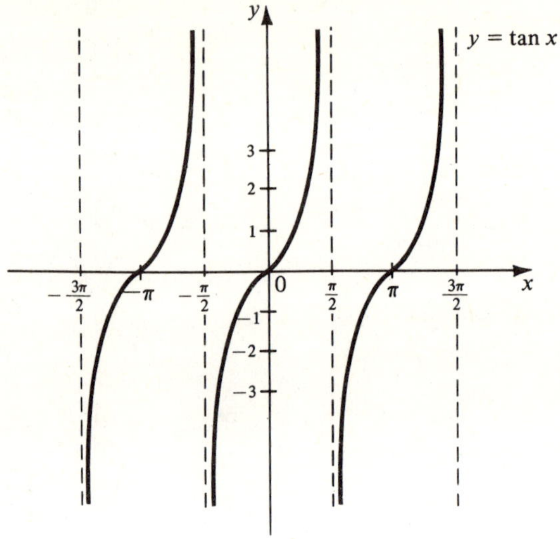

FIGURE 8.9

are vertical asymptotes of the graph of the tangent function, as indicated in Figure 8.9. Since the tangent is not a 1–1 function, it doesn't have an inverse.

However, if we restrict the tangent function to the interval $(-\pi/2, \pi/2)$, then it is strictly increasing and hence has an inverse. Thus, we may make the following definition.

8.21 Definition

The *inverse tangent function*, designated $\tan^{-1}$, is defined by

$$y = \tan^{-1} x \text{ if and only if } x = \tan y \text{ and } -\frac{\pi}{2} < y < \frac{\pi}{2}.$$

By definition, the domain of the inverse tangent function is the set of all real numbers and its range is the open interval $(-\pi/2, \pi/2)$. Its graph is sketched in Figure 8.10. We shall sometimes call the inverse tangent function the *arctangent*. By definition,

8.22

$$\tan(\tan^{-1} x) = \tan(\arctan x) = x, \qquad x \text{ in } (-\infty, \infty).$$

$$\tan^{-1}(\tan y) = \arctan(\tan y) = y, \qquad y \text{ in } \left[-\frac{\pi}{2}, \frac{\pi}{2}\right].$$

Example 2 Find $\tan^{-1}(-1)$, $\arctan\sqrt{3}$, $\tan^{-1} 2$.

Solution: We have

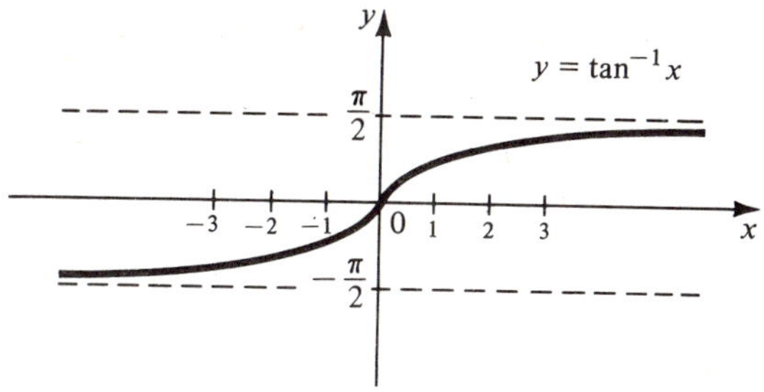

FIGURE 8.10

$$\tan^{-1}(-1) = -\frac{\pi}{4} \text{ because } \tan\left(-\frac{\pi}{4}\right) = -1 \text{ and } -\frac{\pi}{2} < -\frac{\pi}{4} < \frac{\pi}{2}$$

$$\arctan\sqrt{3} = \frac{\pi}{3} \text{ because } \tan\frac{\pi}{3} = \sqrt{3} \text{ and } -\frac{\pi}{2} < \frac{\pi}{3} < \frac{\pi}{2}.$$

We can't tell exactly what $\tan^{-1} 2$ is. From Table IV, page C 5, $\tan 1.10 \doteq 1.96$, $\tan 1.12 \doteq 2.05$, so $\tan 1.11 \doteq 2$. Thus,

$$\tan^{-1} 2 \doteq 1.11 \text{ because } \tan 1.11 \doteq 2 \text{ and } -\frac{\pi}{2} < 1.11 < \frac{\pi}{2}.$$

EXERCISES

Find the following:

1. $\sin^{-1} \frac{1}{2}$
2. $\tan^{-1} 1$
3. $\sin^{-1} \frac{\sqrt{2}}{2}$
4. $\tan^{-1} -\frac{1}{\sqrt{3}}$
5. $\arctan -1$
6. $\arcsin -\frac{\sqrt{3}}{2}$
7. $\sin(\arctan 1)$
8. $\tan\left(\sin^{-1}\frac{\sqrt{3}}{2}\right)$
9. $\csc(\tan^{-1} -\sqrt{3})$
10. $\sin(\cos^{-1} \frac{4}{5})$
11. $\sin^{-1}\left(\cos\frac{\pi}{3}\right)$
12. $\tan(\tan^{-1} 2)$
13. $\sec[\sin^{-1}(-\frac{12}{13})]$
14. $\tan^{-1}\left(\cot\frac{3\pi}{4}\right)$

Simplify the following:

15. $\tan(\sin^{-1} x)$

16. $\cos(2\tan^{-1} x)$

17. $\sin(3\sin^{-1} x)$

18. $\cot(4\cos^{-1} 2x)$

5 DERIVATIVES OF INVERSE TRIGONOMETRIC FUNCTIONS

The derivative of the inverse sine function is as follows.

8.23 Theorem

$$D_x \sin^{-1} x = \frac{1}{\sqrt{1 - x^2}}, \qquad x \text{ in } (-1,1).$$

Proof: If we restrict the sine function to the domain $[-\pi/2, \pi/2]$, then sin and $\sin^{-1}$ are inverse functions. Hence, $\sin^{-1}$ can be differentiated by 3.19. Thus, letting $f = \sin$, $g = \sin^{-1}$,

$$g'(x) = \frac{1}{f'(g(x))}.$$

Hence $\quad D_x \sin^{-1} x = \dfrac{1}{\cos(\sin^{-1} x)} \quad$ if $\cos(\sin^{-1} x) \neq 0$.

Letting $y = \sin^{-1} x$, we have $-\pi/2 \leq y \leq \pi/2$ and $\sin y = x$. Since $\cos y \geq 0$ in $[-\pi/2, \pi/2]$ and $\sin^2 y + \cos^2 y = 1$, evidently

$$\begin{aligned} \cos(\sin^{-1} x) &= \cos y \\ &= \sqrt{1 - \sin^2 y} \\ &= \sqrt{1 - x^2}. \end{aligned}$$

Thus, $\quad D_x \sin^{-1} x = \dfrac{1}{\sqrt{1 - x^2}}, \quad x$ in $(-1,1)$.

The derivative of the arctangent is found similarly.

8.24 Theorem

$$D_x \tan^{-1} x = \frac{1}{1 + x^2}.$$

Proof: If $f(x) = \tan x$, domain $f = (-\pi/2, \pi/2)$ and $g(x) = \tan^{-1} x$, then f and g are inverse functions. Hence, by 3.19,

$$g'(x) = \frac{1}{f'(g(x))}.$$

Thus,
$$D_x \tan^{-1} x = \frac{1}{\sec^2 (\tan^{-1} x)}.$$

If $y = \tan^{-1} x$, then $\tan y = x$ and $\sec^2 y = 1 + \tan^2 y = 1 + x^2$. Thus,

$$D_x \tan^{-1} x = \frac{1}{1 + x^2}.$$

Inspection of 8.23 and 8.24 shows a remarkable feature of the functions $\sin^{-1}$ and $\tan^{-1}$: Their derivatives are *algebraic functions*. Actually, each of the inverse trigonometric functions has this property.

We shall briefly mention one other inverse trigonometric function, namely $\sec^{-1}$, which has domain $(-\infty,-1] \cup [1,\infty)$ and range

$$\left[0, \frac{\pi}{2}\right) \cup \left[\pi, \frac{3\pi}{2}\right)$$

(see Figure 8.11). Its derivative is given by

8.25 $$D_x \sec^{-1} x = \frac{1}{x\sqrt{x^2 - 1}}.$$

The domain of $D_x \sec^{-1}$ is $(-\infty,-1) \cup (1,\infty)$.

Since the inverse trigonometric functions have algebraic derivatives, they will themselves be integrals of algebraic functions. It is convenient to state these integrals in the following forms:

8.26 $$\int \frac{1}{\sqrt{a^2 - x^2}}\, dx = \sin^{-1} \frac{x}{a} + C.$$

8.27 $$\int \frac{1}{a^2 + x^2}\, dx = \frac{1}{a} \tan^{-1} \frac{x}{a} + C.$$

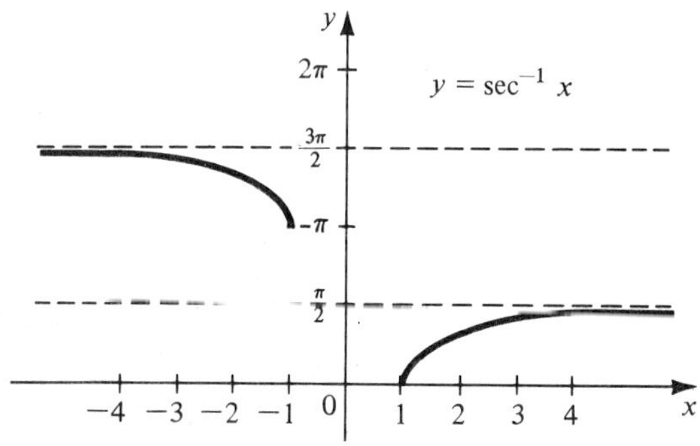

FIGURE 8.11

8.28
$$\int \frac{1}{x\sqrt{x^2 - a^2}}\,dx = \frac{1}{a}\sec^{-1}\frac{x}{a} + C.$$

Proof of 8.26: By 8.23 and the chain rule, we have

$$D_x \sin^{-1}\frac{x}{a} = \frac{1}{\sqrt{1 - (x/a)^2}}\,D_x\frac{x}{a}$$
$$= \frac{1}{a}\frac{1}{\sqrt{1 - (x^2/a^2)}} = \frac{1}{\sqrt{a^2 - x^2}}.$$

The proofs of 8.27 and 8.28 are similar and hence are omitted

Example 1 Find $D_x \tan^{-1}(1/x)$.

Solution: By 8.24 and the chain rule,

$$D_x \tan^{-1}\frac{1}{x} = \frac{1}{1 + (1/x)^2}\,D_x\frac{1}{x} = \frac{1}{1 + (1/x^2)}\left(-\frac{1}{x^2}\right)$$
$$= -\frac{1}{1 + x^2}.$$

Since $D_x \tan^{-1}(1/x) = -D_x \tan^{-1} x$, we must have

$$\tan^{-1}\frac{1}{x} = -\tan^{-1} x + C$$

for some constant C. Since $\tan^{-1} 1 = \pi/4$, we see that $C = \pi/2$. Hence

$$\tan^{-1}\frac{1}{x} = -\tan^{-1} x + \frac{\pi}{2} \qquad \text{for every } x \neq 0.$$

To approximate values of inverse trigonometric functions, we can use the fact that each inverse trigonometric function is an integral of an algebraic function. Consider the following example.

Example 2 Approximate π.

Solution: We know that $\sin^{-1}\frac{1}{2} = \pi/6$, so that

$$\int_0^{1/2} \frac{1}{\sqrt{1 - x^2}}\,dx = \sin^{-1}\tfrac{1}{2} - \sin^{-1} 0 = \frac{\pi}{6}.$$

Also, $\tan^{-1} 1 = \pi/4$, so that

$$\int_0^1 \frac{1}{1 + x^2}\,dx = \tan^{-1} 1 - \tan^{-1} 0 = \frac{\pi}{4}.$$

Thus, if we can approximate one of these integrals, say by Simpson's rule, we can approximate π. Clearly, the inverse tangent integral is easier to approximate, since the integrand does not involve square roots.

Let us approximate $\int_0^1 \frac{1}{1+x^2}\,dx$ by Simpson's rule with $n = 4$. Thus we first make the following table of values.

x	0	$\frac{1}{4}$	$\frac{1}{2}$	$\frac{3}{4}$	1
$1/(1+x^2)$	1	$\frac{16}{17}$	$\frac{4}{5}$	$\frac{16}{25}$	$\frac{1}{2}$

Then, by 6.10,

$$\int_0^1 \frac{1}{1+x^2}\,dx \doteq \tfrac{1}{12}[1 + 4(\tfrac{16}{17}) + 2(\tfrac{4}{5}) + 4(\tfrac{16}{25}) + \tfrac{1}{2}] \doteq .785392.$$

We know that

$$\int_0^1 \frac{1}{1+x^2}\,dx = \tan^{-1} 1 = \frac{\pi}{4}.$$

Thus we have an approximation of π,

$$4(.785392) = 3.141568,$$

accurate, as we know, to four decimal places.

Example 3 Find $\int \frac{x}{\sqrt{4-x^4}}\,dx.$

Solution: If we make a change of variables

$$u = x^2, \qquad du = 2x\,dx,$$

then

$$\int \frac{x}{\sqrt{4-x^4}}\,dx = \frac{1}{2}\int \frac{1}{\sqrt{4-u^2}}\,du\bigg|_{u=x^2}$$

$$= \tfrac{1}{2}\sin^{-1}\frac{u}{2}\bigg|_{u=x^2} + C$$

$$= \tfrac{1}{2}\sin^{-1}\frac{x^2}{2} + C.$$

Example 4 Find $D_x \sec^{-1}\sqrt{x}$.

Solution: Using 8.25 and the chain rule, we have

$$D_x \sec^{-1}\sqrt{x} = \frac{1}{\sqrt{x}\sqrt{(\sqrt{x})^2 - 1}}\,D_x\sqrt{x} = \frac{1}{\sqrt{x}\sqrt{x-1}}\,\frac{1}{2}\,\frac{1}{\sqrt{x}}$$

$$= \frac{1}{2x\sqrt{x-1}}.$$

Example 5 The function cosh is strictly decreasing in the interval $(-\infty,0]$ and strictly increasing in the interval $[0,\infty)$. Find the inverse of the function defined by

$$y = \cosh x, \qquad x \text{ in } [0,\infty).$$

Solution: We have $2y = e^x + e^{-x}$ and

$$e^{2x} - 2ye^x + 1 = 0.$$

We can solve this equation for e^x by the quadratic formula

$$e^x = \frac{2y \pm \sqrt{4y^2 - 4}}{2} = y \pm \sqrt{y^2 - 1}.$$

Since $y - \sqrt{y^2 - 1} \le 1$ whereas $e^x \ge 1$ if x is in $[0,\infty)$, evidently we must have

$$e^x = y + \sqrt{y^2 - 1}$$

and

$$x = \ln(y + \sqrt{y^2 - 1}).$$

Thus

$$\cosh^{-1} y = \ln(y + \sqrt{y^2 - 1})$$

if we take $[0,\infty)$ as the domain of cosh. The domain of $\cosh^{-1}$ clearly is $[1,\infty)$.

If we had taken $(-\infty,0]$ as the domain of cosh, then we would have obtained $\cosh^{-1} y = \ln(y - \sqrt{y^2 - 1})$.

Example 6 Find the derivative of the function $\cosh^{-1}$ of Example 5.

Solution: If we let $f(x) = \cosh x$, domain $f = [0,\infty]$, and $g(x) = \cosh^{-1} x$, then

$$g'(x) = \frac{1}{f'(g(x))}$$

and

$$D_x \cosh^{-1} x = \frac{1}{\sinh(\cosh^{-1} x)}.$$

However, if we let $y = \cosh^{-1} x$, then $x = \cosh y$ and by 7.28

$$\sinh y = \sqrt{\cosh^2 y - 1} = \sqrt{x^2 - 1}.$$

Thus,

$$D_x \cosh^{-1} x = \frac{1}{\sqrt{x^2 - 1}}, \qquad |x| > 1.$$

EXERCISES

1. Define the inverse cosine function with the equation $\cos y = x$, $0 \le y \le \pi$. Sketch its graph and find its derivative.
2. Define the inverse cotangent function with the equation $\cot y = x$, $0 < y < \pi$. Sketch its graph and find its derivative.
3. Define the inverse secant function with the equation $\sec y = x$, where either $0 \le y < \pi/2$ or $\pi \le y < 3\pi/2$. Sketch its graph and verify 8.25.
4. How would you define the function $\csc^{-1}$? Sketch its graph and find its derivative.

In each of Exercises 5 to 30, differentiate.

5. $\sin^{-1} 2x$
6. $\tan^{-1}(x + 1)$
7. $\sec^{-1} 4x$
8. $\sin^{-1} e^x$
9. $(1 + \arcsin 3x)^2$
10. $\arctan \frac{1}{x}$

11. $\dfrac{\arctan e^{2x}}{e^{2x}}$

12. $2x \arctan x - \ln(1 + x^2)$

13. $x \sin^{-1} x + \sqrt{1 - x^2}$

14. $\operatorname{arcsec} \sqrt{x^2 - 1}$

15. $\tan^{-1} \sqrt{x^2 - 1}$

16. $\sin^{-1} \sqrt{1 - x^2}$

17. $\sec^{-1} \sqrt{x}$

18. $\sqrt{\sin^{-1} 3x}$

19. $\ln \arctan x$

20. $\sqrt{1 - x^2} + \arcsin x$

21. $\arcsin x - x\sqrt{1 - x^2}$

22. $\tan^{-1} \dfrac{1 + 2x}{2 - x}$

23. $2x^3 \tan^{-1} x + \ln(1 + x^2) - x^2$

24. $\sec \dfrac{1}{x}$

25. $\ln(\arcsin e^x)$

26. $(\tan^{-1} 3x)^3$

27. $\sec^{-1}(\sinh x)$

28. $\sec^{-1}(\csc x + \sec \sqrt{x})$

29. $\tan^{-1} x^3$

30. $(\sin^{-1} x^2)^\pi$

31. Prove that $\tan^{-1} \dfrac{x + 1}{x - 1} + \tan^{-1} x = c$, a constant, and find c. (*Hint:* Prove that the derivative of the left side is 0.)

32. Prove that $\sec^{-1} x = \pi/2 - \sin^{-1}(1/x)$ for every $x \geq 1$.

Find the following integrals.

33. $\displaystyle\int_0^{\sqrt{3}/2} \frac{1}{\sqrt{1 - x^2}}\, dx$

34. $\displaystyle\int_0^3 \frac{1}{x^2 + 9}\, dx$

35. $\displaystyle\int_{\sqrt{3}}^{3\sqrt{3}} \frac{1}{x^2 + 9}\, dx$

36. $\displaystyle\int \frac{e^x}{e^{2x} + 1}\, dx$

37. $\displaystyle\int_{2/\sqrt{3}}^{\sqrt{2}} \frac{1}{x\sqrt{x^2 - 1}}\, dx$

38. $\displaystyle\int_{-6}^{-3\sqrt{2}} \frac{1}{x\sqrt{x^2 - 9}}\, dx$

39. $\displaystyle\int \frac{\cos x}{1 + \sin^2 x}\, dx$

40. $\displaystyle\int \frac{1}{x\sqrt{1 - \ln^2 x}}\, dx$

41. $\displaystyle\int \frac{1}{\sqrt{3 - 7x^2}}\, dx$

42. $\displaystyle\int \frac{6}{x\sqrt{x^2 - 25}}\, dx$

43. $\displaystyle\int_1^e \frac{1}{x\sqrt{9 - \ln^2 x}}\, dx$

44. $\displaystyle\int_1^\pi \frac{e^{2x}}{7 + e^{2x}}\, dx$

45. $\displaystyle\int \frac{x^2}{16 + x^6}\, dx$

46. $\displaystyle\int_{1/\sqrt{3}}^1 \frac{1}{x\sqrt{9x^2 - 1}}\, dx$

Use implicit differentiation to find dy/dx:

47. $\ln(x^2 + y^2) + 2\tan^{-1}\dfrac{x}{y} = 0$

48. $\sin^{-1} y + x = xy$

49. $\tan^{-1}(x + y) = \sin^{-1}(e^y + x)$

50. $y\cos x + y^3 = \sec^{-1}(x - y)$

51. Find the area of the region bounded by the graph of $y = 1/(1 + x^2)$, the x axis, the y axis, and the line $x = \sqrt{3}$.

52. Show that the hyperbolic sine is an increasing function, and hence that it has an inverse function $\sinh^{-1}$. Thus, $y = \sinh^{-1} x$ if and only if $x = \sinh y$. Find the domain of $\sinh^{-1}$, and prove that $\sinh^{-1} x = \ln(x + \sqrt{x^2 + 1})$. [*Hint:* Solve $x = \sinh y = (e^y - e^{-y})/2$ for y.]

53. Show that the hyperbolic tangent is an increasing function, and hence that it has an inverse function $\tanh^{-1}$. Find the domain of $\tanh^{-1}$, and prove that

$$\tanh^{-1} x = \tfrac{1}{2}\ln\left(\frac{1 + x}{1 - x}\right), \qquad |x| < 1$$

54. Show that the hyperbolic cosine is a decreasing function if $x < 0$ and an increasing function if $x > 0$. If $f(x) = \cosh x$, $x \geq 0$, prove that $f^{-1}(x) = \ln(x + \sqrt{x^2 - 1})$, $x \geq 1$; if $f(x) = \cosh x$, $x \leq 0$, prove that $f^{-1}(x) = -\ln(x + \sqrt{x^2 - 1})$, $x \geq 1$.

55. For the function $\sinh^{-1}$ defined in Exercise 52, show that $D_x \sinh^{-1} x = 1/\sqrt{x^2 + 1}$.

56. For the function $\tanh^{-1}$ defined in Exercise 53, show that $D_x \tanh^{-1} x = 1/(1 - x^2)$.

CHAPTER REVIEW

I

In each of Exercises 1 to 8 find dy/dx.

1. $y = \ln|\sin x|$

2. $y = \sin \ln|x|$

3. $y = \tan \ln x^2$

4. $y = (\sin x^2)^{\exp x^2}$

5. $y = \arcsin(m \sin x)$

6. $y = \arcsin\sqrt{\sin x}$

7. $y = x - \sqrt{1 - x^2}\arcsin x$

8. $y = \frac{2}{3}\arctan x + \frac{1}{3}\arctan\dfrac{x}{1 - x^2}$

In each of Exercises 9 to 14 find the limit:

9. $\lim\limits_{x\to 0} x \sin\dfrac{1}{x}$

10. $\lim\limits_{y\to 0} \dfrac{\sin ay}{\sin by}$

11. $\lim\limits_{x\to 0} \dfrac{\sin x^2}{x}$

12. $\lim\limits_{h\to 0} \dfrac{1 - \cos h}{h^2}$

9

Formal Integration

In this chapter we shall develop methods for finding integrals of some common types of algebraic and transcendental functions.

1 ELEMENTARY INTEGRATION FORMULAS

Several integration formulas were derived in the last four chapters. For example,

$$\int \frac{1}{x}\,dx = \ln |x| + C, \qquad \int \sin x\,dx = -\cos x + C.$$

These and all the others studied so far are tabulated as the first 15 formulas in the Table of Integrals (page B 1).

We are faced with two problems in this chapter. The first is that of extending our basic list of 15 formulas. Although any given (exact) integration formula may be verified by differentiation, we also wish to indicate how the formula might be developed in the first place.

The second problem is that of the use of the formulas. For example, the integral

$$\int \frac{e^x}{e^x + 1}\,dx$$

will not be found in the Table of Integrals. Nevertheless, this integral may easily be evaluated from the table.

The familiar change of variable formula (No. 3 in the table) is one of the important tools available to us:

$$\int f(g(x))g'(x)\,dx = \int f(u)\,du\Big|_{u=g(x)}.$$

It may be used to evaluate the integral above if we let

$$u = e^x + 1, \qquad du = e^x\,dx.$$

Then, using No. 5 of the table, we have

$$\int \frac{e^x}{e^x+1}\,dx = \int \frac{1}{u}\,du\Big|_{u=e^x+1} = \ln|u| + C\Big|_{u=e^x+1}$$
$$= \ln(e^x+1) + C.$$

Another example of the use of the change of variable formula is given below.

Example 1 Find $\int x\sqrt{ax+b}\,dx$, $a \neq 0$.

Solution: If we let

$$u = ax + b, \qquad du = a\,dx,$$

then $x = (u - b)/a$ and

$$\int x\sqrt{ax+b}\,dx = \frac{1}{a^2}\int (u-b)\sqrt{u}\,du\Big|_{u=ax+b}$$
$$= \frac{1}{a^2}\int (u^{3/2} - bu^{1/2})\,du\Big|_{u=ax+b}$$
$$= \frac{1}{a^2}\left(\frac{2}{5}u^{5/2} - \frac{2b}{3}u^{3/2}\right) + C\Big|_{u=ax+b}$$
$$= \frac{1}{a^2}\left[\frac{2}{5}(ax+b)^{5/2} - \frac{2b}{3}(ax+b)^{3/2}\right] + C.$$

Using the same technique, we could evaluate such integrals as

$$\int x^2\sqrt{ax+b}\,dx, \qquad \int x(ax+b)^{3/2}\,dx,$$

and so on.

The trigonometric identities

$$\sin^2\theta = \tfrac{1}{2}(1 - \cos 2\theta)$$
$$\cos^2\theta = \tfrac{1}{2}(1 + \cos 2\theta)$$

are useful in evaluating integrals involving even powers of sine and cosine.

Example 2 Find $\int \sin^2 3x\, dx$.

Solution: Using the first identity above,

$$\int \sin^2 3x\, dx = \tfrac{1}{2}\int (1 - \cos 6x)\, dx.$$

Now

$$\int \cos 6x\, dx = \tfrac{1}{6}\int \cos u\, du\Big|_{u=6x} = \tfrac{1}{6}\sin u + C\Big|_{u=6x}$$
$$= \tfrac{1}{6}\sin 6x + C.$$

Thus,

$$\int \sin^2 3x = \tfrac{1}{2}(x - \tfrac{1}{6}\sin 6x) + C.$$

The quadratic polynomial $x^2 + ax + b$ can be expressed as the sum or difference of two squares by the usual process of the completion of squares. This is useful in evaluating certain integrals, as illustrated in the next examples.

Example 3 Find $\int \frac{1}{x^2 + 4x + 5}\, dx$.

Solution: Since

$$x^2 + 4x + 5 = (x + 2)^2 + 1,$$

we have

$$\int \frac{1}{x^2 + 4x + 5}\, dx = \int \frac{1}{(x + 2)^2 + 1}\, dx$$
$$= \int \frac{1}{u^2 + 1}\, du\Big|_{u=x+2}$$
$$= \tan^{-1} u + C\Big|_{u=x+2} = \tan^{-1}(x + 2) + C.$$

Example 4 Find $\int \frac{x}{\sqrt{7 + 2x - x^2}}\, dx$.

Solution: Since

$$7 + 2x - x^2 = -(x^2 - 2x + 1) + 8 = 8 - (x - 1)^2,$$

we have

$$\int \frac{x}{\sqrt{7 + 2x - x^2}}\, dx = \int \frac{x}{\sqrt{8 - (x - 1)^2}}\, dx$$
$$= \int \frac{u + 1}{\sqrt{8 - u^2}}\, du\Big|_{u=x-1}$$

$$= \left\{\int u(8 - u^2)^{-1/2}\,du + \int \frac{1}{\sqrt{(\sqrt{8})^2 - u^2}}\,du\right\}\Bigg|_{u=x-1}$$

$$= -\sqrt{8 - u^2} + \sin^{-1}\frac{u}{\sqrt{8}} + C\Bigg|_{u=x-1}$$

$$= -\sqrt{7 + 2x - x^2} + \sin^{-1}\frac{x-1}{\sqrt{8}} + C.$$

Formula No. 64 of the Table of Integrals may be developed as in the following example.

Example 5 Find $\int \sec x\,dx$.

Solution: We can change variables by letting

$$u = \sec x + \tan x,$$

in which case

$$\begin{aligned} du &= (\sec x \tan x + \sec^2 x)\,dx \\ &= \sec x(\sec x + \tan x)\,dx \\ &= u \sec x\,dx. \end{aligned}$$

Hence $(1/u)\,du = \sec x\,dx$ and

$$\begin{aligned} \int \sec x\,dx &= \int \frac{1}{u}\,du\Bigg|_{u=\sec x+\tan x} \\ &= \ln|\sec x + \tan x| + C. \end{aligned}$$

EXERCISES

In each of Exercises 1 to 34 find the integral.

1. $\int x(x-2)^{3/2}\,dx$

2. $\int x^2\sqrt{ax+b}\,dx$

3. $\int \frac{2x}{4+\sqrt{x}}\,dx$

4. $\int y(1-y)^{2/3}\,dy$

5. $\int x^3\sqrt{x^2+9}\,dx$

6. $\int \frac{\sqrt{x-4}}{x+4}\,dx$

7. $\int \sin^3 x\, dx$ (*Hint:* Let $\sin^2 x = 1 - \cos^2 x$.)

8. $\int \tan^3 x\, dx$ (*Hint:* Let $\tan^2 x = \sec^2 x - 1$.)

9. $\int \sin^2 x \cos^2 x\, dx$

10. $\int \cos^4 2x\, dx$

11. $\int \frac{x^2}{x + 2}\, dx$

12. $\int \csc x\, dx$

13. $\int \sqrt{x}(2x + 1)^2\, dx$

14. $\int \cos^3 x\, dx$

15. $\int \sin^5 x\, dx$

16. $\int \sin^3 x \cos^3 x\, dx$

17. $\int \frac{\sin^2 x}{\cos^6 x}\, dx$

18. $\int \tan^3 2x \sec^4 2x\, dx$

19. $\int \frac{1}{x^2 - 4x + 13}\, dx$

20. $\int \frac{x - 2}{\sqrt{x^2 - 4x + 13}}\, dx$

21. $\int \frac{1}{\sqrt{2x - x^2}}\, dx$

22. $\int \frac{x}{3x^2 - 2x + 1}\, dx$

23. $\int \frac{x}{3x^4 - 2x^2 + 1}\, dx$

24. $\int x\sqrt[3]{2x + 1}\, dx$

25. $\int \frac{1}{\sqrt{-x^2 - 6x - 6}}\, dx$

26. $\int \frac{1}{3x^2 + 2x - 5}\, dx$

27. $\int \frac{\sin^3 x}{\cos x}\, dx$

28. $\int \sec^2 x\sqrt{\tan x}\, dx$

29. $\int \frac{\tan^3 x}{\sec x}\, dx$

30. $\int \sin^3 x \cos^4 x\, dx$

31. $\int \frac{1}{e^x + 1}\, dx$ [*Hint:* $e^x + 1 = e^x(1 + e^{-x})$]

32. $\int \frac{e^{2x} - 1}{e^{2x} + 1}\, dx$ [*Hint:* $e^{2x} - 1 = e^x(e^x - e^{-x})$, etc.]

33. $\int \cos x\sqrt{1 - \cos x}\, dx$

34. $\int \sin^2 x \cos^2 x\, dx$

35. According to a trigonometric identity, $\sin 2x = 2 \sin x \cos x$. However,

$$\int \sin 2x\, dx = -\tfrac{1}{2} \cos 2x + C_1,$$

whereas

$$\int 2 \sin x \cos x\, dx = \sin^2 x + C_2.$$

Explain the difference in answers.

36. We may evaluate the integral of $\sec^2 x \tan x$ in two ways, namely

$$\int \sec^2 x \tan x\, dx = \int u\, du\Big|_{u=\tan x} = \tfrac{1}{2}\tan^2 x + C_1,$$

$$\int \sec^2 x \tan x\, dx = \int u\, du\Big|_{u=\sec x} = \tfrac{1}{2}\sec^2 x + C_2.$$

Explain the difference in answers.

37. Verify that, for any positive integer n, $\int_0^\pi \sin^2 nx\, dx = \dfrac{\pi}{2}$.

38. Verify that, for any positive integer n, $\int_0^{\pi/n} \sin nx \cos nx\, dx = 0$.

39. Show that $\int_{-1}^{1} \cos m\pi x \sin n\pi x\, dx = 0$ for any positive integers m, n.

40. Find $\int \dfrac{\sqrt{x}}{2 + \sqrt[3]{x}}\, dx$ [*Hint:* Let $x = u^6$.]

2 INTEGRATION BY PARTS

We have not as yet given the integral analog of the product differentiation formula:

$$D_x f(x)g(x) = f(x)g'(x) + g(x)f'(x).$$

This is easily done by integrating each side of the equation above, yielding

$$f(x)g(x) = \int f(x)g'(x)\, dx + \int g(x)f'(x)\, dx,$$

or

9.1 $$\int f(x)g'(x)\, dx = f(x)g(x) - \int g(x)f'(x)\, dx.$$

This formula (No. 16 of the Table of Integrals), called the formula for *integration by parts*, holds for all smooth functions f and g. For definite integrals 9.1 has the form

$$\int_a^b f(x)g'(x)\, dx = f(x)g(x)\Big|_a^b - \int_a^b g(x)f'(x)\, dx.$$

If we let

$$u = f(x), \qquad v = g(x),$$

and

$$du = f'(x)\, dx, \qquad dv = g'(x)\, dx,$$

then 9.1 can be written in the condensed form

9.2 $$\int u\, dv = uv - \int v\, du.$$

The formula for integration by parts allows us to change certain integrals into forms that can be evaluated by previously developed methods. The use of this formula is illustrated by the following examples.

Example 1 Find $\int \ln x\, dx$.

Solution: According to 9.1, we must express the integrand $\ln x$ in the form

$$\ln x = f(x)g'(x)$$

for some functions f and g. The simplest way of doing this is to let

$$f(x) = \ln x \quad \text{and} \quad g'(x) = 1,$$

so that

$$f'(x) = \frac{1}{x} \quad \text{and} \quad g(x) = x.$$

It would not be sensible to let $f(x) = 1$ and $g'(x) = \ln x$, since the problem of finding g is the problem of finding an antiderivative of the logarithmic function, which is equivalent to that of evaluating the given integral.

Furthermore, we note that the logarithm has an algebraic derivative and hence the choice $f(x) = \ln x$ effects a simplification. (Other transcendental functions with algebraic derivatives are the inverse trigonometric functions.)

In the u, v notation, we let

$$u = \ln x, \qquad dv = dx,$$

so that

$$du = \frac{1}{x}\, dx, \qquad v = x.$$

Hence, by 9.2,

$$\int \ln x\, dx = x \ln x - \int x \frac{1}{x}\, dx = x \ln x - x + C.$$

This is essentially No. 72 of the Table of Integrals.

Example 2 Find $\int_0^{\pi} x \sin x\, dx$.

Solution: These are two obvious choices for u and v, namely

(1) $$u = x, \qquad dv = \sin x\, dx,$$

(2) $$u = \sin x, \qquad dv = x\, dx.$$

In case (1) we have

(3) $$du = dx, \qquad v = -\cos x;$$

in case (2),

(4) $$du = \cos x\, dx, \qquad v = \frac{x^2}{2}.$$

Integrating by parts in case (1), we have

$$\int_0^\pi x \sin x \, dx = -x \cos x \Big|_0^\pi - \int_0^\pi (-\cos x) \, dx$$

$$= -\pi \cos \pi + 0 \cos 0 + \sin \pi - \sin 0$$

$$= \pi.$$

Integrating by parts in case (2), we get

$$\int_0^\pi x \sin x \, dx = \frac{x^2}{2} \sin x \Big|_0^\pi - \frac{1}{2} \int_0^\pi x^2 \cos x \, dx.$$

This latter integral certainly is no easier to evaluate than the given one. Clearly case (1) is the better choice of u and v. Note that we have essentially established No. 48 of the Table of Integrals.

Example 3 Find $\int \frac{x^3}{\sqrt{1 + x^2}} \, dx$.

Solution: If we let

$$u = x^3, \qquad dv = \frac{1}{\sqrt{1 + x^2}} \, dx,$$

then the new integral $\int v \, du$ is no easier to evaluate than the given one. However, if we let

$$u = x^2, \qquad dv = \frac{x}{\sqrt{1 + x^2}} \, dx = x(1 + x^2)^{-1/2} \, dx,$$

then

$$du = 2x \, dx, \qquad v = \int x(1 + x^2)^{-1/2} \, dx = \sqrt{1 + x^2},$$

and

$$\int \frac{x^3}{\sqrt{1 + x^2}} \, dx = x^2 \sqrt{1 + x^2} - \int 2x\sqrt{1 + x^2} \, dx$$

$$= x^2 \sqrt{1 + x^2} - \tfrac{2}{3}(1 + x^2)^{3/2} + C = \frac{x^2 - 2}{3} \sqrt{1 + x^2} + C.$$

Example 4 Find $\int e^x \cos x \, dx$.

Solution: If we let

$$u = e^x, \qquad dv = \cos x \, dx,$$

then

$$du = e^x \, dx, \qquad v = \sin x,$$

and

(1) $$\int e^x \cos x\,dx = e^x \sin x - \int e^x \sin x\,dx.$$

Clearly, the new integral is of the same type as the given one, and cannot be evaluated by known methods.

If we try to integrate by parts

$$\int e^x \sin x\,dx$$

by letting

$$u = e^x, \qquad dv = \sin x\,dx,$$

then

$$du = e^x\,dx, \qquad v = -\cos x,$$

and we get

(2) $$\int e^x \sin x\,dx = -e^x \cos x + \int e^x \cos x\,dx.$$

Substituting (2) in (1), we have

$$\int e^x \cos x\,dx = e^x \sin x - \left(-e^x \cos x + \int e^x \cos x\,dx\right)$$

$$= e^x \sin x + e^x \cos x - \int e^x \cos x\,dx.$$

Transposing the latter integral to the other side of the equation, we get

$$2\int e^x \cos x\,dx = e^x \sin x + e^x \cos x,$$

and thus $$\int e^x \cos x\,dx = \frac{e^x}{2}(\sin x + \cos x) + C.$$

This is a special case of No. 71 of the Table of Integrals.

Example 5 Find $\int \sec^3 x\,dx$.

Solution: The easy power of the secant to integrate is $\sec^2 x$; thus, let

$$u = \sec x, \qquad dv = \sec^2 x\,dx.$$

Then

$$du = \sec x \tan x\,dx, \qquad v = \tan x$$

and

$$\int \sec^3 x\,dx = \sec x \tan x - \int \sec x \tan^2 x\,dx.$$

If we replace $\tan^2 x$ by $\sec^2 x - 1$, we get

$$\int \sec^3 x\,dx = \sec x \tan x - \int (\sec^3 x - \sec x)\,dx$$

$$= \sec x \tan x - \int \sec^3 x\,dx + \ln|\sec x + \tan x|.$$

Transposing $-\int \sec^3 x\,dx$, we obtain

$$2\int \sec^3 x\,dx = \sec x \tan x + \ln|\sec x + \tan x|.$$

Thus

$$\int \sec^3 x\,dx = \tfrac{1}{2}(\sec x \tan x + \ln|\sec x + \tan x|) + C.$$

Example 6 Prove that for every integer $n > 1$,

9.3
$$\int \cos^n x\,dx = \frac{1}{n}\sin x \cos^{n-1} x + \frac{n-1}{n}\int \cos^{n-2} x\,dx.$$

Solution: We make the obvious choice for u and dv:

$$u = \cos^{n-1} x, \qquad dv = \cos x\,dx.$$

Then

$$du = (n-1)\cos^{n-2} x \cdot (-\sin x)\,dx, \qquad v = \sin x$$

and

$$\begin{aligned}\int \cos^n x\,dx &= \cos^{n-1} x \sin x + (n-1)\int \cos^{n-2} x \sin^2 x\,dx\\ &= \cos^{n-1} x \sin x + (n-1)\int \cos^{n-2} x(1-\cos^2 x)\,dx\\ &= \cos^{n-1} x \sin x + (n-1)\int \cos^{n-2} x\,dx - (n-1)\int \cos^n x\,dx.\end{aligned}$$

Now transpose the last integral to the left side:

$$\begin{aligned}\int \cos^n x\,dx + (n-1)\int \cos^n x\,dx &= \cos^{n-1} x \sin x + (n-1)\int \cos^{n-2} x\,dx\\ n\int \cos^n x\,dx &= \cos^{n-1} x \sin x + (n-1)\int \cos^{n-2} x\,dx.\end{aligned}$$

On dividing both sides of this equation by n, 9.3 is obtained.

Formula 9.3 (No. 55 of the Table of Integrals) is known as a *reduction formula* for the reason that the integral of a power of the cosine has been expressed in terms of an integral of a lower power of the cosine. Many other reduction formulas are to be found in the table.

With the aid of 9.3, perhaps using it several times, we can integrate any positive integral power of the cosine. For example, letting $n = 3$, we get

$$\begin{aligned}\int \cos^3 x\,dx &= \tfrac{1}{3}\sin x \cos^2 x + \tfrac{2}{3}\int \cos x\,dx\\ &= \tfrac{1}{3}\sin x \cos^2 x + \tfrac{2}{3}\sin x + C;\end{aligned}$$

letting $n = 4$, we get

$$\int \cos^4 x\,dx = \tfrac{1}{4}\sin x \cos^3 x + \tfrac{3}{4}\int \cos^2 x\,dx.$$

Now using 9.3 with $n = 2$,

$$\int \cos^4 x\, dx = \tfrac{1}{4} \sin x \cos^3 x + \tfrac{3}{4}\left[\tfrac{1}{2} \sin x \cos x + \tfrac{1}{2}\int dx\right]$$

$$= \tfrac{1}{4} \sin x \cos^3 x + \tfrac{3}{8} \sin x \cos x + \tfrac{3}{8} x + C.$$

EXERCISES

1. $\int x \ln x\, dx$
2. $\int x^2 \ln x\, dx$
3. $\int_1^2 \sqrt{x} \ln x\, dx$
4. $\int_0^1 \tan^{-1} x\, dx$
5. $\int x \tan^{-1} x\, dx$
6. $\int x^2 \sin x\, dx$
7. $\int x \cos x\, dx$
8. $\int_{-1}^1 xe^x\, dx$
9. $\int x \sin^2 x\, dx$
10. $\int \sec^5 x\, dx$
11. $\int x^2 e^x\, dx$
12. $\int_0^{\sqrt{3}/2} \frac{x^3}{\sqrt{1 - x^2}}\, dx$
13. $\int \frac{x}{\sqrt{2x + 1}}\, dx$
14. $\int \frac{x \ln x}{(x^2 - 1)^{3/2}}\, dx$
15. $\int e^{2x} \sin 3x\, dx$
16. $\int e^{-x} \cos x\, dx$
17. $\int_0^1 x^3 \sqrt{1 - x^2}\, dx$
18. $\int x \sec^2 x\, dx$
19. $\int \ln (x^2 + 1)\, dx$
20. $\int \frac{x^3}{e^{x^2}}\, dx$
21. $\int x^r \ln x\, dx, r \neq -1$
22. $\int x^{-1} \ln x\, dx$
23. $\int_{-\pi/2}^{\pi/2} x^2 \cos 3x\, dx$
24. $\int_0^1 x \tan^{-1} x\, dx$
25. $\int \csc^3 x\, dx$
26. $\int_1^3 \sec^{-1} \sqrt{x}\, dx$

In each of Exercises 27 to 32 integrate by parts.

27. $\int \sin^{-1} ax\, dx$
28. $\int \sinh^{-1} ax\, dx$
29. $\int \tan^{-1} ax\, dx$
30. $\int \tanh^{-1} ax\, dx$
31. $\int x \tanh^{-1} ax\, dx$
32. $\int x \sin^{-1} bx\, dx$

33. Find $\int (\sin^{-1} x)^2 \, dx$ by first making the substitution $y = \sin^{-1} x$.

Find the following integrals.

34. $\int x^2 \tan^{-1} x \, dx$

35. $\int (\ln x)^2 \, dx$

36. $\int x^{3/2}(\ln x)^2 \, dx$

37. $\int \sin (\ln x) \, dx$

38. $\int \frac{\tan^{-1} x}{x^2} \, dx$

Find a reduction formula for each of the following integrals. The exponent n denotes a positive integer.

39. $\int x^n e^x \, dx$

40. $\int x^n \cos x \, dx$

41. $\int \sin^n x \, dx$

42. $\int \cos^n x \, dx$

43. $\int x^n \sin^{-1} x \, dx$

44. $\int x^n \tan^{-1} x \, dx$

45. $\int \sin^m x \cos^n x \, dx$ (No. 52, Table of Integrals). *Hint:* Set $u = \sin^{m-1} x \cos^n x$, $dv = \sin x \, dx$.

3 TRIGONOMETRIC SUBSTITUTIONS

If in the change of variable formula (No. 3 of the Table of Integrals) the function g has an inverse in some interval, then the integral of $f(g(x))g'(x)$ becomes the integral of $f(u)$ if we let $x = g^{-1}(u)$; that is,

$$\int f(u) \, du = \int f(g(x))g'(x) \, dx \Big|_{x=g^{-1}(u)}.$$

For convenience, let us interchange x and u in this formula, obtaining

9.4
$$\int f(x) \, dx = \int f(g(u))g'(u) \, du \Big|_{u=g^{-1}(x)}.$$

Written in this form, the change of variable formula has many uses, as illustrated below.

An integral of an algebraic function involving square roots of the form

$$\sqrt{a^2 - x^2} \quad \text{or} \quad \sqrt{x^2 \pm a^2}, \qquad a > 0,$$

can often be evaluated by changing the integrand into a trigonometric form. This technique is illustrated below.

Example 1 Given a circle of radius r, the shaded region S of Figure 9.1 is called a *sector with central angle* α. Find the area of S, assuming $0 < \alpha \leq \pi/2$.

Solution: The region S is the union of a triangle OPC and the region CPB. Hence $A(S) = A(OPC) + A(CPB)$. Since point P has coordinates $(r \cos \alpha, r \sin \alpha)$, evidently

$$A(OPC) = \tfrac{1}{2}(r \cos \alpha)(r \sin \alpha) = \frac{r^2}{2} \sin \alpha \cos \alpha.$$

The circle has equation $x^2 + y^2 = r^2$, and therefore

$$A(CPB) = \int_{r \cos \alpha}^{r} \sqrt{r^2 - x^2}\, dx.$$

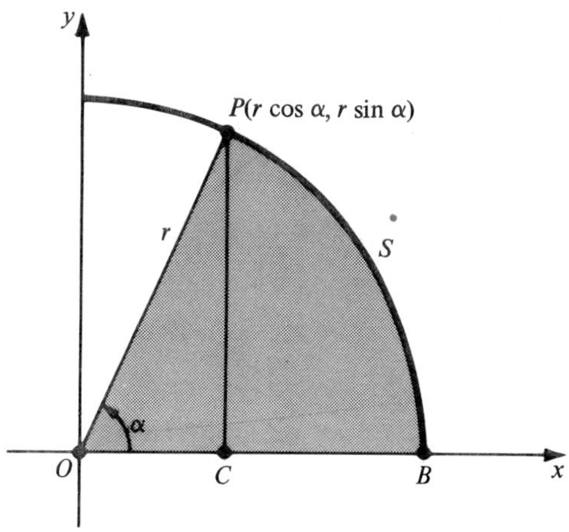

FIGURE 9.1

Hence $$A(S) = \frac{r^2}{2} \sin \alpha \cos \alpha + \int_{r \cos \alpha}^{r} \sqrt{r^2 - x^2}\, dx.$$

We evaluate the above integral by letting

$$x = r \sin u, \qquad u \text{ in } \left[-\frac{\pi}{2}, \frac{\pi}{2}\right].$$

Then $\sqrt{r^2 - x^2} = \sqrt{r^2(1 - \sin^2 u)} = r\sqrt{\cos^2 u} = r|\cos u| = r \cos u.$

We now use 9.4 with

$$x = r \sin u, \qquad dx = r \cos u\, du, \qquad u = \sin^{-1} \frac{x}{r}.$$

Since $u = \sin^{-1} 1 = \pi/2$ when $x = r$ and $u = \sin^{-1} (\cos \alpha) = \pi/2 - \alpha$ [for $\sin (\sin^{-1} (\cos \alpha)) = \cos \alpha = \sin (\pi/2 - \alpha)$], we have

$$A(S) = \frac{r^2}{2} \sin \alpha \cos \alpha + \int_{(\pi/2)-\alpha}^{\pi/2} (r \cos u) r \cos u\, du$$

$$= \frac{r^2}{2} \sin \alpha \cos \alpha + r^2 \int_{(\pi/2)-\alpha}^{\pi/2} \cos^2 u \, du$$

$$= \frac{r^2}{2} \sin \alpha \cos \alpha + \frac{r^2}{2} (\sin u \cos u + u) \Big|_{(\pi/2)-\alpha}^{\pi/2}$$

by No. 56 of the Table of Integrals. Hence

$$A(S) = \frac{r^2}{2} \sin \alpha \cos \alpha + \frac{r^2}{2}\left(\frac{\pi}{2}\right) - \frac{r^2}{2}\left(\cos \alpha \sin \alpha + \frac{\pi}{2} - \alpha\right)$$

$$= \alpha \frac{r^2}{2}.$$

If $\alpha = \pi/2$, then we obtain $\pi r^2/4$ for the area of a quarter-circle, and hence πr^2 for the area of a circle.

The formula $A(S) = \alpha r^2/2$ for the area of sector S with central angle α is the expected result; it says that the area of S is proportional to its central angle; i.e.,

$$\frac{A(S)}{\alpha} = \frac{\pi r^2}{2\pi}.$$

The trigonometric identities

$$\sin^2 \theta + \cos^2 \theta = 1, \qquad \sec^2 \theta = \tan^2 \theta + 1$$

play a basic role in determining the proper substitution for a change of variable. However, since we are employing 9.4 with g as a trigonometric function, we must remember to restrict sufficiently the domain of g so that its inverse g^{-1} exists.

We note in passing that the hyperbolic functions could be used in a similar fashion, using the identity

$$\cosh^2 x - \sinh^2 x = 1.$$

There are three cases when a trigonometric substitution might help solve an integration problem.

Case 1. If the integrand involves

$$\sqrt{a^2 - x^2}, \qquad a > 0,$$

the substitution

$$x = a \sin \theta, \qquad \theta \text{ in } \left[-\frac{\pi}{2}, \frac{\pi}{2}\right]$$

yields

$$\sqrt{a^2 - x^2} = \sqrt{a^2 - a^2 \sin^2 \theta} = a \cos \theta, \qquad \cos \theta \geq 0 \text{ in the range of } \theta.$$

Case 2. If the integrand involves

$$\sqrt{x^2 - a^2}, \qquad a > 0,$$

by substitution

$$x = a \sec \theta, \qquad \theta \text{ in } \left[0, \frac{\pi}{2}\right) \cup \left[\pi, \frac{3\pi}{2}\right)$$

yields

$$\sqrt{x^2 - a^2} = \sqrt{a^2 \sec^2 \theta - a^2} = a \tan \theta, \qquad \tan \theta \geq 0 \text{ in the range of } \theta.$$

Case 3. If the integrand involves

$$\sqrt{x^2 + a^2}, \qquad a > 0,$$

the substitution

$$x = a \tan \theta, \qquad \theta \text{ in } \left(-\frac{\pi}{2}, \frac{\pi}{2}\right),$$

yields

$$\sqrt{x^2 + a^2} = \sqrt{a^2 \tan^2 \theta + a^2} = a \sec \theta, \qquad \sec \theta \geq 1 \text{ in the range of } \theta.$$

Example 2 Find $\displaystyle\int \frac{\sqrt{4 - x^2}}{x^2}\,dx$.

Solution: This is Case 1 with $a = 2$. Thus, we let

$$x = 2 \sin \theta, \qquad dx = 2 \cos \theta \, d\theta, \qquad \theta \text{ in } \left[-\frac{\pi}{2}, \frac{\pi}{2}\right].$$

Then

$$\begin{aligned}
\int \frac{\sqrt{4 - x^2}}{x^2}\,dx &= \int \frac{2 \cos \theta}{4 \sin^2 \theta} \cdot 2 \cos \theta \, d\theta \bigg|_{\theta = \sin^{-1} x/2} \\
&= \int \frac{\cos^2 \theta}{\sin^2 \theta}\, d\theta \bigg|_{\theta = \sin^{-1} x/2} \\
&= \int \frac{1 - \sin^2 \theta}{\sin^2 \theta}\, d\theta \bigg|_{\theta = \sin^{-1} x/2} \\
&= \int (\csc^2 \theta - 1)\, d\theta \bigg|_{\theta = \sin^{-1} x/2} \\
&= (-\cot \theta - \theta) \bigg|_{\theta = \sin^{-1} x/2} + C \\
&= -\frac{\cos \theta}{\sin \theta} - \theta \bigg|_{\theta = \sin^{-1} x/2} + C.
\end{aligned}$$

Since $\sin \theta = x/2$, $\cos \theta = \sqrt{1 - \sin^2 \theta} = \frac{1}{2}\sqrt{4 - x^2}$. Thus,

$$\int \frac{\sqrt{4 - x^2}}{x^2}\,dx = -\frac{\sqrt{4 - x^2}}{x} - \sin^{-1}\frac{x}{2} + C.$$

Example 3 Verify formula No. 29 of the Table of Integrals.

Solution: We evaluate

$$\int \frac{x^2}{\sqrt{x^2 + a^2}}\,dx$$

by letting (Case 3)

$$x = a\tan\theta, \qquad dx = a\sec^2\theta\,d\theta, \qquad -\frac{\pi}{2} < x < \frac{\pi}{2}.$$

Then $\sqrt{x^2 + a^2} = \sqrt{a^2(1 + \tan^2\theta)} = a|\sec\theta| = a\sec\theta,$

and

$$\begin{aligned}
\int \frac{x^2}{\sqrt{x^2 + a^2}}\,dx &= \int \frac{a^2\tan^2\theta}{a\sec\theta}\,a\sec^2\theta\,d\theta\Big|_{\theta=\tan^{-1}x/a} \\
&= a^2\int \tan^2\theta\sec\theta\,d\theta\Big|_{\theta=\tan^{-1}x/a} \\
&= a^2\int (\sec^2\theta - 1)\sec\theta\,d\theta\Big|_{\theta=\tan^{-1}x/a} \\
&= a^2\left\{\int \sec^3\theta\,d\theta - \int \sec\theta\,d\theta\right\}\Big|_{\theta=\tan^{-1}x/a} \\
&= a^2\left\{\tfrac{1}{2}\sec\theta\tan\theta + \tfrac{1}{2}\int \sec\theta\,d\theta - \int \sec\theta\,d\theta\right\}\Big|_{\theta=\tan^{-1}x/a} \\
&= \frac{a^2}{2}(\sec\theta\tan\theta - \ln|\sec\theta + \tan\theta|) + C\Big|_{\theta=\tan^{-1}x/a} \\
&= \frac{x}{2}\sqrt{x^2 + a^2} - \frac{a^2}{2}\ln\left|\frac{\sqrt{x^2 + a^2}}{a} + \frac{x}{a}\right| + C \\
&= \frac{x}{2}\sqrt{x^2 + a^2} - \frac{a^2}{2}\ln|\sqrt{x^2 + a^2} + x| + C',
\end{aligned}$$

where $C' = C + (a^2 \ln a)/2$ is a constant.

If we use the minus sign in No. 29, then we must use the substitution $x = a\sec\theta$ and proceed as above.

We could solve Example 3 by using the hyperbolic functions. Thus, letting

$$x = a\sinh\theta, \qquad dx = a\cosh\theta\,d\theta,$$

we obtain

$$\int \frac{x^2}{\sqrt{x^2 + a^2}}\,dx = \int a^2 \sinh^2 \theta \, d\theta \Big|_{\theta = \sinh^{-1} x/a}$$

$$= \frac{a^2}{2} \int (\cosh 2\theta - 1)\, d\theta \Big|_{\theta = \sinh^{-1} x/a}$$

$$= \frac{a^2}{2} \left(\tfrac{1}{2} \sinh 2\theta - \theta\right) + C \Big|_{\theta = \sinh^{-1} x/a},$$

and so on.

EXERCISES

Find the following integrals.

1. $\int \sqrt{25 - x^2}\,dx$
2. $\int \frac{\sqrt{25 - x^2}}{x}\,dx$
3. $\int \frac{1}{\sqrt{25 - x^2}}\,dx$
4. $\int \sqrt{x^2 - 4}\,dx$
5. $\int \sqrt{9x^2 - 4}\,dx$
6. $\int x\sqrt{9x^2 - 4}\,dx$
7. $\int \frac{1}{x\sqrt{x^2 + 9}}\,dx$
8. $\int \frac{1}{(x^2 + 9)^2}\,dx$
9. $\int \frac{1}{(x^2 - 4)^2}\,dx$
10. $\int \frac{x}{(x^2 - 4)^2}\,dx$

The following integrals are listed in the Table of Integrals. Develop them by the methods of this section.

11. $\int \frac{\sqrt{x^2 - a^2}}{x}\,dx$
12. $\int \frac{\sqrt{x^2 - a^2}}{x^2}\,dx$
13. $\int \frac{\sqrt{a^2 - x^2}}{x^2}\,dx$
14. $\int x^2\sqrt{a^2 - x^2}\,dx$
15. $\int \sqrt{x^2 + a^2}\,dx$
16. $\int \frac{\sqrt{a^2 + x^2}}{x^2}\,dx$
17. $\int \frac{x^2}{\sqrt{x^2 - a^2}}\,dx$
18. $\int \frac{x^2}{\sqrt{a^2 - x^2}}\,dx$
19. $\int \frac{1}{x\sqrt{a^2 - x^2}}\,dx$
20. $\int \frac{1}{x\sqrt{a^2 + x^2}}\,dx$

Find the following integrals.

21. $\int \frac{1}{(x^2 - 4x + 5)^2}\, dx$

22. $\int (x + 3)^2 \sqrt{x^2 + 6x + 8}\, dx$

23. $\int \frac{dx}{x^2\sqrt{16 - x^2}}$

24. $\int \frac{x^2}{(x^2 + 1)^{3/2}}\, dx$

25. $\int \frac{dx}{(x^2 - 2x + 5)^{3/2}}$

26. $\int_1^2 \frac{1}{(9 - x^2)^{3/2}}\, dx$

27. $\int_e^{e^2} \frac{1}{x^3\sqrt{x^2 - 1}}\, dx$

28. $\int_1^2 \frac{1}{x^4\sqrt{x^2 + 5}}\, dx$

4 INTEGRATION OF RATIONAL FUNCTIONS

Let us consider as an example the problem of evaluating the integral

$$\int \frac{2x^4 + 3x^3 - x^2 + x - 1}{x^3 - x}\, dx. \tag{1}$$

By long division we can show that

$$\frac{2x^4 + 3x^3 - x^2 + x - 1}{x^3 - x} = 2x + 3 + \frac{x^2 + 4x - 1}{x^3 - x}, \tag{2}$$

and hence

$$\int \frac{2x^4 + 3x^3 - x^2 + x - 1}{x^3 - x}\, dx = x^2 + 3x + \int \frac{x^2 + 4x - 1}{x^3 - x}\, dx. \tag{3}$$

The integrand of (1) is of the form

$$f(x) = \frac{F(x)}{G(x)},$$

where F and G are polynomial functions. Such a function f is called a *rational function*. Equation (2) illustrates a general theorem which states that

$$\frac{F(x)}{G(x)} = Q(x) + \frac{R(x)}{G(x)},$$

where $Q(x)$ (the quotient) and $R(x)$ (the remainder) are polynomials and $R(x)$ is of degree less than the degree of $G(x)$. If $F(x)$ is of degree less than that of $G(x)$, then $Q(x) = 0$, and $R(x) = F(x)$. Thus the problem of integrating a rational function can always be reduced to one of integrating a quotient of two polynomials where the degree of the numerator is less than the degree of the denominator [as in (3)].

It is easy to verify that

$$(4) \qquad \frac{x^2 + 4x - 1}{x^3 - x} = \frac{1}{x} + \frac{2}{x - 1} - \frac{2}{x + 1},$$

and therefore

$$\begin{aligned}\int \frac{x^2 + 4x - 1}{x^3 - x}\,dx &= \int \frac{1}{x}\,dx + 2\int \frac{1}{x - 1}\,dx - 2\int \frac{1}{x + 1}\,dx \\ &= \ln|x| + 2\ln|x - 1| - 2\ln|x + 1| + C \\ &= \ln\left|\frac{x(x - 1)^2}{(x + 1)^2}\right| + C,\end{aligned}$$

so that the integral (1) has the value

$$x^2 + 3x + \ln\left|\frac{x(x - 1)^2}{(x + 1)^2}\right| + C.$$

In equation (4) we have reduced the quotient $(x^2 + 4x - 1)/(x^3 - x)$ to a sum of *partial fractions*. Obviously, it is this equation which allows us to proceed with the evaluation of (1). It is our purpose in this section to give methods by which these partial fractions may be determined.

Although we shall not give the proof, it can be proved that every polynomial $G(x)$ with real number coefficients can be expressed as a product of linear and quadratic polynomials. For example,

$$\begin{aligned} x^3 - x &= x(x - 1)(x + 1), \\ x^3 + 8 &= (x + 2)(x^2 - 2x + 4), \\ x^4 + 4x^2 + 4 &= (x^2 + 2)^2. \end{aligned}$$

Therefore, starting with a quotient

$$\frac{F(x)}{G(x)}$$

of two polynomials, with the degree of $F(x)$ less than that of $G(x)$, we can first of all factor $G(x)$ into linear and quadratic factors. Having done so, we can hope to express the given quotient as a sum of partial fractions having as denominators factors of $G(x)$.

If $(ax + b)^r$, $r \geq 1$, is the highest power of the linear polynomial $ax + b$ that is a factor of $G(x)$, then included in the sum of partial fractions of $F(x)/G(x)$ will be r terms of the form

$$\frac{A_1}{ax + b} + \frac{A_2}{(ax + b)^2} + \cdots + \frac{A_r}{(ax + b)^r},$$

where $A_1, A_2, \ldots, A_r$ are constants. There will be such a sum associated with each different linear factor of $G(x)$.

If $ax^2 + bx + c$ is a quadratic factor of $G(x)$ that cannot be further factored, and if $(ax^2 + bx + c)^s$, $s \geq 1$, is the highest power of it that is a factor of $G(x)$, then in the sum of partial fractions of $F(x)/G(x)$ there will be included s terms of the form

$$\frac{B_1x + C_1}{ax^2 + bx + c} + \frac{B_2x + C_2}{(ax^2 + bx + c)^2} + \cdots + \frac{B_sx + C_s}{(ax^2 + bx + c)^s},$$

where the B_i and C_i are constants. Such a sum will be associated with each distinct quadratic factor of $G(x)$.

We shall see in the examples below how the numerators of these partial fractions are determined.

Example 1 Establish formula No. 21 of the Table of Integrals.

Solution: According to the discussion above,

$$\frac{1}{x^2 - a^2} = \frac{A}{x - a} + \frac{B}{x + a}$$

for some constants A and B. Adding fractions, we have

$$\frac{1}{x^2 - a^2} = \frac{A(x + a) + B(x - a)}{x^2 - a^2}.$$

Since these fractions have the same denominator, their numerators must be equal,

$$1 = A(x + a) + B(x - a),$$
$$1 = (A + B)x + (A - B)a.$$

The two sides of this equation are equal for every number x if and only if

$$A + B = 0,$$
$$(A - B)a = 1.$$

A simultaneous solution of these two equations is easily found to be

$$A = \frac{1}{2a}, \qquad B = -\frac{1}{2a}.$$

Hence

$$\frac{1}{x^2 - a^2} = \frac{1}{2a}\left(\frac{1}{x - a} - \frac{1}{x + a}\right),$$

and

$$\int \frac{1}{x^2 - a^2}\,dx = \frac{1}{2a}\left(\int \frac{1}{x - a}\,dx - \int \frac{1}{x + a}\,dx\right)$$
$$= \frac{1}{2a}(\ln|x - a| - \ln|x + a|) + C$$
$$= \frac{1}{2a}\ln\left|\frac{x - a}{x + a}\right| + C.$$

This proves No. 21 of the table.

Example 2 Find $\displaystyle\int \frac{x^2 + x + 1}{(2x + 1)(x^2 + 1)}\,dx.$

Solution: We have that

$$\frac{x^2 + x + 1}{(2x + 1)(x^2 + 1)} = \frac{A}{2x + 1} + \frac{Bx + C}{x^2 + 1}$$

for some constants A, B, and C, according to our previous discussion. To determine A, B, and C, we multiply out the right side of the above equation, obtaining

$$\frac{x^2 + x + 1}{(2x + 1)(x^2 + 1)} = \frac{(A + 2B)x^2 + (B + 2C)x + (A + C)}{(2x + 1)(x^2 + 1)}.$$

In order for these fractions to be identically the same, their numerators must be equal:

$$x^2 + x + 1 = (A + 2B)x^2 + (B + 2C)x + (A + C).$$

In turn, these two polynomials are identically the same if the corresponding powers of x have the same coefficients, i.e., if

$$A + 2B = 1, \qquad B + 2C = 1, \qquad A + C = 1.$$

These three equations in the three unknowns A, B, and C may be solved in the usual way to yield

$$A = \tfrac{3}{5}, \qquad B = \tfrac{1}{5}, \qquad C = \tfrac{2}{5}.$$

Hence
$$\frac{x^2 + x + 1}{(2x + 1)(x^2 + 1)} = \frac{1}{5}\left(\frac{3}{2x + 1} + \frac{x + 2}{x^2 + 1}\right),$$

and

$$\int \frac{x^2 + x + 1}{(2x + 1)(x^2 + 1)}\,dx = \frac{1}{5}\left(\int \frac{3}{2x + 1}\,dx + \int \frac{x}{x^2 + 1}\,dx + \int \frac{2}{x^2 + 1}\,dx\right)$$
$$= \tfrac{1}{5}[\tfrac{3}{2}\ln|2x + 1| + \tfrac{1}{2}\ln(x^2 + 1) + 2\tan^{-1} x] + C.$$

Example 3 Find $\displaystyle\int \frac{2x^2 - 3x - 2}{x^3 + x^2 - 2x}\,dx.$

Solution: The denominator factors as $x(x + 2)(x - 1)$; hence

$$\frac{2x^2 - 3x - 2}{x^3 + x^2 - 2x} = \frac{A}{x} + \frac{B}{x + 2} + \frac{C}{x - 1}.$$

On multiplying out the right side of this equation and equating numerators, we get

$$(1) \qquad 2x^2 - 3x - 2 = A(x + 2)(x - 1) + Bx(x - 1) + Cx(x + 2).$$

We can use the method of Example 2 to determine A, B, and C. However, there is an easier way for this example. Since equation (1) is an identity, it holds for every number x. In particular, it holds for $x = 0, 1, -2$. (Note that these are the numbers that make the denominator equal zero.) If we let $x = 0$ in (1), we get

$$-2 = A(2)(-1),$$

and therefore $A = 1$. If we let $x = 1$ in (1), we get

$$-3 = C(1)(3),$$

and hence $C = -1$. When $x = -2$, we have

$$12 = B(-2)(-3),$$

and $B = 2$.

Thus

$$\int \frac{2x^2 - 3x - 2}{x^3 + x^2 - 2x}\,dx = \int \frac{1}{x}\,dx + \int \frac{2}{x+2}\,dx - \int \frac{1}{x-1}\,dx$$

$$= \ln|x| + 2\ln|x+2| - \ln|x-1| + C$$

$$= \ln\left|\frac{x(x+2)^2}{x-1}\right| + C.$$

Example 4 Find $\displaystyle\int \frac{x}{(x-1)^2}\,dx.$

Solution: Here for the first time we have a repeated factor in the denominator. For a repeated linear factor, we have the following sum of partial fractions:

$$\frac{x}{(x-1)^2} = \frac{A}{x-1} + \frac{B}{(x-1)^2}.$$

Equating numerators of each side of this equation, we get

$$x = A(x-1) + B = Ax + (-A + B).$$

Thus we must have

$$A = 1, \qquad -A + B = 0,$$

or $A = 1$, $B = 1$. Hence

$$\int \frac{x}{(x-1)^2}\,dx = \int \frac{1}{x-1}\,dx + \int \frac{1}{(x-1)^2}\,dx$$

$$= \ln|x-1| - \frac{1}{x-1} + C.$$

Example 5 Find $\displaystyle\int \frac{x^3 - 3x^2 + 2x - 3}{(x^2+1)^2}\,dx.$

Solution: In this example there is a repeated quadratic polynomial in the denominator. Hence, according to our previous discussion,

$$\frac{x^3 - 3x^2 + 2x - 3}{(x^2+1)^2} = \frac{A_1x + B_1}{x^2+1} + \frac{A_2x + B_2}{(x^2+1)^2}$$

for some constants A_1, B_1, A_2, and B_2.

An easy way to determine these constants is as follows. By long division,

$$\frac{x^3 - 3x^2 + 2x - 3}{x^2 + 1} = x - 3 + \frac{x}{x^2 + 1},$$

and therefore

$$\frac{x^3 - 3x^2 + 2x - 3}{(x^2 + 1)^2} = \frac{x - 3}{x^2 + 1} + \frac{x}{(x^2 + 1)^2}.$$

Thus $A_1 = 1$, $B_1 = -3$, $A_2 = 1$, and $B_2 = 0$.

We now have

$$\int \frac{x^3 - 3x^2 + 2x - 3}{(x^2 + 1)^2}\, dx = \int \frac{x}{x^2 + 1}\, dx - \int \frac{3}{x^2 + 1}\, dx + \int \frac{x}{(x^2 + 1)^2}\, dx$$

$$= \tfrac{1}{2} \ln (x^2 + 1) - 3 \tan^{-1} x - \frac{1}{2(x^2 + 1)} + C.$$

EXERCISES

Find the following integrals.

1. $\displaystyle\int \frac{x + 1}{x^2 - x}\, dx$

2. $\displaystyle\int \frac{x}{x^2 - 5x + 6}\, dx$

3. $\displaystyle\int \frac{x^3}{x^2 - 2x - 3}\, dx$

4. $\displaystyle\int \frac{6x^2 + 1}{2 - x - 6x^2}\, dx$

5. $\displaystyle\int \frac{3x - 1}{4x^2 - 4x + 1}\, dx$

6. $\displaystyle\int \frac{1}{4x^2 + 12x + 9}\, dx$

7. $\displaystyle\int \frac{x^2 + 1}{x^3 + x^2 - 2x}\, dx$

8. $\displaystyle\int \frac{4x^2 - 3x}{(x + 2)(x^2 + 1)}\, dx$

9. $\displaystyle\int \frac{x^2}{x^4 - 16}\, dx$

10. $\displaystyle\int \frac{1}{x^3 - x^2}\, dx$

11. $\displaystyle\int \frac{x^3 + 1}{x^3 - 4x}\, dx$

12. $\displaystyle\int \frac{x^3 + 1}{x^3 - 1}\, dx$

13. $\displaystyle\int \frac{2x^2 + 1}{(x - 2)^3}\, dx$

14. $\displaystyle\int \frac{x^2 + x + 1}{(x + 1)^3}\, dx$

15. $\displaystyle\int \frac{2x^3 + x^2 + 5x + 4}{x^4 + 8x^2 + 16}\, dx$

16. $\displaystyle\int \frac{x^4 + x^3 + 18x^2 + 10x + 81}{(x^2 + 9)^3}\, dx$

17. $\displaystyle\int \frac{3x + 1}{(x^2 - 4)^2}\, dx$

18. $\displaystyle\int \frac{x^3 + 1}{(4x^2 - 1)^2}\, dx$

The following integrals are listed in the Table of Integrals. Establish them by the methods of this section.

19. $\displaystyle\int \frac{1}{(ax+b)(cx+d)}\,dx$ **20.** $\displaystyle\int \frac{x}{(ax+b)(cx+d)}\,dx$

21. $\displaystyle\int \frac{x}{(ax+b)^2(cx+d)}\,dx$ **22.** $\displaystyle\int \frac{1}{(ax+b)^2(cx+d)}\,dx$

Find the following integrals.

23. $\displaystyle\int \frac{x^3}{x^2-9}\,dx$ **24.** $\displaystyle\int \frac{x}{(x-1)^2(x-2)^2}\,dx$

25. $\displaystyle\int \frac{x^4}{x^2-1}\,dx$ **26.** $\displaystyle\int \frac{x^4}{x^2+1}\,dx$

27. $\displaystyle\int \frac{1}{x^4+x^6}\,dx$ **28.** $\displaystyle\int \frac{x^2}{x^4-1}\,dx$

29. $\displaystyle\int \frac{1}{x^3+1}\,dx$ **30.** $\displaystyle\int \frac{x+1}{x(x^3-1)}\,dx$

31. $\displaystyle\int \frac{e^{2x}}{(e^{2x}+4)^2}\,dx$ **32.** $\displaystyle\int \frac{\sin x}{\cos x+\cos^2 x}\,dx$

5 SEPARABLE DIFFERENTIAL EQUATIONS

Given an equation in x and y such as, for example,

$$\sin x + y^3 = C,$$

C a constant, it is evident that every differentiable function f such that $y = f(x)$ satisfies this equation also satisfies the differential equation

$$\cos x + 3y^2\frac{dy}{dx} = 0.$$

We wish to show in this section that, starting with a differential equation such as the one above, we can find an equation in x and y satisfied by any solution of the differential equation.

The above differential equation is of the type

9.5 $$M(x) + N(y)\frac{dy}{dx} = 0,$$

or, letting $y = f(x)$,

9.6 $$M(x) + N(f(x))f'(x) = 0,$$

where M and N are continuous functions. Such an equation as 9.5 is called a *separable differential equation* (since the variables x and y appear in separate terms). In solving 9.6 we shall seek only smooth solutions f.

If we let

$$F(x) = M(x) + N(f(x))f'(x),$$

then, by 9.6, $F(x) = 0$ and

$$\int F(x)\,dx = C$$

for some constant C. Since

$$\int F(x)\,dx = \int M(x)\,dx + \int N(f(x))f'(x)\,dx,$$

and, by No. 3 of the Table of Integrals,

$$\int N(f(x))f'(x)\,dx = \int N(y)\,dy\bigg|_{y=f(x)},$$

the solution of 9.5 is given by

9.7 $$\int M(x)\,dx + \int N(y)\,dy = C.$$

Thus for every smooth function f such that $y = f(x)$ satisfies 9.5 there is a choice of the constant C such that $y = f(x)$ satisfies 9.7.

For example, the differential equation

$$\cos x + 3y^2 \frac{dy}{dx} = 0$$

has as its solution the equation

$$\int \cos x\,dx + \int 3y^2\,dy = C,$$

or

$$\sin x + y^3 = C.$$

Separable differential equations appear in a natural way in many applications of mathematics. As a matter of fact, the differential equation

$$\frac{dy}{dt} = ky, \qquad y > 0,$$

studied in Section 7 of Chapter 7 is separable, since it can be written in the form

$$k - \frac{1}{y}\frac{dy}{dt} = 0.$$

Let us solve this equation by our present methods.

Example 1 Solve the differential equation

$$k - \frac{1}{y}\frac{dy}{dt} = 0, \qquad y > 0.$$

Solution: By 9.7,

$$\int k\,dt - \int \frac{1}{y}\,dy = C,$$

or

$$kt - \ln y = C.$$

Thus $\ln y = kt - C$, and

$$y = e^{kt-C} = Ae^{kt},$$

where A is a constant (e^{-C}).

Example 2 Solve the differential equation

$$(x + \sec^2 x) + (y - e^y)\frac{dy}{dx} = 0.$$

Solution: For this equation, the functions M and N are defined by

$$M(x) = x + \sec^2 x, \qquad N(y) = y - e^y.$$

By 9.7, its solution is

$$\int (x + \sec^2 x)\,dx + \int (y - e^y)\,dy = C,$$

or

$$\frac{x^2}{2} + \tan x + \frac{y^2}{2} - e^y = C.$$

Example 3 Solve the differential equation

$$\frac{1}{\sqrt{1 - x^2}} + \frac{1}{y}\frac{dy}{dx} = 0, \qquad y > 0.$$

Solution: By 9.7, the solution is

$$\int \frac{1}{\sqrt{1 - x^2}}\,dx + \int \frac{1}{y}\,dy = C,$$

or

$$\sin^{-1} x + \ln y = C.$$

Thus $\ln y = C - \sin^{-1} x$, and

$$y = Ae^{-\sin^{-1} x},$$

where A is a constant (e^C).

In Examples 1 and 3 the solution is given explicitly in the form $y = f(x)$. The solution is given *implicitly* in Example 2 in that the equation is not solved for y in terms of x. It is desirable to give the explicit solution of a differential equation whenever possible. However, the explicit determination of $f(x)$ can offer great difficulties, as it would in the solution of Example 2.

A physical problem that has as its solution a separable differential equation is as follows. Let y be the temperature at time t of a body immersed in a bath of constant temperature a. We shall assume that $y > a$ and hence that the body is being cooled. It is a physical law that the temperature of the body decreases at a rate proportional to the difference between its temperature and the temperature of the surrounding medium (Newton's law of cooling). This law leads to the differential equation

9.8
$$\frac{dy}{dt} = k(y - a), \qquad y > a,$$

for some constant k.

Example 4 A body is immersed in water having a constant temperature of 20°C. The body has initial temperature of 40° and a temperature of 35°C two minutes later. What will its temperature be at the end of 10 minutes?

Solution: Employing 9.8 with $a = 20$, we have the separable differential equation

$$\frac{1}{y - 20}\frac{dy}{dt} = k, \qquad y - 20 > 0,$$

whose solution is

$$\int \frac{1}{y - 20}\, dy = \int k\, dt + C,$$

or

$$\ln (y - 20) = kt + C.$$

Thus

$$y - 20 = e^{kt+C} = Ae^{kt},$$

and

(1)
$$y = Ae^{kt} + 20.$$

We determine A and k as follows. Since $y = 40$ at $t = 0$,

$$40 = A + 20,$$

and $A = 20$. Thus (1) becomes

(2)
$$y = 20e^{kt} + 20.$$

Next, it is given that $y = 35$ when $t = 2$, so that

$$35 = 20e^{2k} + 20.$$

Thus $$e^{2k} = \tfrac{3}{4},$$

and $$k = \tfrac{1}{2} \ln \tfrac{3}{4}.$$

On substituting k in (2), we get

$$y = 20e^{(1/2 \ln 3/4)t} + 20 \tag{3}$$

as the solution of the given problem.

Solution (3) may be written in a more usable form if we observe that

$$e^{\ln 3/4} = \tfrac{3}{4}.$$

Then (3) becomes

$$y = 20(\tfrac{3}{4})^{t/2} + 20. \tag{4}$$

The temperature y when $t = 10$ is given by

$$y = 20(\tfrac{3}{4})^5 + 20 \doteq 24.7°\text{C}.$$

EXERCISES

In each of Exercises 1 to 14 solve the differential equation.

1. $\dfrac{1}{x} + \dfrac{1}{y}\dfrac{dy}{dx} = 0$

2. $\dfrac{1 - x}{x^3} - \dfrac{1}{y^2}\dfrac{dy}{dx} = 0$

3. $\dfrac{dy}{dx} = xy^2$

4. $\dfrac{dy}{dx} = y^2$

5. $\dfrac{y}{y - 1}\dfrac{dy}{dx} - \dfrac{x + 1}{x} = 0$

6. $e^{3x} + 1 + \sin y \dfrac{dy}{dx} = 0$

7. $e^{x-y}\dfrac{dy}{dx} + 1 = 0$

8. $\dfrac{dy}{dx} = \dfrac{1 - x}{1 - y}$

9. $\sec x + \tan y \dfrac{dy}{dx} = 0$

10. $\dfrac{dy}{dx} = \cos^2 y$

11. $\dfrac{dy}{dx} = \dfrac{xy - x}{xy + y}$

12. $\dfrac{dy}{dx} = 1 - y^2$

13. $x\dfrac{dy}{dx} + y = y^2$

14. $\dfrac{dy}{dx} + y = \dfrac{1 + y}{1 - y}$

15. The slope of the tangent line at a point (x,y) on the graph of a function is $3x - 2$. Find the function if the point (4,1) is on its graph.

16. A thermometer reading 80°F is placed in a room whose temperature is 50°F. After 1 min the reading on the thermometer is 70°F.
    - **a.** What is the reading on the thermometer after 3 min?
    - **b.** At what time is the reading on the thermometer 56°F?

17. An object is initially at a temperature of 30°, and is immersed in a bath whose temperature is maintained at 50°. One hour later the temperature of the object is 40°.

a. What is the temperature of the object two hours after being immersed in the bath?

b. When will its temperature be 49°?

18. An object is immersed in a bath which is maintained at 60°. One hour later the temperature of the object is 45°, and two hours later it is 55°. What was its temperature at the instant it was immersed?

19. In a certain chemical reaction two molecules of compound A combine to form one molecule of compound B. The amount x of A present at any time t is given by $dx/dt = -kx^2$, where k is called the rate constant.

a. Initially there are 100 grams of A present and none of B, and one hour later there are 50 grams of A left. Find k.

b. Using the value of k above, how much A will be left after two hours if the amount of A present at the beginning of the reaction was 30 grams?

CHAPTER REVIEW

Find the following integrals.

1. $\int \left(\frac{x-1}{x+1}\right)^3 dx$

2. $\int x^2 \ln (x^3 + 1)\, dx$

3. $\int (\ln x)^3\, dx$

4. $\int \ln (1 + x^2)\, dx$

5. $\int e^{\sqrt{t}}\, dt$

6. $\int \ln (x + \sqrt{x^2 + 1})\, dx$

7. $\int \ln (x + \sqrt{x^2 - a^2})\, dx$

8. $\int \frac{(2 + \sqrt{x})^3}{\sqrt[3]{x}}\, dx$

9. $\int \frac{1}{\sin^2 x + \cos 2x}\, dx$

10. $\int \sin 2x \cos^3 x\, dx$

11. $\int \frac{(\arctan x)^2}{x^2 + 1}\, dx$

12. $\int \frac{1}{(1 - \sin^2 x)\sqrt{1 + \tan x}}\, dx$

13. $\int e^x(\sin e^x)\, dx$

14. $\int \frac{1}{1 + \sin x}\, dx$

15. $\int \cos 2x \cos 3x\, dx$

16. $\int \sin 2x \sin 5x\, dx$

17. $\int x \cdot 5^x\, dx$

18. $\int \arctan \sqrt{x}\, dx$

19. $\int x^2 \ln (1 + x)\, dx$

20. $\int \cos \ln x\, dx$

21. $\int \sqrt{x}\, e^{\sqrt{x}}\, dx$

22. $\int \sin \sqrt[3]{x}\, dx$

23. $\int \sqrt{1 + e^{2x}}\, dx$

24. $\int x \ln (x^3 + 1)\, dx$

25. $\int x^5 e^{-x^2}\, dx$

26. $\int_{-1}^{\sqrt{3}} \frac{x^2}{(25 - x^2)^{3/2}}\, dx$

27. $\int_0^1 \frac{x^2}{x^3 + 4x^2 - 9x - 36}\, dx$

28. $\int_0^{25} \sqrt{25 - \sqrt{x}}\, dx$

29. $\int_0^{\pi/4} \sin^3 2t \cos^3 2t\, dt$

30. $\int_0^2 \frac{x^3}{\sqrt{16 + x^2}}\, dx$

31. $\int_0^1 \sqrt{8y + y^2}\, dy$

32. $\int \frac{1}{\sqrt{6 - 8x - x^2}}\, dx$

33. $\int \cos 6\theta \sin 4\theta\, d\theta$

34. $\int e^{t/3} \cos 3t\, dt$

35. $\int \frac{y + 2}{(y^2 - 4)^2(y^2 + 3)}\, dy$

36. $\int \frac{\sin x}{3 \sin x - 4}\, dx$

Find a reduction formula for each of the following integrals (n denotes a positive integer).

37. $\int (\ln x)^n\, dx$

38. $\int \tan^n x\, dx$

39. $\int x^n \sin^2 x\, dx$

40. $\int \csc^n x\, dx$

41. Show $\int \csc x\, dx = \frac{1}{2} \ln \left| \frac{1 - \cos x}{1 + \cos x} \right| + C.$

42. Find the area of the region bounded by the curve $y = 2 \ln x$, the x axis, and the line $x = e^3$.

43. Find the volume of the solid obtained by revolving about the y axis the region bounded by the curve $y = (x + 1)/(x^2 - 6x + 8)$, the x axis, and the lines $x = 5$, $x = 6$.

44. Find the area of the region bounded by the graph of $f(x) = 2/(x^3 + 1)$, the x axis, the y axis, and the line $x = 2$.

45. Find the area of the region bounded by the catenary $y = 3 \cosh(x/3)$, the x axis, the y axis, and the line $x = a$, $a > 0$.

46. Find the area of the region bounded by the graph of $f(x) = \sin^{-1} 3x$, the x axis, and the line $x = \frac{1}{3}$.

47. The air temperature is 40°. An object in the air cools from 100° to 80° in 20 min. Assuming Newton's law of cooling, find the temperature of the object after 50 min. How long does it take the object to cool down to 68°?

10

Further Applications of the Calculus

Applications of the calculus to such problems as finding the center of gravity of a body and the force of water against a dam will be given in this chapter. Before giving these applications, however, we shall further study conic sections. The conic sections already discussed are the circle and the parabola.

1 THE CENTRAL CONICS; THE ELLIPSE

If a right circular cone of two nappes is cut by a plane not parallel to an edge of the cone and not passing through the vertex of the cone, the resulting curve is either an ellipse (if the plane intersects only one nappe) or a hyperbola (if the plane cuts both nappes). Each of these curves may be defined in an alternate way as follows.

10.1 Definition

An *ellipse* is the set of all points in a plane the sum of whose distances from two fixed points (the foci) in the plane is a constant. A *hyperbola* is the set of all points in a plane the difference of whose distances from two fixed points (the foci) in the plane is a constant.

An ellipse can be constructed from a loop of string in the following way. Place two thumbtacks F and F' (the foci) in the paper and loop the piece of string over them. Then pull the string taut with your pencil point P, as shown in Figure 10.1. Now move the pencil, always keeping the string taut. Since $|FP| + |F'P|$ is always a constant, the curve traced out will be an ellipse, according to 10.1. If F and F' coincide, clearly the ellipse becomes a circle.

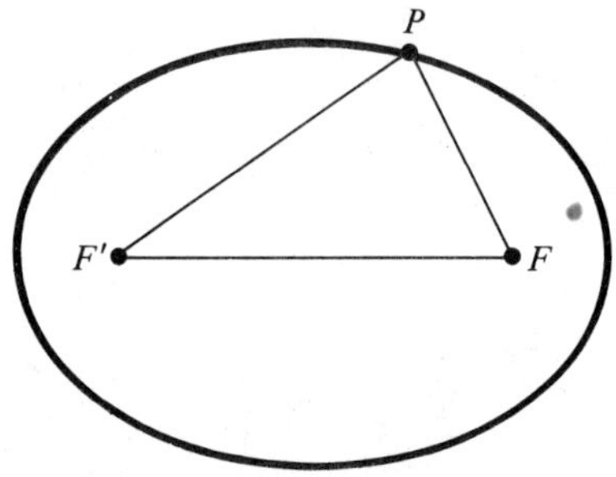

FIGURE 10.1

By placing coordinate axes in the plane, we can find an equation for an ellipse or a hyperbola in much the same way that we found an equation for a parabola. An obvious choice for one axis is the line through the foci and for the other axis is the perpendicular bisector of the segment joining the foci. If the x axis is the axis containing the foci, then the foci will have coordinates $F(c,0)$ and $F'(-c,0)$ for some number $c > 0$.

To find an equation of an ellipse, let $2a$ designate the constant sum of the distances from a point P on the ellipse to the foci F and F'. Evidently, $a > c$. Then a point $P(x,y)$ of the plane is on the ellipse if and only if

$$|FP| + |F'P| = 2a,$$

(see Figure 10.2), i.e., if and only if

$$\sqrt{(x - c)^2 + y^2} + \sqrt{(x + c)^2 + y^2} = 2a.$$

If this equation is rationalized by transposing one of the radicals and then squaring both sides, and similarly for one more step, it reduces to

$$\frac{x^2}{a^2} + \frac{y^2}{a^2 - c^2} = 1.$$

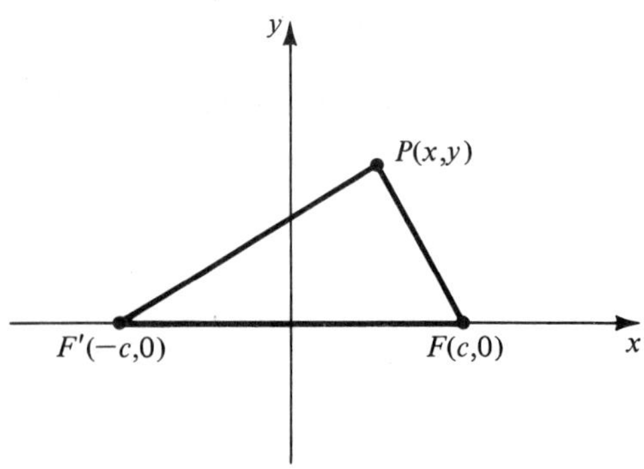

FIGURE 10.2

If we let

10.2 $$b^2 = a^2 - c^2,$$

the above equation becomes

10.3 $$\frac{x^2}{a^2} + \frac{y^2}{b^2} = 1.$$

We have proved that every point $P(x,y)$ on the given ellipse satisfies 10.3. In order to prove that 10.3 is an equation of the ellipse, we must show conversely that every point $P(x,y)$ satisfying 10.3 lies on the given ellipse. The proof of this is left as an exercise for the reader (Exercise 16, page 354).

For each point $P(x,y)$ on the ellipse with 10.3, the points $P_1(-x,y)$, $P_2(x,-y)$, and $P_3(-x,-y)$ are also on the ellipse (Figure 10.3), since their coordinates again are solutions of 10.3. Since P_2 is the image of P in the x axis, evidently the ellipse is symmetric to the x axis. For a similar reason the ellipse is symmetric to the y axis. It follows that the ellipse is also symmetric to the origin; i.e., for each point P on the ellipse the diametrically opposite point P_3 is also on the ellipse. The origin is called the *center* of the ellipse. In 10.3, if $y = 0$, then $x = \pm a$; if $x = 0$, $y = \pm b$. Thus the ellipse cuts the x axis at $(\pm a,0)$ and the y axis at $(0,\pm b)$, as indicated in Figure 10.4. The points $V(a,0)$ and $V'(-a,0)$ are called the *vertices* of the ellipse, the segment VV' is called the *major axis* of the ellipse, and the segment BB' is called the *minor axis* of the ellipse. The major axis of the ellipse contains the foci and is of length $2a$, whereas the minor axis is of length $2b$. From 10.2 it is clear that $a > b$, and therefore the major axis of an ellipse is actually longer than the minor axis (unless the ellipse is a circle).

If the foci of an ellipse are on the y axis and equispaced from the origin, and if $2a$ again designates the sum of the distances of the foci from each point

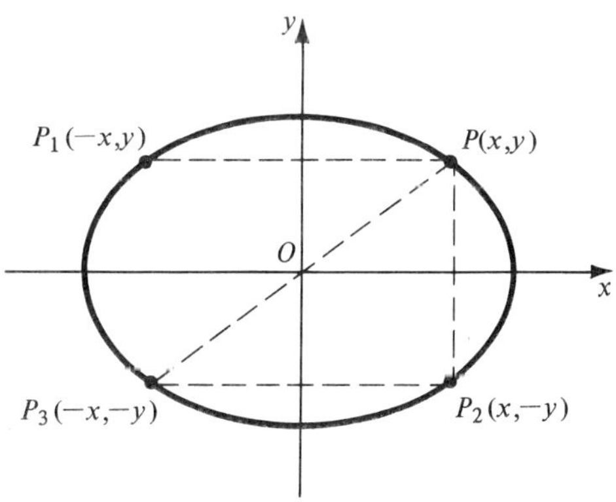

FIGURE 10.3

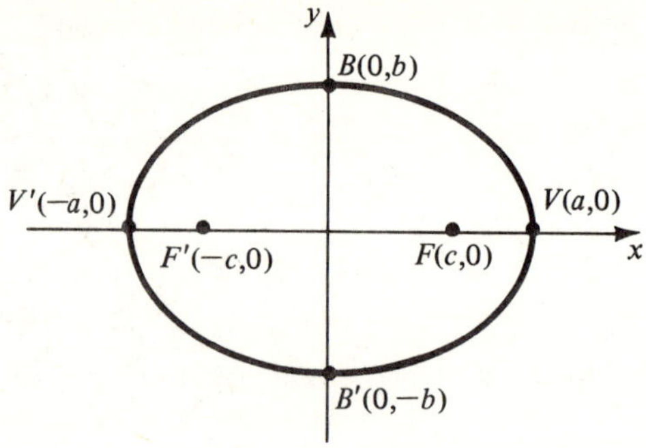

FIGURE 10.4

on the ellipse, then the equation of the ellipse is 10.3 with x and y interchanged; i.e.,

10.4
$$\frac{y^2}{a^2} + \frac{x^2}{b^2} = 1.$$

The vertices and major axis are now on the y axis, as shown in Figure 10.5.

Example 1 Find the equation of the ellipse with foci $(\pm 3,0)$ and vertices $(\pm 5,0)$.

Solution: The ellipse has its center at the origin and its foci on the x axis; therefore its equation is of the form 10.3. Since $c = 3$ and $a = 5$, we have (10.2)

$$b^2 = 5^2 - 3^2 = 16,$$

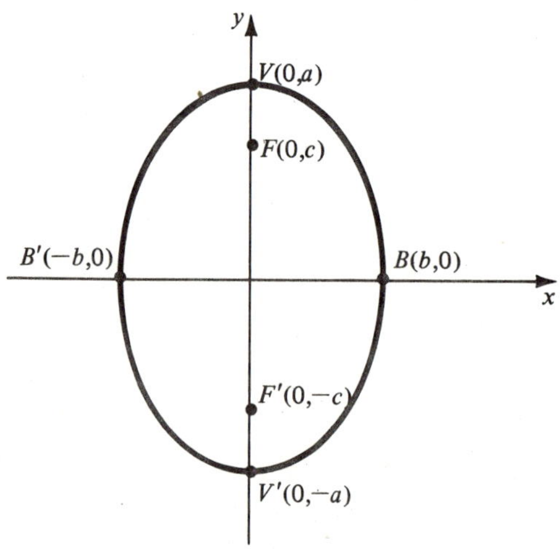

FIGURE 10.5

and $b = 4$. Thus the ellipse has equation

$$\frac{x^2}{25} + \frac{y^2}{16} = 1,$$

or

$$16x^2 + 25y^2 = 400.$$

Example 2 Discuss the graph of the equation

$$4x^2 + y^2 = 4.$$

Solution: If we divide each member of this equation by 4, we get the equation

$$\frac{x^2}{1} + \frac{y^2}{4} = 1.$$

This equation is of the form 10.4 with $a^2 = 4$ and $b^2 = 1$. Thus the graph is an ellipse with major axis on the y axis and minor axis on the x axis. The vertices are $(0, \pm 2)$, and the ends of the minor axis are $(\pm 1, 0)$. Since

$$c^2 = a^2 - b^2 = 3,$$

$c = \sqrt{3}$ and the foci are the points $(0, \pm\sqrt{3})$.

Example 3 Find the area of an elliptical region of the plane.

Solution: Let us find the area of the elliptical region of Figure 10.4. On solving 10.3 for y, we get

$$y = \frac{b}{a}\sqrt{a^2 - x^2}$$

as an equation of the upper half of the ellipse. Hence, by symmetry, the area of the ellipse is given by

$$2\int_{-a}^{a} \frac{b}{a}\sqrt{a^2 - x^2}\,dx \quad \text{or} \quad \frac{2b}{a}\int_{-a}^{a} \sqrt{a^2 - x^2}\,dx.$$

We know that

$$2\int_{-a}^{a} \sqrt{a^2 - x^2}\,dx = \pi a^2$$

is the area of a circle of radius a. Therefore the ellipse with semimajor axis a and semiminor axis b has area $(b/a)(\pi a^2)$, or

$$\pi ab.$$

EXERCISES

I

Discuss and sketch the graph of each of the following equations.

1. $x^2 + 4y^2 = 4$

2. $25x^2 + 16y^2 = 400$

3. $25x^2 + 9y^2 = 225$

4. $4x^2 + 9y^2 = 16$

5. $16x^2 + 25y^2 = 9$
6. $x^2 + 2y^2 = 1$
7. $4x^2 + y^2 = 1$
8. $3x^2 + 4y^2 = 7$

In each of Exercises 9 to 15 find an equation of the ellipse satisfying the given conditions.

9. Foci $(\pm 4,0)$, vertices $(\pm 5,0)$
10. Foci $(0,\pm 2)$, ends of minor axis $(\pm 1,0)$
11. Foci $(0,\pm\sqrt{21})$, end of minor axis $(2,0)$
12. Vertices $(\pm 6,0)$, focus $(-3\sqrt{3},0)$
13. Vertices $(0,\pm 3)$, passing through the point $(\frac{4}{3},2\sqrt{2})$
14. Foci $(\pm\sqrt{5},0)$, passing through the point $(\frac{3}{2},\sqrt{3})$
15. Vertices $(\pm 4,0)$, minor axis of length 5
16. Prove that for every point $P(x,y)$ satisfying 10.3

$$|PF| + |PF'| = 2a$$

(notation of Figure 10.2). [*Hint:* Multiply each side of the equation

$$\sqrt{(x-c)^2 + y^2} + \sqrt{(x+c)^2 + y^2} = 2a$$

by a, then replace a^2y^2 under each radical by $a^2b^2 - b^2x^2$. By 10.2, the resulting equation reduces to $|a^2 - cx| + |a^2 + cx| = 2a^2$. But $x \le a$, etc.]
17. A window arch in the shape of a semi-ellipse is 5 ft wide at the base and 2 ft high. How wide is the window at a height of 1.5 ft above the base?
18. Show that

$$\frac{x_1x}{a^2} + \frac{y_1y}{b^2} = 1$$

is an equation of the tangent line to the ellipse 10.3 at $P(x_1,y_1)$.
19. Define the latus rectum for an ellipse as for a parabola. Show that its length is $2b^2/a$.
20. A diameter of an ellipse is any chord through its center. Show that the minor axis is the shortest and the major axis is the longest diameter if the axes are of unequal lengths.

II

1. Show that there is exactly one square that can be inscribed in a proper ellipse and find its area.
2. What is the maximum area of a rectangle inscribed in the ellipse $x^2/a^2 + y^2/b^2 = 1$?
3. Given the ellipse $x^2/a^2 + y^2/b^2 = 1$ with circumference L, show that

$$L = 4a\int_0^{\pi/2} \sqrt{1 - e^2\sin^2 u}\,du,$$

where $e \doteq \sqrt{a^2 - b^2}/a$ (e is called the eccentricity of the ellipse).

2 THE HYPERBOLA

Let us determine the equation of a hyperbola just as we did that of the ellipse by placing the foci on the x axis equispaced from the origin (Figure 10.6). The point O midway between the foci is called the *center* of the hyperbola. Again, as in the ellipse, we let the coordinates of the foci be $(\pm c,0)$, and we designate the difference of the distances of the foci from a point on the hyperbola by $2a$. Thus the point $P(x,y)$ will be on the hyperbola if and only if

$$|FP| - |F'P| = \pm 2a,$$

(Figure 10.6), i.e., if and only if

$$\sqrt{(x-c)^2+y^2} - \sqrt{(x+c)^2+y^2} = \pm 2a.$$

The plus or minus sign on the right side of the equation is necessitated by the fact that we allow either $|FP|$ or $|F'P|$ to be the larger number, and insist only that the larger number minus the smaller number be $2a$.

In order for the hyperbola to have some point P on it that is not on the x axis, we must have $|F'F| + |FP| > |F'P|$ (i.e., the sum of the lengths of two sides of a triangle exceeds the length of the third side) and $|F'F| + |F'P| > |FP|$. Thus $|F'F| > |F'P| - |FP|$ and $|F'F| > |FP| - |F'P|$, and since $|F'F| = 2c$, $2c > 2a$ and $c > a$. So we insist henceforth that $c > a$.

If we rationalize the above equation of the hyperbola, we obtain the equation

$$\frac{x^2}{a^2} - \frac{y^2}{c^2 - a^2} = 1,$$

or, on letting

10.5 $$b^2 = c^2 - a^2,$$

we get

10.6 $$\frac{x^2}{a^2} - \frac{y^2}{b^2} = 1.$$

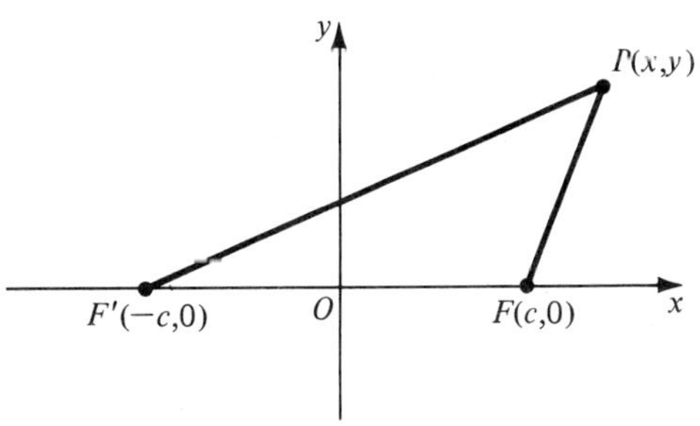

FIGURE 10.6

Every point on the hyperbola satisfies this equation. That, conversely, every point P satisfying 10.6 is on the hyperbola is left as an exercise for the reader. Thus 10.6 is an equation of the given hyperbola.

It is evident from its equation that the hyperbola is symmetric to both axes and its center (the origin in this case). If $y = 0$, then $x = \pm a$, and the hyperbola cuts the x axis at the points $V(a,0)$ and $V'(-a,0)$, called the *vertices* of the hyperbola. The hyperbola does not intersect the y axis. The segment VV' connecting the vertices is called the *transverse axis* of the hyperbola.

Solving 10.6 for y, we obtain

10.7 $$y = \pm \frac{b}{a}\sqrt{x^2 - a^2},$$

from which equation it is evident that the curve does not exist if $x^2 < a^2$. Thus the hyperbola has two branches, one to the right of $x = a$ and the other to the left of $x = -a$. It is sketched in Figure 10.7.

The hyperbola 10.6 has asymptotes, as we shall now show. It is intuitively clear that $\sqrt{x^2 - a^2}$ and x are approximately equal when x is a large positive number. Hence y, given by 10.7, should be approximately equal to

$$y = \pm \frac{b}{a} x.$$

To be more precise, let us prove that

$$\operatorname*{limit}_{x \to \infty} \left(\frac{b}{a} x - \frac{b}{a}\sqrt{x^2 - a^2} \right) = 0,$$

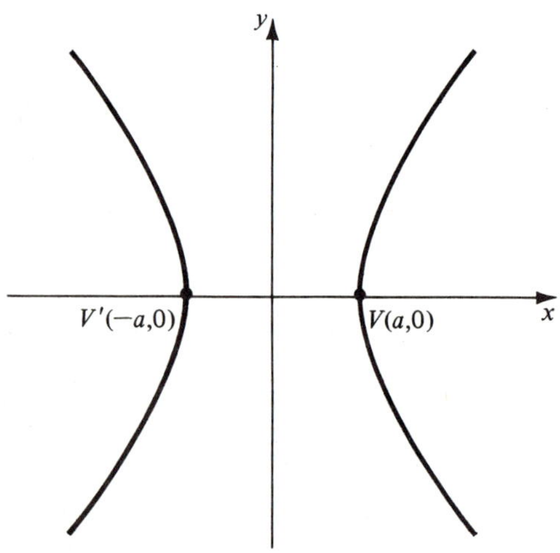

FIGURE 10.7

which will prove that the line $y = bx/a$ is an asymptote of the hyperbola. Since

$$(x - \sqrt{x^2 - a^2})(x + \sqrt{x^2 - a^2}) = a^2,$$

evidently

$$\operatorname*{limit}_{x \to \infty} \frac{b}{a}(x - \sqrt{x^2 - a^2}) = \operatorname*{limit}_{x \to \infty} \frac{ab}{x + \sqrt{x^2 - a^2}} = 0.$$

It can also be shown that

$$\operatorname*{limit}_{x \to -\infty} \left(-\frac{b}{a}x - \frac{b}{a}\sqrt{x^2 - a^2} \right) = 0,$$

which proves that the line $y = -bx/a$ is an asymptote. In a similar manner, it can be shown that the lines $y = \pm bx/a$ are asymptotes of the graph of

$$y = -\frac{b}{a}\sqrt{x^2 - a^2}.$$

Thus the lines

$$y = \pm \frac{b}{a}x$$

are asymptotes of the hyperbola 10.6.

An easy way to construct the asymptotes of a hyperbola is shown in Figure 10.8. The asymptotes are the diagonals of the dotted rectangle. The asymptotes having been drawn in, the hyperbola approaches these lines as indicated in the figure.

If we start with the foci of the hyperbola on the y axis and equispaced from the origin, the equation of the hyperbola takes on the form

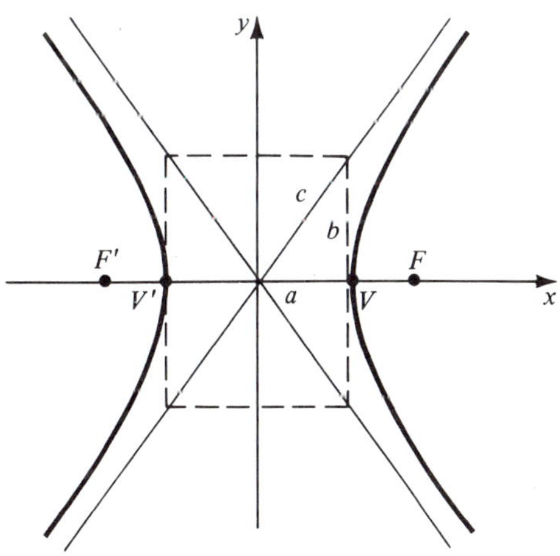

FIGURE 10.8

10.8 $$\frac{y^2}{a^2} - \frac{x^2}{b^2} = 1,$$

where, again, $b^2 = c^2 - a^2$. The vertices of this hyperbola have coordinates $(0, \pm a)$ and the transverse axis is along the y axis. The asymptotes have equations

$$x = \pm \frac{b}{a} y.$$

The hyperbola 10.8 is shown in Figure 10.9.

The ellipse and the hyperbola are called *central conics* for the reason that each has a center of symmetry.

Example 1 Find an equation of the hyperbola with foci $(\pm 5,0)$ and vertices $(\pm 4,0)$.

Solution: The center of this hyperbola is the origin and the transverse axis is along the x axis. Therefore its equation has the form 10.6. Clearly, $c = 5$ and $a = 4$, so that

$$b^2 = c^2 - a^2 = 9.$$

Thus the equation is

$$\frac{x^2}{16} - \frac{y^2}{9} = 1,$$

or $$9x^2 - 16y^2 = 144.$$

The asymptotes of this hyperbola are the lines

$$y = \tfrac{3}{4}x \quad \text{and} \quad y = -\tfrac{3}{4}x.$$

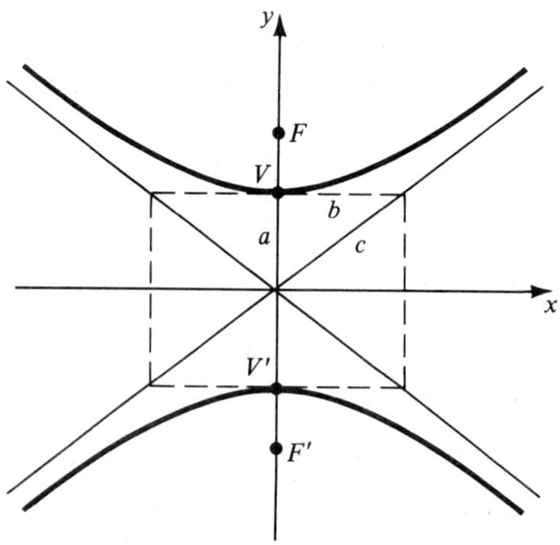

FIGURE 10.9

Example 2 Discuss the graph of the equation

$$4x^2 - 5y^2 + 20 = 0.$$

Solution: If we divide each member of this equation by 20, we obtain the equation

$$\frac{y^2}{4} - \frac{x^2}{5} = 1.$$

This equation has the form 10.8 with $a^2 = 4$ and $b^2 = 5$. Thus

$$c^2 = a^2 + b^2 = 9,$$

and $a = 2$, $b = \sqrt{5}$, and $c = 3$. The transverse axis is along the y axis, the vertices are the points $(0, \pm 2)$, the foci the points $(0, \pm 3)$, and the asymptotes the lines

$$x = \pm \frac{\sqrt{5}}{2} y.$$

EXERCISES

Discuss and sketch the graph of each of the following equations.

1. $x^2 + 4y^2 = 4$
2. $25x^2 - 16y^2 = 400$
3. $9x^2 - 16y^2 + 144 = 0$
4. $16x^2 - 9y^2 + 144 = 0$
5. $x^2 - y^2 = 4$
6. $4x^2 - 4y^2 + 1 = 0$
7. $4x^2 - 9y^2 + 16 = 0$
8. $144x^2 - 25y^2 = 3600$

In each of Exercises 9 to 16, find an equation of the hyperbola satisfying the given conditions.

9. Foci $(\pm 4,0)$, vertices $(\pm 2,0)$
10. Foci $(0,\pm 13)$, vertices $(0,\pm 5)$
11. Foci $(\pm 10,0)$, vertices $(\pm 6,0)$
12. Foci $(0,\pm \sqrt{2})$, vertices $(0,+1)$
13. Vertices $(\pm 2,0)$, asymptotes $y - \pm 2x$
14. Vertices $(0,\pm 4)$, asymptotes $y = \pm 2x/3$
15. Vertices $(0,\pm 2)$, passing through $(3,4)$
16. One focus $(10,0)$, asymptotes $y = \pm 4x$
17. Prove that for every point $P(x,y)$ satisfying Equation 10.6

$$|FP| - |F'P| = \pm 2a$$

(notation of Figure 10.6). (*Hint:* See Exercise I-16, Section 1.)

18. Show that

$$\frac{x_1 x}{a^2} - \frac{y_1 y}{b^2} = 1$$

is an equation of the tangent line to the hyperbola 10.7 at the point $P(x_1,y_1)$.

19. If A and B are the points at which the tangent line to a hyperbola at any point P intersects the asymptotes, show that P is the midpoint of AB.

3 TRANSLATION OF AXES

If, in the coordinate plane with given x and y axes, new coordinate axes are chosen parallel to the given ones, then we shall say that there has been a *translation of axes* in the plane. In Figure 10.10 the given x and y axes have been translated to the x' and y' axes with origin (h,k) relative to the given axes. The positive numbers are assumed to be on the same side of the origin on the new axes as they were on the given axes.

A point P with coordinates (x,y) relative to the given coordinates axes will also have coordinates, say $(x',y')'$, relative to the new axes. These coordinates of P are related to each other by the equations

10.9
$$x' = x - h, \qquad y' = y - k,$$

or

10.10
$$x = x' + h, \qquad y = y' + k.$$

To prove these, let the points O, A, B, O', A', and B' be selected as in Figure 10.10. Then $x = \overline{OA}$, $x' = \overline{O'A'}$, $h = \overline{OB}$, and $k = \overline{OB'}$. Since

$$\overline{OB} = \overline{OA} + \overline{AB} = \overline{OA} + \overline{A'O'} = \overline{OA} - \overline{O'A'},$$

we have

$$h = x - x' \qquad \text{or} \qquad x' = x - h.$$

The other part of 10.9 is proved similarly, whereas 10.10 is just 10.9 written in a slightly different way.

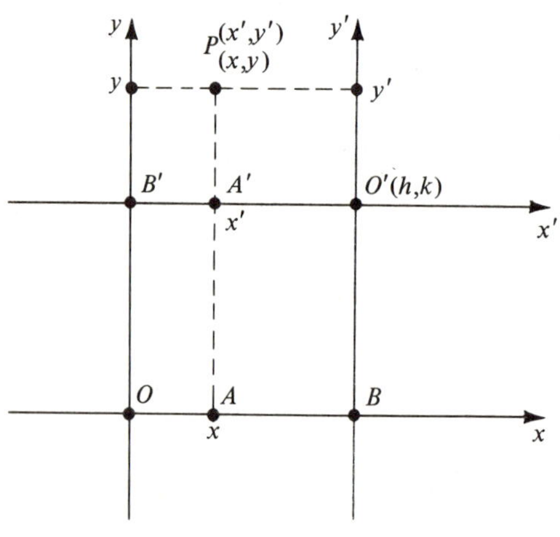

FIGURE 10.10

From a given equation in x and y we may derive an equation in x' and y' simply by replacing x by $x' + h$ and y by $y' + k$ as in 10.10. The graph of the given equation relative to the x and y axes must coincide with the graph of the new equation (in x' and y') relative to the new x' and y' axes, since the point (x,y) satisfies the given equation if and only if $(x',y')'$ satisfies the new equation. We now give an example to illustrate these ideas.

Example 1 Let the given coordinate axes be translated to the new origin (2,3). Find an equation relative to the new x' and y' axes of the graph of the equation

$$x^2 - 4x - 3y + 13 = 0$$

relative to the x and y axes.

Solution: The old and new coordinate axes are shown in Figure 10.11. A point P with coordinates $(x',y')'$ relative to the new coordinate axes has coordinates (x,y) relative to the old axes given by

$$x = x' + 2, \qquad y = y' + 3,$$

according to 10.10. By replacing x by $x' + 2$, and y by $y' + 3$ in the given equation, we obtain the equation

$$(x' + 2)^2 - 4(x' + 2) - 3(y' + 3) + 13 = 0.$$

This simplifies to the equation

$$x'^2 = 3y'.$$

The graph of this equation relative to the x' and y' axes is a parabola with $p = \frac{3}{4}$. It is sketched in Figure 10.11. Thus the graph of the given equation relative to the x and y axes is a parabola with vertex the point (2,3), focus the point $(2,\frac{15}{4})$, and directrix the line $y = \frac{9}{4}$.

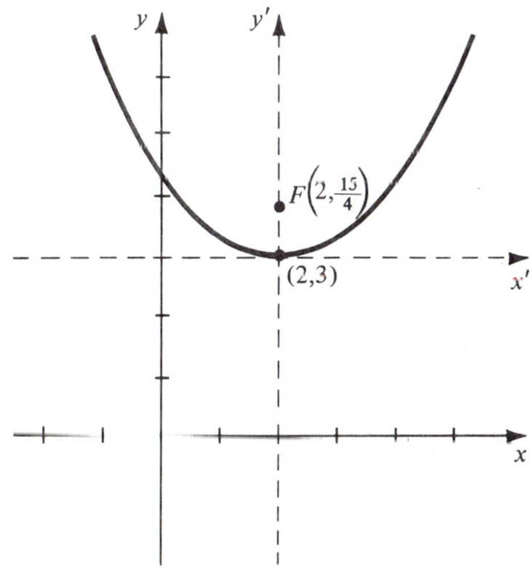

FIGURE 10.11

This example illustrates an important use of translation of axes, namely to reduce an equation to a simpler form so as to facilitate the graphing of the given equation.

The second-degree equation in x and y,

$$x^2 + y^2 + 2 = 0,$$

has no graph at all, since $x^2 \geq 0$, $y^2 \geq 0$, and $x^2 + y^2 + 2 > 0$ for every real number x and y. The equation

$$4(x - 2)^2 + 9(y + 3)^2 = 0$$

has a graph consisting of one point, namely the point $(2, -3)$. The equation

$$y^2 - x^2 = 0$$

has as its graph the two straight lines $y = \pm x$. From these examples it is evident that the graph of a second-degree equation in x and y might not exist, or might consist of just a point or of (one or two) straight lines.

The graph of a second-degree equation of the form

10.11 $$Ax^2 + Cy^2 + Dx + Ey + F = 0,$$

if it exists and is not made up of just a point or straight lines, is necessarily one of the conic sections. A proper translation of axes will reduce the equation into a standard form of one of the conics as given in the previous sections. In Example 1 the second-degree equation reduced to the standard form of the equation of a parabola. More examples of the reduction of equations of the form 10.11 to standard forms are given below.

Example 2 Discuss the graph of the equation

$$9x^2 + 4y^2 - 18x + 16y - 11 = 0.$$

Solution: In order to determine the proper translation of axes to reduce this equation, let us complete the squares on the x and y terms as follows:

$$\begin{aligned} 9(x^2 - 2x) + 4(y^2 + 4y) &= 11, \\ 9(x^2 - 2x + 1) + 4(y^2 + 4y + 4) &= 11 + 9 + 16, \\ 9(x - 1)^2 + 4(y + 2)^2 &= 36, \end{aligned}$$

$$\frac{(x - 1)^2}{4} + \frac{(y + 2)^2}{9} = 1.$$

If we let

$$x' = x - 1, \qquad y' = y + 2,$$

so that $h = 1$ and $k = -2$ in 10.9, the given equation reduces to the standard form of an ellipse (10.4)

$$\frac{x'^2}{4} + \frac{y'^2}{9} = 1.$$

Thus a translation of axes to the new origin $(1,-2)$ shows that the graph of the given equation is an ellipse with axes parallel to the coordinate axes. It is sketched in Figure 10.12.

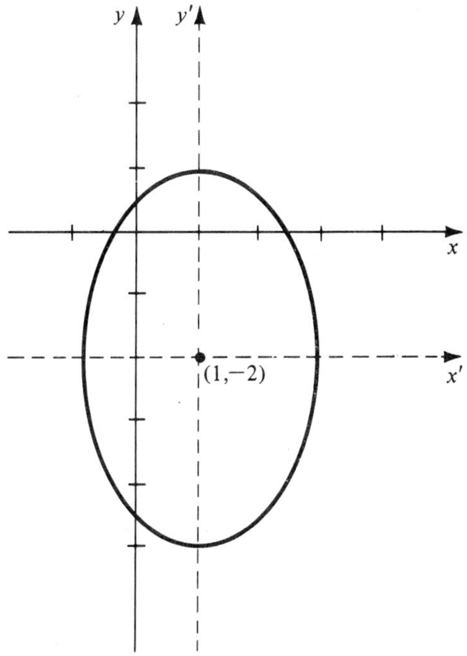

FIGURE 10.12

Example 3 Discuss the graph of the equation

$$x^2 - 4y^2 + 6x + 24y - 31 = 0.$$

Solution: We first complete the squares as follows:

$$(x^2 + 6x + 9) - 4(y^2 - 6y + 9) = 31 + 9 - 36,$$

$$(x + 3)^2 - 4(y - 3)^2 = 4,$$

$$\frac{(x + 3)^2}{4} - \frac{(y - 3)^2}{1} = 1.$$

Letting

$$x' = x + 3, \qquad y' = y - 3,$$

so that $h = -3$ and $k = 3$ in 10.9, the given equation reduces to the standard form of a hyperbola (10.6),

$$\frac{x'^2}{4} - \frac{y'^2}{1} = 1.$$

Hence a translation of axes to the new origin $(-3,3)$ shows that the graph of the given equation is a hyperbola. The graph is sketched in Figure 10.13.

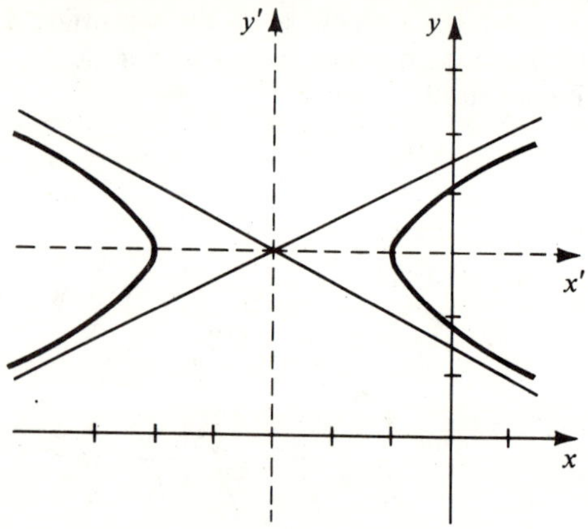

FIGURE 10.13

EXERCISES

I

In each of Exercises 1 to 12, discuss and sketch the graph of the equation.

1. $x^2 - 6x - 4y + 1 = 0$

2. $y^2 + 2y + 8x - 15 = 0$

3. $x^2 + 9y^2 - 4x - 18y + 4 = 0$

4. $2x^2 + 8y^2 - 8x - 16y + 9 = 0$

5. $2x^2 + 3y^2 - 4x + 12y + 8 = 0$

6. $4x^2 - y^2 - 24x - 4y + 36 = 0$

7. $x^2 - y^2 + 4x - 4y + 1 = 0$

8. $x^2 + y^2 + 6x + 5 = 0$

9. $y^2 + 8x - 6 = 0$

10. $x^2 + 2x - 2y + 2 = 0$

11. $9x^2 - 4y^2 - 54x + 45 = 0$

12. $9x^2 - 25y^2 - 90x - 50y - 25 = 0$

13. Find an equation of the parabola with vertex (h,k) and focus $(h + p, k)$.

14. Find an equation of the parabola with vertex (h,k) and focus $(h, k + p)$.

15. Find an equation of the ellipse with center (h,k), foci $(h \pm c, k)$, and vertices $(h \pm a, k)$.

16. Find an equation of the ellipse with center (h,k), foci $(h, k \pm c)$, and vertices $(h, k \pm a)$.

17. Find an equation of the hyperbola with center (h,k), foci $(h \pm c, k)$, and vertices $(h \pm a, k)$.

18. Find an equation of the hyperbola with center (h,k), foci $(h, k \pm c)$, and vertices $(h, k \pm a)$.

19. Find the maximum or minimum point (vertex) of the parabola $y = ax^2 + bx + c$ by a translation of axes.

20. Show how to solve the quadratic equation $ax^2 + bx + c = 0$ by means of Exercise 19.

II

Rotation of Axes

1. If, in a coordinate plane with given x and y axes, new coordinate axes are chosen having the same origin O as the given ones, then we shall say that there has been a *rotation of axes* in the plane. If θ is the angle from the positive half of the x axis to the positive half of the x' axis, as shown in the figure, then we shall say that the new axes are formed by a rotation of axes through an angle θ. If a point P has coordinates (x,y) in the old coordinate system and coordinates $(x',y')'$ in the new, show that

$$\begin{cases} x = x' \cos \theta - y' \sin \theta \\ y = x' \sin \theta + y' \cos \theta \end{cases} \quad \text{and} \quad \begin{cases} x' = x \cos \theta + y \sin \theta \\ y' = -x \sin \theta + y \cos \theta \end{cases}$$

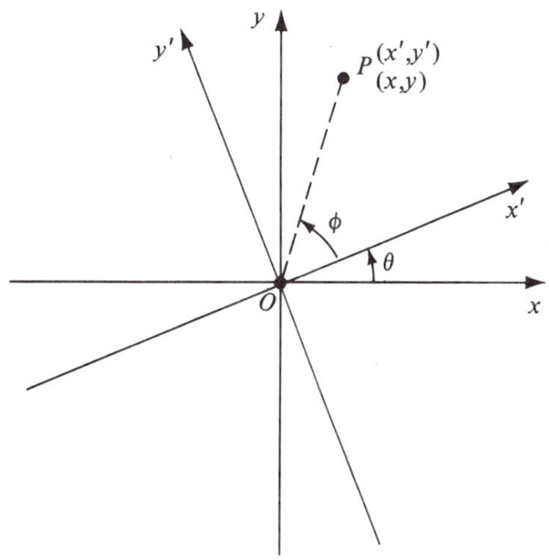

2. Let the given coordinate axes x and y be rotated through an angle of 45°. Find an equation relative to the new axes x' and y' of the graph of the equation $y^2 - x^2 = 4$ in the given coordinate system. Sketch.

3. Show that by a rotation of axes through an angle θ the general second-degree equation $Ax^2 + Bxy + Cy^2 + Dx + Ey + F = 0$ is transformed into a second-degree equation $A'x'^2 + B'x'y' + C'y'^2 + D'x' + E'y' + F' = 0$. Also, show that $B^2 - 4AC$ and $A + C$ are *invariants* of the transformation in the sense that $B^2 - 4AC = B'^2 - 4A'C'$ and $A + C = A' + C'$.

4. Show that the general second-degree equation of Exercise 3 is transformed into an equation in which $B' = 0$ (so that the $x'y'$ term is missing) if θ is chosen so that $\cot 2\theta = (A - C)/B$.

5. Show that the graph of the general second-degree equation $Ax^2 + Bxy + Cy^2 + Dx + Ey + F = 0$ is a hyperbola, a parabola, or an ellipse, depending on whether $B^2 - 4AC$ is positive, zero, or negative, respectively. Under what conditions does the graph degenerate to the empty set, one point, one line, or a pair of lines? (*Hint:* Use Exercise 4.)

6. Given the equation $5x^2 + 4xy + 2y^2 = 6$, show that its graph is an ellipse. Reduce the equation of the ellipse to a standard form by a rotation of axes. Also, reduce the equation to a standard form with the aid of its invariants (see Exercise 3).
7. Find the area of the region bounded by the graph of the equation $2x^2 - \sqrt{3}xy + y^2 = 6$.
8. Find equations of the asymptotes of the graph of the equation $3x^2 - 6xy - 5y^2 + 3 = 0$.

In each of Exercises 9 to 12, reduce the given equation to a standard form by a rotation of axes, and sketch the graph.

9. $x^2 - xy + y^2 - 4x - 4y = 20$
10. $4xy + 3x^2 = 4$
11. $12x^2 + 7xy - 12y^2 - 1 = 0$
12. $9x^2 + 24xy + 16y^2 + 80x - 60y = 0$

4 MOMENTS AND CENTERS OF MASS

The *moment* (of force) of a particle about a line is defined to be the product of the mass of the particle and its distance from the line. We shall find it convenient to consider the particle located on a coordinate plane, and to find the moment of the particle about a coordinate axis (or a line parallel to a coordinate axis). Also, directed distances will be used so that the moment will be positive, negative, or zero, depending on whether the point is on the positive or negative side of the axis, or is on the axis.

In Figure 10.14 the particle of mass m is at the point (x, y) in a coordinate plane. Its moments M_x and M_y about the x and y axes, respectively, are given by

$$M_x = my, \qquad M_y = mx.$$

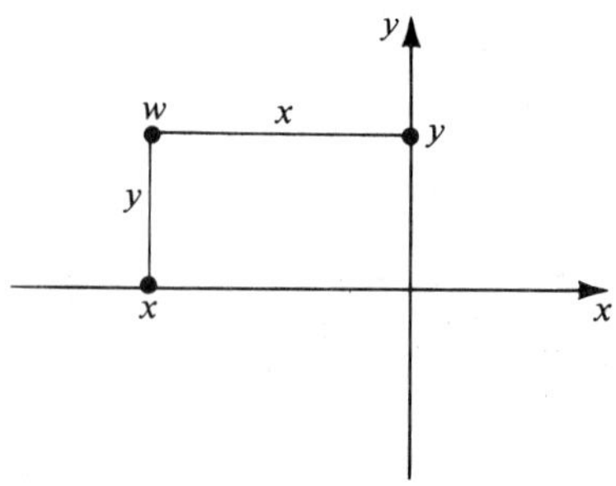

FIGURE 10.14

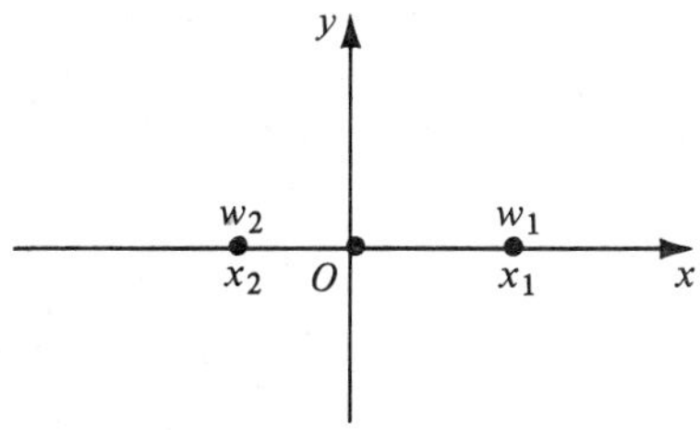

FIGURE 10.15

If the particle is in the second quadrant, as in Figure 10.14, then $M_x > 0$ and $M_y < 0$.

Moments are used to find the *center of mass* of a physical object. As an illustration, consider a seesaw with boys of masses m_1 and m_2 on it. For convenience, we think of the seesaw as being on the x axis with balance point at the origin O (Figure 10.15). The moments of the two boys about the y axis are

$$M_1 = m_1x_1, \qquad M_2 = m_2x_2.$$

The two boys will balance each other provided the sum of their moments is zero, i.e.,

$$m_1x_1 + m_2x_2 = 0.$$

If they balance each other, the balance point O is called the center of mass of the physical system made up of the two boys.

The moments of a system of n particles located in a coordinate plane are defined similarly. Thus if the particles of masses $m_1, m_2, \ldots, m_n$ are at the respective points $(x_1,y_1), (x_2,y_2), \ldots, (x_n,y_n)$ (Figure 10.16), then the moments M_x and M_y of the system of n particles are defined as follows:

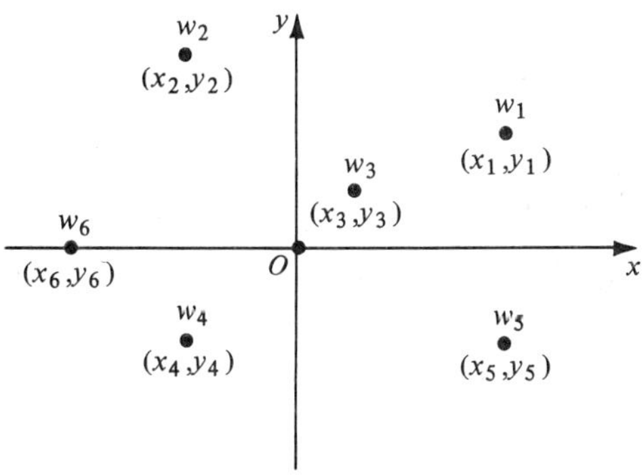

FIGURE 10.16

$$M_x = \sum_{i=1}^{n} m_i y_i, \qquad M_y = \sum_{i=1}^{n} m_i x_i.$$

If $M_x = M_y = 0$, the origin is the center of mass of this system. However, even if $M_x \neq 0$ or $M_y \neq 0$, the given system still has a center of mass, namely the point O' such that relative to the translated axes with center at O', $M_{x'} = M_{y'} = 0$. The following theorem gives the location of the point O'.

10.12 Theorem

Consider a physical system made up of n particles of masses $m_1, m_2, \ldots, m_n$ located at the respective points $(x_1,y_1), (x_2,y_2), \ldots, (x_n,y_n)$. If M_x and M_y are the moments of the system and $m = \sum_{i=1}^{n} m_i$ is its mass, then the center of mass of this system is the point $(\bar{x},\bar{y})$ given by

$$\bar{x} = \frac{M_y}{m}, \qquad \bar{y} = \frac{M_x}{m}.$$

Proof: Let us translate the coordinate axes to a new origin $O'(h,k)$. Relative to the new x' and y' axes, the given particles have coordinates $(x_1',y_1')'$, $(x_2',y_2')', \ldots, (x_n',y_n')'$, where

$$x_i' = x_i - h, \qquad y_i' = y_i - k, \qquad i = 1, 2, \ldots, n.$$

Hence

$$M_x' = \sum_{i=1}^{n} m_i y_i' = \sum_{i=1}^{n} m_i(y_i - k) = \sum_{i=1}^{n} m_i y_i - k \sum_{i=1}^{n} m_i,$$

and therefore $$M_x' = M_x - km.$$

Similarly, $$M_y' = M_y - hm.$$

Now the new origin $O'(h,k)$ will be the center of mass of the given system provided that $M_x' = M_y' = 0$, that is,

$$M_x - km = 0 \qquad \text{and} \qquad M_y - hm = 0.$$

On solving these equations for h and k, we get $h = M_y/m$ and $k = M_x/m$, as stated in the theorem.

An interesting conclusion that can be drawn from this theorem is that the given system of n particles has the same moments relative to the x and y axes as a system made up of one particle of mass $m = m_1 + m_2 + \cdots + m_n$ and located at the point $(\bar{x},\bar{y})$. This follows immediately from the equations

$$M_x = m\bar{y}, \qquad M_y = m\bar{x}$$

of the theorem. Thus each moment of the system may be found by assuming that its mass is concentrated at the center of mass of the system.

The center of mass of a thin homogeneous sheet of substance, called a *lamina*, may be thought of as the balance point of the lamina. If the lamina has a geometric center, then this point will also be the center of mass. For example, the center of mass of a rectangular lamina is the point of intersection of the diagonals of the rectangle.

The moment of a lamina about a line in the plane of the lamina may be defined with the aid of the calculus. We assume that the moment M_L of a lamina of mass m about an axis L is between md_1 and md_2,

$$md_1 \leq M_L \leq md_2,$$

where d_1 is the minimum (directed) distance of any point of the lamina from L, and d_2 is the maximum (directed) distance of any point of the lamina from L (Figure 10.17). We also assume that if the lamina is cut up into pieces, the moment of the lamina is the sum of the moments of its pieces.

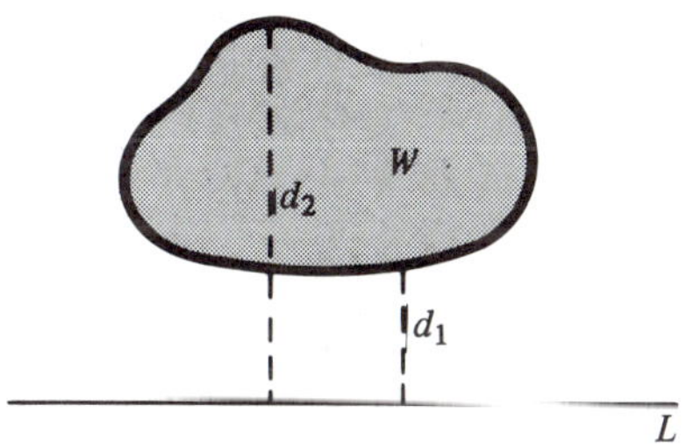

FIGURE 10.17

Let us illustrate the use of these two assumptions in finding the moment of a rectangular lamina of density ρ about an axis parallel to a side of the lamina.* We divide the given rectangle into n congruent rectangles as in Figure 10.18. The height of each smaller rectangle is $\Delta y = b/n$ and its mass is $\rho a \, \Delta y$.

The moment about the x axis of the first rectangle is between $(\rho a \, \Delta y)c$ and $(\rho a \, \Delta y)(c + \Delta y)$; of the second rectangle it is between $(\rho a \, \Delta y)(c + \Delta y)$ and

* The mass of a square unit of a lamina is called its density. The mass of a (homogeneous) lamina is ρA, where ρ is its density and A is its area.

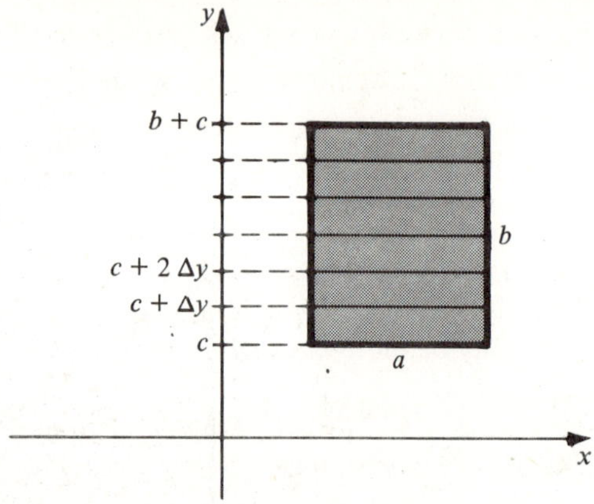

FIGURE 10.18

$(\rho a\ \Delta y)(c + 2\ \Delta y)$; and so on. Hence the moment M_x of the given rectangle satisfies the inequality

$$\sum_{i=1}^{n} (\rho a\ \Delta y)[c + (i - 1)\ \Delta y] \leq M_x \leq \sum_{i=1}^{n} (\rho a\ \Delta y)(c + i\ \Delta y).$$

Using 5.6, we may simplify this as follows ($n\ \Delta y = b$):

$$\sum_{i=1}^{n} (\rho a\ \Delta y)c + \sum_{i=1}^{n} (\rho a\ \Delta y^2)(i - 1) \leq M_x \leq \sum_{i=1}^{n} (\rho a\ \Delta y)c + \sum_{i=1}^{n} (\rho a\ \Delta y^2)i,$$

$$n(\rho a\ \Delta y)c + \rho a\ \Delta y^2\ \frac{(n - 1)n}{2} \leq M_x \leq n(\rho a\ \Delta y)c + \rho a\ \Delta y^2\ \frac{n(n + 1)}{2},$$

$$\rho abc + \rho ab\ \frac{b - \Delta y}{2} \leq M_x \leq \rho abc + \rho ab\ \frac{b + \Delta y}{2}.$$

The limit of each side of this inequality is the same as n approaches ∞ (and Δy approaches 0). Therefore this common limit must be M_x,

$$M_x = \rho ab\left(c + \frac{b}{2}\right).$$

Thus the moment of a rectangular lamina about an axis parallel to a side is the mass ρab of the lamina times the distance $c + b/2$ from the axis to the center of the lamina. In other words, the moment of a rectangular lamina of mass m about an axis parallel to a side is the same as that of a particle of mass m located at the center of mass of the lamina.

Once we have defined the moments of a lamina of mass m about the axes, the center of mass can be defined as the point at which a particle of mass m would be located so that the moments of the particle about the axes would equal the moments of the lamina.

Example

Find the center of mass of a lamina of density ρ lb/in.^{2} having the shape of Figure 10.19.

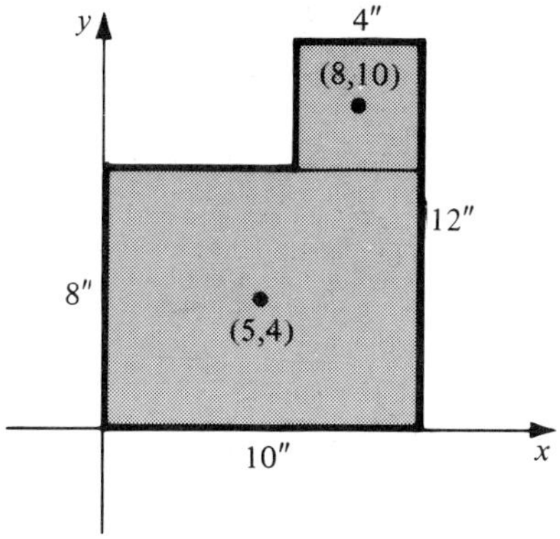

FIGURE 10.19

Solution: The lamina is made up of two rectangles and has a total area of 96 in.^{2}. If we place coordinate axes as indicated in the figure, then the centers of mass of the two rectangles are (5,4) and (8,10). Thus the moments of the lamina about the axes are as follows:

$$M_x = (80\rho)4 + (16\rho)10 = 480\rho.$$
$$M_y = (80\rho)5 + (16\rho)8 = 528\rho.$$

Therefore the center of mass $(\bar{x},\bar{y})$ is given by

$$\bar{x} = \frac{M_y}{m} = \frac{528\rho}{96\rho} = \frac{11}{2}, \qquad y = \frac{M_x}{m} = \frac{480\rho}{96\rho} = 5,$$

that is, by $(\frac{11}{2},5)$.

EXERCISES

I

In the following exercises, the notation $m(x,y)$ signifies that a particle of mass m is located at the point (x,y). Find the center of mass of each of the given systems of particles.

1. 3(2,2), 4(2, −2), 5(−2,2), 2(−2, −2)

2. 8(4,4), 6(4, −4), 3(−4,4), 5(−4, −4)

3. 6(0,0), 6(8,0), 6(0,8), 6(8,8), 3(4,4)

In Exercises 4 and 5 find the center of mass of the lamina of density ρ having the shape in the given figure.

4.

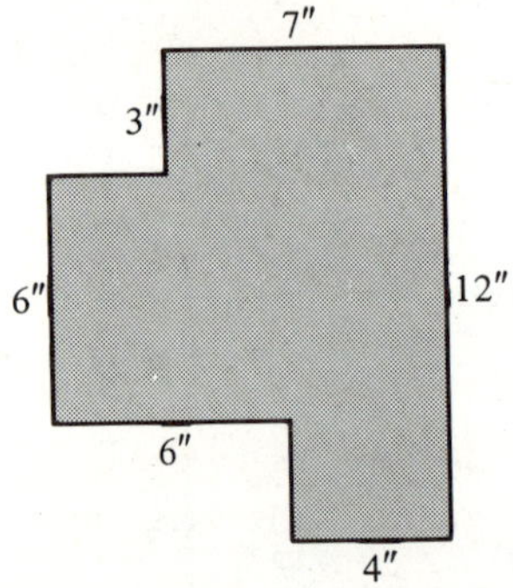

5.

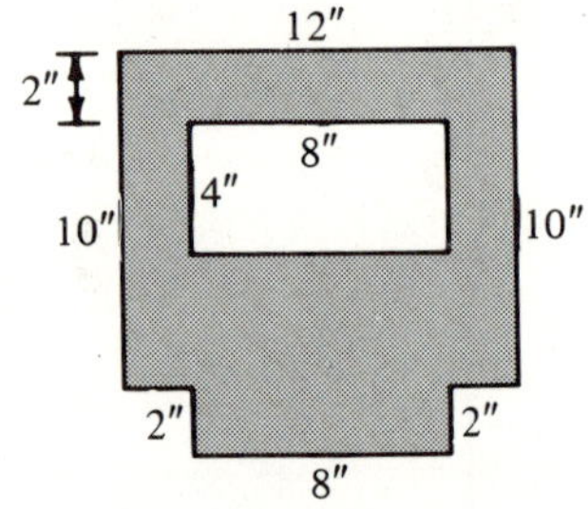

6. Assume that for a given system of n particles $M_x = 0$ and $M_y = 0$. Prove that $M_L = 0$ for every line L passing through the origin.

II

In Exercises 1 and 2, find the center of mass of the lamina of density P having the shape in the given figure.

1.

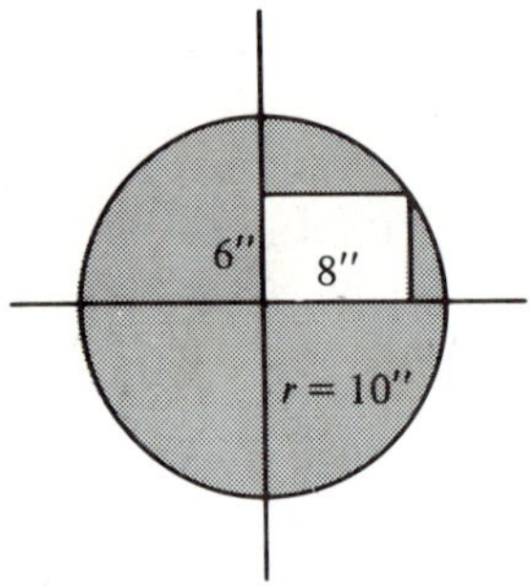

Cut-out circle

2.

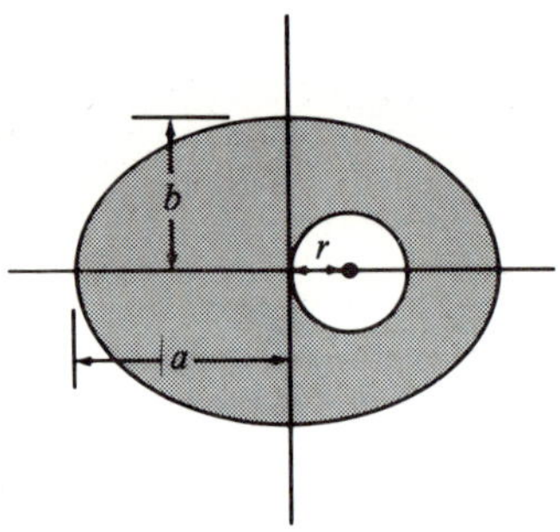

Cut-out ellipse

5 CENTROID OF A PLANE REGION

We turn now to the problem of finding the center of mass of a lamina with a curved boundary. In order to make the problem mathematically solvable, we assume that the boundary of the lamina is made up of graphs of continuous functions.

For simplicity, we first assume that a lamina L of density ρ has the shape of a region bounded by the lines $x = a$ and $x = b$, $a < b$, the x axis, and the graph of a continuous nonnegative function f, as shown in Figure 10.20. We may approximate the moments of L by selecting a partition $P = \{x_0, x_1, \ldots, x_n\}$ of $[a,b]$ and constructing a rectangular polygon relative to P as follows. Let $z_1, z_2, \ldots, z_n$ be the midpoints of segments $[x_0,x_1], [x_1,x_2], \ldots, [x_{n-1}, x_n]$,

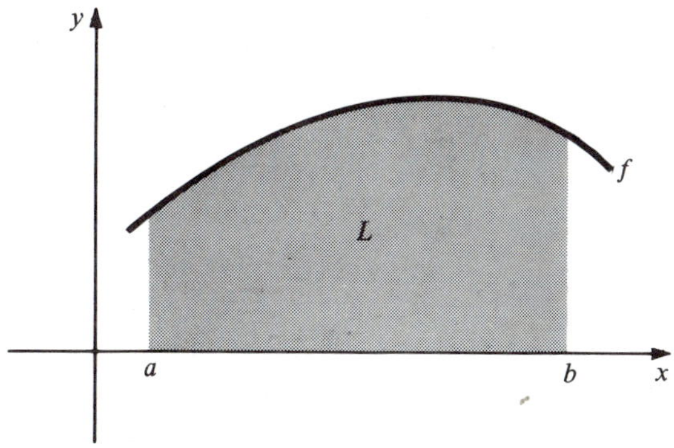

FIGURE 10.20

respectively, and $f(z_1), f(z_2), \ldots, f(z_n)$ be the heights of n rectangles with respective bases $[x_0,x_1], [x_1,x_2], \ldots, [x_{n-1},x_n]$.

The case $n = 3$ is illustrated in Figure 10.21. The first rectangle has area $f(z_1)\,\Delta x_1$ and hence, mass $\rho f(z_1)\,\Delta x_1$. Since the center of this rectangular lamina is the point $(z_1,\frac{1}{2}f(z_1))$, its moment about the x axis is

$$\tfrac{1}{2}f(z_1)\cdot \rho f(z_1)\,\Delta x_1$$

and about the y axis is

$$z_1 \cdot \rho f(z_1)\,\Delta x_1.$$

Using the same procedure for each of the n rectangular laminas and summing, we obtain

$$M_x = \sum_{i=1}^{n} \tfrac{1}{2}f(z_i)\cdot \rho f(z_i)\,\Delta x_i,$$

$$M_y = \sum_{i=1}^{n} z_i \cdot \rho f(z_i)\,\Delta x_i,$$

as the moments of the polygonal lamina about the axes.

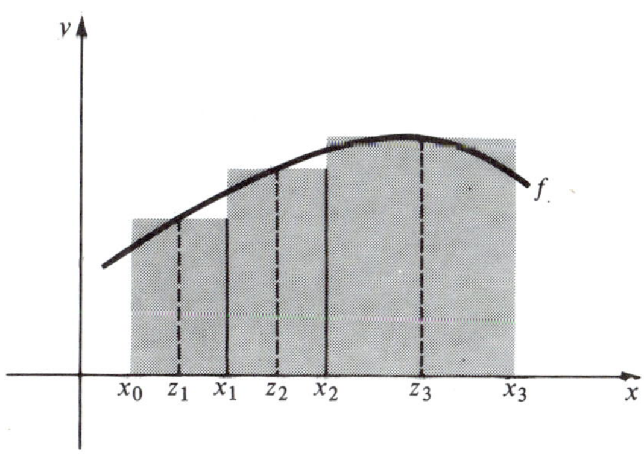

FIGURE 10.21

It is clear that M_x is a Riemann sum, not for the function f, but for the function

$$F(x) = \tfrac{1}{2}f(x) \cdot \rho f(x),$$

and M_y is a Riemann sum for the function

$$G(x) = x \cdot \rho f(x).$$

If we take a sequence of partitions $P_1, P_2, \ldots, P_n, \ldots$ of $[a,b]$ for which

$$\operatorname*{limit}_{n\to\infty} \|P_n\| = 0,$$

and define $(M_n)_x$ and $(M_n)_y$ as above relative to partition P_n, then, by 5.39,

$$\operatorname*{limit}_{n\to\infty} (M_n)_x = \int_a^b \tfrac{1}{2}\rho f^2(x)\,dx$$

$$\operatorname*{limit}_{n\to\infty} (M_n)_y = \int_a^b \rho x f(x)\,dx.$$

Therefore the following definition seems reasonable.

10.13 Definition

The moments of a lamina of density ρ having the shape of the region bounded by the lines $x = a$, $x = b$, and $y = 0$, and by the graph of a continuous nonnegative function f are given by

$$M_x = \frac{\rho}{2}\int_a^b f^2(x)\,dx, \qquad M_y = \rho\int_a^b xf(x)\,dx.$$

Having defined the moments of a lamina, it is natural to define the center of mass of the lamina of mass m to be the point $(\bar{x},\bar{y})$, where

10.14 $$\bar{x} = \frac{M_y}{m}, \qquad \bar{y} = \frac{M_x}{m}.$$

Incidentally, the mass m of the lamina of 10.13 is ρA, where A is the area of the lamina; that is,

$$m = \rho\int_a^b f(x)\,dx.$$

Note that

$$M_x = m\bar{y}, \qquad M_y = m\bar{x},$$

according to 10.14. Thus, again, the moments of the lamina may be found by assuming that the mass of the lamina is concentrated at the center of mass of the lamina.

For the lamina of 10.13 we may write the coordinates of the center of mass in the form

$$\bar{x} = \frac{\int_a^b xf(x)\,dx}{\int_a^b f(x)\,dx}, \qquad \bar{y} = \frac{1}{2}\,\frac{\int_a^b f^2(x)\,dx}{\int_a^b f(x)\,dx}.$$

An interesting feature of the equations for $\bar{x}$ and $\bar{y}$ in this form is that the density factor ρ cancels out, proving that the center of mass of a homogeneous lamina depends only on the shape of the lamina and not on its substance. For this reason we may speak of the center of mass of a plane region in place of the center of mass of a lamina of that shape. We shall use the term *centroid* for the center of mass of a plane region, reserving the term *center of mass* for a material object.

By mathematical arguments similar to those above, we may find the moments and centroids of many different plane regions. If, for example, a region is bounded by the graphs of two functions, then we can imagine the region as being the difference between two regions, one under the graph of each function. Hence its moments will be the difference of two moments, and so on.

A mnemonic device for finding moments of a region such as in Figure 10.22 is as follows. We assume that the region is between the lines $y = a$ and $y = b$, and that for every y in the interval $[a,b]$ the width $W(y)$ of the region is known, W being a continuous function. We imagine the region as being approximated by a polygon made up of n rectangles, a representative one of which, shown in the figure, has the approximate moment $yW(y)\,\Delta y$ about the x axis. By summing up the moments of the n rectangles and taking a limit, we eventually obtain

10.15
$$M_x = \int_a^b yW(y)\,dy$$

for the moment of the given region about the x axis. If the region has area A, then $\bar{y} = M_x/A$ is the y coordinate of its centroid. A similar device may be used to find $\bar{x}$.

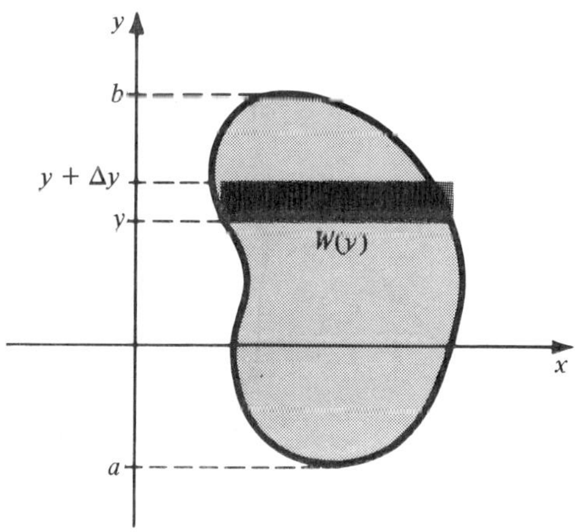

FIGURE 10.22

Example 1 Find the centroid of the region bounded by the lines $x = -1$ and $y = 0$, and the graph of the equation $y = 4 - x^2$.

Solution: For this region, sketched in Figure 10.23, we have, by 10.13 (with $\rho = 1$),

$$M_x = \frac{1}{2}\int_{-1}^{2} (4 - x^2)^2\, dx = \frac{153}{10}, \qquad M_y = \int_{-1}^{2} x(4 - x^2)\, dx = \frac{9}{4}.$$

The area $A = 9$, as may be easily verified. Hence

$$\bar{x} = \frac{M_y}{A} = .25, \qquad \bar{y} = \frac{M_x}{A} = 1.7,$$

and the centroid of the given region is the point (.25,1.7).

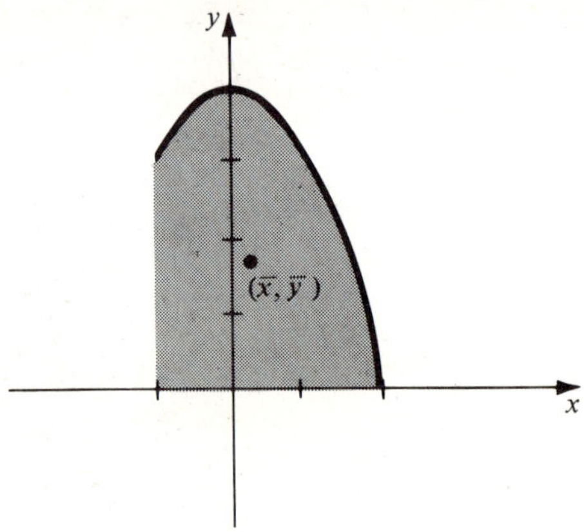

FIGURE 10.23

Example 2 Find the centroid of a triangle.

Solution: Let the coordinate axes be chosen as in Figure 10.24. By similar triangles,

$$\frac{W(y)}{h - y} = \frac{a}{h},$$

or

$$W(y) = \frac{a}{h}(h - y).$$

Hence, by 10.15,

$$M_x = \frac{a}{h}\int_{0}^{h} y(h - y)\, dy = \frac{ah^2}{6}.$$

Since the triangle has area $A = ah/2$,

$$\bar{y} = \frac{M_x}{A} = \frac{h}{3}.$$

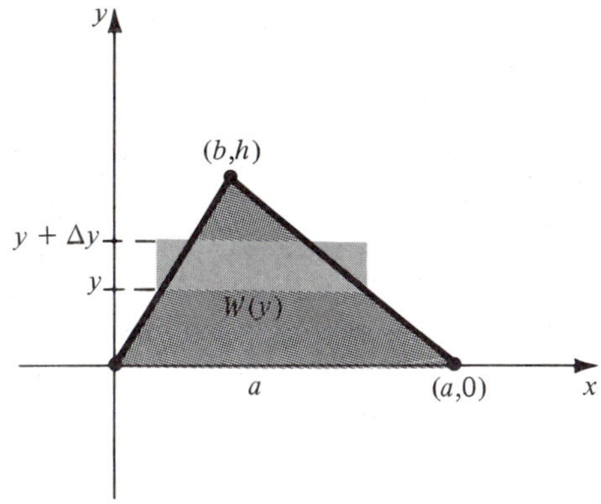

FIGURE 10.24

We could have chosen the x axis on any one of the three sides of the triangle. In each case we would get that the centroid was one-third the altitude above the base. Hence the centroid is the unique point one-third the altitude above each base. By elementary geometry, this is the point of intersection of the medians of the triangle.

EXERCISES

I

In each of Exercises 1 to 14 find the centroid of the region bounded by the graphs of the given equations. Sketch each region.

1. $y = \sqrt{x}, y = 0, x = 4$

2. $y = 1/(x + 1), x = 0, x = 4, y = 0$

3. $y = x^3, x = 0, x = 2, y = 0$

4. $y = \sin x, x = 0, x = \pi, y = 0$

5. $y = \cos x, x = 0, x = \pi/2, y = 0$

6. $y = e^x, x = 0, x = 2, y = 0$

7. $y = \ln x, x = 1, x = e, y = 0$

8. $y = \sec^2 x, x = -\pi/4, x = \pi/4, y = 0$

9. $y = 1/\sqrt{x^2 + 1}, x = 0, x = 1, y = 0$

10. $x^2 - y^2 = 1, x = 3$

11. $y = x^2, y = 6x$

12. $y = \sqrt{9 - x^2}, y = 0$

13. $y = 4 - x^2, y = x^2 - 2x$

14. $y = x^2, y = x^3$ (first quadrant)

15. Find the centroid of a semicircle of radius a.

16. **a.** Find the centroid of the semielliptic region

$$\frac{x^2}{a^2} + \frac{y^2}{b^2} \le 1, \qquad y \ge 0.$$

b. The solution of Exercise 15 can be obtained from this one by merely letting $b = a$. Show how to obtain the solution of Part a from that of Exercise 15.

17. Find the centroid of the region bounded by the x axis, the curve $y = \sinh x$, and the line $x = a$, $a > 0$.

18. Find the centroid of the region bounded by the x axis, the curve $y = \cosh x$, and the lines $x = -a$ and $x = a$.

II

1. Using the result of Exercise I-15, find the centroid of a quarter-circle of radius a.

2. Find the centroid of a circular segment of radius r and height h.

3. Find the centroid of a sector of a circle of radius r and angle 2θ. Find the limit of the position of the centroid as θ approaches zero.

6 CENTROIDS OF SOLIDS OF REVOLUTION

The general problem of finding the center of mass of a solid object will be considered in Chapter 16. However, we can find the center of mass of a homogeneous object having the shape of a solid of revolution, assuming that the center of mass is on the axis of revolution.

Let us assume that a homogeneous object of density* ρ has the shape of a solid generated by rotating about the x axis the region under the graph of a continuous, nonnegative function f between $x = a$ and $x = b$ (Figure 10.25).

We "slice" this object into disks. A representative disk, shown in the figure, has mass $\rho\pi f^2(x)\,\Delta x$, which we imagine as being concentrated at its center. The moment M_y of this disk about the y axis (in reality, with respect to a

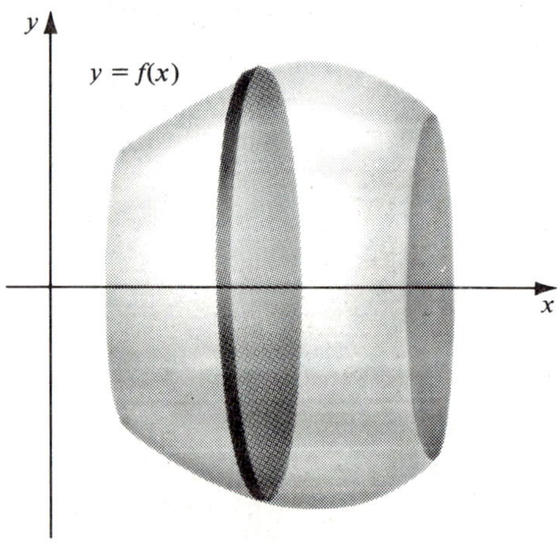

FIGURE 10.25

* Density now means the mass of a cubic unit of the substance.

plane containing the y axis and perpendicular to the x axis) is $x[\rho\pi f^2(x)\,\Delta x]$. Summing up the moments of all the disks and taking a limit, just as in Chapter 5. we finally get

$$M_y = \rho\pi \int_a^b xf^2(x)\,dx.$$

Clearly, the object has mass $m = \rho V$, where V is its volume. Hence the center of mass, $(\bar{x},0)$, of the object is given by

$$\bar{x} = \frac{M_y}{\rho V} = \frac{\int_a^b xf^2(x)\,dx}{\int_a^b f^2(x)\,dx}.$$

We see from the above equation that the center of mass of the homogeneous object depends only on the shape of the object and not on its substance. Thus we may speak of the center of mass of this geometric solid; this point, as in the case of a region, is called the *centroid* of the solid of revolution.

Example Find the centroid of a hemisphere.

Solution: Let the hemisphere be generated by rotating about the x axis the quarter-circle of radius r having equation

$$y = \sqrt{r^2 - x^2}, \qquad 0 \leq x \leq r.$$

Then
$$M_y = \pi \int_0^r x(\sqrt{r^2 - x^2})^2\,dx = \frac{\pi r^4}{4},$$

$$x = \frac{M_y}{V} = \frac{\pi r^4}{4} \cdot \frac{3}{2\pi r^3} = \frac{3r}{8},$$

and the centroid is the point $(3r/8, 0)$.

EXERCISES

I

In each of Exercises 1 to 6, the region bounded by the graphs of the given equations is rotated about the x axis. Find the centroid of the solid generated.

1. $y = \sqrt{x}$, $y = 0$, $x = 4$
2. $y = 1/(x + 1)$, $x = 0$, $x = 4$, $y = 0$
3. $y = \sin x$, $x = 0$, $x = \pi/2$, $y = 0$
4. $y = \sec x$, $x = 0$, $x = \pi/6$, $y = 0$
5. $y = 1/\sqrt{x^2 + 1}$, $x = 0$, $x = 2$, $y = 0$
6. $x^2 - y^2 = 4$, $x = 4$

7. Find the centroid of a right circular cone having radius of base r and altitude h.
8. The region in the first quadrant bounded by the ellipse $x^2/a^2 + y^2/b^2 = 1$ is rotated about the x axis. Find the centroid of the solid generated.
9. The region bounded by a parabola and its latus rectum is rotated about the axis of the parabola. Find the centroid of the solid generated.

Find the centroid of the solid of revolution obtained by revolving the given region about the indicated line. The region bounded by:

10. $y = x^2, x = 1$; about $x = 1$
11. $y = 9x - x^2, y = 0$; about the y axis
12. $y^2 = 6x, y^2 = 24 - 6x$; about the x axis
13. $y = 2x, y = 3x, x + y = 4$; about the y axis

II

1. Prove the following *first theorem of Pappus*: If a plane region, lying on one side of a line L in its plane, is revolved about L, then the volume of the solid generated is equal to the product of the area of the region and the distance traveled by the centroid of the region.

Use Exercise 1 to work each of the following exercises.

2. The volume of a sphere of radius r.
3. The moment of the region bounded by $y = \sqrt{16 - x^2}$ and the x axis with respect to the line $y = -3$.
4. Find the volume of a torus (doughnut-shaped solid) generated by rotating the circle $x^2 + (y - b)^2 = r^2, b > r$, about the x axis.
5. Find the centroid of a semicircular region.
6. Find the volume of a right circular cone of radius r and height h.
7. Find the volume of the ellipsoid of revolution

$$\frac{x^2}{a^2} + \frac{y^2}{b^2} + \frac{z^2}{b^2} = 1.$$

7 FORCE ON A DAM

A liquid in a container exerts a force on the bottom of the container, namely the weight of the liquid therein. The force per square unit of the bottom is called the *pressure* of the liquid at the bottom. Actually, the pressure exerted by a liquid of density ρ at a point d units below the surface is ρd; and this pressure is the same in all directions.

In this section we are interested in finding the total force of a liquid of density ρ against a vertical dam. Let us introduce a coordinate system on a blueprint of the dam, placing the x axis along the line of the surface of the liquid and the positive y axis downward as illustrated in Figure 10.26. In order to make

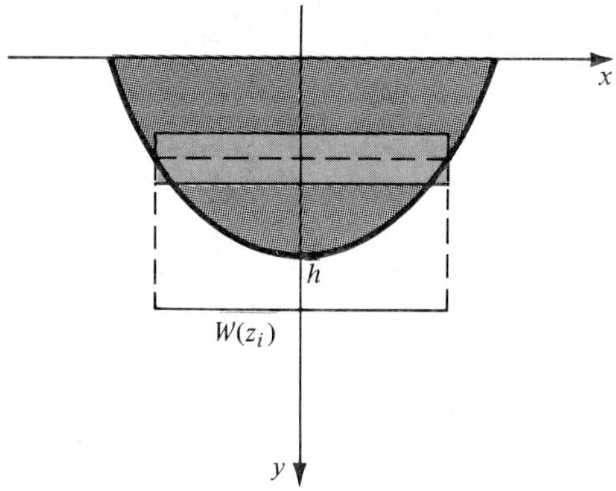

FIGURE 10.26

the problem mathematically solvable, we assume that the width $W(y)$ of the dam at depth y is given by a continuous function W.

For each partition $P = \{y_0, y_1, \ldots, y_n\}$ of the interval $[0,h]$, let us approximate the region of the dam by a polygon made up of rectangles associated with P in the usual way. The force against the ith rectangle (see Figure 10.26) is approximately equal to $\rho z_i W(z_i)\,\Delta y_i$. Hence the force against the dam is approximately equal to

$$\sum_{i=1}^{n} \rho z_i W(z_i)\,\Delta y_i.$$

On taking a sequence of partitions of $[0,h]$ with disappearing norms and finding the limit of the above sum relative to the sequence of partitions, it seems plausible that the force F against the dam is given by the integral

10.16
$$F = \rho \int_0^h yW(y)\,dy.$$

Thus 10.16 is taken to be the definition of the force F. Note that this integral is precisely the moment about the x axis of a lamina of density ρ having the shape of the dam (10.13).

Example 1 Find the force of water against a triangular dam 40 ft wide and 30 ft deep.

Solution: Let the coordinate axes be selected as in Figure 10.27. By similar triangles,

$$\frac{W(y)}{30 - y} = \frac{40}{30} \quad \text{or} \quad W(y) = \tfrac{4}{3}(30 - y).$$

The density of water is 62.5 lb/ft^3. Hence, by 10.16,

$$F = 62.5 \int_0^{30} y \cdot \tfrac{4}{3}(30 - y)\,dy = 3.75 \cdot 10^5 \text{ lb}.$$

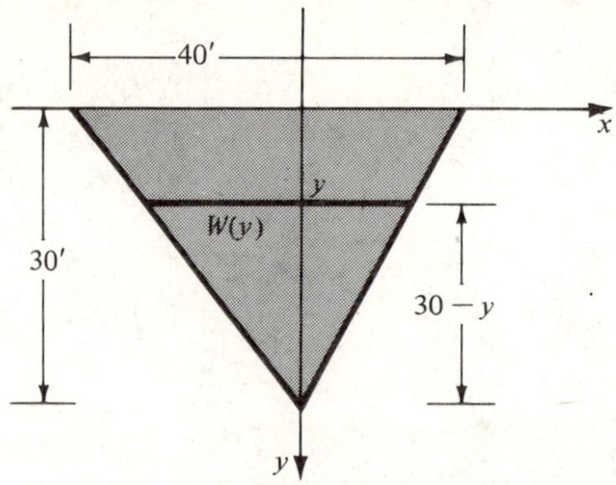

FIGURE 10.27

Example 2 A cylindrical tank 4 ft in diameter and 6 ft long is half full of oil. If the tank is lying on its side, find the force exerted by the oil on one end (assume $\rho = 60$ lb/ft³).

Solution: If the coordinate axes are chosen on an end of the tank as in Figure 10.28, then the semicircle has equation $y = \sqrt{4 - x^2}$ and

$$W(y) = 2x = 2\sqrt{4 - y^2}$$

for each y. Hence, by 10.16,

$$F = 60 \int_0^2 2y\sqrt{4 - y^2}\, dy$$

$$= -40(4 - y^2)^{3/2} \Big|_0^2 = 320 \text{ lb.}$$

Note that the length of the tank is immaterial.

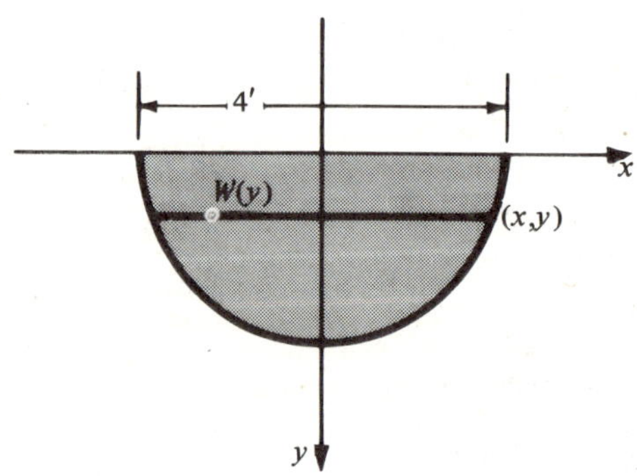

FIGURE 10.28

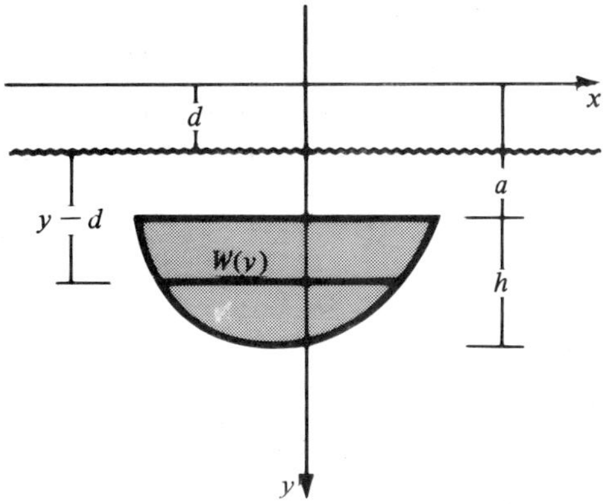

FIGURE 10.29

We need not choose a coordinate axis at the surface of the liquid if it is more convenient to do otherwise. If, for example, we wish to find the force F against the shaded region of Figure 10.29, and if the axes are chosen as in the figure, with the x axis d units above the water level, then by the same reasoning as above,

$$F = \rho \int_a^{a+h} (y - d)W(y)\, dy.$$

EXERCISES

I

1. Find the force of water against an elliptical dam if the major axis of length 50 ft is at the surface of the water and if the semiminor axis has length 20 ft.
2. Find the force of water against a rectangular gate in a dam, if the gate is 6 ft wide and 4 ft high, with the top of the gate parallel to the surface of water and 10 ft below the surface.
3. Find the force of water against the lower half of the gate in Exercise 2.
4. The gate of Exercise 2 is cut into two triangular gates by a diagonal of the gate. Find the force of water against each triangular gate.
5. A dam has a parabolic shape, with the axis of the parabola vertical. If the dam is 50 ft across at the water level and is 60 ft deep, find the force of water against the dam.
6. A trough has a trapezoidal cross section and is 4 ft wide at the top, 2 ft wide at the bottom, and 2 ft deep. Find the force against an end of the trough if it is full of water.
7. A cylindrical tank 6 ft long and 4 ft in diameter is lying on its side. Find the force against the end of the tank if there is 1 ft of water in the tank.

II

1. a. Show that the force against any vertical region under water of a dam is the product of the density of the liquid, the area of the region, and the depth to the centroid of the region.
   b. Use this result to mentally obtain the solution of Exercise I-2.

CHAPTER REVIEW

I

1. Find an equation for the path of a point that moves in the plane in such a way that its distance to the point (3,0) is always equal to e times its distance to the y axis. Sketch the graph of the equation in each of the following cases:
   a. $e = \frac{1}{2}$ b. $e = 1$ c. $e = \frac{3}{2}$
2. Find the length of the latus rectum of the hyperbola
$$\frac{x^2}{a^2} - \frac{y^2}{b^2} = 1.$$
3. If P is any point on an ellipse, prove that the tangent at P makes equal angles with the lines joining P and the two foci.
4. Show that the curves $xy = c$, $x^2 - y^2 = k$ intersect orthogonally, provided $c \neq 0$.

Discuss the graphs of the following equations.

5. $9x^2 - 36x + 16y^2 + 96y + 36 = 0$
6. $x^2 + 2x - 6y + 13 = 0$
7. Find the slope of the tangents to an ellipse at the ends of its latus rectum.
8. Find the centroid of the plane region bounded by $y = 1/(x^2 + a^2)$, $y = 0$, $x = 0$, and $x = a$.
9. The plane region of Exercise 8 is rotated about the x axis. Find the centroid of the solid generated.
10. The surface of the water behind a dam is at the x axis. Find the force on a gate of the dam bounded by the curves $y = -2e^{-x^2}$ and $y = -2/e$.

In each of Exercises 11 to 17, find the centroid of the region with the given boundaries.

11. $y = xe^x$, $y = 0$, $x = 1$
12. $y = xe^x$, $y = 0$, $x = -1$
13. $y = \text{arc sinh } x$, $y = 0$, $x = \pm e$
14. $y = \cosh x$, $y = 0$, $x = -1$, $x = 1$
15. $y = 4x - x^2$, $y = x^2$
16. $y = \sin x$, $y = \cos x$, $x = 0$, $x = \pi/4$

17. One loop of the curve $y^2 = x^2(1 - x^2)$
18. Give an example to show that the centroid of a plane region need not lie within the region.

In each of Exercises 19 to 21, find the volume of the solid generated by revolving the given region about the x axis by the method of the first theorem of Pappus (see Exercise II-1, Section 6).

19. Region of Exercise 11
20. Region of Exercise 12
21. Region of Exercise 14

In each of Exercises 22 to 24, find the centroid of the solid generated by revolving the given region about the x axis.

22. Region of Exercise 11
23. Region of Exercise 12
24. Region of Exercise 14

II

1. Find the shortest distance from the point $(-1, -1)$ to the graph of $9x^2 - 16y^2 + 36x - 32y = 124$.
2. Find the shortest distance from the point $(4, -6)$ to the graph of $81x^2 + 16y^2 - 648x - 32y + 16 = 0$.
3. Show that the two congruent ellipses

$$\frac{x^2}{a^2} + \frac{y^2}{b^2} = 1 \quad \text{and} \quad \frac{(x-h)^2}{a^2} + \frac{(y-k)^2}{b^2} = 1 \qquad h, k \neq 0$$

intersect in at most two points (real).

11

Indeterminate Forms, Improper Integrals, and Taylor's Formula

In this chapter we shall first derive a useful theorem, called l'Hospital's rule, which provides an easy method for finding limits of so-called indeterminate forms. After an introduction to the subject of improper integrals, we then discuss one of the major theorems of the calculus, known as Taylor's theorem. It plays a fundamental role in the numerical computation of functions, since it enables us to approximate the values of a function by calculating the values of a certain polynomial.

1 INDETERMINATE FORMS

The reader will recall that in order to find the derivative of a function ϕ at the number c it is necessary, from the definition of derivative, to find the limit

$$\phi'(c) = \operatorname*{limit}_{x \to c} \frac{\phi(x) - \phi(c)}{x - c}.$$

For example, if $\phi(x) = \tan x$ and $c = \pi/4$, then to find $\phi'(\pi/4)$ from the above definition one has to evaluate

$$\operatorname*{limit}_{x \to (\pi/4)} \frac{\tan x - 1}{x - \dfrac{\pi}{4}}.$$

The attempt to use the theorem on the limit of a quotient, namely

$$\lim_{x \to c} \frac{f(x)}{g(x)} = \frac{\lim_{x \to c} f(x)}{\lim_{x \to c} g(x)}.$$

will obviously fail in this case, since it leads only to the undefined expression 0/0. Indeed, we see from the very definition of the derivative that the attempt at a *direct evaluation* of $\phi'(c)$ always leads to the expression 0/0.

Another important example of a similar type occurred in Chapter 8, Section 2, when we had to find the limit

$$\lim_{x \to 0} \frac{\sin x}{x},$$

which again leads to the expression 0/0. (The reader probably remembers that the determination of this limit was by no means a simple task.)

Both of the above quotients are examples of indeterminate forms. More precisely, we call a quotient f/g of two functions f and g an *indeterminate form* at a number c if either

$$\lim_{x \to c} f(x) = 0 \quad \text{and} \quad \lim_{x \to c} g(x) = 0, \qquad \text{type } \frac{0}{0},$$

or

$$\lim_{x \to c} f(x) = \pm\infty \quad \text{and} \quad \lim_{x \to c} g(x) = \pm\infty, \qquad \text{type } \frac{\infty}{\infty}.$$

If f/g is an indeterminate form at c, then we cannot evaluate its limit by the quotient limit theorem. However, this limit still might exist, as the second example above shows.

The following theorem is of considerable use in evaluating certain limits of indeterminate forms.

11.1 L'Hospital's Rule*

Let the functions f and g be differentiable at every number other than c in some interval, with $g(x) \neq 0$ and $g'(x) \neq 0$ if $x \neq c$. If $\lim_{x \to c} f(x) = \lim_{x \to c} g(x) = 0$, or if $\lim_{x \to c} f(x) = \pm\infty$ and $\lim_{x \to c} g(x) = \pm\infty$, then

$$\lim_{x \to c} \frac{f(x)}{g(x)} = \lim_{x \to c} \frac{f'(x)}{g'(x)},$$

provided this latter limit exists or is infinite.

Proof in the case 0/0: Extend the definition of f and g by defining $f(c) = g(c) = 0$. Then f and g are continuous in the given interval.

Now, for every x in the given interval, define

* Guillaume François Marquis de l'Hospital (1661–1704) was a French mathematician who made several contributions to the calculus in its early formative stage. He wrote the first textbook on the calculus.

$$\phi(t) = f(t) - \frac{f(x)}{g(x)} g(t),$$

whenever t is in the closed interval with endpoints c and x. Note that $\phi(c) = 0$ and $\phi(x) = 0$. Hence by Rolle's Theorem (4.7) there is a number z between c and x such that $\phi'(z) = 0$. Thus

$$0 = f'(z) - \frac{f(x)}{g(x)} g'(z),$$

so that

$$\frac{f(x)}{g(x)} = \frac{f'(z)}{g'(z)}.$$

Clearly the number z depends on x (assuming that c is unchanged throughout our argument), and we indicate this fact by writing

$$\frac{f(x)}{g(x)} = \frac{f'(z(x))}{g'(z(x))}.$$

Since $z(x)$ is always between x and c,

$$\operatorname*{limit}_{x \to c} z(x) = c.$$

If

$$\operatorname*{limit}_{x \to c} \frac{f'(x)}{g'(x)} = k,$$

then also

$$\operatorname*{limit}_{x \to c} \frac{f'(z(x))}{g'(z(x))} = k.$$

For if N is a neighborhood of k and D is a deleted neighborhood of c such that $f'(x)/g'(x)$ is in N for each x in D, then $f'(z(x))/g'(z(x))$ is in N for each x in D, because $z(x)$ is also in D. Thus

$$\operatorname*{limit}_{x \to c} \frac{f(x)}{g(x)} = \operatorname*{limit}_{x \to c} \frac{f'(z)}{g'(z)} = k,$$

and l'Hospital's rule is proved.

If

$$\operatorname*{limit}_{x \to c} \frac{f'(x)}{g'(x)} = \infty \qquad (\text{or } -\infty),$$

then, by the same reasoning as above,

$$\operatorname*{limit}_{x \to c} \frac{f(x)}{g(x)} = \infty \qquad (\text{or } -\infty).$$

This shows that l'Hospital's rule holds for infinite limits also.

Furthermore, the above proof shows that the rule is also valid for *one-sided* limits of $f(x)/g(x)$ as $x \to c^+$ and as $x \to c^-$.

We will not offer a proof here of the more difficult case (∞/∞) when

$$\operatorname*{limit}_{x \to c} f(x) = \pm\infty \qquad \text{and} \qquad \operatorname*{limit}_{x \to c} g(x) = \pm\infty.$$

This may be found in many books on advanced calculus.

Example 1 Find $\operatorname*{limit}_{x \to \pi} \dfrac{\sin x}{x - \pi}$.

Solution: Since

$$\operatorname*{limit}_{x \to \pi} \sin x = \operatorname*{limit}_{x \to \pi} (x - \pi) = 0,$$

we apply l'Hospital's rule to obtain

$$\operatorname*{limit}_{x \to \pi} \frac{\sin x}{x - \pi} = \operatorname*{limit}_{x \to \pi} \frac{\cos x}{1} = -1.$$

Example 2 Find $\operatorname*{limit}_{x \to 0^+} \dfrac{\ln x}{1/x}$.

Solution: Since

$$\operatorname*{limit}_{x \to 0^+} \ln x = -\infty \qquad \text{and} \qquad \operatorname*{limit}_{x \to 0^+} \frac{1}{x} = \infty,$$

we may apply l'Hospital's rule to get

$$\operatorname*{limit}_{x \to 0^+} \frac{\ln x}{1/x} = \operatorname*{limit}_{x \to 0^+} \frac{1/x}{-1/x^2} = \operatorname*{limit}_{x \to 0^+} (-x) = 0.$$

The fraction $(1/x)/-(1/x^2)$ is also an indeterminate form (∞/∞), but we have chosen to reduce the fraction algebraically rather than to employ l'Hospital's rule, which, as is easily seen, would lead us to no conclusion.

In the following example we see that l'Hospital's rule may profitably be applied more than once in certain instances.

Example 3 Find $\operatorname*{limit}_{x \to 0} \dfrac{\sin x - x}{x^3}$.

Solution: We may use l'Hospital's rule, since

$$\operatorname*{limit}_{x \to 0} (\sin x - x) = \operatorname*{limit}_{x \to 0} x^3 = 0.$$

Hence

$$\operatorname*{limit}_{x \to 0} \frac{\sin x - x}{x^3} = \operatorname*{limit}_{x \to 0} \frac{\cos x - 1}{3x^2},$$

if the latter limit exists.

Since

$$\lim_{x\to 0} (\cos x - 1) = \lim_{x\to 0} 3x^2 = 0,$$

we may again use l'Hospital's rule to obtain

$$\lim_{x\to 0} \frac{\cos x - 1}{3x^2} = \lim_{x\to 0} \frac{-\sin x}{6x}.$$

The fraction $-\sin x/6x$ is still indeterminate (0/0), so we reapply the rule to get finally

$$\lim_{x\to 0} \frac{-\sin x}{6x} = \lim_{x\to 0} \frac{-\cos x}{6} = -\frac{1}{6}.$$

We might equally well have recalled in the final step that $\lim_{x\to 0} \sin x/x = 1$ by 8.2. Thus

$$\lim_{x\to 0} \frac{\sin x - x}{x^3} = -\frac{1}{6}.$$

Example 4 Find $\lim_{\theta\to\pi/2^-} \dfrac{\sec\theta}{\tan\theta}$.

Solution: This is an ∞/∞ form to which we apply l'Hospital's rule as follows.

$$\lim_{\theta\to\pi/2^-} \frac{\sec\theta}{\tan\theta} = \lim_{\theta\to\pi/2^-} \frac{\sec\theta\tan\theta}{\sec^2\theta} = \lim_{\theta\to\pi/2^-} \frac{\tan\theta}{\sec\theta}.$$

Reapplying the rule, we get

$$\lim_{\theta\to\pi/2^-} \frac{\tan\theta}{\sec\theta} = \lim_{\theta\to\pi/2^-} \frac{\sec\theta}{\tan\theta}.$$

Clearly, l'Hospital's rule gives us no information in this example. However, since $\sec\theta/\tan\theta = 1/\sin\theta$,

$$\lim_{\theta\to\pi/2^-} \frac{\sec\theta}{\tan\theta} = \lim_{\theta\to\pi/2^-} \frac{1}{\sin\theta} = 1.$$

EXERCISES

I

Evaluate, if possible, each of the following limits.

1. $\lim_{x\to 0} \dfrac{\tan x}{x}$

2. $\lim_{x\to 2} \dfrac{x^2 - 4}{x - 2}$

3. $\lim_{x\to 0} \dfrac{e^x - 1}{x}$

4. $\lim_{y\to 0} \dfrac{\tan^{-1} y}{y}$

5. $\lim_{t\to 0} \dfrac{\sqrt{1+t} - \sqrt{1-t}}{t}$

6. $\lim_{z\to\pi} \dfrac{\ln\cos 2z}{(\pi - z)^2}$

7. $\lim_{\theta \to \pi/2} \frac{\ln |\theta - \pi/2|}{\tan \theta}$

8. $\lim_{x \to 0} \frac{\tan x - x}{x - \sin x}$

9. $\lim_{x \to 0} \frac{e^x - 2 \cos x + e^{-x}}{x \sin x}$

10. $\lim_{x \to 0} \frac{1 - \cos x}{x^2}$

11. $\lim_{x \to -1} \frac{\ln |x|}{x + 1}$

12. $\lim_{t \to 0} \frac{t \sin t}{1 - \cos t}$

13. $\lim_{\theta \to \pi/2} \frac{\ln |\sin \theta|}{\cot \theta}$

14. $\lim_{y \to 0} \frac{\tan^{-1} y - y}{y^3}$

15. $\lim_{y \to 0} \frac{\sin^{-1} y - y}{y^3}$

16. $\lim_{x \to 0} \frac{\sin^{-1} x}{\sin^{-1} 3x}$

17. $\lim_{x \to \pi/2} \frac{\sec^2 3x}{\sec^2 x}$

18. $\lim_{x \to 0} \frac{e^x - 2 + e^{-x}}{1 - \cos 2x}$

19. $\lim_{x \to 0} \frac{\cot 3x}{\cot 2x}$

20. $\lim_{u \to 0} \frac{\tan 2u}{u \sec u}$

21. $\lim_{x \to 0} \frac{10^x - e^x}{x}$

22. $\lim_{x \to 0} \frac{x - \tan^{-1} x}{\sin^{-1} x - x}$

23. $\lim_{x \to 0^+} \frac{e^{-1/x}}{x^2}$

24. $\lim_{x \to 0^+} \frac{\sin x}{\sqrt{x}}$

25. $\lim_{x \to a} \frac{x^n - a^n}{x - a}$

26. $\lim_{x \to 0} \frac{\tanh 3x}{\tanh x}$

27. $\lim_{x \to 0} \frac{\sinh x - \sin x}{\sin^2 x}$

28. $\lim_{x \to 0^+} \frac{b^x - a^x}{x}$

II

Find each of the following limits.

1. $\lim_{x \to 0} \frac{e^x - e^{-x} + 2 \sin x - 4x}{x^5}$

2. $\lim_{x \to 0} \frac{x - \tan^{-1} x}{\tan^3 x}$

3. $\lim_{x \to 1^-} \frac{\ln (1 - x)}{(\ln x)^{-1}}$

4. $\lim_{x \to 0} \frac{2 \sin x - \sin 2x}{\sin^3 x}$

5. $\lim_{x \to 0} \frac{x^3 - 6x + 6 \sin x}{x^5}$

2 FURTHER INDETERMINATE FORMS

L'Hospital's rule may be used in the evaluation of limits of the form

$$\operatorname*{limit}_{x\to\infty} \frac{f(x)}{g(x)},$$

as we shall now prove.

11.2 L'Hospital's Second Rule

Let the functions f and g be differentiable at every x greater than some number a, with $g'(x) \neq 0$. If $\operatorname*{limit}_{x\to\infty} f(x) = \operatorname*{limit}_{x\to\infty} g(x) = 0$, or $\operatorname*{limit}_{x\to\infty} f(x) = \pm\infty$ and $\operatorname*{limit}_{x\to\infty} g(x) = \pm\infty$, then

$$\operatorname*{limit}_{x\to\infty} \frac{f(x)}{g(x)} = \operatorname*{limit}_{x\to\infty} \frac{f'(x)}{g'(x)},$$

provided this latter limit exists or is infinite.

Proof: We shall prove the (0/0) case and omit the proof of the other one. So let us assume that

$$\operatorname*{limit}_{x\to\infty} f(x) = \operatorname*{limit}_{x\to\infty} g(x) = 0.$$

If we let $x = 1/y$, then

$$\operatorname*{limit}_{x\to\infty} \frac{f(x)}{g(x)} = \operatorname*{limit}_{y\to 0+} \frac{f(1/y)}{g(1/y)}.$$

Since

$$D_y f\left(\frac{1}{y}\right) = f'\left(\frac{1}{y}\right) D_y\left(\frac{1}{y}\right) = -\frac{f'(1/y)}{y^2},$$

and similarly for $D_y g\left(\frac{1}{y}\right)$, we may apply the first l'Hospital rule to obtain

$$\operatorname*{limit}_{y\to 0+} \frac{f(1/y)}{g(1/y)} = \operatorname*{limit}_{y\to 0+} \frac{[f'(1/y)]/y^2}{[g'(1/y)]/y^2} = \operatorname*{limit}_{y\to 0+} \frac{f'(1/y)}{g'(1/y)} = \operatorname*{limit}_{x\to\infty} \frac{f'(x)}{g'(x)}.$$

This proves 11.2.

Rule 11.2 also applies to the evaluation of

$$\operatorname*{limit}_{x\to-\infty} \frac{f(x)}{g(x)}.$$

Example 1 Find $\lim_{x\to\infty} \dfrac{\ln x}{x}$.

Solution: This limit is indeterminate of the form (∞/∞). Thus we may use 11.2 to obtain

$$\lim_{x\to\infty} \frac{\ln x}{x} = \lim_{x\to\infty} \frac{1/x}{1} = 0.$$

If

$$\lim_{x\to a} f(x) = 0 \quad \text{and} \quad \lim_{x\to a} g(x) = \pm\infty,$$

then $f(x)g(x)$ is indeterminate of the form $0 \cdot \infty$. We can change it to either the form (0/0) or (∞/∞) as follows:

$$\lim_{x\to a} f(x)g(x) = \lim_{x\to a} \frac{f(x)}{1/g(x)} = \lim_{x\to a} \frac{g(x)}{1/f(x)}.$$

L'Hospital's rule is applicable to these latter two limits. For example,

$$\lim_{x\to 0^+} x \ln x = \lim_{x\to 0^+} \frac{\ln x}{1/x} = 0,$$

by Example 2 of Section 1.

Other indeterminate forms are encountered in evaluating limits of the form

$$\lim\, [f(x)]^{g(x)}.$$

If $\lim f(x) = \lim g(x) = 0$, we have the 0^0 form; other forms are 1^∞ and ∞^0. Each of these types is attacked by writing

$$[f(x)]^{g(x)} = \exp\,[g(x) \ln f(x)].$$

By the continuity of the exponential function (and 2.28), we have, for instance, that

$$\lim_{x\to a} \exp\,[g(x) \ln f(x)] = \exp\,[\lim_{x\to a} g(x) \ln f(x)].$$

Thus we have only to evaluate

$$\lim_{x\to a} g(x) \ln f(x)$$

in order to find the desired limit.

Example 2 Find $\lim_{x\to 0^+} x^x$.

Solution: This 0^0 indeterminate form is evaluated as follows:

$$\lim_{x\to 0^+} x^x = \lim_{x\to 0^+} \exp\,(x \ln x)$$

$$= \exp\,(\lim_{x\to 0^+} x \ln x) = e^0 = 1.$$

Example 3 Find $\lim\limits_{h\to 0^+} (1 + ah)^{1/h}$.

Solution: Since $1 + ah$ has limit 1 and $1/h$ has limit ∞ as h approaches 0^+, we have the indeterminate form 1^∞. We evaluate this limit as follows:

$$\lim_{h\to 0^+} (1 + ah)^{1/h} = \lim_{h\to 0^+} \exp\left[\frac{1}{h}\ln(1 + ah)\right] = \exp \lim_{h\to 0^+} \frac{\ln(1 + ah)}{h}.$$

By l'Hospital's first rule,

$$\lim_{h\to 0^+} \frac{\ln(1 + ah)}{h} = \lim_{h\to 0^+} \frac{a/(1 + ah)}{1} = a.$$

Hence

$$\lim_{h\to 0^+} (1 + ah)^{1/h} = e^a.$$

Example 4 Sketch the graph of the function f defined by

$$f(x) = x \ln x.$$

Solution: The domain of f is the set of positive real numbers. Since

$$\lim_{x\to 0^+} x \ln x = 0,$$

the graph approaches the origin when x approaches zero. Now

$$f'(x) = 1 + \ln x,$$

and $f'(x) = 0$ if and only if $\ln x = -1$. Thus $e^{-1} = 1/e$ is the only critical number of f. Since $f''(x) = 1/x$, $f''(1/e) > 0$, and the point $(1/e, -1/e)$ is a minimum point on the graph. Clearly, $f(1) = 0$. The graph is shown in Figure 11.1.

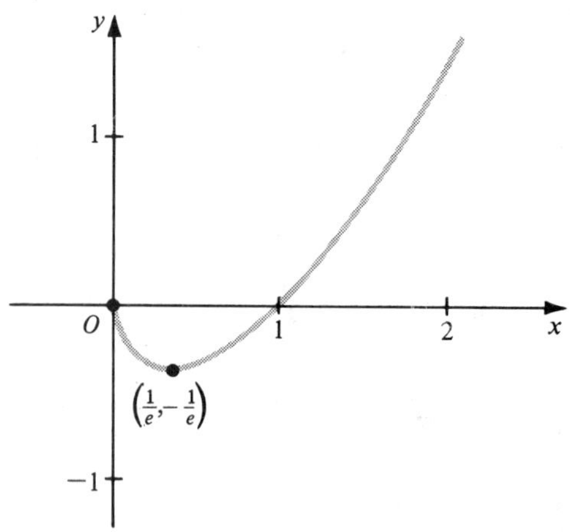

FIGURE 11.1

EXERCISES

I

In each of Exercises 1 to 18, evaluate the limit, if possible.

1. $\lim_{x \to \infty} xe^{-x}$

2. $\lim_{x \to \infty} \frac{\ln x}{\sqrt{x}}$

3. $\lim_{x \to \infty} \frac{x^n}{e^x}$, n a positive integer

4. $\lim_{x \to -\infty} x^2 e^x$

5. $\lim_{x \to \infty} \frac{\ln x}{x^a}$, $a > 0$

6. $\lim_{x \to 0^+} \left(1 + \frac{1}{x}\right)^x$

7. $\lim_{x \to 0} (1 - x)^{1/x}$

8. $\lim_{x \to \infty} x^{1/x}$

9. $\lim_{x \to 0^+} \left(\frac{1}{\sin x} - \frac{1}{x}\right)$

10. $\lim_{x \to \pi/2} (\sec x - \tan x)$

11. $\lim_{x \to \pi/2} (\sin x)^{\tan x}$

12. $\lim_{x \to 0} (1 + 2 \sin x)^{\cot x}$

13. $\lim_{x \to 0} (x + e^{2x})^{1/x}$

14. $\lim_{x \to 0^+} (\sin x)^x$

15. $\lim_{x \to 0^+} x^{\sin x}$

16. $\lim_{x \to 0} \left(\frac{1}{x} - \frac{1}{\tan^{-1} x}\right)$

17. $\lim_{x \to 0^+} x^{a/\ln x}$

18. $\lim_{x \to 1} x^{1/(1-x)}$

19. $\lim_{x \to 0} (e^{ax} + bx)^{c/x}$.

20. Prove that $\lim_{x \to \infty} xb^x = 0$ for $0 < b < 1$.

21. Prove that $\lim_{x \to 0^+} x^p \ln^2 x = 0$ for $p > 0$.

II

Find each of the following limits.

1. $\lim_{x \to (\pi/2)^-} (\tan x)^{\sin 2x}$.

2. $\lim_{x \to 0} \left(\frac{1}{\sin^2 x} - \frac{1}{x^2}\right)$.

3. $\lim_{x \to \infty} \frac{\ln (\ln x)}{\ln (x - \ln x)}$.

4. $\lim_{x \to 0^+} \left(\frac{\tan x}{x}\right)^{1/x^2}$.

5. $\lim_{x \to 1^-} (1 - x)^{\tan \pi x}$

3 IMPROPER INTEGRALS

The reader will recall that the definition of $\int_a^b f(x)\,dx$ in Chapter 5 applied only to the case when the function f is continuous in the closed interval $[a,b]$. We shall now take up the question of extending the definition of $\int_a^b f(x)\,dx$ to the case when f fails to be continuous at one of the endpoints of $[a,b]$, say the point b.

We make the following definition.

11.3 Definition

(1) If the function f is continuous in the half-open interval $[a,b)$, then

$$\int_a^b f(x)\,dx = \lim_{t\to b^-} \int_a^t f(x)\,dx, \qquad \text{if this limit exists.}$$

(2) If f is continuous in $(a,b]$, then

$$\int_a^b f(x)\,dx = \lim_{t\to a^+} \int_t^b f(x)\,dx, \qquad \text{if this limit exists.}$$

An integral that exists in the sense of Definition 11.3 is called an *improper* integral (as cpposed to integrals in the usual sense, which we might call "proper"). Improper integrals will be of use to us in our study of infinite series in Chapter 12.

Example 1 Find $\int_0^4 \frac{1}{\sqrt{x}}\,dx$.

Solution: Evidently the function $f(x) = 1/\sqrt{x}$ is unbounded in $(0,4]$, but is continuous in $[t,4]$ for $0 < t < 4$. This is an improper integral of type 11.3(2). Hence

$$\int_0^4 \frac{1}{\sqrt{x}}\,dx = \lim_{t\to 0^+} \int_t^4 x^{-1/2}\,dx$$

$$= \lim_{t\to 0^+} 2x^{1/2}\Big|_t^4 = \lim_{t\to 0^+} (4 - 2\sqrt{t}) = 4.$$

Geometrically, we have shown that the shaded region of Figure 11.2, a region of infinite extent, has a finite area of 4 square units.

Example 2 Find $\int_1^0 x \ln x\,dx$. (This is the area of the region bounded by the graph of $y = x \ln x$ and the x axis sketched in Figure 11.1.)

Solution: The integrand has a discontinuity at 0 since ln 0 does not exist. Hence

$$\int_1^0 x \ln x\,dx = \lim_{t\to 0^+} \int_1^t x \ln x\,dx$$

$$= \lim_{t\to 0^+} \tfrac{1}{4}(2x^2 \ln x - x^2)\Big|_1^t$$

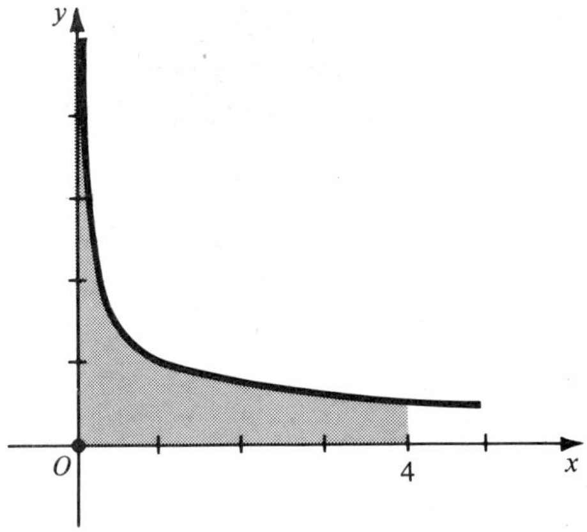

FIGURE 11.2

$$= \operatorname*{limit}_{t \to 0^+} \tfrac{1}{4}(2t^2 \ln t - t^2 + 1).$$

By l'Hospital's first rule,

$$\operatorname*{limit}_{t \to 0^+} t^2 \ln t = \operatorname*{limit}_{t \to 0^+} \frac{\ln t}{1/t^2}$$

$$= \operatorname*{limit}_{t \to 0^+} \frac{1/t}{-2/t^3} = \operatorname*{limit}_{t \to 0^+} \left(-\frac{t^2}{2}\right) = 0.$$

Thus $$\operatorname*{limit}_{t \to 0^+} \tfrac{1}{4}(2t^2 \ln t - t^2 + 1) = \tfrac{1}{4},$$

and $$\int_1^0 x \ln x \, dx = \tfrac{1}{4}.$$

A second type of improper integral arises when we wish to consider an integral over an *infinite* interval. Suppose, for example, that the function f is continuous in the interval $[a,\infty)$. Then we *define*

11.4 $$\int_a^\infty f(x)\,dx = \operatorname*{limit}_{t \to \infty} \int_a^t f(x)\,dx,$$

provided that this limit exists. Likewise, if f is continuous in $(-\infty,b]$, we *define*

11.5 $$\int_{-\infty}^b f(x)\,dx = \operatorname*{limit}_{t \to -\infty} \int_t^b f(x)\,dx,$$

provided that the limit exists.

Example 3 Find $\int_0^\infty xe^{-x}\,dx$.

Solution: The part of the graph of f in the first quadrant is sketched in Figure 11.3. The function f is continuous in $[0,\infty)$, For each $t > 0$,

$$\int_0^t xe^{-x}\,dx = (-xe^{-x} - e^{-x})\Big|_0^t$$

$$= -te^{-t} - e^{-t} + 1.$$

Hence $\int_0^\infty xe^{-x}\,dx = \lim_{t\to\infty}(-te^{-t} - e^{-t} + 1) = 1$. We see, therefore, that the region R between the graph of f and the x axis (Figure 11.3) may be assigned the value 1 for its area.

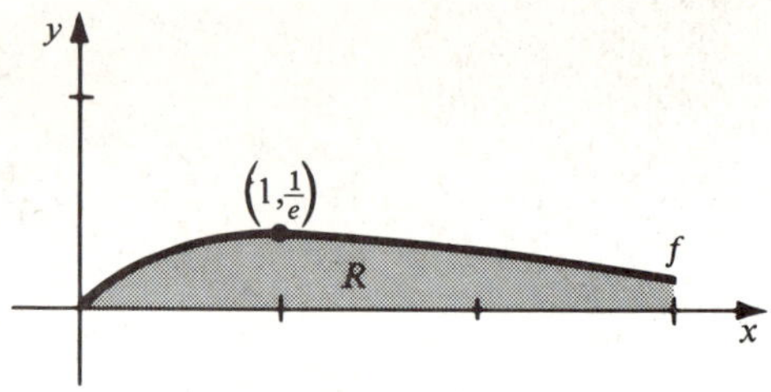

FIGURE 11.3

Example 4 The graph of the function f defined by

$$f(x) = \frac{1}{x}$$

is sketched in the first quadrant in Figure 11.4. Can an area be assigned to the shaded region R? If the region R is rotated about the x axis, can a volume be assigned to the solid so formed?

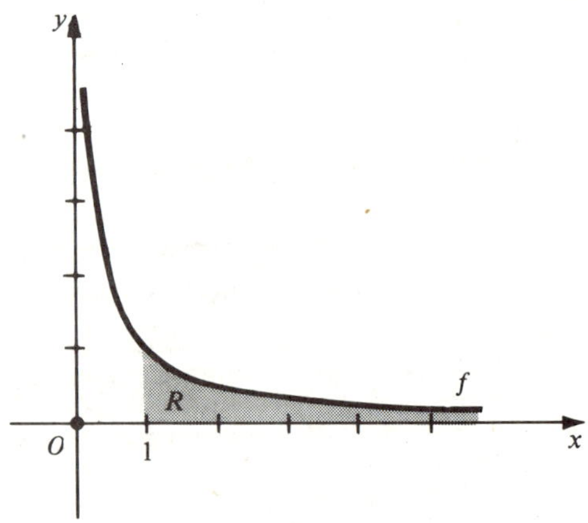

FIGURE 11.4

Solution: If R is to have an area consistent with the area of a finite region, it must be given by

$$\int_1^{\infty} \frac{1}{x}\,dx.$$

However $$\int_1^{\infty} \frac{1}{x}\,dx = \lim_{t\to\infty} \int_1^t \frac{1}{x}\,dx = \lim_{t\to\infty} \ln t = \infty.$$

Thus R does not have a finite area.

The volume of the solid obtained by rotating R about the x axis must be given by

$$\pi \int_1^{\infty} \frac{1}{x^2}\,dx.$$

Since

$$\int_1^{\infty} \frac{1}{x^2}\,dx = \lim_{t\to\infty} \int_1^t \frac{1}{x^2}\,dx$$

$$= \lim_{t\to\infty} \left(-\frac{1}{t} + 1\right) = 1,$$

the volume assigned to this solid is π.

EXERCISES

I

In each of Exercises 1 to 16 find the value of the improper integral, if it exists.

1. $\int_1^{\infty} \frac{1}{x\sqrt{x}}\,dx$

2. $\int_{-\infty}^{4} \frac{1}{(5-x)^2}\,dx$

3. $\int_{-\infty}^{3} \frac{1}{\sqrt{7-x}}\,dx$

4. $\int_0^{\infty} \frac{x}{1+x^2}\,dx$

5. $\int_3^{4} \frac{1}{\sqrt{x-3}}\,dx$

6. $\int_{-2}^{0} \frac{1}{\sqrt{4-x^2}}\,dx$

7. $\int_0^{\infty} \frac{1}{\sqrt{e^x}}\,dx$

8. $\int_0^{\infty} e^{-ax}\,dx,\ a > 0$

9. $\int_0^{1} \frac{1}{\sqrt[3]{x}}\,dx$

10. $\int_0^{4} \frac{1}{x\sqrt{x}}\,dx$

11. $\int_1^{\infty} \frac{1}{1 + x^2}\, dx$

12. $\int_3^{\infty} \frac{1}{x^2 - 2x}\, dx$

13. $\int_{\pi/4}^{\pi/2} \sec x\, dx$

14. $\int_0^{\pi/2} \frac{1}{1 - \sin x}\, dx$

15. $\int_{-1}^{\infty} \frac{x}{e^{x^2}}\, dx$

16. $\int_2^{\infty} \frac{1}{x^2 - 1}\, dx$

17. If f is continuous at each real number x, define

(1) $$\int_{-\infty}^{\infty} f(x)\, dx = \int_{-\infty}^{0} f(x)\, dx + \int_0^{\infty} f(x)\, dx$$

provided each integral on the right side of (1) exists.

a. Show that if the integral (1) exists then

$$\int_{-\infty}^{\infty} f(x)\, dx = \int_{-\infty}^{a} f(x)\, dx + \int_a^{\infty} f(x)\, dx$$

for every real number a.

b. Show by example that $\lim_{t \to \infty} \int_{-t}^{t} f(x)\, dx$ can exist even though $\int_{-\infty}^{\infty} f(x)\, dx$ does not exist.

Find the integrals in Exercises 18 and 19.

18. $\int_{-\infty}^{\infty} \frac{1}{1 + x^2}\, dx$

19. $\int_{-\infty}^{\infty} \frac{1}{e^x + e^{-x}}\, dx$

20. If f is continuous at each number in the closed interval $[a,b]$ except c, $a < c < b$, then define

(1) $$\int_a^b f(x)\, dx = \int_a^c f(x)\, dx + \int_c^b f(x)\, dx,$$

provided each integral on the right side of (1) exists.

a. Does

$$\int_a^b f(x)\, dx = \lim_{\varepsilon \to 0^+} \left[\int_a^{c-\varepsilon} f(x)\, dx + \int_{c+\varepsilon}^{b} f(x)\, dx \right] ?$$

b. What is wrong with the reasoning,

$$\int_{-1}^{1} \frac{1}{x}\, dx = \ln |x| \Big|_{-1}^{1} = \ln 1 - \ln 1 = 0?$$

Find the integrals in Exercises 21 and 22, if they exist.

21. $\int_{-2}^{1} \frac{1}{\sqrt[3]{x^2}}\, dx$

22. $\int_0^{4} \frac{1}{(x - 2)^2}\, dx$

23. Show that $\int_0^1 \frac{1}{\sqrt[3]{x^2}}\,dx$ exists but that $\int_0^1 \frac{1}{\sqrt[3]{x^4}}\,dx$ does not, and interpret the result geometrically.

24. Show that

$$\int_0^4 \frac{1}{(x+4)\sqrt{x}}\,dx = \frac{\pi}{4}, \qquad \int_4^\infty \frac{1}{(x+4)\sqrt{x}}\,dx = \frac{\pi}{4}.$$

Hence find

$$\int_0^\infty \frac{1}{(x+4)\sqrt{x}}\,dx.$$

Find the following integrals.

25. $\int_1^\infty \frac{\ln x}{x^2}\,dx$

26. $\int_1^{\pi/2} \left(\frac{1}{x^2} - \csc x \cot x\right) dx$

27. $\int_0^{\pi/2} (\sec x \tan x - \sec^2 x)\,dx$

28. $\int_0^e x^2 \ln x\,dx$

29. Prove that $\int_0^\infty x^n e^{-x}\,dx = n!$, for n a positive integer.

30. For what values of p do the following integrals exist?

a. $\int_0^1 x^p\,dx$,

b. $\int_1^\infty x^p\,dx$.

II

Find the following integrals.

1. $\int_1^\infty \frac{dx}{x(x+1)(x+2)}$.

2. $\int_0^1 x^2 \ln^2 x\,dx$.

3. Prove that $\int_0^1 \left(\ln \frac{1}{x}\right)^n dx = n!$ for n a positive integer. *Suggestion:* Set $t = \ln(1/x)$ and refer to Exercise I–29.

4 TAYLOR'S FORMULA

Let f be a function for which the nth derivative $f^{[n]}$ exists at some number c. The polynomial*

11.6 $$P_n(x) = f(c) + f'(c)(x-c) + \frac{f''(c)}{2!}(x-c)^2 + \cdots + \frac{f^{[n]}(c)}{n!}(x-c)^n,$$

of degree n in x, is called the nth-degree *Taylor's*† *polynomial* of f at c. Using the $\sum$ notation, with $0! = 1$ and $f^{[0]} = f$,

$$P_n(x) = \sum_{k=0}^{n} \frac{f^{[k]}(c)}{k!}(x - c)^k.$$

Since

$$D_x \frac{f^{[k]}(c)}{k!}(x - c)^k = \frac{kf^{[k]}(c)}{k!}(x - c)^{k-1} = \frac{f^{[k]}(c)}{(k-1)!}(x - c)^{k-1},$$

evidently

$$P_n'(x) = \sum_{k=1}^{n} \frac{f^{[k]}(c)}{(k-1)!}(x - c)^{k-1},$$

and, in general,

11.7 $$P_n^{[j]}(x) = \sum_{k=j}^{n} \frac{f^{[k]}(c)}{(k-j)!}(x - c)^{k-j}, \qquad 0 \le j \le n.$$

Since $P_n(x)$ is a polynomial of degree n [or less if $f^{[n]}(c) = 0$], clearly

$$P_n^{[j]}(x) = 0, \qquad j > n.$$

If we replace x with c in 11.6, each term becomes zero with the exception of the first one; thus

$$P_n(c) = f(c).$$

Similarly, each term of the right side of 11.7 after the first one becomes zero if we replace x with c; that is,

11.8 $$P_n^{[j]}(c) = f^{[j]}(c), \qquad 0 \le j \le n.$$

Thus not only does the Taylor's polynomial $P_n(x)$ have the same value as f at c, but also each derivative of P_n through the nth has the same value as the corresponding derivative of f at c.

Example 1 Find the fourth-degree Taylor's polynomial of the sine function at $\pi/2$.

Solution: If $f(x) = \sin x$, then, by 11.6,

$$P_4(x) = f\left(\frac{\pi}{2}\right) + f'\left(\frac{\pi}{2}\right)\left(x - \frac{\pi}{2}\right) + \frac{f''(\pi/2)}{2}\left(x - \frac{\pi}{2}\right)^2$$
$$+ \frac{f'''(\pi/2)}{3!}\left(x - \frac{\pi}{2}\right)^3 + \frac{f^{[4]}(\pi/2)}{4!}\left(x - \frac{\pi}{2}\right)^4.$$

* $n!$, read "n factorial," equals $n(n-1)(n-2)\cdots 1$.

† Brook Taylor (1685–1731) gave methods of expanding a function in a series in his book *Methodus Incrementorum Directa et Inversa*, published in 1715. His methods were made mathematically rigorous a century later, notably by Gauss and Cauchy.

Since

$$f(x) = \sin x, \qquad f\left(\frac{\pi}{2}\right) = 1,$$

$$f'(x) = \cos x, \qquad f'\left(\frac{\pi}{2}\right) = 0,$$

$$f''(x) = -\sin x, \qquad f''\left(\frac{\pi}{2}\right) = -1,$$

$$f'''(x) = -\cos x, \qquad f'''\left(\frac{\pi}{2}\right) = 0,$$

$$f^{[4]}(x) = \sin x, \qquad f^{[4]}\left(\frac{\pi}{2}\right) = 1,$$

we obtain

$$P_4(x) = 1 - \frac{1}{2}\left(x - \frac{\pi}{2}\right)^2 + \frac{1}{24}\left(x - \frac{\pi}{2}\right)^4.$$

We come now to the fundamental Taylor's formula, relating a function to its Taylor's polynomials.

11.9 Taylor's Formula

Let f be a function and n an integer such that $f^{[n+1]}(x)$ exists for every number x in a closed interval $[a,b]$ containing the number c. Then if $P_n(x)$ is the nth-degree Taylor's polynomial of f at c, there exists some number z (depending on x) between x and c such that

$$f(x) = P_n(x) + \frac{f^{[n+1]}(z)}{(n+1)!}(x - c)^{n+1}, \qquad a \le x \le b.$$

Proof: For the purpose of this proof we consider x as a *constant*, and define a number K by the equation

(1)
$$f(x) = f(c) + f'(c)(x - c) + \cdots + \frac{f^{(n)}(c)}{n!}(x - c)^n + \frac{K(x - c)^{n+1}}{(n+1)!}.$$

We could, of course, solve this equation for K in terms of x and c, but this would not be very useful. Instead, we shall find K by a rather clever trick. Replace the constant c by a *variable* t; then transpose $f(x)$ to the right-hand side and define a function $h(t)$ by

(2)
$$h(t) = -f(x) + f(t) + f'(t)(x - t) + \frac{f''(t)}{2!}(x - t)^2 + \cdots + \frac{f^{(n)}(t)}{n!}(x - t)^n + \frac{K(x - t)^{n+1}}{(n+1)!}.$$

The reader may verify immediately that $h(x) = 0$ and, by using (1), that $h(c) = 0$. Hence by Rolle's Theorem (4.7), there is a number z between x and c such that $h'(z) = 0$. Let us differentiate equation (2) to find a formula for $h'(t)$. Note that in this differentiation x is a *constant*, and all of the terms of (2), except the first, are functions of t. We obtain

$$h'(t) = f'(t) - f'(t) + f''(t)(x - t) - f''(t)(x - t) + \frac{f'''(t)}{2!}(x - t)^2 - \cdots - \frac{f^{(n)}(t)}{(n - 1)!}(x - t)^{n-1} + \frac{f^{(n+1)}(t)}{n!}(x - t)^n - \frac{K(x - t)^n}{n!}.$$

Observe that most of the terms drop out, and we find that

$$h'(t) = \frac{f^{(n+1)}(t)}{n!}(x - t)^n - \frac{K(x - t)^n}{n!}.$$

But for the above number z between x and c we have $h'(z) = 0$. Hence

$$\frac{f^{(n+1)}(z)}{n!}(x - z)^n - \frac{K(x - z)^n}{n!} = 0,$$

from which we obtain $K = f^{(n+1)}(z)$. The proof of Taylor's formula is completed by putting this value for K in equation (1).

The term

11.10 $$R_n(x) = \frac{f^{[n+1]}(z)}{(n + 1)!}(x - c)^{n+1}$$

appearing in Taylor's formula is called the nth *remainder term* of f at c. Using 11.10, we may write Taylor's formula in the form

$$f(x) = P_n(x) + R_n(x).$$

We shall soon show that under certain conditions the remainder term is very small; hence

$$f(x) \doteq P_n(x).$$

Example 2 Write Taylor's formula for the sine function of Example 1.

Solution: Since $f^{[5]}(x) = \cos x$, we have

$$R_4(x) = \frac{f^{[5]}(z)}{5!}\left(x - \frac{\pi}{2}\right)^5 = \frac{\cos z}{120}\left(x - \frac{\pi}{2}\right)^5.$$

Hence, using P_4 from Example 1,

$$\sin x = 1 - \frac{1}{2}\left(x - \frac{\pi}{2}\right)^2 + \frac{1}{24}\left(x - \frac{\pi}{2}\right)^4 + \frac{\cos z}{120}\left(x - \frac{\pi}{2}\right)^5$$

for some number z between x and $\pi/2$.

If $c = 0$, Taylor's formula becomes

11.11 $$f(x) = f(0) + f'(0)x + \frac{f''(0)}{2!}x^2 + \cdots + \frac{f^{[n]}(0)}{n!}x^n + \frac{f^{[n+1]}(z)}{(n+1)!}x^{n+1}$$

with z a number between 0 and x. This form of Taylor's formula is called *Maclaurin's formula.**

Example 3 Write Maclaurin's formula for the exponential function.

Solution: If $f(x) = e^x$, then $f^{[k]}(x) = e^x$ for each k and therefore $f^{[k]}(0) = e^0 = 1$. Thus Maclaurin's formula becomes

$$e^x = 1 + x + \frac{x^2}{2!} + \frac{x^3}{3!} + \cdots + \frac{x^n}{n!} + \frac{e^z}{(n+1)!}x^{n+1},$$

with z a number between 0 and x.

Example 4 Prove that e is an irrational number.

Solution: We will show the assumption that e is a rational number leads to a contradiction. We know that $2 < e < 3$, so e is not an integer. Therefore, let us assume that $e = k/m$, where k and m are integers with $m \geq 2$. In the formula for e^x in Example 3, set $x = 1$ and $n = m$ to obtain

$$e = 1 + 1 + \frac{1}{2!} + \frac{1}{3!} + \cdots + \frac{1}{m!} + \frac{e^z}{(m+1)!},$$

where $0 < z < 1$. Hence

$$e \cdot m! = p + \frac{e^z}{m+1},$$

where p is an integer. Since $e \cdot m! = (k/m) \cdot m!$ is also an integer, it follows that $e^z/(m+1)$ must be an integer. But this is impossible, because $e^z < e < 3$ and $m + 1 \geq 3$. Thus, our assumption that e is rational must be false.

* Colin Maclaurin (1698–1746), a Scottish mathematician and contemporary of Taylor, gave this formula in his *Treatise of Fluxions* in 1742. However, this formula had appeared 25 years earlier in a publication by Stirling.

EXERCISES

In each of the following exercises, find $P_n(x)$, $R_n(x)$, and write the Taylor's formula for the given function, value of n, and value of c.

1. $f(x) = \cos x;\ n = 5,\ c = \pi/2$
2. $g(x) = \ln x;\ n = 5,\ c = 1$
3. $F(x) = \tan^{-1} x;\ n = 3,\ c = 1$
4. $G(x) = e^x;\ n = 4,\ c = -1$
5. $f(x) = \sec x;\ n = 3,\ c = \pi/4$
6. $h(x) = 1/(x + 1);\ n = 6,\ c = -2$

In each of the following exercises, write the Maclaurin's formula for the given function and value of n.

7. $f(x) = \sin x,\ n = 6$
8. $f(x) = \cos x,\ n = 5$
9. $g(x) = \ln(x + 1),\ n = 5$
10. $F(x) = \tan^{-1} x,\ n = 4$
11. $F(x) = \sin^{-1} x,\ n = 3$
12. $g(x) = \sqrt{1 + x},\ n = 5$
13. $f(x) = \dfrac{1}{\sqrt{1 - x}},\ n = 4$
14. $f(x) = e^{-x},\ n = 5$
15. $g(x) = \tan x,\ n = 4$
16. $g(x) = \dfrac{1}{(1 - x)^2},\ n = 4$
17. $F(x) = \dfrac{1}{1 + e^x},\ n = 3$
18. $g(x) = x^5,\ n = 6$
19. $f(x) = (1 - x)^{-1/2},\ n = 4$
20. $g(x) = e^{-x^2},\ n = 3$
21. Find the Maclaurin expansion of y up to the x^3 term if
$$y^3 + xy = 1 \quad \text{and} \quad y = 1 \quad \text{for } x = 0.$$
(*Suggestion:* Use implicit differentiation.)
22. Same as Exercise 21, except that
$$e^x - e^y = xy \quad \text{and} \quad y = 0 \quad \text{for } x = 0.$$

5 APPROXIMATIONS BY TAYLOR'S POLYNOMIALS

One of the easiest functions with which to compute values is the polynomial function. Thus it is an easy matter to find values of

$$f(x) = x - \frac{x^3}{6}$$

corresponding to various choices of x. For example,

$$f(.2) = .2 - \frac{.008}{6} \doteq .19867.$$

The theory of Section 4 allows us to approximate many functions by polynomial functions. Thus, if $P_n(x)$ and $R_n(x)$ are the Taylor polynomial and the remainder term for a function f at c,

$$f(x) = P_n(x) + R_n(x), \qquad \text{and} \qquad |f(x) - P_n(x)| = |R_n(x)|.$$

If we can find a number d such that

$$|R_n(x)| \leq d,$$

then we will have $|f(x) - P_n(x)| \leq d$, or

$$P_n(x) - d \leq f(x) \leq P_n(x) + d.$$

We shall write

$$f(x) = P_n(x) \pm d$$

in this case, with the understanding that this equation means $f(x) \doteq P_n(x)$, with an error of not more than d units.

Example 1 Approximate $\sin x$ by a fourth-degree polynomial in x, if $0 \leq x \leq .2$.

Solution: If $f(x) = \sin x$, then

$$f(0) = 0, \quad f'(0) = 1, \quad f''(0) = 0, \quad f'''(0) = -1, \quad f^{[4]}(0) = 0, \quad f^{[5]}(x) = \cos x.$$

Hence the fourth-degree Taylor's polynomial of f at 0 is given by

$$P_4(x) = x - \frac{x^3}{3!}.$$

Note that this polynomial is actually of the third degree, since the coefficient $f^{[4]}(0)$ of the term of fourth degree is zero. Also,

$$R_4(x) = \frac{\cos z}{5!} x^5,$$

where $0 < z < .2$.

Since $\cos z \leq 1$,

$$0 < R_4(x) \leq \frac{(.2)^5}{5!} < .000003$$

in the given range $0 \leq x \leq .2$. Thus

$$\sin x = x - \frac{x^3}{6} \pm .000003, \qquad 0 \leq x \leq .2.$$

For example,

$$\sin .2 \doteq .19867,$$

accurate to five decimal places.

If we increase the range of x in the above example to $0 \leq x \leq .5$, then it is easily shown that $R_4(x) < .0003$. Hence

$$\sin x = x - \frac{x^3}{3!} \pm .0003, \qquad 0 \le x \le .5.$$

We still have accuracy to three decimal places, allowing x to range up to .5 radian (about 28°). Thus

$$\sin .5 \doteq .479,$$

accurate to three decimal places. The accuracy could be increased by choosing a Taylor's polynomial of higher degree.

Example 2 Approximate $\ln x$ by a fifth-degree polynomial in x, if $1 \le x \le 1.2$.

Solution: Let us find $P_5(x)$ and $R_5(x)$ for $f(x) = \ln x$ at $c = 1$. Since

$$f'(x) = \frac{1}{x},\ f''(x) = -\frac{1}{x^2},\ f'''(x) = \frac{2}{x^3},\ f^{[4]}(x) = -\frac{6}{x^4},\ f^{[5]}(x) = \frac{24}{x^5},$$

we have

$$f(1) = 0,\ \ f'(1) = 1,\ \ f''(1) = -1,\ \ f('''1) = 2,\ \ f^{[4]}(1) = -6,\ \ f^{[5]}(1) = 24.$$

Thus

$$P_5(x) = (x-1) - \frac{1}{2!}(x-1)^2 + \frac{2}{3!}(x-1)^3 - \frac{6}{4!}(x-1)^4 + \frac{24}{5!}(x-1)^5,$$

or $P_5(x) = (x-1) - \frac{1}{2}(x-1)^2 + \frac{1}{3}(x-1)^3 - \frac{1}{4}(x-1)^4 + \frac{1}{5}(x-1)^5.$

The remainder term is given by

$$R_5(x) = \frac{f^{[6]}(z)}{6!}(x-1)^6 = -\frac{(x-1)^6}{6z^6}$$

for some number z, $1 < z < 1.2$. Since $1/z < 1$,

$$|R_5(x)| < \frac{(.2)^6}{6} < .000011.$$

Thus

$$\ln x = (x-1) - \tfrac{1}{2}(x-1)^2 + \tfrac{1}{3}(x-1)^3 - \tfrac{1}{4}(x-1)^4 + \tfrac{1}{5}(x-1)^5 \pm .000011, \qquad \text{if } 1 \le x \le 1.2.$$

For example,

$$\ln 1.2 = .2 - \frac{(.2)^2}{2} + \frac{(.2)^3}{3} - \frac{(.2)^4}{4} + \frac{(.2)^5}{5} \pm .000011,$$

and

$$\ln 1.2 \doteq .18233,$$

with a possible error of 1 in the fifth decimal place.

Example 3 Find $\sqrt[3]{e}$ to an accuracy of five decimal places.

Solution: By Example 3 of Section 4,

$$e^x = 1 + x + \frac{x^2}{2!} + \frac{x^3}{3!} + \cdots + \frac{x^n}{n!} + R_n(x),$$

where

$$R_n(x) = \frac{e^z}{(n+1)!}\,x^{n+1}, \qquad 0 < z < x \text{ if } x > 0.$$

We wish to choose n so that $|R_n(\frac{1}{3})| < 10^{-5}$. Clearly $e^{1/3} < 2$; therefore

$$\left|R_n\left(\frac{1}{3}\right)\right| < \frac{2}{3^{n+1}(n+1)!}.$$

If $n = 5$,

$$\left|R_5\left(\frac{1}{3}\right)\right| < \frac{2}{3^6 6!} = \frac{1}{729 \times 360} < 10^{-5}.$$

Thus

$$\sqrt[3]{e} \doteq 1 + \tfrac{1}{3} + \tfrac{1}{2}(\tfrac{1}{3})^2 + \tfrac{1}{6}(\tfrac{1}{3})^3 + \tfrac{1}{24}(\tfrac{1}{3})^4 + \tfrac{1}{120}(\tfrac{1}{3})^5,$$

or

$$\sqrt[3]{e} \doteq 1.39561.$$

EXERCISES

I

In each of Exercises 1 to 10 approximate the given function by a polynomial of given degree n in the given interval. State the error of approximation and compute the value of the function at the given number d as accurately as possible with the polynomial at hand.

1. $f(x) = \sin x,\ 0 \le x \le .5,\ n = 6,\ d = .5$
2. $g(x) = \cos x,\ 0 \le x \le .2,\ n = 3,\ d = .2$
3. $F(x) = e^x,\ 0 \le x \le 1,\ n = 5,\ d = 1$
4. $G(x) = e^{-x},\ 0 \le x \le 1,\ n = 6,\ d = 1$
5. $f(x) = \ln(1 + x),\ 0 \le x \le \frac{1}{2},\ n = 6,\ d = \frac{1}{2}$
6. $g(x) = \sqrt{x},\ 1 \le x \le 1.2,\ n = 5,\ d = 1.2$
7. $F(x) = \sinh x = \dfrac{e^x - e^{-x}}{2},\ 0 \le x \le 1,\ n = 6,\ d = 1$
8. $G(x) = \cosh x = \dfrac{e^x + e^{-x}}{2},\ 0 \le x \le \frac{1}{2},\ n = 4,\ d = \frac{1}{2}$
9. $g(x) = \tan^{-1} x,\ 0 \le x \le .2,\ n = 3,\ d = .2$
10. $f(x) = \sin^{-1} x,\ 0 \le x \le \frac{1}{2},\ n = 4,\ d = \frac{1}{2}$
11. Show that if $F(x)$ is a polynomial in x of degree n then $F(x) = P_n(x)$, the Taylor's polynomial of F at any number c.
12. Write the general Maclaurin's formula for $\ln(1 - x)$ and for $\ln(1 + x)$.

In Exercises 13 to 22, use an appropriate Taylor polynomial to compute the indicated values to three decimal places.

13. $\cos 46°$

14. $\sin 47°$

15. $\tan 43°$

16. $\cos 58°$

17. $\ln 2.1$, given that $\ln 2 \doteq .69315$

18. $\ln 3$

19. $\sqrt{e}$

20. $\sqrt[3]{26}$

21. $\sqrt[5]{28}$ (take $f(x) = \sqrt[5]{x}$, $c = 32$)

22. $\sqrt[5]{1.08}$

II

In Exercises 1–3, find the Taylor polynomial $P_n(x)$ in powers of x for the given function and value of n.

1. $\sec x$, $n = 4$.
2. $\tan x$, $n = 5$.
3. $\sin^2 x$, $n = 4$.
4. Suppose f is a function for which $f''(x)$ exists and is ≥ 0 for all x in an interval I. Prove that for any point c in I, the graph of f on I is never below the tangent line at the point $(c,f(c))$. *Suggestion:* Write Taylor's formula for f with $n = 1$.
5. If k is a small positive number, the straight line $y = kx$ intersects the graph of $y = \cot x$ near $x = \pi/2$. Use Taylor's formula to find an approximate value of the x-coordinate of this point of intersection. *Suggestion:* Consider the function $f(x) = \cot x - kx$, and find the polynomial $P_2(x)$ for f with $c = \pi/2$. Use $P_2(x)$ as an approximation to $f(x)$, neglecting $R_2(x)$.
6. Suppose f and g are functions for which $f^{(n)}(x)$ and $g^{(n)}(x)$ exist and are continuous throughout an interval I containing the number c. Suppose, furthermore, that

$$f(c) = f'(c) = \cdots = f^{(n-1)}(c) = 0,$$
$$g(c) = g'(c) = \cdots = g^{(n-1)}(c) = 0,$$

but that $g^{(n)}(x) \neq 0$ for all x in I. By applying Taylor's formula to both f and g, prove that

$$\operatorname*{limit}_{x \to c} \frac{f(x)}{g(x)} = \frac{f^{(n)}(c)}{g^{(n)}(c)}.$$

Compare this result with l'Hospital's rule.

CHAPTER REVIEW

Find each of the following limits.

1. $\operatorname*{limit}_{x \to 1} \dfrac{\ln x}{x^2 - x}$

2. $\operatorname*{limit}_{x \to 0} \dfrac{e^x - e^{-x}}{\sin x}$

3. $\lim_{\theta\to\pi/2^-} (\sec\theta)^{\cos\theta}$

4. $\lim_{\alpha\to\pi/2} (1 - \sin\alpha)\tan\alpha$

5. $\lim_{x\to 0^+} x \ln(\sin x)$

6. $\lim_{z\to 0^+} \dfrac{\ln z}{\cot z}$

7. $\lim_{x\to\infty} (\sqrt{x^2+2} - \sqrt{x^2-2})$

8. $\lim_{x\to\infty} x^{3/2}(\sqrt{x^3+a} - \sqrt{x^3-a})$

9. $\lim_{x\to\pi/2} \left(\dfrac{2}{1-\sin x} - \dfrac{1}{\cos^2 x}\right)$

10. $\lim_{x\to 0} (1 + \sin x)^{1/x}$

11. $\lim_{x\to 0^+} \left(\dfrac{1}{x^2}\right)^{\sin x}$

12. $\lim_{x\to 0} \left(\dfrac{\sin 3x}{3x}\right)^{1/x}$

13. $\lim_{x\to\infty} x \ln\left(\dfrac{x-1}{x+1}\right)$

14. $\lim_{x\to 3} \dfrac{3-x}{\ln(3-x)}$

In each of Exercises 15 to 24, find the integral if it exists.

15. $\int_0^\infty e^{-ax} \sin bx\, dx,\ a > 0$

16. $\int_0^\infty e^{-ax} \cos bx\, dx,\ a > 0$

17. $\int_0^1 \dfrac{1}{e^x - e^{-x}}\, dx$

18. $\int_{-\infty}^\infty \dfrac{1}{e^x - e^{-x}}\, dx$

19. $\int_0^\infty \dfrac{1}{e^x + e^{-x}}\, dx$

20. $\int_0^a \dfrac{1}{\sqrt{ax - x^2}}\, dx,\ a > 0$

21. $\int_3^4 \dfrac{x}{\sqrt{x-3}}\, dx$

22. $\int_1^\infty \dfrac{1}{\sqrt{x-1}}\, dx$

23. $\int_{-\infty}^\infty \dfrac{1}{3x^2 + 3x + 4}\, dx$

24. $\int_{-1}^1 \dfrac{x^2}{\sqrt{1-x}}\, dx$

In each of Exercises 25 to 28, find $P_n(x)$, $R_n(x)$ and write the Taylor's formula for the given function and given values of n and c.

25. $F(x) = \dfrac{1}{2+x};\ n = 4,\ c = -1$

26. $F(x) = \begin{cases} \dfrac{e^x - 1}{x} & \text{for } x \neq 0 \\ 1 & \text{for } x = 0, \end{cases} \quad n = 4,\ c = 0$

27. $f(x) = (2 + x^2)^{-1};\ n = 3,\ c = 1$

28. $f(x) = 4x^4 + 3x^3 + x^2 + x - 1;\ n = 4,\ c = 2$

29. Prove $\lim_{x\to\infty} \dfrac{x^n}{e^x} = 0;\ n = 1, 2, 3, \ldots$

30. Find $f'(0)$, where $f(x) = \begin{cases} \dfrac{e^x - 1}{x} & \text{if } x \neq 0 \\ 1 & \text{if } x = 0 \end{cases}$

Write the Maclaurin's formula for the given function and value of n.

31. $\sin 2x$, $n = 6$

32. e^{x^2}, $n = 4$

33. $\tan 3x$, $n = 4$

34. $\sin x^2$, $n = 3$

35. Calculate $\int_0^{1/4} \dfrac{2e^{-x^2}}{\sqrt{\pi}}\, dx$ accurate to five decimal places.

36. Calculate the smallest positive root of the equation $e^x + 10x = 2.216$ accurate to three decimal places by use of Maclaurin's formula.

12

Infinite Series

An infinite series is a certain special kind of sequence which is of great importance in both pure and applied mathematics. Of particular interest to us will be the so-called Taylor series, which is intimately related to the Taylor polynomials studied in the previous chapter.

1 PARTIAL SUMS AND TERMS

The reader should at this point review the basic properties of sequences discussed in Chapter 5, Section 8. In particular, it will be recalled that a sequence is a function whose domain is the set of positive integers. If a is a sequence, its values or members are denoted by

$$a_1, a_2, a_3, \ldots, a_n, \ldots.$$

A somewhat more concise notation for the above sequence is obtained simply by writing the formula for the nth member in brackets; thus the above sequence may be denoted by $\{a_n\}$. For example, $\{n^2\}$ denotes the sequence

$$1^2, 2^2, 3^2, \ldots, n^2, \ldots,$$

and $\left\{\frac{1}{n}\right\}$ denotes the sequence

$$1, \frac{1}{2}, \frac{1}{3}, \ldots, \frac{1}{n}, \ldots.$$

We shall make frequent use of this notation in this chapter.

Now, given any sequence $\{a_n\}$ of real numbers, we can form a new sequence $\{S_n\}$ as follows. We define

12.1

$$\begin{aligned}
S_1 &= a_1,\\
S_2 &= a_1 + a_2,\\
S_3 &= a_1 + a_2 + a_3,\\
&\cdots\cdots\cdots\cdots\cdots\\
S_n &= a_1 + a_2 + \cdots + a_n = \sum_{k=1}^{n} a_k,\\
&\cdots\cdots\cdots\cdots\cdots
\end{aligned}$$

The new sequence $\{S_n\}$ is called an *infinite series*. The members $a_1, a_2, \ldots, a_n, \ldots$ of the original sequence are called the *terms* of the series, while the numbers $S_1, S_2, \ldots, S_n, \ldots$ are called its *partial sums*.

Thus, in essence, an infinite series is a *pair* of sequences: a sequence of *terms* $\{a_n\}$ and a sequence of *partial sums* $\{S_n\}$. The terms and the partial sums are of course related by the equations 12.1. The student will avoid much confusion if he keeps clearly in mind which of the two sequences is being discussed in any particular theorem or problem.

Suppose, for example, we are given the sequence

$$\frac{1}{2}, \frac{1}{4}, \frac{1}{8}, \ldots, \frac{1}{2^n}, \ldots$$

as the sequence of terms $\{a_n\}$ for an infinite series. Then the sequence of partial sums $\{S_n\}$ is given by

$$\begin{aligned}
S_1 &= \frac{1}{2},\\
S_2 &= \frac{1}{2} + \frac{1}{4} = \frac{3}{4},\\
S_3 &= \frac{1}{2} + \frac{1}{4} + \frac{1}{8} = \frac{7}{8},\\
&\cdots\cdots\cdots\cdots\cdots\\
S_n &= \frac{1}{2} + \frac{1}{4} + \cdots + \frac{1}{2^n} = 1 - \frac{1}{2^n},\\
&\cdots\cdots\cdots\cdots\cdots
\end{aligned}$$

Usually an infinite series is specified by giving its *terms*; i.e., by giving the sequence $\{a_n\}$. Thus, if we wanted to talk about the infinite series (i.e., the sequence of partial sums) formed from the sequence $\{a_n\}$, some sort of notation such as $\left\{\sum_{k=1}^{n} a_k\right\}$ would be logical and appropriate. For the sequence of partial sums in the above example we would then write $\left\{\sum_{k=1}^{n} \frac{1}{2^k}\right\}$, which of course is the same as the sequence $\left\{1 - \frac{1}{2^n}\right\}$.

Unfortunately, in mathematical literature the traditional notation for infinite series is not as clear or logical as that suggested above. An infinite series with terms $\{a_n\}$ is frequently indicated by

$$a_1 + a_2 + \cdots + a_n + \cdots.$$

In this notation, for instance, the example above would be written

$$\frac{1}{2} + \frac{1}{4} + \cdots + \frac{1}{2^n} + \cdots.$$

This notation is open to objection because it seems to suggest that some sort of infinite "sum" is to be taken, but these symbols are intended only to denote the sequence $\left\{\sum_{k=1}^{n} \frac{1}{2^k}\right\}$.

Once the notation

$$a_1 + a_2 + \cdots + a_n + \cdots$$

is accepted, however, then we might just as well write this in the sigma notation as

12.2 $$\sum_{n=1}^{\infty} a_n,$$

and this again just denotes the infinite series with terms $\{a_n\}$. The wide prevalence of this notation in mathematical texts induces us to use it in this book, but not without some reluctance. Following common usage, we shall also frequently omit the limits on the sigma sign in 12.2, and simply denote the infinite series 12.2 by Σa_n or Σa_k.

We have indicated that an infinite series is usually specified by giving its terms. We may then compute its partial sums from equations 12.1. However, knowing the partial sums $\{S_n\}$ of an infinite series also enables us to compute its terms $\{a_n\}$, because we can "invert" equations 12.1 and write

12.3
$$\begin{aligned}
a_1 &= S_1,\\
a_2 &= S_2 - a_1 = S_2 - S_1,\\
a_3 &= S_3 - (a_1 + a_2) = S_3 - S_2,\\
&\cdots\cdots\cdots\cdots\cdots\cdots\\
a_n &= S_n - (a_1 + \cdots + a_{n-1}) = S_n - S_{n-1}, \qquad \text{for } n > 1.\\
&\cdots\cdots\cdots\cdots\cdots\cdots
\end{aligned}$$

For example,

$$\frac{1}{2}, \frac{2}{3}, \ldots, \frac{n}{n+1}, \ldots$$

is the sequence of partial sums of the infinite series

$$\frac{1}{2} + \left(\frac{2}{3} - \frac{1}{2}\right) + \cdots + \left(\frac{n}{n+1} - \frac{n-1}{n}\right) + \cdots,$$

or
$$\frac{1}{1 \cdot 2} + \frac{1}{2 \cdot 3} + \cdots + \frac{1}{n(n+1)} + \cdots.$$

In fact, as the reader can see from equations 12.3, any sequence is the sequence of partial sums corresponding to some infinite series.

The reader should now recall the definition of limit of a sequence in Chapter 5, Section 8. Thus,

$$\operatorname*{limit}_{n \to \infty} a_n = b$$

means that for every neighborhood N of b, there exists a positive number k such that a_n is in N for every $n > k$. Or, stated another way, the sequence $\{a_n\}$ has limit b if for every number $\varepsilon > 0$, there is a number $k > 0$ such that $|a_n - b| < \varepsilon$ for all $n > k$. If a sequence has a limit, we say that it is *convergent*; otherwise, it is *divergent*.

The reader should now also review the basic properties of limits of sequences, which were presented in Chapter 5. Also, it is an immediate consequence of the definition of limit of a sequence that if the sequence

$$S_1, S_2, \ldots, S_n, \ldots$$

converges to the limit S, then so does the sequence

$$S_m, S_{m+1}, \ldots, S_{m+n}, \ldots$$

obtained by deleting the first $m - 1$ members of the given sequence.

Given an infinite series with terms $\{a_n\}$, it may happen that its sequence of partial sums $\{S_n\}$ converges to a limit S. In this case S is called the *sum* of the infinite series, and according to our previous terminology we say that the series Σa_n *converges* to S. If $\{S_n\}$ has no limit, we say that Σa_n *diverges*. When Σa_n converges to S, this is indicated in traditional notation by writing

$$S = a_1 + a_2 + \cdots + a_n + \cdots,$$

or
$$S = \sum_{n=1}^{\infty} a_n.$$

The difficulty with this notation is that sometimes the symbols

$$a_1 + a_2 + \cdots + a_n + \cdots$$

denote the infinite series itself (i.e., the sequence of partial sums) and sometimes the sum S to which the series converges. It is often necessary to rely on the context to determine which usage is intended. Thus,

$$\frac{1}{2} + \frac{1}{4} + \cdots + \frac{1}{2^n} + \cdots$$

denotes the infinite series whose sequence of terms is $\left\{\frac{1}{2^n}\right\}$. The sequence of partial sums of this series is $\left\{1 - \frac{1}{2^n}\right\}$, which converges to the limit 1. Thus it is also customary to write

$$1 = \frac{1}{2} + \frac{1}{4} + \cdots + \frac{1}{2^n} + \cdots.$$

The student should understand clearly that the sum S of an infinite series Σa_n is the limit of the sequence of partial sums $\{S_n\}$ and not the limit of the sequence of terms $\{a_n\}$. However, if the sequence $\{S_n\}$ converges, then so does $\{a_n\}$, according to the following theorem.

12.4 Theorem

If the infinite series Σa_n converges to the sum S, then $\lim_{n\to\infty} a_n$ exists and $= 0$.

Proof: From 12.3 we have $a_n = S_n - S_{n-1}$, for $n = 2, 3, 4, \ldots$. Hence $\lim_{n\to\infty} a_n = \lim_{n\to\infty} S_n - \lim_{n\to\infty} S_{n-1}$. But, since $n \geq 2$, the sequence $\{S_n\}$ has members

(1) $$S_2, S_3, S_4, \ldots,$$

while the sequence $\{S_{n-1}\}$ has members

(2) $$S_1, S_2, S_3, \ldots.$$

Sequence (2) has limit S, and hence it is obvious that sequence (1) also has limit S. Hence $\lim_{n\to\infty} a_n = S - S = 0$.

Theorem 12.4 can also be stated in a useful *contrapositive* form: *If $\lim_{n\to\infty} a_n \neq 0$, then the infinite series Σa_n diverges.*

It is important to emphasize that the *converse* of Theorem 12.4 is not true; i.e., we may have $\lim_{n\to\infty} a_n = 0$, but the series Σa_n may fail to converge. For example, the so called *harmonic series*,

$$\sum_{n=1}^{\infty} \frac{1}{n} = 1 + \frac{1}{2} + \cdots + \frac{1}{n} + \cdots,$$

clearly has $\lim_{n\to\infty} a_n = 0$. However, we shall show in the next section that this series diverges (also see Exercise II-2 of this section).

An important type of infinite series is the *geometric series*,

$$\sum_{k=1}^{\infty} ar^{k-1} = a + ar + ar^2 + \cdots + ar^{k-1} + \cdots,$$

where a and r are constants. From the identity

$$1 - r^n = (1 - r)(1 + r + r^2 + \cdots + r^{n-1}),$$

we easily deduce that the nth partial sum S_n of the geometric series is given by

$$S_n = a \cdot \sum_{k=1}^{n} r^{k-1} = \frac{a(1 - r^n)}{1 - r},$$

or

$$S_n = \frac{a}{1 - r} - \frac{ar^n}{1 - r}, \qquad r \neq -1.$$

Since

$$\operatorname*{limit}_{n\to\infty} r^n = 0 \qquad \text{if } |r| < 1,$$

evidently

$$\operatorname*{limit}_{n\to\infty} S_n = \frac{a}{1 - r} - \frac{a}{1 - r} \operatorname*{limit}_{n\to\infty} r^n = \frac{a}{1 - r}$$

if $|r| < 1$. This proves that the geometric series converges and has sum $a/(1 - r)$ if $|r| < 1$; that is,

12.5 $$\sum_{k=1}^{\infty} ar^{k-1} = \frac{a}{1 - r}, \qquad |r| < 1.$$

If $|r| \geq 1$, then $\operatorname*{limit}_{n\to\infty} r^n \neq 0$; and hence in this case the geometric series diverges by Theorem 12.4.

For example,

$$1 + \frac{1}{2} + \frac{1}{4} + \cdots + \frac{1}{2^{n-1}} + \cdots = \frac{1}{1 - \frac{1}{2}} = 2,$$

$$.3 + .03 + .003 + \cdots + \frac{3}{10^n} + \cdots = \frac{.3}{1 - 10^{-1}} = \frac{1}{3}.$$

We recognize this latter series to be $.3333\cdots$, the infinite decimal representation of $\frac{1}{3}$.

We can, conversely, use 12.5 to obtain an infinite series expansion for fractions such as $1/(c - x)$. For example, if we let $a = 1$ and $r = -x^2$ in 12.5, we obtain

$$\frac{1}{1+x^2} = \frac{1}{1-(-x^2)} = 1 + (-x^2) + (-x^2)^2 + (-x^2)^3 + \cdots$$
$$= 1 - x^2 + x^4 - x^6 + \cdots$$
$$= \sum_{k=1}^{\infty} (-1)^{k-1} x^{2k-2}$$
$$= \sum_{k=0}^{\infty} (-1)^k x^{2k}.$$

The series converges if $x^2 < 1$, that is, $|x| < 1$.

EXERCISES

I

Find the sum of each of the following convergent series.

1. $5 + \frac{5}{7} + \cdots + \frac{5}{7^{n-1}} + \cdots$

2. $5 - \frac{5}{7} + \cdots + \frac{(-1)^{n-1}5}{7^{n-1}} + \cdots$

3. $\pi + \frac{\pi}{\sqrt{2}} + \cdots + \frac{\pi}{\sqrt{2^{n-1}}} + \cdots$

4. $.\overline{23}2323\cdots$ (the bar over the number 23 means that 23 is to be repeated indefinitely)

5. $.\overline{612}612612\cdots$

6. $1 - \frac{1}{4} + \frac{1}{16} - \frac{1}{64} + \cdots + \frac{(-1)^{n-1}}{4^{n-1}} + \cdots$

7. $\sum_{n=1}^{\infty} \frac{2^{n+1}}{3^n}$

8. $\sum_{n=1}^{\infty} 6(\frac{1}{4})^n$

Each of the following is the nth partial sum of an infinite series. Find the infinite series, tell whether it is convergent or divergent, and if convergent, find its sum.

9. $S_n = \frac{n}{2n+1}$

10. $S_n = \frac{3n}{4n+1}$

11. $S_n = \frac{n^2}{n+1}$

12. $S_n = \ln(n+1)$

13. $S_n = \frac{1}{2^n}$

14. $S_n = 2^n$

15. $S_n = \dfrac{n^2}{n^2 + 1}$

16. $S_n = \dfrac{n^3 - 1}{n^3}$

In each of Exercises 17 to 22, show that the infinite series diverges.

17. $\dfrac{1}{2} + \dfrac{2}{3} + \cdots + \dfrac{n}{n + 1} + \cdots$

18. $-1 + 1 + \cdots + (-1)^n + \cdots$

19. $\frac{3}{2} + \frac{9}{4} + \cdots + (\frac{3}{2})^n + \cdots$

20. $\displaystyle\sum_{n=1}^{\infty} \sin \pi n$

21. $\displaystyle\sum_{n=1}^{\infty} [1 + (-1)^{n-1}]$

22. $\displaystyle\sum_{n=1}^{\infty} \ln\left(\frac{3}{n}\right)$

23. A ball is dropped from a height of 20 ft. Each time it hits the ground it bounces back to three-fourths of its previous height. Find the total distance the ball travels before coming to rest.

24. If Σa_k is a convergent infinite series and

$$R_n = a_{n+1} + a_{n+2} + \cdots,$$

the *remainder* of the series after n terms, then prove that

$$\operatorname*{limit}_{n \to \infty} R_n = 0.$$

25. Let $b_1, b_2, \ldots, b_n, \ldots$ be a sequence of nonzero numbers such that $\operatorname*{limit}_{n \to \infty} b_n = \pm\infty$. Prove that the infinite series $\Sigma(b_{k+1} - b_k)$ diverges whereas the series $\Sigma(1/b_k - 1/b_{k+1})$ converges.

26. Using Exercise 25, show that the infinite series

$$\sum_{k=1}^{\infty} \ln\left(1 + \frac{1}{k}\right)$$

diverges.

In each of the following series, find for what x the series converges, and then find the sum of the series.

27. $\displaystyle\sum_{n=1}^{\infty} x^{3n-1}$

28. $\displaystyle\sum_{n=0}^{\infty} (2x)^{2n+1}$

29. $\displaystyle\sum_{j=2}^{\infty} (-1)^j(2x - 1)^{2j-4}$

30. $\displaystyle\sum_{m=0}^{\infty} \left(\frac{x^2 - 1}{2}\right)^m$

31. Find the sum of the series

$$\frac{1}{1 \cdot 2} + \frac{1}{2 \cdot 3} + \cdots + \frac{1}{k(k + 1)} + \cdots$$

$$\left[\textit{Hint:}\ \frac{1}{k(k + 1)} = \frac{1}{k} - \frac{1}{k + 1}.\right]$$

II *Exercises on Sequences*

1. A sequence $\{S_n\}$ is called a *Cauchy* sequence if for every number $\varepsilon > 0$ there is a number $k > 0$ such that $|S_m - S_n| < \varepsilon$ for all $m > k$, $n > k$. Prove that any convergent sequence is a Cauchy sequence. (It is also true that any Cauchy sequence is a convergent sequence, but this is more difficult to prove.)

2. Use Exercise 1 above to prove that the harmonic series $\sum_{n=1}^{\infty} \frac{1}{n}$ is divergent; i.e., prove that the sequence of partial sums $\{S_n\}$ is not a Cauchy sequence. (*Hint:* Show that $S_{2n} - S_n > \frac{1}{2}$ for all n.)

3. Suppose that the sequence $\{S_n\}$ converges to 0. Prove that the sequence $\{\sigma_n\}$ of arithmetic means defined by

$$\sigma_n = \frac{S_1 + S_2 + \cdots + S_n}{n}$$

also converges to 0. *Suggestion:* Given any $\varepsilon > 0$, choose n_0 so that $|S_n| < \varepsilon/2$ for all $n > n_0$. Then

$$\begin{aligned} |\sigma_n| &\leq \frac{|S_1| + \cdots + |S_{n_0}|}{n} + \frac{|S_{n_0+1}| + \cdots + |S_n|}{n} \\ &\leq \frac{|S_1| + \cdots + |S_{n_0}|}{n} + \frac{(n - n_0)}{n}\left(\frac{\varepsilon}{2}\right). \end{aligned}$$

Now choose k so that $(|S_1| + \cdots + |S_{n_0}|)/n < \varepsilon/2$ for all $n > k$.

4. Prove that if $\{S_n\}$ converges to S, then the sequence $\{\sigma_n\}$ defined by

$$\sigma_n = \frac{S_1 + S_2 + \cdots + S_n}{n}$$

also converges to S. *Hint:* Use Exercise 3.

5. If $\{x_n\}$ is a sequence of positive numbers converging to the positive number x, prove that the sequence $\{y_n\}$ of geometric means defined by

$$y_n = \sqrt[n]{x_1 x_2 \ldots x_n}$$

also converges to x. *Hint:* Let $S_n = \ln x_n$ and use Exercise 4.

6. Use Exercise 5 to prove that $\lim_{n\to\infty} \sqrt[n]{n} = 1$ *Hint:* Write $n = 1 \cdot \frac{2}{1} \cdot \frac{3}{2} \cdot \cdots \cdot \frac{n}{n-1}$ and observe that $\lim_{n\to\infty} \frac{n}{n-1} = 1$.

7. If $\{x_n\}$ is a sequence with $\lim_{n\to\infty} x_{2n} = b$ and $\lim_{n\to\infty} x_{2n+1} = b$, prove that $\lim_{n\to\infty} x_n = b$.

8. If the sequence $\{x_n\}$ has a limit b, prove that the sequence is *bounded*: that is, there exists a positive number M such that $|x_n| \leq M$ for all n.

2 POSITIVE TERM SERIES

It will be helpful, at this point, to mention some further elementary properties of infinite series. The following theorem, for example, is a direct application of the results of Chapter 5, Section 8, on the algebra of limits of sequences.

12.6 Theorem

Let Σa_k and Σb_k be infinite series which converge to the sums S and T, respectively. Then:

(a) *The series $\Sigma(a_k + b_k)$ converges to the sum $S + T$.*
(b) *For any constant c, Σca_k converges to cS.*

Proof: Let $\{S_n\}$ and $\{T_n\}$ be the sequences of partial sums for Σa_k and Σb_k, respectively. Then $\{S_n + T_n\}$ is the sequence of partial sums for $\Sigma(a_k + b_k)$, and $\operatorname{limit}_{n\to\infty} (S_n + T_n) = \operatorname{limit}_{n\to\infty} S_n + \operatorname{limit}_{n\to\infty} T_n = S + T$, by 5.31. The proof of (b) follows in a similar way from 5.30 and 5.32.

The convergence or divergence of an infinite series Σa_k is not changed by modifying a finite number of its terms. Thus, if the two series

$$\sum_{k=1}^{\infty} a_k, \qquad \sum_{k=1}^{\infty} b_k,$$

with nth partial sums S_n and T_n, respectively, differ only in their first m terms (that is, $a_k = b_k$ if $k > m$), then

$$S_n - T_n = S_m - T_m$$

for every integer $n \geq m$. Hence

$$\operatorname{limit}_{n\to\infty} S_n = (S_m - T_m) + \operatorname{limit}_{n\to\infty} T_n,$$

and either both limits exist or both limits do not exist; i.e., the series Σa_k and Σb_k either both converge or both diverge. If the series converge, their sums differ by $S_m - T_m$. In particular, if $b_1 = b_2 = \cdots = b_m = 0$, then

$$\sum_{k=1}^{\infty} a_k \quad \text{and} \quad \sum_{k=m+1}^{\infty} a_k = \sum_{k=1}^{\infty} b_k;$$

either both converge or both diverge.

An infinite series Σa_k with each $a_k > 0$ is called a *positive term series.* For such a series, the partial sums $\{S_n\}$ form an *increasing* sequence: in fact, we have

$$S_1 < S_2 < \cdots < S_n < \cdots.$$

According to Theorem 5.36, the sequence $\{S_n\}$ therefore converges if it has an upper bound, that is, if there exists a number k such that $S_n < k$ for all n. This gives the following result.

12.7 Theorem

Let Σa_k be a positive term series with sequence of partial sums $\{S_n\}$. The series Σa_k converges if the sequence $\{S_n\}$ is bounded and diverges otherwise. Furthermore, if Σa_k converges to the sum S, then $S_n < S$ for every n.

Proof: If every $S_n < K$, then $\lim_{n\to\infty} S_n$ exists by Theorem 5.36 and hence Σa_k is convergent. If, on the other hand, $\{S_n\}$ has no upper bound, then $\lim_{n\to\infty} S_n = \infty$, and hence Σa_k diverges. To prove the last assertion, recall from the proof of 5.36 that, if the increasing sequence $\{S_n\}$ has an upper bound, then $\{S_n\}$ converges to its *least upper bound.*

Example 1 Prove that the harmonic series $\sum_{k=1}^{\infty} \frac{1}{k}$ diverges.

Solution: Writing $S_n = 1 + \frac{1}{2} + \cdots + 1/n$, observe that

$$S_{2n} - S_n = \frac{1}{n+1} + \frac{1}{n+2} + \cdots + \frac{1}{2n} > \frac{1}{2n} + \frac{1}{2n} + \cdots + \frac{1}{2n} = \frac{1}{2}.$$

Thus $S_{2n} - S_n > \frac{1}{2}$ for every n. In particular,

$$\begin{aligned} S_2 &> S_1 + \tfrac{1}{2} = \tfrac{3}{2}, \\ S_4 &> S_2 + \tfrac{1}{2} > \tfrac{3}{2} + \tfrac{1}{2} = 2, \\ S_8 &> S_4 + \tfrac{1}{2} > 2 + \tfrac{1}{2} = \tfrac{5}{2}, \end{aligned}$$

and by induction we can show that

$$S_{2^p} > \frac{p+2}{2} \qquad \text{for every positive integer } p.$$

This means that the sequence $\{S_n\}$ cannot have an upper bound, and hence $\sum \frac{1}{k}$ diverges by 12.7.

From a given positive term series Σa_k, we may form other series by grouping the terms in some order. For example, we might form the series

$$a_1 + (a_2 + a_3) + (a_4 + a_5 + a_6) + (a_7 + a_8 + a_9 + a_{10}) + \cdots.$$

Designating this series by Σb_k, we have

$$b_1 = a_1, \quad b_2 = a_2 + a_3, \quad b_3 = a_4 + a_5 + a_6, \quad \cdots.$$

It is evident that each partial sum of the series Σb_k is also a partial sum of the series Σa_k. Thus, if the series Σa_k converges, so does the series Σb_k and their sums are equal. Similar remarks hold for any grouping of the terms of Σa_k.

We may also form a new series from a series Σa_k by rearranging the terms. For example, we might form the sum

$$a_3 + a_1 + a_5 + a_2 + a_7 + a_4 + \cdots.$$

If the series Σa_k is convergent with sum S and if T_n is the nth partial sum of a series Σb_k formed by rearranging the terms of Σa_k, then $T_n < S$ since T_n is a sum of terms of the series Σa_k. Hence Σb_k converges and has a sum $T \leq S$ by 12.7. Since the series Σa_k may be obtained by rearranging the terms of Σb_k, we have $S \leq T$ by the same reasoning. Thus $S = T$, and we have proved that each rearrangement of a positive term convergent series is a convergent series having the same sum. (This result, however, need not hold for series whose terms are not all positive. See Section 5, Exercise II-3.)

The theory of improper integrals may be used to give the following test for convergence of an infinite series.

12.8 Integral Test

If f is a positive-valued, continuous, and decreasing function, then the infinite series

$$f(1) + f(2) + \cdots + f(n) + \cdots$$

converges or diverges according as the improper integral

$$\int_1^\infty f(x)\,dx$$

exists or is infinite.

Proof: It is clear from Figure 12.1 that

$$f(2) + f(3) + \cdots + f(n)$$

is a lower sum and

$$f(1) + f(2) + \cdots + f(n-1)$$

is an upper sum of f over the interval $[1,n]$. Hence

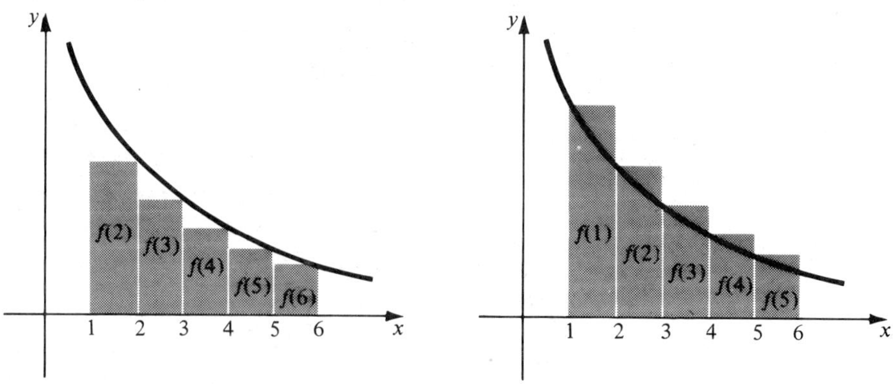

FIGURE 12.1

$$(1) \qquad \sum_{k=2}^{n} f(k) \leq \int_1^n f(x)\,dx \leq \sum_{k=1}^{n-1} f(k).$$

If the given improper integral exists, then

$$\sum_{k=2}^{n} f(k) \leq \int_1^n f(x)\,dx < \int_1^\infty f(x)\,dx$$

and the infinite series $\Sigma f(k)$ converges by 12.7. If the given improper integral is infinite, then clearly $\Sigma f(k) = \infty$ also by (1) above. This proves the integral test.

If we let

$$f(x) = \frac{1}{x^p}, \qquad p > 0,$$

then the series $\Sigma f(k)$ is simply the *p series*

$$1 + \frac{1}{2^p} + \frac{1}{3^p} + \cdots + \frac{1}{n^p} + \cdots.$$

We easily verify that

$$\begin{aligned}\int_1^\infty \frac{1}{x^p}\,dx &= \operatorname*{limit}_{t\to\infty} \int_1^t x^{-p}\,dx \\ &= -\frac{1}{p-1} \operatorname*{limit}_{t\to\infty} \left(\frac{1}{t^{p-1}} - 1\right) \\ &= \frac{1}{p-1} \qquad \text{if } p > 1, \text{ or } \infty \text{ if } p < 1.\end{aligned}$$

Hence the p series converges by the integral test if $p > 1$ and diverges if $p < 1$. If $p = 1$, the series is the harmonic series which we previously showed diverged. Thus, we have the following result.

12.9 Theorem

The p series $\Sigma 1/k^p$ converges if $p > 1$ and diverges if $p \leq 1$.

Example 2 Test the series

$$\sum_{k=1}^{\infty} \frac{k}{e^k}$$

for convergence.

Solution: If we let $f(x) = x/e^x = xe^{-x}$, then $f(x) > 0$ and $f'(x) = e^{-x} - xe^{-x} < 0$ if $x > 1$. Hence f is a decreasing function in $[1,\infty)$. By formula 6.8 of the Table of Integrals,

$$\int_1^t xe^{-x}\,dx = 2e^{-1} - (t+1)e^{-t}.$$

Since

$$\int_1^\infty xe^{-x}\,dx = \lim_{t\to\infty} [2e^{-1} - (t+1)e^{-t}] = 2e^{-1}$$

the given series converges by the integral test.

Example 3 Test the series

$$\sum_{k=2}^{\infty} \frac{1}{k \ln k}$$

for convergence.

Solution: If we let $f(x) = 1/x \ln x$, then

$$f'(x) = -\frac{1 + \ln x}{(x \ln x)^2}.$$

Since $f(x) > 0$ and $f'(x) < 0$ if $x > 1$, f is decreasing in the interval $[2,\infty)$ and the integral test can be applied to $\Sigma 1/k \ln k$. Now

$$\int_2^t \frac{1}{x \ln x}\,dx = \ln(\ln t) - \ln(\ln 2)$$

and therefore

$$\int_2^\infty \frac{1}{x \ln x}\,dx = \lim_{t\to\infty} [\ln(\ln t) - \ln(\ln 2)] = \infty.$$

Hence the series $\Sigma 1/k \ln k$ diverges according to the integral test.

EXERCISES

I

In each of Exercises 1 to 1 , determine whether the series is convergent or divergent.

1. $1 + \dfrac{1}{2^3} + \dfrac{1}{3^3} + \cdots + \dfrac{1}{n^3} + \cdots$

2. $1 + \dfrac{1}{2\sqrt{2}} + \dfrac{1}{3\sqrt{3}} + \cdots + \dfrac{1}{n\sqrt{n}} + \cdots$

3. $\dfrac{1}{3} + \dfrac{1}{5} + \dfrac{1}{11} + \cdots + \dfrac{1}{3^{n-1} + 2} + \cdots$

4. $\frac{1}{4} + \frac{1}{7} + \frac{1}{10} + \cdots + \frac{1}{3n+1} + \cdots$

5. $\sum_{k=1}^{\infty} \frac{k}{k^2 + 10^6}$

6. $\sum_{k=1}^{\infty} \frac{\ln k}{k}$

7. $\sum_{i=3}^{\infty} \frac{1}{\sqrt[i]{2}}$

8. $\sum_{j=4}^{\infty} \frac{\sqrt{j}}{j^2 - 4}$

9. $\sum_{k=3}^{\infty} \frac{\ln k}{k^2}$

10. $\sum_{k=2}^{\infty} \frac{1}{\ln k}$

11. $\sum_{n=1}^{\infty} n^2 e^{-n}$

12. $\sum_{r=1}^{\infty} \frac{\tan^{-1} r}{r^2 + 1}$

13. $\sum_{k=1}^{\infty} \frac{1}{(2k+1)^2}$

14. $\sum_{k=1}^{\infty} \frac{k}{1 + k^4}$

II

1. Let $\sum_{n=1}^{\infty} f(n)$ be a positive term series. Suppose there is a number $r > 1$ and a number B such that $f(n) \cdot n^r \le B$ for all $n > a$. Prove that $\Sigma f(n)$ converges. $\left(\textit{Hint:}\ 0 \le \int_a^{\infty} f(x)\,dx \le \int_a^{\infty} \frac{B}{x^r}\,dx.\right)$

2. To which exercises of Part I does the preceding exercise immediately apply?

3. For what values of r does $\sum_{k=2}^{\infty} \frac{1}{k(\ln k)^r}$ converge? Use the integral test.

4. Use the methods of the proof of the integral test to show that

$$\frac{1}{(p-1)(n+1)^{p-1}} < \sum_{k=1}^{\infty} \frac{1}{k^p} - \sum_{k=1}^{n} \frac{1}{k^p} < \frac{1}{(p-1)n^{p-1}}$$

if $p > 1$.

a. How good an approximation of the infinite series $\Sigma 1/k^2$ is its nth partial sum when $n = 10$? $n = 1000$?

b. How good an approximation of the infinite series $\Sigma 1/k^3$ is its nth partial sum when $n = 10$? $n = 1000$?

5. (*Cauchy's Condensation Theorem.*) Let $\sum_{k=1}^{\infty} a_k$ be any series of positive *decreasing* terms. From this series we may form a new series

$$\sum_{k=0}^{\infty} 2^k a_{2^k} = a_1 + 2a_2 + 4a_4 + 8a_8 + \cdots,$$

which is called the *condensed series*. Prove that the given series Σa_k and its condensed series either both converge or both diverge. *Suggestion:* Let T_m denote the mth partial sum of the condensed series; that is,

$$T_m = a_1 + 2a_2 + \cdots + 2^m a_{2^m}.$$

If $n = 2^m$, and if S_n is the nth partial sum of the given series, prove that $S_n \le T_m \le 2S_n$. This shows that the two sequences $\{S_n\}$ and $\{T_m\}$ are either both bounded or both unbounded, so that the conclusion follows from 12.7.

The condensation theorem is remarkable in that it shows that the convergence or divergence of a series may be determined by considering only a small number of its terms, namely, $a_1, a_2, a_4, a_8, \ldots$.

6. Use the condensation theorem to prove that the p-series $\Sigma 1/k^p$ converges for $p > 1$ and diverges for $p \le 1$.
7. Do Exercise 3 above by the condensation theorem.
8. Prove that the series $\sum_{k=3}^{\infty} \dfrac{1}{k \cdot \ln k \cdot \ln(\ln k)}$ diverges. *Hint:* Prove that the condensed series has larger terms than $\sum_{k=3}^{\infty} 1/k \ln k$ and use Example 3 of this section.

3 COMPARISON TEST

Another useful test for convergence and divergence of a positive term series is as follows.

12.10 Comparison Test I

Let Σa_k and Σb_k be positive term series.

(1) *If Σa_k converges and if $b_k \le a_k$ for every integer k, then the series Σb_k converges.*

(2) *If Σa_k diverges and if $b_k \ge a_k$ for every integer k, then the series Σb_k diverges.*

Proof: Designate the nth partial sums of Σa_k and Σb_k by S_n and T_n, respectively. If Σa_k converges and every $b_k \le a_k$, then

$$T_n \le S_n < \sum_{k=1}^{\infty} a_k$$

for every integer n. Hence the series Σb_k converges by 12.7. We omit the proof of (2) because of its similarity to that of (1).

If $0 < b_k \le a_k$ for every integer k, then the series Σa_k is said to *dominate* the series Σb_k. According to the theorem above, every infinite series dominated by a convergent series also is convergent.

In view of the fact that a finite number of terms of a series may be changed without affecting the convergence or divergence of the series, it is clear that comparison test I need only hold from the mth term on, where m is any positive integer.

Example 1 Test the series

$$\frac{1}{2} + \frac{1}{3} + \frac{1}{5} + \cdots + \frac{1}{2^{n-1} + 1} + \cdots$$

for convergence.

Solution: We compare the given series with the convergent geometric series $\sum \frac{1}{2^{k-1}}$. Thus

$$\frac{1}{2^{k-1} + 1} < \frac{1}{2^{k-1}}$$

for every integer k, and the given series must converge.

Example 2 Test the series

$$\sum_{k=2}^{\infty} \frac{1}{\ln k}$$

for convergence.

Solution: Since $\ln k < k$ for every $k \geq 2$,

$$\frac{1}{\ln k} > \frac{1}{k} \qquad \text{for every } k \geq 2.$$

Hence, $\Sigma 1/\ln k$ diverges by 12.10(2).

The following test is essentially a corollary of comparison test I.

12.11 Comparison Test II

Let Σa_k and Σb_k be positive term series, and let $c_k = b_k/a_k$. If $\lim_{k\to\infty} c_k = c$, where $c > 0$, then the two series Σa_k and Σb_k either both converge or both diverge.

Proof: Since $\lim_{k\to\infty} c_k = c > 0$, there exists an integer k_0 such that

$$\frac{c}{2} < c_k < \frac{3c}{2} \qquad \text{for all } k \geq k_0.$$

Hence for $k \geq k_0$ we have

$$\frac{c}{2} a_k < c_k a_k < \frac{3c}{2} a_k,$$

or

$$\frac{c}{2} a_k < b_k < \frac{3c}{2} a_k.$$

If the series Σa_k converges, then so does the series Σb_k because it is dominated by a convergent series $\Sigma(3c/2)a_k$. Conversely, if the series Σb_k is convergent, then so is the series $\Sigma(c/2)a_k$, and in turn so is $\Sigma a_k = (2/c)\Sigma(c/2)a_k$. This proves that Σa_k converges if and only if Σb_k converges. But this implies that Σa_k diverges if and only if Σb_k diverges. Thus the theorem is proved.

Example 3 Test the series

$$\sum_{k=1}^{\infty} \frac{k+1}{k(2k-1)}$$

for convergence.

Solution: We may put the given series in the form

$$\sum_{k=1}^{\infty} \left(\frac{1+1/k}{2-1/k}\right)\frac{1}{k}.$$

Since

$$\underset{n\to\infty}{\text{limit}} \frac{1+1/n}{2-1/n} = \frac{1}{2}$$

and the harmonic series $\Sigma 1/k$ diverges, the given series diverges by 12.11.

Example 4 Test the series

$$\sum_{k=1}^{\infty} \frac{k}{\sqrt{k^5+1}}$$

for convergence.

Solution: The series can be put in the form

$$\sum_{k=1}^{\infty} \frac{1}{\sqrt{1+1/k^5}}\frac{1}{k^{3/2}}.$$

Since

$$\underset{n\to\infty}{\text{limit}} \frac{1}{\sqrt{1+1/n^5}} = 1,$$

and the p series $\Sigma 1/k^{3/2}$ converges by 12.9, the given series converges by 12.11.

EXERCISES

I

In each of Exercises 1 to 22 determine whether the series is convergent or divergent, using the methods of this section.

1. $\displaystyle\sum_{k=0}^{\infty} \frac{k+1}{(k+2)2^k}$

2. $\displaystyle\sum_{k=1}^{\infty} \frac{2+10^6\sin^2 3k}{k^2}$

3. $\sum_{k=1}^{\infty} \frac{k+1}{\ln (k+2)}$

4. $\sum_{k=1}^{\infty} \frac{k^2}{k^3+1000}$

5. $\sum_{k=1}^{\infty} ke^{-k^2}$

6. $\sum_{k=1}^{\infty} \frac{1}{\sqrt[3]{k(k+1)(k+2)}}$

7. $\sum_{k=1}^{\infty} \frac{1}{\sqrt{k(k+1)(k+2)}}$

8. $\sum_{k=1}^{\infty} k^2 2^{-k}$

9. $\sum_{k=2}^{\infty} \frac{\sqrt{k}}{k^2 - \sin^2 100k}$

10. $\sum_{k=1}^{\infty} \frac{\tan^{-1} k}{k^2+1}$

11. $\sum_{k=2}^{\infty} \frac{\sqrt{k^4+1}}{k^3 \ln k}$

12. $\sum_{k=1}^{\infty} \frac{2^k}{3^k+1}$

13. $\sum_{n=1}^{\infty} \frac{n^3}{n!}$

14. $\sum_{n=1}^{\infty} \frac{n!}{(3n)!}$

15. $\sum_{n=1}^{\infty} \frac{|\sec n|}{\sqrt{n}}$

16. $\sum_{n=1}^{\infty} \frac{1}{1+\sqrt{n}}$

17. $\sum_{n=1}^{\infty} \frac{1}{1+3^{1/n}}$

18. $\sum_{n=3}^{\infty} \frac{1}{(\ln n)^{1/n}}$

19. $\sum_{n=1}^{\infty} \frac{n!}{n^n}$

20. $\sum_{n=4}^{\infty} \frac{1}{\sqrt[3]{n^2+4}}$

21. $\sum_{n=1}^{\infty} \frac{1}{n^{1+1/n}}$

22. $\sum_{n=1}^{\infty} \frac{\sqrt{n+1}-\sqrt{n}}{n}$

II

Suppose Σa_k is a divergent positive term series. State what you can about the behavior of each of the series in Exercises 1–5.

1. $\sum \frac{a_k}{1+a_k}$

2. $\sum \frac{a_k}{1+a_k \ln k}$

3. $\sum \frac{a_k}{1+ka_k}$

4. $\sum \frac{a_k}{1+k^2 a_k}$

5. $\sum \frac{a_k}{1+a_k^2}$

6. Prove that if $a_n \geq 0$ and $\sum_{n=1}^{\infty} a_n$ converges, then so does $\sum_{n=1}^{\infty} a_n^2$.

7. Prove that if $a_n \geq 0$ and $\sum_{n=1}^{\infty} a_n^2$ converges, then so does $\sum_{n=1}^{\infty} \frac{a_n}{n}$.

8. Suppose $\{b_k\}$ is any sequence of positive numbers with $b_k \to \infty$. Determine the behavior of the series

$$\sum_{k \to 1}^{\infty} \frac{1}{b_k^{\ln k}}.$$

9. (*Cauchy's Root Test.*) Let Σa_n be a positive term series. If $\sqrt[n]{a_n} \le c < 1$ for all sufficiently large n, prove that Σa_n converges. On the other hand, if $\sqrt[n]{a_n} \ge c > 1$ for all sufficiently large n, prove that Σa_n diverges. *Hint:* Compare with the geometric series Σc^n.

10. (*Another form of Cauchy's Root Test.*) Let Σa_n be a positive term series and let $\lim_{n\to\infty} \sqrt[n]{a_n} = b$. *Prove:* if $b < 1$ the series converges, and if $b > 1$ the series diverges. Use Exercise II-9.

4 ALTERNATING SERIES

An infinite series of the form

$$a_1 - a_2 + a_3 - a_4 + \cdots + (-1)^{n-1}a_n + \cdots,$$

each $a_n > 0$, having alternating positive and negative terms is called an *alternating series*. A classical example of an alternating series is the alternating harmonic series

$$1 - \frac{1}{2} + \frac{1}{3} - \frac{1}{4} + \cdots + \frac{(-1)^{n-1}}{n} + \cdots,$$

which we shall show presently to be convergent with sum ln 2. A simple test for the convergence of an alternating series is as follows.

12.12 Alternating Series Test

If (1) $a_{n+1} \le a_n$ *for each* n, *and* (2) $\lim_{n\to\infty} a_n = 0$, *then the alternating series* $\Sigma(-1)^{k-1}a_k$ *is convergent.*

Proof: The partial sums of $\Sigma(-1)^{k-1}a_k$ having an even number of terms may be written as follows:

$$S_2 = (a_1 - a_2), \qquad S_4 = (a_1 - a_2) + (a_3 - a_4),$$

and, in general,

$$S_{2n} = (a_1 - a_2) + (a_3 - a_4) + \cdots + (a_{2n-1} - a_{2n}).$$

By assumption, $a_1 - a_2 \ge 0$, $a_3 - a_4 \ge 0$, and, in general, $a_{2n-1} - a_{2n} \ge 0$. Thus it is clear that

$$0 \le S_2 \le S_4 \le \cdots \le S_{2n} \le \cdots.$$

On the other hand, S_{2n} may also be written in the form

$$S_{2n} = a_1 - (a_2 - a_3) - (a_4 - a_5) - \cdots - (a_{2n-2} - a_{2n-1}) - a_{2n},$$

from which it is clear that, for every integer n,

$$S_{2n} \leq a_1.$$

That

$$\lim_{n\to\infty} S_{2n}$$

exists and is less than or equal to a_1 now follows from 5.37.

Since $S_{2n+1} = S_{2n} + a_{2n+1}$,

$$\lim_{n\to\infty} S_{2n+1} = \lim_{n\to\infty} S_{2n} + \lim_{n\to\infty} a_{2n+1} = \lim_{n\to\infty} S_{2n}.$$

Thus

$$\lim_{n\to\infty} S_n = \lim_{n\to\infty} S_{2n} = \lim_{n\to\infty} S_{2n+1} = S,$$

and $S \leq a_1$. This proves the test.

If $\Sigma(-1)^{k-1}a_k$ is an alternating series with the properties of 12.12, then the remainder R_n of this series after n terms,

$$R_n = (-1)^n(a_{n+1} - a_{n+2} + a_{n+3} - \cdots),$$

is again an alternating series with these same properties. Clearly,

$$|R_n| = a_{n+1} - a_{n+2} + a_{n+3} - \cdots,$$

and therefore

$$|R_n| \leq a_{n+1}$$

as shown in the proof of 12.12. If S is the sum of $\Sigma(-1)^{k-1}a_k$, then $S = S_n + R_n$ and

$$|S - S_n| = |R_n| \leq a_{n+1}.$$

This proves the following result.

12.13 Theorem

If $\Sigma(-1)^{k-1}a_k$ is an alternating series having the properties stated in 12.12, and if S_n and S are, respectively, the nth partial sum and the sum of this series, then $S \doteq S_n$ with an error of no more than a_{n+1}.

Example 1 Show that the alternating series

$$1 - \frac{1}{1!} + \frac{1}{2!} - \frac{1}{3!} + \cdots + \frac{(-1)^n}{n!} + \cdots$$

is convergent. Approximate its sum S to three decimal places.

Solution: Since

$$\frac{1}{n!} > \frac{1}{(n+1)!} \quad \text{and} \quad \lim_{n\to\infty} \frac{1}{n!} = 0,$$

the series converges by 12.12. We easily verify that

$$\frac{1}{7!} < .0002.$$

Hence

$$S \doteq 1 - 1 + \tfrac{1}{2} - \tfrac{1}{6} + \tfrac{1}{24} - \tfrac{1}{120} + \tfrac{1}{720} \doteq .368,$$

accurate to three decimal places. We shall show presently that this alternating series has sum $1/e$, so $1/e \doteq .368$.

Example 2 Test the alternating series

$$\sum_{n=1}^{\infty} \frac{(-1)^n \ln^2 n}{n}$$

for convergence.

Solution: In checking condition (1) of 12.12, it is sometimes convenient to think of n as a *continuous* variable for $n \geq 1$ and examine the sign of $d(a_n)/dn$. Thus if $d(a_n)/dn < 0$ for all n (or all sufficiently large n), then a_n is a decreasing function of n and 12.12(1) is satisfied. In this case,

$$\frac{d}{dn}\left(\frac{\ln^2 n}{n}\right) = \frac{2 \ln n - \ln^2 n}{n^2} = \frac{\ln n}{n^2}(2 - \ln n).$$

But $2 < \ln n$ for $n > e$, so that a_n is a decreasing function for all such n. By l'Hospital's rule we have $\lim_{n\to\infty} a_n = 0$, and hence the series converges by 12.12.

Example 3 Test the alternating series

$$\sum_{n=1}^{\infty} \frac{(-1)^{n-1} n}{2n - 1}$$

for convergence.

Solution: The reader may easily verify, using the method of Example 2, that $n/(2n - 1)$ is a decreasing function of n and hence 12.12(1) is satisfied. However,

$$\lim_{n\to\infty} \frac{n}{2n - 1} = \lim_{n\to\infty} \frac{1}{2 - 1/n} = \frac{1}{2},$$

so that 12.12(2) is not satisfied. Since the nth term of the series does not approach 0, the series diverges by 12.4.

EXERCISES

Test each of the following alternating series for convergence.

1. $1 - \frac{1}{2} + \frac{1}{3} - \cdots + \frac{(-1)^{n-1}}{n} + \cdots$

2. $\dfrac{2}{\ln 2} - \dfrac{2}{\ln 3} + \cdots + \dfrac{2(-1)^{n-1}}{\ln (n+1)} + \cdots$

3. $\dfrac{2}{3} - \dfrac{1}{2} + \dfrac{4}{9} - \cdots + \dfrac{(n+1)(-1)^{n-1}}{3n} + \cdots$

4. $-\dfrac{1}{3} + \dfrac{1}{3} - \dfrac{3}{11} + \cdots + \dfrac{(-1)^n n}{n^2+2} + \cdots$

5. $\displaystyle\sum_{k=1}^{\infty} \frac{(-1)^k}{2k-1}$

6. $\displaystyle\sum_{k=1}^{\infty} \frac{k(-1)^{k-1}}{2^k}$

7. $\displaystyle\sum_{k=2}^{\infty} \frac{k(-1)^{k-1}}{\ln k}$

8. $\displaystyle\sum_{k=1}^{\infty} \frac{(2k-1)(-1)^k}{5k+1}$

9. $\displaystyle\sum_{k=1}^{\infty} \frac{\sqrt{k}(-1)^{k-1}}{2k+1}$

10. $\displaystyle\sum_{k=1}^{\infty} \frac{(-1)^{k-1} \ln k}{k}$

11. $\displaystyle\sum_{n=1}^{\infty} \frac{(-1)^n n}{10n-1}$

12. $\displaystyle\sum_{n=1}^{\infty} (-1)^{n+1} \cos \frac{\pi}{n}$

13. Consider the alternating series $\displaystyle\sum_{k=1}^{\infty} (-1)^k a_k$, where

$$\left.\begin{aligned} a_{2k-1} &= \frac{1}{k} \\ a_{2k} &= \frac{1}{k^2} \end{aligned}\right\} \quad k = 1, 2, 3, \ldots .$$

It is obvious that $\lim_{k \to \infty} a_k = 0$ and $a_{2k} \leq a_{2k-1}$ for all k. Show that the series does not converge and explain why the alternating series test does not apply.

14. If $\displaystyle\sum_{n=1}^{\infty} (-1)^{n+1}(a_n - b_n)$ is an alternating decreasing series, which converges, what can be said about the convergence of $\displaystyle\sum_{n=1}^{\infty} (-1)^{n+1} a_n$? Illustrate by specific examples.

Approximate the sum of each of the following series to three decimal places.

15. $1 - \dfrac{1}{2^2} + \dfrac{1}{2^4} - \cdots + \dfrac{(-1)^{n-1}}{2^{2(n-1)}} + \cdots$

16. $1 - \dfrac{1}{2!} + \cdots + \dfrac{(-1)^{n-1}}{(2n-2)!} + \cdots$

17. $1 - \dfrac{1}{3^3} + \cdots + \dfrac{(-1)^{n-1}}{(2n-1)^3} + \cdots$

18. $\dfrac{1}{3} - \dfrac{1}{2 \cdot 3^2} + \cdots + \dfrac{(-1)^{n-1}}{n \cdot 3^n} + \cdots$

Test the following alternating series for convergence.

19. $\sum_{k=1}^{\infty} \frac{(-1)^k k}{k^2+1}$

20. $\sum_{k=1}^{\infty} \frac{(-1)^{k-1} k^2}{k^3 - k + 5}$

21. $\sum_{k=1}^{\infty} (-1)^k k \arctan\left(\frac{1}{k^2}\right)$

22. $\sum_{k=1}^{\infty} (-1)^k k^9 e^{-k}$

5 ABSOLUTE CONVERGENCE

We shall consider in this section series having negative as well as positive terms. Associated with each infinite series Σa_k is its series of absolute values

$$\sum_{k=1}^{\infty} |a_k| = |a_1| + |a_2| + \cdots + |a_n| + \cdots.$$

A series Σa_k is called *absolutely convergent* if its series of absolute values $\Sigma |a_k|$ converges. Let us prove the following theorem on absolutely convergent series.

12.14 Theorem

Every absolutely convergent series Σa_k is convergent, and

$$\left|\sum_{k=1}^{\infty} a_k\right| \leq \sum_{k=1}^{\infty} |a_k|.$$

Proof: Consider the three infinite series Σa_k, $\Sigma |a_k|$, and $\Sigma(a_k + |a_k|)$ having respective nth partial sums S_n, T_n, and U_n. Clearly,

$$S_n + T_n = U_n.$$

Since $0 \leq a_k + |a_k| \leq 2|a_k|$,

$$0 \leq U_n \leq 2T_n \leq 2T$$

for every integer n, where T is the sum of $\Sigma |a_k|$. Since $U_1 \leq U_2 \leq \cdots \leq U_n \leq \cdots \leq 2T$, we have by 5.37 that

$$\operatorname*{limit}_{n\to\infty} U_n = U \leq 2T.$$

Thus

$$\operatorname*{limit}_{n\to\infty} S_n = \operatorname*{limit}_{n\to\infty} U_n - \operatorname*{limit}_{n\to\infty} T_n = U - T,$$

and the series Σa_k is convergent with sum $S = U - T \leq T$.

The series $\Sigma(-a_k)$ has sum $-S$, and, since $T = \Sigma |-a_k|$, we have $-S \leq T$ by the same argument as above with the series Σa_k replaced by $\Sigma(-a_k)$. Since both S and $-S$ are less than or equal to T, then $|S| \leq T$ and the theorem is proved.

Example 1 Test the series

$$\frac{\sin 1}{1^2} + \frac{\sin 2}{2^2} + \frac{\sin 3}{3^2} + \cdots + \frac{\sin n}{n^2} + \cdots$$

for absolute convergence.

Solution: Note that this is not a series of positive terms. However, for its series of absolute values,

$$\sum_{k=1}^{\infty} \frac{|\sin k|}{k^2},$$

the comparison test (Theorem 12.10) applies. The positive term series $\Sigma 1/k^2$ is known to be convergent, and

$$\frac{|\sin k|}{k^2} \leq \frac{1}{k^2}$$

for all k. Hence by 12.10 the given series is absolutely convergent. By Theorem 12.14, the given series is also convergent.

Not every convergent series is absolutely convergent. Thus, by the alternating series test, the alternating harmonic series

$$1 - \tfrac{1}{2} + \tfrac{1}{3} - \tfrac{1}{4} + \tfrac{1}{5} - \cdots$$

is convergent, whereas its series of absolute values is the divergent harmonic series. A series such as the alternating harmonics series that is convergent without being absolutely convergent is called *conditionally convergent.*

We may easily prove the following results on absolutely convergent series. (Compare these results with 12.6.)

12.15 Theorem

If the series Σa_k and Σb_k are absolutely convergent, then so are the series $\Sigma(a_k + b_k)$ and $\Sigma c a_k$ for any constant c.

Proof: These follow readily from the inequalities

$$\sum_{k=1}^{n} |a_k \pm b_k| \leq \sum_{k=1}^{n} |a_k| + \sum_{k=1}^{n} |b_k| \leq \sum_{k=1}^{\infty} |a_k| + \sum_{k=1}^{\infty} |b_k|,$$

$$\sum_{k=1}^{\infty} |c a_k| = |c| \sum_{k=1}^{\infty} |a_k|,$$

and from 12.7.

A very useful test for absolute convergence of a series is the ratio test that follows.

12.16 Ratio Test

The series Σa_k is:

(1) *Absolutely convergent if* $\lim_{n\to\infty}\left|\dfrac{a_{n+1}}{a_n}\right| = L < 1.$

(2) *Divergent if* $\lim_{n\to\infty}\left|\dfrac{a_{n+1}}{a_n}\right| = \begin{cases} L > 1 \\ \infty. \end{cases}$

Proof: To prove (1), let r be any number such that

$$0 \le L < r < 1.$$

Since the limit of $|a_{n+1}/a_n|$ as n approaches ∞ is L, there exists a positive integer N such that

$$\left|\frac{a_{n+1}}{a_n}\right| < r \qquad \text{for every integer } n \ge N.$$

Thus

$$\begin{aligned} |a_{N+1}| &< r|a_N|, \\ |a_{N+2}| &< r|a_{N+1}| < r^2|a_N|, \\ |a_{N+3}| &< r|a_{N+2}| < r^2|a_{N+1}| < r^3|a_N|, \end{aligned}$$

and, in general,

$$|a_{N+k}| < r^k|a_N| \qquad \text{for every integer } k > 0.$$

Hence the series

$$|a_{N+1}| + |a_{N+2}| + \cdots + |a_{N+n}| + \cdots$$

is convergent, since it is dominated by the convergent geometric series

$$|a_N|r + |a_N|r^2 + \cdots + |a_N|r^n + \cdots.$$

Therefore the series

$$\sum_{k=1}^{\infty} |a_k| = \sum_{k=1}^{N} |a_k| + \sum_{k=1}^{\infty} |a_{N+k}|$$

is convergent, and the given series Σa_k is absolutely convergent.

To prove (2), if $L > 1$, then there exists a number N such that

$$\left|\frac{a_{n+1}}{a_n}\right| > 1 \qquad \text{for every integer } n > N.$$

Thus $|a_{n+1}| > |a_n|$ for every $n > N$, and

$$\lim_{n\to\infty} |a_n|$$

cannot equal zero. Hence, by 5.35,

$$\lim_{n\to\infty} a_n \ne 0,$$

and therefore the series Σa_k is divergent by 12.4. An analogous argument proves the other part of (2).

If

$$\lim_{n\to\infty} \left|\frac{a_{n+1}}{a_n}\right| = 1,$$

nothing can be said directly about the convergence of the series Σa_k. For example,

$$\lim_{n\to\infty} \frac{1/(n+1)^p}{1/n^p} = \lim_{n\to\infty} \left(\frac{n}{n+1}\right)^p = 1$$

for every p series $\Sigma 1/k^p$. We have already shown that the p series converges if $p > 1$ and diverges if $p \leq 1$. The ratio test is of no help in testing series similar to the p series for convergence.

Example 2 Test the convergence of the series

$$1 + \frac{1}{2!} + \frac{1}{3!} + \cdots + \frac{1}{n!} + \cdots.$$

Solution: Since

$$\lim_{n\to\infty} \frac{1/(n+1)!}{1/n!} = \lim_{n\to\infty} \frac{1}{n+1} = 0,$$

the series converges by the ratio test.

Example 3 Test the convergence of the series

$$\frac{1}{5} - \frac{2}{5^2} + \frac{3}{5^3} - \cdots + \frac{n(-1)^{n-1}}{5^n} + \cdots.$$

Solution: We have

$$\lim_{n\to\infty} \left|\frac{(n+1)(-1)^n/5^{n+1}}{n(-1)^{n-1}/5^n}\right| = \lim_{n\to\infty} \frac{n+1}{5n} = \frac{1}{5},$$

and therefore the series converges absolutely by the ratio test.

Example 4 Test the convergence of the series

$$1 + \frac{2^2}{2!} + \frac{3^3}{3!} + \cdots + \frac{n^n}{n!} + \cdots.$$

Solution: The ratio a_{n+1}/a_n for this series is given by

$$\frac{a_{n+1}}{a_n} = \frac{(n+1)^{n+1}}{(n+1)!} \cdot \frac{n!}{n^n} = \frac{(n+1)^n \cdot (n+1) \cdot n!}{(n+1)! \cdot n^n} = \left(\frac{n+1}{n}\right)^n,$$

or

$$\frac{a_{n+1}}{a_n} = \left(1 + \frac{1}{n}\right)^n.$$

Since

$$\lim_{n\to\infty} \left(1 + \frac{1}{n}\right)^n = e$$

and $e > 1$ (see Chapter 7, Section 3) the series diverges by the ratio test.

EXERCISES

I

In each of Exercises 1 to 24 test the given series for convergence or divergence. In case of convergence, determine whether the series converges absolutely.

1. $1 - \dfrac{1}{3!} + \cdots + \dfrac{(-1)^{n-1}}{(2n-1)!} + \cdots$

2. $2 - \dfrac{2^2}{2!} + \cdots - \dfrac{(-2)^n}{n!} + \cdots$

3. $\dfrac{1}{2} + \dfrac{3}{5} + \dfrac{9}{10} + \cdots + \dfrac{3^{n-1}}{n^2+1} + \cdots$

4. $\dfrac{1}{9} - \dfrac{2}{81} + \dfrac{6}{729} - \cdots + \dfrac{n!\,(-1)^{n-1}}{9^n} + \cdots$

5. $\displaystyle\sum_{k=1}^{\infty} \frac{(-1)^{k+1}}{\sqrt[3]{k}}$

6. $\displaystyle\sum_{k=1}^{\infty} \frac{(-1)^k(k+1000)}{k^3}$

7. $\displaystyle\sum_{k=1}^{\infty} \frac{(-1)^k k^{1000}}{(k+2)!}$

8. $\displaystyle\sum_{k=1}^{\infty} (-1)^k \left(\frac{1}{k}\right)^{1/k}$

9. $\displaystyle\sum_{k=3}^{\infty} \frac{(-1)^k}{\ln k}$

10. $\displaystyle\sum_{k=1}^{\infty} \frac{(-1)^k k!}{100^k}$

11. $\displaystyle\sum_{k=2}^{\infty} \frac{(-1)^k(4k+1)}{7k^2-1}$

12. $\displaystyle\sum_{k=1}^{\infty} \frac{(-1)^{k+1}3^k}{k^3 \cdot 2^{k+3}}$

13. $\displaystyle\sum_{k=1}^{\infty} \frac{(-1)^k k^7 \cdot 7^{k+3}}{2^{3k}}$

14. $\displaystyle\sum_{k=1}^{\infty} \frac{(-1)^k k!}{1 \cdot 3 \cdot 5 \cdots (2k-1)}$

15. $\displaystyle\sum_{k=2}^{\infty} \frac{(-1)^k}{k \ln^2 k}$

16. $\displaystyle\sum_{k=1}^{\infty} (-1)^k \frac{\sin 10k}{\sqrt{k^3}}$

17. $\displaystyle\sum_{n=1}^{\infty} \frac{\cos (n\pi/4)}{n!}$

18. $\displaystyle\sum_{n=1}^{\infty} \frac{(-1)^{n+1}\sqrt{3n-1}}{n}$

19. $\displaystyle\sum_{n=1}^{\infty} (-1)^{n+1}3^{1/n}$

20. $\displaystyle\sum_{n=1}^{\infty} \frac{(-1)^{n+1}}{(2n-1)!}$

21. $\displaystyle\sum_{k=1}^{\infty} \frac{(-1)^k \tan^{-1} k}{k}$

22. $\displaystyle\sum_{n=1}^{\infty} \frac{n!}{2n^{n^2}}$

23. $\displaystyle\sum_{n=1}^{\infty} \frac{2^n \cdot n!}{n^n}$

24. $\displaystyle\sum_{n=1}^{\infty} \frac{3^n \cdot n!}{n^n}$

25. Prove that if Σa_k^2 and Σb_k^2 both converge, then $\Sigma a_k b_k$ converges absolutely.

II

1. a. Prove that if $\sum_{n=1}^{\infty} a_n$ converges absolutely, then $\sum_{n=1}^{\infty} a_n^2$ converges.
   b. Give an example to show that the converse of (a) is not true.
2. Let Σa_n be any conditionally convergent series. From this series we form two new series, Σa_n^+ and Σa_n^-, by taking respectively all the positive terms alone and all the negative terms alone. To be more precise, define

$$a_n^+ = \frac{a_n + |a_n|}{2}, \qquad a_n^- = \frac{a_n - |a_n|}{2},$$

for all n. (Note that when $a_n > 0$, then $a_n^+ = a_n$ and $a_n^- = 0$; and when $a_n < 0$, then $a_n^- = a_n$ and $a_n^+ = 0$.) Prove that Σa_n^+ and Σa_n^- both diverge.
3. (*Riemann's rearrangement theorem.*) Let Σa_n be any conditionally convergent series, and let T be any real number. Prove that there is a rearrangement of the terms of Σa_n which converges to the sum T. (*Hint:* Define a_n^+ and a_n^- as in Exercise 8 above. Take just enough terms of Σa_n^+ so that their sum exceeds T. To this sum add just enough terms of Σa_n^- so that the combined sum is less than T. Now continue this process.)

6 POWER SERIES

An infinite series of the form

$$\sum_{k=0}^{\infty} a_k x^k = a_0 + a_1 x + a_2 x^2 + \cdots + a_k x^k + \cdots$$

is called a *power series* in x. The coefficients $a_0, a_1, \ldots, a_k, \ldots$ are to be regarded as constants, while x is a *variable* for which we may substitute real numbers from some specified domain. More generally, an infinite series of the form

$$\sum_{k=0}^{\infty} a_k(x-c)^k = a_0 + a_1(x-c) + a_2(x-c)^2 + \cdots + a_k(x-c)^k + \cdots$$

is called a power series in $x - c$.

Our first consideration, in dealing with a power series, is to determine the set of all those values of x for which the given series converges. The following basic theorem will enable us to solve this type of problem rather easily. We shall state this theorem for a power series of the form $\Sigma a_k x^k$, and leave it to the reader to formulate the corresponding result for a series of the form $\Sigma a_k(x-c)^k$.

12.17 Theorem

(i) *If $\Sigma a_k x^k$ converges for $x = t$, where $t \neq 0$, then it converges absolutely for all x with $|x| < |t|$.*
(ii) *If $\Sigma a_k x^k$ diverges for $x = t$, then it diverges for all x with $|x| > |t|$.*

Proof: (i) If $\Sigma a_k t^k$ converges, then $\lim_{k\to\infty} |a_k t^k| = 0$ by 12.4. This implies that the sequence $\{a_k t^k\}$ is *bounded* (see Exercise II-8, Section 1 of this chapter). Thus, there is some positive number M such that $|a_k t^k| \leq M$ for all k. Since $|x| < |t|$, the number

$$s = \frac{|x|}{|t|}$$

is less than 1. Clearly

$$|a_k x^k| = |a_k t^k s^k| \leq Ms^k$$

for every k; and, therefore, the series $\Sigma|a_k x^k|$ is dominated by the convergent geometric series ΣMs^k. It follows that $\Sigma|a_k x^k|$ converges, by Comparison Test I, which completes the proof of (i).

(ii) Suppose $\Sigma a_k t^k$ diverges. Then it is impossible for $\Sigma a_k x^k$ to converge if $|x| > |t|$: because in this case we would have $\Sigma a_k x^k$ convergent, $|t| < |x|$, and $\Sigma a_k t^k$ divergent, contradicting what we just proved in (i) above.

Clearly any power series $\Sigma a_k x^k$ converges for $x = 0$. Theorem 12.17 suggests that if $\Sigma a_k x^k$ also converges for some $t \neq 0$, then the set of numbers for which it converges might form an *interval*. This leads us to make the following definition.

12.18 Definition

The *radius of convergence* r of a power series $\Sigma a_k x^k$ is the least upper bound of the set of all numbers x such that $\Sigma a_k x^k$ is convergent. If $\Sigma a_k x^k$ converges for all x, we define $r = \infty$.

We immediately obtain the following theorem.

12.19 Theorem

If r is the radius of convergence of the power series $\Sigma a_k x^k$, then the series converges absolutely if $|x| < r$ and diverges if $|x| > r$.

Proof: If $r = 0$, then the series converges only for $x = 0$ and there is nothing to prove. So suppose $r > 0$ and $|x| < r$. Then there is some positive number t between $|x|$ and r for which $\Sigma a_k t^k$ converges; for otherwise, r would not be the *least* upper bound of the set of numbers for which the series converges. Since $|x| < |t|$, it follows from 12.17(i) that $\Sigma a_k x^k$ converges absolutely. If $|x| > r$, then clearly $\Sigma a_k x^k$ must diverge by the very definition of r.

To summarize the above results, a power series $\Sigma a_k x^k$ might converge

(i) only for $x = 0$ (in this case $r = 0$),
(ii) for all x (in this case $r = \infty$),

(iii) for all x inside some interval $(-r,r)$, and diverge outside this interval.

In case (iii) the behavior of the series when $x = r$ and when $x = -r$ is still not determined. Thus the set of numbers for which $\Sigma a_k x^k$ converges could consist of an open interval $(-r,r)$, a half-closed interval $[-r,r)$ or $(-r,r]$, or a closed interval $[-r,r]$. In each of these cases, the set of numbers for which $\Sigma a_k x^k$ converges is called the *interval of convergence* of the power series.

Similar results hold for a power series of the more general form

$$\sum_{k=0}^{\infty} a_k(x-c)^k = a_0 + a_1(x-c) + a_2(x-c)^2 + \cdots + a_k(x-c)^k + \cdots.$$

Its interval of convergence can be found by determining the interval of convergence of the corresponding series $\Sigma a_k x^k$ with the same coefficients $a_0, a_1, \ldots, a_k, \ldots$. If, for example, $\Sigma a_k x^k$ has interval of convergence $[-r,r)$ then the series $\Sigma a_k(x-c)^k$ converges for $-r \le x - c < r$, that is, for every x in the interval $[c - r, c + r)$. This interval is called the *interval of convergence* of the series $\Sigma a_k(x-c)^k$, and the number r is its *radius of convergence*.

The interval of convergence of a power series can often be found by use of the ratio test.

Example 1 Find the interval of convergence of each of the following power series:

(a) $1 + x + \dfrac{x^2}{2!} + \cdots + \dfrac{x^n}{n!} + \cdots.$

(b) $1 - x + \dfrac{x^2}{2} - \cdots + \dfrac{(-1)^{n-1}x^n}{n} + \cdots.$

(c) $1 + x + 2x^2 + \cdots + nx^n + \cdots.$

(d) $\dfrac{1}{3} + \dfrac{(x-2)}{36} + \dfrac{(x-2)^2}{243} + \cdots + \dfrac{(x-2)^n}{3^n n^2} + \cdots.$

(e) $1 + x + 2!\,x^2 + 3!\,x^3 + \cdots + n!\,x^n + \cdots.$

Solution: (a) Since

$$\operatorname*{limit}_{n\to\infty} \left|\frac{x^{n+1}/(n+1)!}{x^n/n!}\right| = \operatorname*{limit}_{n\to\infty} \frac{|x|}{n+1} = 0$$

for every number x, this series converges for every number x by the ratio test. Its interval of convergence is $(-\infty,\infty)$.

(b) We have

$$\operatorname*{limit}_{n\to\infty} \left|\frac{(-1)^n x^{n+1}/(n+1)}{(-1)^{n-1}x^n/n}\right| = \operatorname*{limit}_{n\to\infty} \frac{n}{n+1}|x| = |x|,$$

and therefore the series converges if $|x| < 1$ and diverges if $|x| > 1$ by the ratio test. Hence $r = 1$ for this series. Clearly, the series converges if $x = 1$ and diverges if $x = -1$. Hence $(-1,1]$ is its interval of convergence.

(c) Since

$$\lim_{n\to\infty} \left|\frac{(n+1)x^{n+1}}{nx^n}\right| = \lim_{n\to\infty} \frac{n+1}{n}|x| = |x|,$$

again $r = 1$ by the ratio test. This series evidently diverges if $x = \pm 1$. Hence its interval of convergence is the open interval $(-1,1)$.

(d) We have

$$\lim_{n\to\infty} \left|\frac{(x-2)^{n+1}}{3^{n+1}(n+1)^2} \cdot \frac{3^n n^2}{(x-2)^n}\right| = \lim_{n\to\infty} \frac{n^2}{3(n+1)^2}|x-2| = \frac{|x-2|}{3}.$$

Hence the series converges if $|x - 2|/3 < 1$, or if

$$|x - 2| < 3.$$

If $|x - 2| = 3$, the series is just the p series (or alternating p series) for $p = 2$. Since these series also converge, the interval of convergence of the given series is the set of all x such that $|x - 2| \le 3$, that is, the closed interval $[-1,5]$.

(e) For $x \ne 0$, we have

$$\lim_{n\to\infty} \left|\frac{(n+1)!\, x^{n+1}}{n!\, x^n}\right| = \lim_{n\to\infty} (n+1)|x| = \infty.$$

Hence this series diverges whenever $x \ne 0$. Its interval of convergence consists only of the point 0.

Similar methods may sometimes be applied to the study of an infinite series of variable terms that are not of the form $a_k(x - c)^k$. The set of numbers for which such a series converges need not necessarily be an interval, and hence we shall instead refer to this set of numbers as the *region of convergence* of the series. It can frequently be found by the ratio test.

Example 2 Find the region of convergence of the series

$$\sum_{k=1}^{\infty} \frac{\ln k}{3^k x^k}.$$

Solution: Using the ratio test,

$$\lim_{k\to\infty} \left|\frac{\ln (k+1)/3^{k+1}x^{k+1}}{\ln k/3^k x^k}\right| = \lim_{k\to\infty} \left|\frac{\ln (k+1)}{\ln k}\right| \cdot \left|\frac{1}{3x}\right| = \frac{1}{3|x|}.$$

Here we used l'Hospital's rule to determine that

$$\lim_{k\to\infty} \frac{\ln (k+1)}{\ln k} = 1.$$

Thus the given series converges if $1/(3|x|) < 1$, or $|x| > \frac{1}{3}$. The series diverges if $|x| < \frac{1}{3}$. It remains to check the behavior of the series for $x = \frac{1}{3}$ and $x = -\frac{1}{3}$. For $x = \frac{1}{3}$ the series reduces to $\sum_{k=1}^{\infty} \ln k$; and for $x = -\frac{1}{3}$ the series becomes $\sum_{k=1}^{\infty} (-1)^k \ln k$. Both of these series diverge. Hence the region of convergence of the given series consists of all those numbers x with $x > \frac{1}{3}$ or $x < -\frac{1}{3}$.

The multiplication of a series termwise by a nonzero number does not alter the convergence or divergence of a series (12.6). Thus, for example, the power series

$$a_0 x + a_1 x^2 + \cdots + a_n x^{n+1} + \cdots$$

has the same interval of convergence as $\Sigma a_k x^k$, since it is obtained by multiplying $\Sigma a_k x^k$ termwise by x. If we multiply $\Sigma a_k x^k$ termwise by $1/x$ and omit the first term of the resulting series, we obtain the series

$$a_1 + a_2 x + \cdots + a_n x^{n-1} + \cdots,$$

which again has the same interval of convergence as $\Sigma a_k x^k$.

EXERCISES

I

In each of Exercises 1 to 20 find the interval of convergence of the power series.

1. $x - \dfrac{x^3}{3!} + \cdots + \dfrac{(-1)^{n-1}x^{2n-1}}{(2n-1)!} + \cdots$

2. $1 - x + \cdots + (-x)^{n-1} + \cdots$

3. $1 + x + \dfrac{x^2}{\sqrt{2}} + \dfrac{x^3}{\sqrt{3}} + \cdots + \dfrac{x^n}{\sqrt{n}} + \cdots$

4. $1 + \dfrac{x}{2} + \dfrac{2x^2}{2^2} + \dfrac{3x^3}{2^3} + \cdots + \dfrac{nx^n}{2^n} + \cdots$

5. $\displaystyle\sum_{k=0}^{\infty} \frac{x^{2k}}{k!}$

6. $\displaystyle\sum_{k=0}^{\infty} \frac{(-1)^k x^k}{(k+1)^2}$

7. $\displaystyle\sum_{k=0}^{\infty} \frac{(3x)^k}{2^{k+1}}$

8. $\displaystyle\sum_{k=0}^{\infty} \frac{k!\, x^k}{10^k}$

9. $\displaystyle\sum_{k=1}^{\infty} (-1)^k k^2 x^k$

10. $\displaystyle\sum_{k=1}^{\infty} \frac{(-1)^{k-1}x^{2k-1}}{k+1}$

11. $\sum_{k=1}^{\infty} \frac{k(x-1)^{k-1}}{3^k}$

12. $\sum_{k=0}^{\infty} \frac{(x+2)^k}{(k+1)2^k}$

13. $\sum_{k=0}^{\infty} \frac{(-1)^{k+1}(x+1)^{2k}}{(k+1)^2 5^k}$

14. $\sum_{k=1}^{\infty} \frac{(-1)^k(2x-1)^k}{k!}$

15. $\sum_{k=1}^{\infty} \frac{k!\,k!}{(2k)!} x^k$

16. $\sum_{k=1}^{\infty} \frac{x^k}{\ln(k+1)}$

17. $\sum_{k=2}^{\infty} \frac{(-1)^k x^k}{k(\ln k)^2}$

18. $\sum_{k=0}^{\infty} \frac{(2x+1)^k}{3^k}$

19. $\sum_{k=0}^{\infty} \frac{(x-2)^k}{2^k\sqrt{k+1}}$

20. $\sum_{k=0}^{\infty} \frac{kx^k}{(k+1)(k+2)2^k}$

In each of Exercises 21 to 28, find the set of all x for which the given series converges.

21. $\sum_{n=1}^{\infty} \frac{n^2}{x^n}$

22. $\sum_{n=0}^{\infty} e^{-nx^2}$

23. $\sum_{n=1}^{\infty} \frac{1}{n^2}\left(\frac{x}{1+x}\right)^n$

24. $\sum_{n=1}^{\infty} \frac{(-1)^{n+1}(x-3)^n}{n}$

25. $\sum_{n=1}^{\infty} (\ln x)^n$

26. $\sum_{n=1}^{\infty} \frac{(x/3)^n}{n!}$

27. $\sum_{n=1}^{\infty} \frac{1}{x^{2n}+5}$

28. $\sum_{n=1}^{\infty} \frac{(-1)^{n-1}}{n^2}\left(\frac{x-6}{5}\right)^n$

II

1. If $a_k = 2^{-k}$, k an even integer, and $a_k = 2^{-k+1}$, k an odd integer, show that the power series $\Sigma a_k x^k$ has the open interval $(-2,2)$ as its interval of convergence. Note that $\lim_{n\to\infty} (a_{n+1}/a_n)$ does not exist.

2. Prove that if $\lim_{n\to\infty} \sqrt[n]{|a_n|} = r > 0$, then the power series $\Sigma a_k x^k$ has radius of convergence $1/r$. Use Exercise II-10, Section 3.

7 OPERATIONS ON POWER SERIES

Let $\Sigma a_k x^k$ be a power series and I its interval of convergence. Corresponding to each x in I the series converges to a sum that obviously depends on the choice of x. Thus each power series $\Sigma a_k x^k$ defines a function f, where $f(x)$ is the sum of the series $\Sigma a_k x^k$:

$$f(x) = \sum_{i=0}^{\infty} a_k x^k = a_0 + a_1 x + a_2 x^2 + \cdots + a_k x^k + \cdots .$$

The domain of f is the interval of convergence of the given series.

It might be asked whether this function f is differentiable. If so, could we find its derivative $f'(x)$ by differentiating the above series term by term? In other words, would it be true that the power series

$$\sum_{k=1}^{\infty} ka_k x^{k-1} = a_1 + 2a_2 x + 3a_3 x^2 + \cdots + ka_k x^{k-1} + \cdots$$

converges to $f'(x)$ in some interval of convergence?

Before giving the complete answer to this question, we must first prove the following theorem.

12.20 Theorem

If the series $\Sigma a_k x^k$ converges for all x in the interval $(-r,r)$, then so does the series $\Sigma k a_k x^{k-1}$.

Proof: Let x_0 be any number in $(-r,r)$ with $x_0 \neq 0$. We will show that the series $\Sigma k a_k x_0^{k-1}$ converges.

Choose another number x_1 in $(-r,r)$ with $|x_0| < |x_1| < r$, and set $b = |x_0|/|x_1|$. Recall from Exercise I-20, Chapter 11, Section 2, that $\lim_{k\to\infty} kb^k = 0$, since we have $0 < b < 1$. Hence there is a number k_0 such that $kb^k < |x_0|$ for all $k > k_0$, or

$$\frac{kb^k}{|x_0|} < 1 \qquad \text{for all } k > k_0.$$

Then we have

$$|ka_k x_0^{k-1}| = \left|\frac{ka_k x_1^k}{x_0}\left(\frac{x_0}{x_1}\right)^k\right| = \frac{kb^k}{|x_0|}|a_k x_1^k| < |a_k x_1^k|$$

for all $k > k_0$. Since $\Sigma |a_k x_1^k|$ is a convergent positive term series, comparison test I (12.10) can be applied to conclude that $\Sigma k a_k x_0^{k-1}$ converges.

We may now state the desired theorem on differentiation of power series.

12.21 Theorem

Let $\Sigma a_k x^k$ be a power series with nonzero radius of convergence r, and let the function f be defined by

$$f(x) = \sum_{k=0}^{\infty} a_k x^k.$$

Then for every x in $(-r,r)$ the derivative of f is given by

$$f'(x) = \sum_{k=1}^{\infty} ka_k x^{k-1}.$$

A corresponding result holds for the integration of power series.

12.22 Theorem

Let $\Sigma a_k x^k$ be a power series converging to the sum $f(x)$ in the open interval $(-r,r)$. If $[a,b]$ is any closed interval contained within $(-r,r)$, then

$$\int_a^b f(x)\,dx = \sum_{k=0}^{\infty}\left[\int_a^b a_k x^k\,dx\right] = \sum_{k=0}^{\infty}\frac{a_k}{k+1}(b^{k+1} - a^{k+1}).$$

A special case of 12.22 arises when we integrate f from 0 to a variable value x. Note that here we denote the variable of integration by t.

12.23 Corollary

If $\Sigma a_k x^k$ is a power series converging to the sum $f(x)$ in $(-r,r)$, then

$$\int_0^x f(t)\,dt = \sum_{k=0}^{\infty}\left[\int_0^x a_k t^k\,dt\right] = \sum_{k=0}^{\infty}\frac{a_k x^{k+1}}{k+1},$$

for every x in $(-r,r)$.

We shall give some examples to show how the above theorems are used. The proofs of the theorems will be given in the appendix to this chapter (Section 9).

Example 1 Show that for all x in $(-1,1)$,

$$\frac{2}{(1+x)^3} = 1\cdot 2 - 2\cdot 3x + 3\cdot 4x^2 - 4\cdot 5x^3 + \cdots.$$

Solution: We know from 12.5 that

$$\frac{1}{1+x} = 1 - x + x^2 - x^3 + \cdots$$

for all x in $(-1,1)$. Applying 12.21 we can conclude that

$$-\frac{1}{(1+x)^2} = -1 + 2x - 3x^2 + 4x^3 - \cdots$$

for all x in $(-1,1)$. The desired result follows by applying 12.21 again.

Example 2 Show that

$$\tan^{-1} x = x - \frac{x^3}{3} + \frac{x^5}{5} - \frac{x^7}{7} + \cdots, \qquad |x| < 1.$$

Solution: The geometric series

$$\frac{1}{1+x^2} = 1 - x^2 + x^4 - x^6 + \cdots$$

has radius of convergence $r = 1$. Hence we may integrate the series termwise by 12.23 to obtain

$$\int_0^x \frac{1}{1+t^2}\,dt = \tan^{-1} x = x - \frac{x^3}{3} + \frac{x^5}{5} - \frac{x^7}{7} + \cdots, \qquad |x| < 1.$$

Example 3 Approximate $\tan^{-1} \frac{1}{2}$ to three decimal places.

Solution: By Example 2,

$$\tan^{-1} \tfrac{1}{2} \doteq \tfrac{1}{2} - \tfrac{1}{3}(\tfrac{1}{2})^3 + \tfrac{1}{5}(\tfrac{1}{2})^5 - \tfrac{1}{7}(\tfrac{1}{2})^7 \doteq .463.$$

By 12.13, the error does not exceed

$$\tfrac{1}{9}(\tfrac{1}{2})^9 < 3 \times 10^{-4}.$$

Similar results also hold for power series of the form $\Sigma a_k(x - c)^k$. For if

$$f(x) = \sum_{k=0}^{\infty} a_k(x - c)^k, \qquad g(x) = \sum_{k=0}^{\infty} a_k x^k,$$

then $f(x) = g(x - c)$, and by the chain rule

$$f'(x) = g'(x - c)D_x(x - c) = g'(x - c).$$

Hence

$$f'(x) = \sum_{k=1}^{\infty} ka_k(x - c)^{k-1},$$

and the power series $\Sigma a_k(x - c)^k$ is differentiated just as the power series $\Sigma a_k x^k$. The integral of $f(x)$ is also found by integrating the series termwise.

Power series possess many of the properties of polynomials. In particular, power series may be multiplied together in much the same way as polynomials. Suppose, for example, we multiply together the two polynomials

$$f(x) = a_0 + a_1 x + a_2 x^2,$$
$$g(x) = b_0 + b_1 x + b_2 x^2.$$

For the product we of course obtain the polynomial

$$f(x) \cdot g(x) = c_0 + c_1 x + c_2 x^2 + c_3 x^3 + c_4 x^4,$$

where

$$\begin{aligned} c_0 &= a_0 b_0, \\ c_1 &= a_1 b_0 + a_0 b_1, \\ c_2 &= a_2 b_0 + a_1 b_1 + a_0 b_2, \\ c_3 &= a_3 b_0 + a_2 b_1 + a_1 b_2 + a_0 b_3, \\ c_4 &= a_4 b_0 + a_3 b_1 + a_2 b_2 + a_1 b_3 + a_0 b_4. \end{aligned}$$

A similar result holds for power series.

12.24 Theorem

Suppose the two power series,

$$\sum_{k=0}^{\infty} a_k x^k, \qquad \sum_{j=0}^{\infty} b_j x^j,$$

converge respectively to the functions $f(x)$ *and* $g(x)$, *for* $|x| < r$. *Then the power series*

$$\sum_{n=0}^{\infty} c_n x^n,$$

where

$$c_n = \sum_{k+j=n} a_k b_j = a_n b_0 + a_{n-1} b_1 + \cdots + a_1 b_{n-1} + a_0 b_n,$$

converges to the function $f(x) \cdot g(x)$ *for* $|x| < r$.

We shall not give a formal proof of 12.24. In many specific problems, since the form of the coefficients a_k and b_j might be rather complicated, it is often best simply to "multiply out" the two series. We illustrate this in the following example.

Example 4 Find a series of powers of x that converges to the function $\dfrac{\ln(1+x)}{1+x^2}$.

Solution: By integrating the power series for $1/(1+x)$ (Exercise I-1 below), we obtain, for $|x| < 1$,

$$\ln(1+x) = x - \frac{x^2}{2} + \frac{x^3}{3} - \frac{x^4}{4} + \cdots.$$

The series for $1/(1+x^2)$, which was given in Example 2 above, also converges for $|x| < 1$. We "multiply out," arranging the work as follows.

$$\ln(1+x) = x - \frac{x^2}{2} + \frac{x^3}{3} - \frac{x^4}{4} + \frac{x^5}{5} - \cdots$$

$$\frac{1}{1+x^2} = 1 - x^2 + x^4 + \cdots$$

$$x - \frac{x^2}{2} + \frac{x^3}{3} - \frac{x^4}{4} + \frac{x^5}{5} - \cdots$$

$$- x^3 + \frac{x^4}{2} - \frac{x^5}{3} + \cdots$$

$$x^5 - \cdots$$

$$\frac{\ln(1+x)}{1+x^2} = x - \frac{x^2}{2} - \frac{2}{3}x^3 + \frac{x^4}{4} + \frac{13x^5}{15} - \cdots$$

The product series also converges for $|x| < 1$, according to Theorem 12.24.

A power series may also be divided by another power series, in a manner analogous to the "long division" process for polynomials. We shall not state a formal theorem on division, but shall illustrate the procedure by an example.

Example 5 Find a series of powers of x that converges to $\tan x = \dfrac{\sin x}{\cos x}$.

Solution: In the next section it will be shown that the following series converge to $\sin x$ and $\cos x$, respectively, for all values of x.

$$\sin x = x - \frac{x^3}{3!} + \frac{x^5}{5!} - \cdots,$$

$$\cos x = 1 - \frac{x^2}{2!} + \frac{x^4}{4!} - \cdots.$$

We arrange the work as follows.

$$\begin{array}{r|l}
 & x + \dfrac{x^3}{3} + \dfrac{2x^5}{15} + \cdots \\ \hline
1 - \dfrac{x^2}{2} + \dfrac{x^4}{24} - \cdots & x - \dfrac{x^3}{6} + \dfrac{x^5}{120} - \cdots \\
 & x - \dfrac{x^3}{2} + \dfrac{x^5}{24} - \cdots \\ \cline{2-2}
 & \qquad \dfrac{x^3}{3} - \dfrac{x^5}{30} + \cdots \\
 & \qquad \dfrac{x^3}{3} - \dfrac{x^5}{6} + \cdots \\ \cline{2-2}
 & \qquad\qquad \dfrac{2x^5}{15} + \cdots
\end{array}$$

It is clear that we may compute as many terms of the quotient series as we desire. Through the terms of fifth degree,

$$\tan x = x + \frac{x^3}{3} + \frac{2x^5}{15} + \cdots.$$

EXERCISES

1. Show that

$$\ln(1 + x) = x - \frac{x^2}{2} + \frac{x^3}{3} - \frac{x^4}{4} + \cdots, \qquad |x| < 1.$$

Approximate ln (1.2) to three decimal places.

2. Show that

$$\ln(1-x) = -x - \frac{x^2}{2} - \frac{x^3}{3} - \frac{x^4}{4} - \cdots, \qquad |x| < 1.$$

Approximate $\ln \frac{3}{4}$ to three decimal places.

3. By combining the series in Exercises 1 and 2, show that

$$\ln\left(\frac{1+x}{1-x}\right) = 2\left(x + \frac{x^3}{3} + \frac{x^5}{5} + \cdots\right), \qquad |x| < 1.$$

Approximate ln 3 to three decimal places.

4. Approximate $\int_0^{1/2} \frac{x}{1+x^3}\,dx$ to four decimal places.

5. Find an infinite series for the improper integral

$$\int_0^x \frac{\ln(1+t)}{t}\,dt.$$

6. Find an infinite series for the improper integral

$$\int_0^x \frac{\tan^{-1} t}{t}\,dt.$$

7. Find the sum of the series

$$s(x) = \frac{x^2}{1\cdot 2} + \frac{x^3}{2\cdot 3} + \cdots + \frac{x^n}{(n-1)(n)} + \cdots$$

by first obtaining $s'(x)$ and using Exercise 2.

8. Find the sum of the series

$$s(x) = \frac{x^2}{1\cdot 2} - \frac{x^4}{3\cdot 4} + \cdots + \frac{(-1)^{n-1}x^{2n-2}}{(2n-3)(2n-2)} + \cdots$$

by first obtaining $s''(x)$.

9. Find the sum of the series

$$\sum_{n=2}^{\infty} \frac{x^{3n}}{2n}.$$

In each of the following problems, find a series in powers of x that converges to the given function, and determine the radius of convergence.

10. $\dfrac{1}{(1-x)^2}$

11. $\dfrac{1}{(1-x)^3}$

12. $\sin^2 x$

13. $\dfrac{\cos x}{1+x}$

14. $\dfrac{x^2}{(1 - x)(1 + x^2)}$

15. $\sec x$ (use division)

16. $\dfrac{1}{1 + x + x^2}$ (use division)

17. $\dfrac{1 - x}{1 - x + x^2}$

18. $\dfrac{x}{(1 - x)^2}$

19. $\dfrac{\ln (1 + x)}{2 + x}$

20. $\dfrac{x^2}{\cos^2 x}$

Approximate each of the following to four decimal places.

21. $\displaystyle\int_0^{1/2} \frac{dx}{1 + x^4}$

22. $\displaystyle\int_0^{.2} \frac{\sin x}{1 - x^2}\, dx$

23. Prove the converse of Theorem 12.20: if the series $\Sigma k a_k x^{k-1}$ converges for all x in $(-r,r)$, then so does the series $\Sigma a_k x^k$. *Hint:* For a given x_0 in $(-r,r)$, we have $|x_0| < k$ for all sufficiently large k. Then use comparison test I.

8 TAYLOR'S SERIES

A function defined by a power series in x possesses derivatives of all orders, obtainable by differentiating the power series termwise according to 12.21. We might well ask whether, conversely, if a function has derivatives of all orders, it may then be represented by a power series. While this is not true in general, it is true for most of the elementary functions studied in this book.

If

$$f(x) = \sum_{k=0}^{\infty} a_k(x - c)^k$$

has an open interval containing c as its domain, then

$$f'(x) = \sum_{k=1}^{\infty} k a_k(x - c)^{k-1}, \qquad f''(x) = \sum_{k=2}^{\infty} k(k - 1)a_k(x - c)^{k-2},$$

and, in general,

$$f^{[n]}(x) = \sum_{k=n}^{\infty} k(k - 1)\cdots(k - n + 1)a_k(x - c)^{k-n}.$$

According to 12.20 and 12.21, each of these differentiated series converges to the corresponding derivative whenever the series for $f(x)$ converges. Evaluating the function f and its derivatives at the number c, we get

$$f(c) = a_0, \quad f'(c) = a_1, \quad f''(c) = 2a_2,$$

and, in general,

$$f^{[n]}(c) = n!\, a_n.$$

Thus $$a_n = \frac{f^{[n]}(c)}{n!}$$

for each integer n, and the power series for f is given by

12.25 $$f(x) = f(c) + f'(c)(x - c) + \frac{f''(c)}{2!}(x - c)^2 + \cdots + \frac{f^{[n]}(c)}{n!}(x - c)^n + \cdots.$$

Series 12.25 is called the *Taylor's series of f at c*.

If $c = 0$, then we get the following *Maclaurin's series* (11.11) for $f(x)$.

12.26 $$f(x) = f(0) + f'(0)x + \frac{f''(0)}{2!}x^2 + \cdots + \frac{f^{[n]}(0)}{n!}x^n + \cdots.$$

The nth partial sum of 12.25 is a Taylor's polynomial $P_n(x)$ as defined in 11.6. The nth-degree remainder term $R_n(x)$ is given by (11.10)

$$R_n(x) = \frac{f^{[n+1]}(z)}{(n + 1)!}(x - c)^{n+1},$$

where z is some number between x and c, and

$$f(x) = P_n(x) + R_n(x).$$

The fundamental theorem for representing a function by a power series may be stated as follows.

12.27 Theorem

If the function f has derivatives of all orders in an interval containing the number c and if

$$\lim_{n\to\infty} R_n(x) = 0$$

for every x in the interval, then f is given by 12.25 *in this interval.*

Proof: Since $P_n(x) = f(x) - R_n(x)$, we have

$$\lim_{n\to\infty} P_n(x) = f(x) - \lim_{n\to\infty} R_n(x) = f(x),$$

and the theorem is proved.

The power series $\Sigma z^k/k!$ converges for every number z. Hence the limit of its nth term must be 0,

12.28 $$\lim_{n\to\infty} \frac{z^n}{n!} = 0.$$

This fact will be used in some of the following examples.

Example 1 Find the Maclaurin's series for e^x, and prove that it represents e^x for all x.

Solution: If $f(x) = e^x$, then $f^{[n]}(x) = e^x$ for every n. Thus $f^{[n]}(0) = 1$ and the Maclaurin's series for e^x is as follows:

12.29
$$e^x = 1 + x + \frac{x^2}{2!} + \frac{x^3}{3!} + \cdots + \frac{x^n}{n!} + \cdots.$$

Let us prove that 12.29 is valid for every number x. Evidently,

$$R_n(x) = \frac{e^{z_n}x^{n+1}}{(n+1)!},$$

where each z_n is between 0 and x. If $x > 0$, $e^{z_n} < e^x$ for every integer n, and therefore

$$0 < R_n(x) < \frac{e^x x^{n+1}}{(n+1)!}.$$

Since

$$\operatorname*{limit}_{n\to\infty} e^x \frac{x^{n+1}}{(n+1)!} = 0$$

by 12.28, also

$$\operatorname*{limit}_{n\to\infty} R_n(x) = 0.$$

If $x < 0$, each $e^{z_n} < 1$, and the same conclusion holds. Thus 12.29 is valid for every real number x.

If $x = 1$, 12.29 yields the following expression for e:

$$e = 1 + 1 + \frac{1}{2} + \frac{1}{6} + \cdots + \frac{1}{n!} + \cdots.$$

Example 2 Find the Maclaurin's series for $\sin x$, and prove that it represents $\sin x$ for all x.

Solution: If $f(x) = \sin x$, then $f(0) = 0$ and

$$\begin{aligned} f'(x) &= \cos x, & f'(0) &= 1, \\ f''(x) &= -\sin x, & f''(0) &= 0, \\ f'''(x) &= -\cos x, & f'''(0) &= -1, \\ f^{[4]}(x) &= \sin x, & f^{[4]}(0) &= 0, \end{aligned}$$

and so on. Thus

$$\sin x = x - \frac{x^3}{3!} + \frac{x^5}{5!} - \frac{x^7}{7!} + \cdots.$$

Since $|f^{[n]}(x)|$ is either $|\sin x|$ or $|\cos x|$, evidently $|f^{[n]}(x)| \le 1$ and

$$\operatorname*{limit}_{n\to\infty} |R_n(x)| = \operatorname*{limit}_{n\to\infty} |f^{[n+1]}(z_n)| \left|\frac{x^{n+1}}{(n+1)!}\right| = 0.$$

Hence the sine series is valid for every number x.

Example 3 Find the Taylor's series for $\ln x$ in powers of $x - 1$.

Solution: If $f(x) = \ln x$, then $f(1) = 0$ and

$$f'(x) = \frac{1}{x}, \qquad f'(1) = 1,$$

$$f''(x) = -\frac{1}{x^2}, \qquad f''(1) = -1,$$

$$f'''(x) = \frac{2}{x^3}, \qquad f'''(1) = 2,$$

and, in general,

$$f^{[n+1]}(x) = \frac{(-1)^n n!}{x^{n+1}}, \qquad f^{[n+1]}(1) = (-1)^n n!.$$

The Taylor's series for $\ln x$ thus has the form

$$\ln x = (x-1) - \frac{(x-1)^2}{2} + \frac{(x-1)^3}{3} - \cdots + \frac{(-1)^{n-1}(x-1)^n}{n} + \cdots.$$

This series converges for each x in the interval $(0,2]$ and diverges elsewhere. Evidently,

$$R_n(x) = \frac{(-1)^n n!}{(z_n)^{n+1}} \frac{(x-1)^{n+1}}{(n+1)!} = \frac{(-1)^n}{n+1}\left(\frac{x-1}{z_n}\right)^{n+1},$$

where z_n is between 1 and x. If $x > 1$, then $1 < z_n < x \leq 2$ and $0 < x - 1 \leq 1 < z_n$. Hence $(x-1)/z_n < 1$ and

$$\lim_{n\to\infty} |R_n(x)| = \lim_{n\to\infty} \frac{1}{n+1} = 0.$$

Thus this Taylor's series for $\ln x$ has $\ln x$ as its sum if $1 < x \leq 2$. We omit the details, but it may be shown that $\ln x$ is also given by this series if $0 < x \leq 1$.

The reader will recall the binomial theorem,

$$(x+y)^m = x^m + mx^{m-1}y + \frac{m(m-1)}{2!}x^{m-2}y^2 + \cdots + \frac{m(m-1)\cdots(m-k+1)}{k!}x^{m-k}y^k + \cdots + y^m,$$

which holds when m is a positive integer. Using Taylor's series, it is possible to extend the above formula to the case when the exponent m is replaced by an arbitrary real number a. In this general case, the expression on the right becomes an infinite series.

Let us consider the function $f(x) = (1+x)^a$, where a is arbitrary, and let us compute its Maclaurin's series. We find that the kth derivative of f is given by

$$f^{[k]}(x) = a(a-1)\cdots(a-k+1)(1+x)^{a-k}.$$

Hence the Maclaurin series for this function, which is called the *binomial series*, is given by

12.30 $$\sum_{k=0}^{\infty} c_k x^k = 1 + ax + \frac{a(a-1)}{2!}x^2 + \cdots$$

$$+ \frac{a(a-1)\cdots(a-n+1)}{n!}x^n + \cdots,$$

where

$$c_k = \frac{a(a-1)(a-2)\cdots(a-k+1)}{k!}, \qquad k = 1, 2, \ldots.$$

The binomial series is finite if and only if a is a nonnegative integer.

We can find the radius of convergence of 12.30 by the ratio test. Thus

$$\lim_{n\to\infty} \left|\frac{c_{n+1}x^{n+1}}{c_n x^n}\right| = \lim_{n\to\infty} \left|\frac{a-n}{n+1}\right| |x| = |x|,$$

and the series is convergent if $|x| < 1$ and divergent if $|x| > 1$. Hence its radius of convergence is 1.

It is possible to prove that the series 12.30 actually converges to the function $(1 + x)^a$, but the details are somewhat more difficult and will be omitted. However, we shall use this result and illustrate it with some examples.

Note that the expansion for the function $(b + x)^a$ may be obtained by factoring out the number b and thus reducing the expression to the form $(1 + x)^a$. For example, we may write $\sqrt{9 + x} = 3\sqrt{1 + x/9}$, and then apply 12.30 with x replaced by $x/9$.

Example 4 Find the series expansion of $\sqrt{1 + x}$.

Solution: Using the series 12.30 with $a = \frac{1}{2}$, we have

$$\sqrt{1+x} = 1 + \tfrac{1}{2}x + \frac{\frac{1}{2}(\frac{1}{2}-1)}{2!}x^2 + \cdots + \frac{\frac{1}{2}(\frac{1}{2}-1)\cdots(\frac{1}{2}-n+1)}{n!}x^n + \cdots$$

$$= 1 + \tfrac{1}{2}x - \frac{1}{2^2 2!}x^2 + \cdots + \frac{(-1)^{n+1}1\cdot 3\cdots(2n-3)}{2^n n!}x^n + \cdots,$$

if $|x| < 1$.

Example 5 Find the series expansion of $\ln(x + \sqrt{1 + x^2})$.

Solution: If $f(x) = \ln(x + \sqrt{1 + x^2})$, then

$$f'(x) = \frac{1 + x/\sqrt{1+x^2}}{x + \sqrt{1+x^2}} = \frac{1}{\sqrt{1+x^2}}.$$

Thus to find the series expansion of f we need only find the series expansion of $(1 + t^2)^{-1/2}$ and then integrate it termwise. Using 12.30,

$$(1+t^2)^{-1/2} = 1 - \tfrac{1}{2}t^2 + \frac{1\cdot 3}{2^2\cdot 2!}t^4 + \cdots + (-1)^n\frac{1\cdot 3\cdots(2n-1)}{2^n\cdot n!}t^{2n} + \cdots$$

if $t^2 < 1$. Hence, by 12.23,

$$\ln(x + \sqrt{1+x^2}) = \int_0^x \frac{1}{\sqrt{1+t^2}}\,dt$$

$$= x - \frac{1}{2 \cdot 3} x^3 + \frac{1 \cdot 3}{2^2 \cdot 5 \cdot 2!} x^5 + \cdots$$

$$+ (-1)^n \frac{1 \cdot 3 \cdots (2n-1)}{2^n \cdot (2n+1) \cdot n!} x^{2n+1} + \cdots,$$

if $|x| < 1$.

We remark in passing that it is not true in general that the sum of the Taylor's series of f at c (12.25) necessarily equals $f(x)$ for every x in the interval of convergence of this series. Exercise II-1 at the end of this section illustrates this point.

The infinite series expansion of a function may be used to approximate values of the function, as we have illustrated previously in this chapter. Some more examples of this are given below.

Example 6 Approximate $\int_0^1 e^{-x^2}\,dx$ to three decimal places.

Solution: Replacing x by $-x^2$ in 12.29, we get

$$e^{-x^2} = 1 - x^2 + \frac{x^4}{2!} - \frac{x^6}{3!} + \cdots + \frac{(-1)^n x^{2n}}{n!} + \cdots.$$

We may integrate this series termwise according to 12.23, getting

$$\int_0^1 e^{-x^2}\,dx = 1 - \frac{1}{3} + \frac{1}{10} - \frac{1}{42} + \cdots + \frac{(-1)^n}{(2n+1)n!} + \cdots.$$

Using five terms, we get

$$\int_0^1 e^{-x^2}\,dx \doteq .747,$$

with an error less than the next term, $\frac{1}{1320}$, by 12.13.

Example 7 Approximate $\int_0^{1/2} \cos\sqrt{x}\,dx$ to four significant digits.

Solution: We may easily verify that

$$\cos x = 1 - \frac{x^2}{2!} + \frac{x^4}{4!} - \frac{x^6}{6!} + \cdots$$

for every number x. Hence

$$\cos\sqrt{x} = 1 - \frac{x}{2!} + \frac{x^2}{4!} - \frac{x^3}{6!} + \cdots$$

for every nonnegative number x, and

$$\int_0^{1/2} \cos\sqrt{x}\,dx = \frac{1}{2} - \frac{1}{2 \cdot 2!}\left(\frac{1}{2}\right)^2 + \frac{1}{3 \cdot 4!}\left(\frac{1}{2}\right)^3 - \frac{1}{4 \cdot 6!}\left(\frac{1}{2}\right)^4 + \cdots \doteq .4392$$

with an error less than the fifth term $1/(2^6 \times 6!)$ by 12.13.

Example 8 Find $\lim_{x\to 0} \dfrac{e^x - x - \cos x}{x \sin x}$.

Solution: Although this problem is of a type to which l'Hospital's rule applies, it is simpler to handle by using infinite series. The first step is to expand numerator and denominator in MacLaurin's series.

$$\frac{e^x - x - \cos x}{x \sin x} = \frac{\left(1 + x + \dfrac{x^2}{2!} + \dfrac{x^3}{3!} + \dfrac{x^4}{4!} + \cdots\right) - x - \left(1 - \dfrac{x^2}{2!} + \dfrac{x^4}{4!} - \cdots\right)}{x\left(x - \dfrac{x^3}{3!} + \dfrac{x^5}{5!} - \cdots\right)}$$

$$= \frac{x^2 + \dfrac{x^3}{3!} + \cdots}{x^2 - \dfrac{x^4}{3!} + \dfrac{x^6}{5!} - \cdots} = \frac{1 + \dfrac{x}{3!} + \cdots}{1 - \dfrac{x^2}{3!} + \cdots}.$$

In the last step we were able to cancel x^2 from both numerator and denominator because we do not allow $x = 0$. Since now the limit of the denominator is not 0, we may let x approach 0 to obtain

$$\lim_{x\to 0} \frac{e^x - x - \cos x}{x \sin x} = \frac{1}{1} = 1.$$

EXERCISES

I

Find an infinite series expansion for each of the following functions. State its radius of convergence.

1. $\cos x$ in powers of $x - \pi/4$

2. xe^x in powers of x

3. $\ln |x|$ in powers of $x + 1$

4. $\dfrac{\sin x}{x}$ in powers of x

5. $\sinh x$ in powers of x

6. $\cosh x$ in powers of x

7. $\dfrac{1 - \cos x}{x}$ in powers of x

8. $\sin x$ in powers of $x + \pi/3$

9. $\sin^2 x$ in powers of x. (*Hint:* Use double-angle formula.)

10. $\cos^2 x$ in powers of x

11. $4x^4 - 15x^3 + 20x^2 - 10x + 14$ in powers of $x + 1$

12. 2^x in powers of x

Use the binomial series to find expansions of the following functions in powers of x. Determine the radius of convergence.

13. $\dfrac{1}{(1 + x)^2}$

14. $\dfrac{z}{\sqrt{1 - z^3}}$

15. $\dfrac{1}{\sqrt{1 - x^2}}$

16. $(1 + 2x)^{-3}$

17. $x(4 - x)^{3/2}$. (*Hint:* Factor out the number 4.)

18. $\sqrt[3]{x + 8}$

19. $\dfrac{1}{\sqrt{16 + x^4}}$

20. $\sqrt{2 + z}$

21. $\sin^{-1} x$

22. $\ln(\sqrt{1 + x^2} - x)$

23. $\sec^{-1} \dfrac{1}{x}$

24. Show that

$$\frac{\sin^{-1} x}{\sqrt{1 - x^2}} = x + \frac{2}{3}x^3 + \frac{2 \cdot 4}{3 \cdot 5}x^5 + \frac{2 \cdot 4 \cdot 6}{3 \cdot 5 \cdot 7}x^7 + \cdots = \sum_{n=0}^{\infty} \frac{2^{2n}(n!)^2 x^{2n+1}}{(2n + 1)!}.$$

Approximate each of the following to the stated degree of accuracy.

25. $\sqrt[3]{e}$, five decimal places

26. $\sin 10^0$, four decimal places. (*Hint:* $1° \doteq .017453$ radian.)

27. $\int_0^{.4} \sin x^2 \, dx$, six decimal places

28. $\sinh \frac{1}{2}$, five decimal places

Approximate each of the following to four decimal places.

29. $\int_0^1 \sqrt{1 - x^3} \, dx$

30. $\int_0^{1/2} \dfrac{1 - e^{-x^2}}{x} \, dx$

31. $\cos 61°$

32. $\tan 47°$

Use infinite series to find the following limits.

33. $\lim\limits_{x \to 0} \dfrac{\sin x - x \cos x}{x^2 \tan x}$

34. $\lim\limits_{x \to 0} \dfrac{x \ln(1 + x)}{\sin^2 x}$

35. $\lim\limits_{x \to 0} \dfrac{2 - \cos x - \sqrt{1 + x^2}}{x^4}$

36. $\lim\limits_{x \to 0} \dfrac{1 + x^2 - e^{x^2}}{x^3 \sin x}$

II

1. Define the function f by $f(x) = e^{-1/x^2}$ if $x \neq 0$, $f(0) = 0$. Prove that $f^{[n]}(0)$ exists and $= 0$ for every positive integer n. Also prove that MacLaurin's series for f does not converge to f in any neighborhood of 0.

9 APPENDIX: DIFFERENTIATION AND INTEGRATION OF POWER SERIES

Our purpose in this section is to give proofs of Theorems 12.21 and 12.22. Before doing this, however, it is necessary to establish a preliminary theorem that assures us that any function defined by a power series is *continuous*.

12.31 Theorem

If $\Sigma a_k x^k$ has non-zero radius of convergence r, and if

$$f(x) = \sum_{k=0}^{\infty} a_k x^k ,$$

then f is continuous for every number t in $(-r,r)$.

Proof: We will prove the continuity of f at each t in $(-r,r)$ by showing that

$$\lim_{x \to t} |f(x) - f(t)| = 0.$$

First we write

$$|f(x) - f(t)| = \left| \sum_{k=0}^{\infty} a_k x^k - \sum_{k=0}^{\infty} a_k t^k \right|$$

$$= \left| \sum_{k=0}^{\infty} a_k (x^k - t^k) \right| \leq \sum_{k=0}^{\infty} |a_k| \, |x^k - t^k|,$$

where the above inequality follows from 12.14.

Now apply the Mean Value Theorem (4.8) to the function x^k on the interval from t to x. We can then state that

$$x^k - t^k = kz^{k-1}(x - t)$$

for some number z between t and x. Although this number z depends on the particular value of k, nevertheless z always lies between t and x, and x may be restricted to lie in a small neighborhood of t. Hence there is certainly some number s in $(-r,r)$ such that $|z| \leq |s|$, where s is independent of k. We then have

$$|f(x) - f(t)| \leq \sum_{k=0}^{\infty} k|a_k| \cdot |z^{k-1}| \cdot |x - t|$$

$$\leq |x - t| \sum_{k=0}^{\infty} k|a_k| \, |s|^{k-1}$$

Since s is in $(-r,r)$, the above series $\Sigma k|a_k| \, |s|^{k-1}$ converges by Theorem 12.20. If we denote its sum by B, then we obtain

$$|f(x) - f(t)| \leq |x - t| \cdot B .$$

Letting x approach t, it follows that $|f(x) - f(t)|$ approaches 0, which completes the proof of 12.31.

Next we prove Theorem 12.22 on the integration of power series.

Proof of 12.22: Let r be the radius of convergence of $\Sigma a_k x^k$, $r \neq 0$. Let $[a,b]$ be a closed interval contained in $(-r,r)$. Note that if

$$f(x) = \sum_{k=0}^{\infty} a_k x^k ,$$

then f is continuous in $(-r,r)$ by 12.31 and hence we know $\int_a^b f(x)\,dx$ exists.

We again use our previous notation for the remainder after n terms of a power series:

$$R_n(x) = f(x) - \sum_{k=0}^{n} a_k x^k = \sum_{k=n+1}^{\infty} a_k x^k .$$

Let S_n denote the nth partial sum of the integrated series; that is,

$$S_n = \sum_{k=0}^{n} \frac{a_k}{k+1} (b^{k+1} - a^{k+1}).$$

Since the integral of a finite sum can be found by term-by-term integration, we also have

$$S_n = \sum_{k=0}^{n} \int_a^b a_k x^k \, dx = \int_a^b \left(\sum_{k=0}^{n} a_k x^k \right) dx.$$

We want to prove that $\lim_{n\to\infty} S_n = \int_a^b f(x)\,dx$.

Let us write

$$\left| \int_a^b f(x)\,dx - S_n \right| = \left| \int_a^b \left[f(x) - \sum_{k=0}^{n} a_k x^k \right] dx \right|$$
$$= \left| \int_b^b R_n(x)\,dx \right| \le \int_b^b |R_n(x)|\,dx.$$

The proof will be completed if we can show that

$$\lim_{n\to\infty} \int_a^b |R_n(x)|\,dx = 0.$$

To do this, let $d = \max(|a|, |b|)$. Then for $a \le x \le b$

$$(1) \qquad |R_n(x)| = \left| \sum_{k=n+1}^{\infty} a_k x^k \right| \le \sum_{k=n+1}^{\infty} |a_k|\,|x^k| \le \sum_{k=n+1}^{\infty} |a_k| d^k,$$

where the first of the above inequalities follows by 12.14. Now $\Sigma|a_k|d^k$ is a convergent series of positive constants, and the quantity on the right of (1) represents its *remainder* after the nth term. This quantity must therefore approach 0 as n approaches ∞. Hence, given any $\varepsilon > 0$, we can find n_0 such that $|R_n(x)| < \varepsilon/(b-a)$ for all $n > n_0$ and all x in $[a,b]$. Thus, corresponding to any given $\varepsilon > 0$, for $n > n_0$ we have

$$\int_a^b |R_n(x)|\,dx \le \frac{\varepsilon}{b-a}(b-a) = \varepsilon.$$

But this means that $\lim_{n\to\infty} \int_a^b |R_n(x)|\,dx = 0$.

It is clear that Corollary 12.23 follows immediately from 12.22 by setting $a = 0, b = x$. We now use 12.23 to prove Theorem 12.21 on differentiation of power series.

Proof of 12.21: We want to prove that the series $\Sigma k a_k x^{k-1}$ converges to $f'(x)$ for all x in $(-r,r)$. We already know from 12.20 that $\Sigma k a_k x^{k-1}$ converges for all x in $(-r,r)$. So let

$$g(x) = \sum_{k=1}^{\infty} k a_k x^{k-1}.$$

Then by 12.23 we have

$$\int_0^x g(t)\,dt = \sum_{k=1}^{\infty} a_k x^k = f(x) - a_0.$$

Now differentiate both sides of the above equation with respect to x. By the Fundamental Theorem of Calculus (5.20), the derivative of the left-hand side is $g(x)$. Hence $g(x) = f'(x)$, completing the proof of 12.21.

CHAPTER REVIEW

In each of Exercises 1 to 16, test the series whose nth term is specified for convergence or divergence. In case of convergence, determine whether or not the series converges absolutely.

1. $\dfrac{\tan^{-1} n}{n^{3/2}}$

2. $\dfrac{\sqrt{n} - 7}{n(n+3)}$

3. $\dfrac{n!}{(2n)!}$

4. $\dfrac{(-1)^n 3^n}{n^2 \cdot 2^n}$

5. $\dfrac{(-1)^n n^2}{a^n}\ (a > 1)$

6. $\dfrac{\sqrt{n^4 + 2}}{n^3 \ln n}$

7. $\dfrac{(-1)^n \ln n}{n^{3/2}}$

8. $\dfrac{(-1)^n (\ln n)^{10}}{n^p}\ (p > 1)$

9. $\left(1 + \dfrac{1}{n}\right)^n \dfrac{1}{n^2}$

10. $(-1)^n \ln\left(1 + \dfrac{1}{n}\right)$

11. $\dfrac{(-1)^n 4^{3n+1}}{(3n+1)!}$

12. $\dfrac{4 + \sin n}{n^3}$

13. $\sin \dfrac{\pi}{2n^2 - 1}$

14. $\dfrac{2^n + 4^n}{6^n}$

15. $\dfrac{\sqrt{n+1}-\sqrt{n}}{\sqrt{n^2+1}}$

16. $\dfrac{(-1)^n n}{\sqrt{n^2+1}}$

In Exercises 17–22, determine the set of all x for which the given series converges.

17. $\sum_{n=0}^{\infty} (-1)^n \dfrac{(2x+3)^n}{2^n}$

18. $\sum_{n=0}^{\infty} \dfrac{(-1)^n 5^n}{\sqrt{n+3}\, x^{2n}}$

19. $\sum_{n=1}^{\infty} \dfrac{1}{n}\left(\dfrac{x-1}{x}\right)^n$

20. $\sum_{n=1}^{\infty} \dfrac{n!}{x^n}$

21. $\sum_{n=1}^{\infty} \dfrac{n!\, x^n}{n^n}$

22. $\sum_{n=1}^{\infty} \dfrac{\sec^2 nx}{n}$

In each of Exercises 23–26, expand the given function into a Taylor series about the given point.

23. a^x, $x = 0$

24. $\dfrac{1}{x}$, $x = 2$

25. e^{x+3}, $x = 2$

26. $\dfrac{1}{x^2+4x+2}$, $x = -2$

In Exercises 27–32, find the specified number of terms of the series expansion of the given function in powers of x.

27. $e^x \sin x$, four terms

28. $\dfrac{2}{e^x + e^{-x}}$, three terms

29. $\csc x$, three terms

30. $\sec^2 x$, four terms. (Use the series of Example 5, Section 7.)

31. $\ln \sec x$, three terms

32. $\cot x$, four terms

In Exercises 33–36, use infinite series to find the indicated limits.

33. $\lim\limits_{x\to 0} \dfrac{e^x - 1 - x}{x^2 \cos x}$

34. $\lim\limits_{x\to 0} \dfrac{x - e^x \sin x}{x \ln(1+x)}$

35. $\lim\limits_{x\to 0} \dfrac{(1-\cos x)^2}{x^3 \sin x}$

36. $\lim\limits_{x\to 0} \dfrac{x^2 e^x}{1 - \cos 3x}$

Compute the following to four decimal places.

37. $\int_0^1 \sin x^3 \, dx$

38. $\int_0^{.1} \dfrac{\sin x}{1-x^2}\, dx$

39. *Computation of π.* In Example 2, Section 7, we obtained the series

$$\tan^{-1} x = x - \frac{x^3}{3} + \frac{x^5}{5} - \frac{x^7}{7} + \cdots, \qquad |x| < 1. \tag{12.32}$$

This series also converges for $x = 1$ (by the alternating series test), which gives

$$\frac{\pi}{4} = \tan^{-1} 1 = 1 - \frac{1}{3} + \frac{1}{5} - \frac{1}{7} + \cdots.$$

However, it is difficult to compute $\pi/4$ from this series to an accuracy of, say, four decimal places, because a large number of terms would be needed. A better way to compute $\pi/4$ is the following.

a. Use an appropriate number of terms of series (12.32) to obtain the values $\tan^{-1} \frac{1}{3} \doteq 0.32175$, $\tan^{-1} \frac{1}{7} \doteq 0.14190$.

b. Use the formula $\tan (A + B) = (\tan A + \tan B)/(1 - \tan A \tan B)$ to prove that $\tan (2 \tan^{-1} \frac{1}{3} + \tan^{-1} \frac{1}{7}) = 1$.

c. From part (a) and the fact that $\pi/4 = 2 \tan^{-1} \frac{1}{3} + \tan^{-1} \frac{1}{7}$, deduce that $\pi/4 \doteq 0.7854$. This gives π accurate to four decimal places.

40. Find an approximate value of π by setting $x = \frac{1}{2}$ in the MacLaurin series for $\sin^{-1} x$ (Exercise I-21, Section 8). Use the first four terms of the series.

13

Plane Curves, Vectors, and Polar Coordinates

Heretofore we have considered only real functions, i.e., functions having domains and ranges in the set of real numbers. In this chapter we shall study functions, called curves, that map intervals into sets of points in a coordinate plane. This will allow us to study the motion of particles along much more complicated paths than graphs of functions, such as that shown in Figure 13.1, where a particle might move along the path from time $t = a$ to time $t = b$.

We shall also introduce two-dimensional vectors and discuss tangent vectors to curves. The useful polar coordinate system in the plane will be described and equations of the conic sections in polar coordinates will be developed. Finally, arc length of a curve and area of a surface of revolution will be discussed.

1 PLANE CURVES

For convenience we shall henceforth designate the set of all real numbers by

$$\mathsf{R}$$

and the set of all points in a rectangular coordinate plane by

$$\mathsf{R}^2.$$

Thus R^2 is the cartesian product $\mathsf{R} \times \mathsf{R} = \{(a,b) \mid a,b \text{ in } \mathsf{R}\}$.

A mapping λ (lambda) having an interval I of R as its domain and a subset of R^2 as its range is called a *plane curve*, or simply a *curve*. Space curves will be discussed in Chapter 14. The range of a curve λ is called the *trace* of λ.

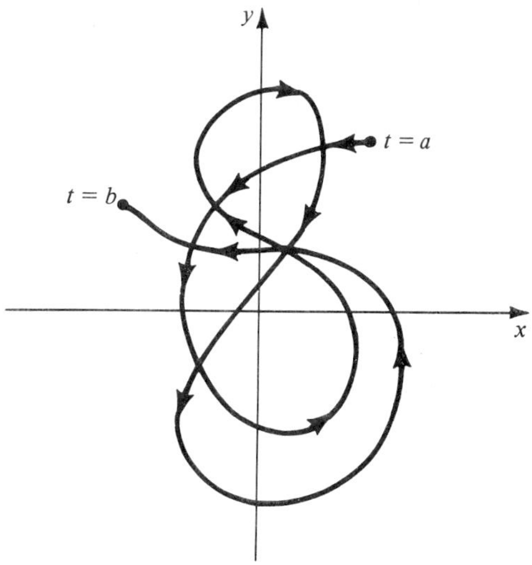

FIGURE 13.1

Example 1 Discuss the plane curve λ defined by

$$\lambda(t) = (\cos t, \sin t)$$

when (1) domain $\lambda = [0,\pi]$; (2) domain $\lambda = [-\pi,\pi]$; (3) domain $\lambda = [-2\pi,2\pi]$.

Solution: Since

$$\cos^2 t + \sin^2 t = 1$$

for every number t, evidently the trace of λ lies on the unit circle of Figure 13.2. In

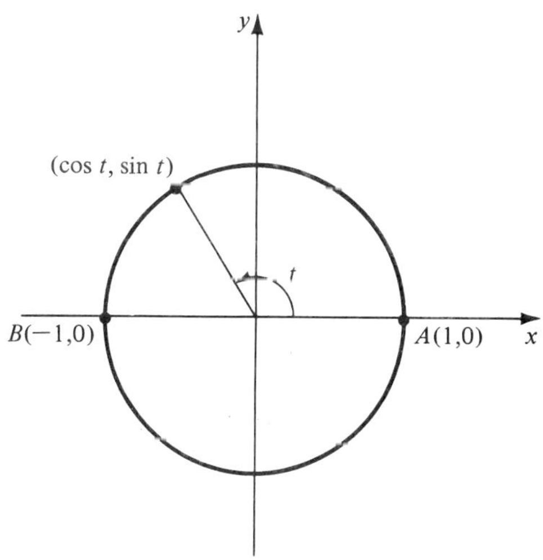

FIGURE 13.2

case (1), as t varies from 0 to π, clearly $\cos t$ varies from 1 to -1 and $\sin t$ varies from 0 to 1 and back again to 0. Thus the point $\lambda(t)$ moves from A to B on the circle in a counterclockwise direction. In case (2) the point $\lambda(t)$ starts at B when $t = \pi$, moves to A when $t = 0$, and then moves back to B when $t = \pi$. That is, $\lambda(t)$ makes one complete revolution of the circle in a counterclockwise direction starting from B. It is easily seen that in case (3) the point $\lambda(t)$ starts at A when $t = -2\pi$ and makes two complete revolutions of the circle in a counterclockwise direction, ending up at point A when $t = 2\pi$. It is worthwhile noting that the trace of λ is the same in (2) and (3), namely the circle, although the curves clearly are different.

Each curve λ with domain I has the form

$$\lambda(t) = (f(t),g(t)), \qquad t \text{ in } I,$$

for some real-valued functions f and g. Thus in Example 1 we have $f(t) = \cos t$ and $g(t) = \sin t$. That is, each point (x,y) of the range of λ is given by

13.1 $$x = f(t), \qquad y = g(t)$$

for some t in I. These equations are usually called *parametric equations* of curve λ and t is called the *parameter*.

Example 2 Discuss the curve with parametric equations

$$x = \sin t, \qquad y = \cos^2 t$$

and domain $[0,2\pi]$.

Solution: In our previous notation the curve λ is defined by

$$\lambda(t) = (\sin t, \cos^2 t), \qquad \text{domain } \lambda = [0,2\pi].$$

Since $\sin^2 t + \cos^2 t = 1$, it is evident that the trace of λ lies on the parabola

(1) $$x^2 + y = 1.$$

While the trace of λ is part of the parabola (1), it is not true that the trace of λ is the complete parabola (1). Thus, since $x = \sin t$ and $y = \cos^2 t$, necessarily $-1 \le x \le 1$ and $0 \le y \le 1$ for every point (x,y) in the trace of λ. Hence the trace of λ consists of at most the part of the parabola (1) on and above the x axis (Figure 13.3).

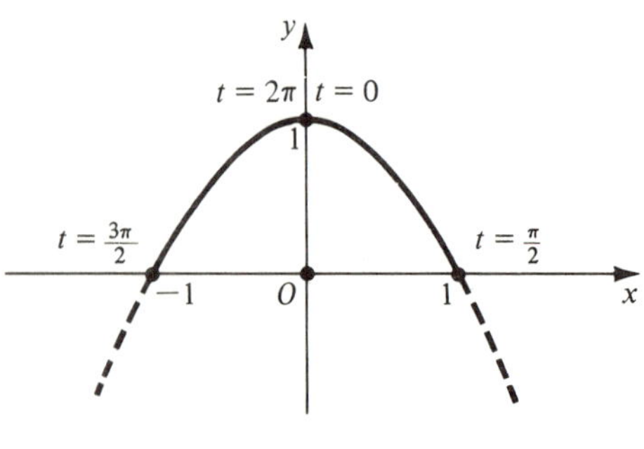

FIGURE 13.3

That the trace of λ is precisely this part of the parabola is seen as follows. Since the sine is continuous in $[0,2\pi]$, for each number x between 1 ($=\sin \pi/2$) and -1 ($=\sin 3\pi/2$) there exists a number t in $[\pi/2, 3\pi/2]$ such that $\sin t = x$. Hence the point $(x, 1 - x^2)$ is in the trace of λ for each x in $[-1,1]$.

The point $\lambda(t)$ is at $(0,1)$ when $t = 0$, moves to $(1,0)$ when $t = \pi/2$, moves back to $(0,1)$ when $t = \pi$, moves to $(-1,0)$ when $t = 3\pi/2$, and finally moves back to $(0,1)$ when $t = 2\pi$.

Example 3 Describe the curve λ defined by

$$\lambda(t) = (1 + 4t, 3 - 2t), \qquad \text{domain } \lambda = \mathsf{R}.$$

Solution: The curve λ has parametric equations

$$x = 1 + 4t, \qquad y = 3 - 2t.$$

If we solve the first equation for t, obtaining

(1) $$t = \tfrac{1}{4}(x - 1),$$

and substitute this value of t in the second equation, we get the equation

(2) $$y - 3 = -\tfrac{1}{2}(x - 1).$$

Hence the trace of λ is on the line L with equation (2). Since every point (x,y) on L has the form $\lambda(t)$ if we choose t as in (1), it is clear that the trace of λ is the whole line L sketched in Figure 13.4. Furthermore, the line L is traced out only once by λ. As t varies from negative numbers, through 0, and on to positive numbers, the point $\lambda(t)$ moves from left to right on L, being at the point $(1,3)$ when $t = 0$. For example, $\lambda(t)$ is on the y axis when $t = -\frac{1}{4}$ and on the x axis when $t = \frac{3}{2}$.

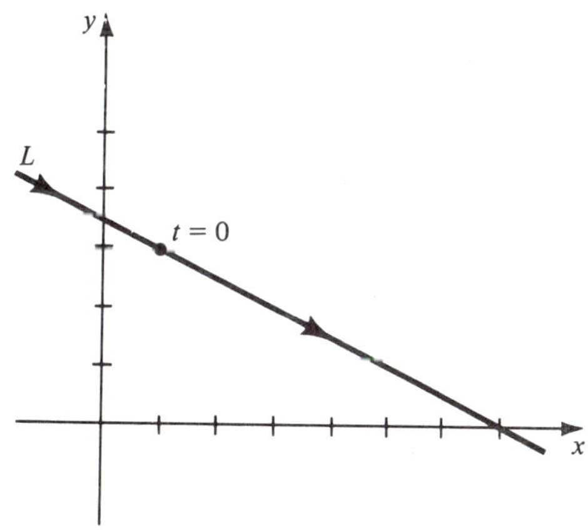

FIGURE 13.4

As the three examples above show, a curve λ has *direction* determined by the order in I, the domain of λ. If I is a closed interval $[a,b]$, then $\lambda(a)$ is called the *initial point* and $\lambda(b)$ the *terminal point* of λ. We think of point $\lambda(t_1)$ preceding point $\lambda(t_2)$ if and only if $t_1 < t_2$. If curve λ has domain $[a,b]$ and $\lambda(a) = \lambda(b)$, then λ is called a *closed curve.* Example 1, cases (2) and (3), and Example 2 are closed curves. A curve λ with domain I is said to be a *one-to-one curve* (or a 1–1 *curve*) if $\lambda(t_1) \neq \lambda(t_2)$ whenever $t_1 \neq t_2$ in I. Finally, a closed curve λ with domain $[a,b]$ is called a *simple closed curve* if $\lambda(t_1) \neq \lambda(t_2)$ whenever $t_1 \neq t_2$, where t_1 is in (a,b) and t_2 in $[a,b]$. Thus the trace of λ is covered only once as t varies from a to b if λ is a simple closed curve, except that $\lambda(a) = \lambda(b)$. Example 1, case (2), is a simple closed curve, whereas Example 3 is a 1–1 curve.

Example 4 As a circle rolls along a straight line in a plane, the curve described by a fixed point P on the circle is called a *cycloid.* Find parametric equations for the cycloid.

Solution: We shall assume that the circle has radius r, that it rolls along the x axis and above it, and that the origin is so chosen on the x axis that the point P makes contact with the x axis at the origin. Let us choose as parameter the angle t (in radians) of rotation of the circle from the position when P is at the origin (Figure 13.5).

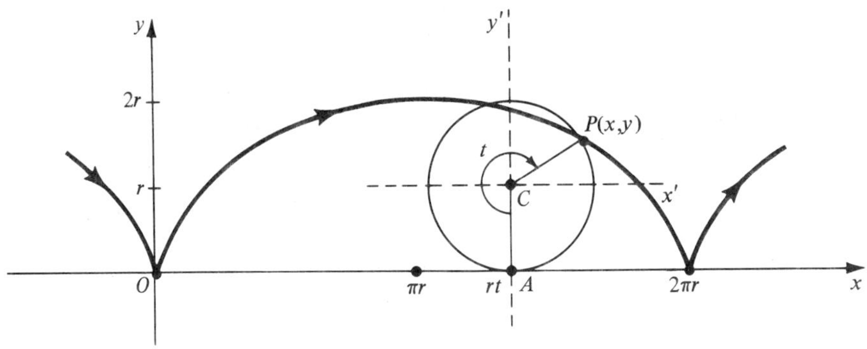

FIGURE 13.5

When the circle has revolved through an angle of t radians, it will have rolled a distance $\overline{OA} = rt$ from the origin. Hence the center of the circle will be the point $C(rt,r)$. If we translate axes to the new origin C, then the coordinates of P relative to the new axes will be

(1) $$x' = x - rt, \qquad y' = y - r.$$

A close-up view of the new coordinate axes, given in Figure 13.6, shows us that

(2) $$x' = r\cos\left(\frac{3\pi}{2} - t\right) = -r\sin t,$$
$$y' = r\sin\left(\frac{3\pi}{2} - t\right) = -r\cos t.$$

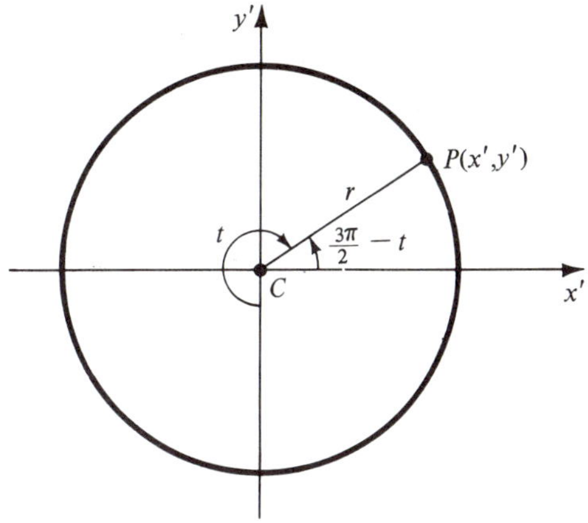

FIGURE 13.6

Finally, on combining (1) and (2), we get

$$x = r(t - \sin t), \qquad y = r(1 - \cos t)$$

as parametric equations of the cycloid.

If we think of the cycloid as a curve λ, then

$$\lambda(t) = (r(t - \sin t), r(1 - \cos t)), \qquad \text{domain } \lambda = \mathsf{R}.$$

The curve is directed from left to right in Figure 13.5.

2 CONTINUITY OF A CURVE

If P and Q are points in R^2, then let us designate the *distance* between P and Q by $d(P,Q)$. We know that if $P = (x_1,y_1)$ and $Q = (x_2,y_2)$ then

$$d(P,Q) = \sqrt{(x_2 - x_1)^2 + (y_2 - y_1)^2}.$$

For each positive number r and each point A in R^2, let us define the *open disc* with center A and radius r to be

$$B(A,r) = \{P \text{ in } \mathsf{R}^2 \mid d(A,P) < r\},$$

and the *closed disc* with center A and radius r to be

$$B[A,r] = \{P \text{ in } \mathsf{R}^2 \mid d(A,P) \leq r\}.$$

A subset S of R^2 is called *open* if for each point A in S there exists a positive number r such that the open disc $B(A,r)$ is contained in S. Naturally, R^2 is itself an open set. Another example of an open set in R^2 is

$$H = \{(x,y) \mid x > 0\},$$

the open half-plane to the right of the y axis.

13.2 Definition

If λ is a curve with domain I, then λ is said to be *continuous at* c *in* I if for each open disc $B(\lambda(c),\varepsilon)$ there exists a neighborhood N of c such that $\lambda(t)$ is in $B(\lambda(c),\varepsilon)$ for every t in $N \cap I$.

As usual, we call a curve λ *continuous* if λ is continuous at each number in its domain. The following result gives us an easy test for continuity of curves.

13.3 Theorem

If λ is a curve given by

$$\lambda(t) = (f(t),g(t)), \qquad \textit{domain } \lambda = I,$$

then λ is continuous at c in I if and only if functions f and g are continuous at c.

Proof: Assume that λ is continuous at c. Then for every number $\varepsilon > 0$ there exists a neighborhood N of c such that $\lambda(t)$ is in $B(\lambda(c),\varepsilon)$ for every t in $N \cap I$. Hence $|f(t) - f(c)| < \varepsilon$ for every t in $N \cap I$ (see Figure 13.7). Thus f is continuous at c. A similar argument proves that g is continuous at c.

Conversely, assume that f and g are continuous at c. Then for each number $\varepsilon > 0$ there exists a neighborhood N of c such that

$$|f(t) - f(c)| < \frac{\varepsilon}{\sqrt{2}} \quad \text{and} \quad |g(t) - g(c)| < \frac{\varepsilon}{\sqrt{2}} \quad \text{for every } t \text{ in } N \cap I.$$

Hence $$[f(t) - f(c)]^2 + [g(t) - g(c)]^2 < \frac{\varepsilon^2}{2} + \frac{\varepsilon^2}{2} = \varepsilon^2$$

for every t in $N \cap I$; that is,

$$\lambda(t) \text{ is in } B(\lambda(c),\varepsilon) \qquad \text{for every } t \text{ in } N \cap I.$$

This proves that λ is continuous at c and completes the proof of 13.3.

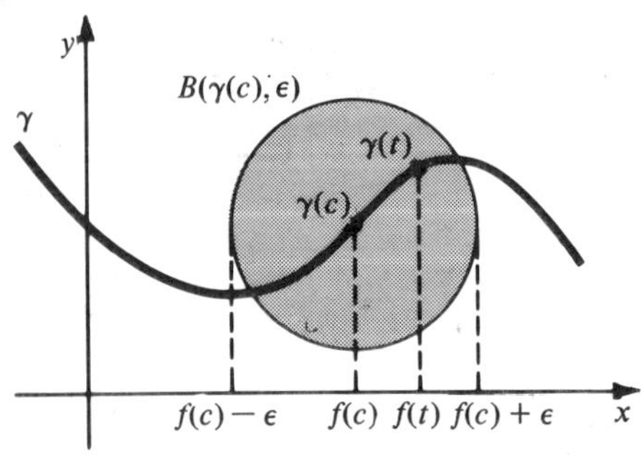

FIGURE 13.7

According to 13.3, each of the curves of Examples 1 to 4 of Section 1 is continuous, since the functions f and g are continuous in every case.

Given a curve λ and a number c, we can define

$$\lim_{t \to c} \lambda(t)$$

in an obvious way. That is,

$$\lim_{t \to c} \lambda(t) = P$$

if λ is continuous at c when we define (or redefine) $\lambda(c) = P$. In view of 13.3, we have that if

$$\lambda(t) = (f(t), g(t))$$

then

$$\lim_{t \to c} \lambda(t) = (\lim_{t \to c} f(t), \lim_{t \to c} g(t)),$$

provided the latter two limits exist.

EXERCISES

I

In each of Exercises 1 to 10, discuss the curve λ, describe its direction, and sketch its trace.

1. $\lambda(t) = (3t, 2 - t)$, domain $\lambda = \mathrm{R}$
2. $\lambda(t) = (t^2, t + 1)$, domain $\lambda = \mathrm{R}$
3. $\lambda(t) = (\sin 2t, 2 \sin^2 t)$, domain $\lambda = [0,\pi]$
4. $\lambda(t) = (\sin 2t, 2 \sin^2 t)$, domain $\lambda = [0,2\pi]$
5. $\lambda(t) = (2 \sin t, 3 \cos t)$, domain $\lambda = [0,\pi]$
6. $\lambda(t) = (2 \sin t, 3 \cos t)$, domain $\lambda = [0,2\pi]$
7. $\lambda(t) = (2 \cos t, 3 \sin t)$, domain $\lambda = [0,2\pi]$
8. $\lambda(t) = (\sinh t, \cosh t)$, domain $\lambda = \mathrm{R}$
9. $\lambda(t) = (h + a \cosh t, k + b \sinh t)$ domain $\lambda = \mathrm{R}$
10. $\lambda(t) = (1 + \cos t, 1 - \sin t)$, domain $\lambda = [-\pi,\pi]$

In each of Exercises 11 to 18 discuss the curve with the given parametric equations. The domain of each curve is taken to be the set of all real numbers for which the equations are defined.

11. $x = t^3, y = t^2$
12. $x = \tan s, y = \sec s$
13. $x = \tan^2 r, y = \sec r$
14. $x = 3e^t, y = 1 - e^t$
15. $x = 5 \cos s, y = 3 \sin s$
16. $x = 2 + \cos t, y = -1 + \sin t$
17. $x = a \sec t, y = b \tan t$
18. $x = 1 + \dfrac{1}{t}, y = t - \dfrac{1}{t}$

19. Show that the trace of each of the following curves is part or all of the parabola $y = x^2$. Describe the trace in each case.

a. $\lambda(t) = (t,t^2)$, domain $\lambda = \mathsf{R}$

b. $\lambda(t) = (t^2,t^4)$, domain $\lambda = \mathsf{R}$

c. $\lambda(t) = (|t|,t^2)$, domain $\lambda = (-\infty,0]$

d. $\lambda(t) = (e^t,e^{2t})$, domain $\lambda = [0,\infty)$

e. $\lambda(t) = (1 - 1/t^2, 1 - 2/t^2 + 1/t^4)$, domain $\lambda = (-\infty,0) \cup (0,\infty)$

f. $\lambda(t) = (\sec t, 1 + \tan^2 t)$, domain $\lambda = \left[0, \frac{\pi}{2}\right) \cup \left(\frac{\pi}{2}, \pi\right]$

20. Show that $x = x_1 + (x_2 - x_1)t, y = y_1 + (y_2 - y_1)t$ are parametric equations of the line passing through the distinct points (x_1,y_1) and (x_2,y_2).

21. Find parametric equations of the circle with center (h,k) and radius r.

22. Show that the curve λ defined by $\lambda(t) = (h + a \cos t, k + b \sin t)$, domain $\lambda = [0,2\pi]$, is an ellipse.

23. Show that the curve λ defined by $\lambda(t) = (h + a \sec t, k + b \tan t)$, domain

$$\lambda = \left[0, \frac{\pi}{2}\right) \cup \left(\frac{\pi}{2}, \frac{3\pi}{2}\right) \cup \left(\frac{3\pi}{2}, 2\pi\right],$$ is a hyperbola.

24. A bicycle is proceeding along a straight road at a constant speed of v ft/sec. If each wheel has a radius of r ft, find parametric equations of the motion of a point P on a tire in terms of time t.

II

In Exercises 1–3, find parametric equations for the graph of the given equation by setting $y = tx$ and solving the equation for x in terms of t. Then sketch the graph.

1. $x^4 - 3xy^2 + 2y^3 = 0$

2. $x^4 - 4x^2y + y^3 = 0$

3. $(x^2 + y^2)^2 = 16x^2y$

4. Show how to construct, using only straightedge and compass, any number of points on the ellipse $(x^2/a^2) + (y^2/b^2) = 1$. *Hint:* Draw circles of radii a and b centered at the origin and use Exercise I-22.

5. (*Hyperbolic functions and the unit hyperbola.*) Let us represent the unit circle $x^2 + y^2 = 1$ by the parametric equations $x = \cos t$, $y = \sin t$, as in Example 1 of this section. In Figure 13.2, if we let P be the point $(\cos t, \sin t)$, then the area of the circular sector AOP is $\frac{1}{2}t$. (Chapter 9, Section 1, Example 1). Hence the parameter t, in addition to denoting the angle between the positive x-axis and the radius OP, has another geometric significance; namely, t is twice the area of the sector AOP.

Now the unit hyperbola $x^2 - y^2 = 1$ may be represented by the parametric equations $x = \cosh t$, $y = \sinh t$ (Exercise I-8 above). Let $A = (1,0)$ and let P be the point $(\cosh t, \sinh t)$ on the hyperbola (see figure below). Prove that in this case also, just as for the unit circle, the parameter t is twice the area of the sector AOP.

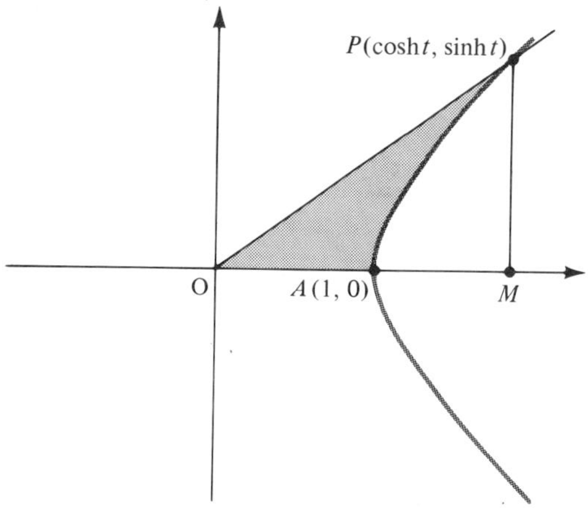

Suggestion: Referring to the above figure, first compute the area of the sector APM, which is given by

$$\text{Area }(APM) = \int_1^{\cosh t} y\,dx = \int_1^{\cosh t} \sqrt{x^2 - 1}\,dx = \int_0^t \sinh^2 t\,dt.$$

3 TWO-DIMENSIONAL VECTOR ALGEBRA

In the representation of many physical concepts, it is not sufficient merely to specify the magnitude of some quantity. For example, in physics the notions of *displacement*, *velocity*, *acceleration*, and *force* all require that we indicate a *direction* as well as a magnitude. To provide a mathematical description of such quantities, we now introduce the concept of a *vector*.

A *vector* may be defined simply as a directed line segment. If A and B are two points in the plane, then the directed line segment from A to B is a vector, which we denote by

$$\overrightarrow{AB}.$$

The vector is represented geometrically by drawing an "arrow" from A to B (Figure 13.8). We say that A is the initial point and B is the *terminal point* of the vector $\overrightarrow{AB}$.

It might be argued that the notion of a "directed line segment" is based only on geometric intuition and hence is somewhat vague from a strictly "analytic" point of view. This objection could be overcome simply by defining a vector as an *ordered pair of points* (A,B) (in which, of course, A is the "first" point and B is the "second" point). However, our intuition is certainly improved

FIGURE 13.8

by visualizing the vector $\overrightarrow{AB}$ as an "arrow" originating from A with its "head" at B.

In this section, all vectors are assumed to lie in a rectangular coordinate plane. Hence, each vector $\overrightarrow{AB}$ has a *length* or *magnitude*, denoted by $|\overrightarrow{AB}|$, and defined to be the distance between A and B,

$$|\overrightarrow{AB}| = d(A,B).$$

If A and B have coordinates (a_1,a_2) and (b_1,b_2) respectively, then the numbers $b_1 - a_1$ and $b_2 - a_2$ are called the *components* of the vector $\overrightarrow{AB}$; $b_1 - a_1$ is the x *component* and $b_2 - a_2$ is the y *component*.

We now raise the question: Under what conditions shall two vectors be regarded as "equal"? Any definition of equality of vectors should, from physical considerations, have the property that two vectors are "equal" if they have the same magnitude and the same direction. This result is accomplished by a very simple definition: Two vectors are *equal* if and only if they have the same x component and the same y component. More precisely, if the points A, B, C, and D have respective coordinates (a_1,a_2), (b_1,b_2), (c_1,c_2), and (d_1,d_2), then

$$\overrightarrow{AB} = \overrightarrow{CD} \text{ if and only if } b_1 - a_1 = d_1 - c_1 \text{ and } b_2 - a_2 = d_2 - c_2.$$

Thus in Figure 13.9, the vectors $\overrightarrow{AB}$ and $\overrightarrow{CD}$ are equal if and only if $\overline{AE} = \overline{CF}$ and $\overline{EB} = \overline{FD}$.

From the definition of equality for vectors, it follows that a vector is unchanged if it is moved parallel to itself. For many considerations, the location of the initial point of a vector is immaterial.

The notation

$$\langle h_1,h_2 \rangle$$

will be used for the components of a vector, where h_1 is the x component and h_2 is the y component. Furthermore, we actually identify a vector with its

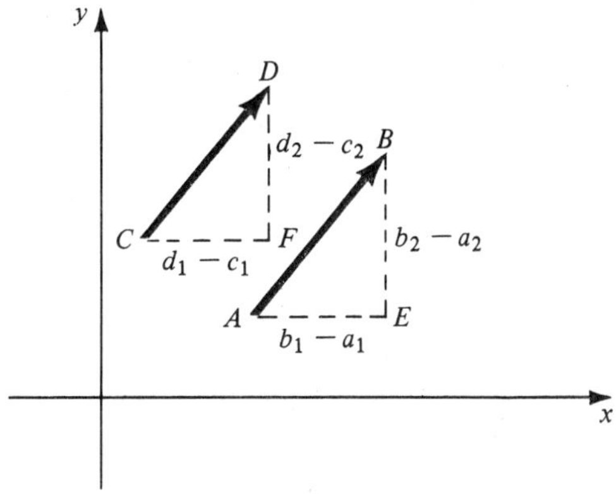

FIGURE 13.9

components. Thus, if $\overrightarrow{AB}$ has components h_1 and h_2, we shall write

$$\overrightarrow{AB} = \langle h_1,h_2\rangle.$$

In terms of components, the length of $\overrightarrow{AB}$ is

$$|\overrightarrow{AB}| = d(A,B) = \sqrt{h_1^2 + h_2^2}.$$

We shall often use a single letter to designate a vector. When we do, the letter will be printed in boldface, as for example **u**, **v**, and **w**. (Since it is difficult to write boldface letters, the reader may wish to use an arrow above a letter to indicate it denotes a vector.) The length of a vector **u** will be denoted by $|\mathbf{u}|$.

Operations of addition and subtraction can be defined for vectors in a natural way. Thus, we define

$$\langle h_1,h_2\rangle + \langle k_1,k_2\rangle = \langle h_1 + k_1, h_2 + k_2\rangle,$$
$$\langle h_1,h_2\rangle - \langle k_1,k_2\rangle = \langle h_1 - k_1, h_2 - k_2\rangle.$$

Geometrically, vector addition is the familiar triangular addition shown in Figure 13.10. Thus, if $\mathbf{u} = \langle h_1,h_2\rangle$, $\mathbf{v} = \langle k_1,k_2\rangle$, and $A = (a_1,a_2)$, then $\mathbf{u} = \overrightarrow{AB}$ where $B = (a_1 + h_1, a_2 + h_2)$ and $\mathbf{v} = \overrightarrow{BC}$ where $C = ((a_1 + h_1) + k_1, (a_2 + h_2) + k_2)$. On the other hand, $\mathbf{u} + \mathbf{v} = \langle h_1 + k_1, h_2 + k_2\rangle$ and hence $\mathbf{u} + \mathbf{v} = \overrightarrow{AC}$.

The usual rules for addition of numbers hold for the addition of vectors. Thus, for all vectors **u**, **v**, **w**,

$$\mathbf{u} + \mathbf{v} = \mathbf{v} + \mathbf{u}, \qquad \textit{commutative law,}$$

$$\mathbf{u} + (\mathbf{v} + \mathbf{w}) = (\mathbf{u} + \mathbf{v}) + \mathbf{w}, \qquad \textit{associative law.}$$

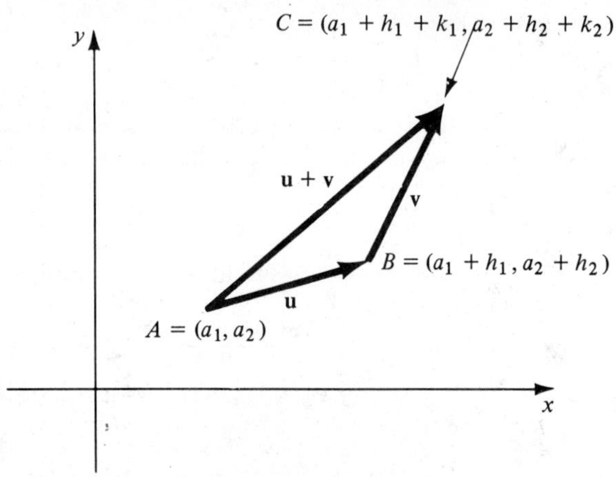

FIGURE 13.10

The vector $\langle 0,0 \rangle$ is called the *zero vector* and is designated by $\mathbf{0}$,

$$\mathbf{0} = \langle 0,0 \rangle.$$

Geometrically, $\mathbf{0} = \overrightarrow{AA}$ for any point A. Clearly

$$\mathbf{v} + \mathbf{0} = \mathbf{v}$$

for every vector $\mathbf{v}$. Each vector $\mathbf{v} = \langle k_1,k_2 \rangle$ has a *negative* $-\mathbf{v} = \langle -k_1,-k_2 \rangle$ such that

$$\mathbf{v} + (-\mathbf{v}) = \mathbf{0}.$$

We see that

$$\mathbf{u} + (-\mathbf{v}) = \mathbf{u} - \mathbf{v}$$

for all vectors $\mathbf{u}$ and $\mathbf{v}$.

The rules stated above for addition of vectors may be proved from corresponding rules for addition of numbers. For example, if $\mathbf{u} = \langle h_1,h_2 \rangle$ and $\mathbf{v} = \langle k_1,k_2 \rangle$, then

$$\mathbf{u} + \mathbf{v} = \langle h_1 + k_1, h_2 + k_2 \rangle = \langle k_1 + h_1, k_2 + h_2 \rangle = \mathbf{v} + \mathbf{u}$$

since $h_1 + k_1 = k_1 + h_1$ and $h_2 + k_2 = k_2 + h_2$ by the commutative law for the real number system. This proves the commutative law for vector addition.

Another useful operation with vectors is that of *scalar multiplication.* For each vector $\mathbf{v} = \langle h,k \rangle$ and each number c (called a *scalar*), the scalar multiple of $\mathbf{v}$ by c is the vector $c\mathbf{v}$ defined as follows:

$$c\langle h,k \rangle = \langle ch,ck \rangle.$$

For example,

$$3\langle 2,-1\rangle = \langle 6,-3\rangle, \qquad (-2)\langle -5,4\rangle = \langle 10,-8\rangle.$$

Geometrically, the vector $c\mathbf{v}$ has the same direction as $\mathbf{v}$ if $c > 0$, and the opposite direction if $c < 0$. The length of $c\mathbf{v}$ is simply the absolute value of c times the length of $\mathbf{v}$,

$$|c\mathbf{v}| = |c||\mathbf{v}|.$$

The following rules hold for scalar multiplication. For all vectors $\mathbf{u}$ and $\mathbf{v}$ and scalars c and d,

$$c(\mathbf{u} + \mathbf{v}) = c\mathbf{u} + c\mathbf{v}, \qquad (c + d)\mathbf{v} = c\mathbf{v} + d\mathbf{v},$$
$$(cd)\mathbf{v} = c(d\mathbf{v}), \qquad (-c)\mathbf{v} = -(c\mathbf{v}),$$
$$1\mathbf{v} = \mathbf{v}, \qquad 0\mathbf{v} = \mathbf{0}, \qquad c\mathbf{0} = \mathbf{0}.$$

Proof of $c(\mathbf{u} + \mathbf{v}) = c\mathbf{u} + c\mathbf{v}$: We have

$$\begin{aligned} c(\langle h_1,h_2\rangle + \langle k_1,k_2\rangle) &= c\langle h_1 + k_1, h_2 + k_2\rangle \\ &= \langle c(h_1 + k_1), c(h_2 + k_2)\rangle \\ &= \langle ch_1 + ck_1, ch_2 + ck_2\rangle \\ &= \langle ch_1,ch_2\rangle + \langle ck_1,ck_2\rangle \\ &= c\langle h_1,h_2\rangle + c\langle k_1,k_2\rangle. \end{aligned}$$

The proofs of the other laws are left to the reader.

The *inner product* of two vectors $\mathbf{u}$ and $\mathbf{v}$ is a number denoted by $\mathbf{u}\cdot\mathbf{v}$ and defined as follows. If $\mathbf{u} = \langle h_1,h_2\rangle$ and $\mathbf{v} = \langle k_1,k_2\rangle$, then

$$\mathbf{u}\cdot\mathbf{v} = \langle h_1,h_2\rangle\cdot\langle k_1,k_2\rangle = h_1k_1 + h_2k_2.$$

We emphasize that the inner product of two vectors is a number and not a vector.

For example,

$$\langle 2,\ \ 2\rangle\cdot\langle 4,3\rangle - 2\cdot 4 + (-2)\cdot 3 = 2,$$
$$\langle 3,-7\rangle\cdot\langle 7,3\rangle = 3\cdot 7 + (-7)\cdot 3 = 0.$$

The geometric meaning of the inner product is as follows. If $\mathbf{u}$ or $\mathbf{v}$ is $\mathbf{0}$, then $\mathbf{u}\cdot\mathbf{v} = 0$. Otherwise, if $\mathbf{u} = \langle h_1,h_2\rangle \neq \mathbf{0}$ and $\mathbf{v} - \langle k_1k_2\rangle \neq \mathbf{0}$, then take $\mathbf{u}$ and $\mathbf{v}$ to have the same initial point and let θ be the least angle between $\mathbf{u}$ and $\mathbf{v}$ (Figure 13.11). By the law of cosines (page 752),

$$|BC|^2 = |\mathbf{u}|^2 + |\mathbf{v}|^2 - 2|\mathbf{u}||\mathbf{v}|\cos\theta.$$

Hence

$$(k_1 - h_1)^2 + (k_2 - h_2)^2 = (h_1^2 + h_2^2) + (k_1^2 + k_2^2) - 2|\mathbf{u}||\mathbf{v}|\cos\theta.$$

From this equation and the definition of $\mathbf{u}\cdot\mathbf{v} = h_1k_1 + h_2k_2$, we easily obtain

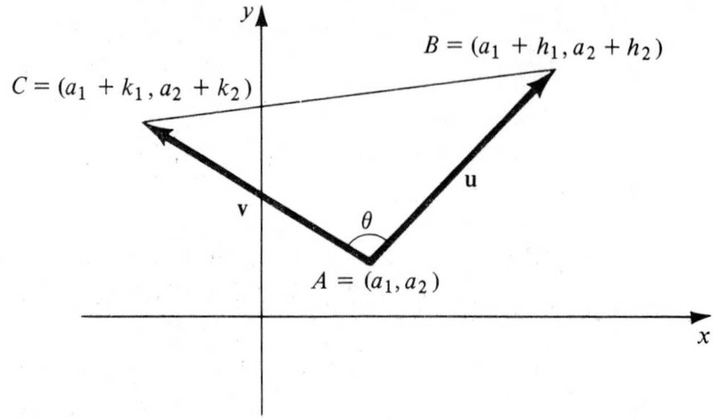

FIGURE 13.11

13.4 $$\mathbf{u} \cdot \mathbf{v} = |\mathbf{u}||\mathbf{v}| \cos \theta.$$

We can use 13.4 to find the angle between two nonzero vectors. For example, if $\mathbf{u} = \langle 1,-1 \rangle$ and $\mathbf{v} = \langle 7,1 \rangle$, then by 13.4

$$1 \cdot 7 + (-1) \cdot 1 = \sqrt{1^2 + (-1)^2}\sqrt{7^2 + 1^2} \cos \theta,$$

where θ is the least angle between $\mathbf{u}$ and $\mathbf{v}$. On solving this equation for $\cos \theta$, we obtain

$$\cos \theta = .6 \qquad \text{and hence } \theta \doteq .53^\circ.$$

Since $0 \le \theta \le \pi$, $\cos \theta = 0$ if and only if $\theta = \pi/2$. This proves the following theorem.

13.5 **Theorem**

The nonzero vectors $\mathbf{u}$ *and* $\mathbf{v}$ *are perpendicular if and only if* $\mathbf{u} \cdot \mathbf{v} = 0$.

Some of the useful properties of the inner product are listed below. For all vectors $\mathbf{u}$, $\mathbf{v}$, and $\mathbf{w}$ and each scalar c,

$$\mathbf{u} \cdot \mathbf{v} = \mathbf{v} \cdot \mathbf{u}, \qquad (\mathbf{u} + \mathbf{v}) \cdot \mathbf{w} = \mathbf{u} \cdot \mathbf{w} + \mathbf{v} \cdot \mathbf{w},$$
$$(c\mathbf{u}) \cdot \mathbf{v} = \mathbf{u} \cdot (c\mathbf{v}) = c(\mathbf{u} \cdot \mathbf{v}),$$
$$\mathbf{0} \cdot \mathbf{v} = 0, \qquad \mathbf{v} \cdot \mathbf{v} > 0 \quad \text{if } \mathbf{v} \ne \mathbf{0}.$$

Proof of $(\mathbf{u} + \mathbf{v}) \cdot \mathbf{w} = \mathbf{u} \cdot \mathbf{w} + \mathbf{v} \cdot \mathbf{w}$: If $\mathbf{u} = \langle h_1 h_2 \rangle$, $\mathbf{v} = \langle k_1 k_2 \rangle$, $\mathbf{w} = \langle l_1, l_2 \rangle$, then

$$\begin{aligned}(\mathbf{u} + \mathbf{v}) \cdot \mathbf{w} &= \langle h_1 + k_1, h_2 + k_2 \rangle \cdot \langle l_1, l_2 \rangle \\ &= (h_1 + k_1)l_1 + (h_2 + k_2)l_2 \\ &= (h_1 l_1 + h_2 l_2) + (k_1 l_1 + k_2 l_2) \\ &= \mathbf{u} \cdot \mathbf{w} + \mathbf{v} \cdot \mathbf{w}.\end{aligned}$$

The proofs of the other properties of the inner product are similar and hence are omitted.

Some other useful properties of the operations with vectors are:

$$|\mathbf{u} \cdot \mathbf{v}| \le |\mathbf{u}||\mathbf{v}|, \qquad \textit{Cauchy's inequality,}$$
$$|\mathbf{u} + \mathbf{v}| \le |\mathbf{u}| + |\mathbf{v}|, \qquad \textit{triangle inequality.}$$

These follow geometrically from 13.4 ($|\cos \theta| \le 1$) and the fact that the length of a side of a triangle is less than the sum of the lengths of the other two sides. Purely algebraic proofs are as follows.

Proof of Cauchy's inequality: If $\mathbf{u} = \mathbf{0}$, then $|\mathbf{u} \cdot \mathbf{v}| = |\mathbf{u}||\mathbf{v}| = 0$ and the inequality holds. If $\mathbf{u} \ne \mathbf{0}$, define the vector $\mathbf{w}$ as

$$\mathbf{w} = a\mathbf{v} - b\mathbf{u},$$

where $a = \mathbf{u} \cdot \mathbf{u}$ and $b = \mathbf{u} \cdot \mathbf{v}$. Then

$$\begin{aligned}\mathbf{w} \cdot \mathbf{w} &= (a\mathbf{v}) \cdot (a\mathbf{v}) + (a\mathbf{v}) \cdot (-b\mathbf{u}) + (-b\mathbf{u}) \cdot (a\mathbf{v}) + (-b\mathbf{u}) \cdot (-b\mathbf{u}) \\ &= a^2(\mathbf{v} \cdot \mathbf{v}) - 2ab(\mathbf{u} \cdot \mathbf{v}) + b^2(\mathbf{u} \cdot \mathbf{u}) = a^2(\mathbf{v} \cdot \mathbf{v}) - ab^2 \\ &= a[a(\mathbf{v} \cdot \mathbf{v}) - b^2].\end{aligned}$$

Since $a > 0$ and $\mathbf{w} \cdot \mathbf{w} \ge 0$, we have $a(\mathbf{v} \cdot \mathbf{v}) - b^2 \ge 0$, $a(\mathbf{v} \cdot \mathbf{v}) \ge b^2$, and $\sqrt{a}\sqrt{\mathbf{v} \cdot \mathbf{v}} \ge |b|$. Thus $|\mathbf{u}||\mathbf{v}| \ge |\mathbf{u} \cdot \mathbf{v}|$, as desired.

Proof of the triangle inequality: Using the Cauchy inequality, we have

$$\begin{aligned}|\mathbf{u} + \mathbf{v}|^2 = (\mathbf{u} + \mathbf{v}) \cdot (\mathbf{u} + \mathbf{v}) &= \mathbf{u} \cdot \mathbf{u} + 2\mathbf{u} \cdot \mathbf{v} + \mathbf{v} \cdot \mathbf{v} \\ &\le |\mathbf{u}|^2 + 2|\mathbf{u}||\mathbf{v}| + |\mathbf{v}|^2 = (|\mathbf{u}| + |\mathbf{v}|)^2.\end{aligned}$$

Hence $|\mathbf{u} + \mathbf{v}| \le |\mathbf{u}| + |\mathbf{v}|$, as desired.

A vector of length 1 is called a *unit vector.* Special unit vectors are $\mathbf{i}$ and $\mathbf{j}$ defined by

$$\mathbf{i} = \langle 1,0 \rangle, \qquad \mathbf{j} = \langle 0,1 \rangle.$$

Every vector $\mathbf{v} = \langle h_1,h_2 \rangle$ is a *linear combination* of $\mathbf{i}$ and $\mathbf{j}$. Thus $\mathbf{v} = \langle h_1,0 \rangle + \langle 0,h_2 \rangle = h_1\langle 1,0 \rangle + h_2\langle 0,1 \rangle$, or

$$\langle h_1,h_2 \rangle = h_1\mathbf{i} + h_2\mathbf{j}.$$

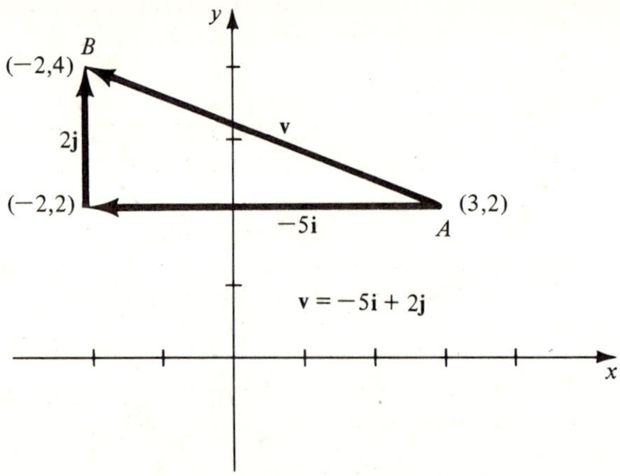

FIGURE 13.12

Geometrically, **v** is resolved into *horizontal* and *vertical components.* This is illustrated in Figure 13.12, where

$$\mathbf{v} = \overrightarrow{AB} = \langle -5,2\rangle = -5\mathbf{i} + 2\mathbf{j}.$$

Every vector **v** can be written as a scalar multiple of a unit vector **u**. For if $c = |\mathbf{v}|$, then the vector **u** defined by

$$\mathbf{u} = \frac{1}{c}\mathbf{v}$$

is a unit vector, and $\mathbf{v} = c\mathbf{u}$.

Example 1 Find a unit vector **v** that is perpendicular to the vector $\mathbf{u} = \langle -2,3\rangle$.

Solution: If the vector $\langle h_1,h_2\rangle$ is perpendicular to **u**, then by 13.5 we must have $-2h_1 + 3h_2 = 0$. An obvious solution of this equation is $h_1 = 3$, $h_2 = 2$, and so the vector $\mathbf{w} = \langle 3,2\rangle$ is perpendicular to **u**. Since $|\mathbf{w}| = \sqrt{13}$, a unit vector **v** in the direction of **w** is $\left\langle \frac{3}{\sqrt{13}}, \frac{2}{\sqrt{13}}\right\rangle$.

Example 2 Let $\mathbf{v}_1 = \langle -1,1\rangle$, $\mathbf{v}_2 = \langle 4,2\rangle$, and $\mathbf{u} = \langle 6,-5\rangle$. Express **u** as a *linear combination* of $\mathbf{v}_1$ and $\mathbf{v}_2$: that is, find scalars c_1 and c_2 such that $\mathbf{u} = c_1\mathbf{v}_1 + c_2\mathbf{v}_2$.

Solution: The scalars c_1 and c_2 must satisfy the vector equation

$$\langle 6,-5\rangle = c_1\langle -1,1\rangle + c_2\langle 4,2\rangle = \langle -c_1 + 4c_2, c_1 + 2c_2\rangle.$$

Setting corresponding components equal to each other,

$$6 = -c_1 + 4c_2,$$
$$-5 = c_1 + 2c_2.$$

Solving for c_1 and c_2, we obtain $c_1 = -16/3$, $c_2 = \frac{1}{6}$.

EXERCISES

I

1. If $\mathbf{u} = \langle 2,3 \rangle$, $\mathbf{v} = \langle 4,6 \rangle$, $\mathbf{w} = \langle -6,4 \rangle$, $\mathbf{z} = \langle 1,-1 \rangle$, determine the following:
   a. $2\mathbf{u} - \mathbf{v}$
   b. $\mathbf{u} + \mathbf{v} + 2\mathbf{z}$
   c. $\mathbf{u}^2 + \mathbf{v}^2 + \mathbf{w}^2 + \mathbf{z}^2$
   (*Note:* By $\mathbf{u}^2$, we mean $\mathbf{u} \cdot \mathbf{u}$.)
   d. $\mathbf{u} \cdot \mathbf{w} + \mathbf{v} \cdot \mathbf{w}$
   e. $(\mathbf{u} + \mathbf{v} - \mathbf{z}) \cdot (\mathbf{u} - \mathbf{v} + \mathbf{z})$
2. Determine the angle between the two vectors $\mathbf{u}$ and $\mathbf{v}$ if
   a. $\mathbf{u} = \langle 3,4 \rangle$, $\mathbf{v} = \langle -4,3 \rangle$
   b. $\mathbf{u} = \langle 1,1 \rangle$, $\mathbf{v} = \langle \sqrt{3},1 \rangle$
   c. $\mathbf{u} = \langle 1,\sqrt{3} \rangle$, $\mathbf{v} = \langle 2,2 \rangle$
3. Determine a unit vector perpendicular to $\mathbf{u}$ if
   a. $\mathbf{u} = \langle 3,4 \rangle$
   b. $\mathbf{u} = \langle 4,3 \rangle$
   c. $\mathbf{u} = \langle 2,2 \rangle$
   d. $\mathbf{u} = \langle a \cos \theta, a \sin \theta \rangle$
   e. $\mathbf{u} = \langle a,b \rangle$
4. When does $|\mathbf{u} \cdot \mathbf{v}| = |\mathbf{u}||\mathbf{v}|$?
5. Prove that $|\mathbf{u}_1| + |\mathbf{u}_2| + \cdots + |\mathbf{u}_n| \geq |\mathbf{u}_1 + \mathbf{u}_2 + \cdots + \mathbf{u}_n|$ and give a geometric interpretation.
6. If $\mathbf{u} \cdot \mathbf{v} = \mathbf{u} \cdot \mathbf{w}$ and $\mathbf{u} \neq \mathbf{0}$, must $\mathbf{v} = \mathbf{w}$?
7. If $A = (1,6)$ and $B = (4,-1)$, find a point C such that
   a. $\overrightarrow{AB} = \overrightarrow{BC}$
   b. $\overrightarrow{CA} = \overrightarrow{AB}$
8. Find the components of the vector $\mathbf{u}$ if $\mathbf{u}$ is parallel to the line $4x + 3y = 0$ and $|\mathbf{u}| = 10$.
9. If $A = (-1,2)$, $B = (2,7)$, and $C = (1,-6)$, find a point D such that $\overrightarrow{AB} = \overrightarrow{CD}$.
10. Let $\mathbf{v}_1 = \langle 4,2 \rangle$, $\mathbf{v}_2 = \langle 5,-1 \rangle$, and $\mathbf{u} = \langle 1,1 \rangle$. Can you find scalars c_1 and c_2 such that $\mathbf{u} = c_1\mathbf{v}_1 + c_2\mathbf{v}_2$?
11. Suppose that $\mathbf{v}_1$ and $\mathbf{v}_2$ are vectors in the plane such that neither one is a scalar multiple of the other. Prove that if $b_1\mathbf{v}_1 + b_2\mathbf{v}_2 = c_1\mathbf{v}_1 + c_2\mathbf{v}_2$, then $b_1 = c_1$ and $b_2 = c_2$.
12. Show that the vectors $\mathbf{u} = (\cos \theta)\mathbf{i} + (\sin \theta)\mathbf{j}$ and $\mathbf{v} = (\cos \phi)\mathbf{i} + (\sin \phi)\mathbf{j}$ are unit vectors. Derive the addition formula for $\cos (\theta - \phi)$ by expressing the angle between $\mathbf{u}$ and $\mathbf{v}$ in terms of $\theta - \phi$ and using the inner product.
13. Suppose that A, B, C, D are vertices of a parallelogram. Let $\mathbf{u} = \overrightarrow{AB} = \overrightarrow{CD}$, $\mathbf{v} = \overrightarrow{AC}$, $\overrightarrow{DE} = \frac{1}{3}\overrightarrow{DB}$, and $\overrightarrow{DF} = \frac{1}{4}\overrightarrow{DA}$. Express $\overrightarrow{CE}$ and $\overrightarrow{CF}$ in terms of $\mathbf{u}$ and $\mathbf{v}$, and show that the points C, F, and E lie on a straight line.
14. If $\mathbf{u} \cdot \mathbf{v} = \mathbf{u} \cdot \mathbf{w}$ for every vector $\mathbf{u}$, prove that $\mathbf{v} = \mathbf{w}$.
15. Given two nonzero vectors $\mathbf{u}$ and $\mathbf{v}$, the *component* of $\mathbf{u}$ on $\mathbf{v}$ is defined to be the number $|\mathbf{u}| \cos \theta$, where θ is the angle between $\mathbf{u}$ and $\mathbf{v}$. The *projection* of $\mathbf{u}$ on $\mathbf{v}$ is the vector $\mathbf{w} = c\mathbf{v}_1$, where $c = |\mathbf{u}| \cos \theta$ and $\mathbf{v}_1$ is a unit vector in the direction of $\mathbf{v}$. Show that the projection of $\mathbf{u}$ on $\mathbf{v}$ is given by the formula

$$\mathbf{w} = \left(\frac{\mathbf{u} \cdot \mathbf{v}}{|\mathbf{v}|^2}\right)\mathbf{v}.$$

Draw diagrams to illustrate the cases that occur when $c > 0$ and $c < 0$.

In Exercises 16 to 18, use the result of Exercise 15 to find the projection of the first vector on the second.

16. $5\mathbf{i} - 2\mathbf{j}$ on $\mathbf{i} + \mathbf{j}$

17. $-2\mathbf{i} + 3\mathbf{j}$ on $\mathbf{i} + 2\mathbf{j}$

18. $3\mathbf{i} + 4\mathbf{j}$ on $-2\mathbf{i} - \mathbf{j}$

19. Show that $\mathbf{u} = \frac{1}{2}\mathbf{i} + (\sqrt{3}/2)\mathbf{j}$ and $\mathbf{v} = (\sqrt{3}/2)\mathbf{i} - \frac{1}{2}\mathbf{j}$ are perpendicular unit vectors, and express the vector $\mathbf{r} = 2\mathbf{i} + 3\mathbf{j}$ as a linear combination of $\mathbf{u}$ and $\mathbf{v}$. [*Suggestion:* Find the components (Exercise 15) of $\mathbf{r}$ on $\mathbf{u}$ and on $\mathbf{v}$.]

20. Show that $\mathbf{u} = \frac{4}{5}\mathbf{i} - \frac{3}{5}\mathbf{j}$ and $\mathbf{v} = \frac{3}{5}\mathbf{i} + \frac{4}{5}\mathbf{j}$ are perpendicular unit vectors, and express $3\mathbf{i} - 2\mathbf{j}$ as a linear combination of $\mathbf{u}$ and $\mathbf{v}$.

21. If ship A is moving 10 mph to the north, and ship B is moving 20 mph to the east, how fast and in what direction does ship B appear to be moving from ship A?

22. A man wishes to swim directly across (easterly in a straight line) a river that is flowing south at a rate of $\frac{1}{2}$ mph. If the man's swimming speed in still water is 1 mph, what direction should he swim? Also, if the river is 2 miles wide, how long does it take the man to swim across?

II

1. Give an alternate proof of Cauchy's inequality by showing directly that $(a_1b_1 + a_2b_2)^2 \le (a_1^2 + a_2^2)(b_1^2 + b_2^2)$.

Prove that the following identities hold for all vectors $\mathbf{u}$ and $\mathbf{v}$.

2. $\mathbf{u}\cdot\mathbf{v} = \frac{1}{2}(|\mathbf{u} + \mathbf{v}|^2 - |\mathbf{u}|^2 - |\mathbf{v}|^2)$

3. $\mathbf{u}\cdot\mathbf{v} = \frac{1}{4}|\mathbf{u} + \mathbf{v}|^2 - \frac{1}{4}|\mathbf{u} - \mathbf{v}|^2$

4. $|\mathbf{u} - \mathbf{v}|^2 + |\mathbf{u} + \mathbf{v}|^2 = 2(|\mathbf{u}|^2 + |\mathbf{v}|^2)$

4 VECTOR METHODS IN GEOMETRICAL PROOFS

In the previous section it was pointed out that a vector is unchanged when it is moved parallel to itself. In many problems it is desirable to consider all the vectors involved as emanating from the same initial point. The most convenient point for this purpose is of course the origin O of our coordinate system. Therefore we define a *position vector* as a vector whose initial point is the origin O.

Every point $P = (a,b)$ in the plane may thus be regarded as determining a unique vector, namely the position vector $\overrightarrow{OP}$. Conversely, given any vector $\mathbf{v}$, we could consider $\mathbf{v}$ as a position vector $\overrightarrow{OP}$ and associate with $\mathbf{v}$ its terminal point P. In this way we see that there is a perfect one-to-one correspondence between the set R^2 of all points in the plane and the set of all two-dimensional vectors. As a matter of fact, for this reason the distinction between the point (a,b) and the vector $\langle a,b\rangle$ is frequently ignored.

We shall use a capital boldface letter $\mathbf{A}$ to denote the position vector determined by the point A; that is, $\mathbf{A} = \overrightarrow{OA}$. More generally, **A, B, C**, . . . , **P**, . . . will denote position vectors of points $A, B, C, \ldots, P, \ldots$.

The following theorem is frequently used in geometrical problems.

13.6 Theorem

Let A and B be two distinct points in the plane. Let P be the point on the segment AB such that

$$\frac{d(A,P)}{d(P,B)} = \frac{r}{s}.$$

Then the position vector of the point P is given by

$$\mathbf{P} = \left(\frac{s}{r+s}\right)\mathbf{A} + \left(\frac{r}{r+s}\right)\mathbf{B}.$$

Proof: Referring to Figure 13.13, observe that the vector $\overrightarrow{AB} = \mathbf{B} - \mathbf{A}$. We then have

$$\mathbf{P} = \mathbf{A} + \overrightarrow{AP} = \mathbf{A} + \left(\frac{r}{r+s}\right)\overrightarrow{AB},$$

$$\mathbf{P} = \mathbf{A} + \left(\frac{r}{r+s}\right)(\mathbf{B} - \mathbf{A}) = \left(1 - \frac{r}{r+s}\right)\mathbf{A} + \left(\frac{r}{r+s}\right)\mathbf{B},$$

from which the theorem follows.

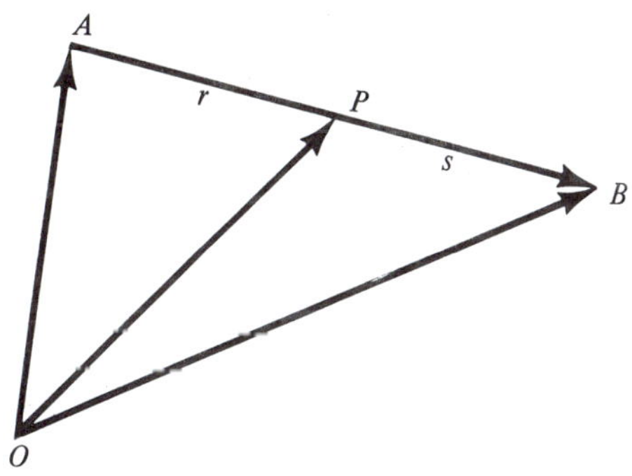

FIGURE 13.13

As special cases of Theorem 13.6, note that the position vector of the midpoint of AB is

$$\mathbf{P} = \tfrac{1}{2}\mathbf{A} + \tfrac{1}{2}\mathbf{B},$$

while the position vector of the point P two-thirds of the way from A to B is given by

$$\mathbf{P} = \tfrac{1}{3}\mathbf{A} + \tfrac{2}{3}\mathbf{B}.$$

We shall now give some examples of the use of vector methods in proving geometrical theorems.

Example 1 Prove that the line segment joining the midpoints of two sides of a triangle is parallel to, and half the length of, the third side.

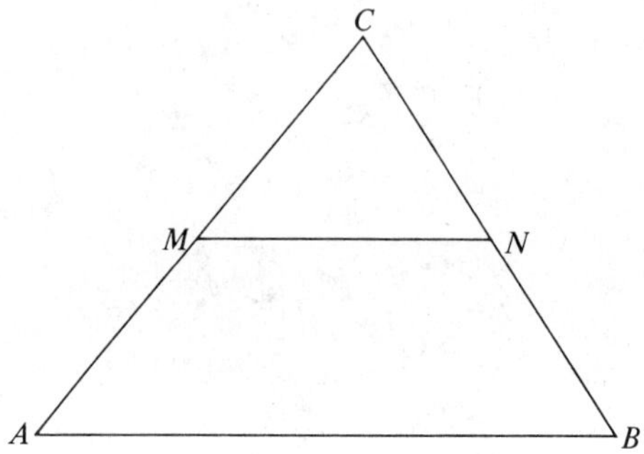

FIGURE 13.14

Solution: Let the vertices of the triangle be A, B, and C, and let M and N be the midpoints of AC and BC, respectively (Figure 13.14). By 13.6, the position vectors of M and N are

$$\mathbf{M} = \tfrac{1}{2}\mathbf{A} + \tfrac{1}{2}\mathbf{C}, \qquad \mathbf{N} = \tfrac{1}{2}\mathbf{B} + \tfrac{1}{2}\mathbf{C}.$$

The vector $\overrightarrow{MN}$ is given by

$$\overrightarrow{MN} = \mathbf{N} - \mathbf{M} = \tfrac{1}{2}[\mathbf{B} + \mathbf{C} - (\mathbf{A} + \mathbf{C})],$$
$$\overrightarrow{MN} = \tfrac{1}{2}(\mathbf{B} - \mathbf{A}) = \tfrac{1}{2}\overrightarrow{AB}.$$

Thus $\overrightarrow{MN}$ has half the length of $\overrightarrow{AB}$ and is parallel to $\overrightarrow{AB}$.

Example 2 Prove that the three medians in any triangle intersect at a single point, which is two-thirds of the distance from each vertex to the opposite side.

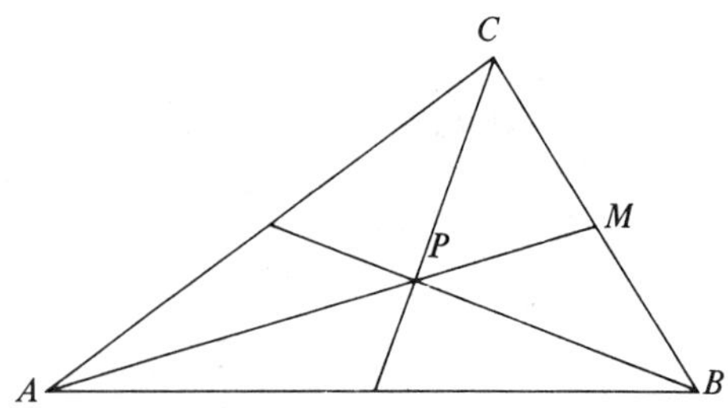

FIGURE 13.15

Solution: Recall that a median of a triangle is a line from a vertex to the midpoint of the opposite side. Let the vertices of the triangle be A, B, and C, and let M be the midpoint of the side BC (Figure 13.15). Then, by 13.6, the position vector of M is

$$\mathbf{M} = \tfrac{1}{2}\mathbf{B} + \tfrac{1}{2}\mathbf{C}.$$

Let P be the point on the median AM that is two-thirds of the way from A to M. By 13.6 the position vector of P is given by

$$\mathbf{P} = \tfrac{1}{3}\mathbf{A} + \tfrac{2}{3}\mathbf{M},$$
$$\mathbf{P} = \tfrac{1}{3}\mathbf{A} + \tfrac{2}{3}(\tfrac{1}{2}\mathbf{B} + \tfrac{1}{2}\mathbf{C}) = \tfrac{1}{3}(\mathbf{A} + \mathbf{B} + \mathbf{C}).$$

This formula for $\mathbf{P}$ is symmetric in $\mathbf{A}$, $\mathbf{B}$, and $\mathbf{C}$, and hence exactly the same result will be obtained for those points that are two-thirds of the way along the medians from B and from C. This proves that the point whose position vector is $\tfrac{1}{3}(\mathbf{A} + \mathbf{B} + \mathbf{C})$ lies on all three medians, and hence is their single point of intersection.

EXERCISES

I

1. Let $\mathbf{A}$, $\mathbf{B}$, and $\mathbf{C}$ be position vectors of points A, B, and C. If D is the point on BC one-fourth of the way from B to C, express the vector $\overrightarrow{AD}$ in terms of $\mathbf{A}$, $\mathbf{B}$, and $\mathbf{C}$.
2. If D is the point on the segment AB two-fifths of the way from A to B, and E is the midpoint of AC, express the vector DE in terms of $\mathbf{A}$, $\mathbf{B}$, and $\mathbf{C}$.
3. Let $A = (1,5)$ and $B = (2,1)$. Suppose that P lies on the line through A and B, on the opposite side of A from B, and that $|\overrightarrow{AB}| = 2|\overrightarrow{AP}|$. Find the coordinates of P.
4. Use the result of Example 2 to find the coordinates of the intersection of the medians of the triangle ABC, where $A = (-2,-1)$, $B = (3,7)$, $C = (5,2)$.
5. Let $\mathbf{u}$, $\mathbf{v}$, and $\mathbf{w}$ be the vectors from the vertices of a triangle to the midpoints of the opposite sides. Show that $\mathbf{u} + \mathbf{v} + \mathbf{w} = \mathbf{0}$.
6. In a triangle ABC, let D be the midpoint of AB, and let E be the point on BC that is two-thirds of the way from B to C. Prove that the segment AE bisects the segment CD.

Use vector methods to prove the following theorems.

7. Four points A, B, C, D form the consecutive vertices of a parallelogram if and only if the diagonals AC and BD bisect each other.
8. The midpoints of the sides of any quadrilateral form the vertices of a parallelogram.
9. The line segment joining the midpoints of the nonparallel sides of a trapezoid is parallel to the other two sides, and its length is one-half the sum of the lengths of those sides.

10. The diagonals of a rhombus are perpendicular to each other.
11. In a parallelogram, the sum of the squares of the sides is equal to the sum of the squares of the diagonals.

II

Prove the following theorems by vector methods.

1. The altitudes of a triangle intersect in a single point.
2. An angle inscribed in a semicircle is a right angle.
3. The sum of the squares of the sides of a quadrilateral equals the sum of the squares of the diagonals plus four times the square of the line segment joining the midpoints of the diagonals.

5 VECTOR-VALUED FUNCTIONS

In our discussion of curves in Sections 1 and 2, we regarded a curve as a function λ whose domain is a set of real numbers and whose range is a set of points in the plane. Thus, for each t in the domain of λ, $\lambda(t)$ is a point that can be designated as $(x(t),y(t))$, where $x(t)$ and $y(t)$ are real-valued functions of t.

It should be clear that instead of considering the *point* $\lambda(t) = (x(t),y(t))$, we might just as well consider the *position vector* of this point, namely the vector from 0 to $\lambda(t)$. This position vector will be denoted, using boldface type, by $\boldsymbol{\lambda}(t)$. The components of $\boldsymbol{\lambda}(t)$ of course are given by $\langle x(t),y(t)\rangle$. In this way, we are simply thinking of the curve λ as being "traced out" by the tip of the position vector from the origin to $\lambda(t)$. This representation of a curve will have several advantages.

A function such as that defined by $\boldsymbol{\lambda}(t) = \langle x(t),y(t)\rangle$ is an example of a *vector-valued* function; i.e., a function whose domain is a set of real numbers and whose range is a set of vectors.

The equation of a straight line may be put into a very simple form using a vector-valued function. Suppose it is required to find a representation for the line L through the point $P = (a,b)$ in the direction of a given nonzero vector $\mathbf{v}$. Let $\mathbf{P}$ denote the position vector of the point P, and let us consider $\mathbf{v}$ as having its initial point at P (Figure 13.16). Then a point Q is on L if and only if the vector $\overrightarrow{PQ}$ is a scalar multiple of $\mathbf{v}$, say $\overrightarrow{PQ} = t\mathbf{v}$. If the position vector $\overrightarrow{OQ}$ is denoted by $\boldsymbol{\lambda}(t)$, then we have

13.7
$$\boldsymbol{\lambda}(t) = \mathbf{P} + t\mathbf{v}.$$

As t runs through all real numbers, the position vector $\boldsymbol{\lambda}(t)$ traces out the line L. Thus 13.7 can be considered as a *vector equation* of L.

We may also obtain parametric equations for the line L from 13.7. If the components of $\mathbf{v}$ are $\langle h,k\rangle$, then

$$\boldsymbol{\lambda}(t) = \langle x(t),y(t)\rangle = \mathbf{P} + t\mathbf{v} = \langle a,b\rangle + t\langle h,k\rangle,$$

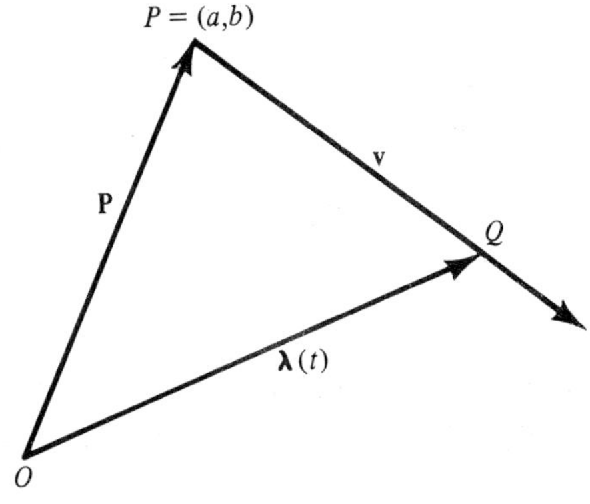

FIGURE 13.16

and hence

13.8 $$x(t) = a + th, \qquad y(t) = b + tk.$$

By eliminating the parameter t, it can be verified that the graph of the parametric equations 13.8 is indeed the line L. Thus, if $h \neq 0$, we have $t = (x - a)/h$, and it follows that

$$y - b = \frac{k}{h}(x - a).$$

We recognize the graph of this equation to be the line through (a,b) with slope k/h.

The line L described by 13.7 has a natural direction assigned to it, namely the direction of the vector $\mathbf{v}$. The points traced out on L by $\boldsymbol{\lambda}(t)$ when $t > 0$ will be called the *positive half* of L. Likewise, those points on L corresponding to $\boldsymbol{\lambda}(t)$ when $t < 0$ will be called the *negative half* of L. Suppose, for example, that $P = (2,-1)$ and $\mathbf{v} = \langle 1,3 \rangle$. Then the positive and negative halves of L, indicated by L_1 and L_2, respectively, are shown in Figure 13.17.

We consider next the definition of *limit* for a vector-valued function. Our discussion of curves in Section 2 showed that for a curve λ,

$$\operatorname*{limit}_{t \to c} \lambda(t) = (\operatorname*{limit}_{t \to c} x(t), \operatorname*{limit}_{t \to c} y(t)).$$

We shall simply take this "componentwise" limit as our *definition* in the case of vector-valued functions. This definition may be stated more precisely as follows:

13.9 Definition

Let $\mathbf{F}$ be a vector-valued function defined by

$$\mathbf{F}(t) = \langle g(t), h(t) \rangle$$

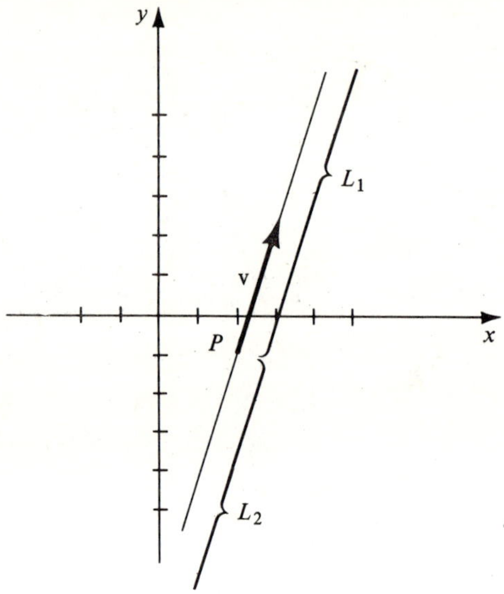

FIGURE 13.17

for all real numbers in an interval I. If c is a number in I, then

$$\lim_{t \to c} \mathbf{F}(t) = \langle \lim_{t \to c} g(t), \lim_{t \to c} h(t) \rangle,$$

provided that the limits of the component functions exist.

Now let λ be a curve traced out by the position vector defined by

$$\boldsymbol{\lambda}(t) = \langle x(t), y(t) \rangle,$$

for each t in an interval I. Making use of the above concept of limit, it is now possible to "differentiate" this vector-valued function. If c is a number in I, then we define

13.10 $$\boldsymbol{\lambda}'(c) = \lim_{h \to 0} \frac{1}{h} [\boldsymbol{\lambda}(c + h) - \boldsymbol{\lambda}(c)].$$

The reader should note that 13.10 is the exact formal analogue of the definition of derivative for a real-valued function. The vector $\boldsymbol{\lambda}'(c)$ defined by 13.10 is called the *tangent vector* to the curve λ at the point $\lambda(c)$.

The geometrical significance of the tangent vector is illustrated in Figure 13.18. Writing

$$\mathbf{s}(h) = \boldsymbol{\lambda}(c + h) - \boldsymbol{\lambda}(c),$$

we see that the initial and terminal points of the vector $\mathbf{s}(h)$ are $\lambda(c)$ and $\lambda(c + h)$, respectively. The vector $\mathbf{s}(h)$ is called the *secant vector* determined

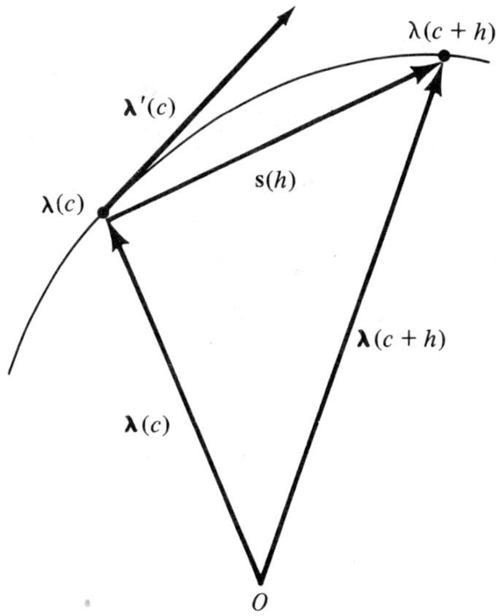

FIGURE 13.18

by these two points. As h approaches 0, the point $\lambda(c + h)$ moves along the curve toward $\lambda(c)$. The secant vector thus changes its direction as h approaches 0, and assumes the direction of $\boldsymbol{\lambda}'(c)$ as a "limiting position." It is customary to regard the tangent vector $\boldsymbol{\lambda}'(c)$ as having $\lambda(c)$ as its initial point.

The components of $\mathbf{s}(h)$ are given by

$$\mathbf{s}(h) = \langle x(c + h) - x(c), y(c + h) - y(c)\rangle,$$

so that

$$\frac{1}{h}\,\mathbf{s}(h) = \left\langle \frac{x(c + h) - x(c)}{h}, \frac{y(c + h) - y(c)}{h} \right\rangle,$$

and by 13.10,

$$\begin{aligned}\boldsymbol{\lambda}'(c) &= \operatorname*{limit}_{h\to 0} \frac{1}{h}\,\mathbf{s}(h)\\ &= \left\langle \operatorname*{limit}_{h\to 0} \frac{x(c + h) - x(c)}{h}, \operatorname*{limit}_{h\to 0} \frac{y(c + h) - y(c)}{h} \right\rangle\\ &= \langle x'(c), y'(c)\rangle.\end{aligned}$$

Hence, to compute the tangent vector we simply differentiate "componentwise." In other words, for all t in I:

13.11 $$\text{If } \boldsymbol{\lambda}(t) = \langle x(t), y(t)\rangle, \text{ then } \boldsymbol{\lambda}'(t) = \langle x'(t), y'(t)\rangle.$$

Thus there is associated with the function $\boldsymbol{\lambda}(t)$ another vector-valued function $\boldsymbol{\lambda}'(t)$.

We shall call λ a *differentiable* curve if $\boldsymbol{\lambda}'(t)$ exists for all t in the domain of λ. It is clear that if λ is differentiable, then the component functions $x(t)$ and $y(t)$ must be continuous, and hence λ is continuous. If the functions $x'(t)$ and $y'(t)$ are continuous, then we say that λ is a *smooth curve*.

Example 1 If $\boldsymbol{\lambda}(t) = \langle \cos t, \sin t \rangle$, find the function $\boldsymbol{\lambda}'(t)$ and show that each tangent vector lies along the tangent line in the previously defined sense (Figure 13.19).

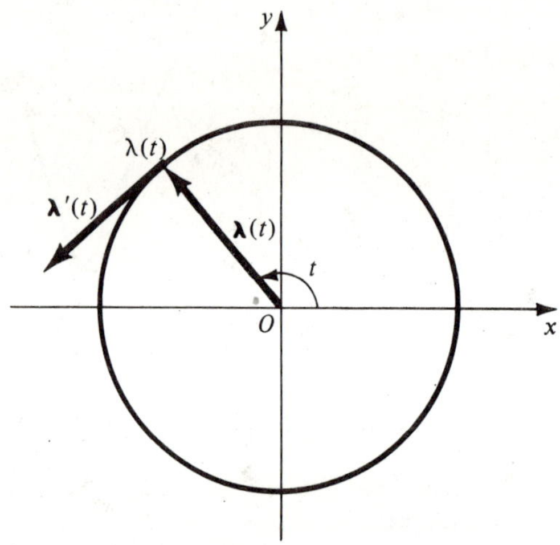

FIGURE 13.19

Solution: The trace of λ is the unit circle centered at the origin. By 13.11,

$$\boldsymbol{\lambda}'(t) = \langle -\sin t, \cos t \rangle.$$

To prove that $\boldsymbol{\lambda}'(t)$ lies along the tangent line to the circle in the usual sense at the point $\lambda(t)$, it is sufficient to show that $\boldsymbol{\lambda}'(t)$ and $\boldsymbol{\lambda}(t)$ are perpendicular. But this follows from 13.5, since

$$\boldsymbol{\lambda}'(t) \cdot \boldsymbol{\lambda}(t) = -\sin t \cos t + \cos t \sin t = 0.$$

Also note that $\boldsymbol{\lambda}'(t)$ is a unit vector, since

$$|\boldsymbol{\lambda}'(t)| = \sqrt{(-\sin t)^2 + (\cos t)^2} = 1.$$

The *tangent line* to a curve λ at the point $\lambda(c)$ will now be defined, in the obvious way, as the line through the tangent vector $\boldsymbol{\lambda}'(c)$. More precisely, let λ be a differentiable curve with domain I, and let c be a number in I with $\boldsymbol{\lambda}'(c) \neq \mathbf{0}$. Then the line L traced out by the position vector

$$\mathbf{Q}(t) = \boldsymbol{\lambda}(c) + t\boldsymbol{\lambda}'(c)$$

passes through the point $\lambda(c)$ and has the direction of the tangent vector $\lambda'(c)$, according to 13.7. The line L is defined to be the *tangent line* to λ at the point $\lambda(c)$.

If the curve λ defined by

$$\lambda(t) = \langle x(t), y(t)\rangle, \qquad \text{domain } \lambda = I,$$

has a nonzero tangent vector at some t,

$$\lambda'(t) = \langle x'(t), y'(t)\rangle \neq 0,$$

then either $x'(t) = 0$, in which case the tangent line $L(\lambda,t)$ is vertical, or $x'(t) \neq 0$, in which case $L(\lambda,t)$ has slope $m(t)$ given by

13.12 $$m(t) = \frac{y'(t)}{x'(t)} \qquad \text{if } x'(t) \neq 0.$$

It might happen that in some interval I the trace of a differentiable curve λ coincides with the graph of a differentiable function f. If $\lambda(t) = \langle x(t), y(t)\rangle$, then this means that

$$y(t) = f(x(t))$$

for every t in I. Hence, by the chain rule,

$$y'(t) = f'(x(t))x'(t)$$

and

13.13 $$f'(x(t)) = \frac{y'(t)}{x'(t)} \qquad \text{if } x'(t) \neq 0.$$

This equation is often written in the abbreviated form

$$\frac{dy}{dx} = \frac{dy/dt}{dx/dt} \qquad \text{if } \frac{dx}{dt} \neq 0.$$

Since $f'(x(t))$ is the slope of the tangent line to the graph of f at the point $(x(t), y(t))$, it is clear that the tangent line to the curve λ at t (with slope given by 13.12) coincides with the tangent line to the graph of f at the point $(x(t), y(t))$.

Example 2 Describe the tangent lines to the cycloid λ defined by

$$\lambda(t) = \langle r(t - \sin t), r(1 - \cos t)\rangle.$$

Solution: We have

$$\lambda'(t) = \langle r(1 - \cos t), r \sin t\rangle$$

for every t in R. Thus the tangent vector exists at each number t. However, we see that

$$\lambda'(t) = \mathbf{0} \qquad \text{if } t = 2n\pi,\ n \text{ an integer}.$$

That is, the tangent vectors are zero at each point of the cycloid on the x axis. Hence the slope of the tangent line to the cycloid is given by (13.12)

$$m(t) = \frac{\sin t}{1 - \cos t} \qquad \text{if } t \neq 2n\pi,\ n \text{ an integer.}$$

Although $\lambda'(2n\pi) = \mathbf{0}$, n an integer, this does not necessarily mean that the cycloid has no tangent line at the point $(2n\pi r,0)$. As usual, we say that the curve has a tangent line at $(2n\pi r,0)$ provided the limit of the slope of the line joining this point to a point $(x(t), y(t))$ as t approaches 0 exists or is infinite. Since

$$\begin{aligned}\operatorname*{limit}_{t\to 2n\pi} \frac{y(t)}{x(t) - 2n\pi r} &= \operatorname*{limit}_{t\to 2n\pi} \frac{r(1 - \cos t)}{r(t - \sin t - 2n\pi)} \\ &= \operatorname*{limit}_{t\to 2n\pi} \frac{\sin t}{(1 - \cos t)} \\ &= \operatorname*{limit}_{t\to 2n\pi} \frac{\cos t}{\sin t} = \infty,\end{aligned}$$

by l'Hospital's rule (11.1), the line

$$x = 2n\pi r$$

is tangent to the cycloid at the point $(2n\pi r,0)$.

EXERCISES

I

1. Find parametric equations for the following straight lines:
   a. Through $P = (2,3)$ in the direction of $\mathbf{v} = \langle 3,4 \rangle$
   b. Through $P = (1,2)$ in the direction of $\mathbf{v} = \langle 1,0 \rangle$
   c. Through the points $(-1,5)$ and $(3,-4)$
2. Find the vector form of the equation of the straight line passing through:
   a. Points (x_1,y_1) and (x_2,y_2)
   b. Point (x_1,y_1) with slope m

In each of Exercises 3 to 6, find the tangent vector to the curve at the specified value of the parameter.

3. $\lambda(t) = \langle t^2,t^3 \rangle$, $t = 2$
4. $\lambda(t) = \langle e^t,e^{-t} \rangle$, $t = 1$
5. $\lambda(r) = \langle \tan r, \cot r \rangle$, $r = \frac{3}{4}\pi$
6. $\lambda(m) = \langle 3 \cos m, 2 \sin m \rangle$, $m = \frac{1}{3}\pi$

In each of Exercises 7 to 10, sketch the trace of the curve and find on it each point at which the tangent vector is either horizontal or vertical.

7. $\lambda(t) = \langle \cos 2t, \cos t \rangle$
8. $\lambda(s) = \langle 2s - \sin s, 2 - \cos s \rangle$
9. $\lambda(r) = \langle a \cos^3 r, a \sin^3 r \rangle$
10. $\lambda(t) = \langle 2 - 3 \cos t, -1 + 2 \sin t \rangle$

In Exercises 11 to 14, consider $\lambda'(t)$ as the *position vector* of a curve, and sketch the trace of the curve it defines for each of the given functions $\lambda(t)$.

11. $\lambda(t) = \langle 3t, t^3 \rangle$

12. $\lambda(t) = \langle e^t, te^{-t} \rangle$

13. $\lambda(r) = \langle a(r - \sin r), a(1 - \cos r) \rangle$

14. $\lambda(s) = \langle a \cos^3 s, b \sin^3 s \rangle$

15. If the curve with parametric equations

$$x = x(t), \qquad y = y(t)$$

also has equation $y = f(x)$, then show that

$$\frac{d^2y}{dx^2} = \frac{(d/dt)(dy/dx)}{dx/dt} = \frac{x'(t)y''(t) - y'(t)x''(t)}{(x'(t))^3}.$$

[*Hint:* The derivative f' is given parametrically by $x = x(t)$, $\dfrac{dy}{dx} = \dfrac{y'(t)}{x'(t)}$.]

In each of Exercises 16 to 20 assume that the curve given parametrically is also the graph of an equation $y = f(x)$. Then use Exercise 15 to find dy/dx and d^2y/dx^2 for each curve.

16. $x = 3t$, $y = 2t^3$

17. $x = e^t$, $y = te^{-t}$

18. $x = r(\theta - \sin \theta)$, $y = r(1 - \cos \theta)$

19. $x = a \cos^3 \theta$, $y = a \sin^3 \theta$

20. $x = \ln t$, $y = t^3$

II

1. Prove that if $x'(a)$ and $y'(a)$ are not both zero then $x = x(a) + x'(a)t$, $y = y(a) + y'(a)t$ are parametric equations of the tangent line to the curve $\lambda(t) = (x(t), y(t))$ at the point $t = a$.
2. Show that the trace of the curve $\lambda(t) = \langle a \cos^4 t, a \sin^4 t \rangle$, $a > 0$, is the same as the graph of the equation $\sqrt{x} + \sqrt{y} = \sqrt{a}$. Sketch this graph.
3. Show that for the curve in Exercise 2 the sum of the x and y intercepts of every tangent line is the constant a.
4. Show that the length of the segment of each tangent line of the curve $\lambda(t) = \langle a \cos^3 t, a \sin^3 t \rangle$ cut off by the coordinate axes is constant.
5. Show that the tangent and normal lines to the cycloid in Example 4, Section 1, pass through the highest and lowest points of the rolling circle at each position.
6. Referring to Exercise I-15, find a formula for d^3y/dx^3 in terms of derivatives with respect to t of the functions $x(t)$ and $y(t)$.

6 PLANE MOTION

One possible interpretation of a curve λ is as the *motion of a particle* in a coordinate plane. The parameter t is taken to be in time units, so that $\lambda(t)$ is the position vector of the moving particle at time t.

The tangent vector $\lambda'(t)$ is called the *velocity* of the particle at time t. Its length $|\lambda'(t)|$ is commonly called the *speed* of the particle at time t. Thus the velocity is a vector and the speed is a scalar. If

$$\lambda(t) = \langle x(t), y(t) \rangle,$$

then $\quad \lambda'(t) = \langle x'(t), y'(t) \rangle \quad$ and $\quad |\lambda'(t)| = \sqrt{x'^2(t) + y'^2(t)}.$

By 13.11, the derivative of the vector-valued function λ' is given by

$$\lambda''(t) = \langle x''(t), y''(t) \rangle.$$

The vector $\lambda''(t)$ is called the *acceleration* of the particle at time t.

Example 1 Describe the motion of a particle for which

$$\lambda(t) = \langle 2 - t, t^2 - 1 \rangle.$$

Sketch the velocity and acceleration vectors at $t = 0$ and $t = 1$.

Solution: If we eliminate t from the parametric equations

$$x = 2 - t, \qquad y = t^2 - 1,$$

we see that the path of the particle is the parabola

$$y + 1 = (x - 2)^2$$

sketched in Figure 13.20. The velocity and acceleration vectors are given by

$$\lambda'(t) = \langle -1, 2t \rangle, \qquad \lambda''(t) = \langle 0,2 \rangle.$$

Hence $\quad \lambda'(0) = \langle -1,0 \rangle, \qquad \lambda''(0) = \langle 0,2 \rangle$

and $\quad \lambda'(1) = \langle -1,2 \rangle, \qquad \lambda''(1) = \langle 0,2 \rangle$

are the velocity and acceleration vectors at $t = 0$ and $t = 1$, respectively. These are also sketched in Figure 13.20. We note that the acceleration vector is constant. As t increases, the particle moves from right to left along the parabola.

If λ is the motion of a particle of constant mass m, then the vector

$$\mathbf{F}(t) = m\lambda''(t)$$

is called the *force* acting on the particle at time t.

Example 2 Show that if a particle of mass m is moving in a circular path at a constant angular speed of ω revolutions per second, then the (centripetal) force acting on the particle is directed towards the center of the circle.

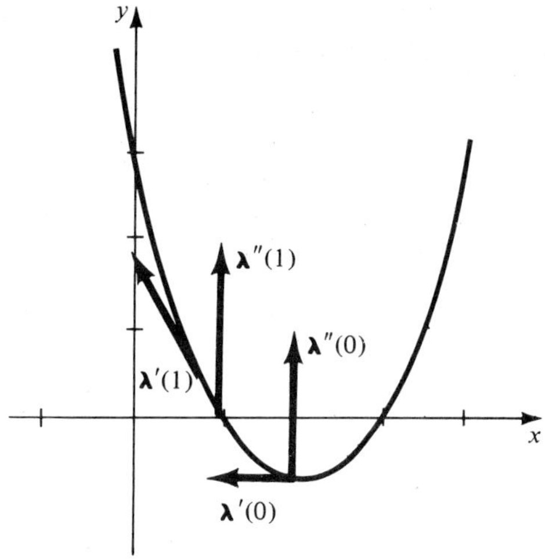

FIGURE 13.20

Solution: The particle sweeps out an angle of $\alpha = 2\pi\omega t$ radians in t seconds (Figure 13.21). Hence the motion function λ is given by

$$\lambda(t) = \langle r\cos 2\pi\omega t, r\sin 2\pi\omega t\rangle,$$

assuming that $\lambda(0) = \langle r,0\rangle$. We easily compute

$$\lambda'(t) = \langle -2\pi\omega r\sin 2\pi\omega t, 2\pi\omega r\cos 2\pi\omega t\rangle,$$
$$\lambda''(t) = -k\langle r\cos 2\pi\omega t, r\sin 2\pi\omega t\rangle,$$

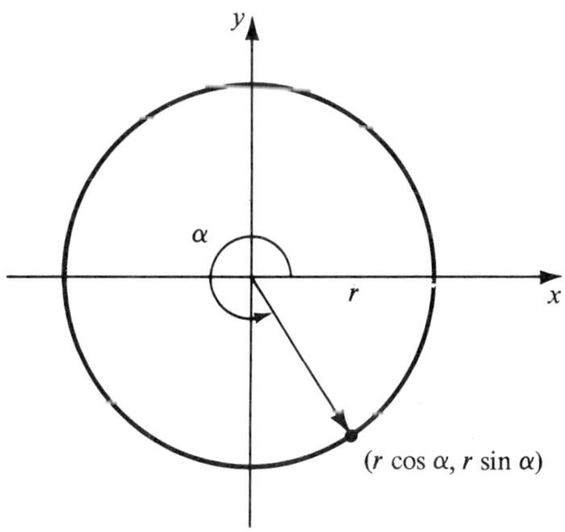

FIGURE 13.21

where $k = 4\pi^2\omega^2$. Since $\boldsymbol{\lambda}''(t) = -k\boldsymbol{\lambda}(t)$ and $k > 0$, evidently the force vector

$$\mathbf{F}(t) = m\boldsymbol{\lambda}''(t)$$

has the opposite direction to $\boldsymbol{\lambda}(t)$. Therefore $\mathbf{F}(t)$ acts on the particle in the direction of the center O of the circle.

Example 3 A projectile is fired in the air at an angle α with the ground, and with an initial speed of v_0 ft/sec. Find its position $\boldsymbol{\lambda}(t)$ at any time t. Neglect air resistance, and assume a constant gravitational deceleration of 32 ft/sec^2.

Solution: Set up coordinate axes with the origin at the point on the ground at which the projectile is fired, and let this point correspond to time $t = 0$ (Figure 13.22).

From the given conditions, the initial velocity vector $\boldsymbol{\lambda}'(0)$ may be written as

$$\boldsymbol{\lambda}'(0) = v_0\langle\cos\alpha, \sin\alpha\rangle.$$

If m is the mass of the projectile, then the force $\mathbf{F}(t)$ acting on the projectile at time t is

$$\mathbf{F}(t) = m\boldsymbol{\lambda}''(t) = m\langle 0, -32\rangle.$$

Integrating this equation, and making use of the initial conditions, we obtain

$$\boldsymbol{\lambda}'(t) = \langle v_0\cos\alpha, v_0\sin\alpha - 32t\rangle.$$

Integrating again,

$$\boldsymbol{\lambda}(t) = \langle v_0 t\cos\alpha, v_0 t\sin\alpha - 16t^2\rangle.$$

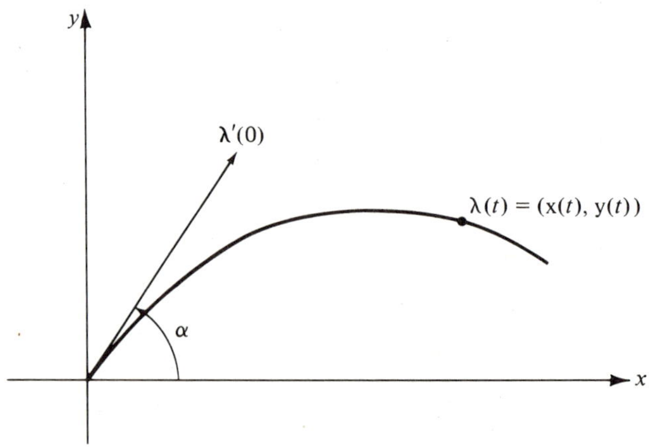

FIGURE 13.22

EXERCISES

In each of Exercises 1 to 11, $\lambda(t)$ is the position vector of a moving particle at time t. Sketch the trace of the particle and find its velocity and acceleration. Sketch the velocity and acceleration of the particle and find its speed at the given time t.

1. $\lambda(t) = \langle t^2, 2t \rangle$, $t = 2$
2. $\lambda(t) = \langle 2t, t^4 \rangle$, $t = 2$
3. $\lambda(t) = \langle 2 \sin t, 2 \cos t \rangle$, $t = 4\pi$
4. $\lambda(t) = \langle \tan t, \cot t \rangle$, $t = \frac{3}{4}\pi$
5. $\lambda(t) = \langle \cos t, \cos 2t \rangle$, $t = \frac{1}{2}\pi$
6. $\lambda(t) = \langle 2t, e^{-t} \rangle$, $t = 0$
7. $\lambda(t) = \langle e^t, e^{-t} \rangle$, $t = 0$
8. $\lambda(t) = \langle \sin t, \cos^2 t \rangle$, $t = \frac{1}{2}\pi$
9. $\lambda(t) = \langle 2 \sin t, 2(1 - \cos t) \rangle$, $t = \frac{1}{6}\pi$
10. $\lambda(t) = \langle \sqrt{t}, \sqrt{2 - t} \rangle$, $t = 1$
11. $\lambda(t) = \langle a \cos bt, a \sin bt \rangle$, $a > 0$, $b > 0$, at time t.
12. Describe the motion of the particle during the time interval $[-2,2]$ if $\lambda(t) = \langle t^2/2, t^4/4 \rangle$. Find $\lambda'(t)$ and $\lambda''(t)$ when $t = -2, -1, 0, 1, 2$.
13. What is the significance of the fact that $\langle 2 \sin t, 2 \cos t \rangle \cdot \lambda'(t) = 0$ in Exercise 3?
14. Show that if $\lambda(t) = \langle e^t \cos t, e^t \sin t \rangle$ then the acceleration vector $\lambda''(t)$ is perpendicular to the vector $\langle e^t \cos t, e^t \sin t \rangle$ at each time t.

Exercises 15–17 refer to the motion of the projectile described in Example 3.

15. **a.** What is the range of the projectile (maximum value of x)?
    b. What is the maximum height reached?
    c. What is the total time of flight?
    d. Find an equation in x and y for its trajectory.
    e. Determine the angle α for which the range is a maximum, and find this range.
    f. If two projectiles are fired with the same initial speed at complementary angles with the horizontal, show that their ranges are the same.
16. When half the time required to reach its maximum height has elapsed, what fraction of its maximum height has the projectile attained?
17. A shell can be fired from a certain gun with an initial speed of 1000 ft/sec. Find two values of the angle of elevation α if a target on the ground 20,000 feet away is to be hit.
18. A ball is thrown in the air from a point O on the side of a hill with an initial speed of v_0 ft/sec and at an angle of 45° with the horizontal. If β is the angle of elevation of the hillside, determine the point at which the ball hits the ground (see figure below).

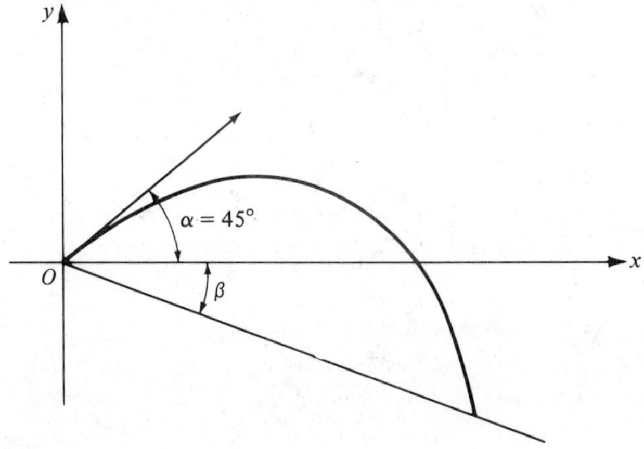

19. If **u** and **v** are vector-valued functions of a real variable t, prove that $D_t(\mathbf{u} \cdot \mathbf{v}) = \mathbf{u} \cdot D_t\mathbf{v} + D_t\mathbf{u} \cdot \mathbf{v}$.

20. A particle moves in the plane so that its speed is constant. Prove that its acceleration vector is always perpendicular to its velocity vector. (*Hint:* The velocity vector **v** satisfies $\mathbf{v} \cdot \mathbf{v} = constant$. Apply Exercise 19).

7 POLAR COORDINATE SYSTEMS

Another convenient way of introducing coordinates in a plane is by the polar coordinate system described below.

We start off with a fixed point O, called the *pole*, and the positive half of a coordinate line, called the *polar axis*, emanating from O. The presence of the polar axis allows us to assign a length to each line segment in the plane.

Each point P in the plane may now be assigned coordinates (r,θ), where r is the length of the segment OP and θ is the measure of an angle with initial side along the polar axis and terminal side along OP (Figure 13.23). We shall also allow the coordinates $(-r,\theta)$ for the point P if $|OP| = r$ and θ is the measure of an angle with initial side along the polar axis and terminal side along the extension of OP through the pole (Figure 13.24). The pole O is assigned

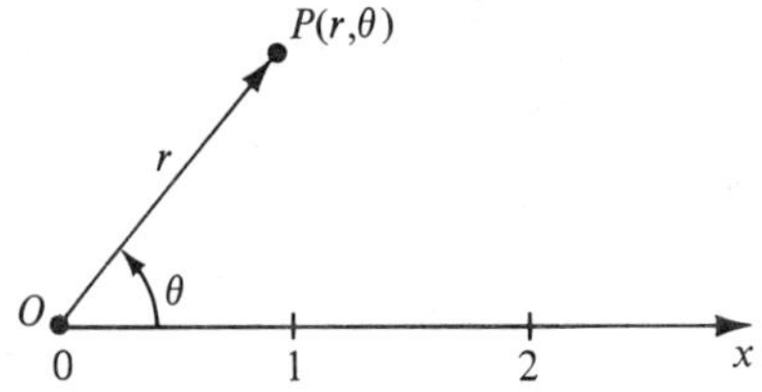

FIGURE 13.23

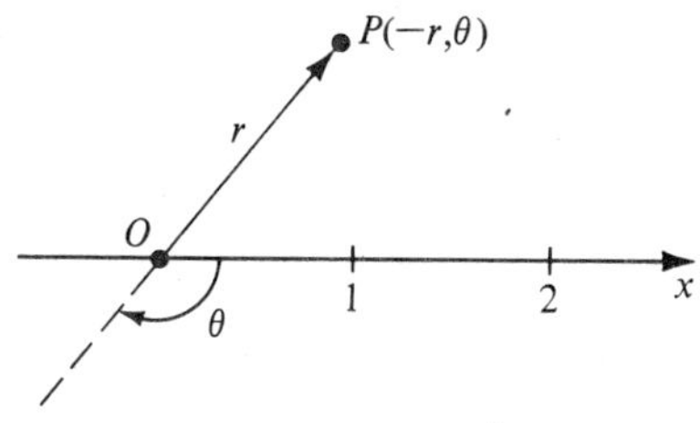

FIGURE 13.24

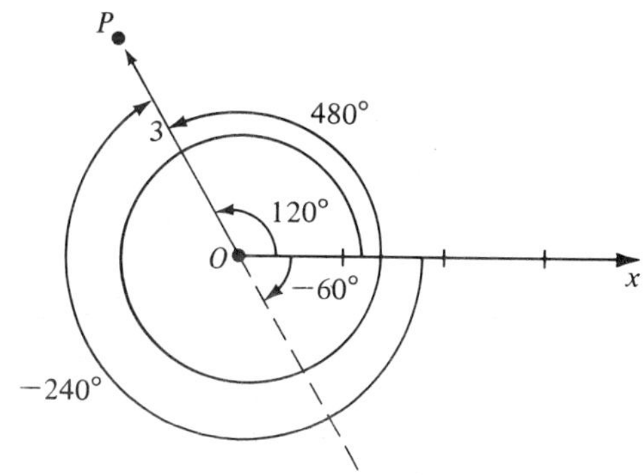

FIGURE 13.25

coordinates $(0,\theta)$, where θ is any real number.

There certainly is nothing unique about the polar coordinates of a point. The point P of Figure 13.25, for example, has as possible polar coordinates

$$(3, 120^\circ),\ (-3, -60^\circ),\ (3, 480^\circ),\ (3, -240^\circ),\ (-3, 300^\circ),$$

and so on. In general, if P has polar coordinates (r,θ), then P also has polar coordinates $(r, \theta \pm 2n\pi)$ and $(-r, \theta \pm (2n - 1)\pi)$ for every integer n. However, if P is not the pole, P does have a unique set of coordinates (r,θ) where $r > 0$ and $0^\circ \leq \theta < 360^\circ$. The only coordinates of P in Figure 13.25 satisfying these restrictions are $(3, 120^\circ)$.

Every pair of numbers (r,θ) determines a unique point P such that $|OP| = |r|$ and θ is the (radian) measure of an angle having initial side along the polar axis and terminal side along OP if $r > 0$ and along OP extended through the origin if $r < 0$. This association of pairs of numbers with points is called a *polar coordinate system* in the plane.

Just as an equation in x and y has a graph in a rectangular coordinate plane, so does an equation in r and θ have a graph in a polar coordinate plane.

Thus the graph of an equation in r and θ consists of those and only those points P having some pair of coordinates satisfying the given equation.

The graph of the equation

$$r = c,$$

c a constant, is a circle of radius $|c|$ having its center at the pole, since each point P on this circle has a pair of coordinates of the form (c,θ) for some θ, whereas each point $P(r,\theta)$ off this circle has $|r| \neq |c|$. The graph of $r = -c$ is the same circle.

The graph of the equation

$$\theta = c,$$

c a constant, is a straight line passing through the pole and making an angle of measure c with the polar axis. The graph of $\theta = c \pm n\pi$, n any integer, is the same line.

It is natural to construct coordinate paper for a polar coordinate system, as indicated in Figure 13.26, with each circle having a constant value of r and each line a constant value of θ. Some points are plotted in the figure to indicate how the paper is used.

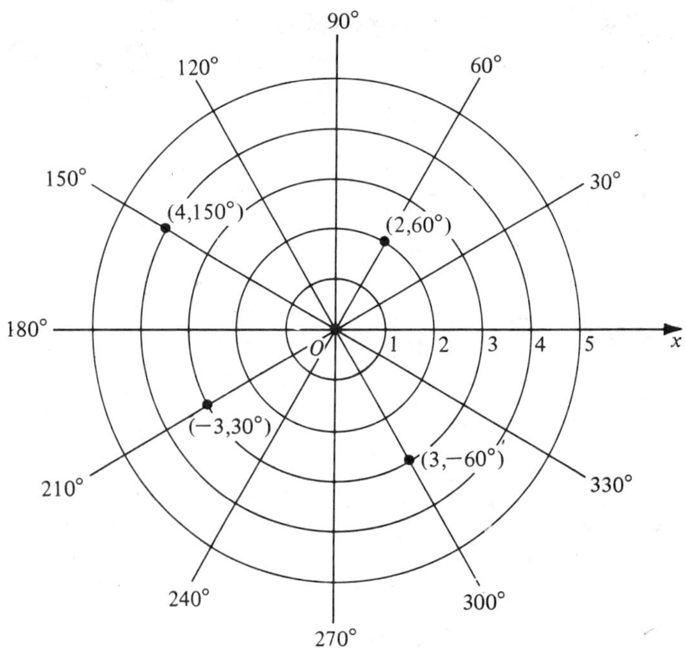

FIGURE 13.26

Example 1 Sketch the graph of the equation

$$r = 4 \sin \theta.$$

Solution: We need graph this equation only for $0 \le \theta \le 2\pi$, since the sine function has period 2π. Using the accompanying table of values, we sketch the graph as shown in Figure 13.27. The graph is traced out twice, once as θ ranges from 0 to π and again as θ ranges from π to 2π. Starting from the pole, the graph is traced out as indicated by the arrowhead. The graph is a circle, as we shall presently show.

θ	0	$\frac{\pi}{6}$	$\frac{\pi}{4}$	$\frac{\pi}{3}$	$\frac{\pi}{2}$	$\frac{2\pi}{3}$	$\frac{3\pi}{4}$	$\frac{5\pi}{6}$
r	0	2	$2\sqrt{2}$	$2\sqrt{3}$	4	$2\sqrt{3}$	$2\sqrt{2}$	2

π	$\frac{7\pi}{6}$	$\frac{5\pi}{4}$	$\frac{4\pi}{3}$	$\frac{3\pi}{2}$	$\frac{5\pi}{3}$	$\frac{7\pi}{4}$	$\frac{11\pi}{6}$	2π
0	-2	$-2\sqrt{2}$	$-2\sqrt{3}$	-4	$-2\sqrt{3}$	$-2\sqrt{2}$	-2	0

Example 2 Sketch the graph of the equation

$$r = 2(1 - 2 \sin \theta).$$

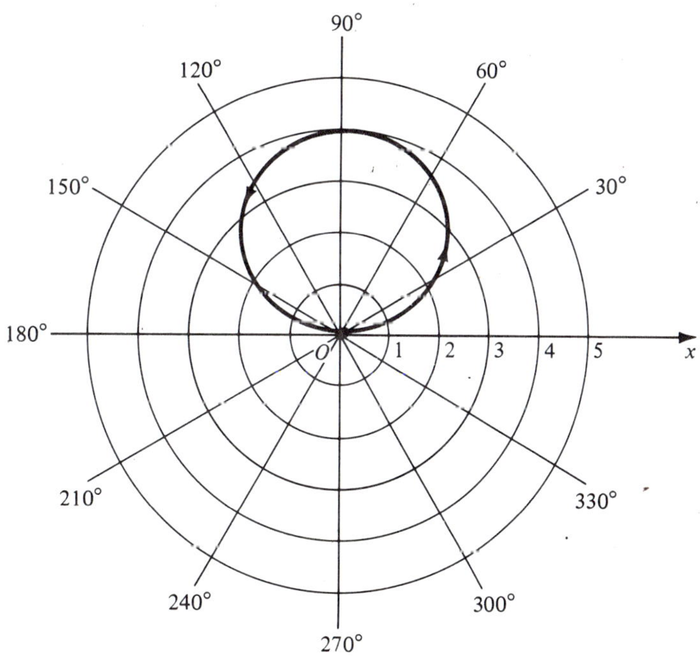

FIGURE 13.27

Solution: We may again limit the range of θ to $0 \le \theta \le 2\pi$. The graph is sketched in Figure 13.28 from the accompanying table of values, with r approximated to one decimal place. This curve is called a limacon.

θ	0	$\frac{\pi}{6}$	$\frac{\pi}{4}$	$\frac{\pi}{3}$	$\frac{\pi}{2}$	$\frac{2\pi}{3}$	$\frac{3\pi}{4}$	$\frac{5\pi}{6}$	π	$\frac{7\pi}{6}$	$\frac{5\pi}{4}$	$\frac{4\pi}{3}$	$\frac{3\pi}{2}$	$\frac{5\pi}{3}$	$\frac{7\pi}{4}$	$\frac{11\pi}{6}$	2π
r	2	0	−.8	−1.5	−2	−1.5	−.8	0	2	4	4.8	5.5	6	5.5	4.8	4	2

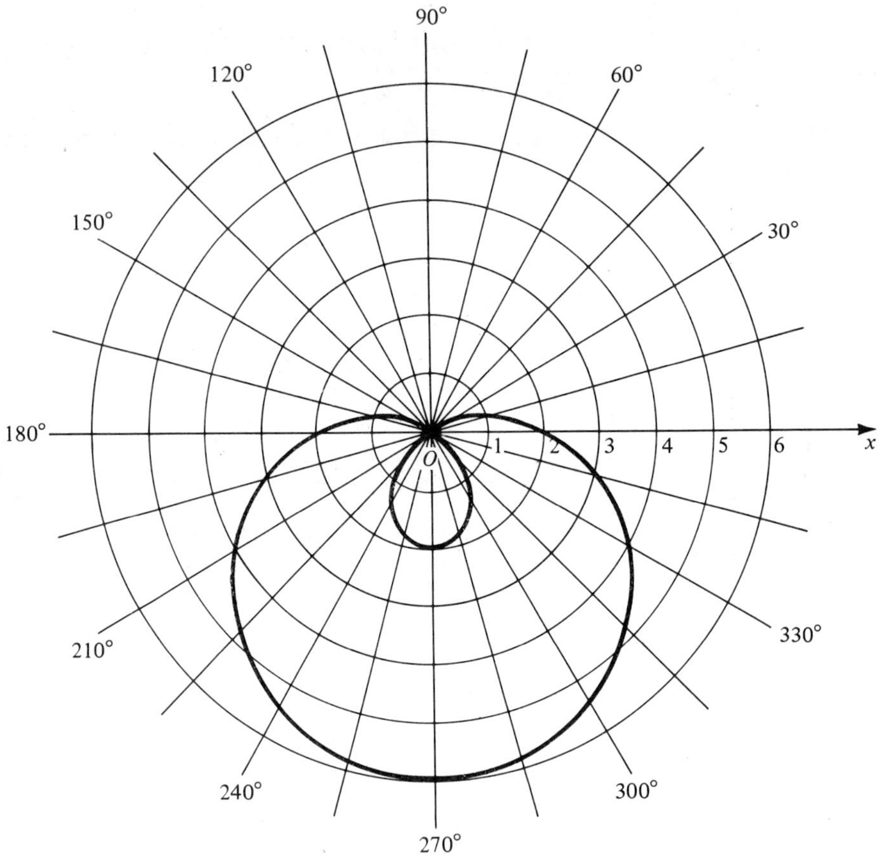

FIGURE 13.28

If a rectangular and a polar coordinate system are placed in the same plane, with the positive x axis of the first the polar axis of the second (Figure 13.29), then each point P in the plane has both rectangular coordinates (x,y) and polar coordinates (r,θ). If $r > 0$, then, according to the definition of the trigonometric functions,

$$\sin\theta = \frac{y}{r}, \qquad \cos\theta = \frac{x}{r},$$

or

13.14 $$x = r\cos\theta, \qquad y = r\sin\theta.$$

It may be verified that even if $r \le 0$ 13.14 still holds. Thus the rectangular and polar coordinates of each point in the plane are related by 13.14. It is clear from 13.14 that x, y, and r are related by the equation

13.15 $$x^2 + y^2 = r^2.$$

In view of 13.14, the graph of the equation

$$r = f(\theta)$$

in polar coordinates is the same as the graph of the parametric equations (with parameter θ)

$$x = f(\theta)\cos\theta, \qquad y = f(\theta)\sin\theta$$

in rectangular coordinates. Conversely, the graph of a given equation in x and y is the same as the graph of the equation in r and θ obtained by replacing x by $r\cos\theta$ and y by $r\sin\theta$.

Example 3 Find an equation in polar coordinates of the hyperbola

$$x^2 - y^2 = 1.$$

Solution: Using 13.14, we get

$$r^2\cos^2\theta - r^2\sin^2\theta = 1,$$

or, since $\cos^2\theta - \sin^2\theta = \cos 2\theta$,

$$r^2 = \sec 2\theta.$$

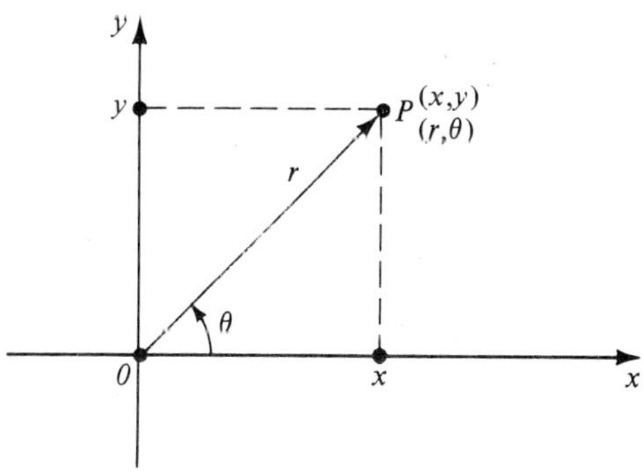

FIGURE 13.29

Example 4 Show that the graph of either $r = a \sin \theta$ or $r = a \cos \theta$, $(a > 0)$, is a circle of diameter a.

Solution: The equation $r = a \sin \theta$ has the same graph as the equation

$$r^2 = ar \sin \theta,$$

since, if $r \neq 0$, we may cancel out an r in the equation above to obtain $r = a \sin \theta$, while the pole is on both graphs. Hence, by 13.14 and 13.15,

$$x^2 + y^2 = ay$$

is a rectangular-coordinate equation of this graph. This latter equation may be put in the form

$$x^2 + \left(y - \frac{a}{2}\right)^2 = \left(\frac{a}{2}\right)^2,$$

which we recognize as the equation of a circle of radius $a/2$ and with center $(0, a/2)$. Thus the graph of $r = a \sin \theta$ is a circle of diameter a and with center $(a/2, \pi/2)$. By similar arguments, the graph of $r = a \cos \theta$ is a circle of diameter a and with center $(a/2, 0)$.

EXERCISES

I

Sketch the graph of each of the following equations.

1. $r = 3$

2. $\theta = 2\pi/3$

3. $\theta = -\pi/4$

4. $r = -4$

5. $r = 6 \cos \theta$

6. $r = -2 \sin \theta$

7. $r = \theta$

8. $r = 1/\theta$

9. $r = 2(1 - \cos \theta)$ (cardioid)

10. $r = 2 - \sin \theta$ (limacon)

11. $r = 4 \sin 3\theta$ (three-leaved rose)

12. $r = 2 \cos 2\theta$ (four-leaved rose)

13. $r = 1 + \sin \theta$ (cardioid)

14. $r^2 = a^2 \cos 2\theta$ (lemniscate)

15. $r = 2 \tan \theta$

16. $r \cos \theta = 3$

17. $r = 2 \sec \theta + 1$ (conchoid)

18. $r = a(1 + \sin^2 \theta)$

19. $r = a \csc \theta$

20. $r = \sin 4\theta$

Find an equation in polar coordinates of the graph of each of the following rectangular equations.

21. $x^2 + y^2 = 9$

22. $x = 4$

23. $xy = 1$

24. $y^2 = 8x$

25. $x^2 + y^2 + 4x = 0$

26. $x^2 + 4y^2 = 4$

Find an equation in rectangular coordinates of the graph of each of the following polar equations.

27. $r = 2 \sin \theta$

28. $r = 4$

29. $r = 1 - \sin \theta$

30. $r = \sec \theta$

31. $r = 2 \csc \theta$

32. $r = 3 \tan \theta$

33. Use the law of cosines (Appendix A) to find a formula in polar coordinates for the distance between the points $P(r_1,\theta_1)$ and $Q(r_2,\theta_2)$.

34. Prove that the graph of $r = a \cos \theta + b \sin \theta$ is a circle, and find its center and radius.

II

1. Transform the equation

$$(x^2 + y^2)^2 = 4x(x^2 - 3y^2)$$

to polar coordinates, and show that its graph is a three-leaved "rose curve" (see Exercise I-11).

2. Find parametric equations for the curve of Exercise II-1 by setting $y = tx$.

In each of Exercises 3 to 10, sketch the graph of the equation.

3. $r = \sqrt{1 - \theta}$

4. $r(1 - \theta) = 4$

5. $r = 2 + \sin 2\theta$

6. $r = 1 + \sin 2\theta$

7. $r = 1 + 2 \sin 2\theta$

8. $r \sin 2\theta = 1$

9. $r = \dfrac{3 \sin 2\theta}{\sin^3 \theta + \cos^3 \theta}$

10. $r^2 \sin 2\theta = 1$

In each of Exercises 11 to 15, find all points of intersection of the given pair of equations. (*Hint:* Remember that the polar coordinate representation of a point is not unique.)

11. $r = 2 \sin 2\theta,\ r = 2 \sin \theta$

12. $r = 2 \sin \theta,\ r = \cos \theta - 1$

13. $r = 2(1 + \cos \theta),\ r = 2 \cos 2\theta$

14. $r = 1 - \sin \theta,\ r = 1 - \cos \theta$

15. $r = 4(1 + \sin \theta),\ r(1 - \sin \theta) = 3$

8 THE CONIC SECTIONS

We recall that the parabola was defined to be the set of all points in a plane equidistant from a fixed point (the focus) and a fixed line (the directrix) of the plane. In like manner, every conic section (other than the circle) may be defined to be the set of all points P in a plane such that the ratio of the distance between P and a fixed point F (the focus) to the distance between P and a fixed line L (the directrix) is a positive constant e, called the *eccentricity* of the conic section.

In order to find an equation in polar coordinates of a conic section defined as above, let us place the focus F at the pole and the directrix L perpendicular to the polar axis, as in Figure 13.30. We let $2p$, $p > 0$, be the distance between the directrix and the focus. We limit our discussion to the case

$$e \leq 1,$$

leaving the case $e > 1$ for the reader.

By definition, the point P is on the conic section if and only if (Figure 13.30)

$$\frac{|FP|}{|PQ|} = e.$$

Since $e \leq 1$ by assumption, the point P is necessarily on the same side of the directrix as the focus. If (r,θ) is any pair of coordinates of P with $r > 0$, then $|FP| = r$, $|PQ| = 2p + r\cos\theta$, and r and θ satisfy the equation

$$\frac{r}{2p + r\cos\theta} = e.$$

On solving this equation for r, we get

13.16 $$r = \frac{2ep}{1 - e\cos\theta}$$

as an equation satisfied by every point $P(r,\theta)$, $r > 0$, on the conic section.

Conversely, for each point $P(r,\theta)$ satisfying 13.16, necessarily $r > 0$ (since $1 - e\cos\theta \geq 0$) and a reversal of the argument above proves that P is on the conic section. Thus 13.16 is an equation in polar coordinates of the conic section as defined above. Even if $e > 1$, it may be proved that 13.16 is an equation of a conic section, although in this case $r < 0$ for the points $P(r,\theta)$ on the opposite side of the directrix from the focus.

Let us prove that 13.16 actually is an equation of a conic section by finding an equation in rectangular coordinates of the graph of 13.16. We shall still assume that $e \leq 1$.

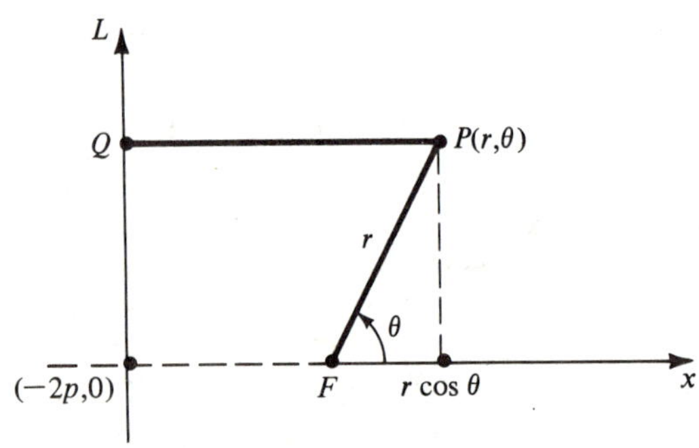

FIGURE 13.30

We may write 13.16 in the form

$$r = e(r\cos\theta + 2p),$$

and since $r > 0$, we have by 13.14 and 13.15 that the graph of this equation has equation

$$\sqrt{x^2 + y^2} = e(x + 2p)$$

in rectangular coordinates. In turn, the graph of this equation is the same as the graph of

$$(1) \qquad x^2 + y^2 = e^2(x^2 + 4px + 4p^2).$$

If $e = 1$, (1) becomes

$$y^2 = 4p(x + p),$$

the equation of a parabola with focus at the origin and directrix $x = -2p$.

If $e < 1$, we may complete squares and put (1) in the form

$$(2) \qquad \left(x - \frac{2e^2p}{1 - e^2}\right)^2 + \frac{y^2}{1 - e^2} = \frac{4e^2p^2}{(1 - e^2)^2}.$$

We recognize (2) as an equation of an ellipse with foci on the x axis and with

$$a^2 = \frac{4e^2p^2}{(1 - e^2)^2}, \qquad b^2 = \frac{4e^2p^2}{1 - e^2}, \qquad c^2 = \frac{4e^4p^2}{(1 - e^2)^2}.$$

Since $c = 2e^2p/(1 - e^2)$, the origin is at a focus. The given directrix has equation $x = -2p$. We note incidently that $e = c/a$.

If $e > 1$, 13.16 may be shown to be an equation of a hyperbola with a focus at the pole.

If the focus is kept at the origin but the directrix is varied to either side of the focus or parallel to the polar axis, either above it or below, the conic section may have an equation of the form

$$r = \frac{2ep}{1 \pm e\cos\theta}, \qquad r = \frac{2ep}{1 \pm e\sin\theta}.$$

The conic section is an ellipse if $0 < e < 1$, a parabola if $e = 1$, and a hyperbola if $e > 1$.

Example 1 Find an equation of the ellipse with focus at the pole, eccentricity $e = \frac{1}{2}$ and directrix perpendicular to the polar axis at the point $(-4,0)$.

Solution: We let $e = \frac{1}{2}$ and $p = 2$ in 13.16, obtaining

$$r = \frac{4}{2 - \cos\theta}$$

as the desired equation.

Example 2 Find an equation of the parabola with focus at the pole and directrix perpendicular to the polar axis at the point $(-3,0)$.

Solution: We let $e = 1$ and $p = \frac{3}{2}$ in 13.16, getting

$$r = \frac{3}{1 - \cos \theta}$$

as an equation of the parabola.

Example 3 Describe and sketch the graph of the equation

$$r = \frac{16}{5 + 3 \sin \theta}.$$

Solution: We may put this equation in the form

$$r = \frac{\frac{16}{5}}{1 + \frac{3}{5} \sin \theta},$$

which is an equation of an ellipse with focus at the pole and major axis perpendicular to the polar axis. By giving θ the values $\pi/2$ and $3\pi/2$, we find the ends of the major axis to be $(2, \pi/2)$ and $(8, 3\pi/2)$. Thus the length of the major axis is 10, and $a = 5$. The center of the ellipse is the point $(3, 3\pi/2)$, and $c = 3$. Hence $b^2 = a^2 - c^2 = 16$, and $b = 4$. The ellipse is sketched in Figure 13.31.

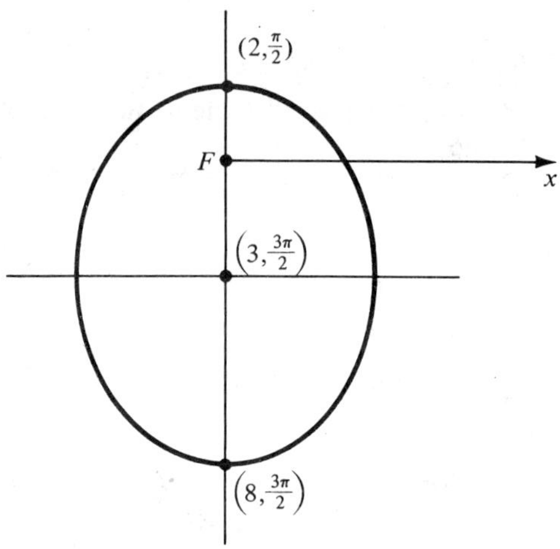

FIGURE 13.31

EXERCISES

I

Describe and sketch the graph of each of the following equations.

1. $r = \dfrac{2}{1 - \cos \theta}$ **2.** $r = \dfrac{4}{1 - \sin \theta}$

3. $r = \dfrac{12}{3 + \cos\theta}$

4. $r = \dfrac{12}{1 - 3\cos\theta}$

II

In Exercises 1 and 2, describe and sketch the graph of the given equation.

1. $r = \dfrac{2}{1 - \cos\theta - \sin\theta}$

2. $r = \dfrac{2}{2 + \cos\theta - \sin\theta}$

3. Show that if

$$r = \frac{ae}{1 - e\cos\theta}$$

is an equation of a hyperbola, then the inclination (with the horizontal) of the asymptotes is given by $\cos\theta = 1/e$.

4. Find a polar equation of an ellipse whose center is at the pole.
5. Find a polar equation of a hyperbola whose center is at the pole.

9 TANGENT LINES IN POLAR COORDINATES

The graph of the equation

$$r = f(\theta)$$

in polar coordinates is the same as the trace of the curve λ defined by

13.17 $$\lambda(\theta) = (f(\theta)\cos\theta, f(\theta)\sin\theta)$$

in the associated rectangular coordinate system. Thus we may study tangent lines to the graph of f in polar coordinates by looking at the tangent lines of λ in rectangular coordinates.

If the tangent line to the curve λ of 13.17 at point $\lambda(\theta)$ has slope $m(\theta)$, then

$$m(\theta) = \frac{y'(\theta)}{x'(\theta)} = \frac{f(\theta)\cos\theta + f'(\theta)\sin\theta}{-f(\theta)\sin\theta + f'(\theta)\cos\theta},$$

according to 13.12. Hence, if α is the inclination of the tangent line, then $\tan\alpha = m(\theta)$, and if $\cos\theta \neq 0$, then

$$\tan\alpha = \frac{f(\theta) + f'(\theta)\tan\theta}{-f(\theta)\tan\theta + f'(\theta)}.$$

We may solve this equation for $f'(\theta)$ (if $\tan\alpha \neq \tan\theta$), obtaining

$$f'(\theta) = \frac{1 + \tan\alpha\tan\theta}{\tan\alpha - \tan\theta} f(\theta).$$

This may be put in the form

13.18
$$f'(\theta) = f(\theta) \cot (\alpha - \theta).$$

Under certain conditions, as indicated in Figure 13.32, $\alpha - \theta = \psi$, an angle between the position vector and tangent line at P. Hence ψ may be computed from the equation

$$\cot \psi = \frac{f'(\theta)}{f(\theta)}.$$

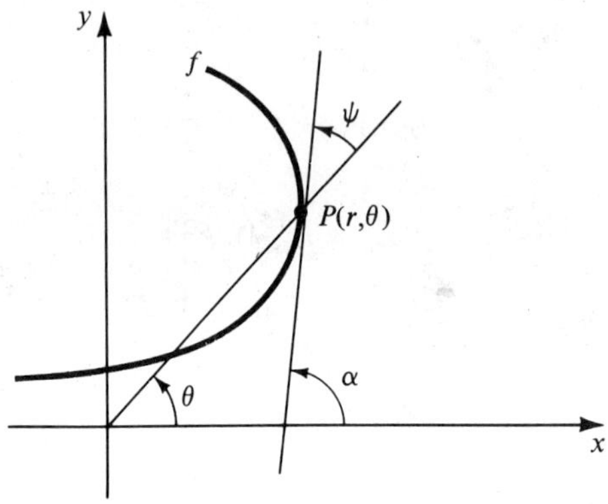

FIGURE 13.32

Example

Find the angle ψ for the limacon

$$r = 2(1 - 2 \sin \theta),$$

shown in Figure 13.28, at the point (2,0).

Solution: We have $r = f(\theta)$, and

$$\cot (\alpha - \theta) = -\frac{4 \cos \theta}{2(1 - 2 \sin \theta)},$$

by 13.18. If $\theta = 0$, $\cot (\alpha - \theta) = \cot \alpha = -2$. In this case

$$\alpha = \psi = \tan^{-1} (-\tfrac{1}{2}) \doteq 153°26'.$$

EXERCISES

1. Show that the angle ψ is a constant for the logarithmic spiral $r = e^{a\theta}$. Sketch.

In each of Exercises 2 to 6 find the angle of intersection of the graphs of the given pair of equations. (The angle between the graphs is defined to be the angle between their tangent lines at a point of intersection.)

2. $r \cos \theta = 4$, $r = 10 \sin \theta$

3. $r = a \cos \theta$, $r = -a \sin 2\theta$

4. $r = 2(1 + \cos \theta)$, $r = -2 \sin \theta$

5. $r = \sec \theta$, $r \sin 2\theta = 2$

6. $r = a(1 + \sin \theta)$, $r = a(1 - \sin \theta)$

In Exercises 7 to 10, find the angle of inclination α of the tangent line to the graph of the given equation at the indicated point P (in polar coordinates).

7. $r = e^{\theta}$, $P = (1,0)$

8. $r = \cos 2\theta$, $P = \left(\frac{1}{2}, \frac{\pi}{6}\right)$

9. $r = 1 - \cos \theta$, $P = \left(\frac{1}{2}, \frac{\pi}{3}\right)$

10. $r = \theta$, $P = \left(\frac{\pi}{2}, \frac{\pi}{2}\right)$

In Exercises 11 to 14, find an equation (in polar coordinates) of a curve that satisfies the given formula for cotan ψ at all points (r,θ) on the curve, and which contains the given point P. Note that it will be necessary to solve a separable differential equation.

11. $\text{cotan } \psi = \frac{1}{\theta}$, $P = (2,1)$

12. $\text{cotan } \psi = \frac{1}{r}$, $P = (0,0)$

13. $\text{cotan } \psi = -r \cos \theta$, $P = \left(1, \frac{\pi}{2}\right)$

14. $\text{cotan } \psi = \frac{\sin \theta}{r}$, $P = (0,0)$

10 AREAS IN POLAR COORDINATES

We may find the area of a region bounded by the graph of a function in polar coordinates and two radius vectors much as we found areas in rectangular coordinates.

Let f be a continuous, nonnegative function in an interval $[a,b]$, and let R be the region bounded by the graph of f and the lines $\theta = a$ and $\theta = b$ (Figure 13.33). Let $P = \{\theta_0, \theta_1, \ldots, \theta_n\}$ be a partition of $[a,b]$, and, as usual, let $f(u_i)$ be the minimum value and $f(v_i)$ the maximum value of f in the subinterval $[\theta_{i-1},\theta_i]$.

The region bounded by the lines $\theta = \theta_{i-1}$ and $\theta = \theta_i$ and the graph of f contains the sector of a circle with radius $f(u_i)$ and central angle $\Delta\theta_i = \theta_i - \theta_{i-1}$, and in turn is contained in the sector with radius $f(v_i)$ and central angle $\Delta\theta_i$ (Figure 13.33); thus

$$\tfrac{1}{2}f^2(u_i)\,\Delta\theta_i \le \Delta A_i \le \tfrac{1}{2}f^2(v_i)\,\Delta\theta_i,$$

where ΔA_i designates the area of this region. Since the sum of the ΔA_i, $i = 1, 2, \ldots, n$, is $A(R)$, the area of R, we have

$$\sum_{i=1}^{n} \tfrac{1}{2}f^2(u_i)\,\Delta\theta_i \le A(R) \le \sum_{i=1}^{n} \tfrac{1}{2}f^2(v_i)\,\Delta\theta_i.$$

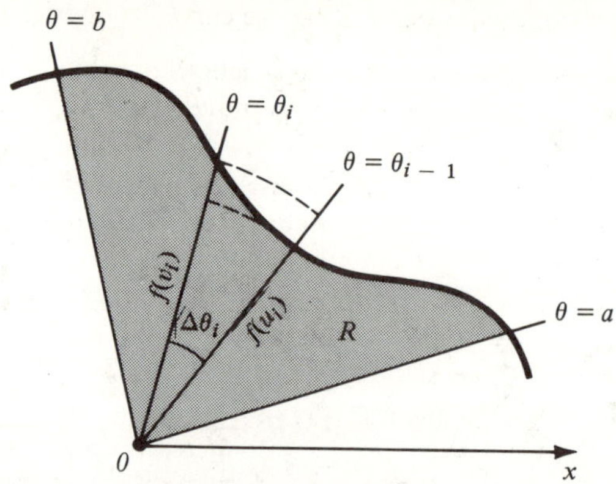

FIGURE 13.33

If now we take a sequence of partitions with norms having limit zero, then each of the above sums approaches the same definite integral, and $A(R)$ must equal this integral; thus the area $A(R)$ of the region R bounded by the graph of f and the lines $\theta = a$ and $\theta = b$ is given by

13.19
$$A(R) = \tfrac{1}{2}\int_a^b f^2(\theta)\, d\theta.$$

Example 1 Find the area of the region R bounded by the graph of $r = 1 + \cos\theta$ and the lines $\theta = 0$ and $\theta = \pi/2$ (Figure 13.34).

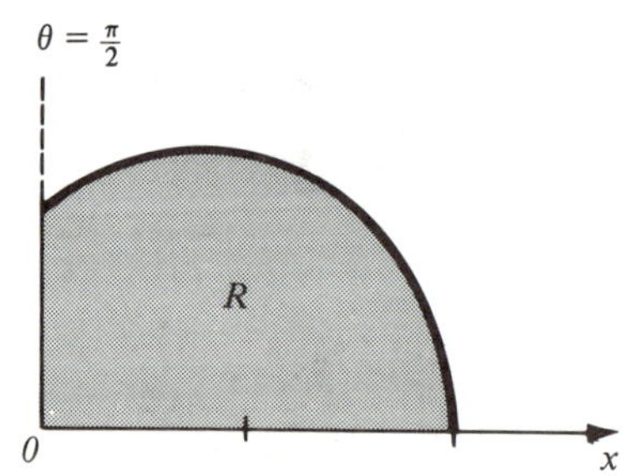

FIGURE 13.34

Solution: By 13.19,

$$A(R) = \tfrac{1}{2}\int_0^{\pi/2} (1 + \cos\theta)^2\, d\theta = \tfrac{1}{2}\int_0^{\pi/2} (1 + 2\cos\theta + \cos^2\theta)\, d\theta$$

$$= \frac{1}{2}\left(\theta + 2\sin\theta + \frac{\theta}{2} + \frac{\sin 2\theta}{4}\right)\Bigg|_0^{\pi/2} = 1 + \frac{3\pi}{8}.$$

Example 2 Find the area of one loop of the curve $r = 2 \sin 3\theta$.

Solution: The least positive angle θ for which $r = 0$ is $\pi/3$. Thus there is a loop of the curve between $\theta = 0$ and $\theta = \pi/3$ whose area A is given by

$$A = \tfrac{1}{2}\int_0^{\pi/3} (2 \sin 3\theta)^2 \, d\theta = 2\int_0^{\pi/3} \sin^2 3\theta \, d\theta$$

$$= \frac{2}{3}\left(\frac{3\theta}{2} - \frac{\sin 6\theta}{4}\right)\Bigg|_0^{\pi/3} = \frac{\pi}{3}.$$

EXERCISES

I

In each of Exercises 1 to 8 find the area of the region bounded by the graphs of the given equations. Sketch each region.

1. $r = \theta;\ \theta = 0,\ \theta = \dfrac{\pi}{2}$

2. $r = \tan\theta;\ \theta = \dfrac{\pi}{6},\ \theta = \dfrac{\pi}{4}$

3. $r = \dfrac{1}{\cos\theta};\ \theta = -\dfrac{\pi}{4},\ \theta = \dfrac{\pi}{4}$

4. $r = a\sec^2\dfrac{\theta}{2};\ \theta = 0,\ \theta = \dfrac{\pi}{2}$

5. $r = e^{\theta};\ \theta = 0,\ \theta = \pi$

6. $r = \sqrt{\sin\theta};\ \theta = \dfrac{\pi}{6},\ \theta = \dfrac{\pi}{2}$

7. $r = \sqrt{1 - \cos\theta};\ \theta = \dfrac{\pi}{2},\ \theta = \pi$

8. $r = \sin\theta + \cos\theta;\ \theta = -\dfrac{\pi}{4},\ \theta = 0$

In each of Exercises 9 to 18 find the area of the region bounded by the graph of the equation. (Find the area of just one loop if there is more than one loop.) Sketch each region.

9. $r = 10\cos\theta$

10. $r = 3\sin\theta$

11. $r = 1 - \cos\theta$

12. $r = 2(1 + \sin\theta)$

13. $r = 2\sin 2\theta$

14. $r = \cos 4\theta$

15. $r^2 = \cos 2\theta$

16. $r^2 = \sin\theta$

17. $r = a\sin n\theta$; n a positive integer, $a > 0$

18. $r^2 = a\cos n\theta$; n a positive integer, $a > 0$

In each of Exercises 19 to 22, find the area of the region common to the two given regions.

19. $r = \cos\theta,\ r = \sin\theta$

20. $r = 3\cos\theta,\ r = 1 + \cos\theta$

21. $r = 4(1 + \cos\theta),\ r = -4\sin\theta$

22. Find the area of the region between the two loops of the limacon $r = 2 + 4\cos\theta$.

II

1. Find the area of the region inside the graph of the equation $r^2 = 2a^2 \cos 2\theta$ and outside the circle $r = a$. Sketch the region.
2. Find the area of the region inside the graph of the equation $r = 3a \sin \theta$ and outside the graph of $r = a(1 + \sin \theta)$. Sketch the region.
3. Find the area of the loop of the graph of the equation $r \cos \theta = a \cos 2\theta$.
4. Find the area of the loop of the curve in Exercise II-9, Section 7.
5. Find the area of the region bounded by the curve in Exercise 4 and its asymptote.

11 ARC LENGTH OF A CURVE

In Chapter 6, Section 4, we discussed how we can assign a "length" to an arc of the graph of a smooth function. Our purpose in this section will be to extend this discussion to the more general case of an arc of a smooth curve. To accomplish this it will be necessary to examine these concepts more carefully and, in particular, to give a precise *definition* of "arc length."

It is natural to start with some observations about "polygonal" curves. Suppose we are given $n + 1$ points $P_0, P_1, P_2, \ldots, P_n$ in the plane, with no two consecutive points the same. A curve that has the set $P_0P_1 \cup P_1P_2 \cup \cdots \cup P_{n-1}P_n$ as its trace is called a *polygonal curve*, and the points $P_0, P_1, P_2, \ldots, P_n$ are called its *vertices*. An example of a polygonal curve, with $n = 4$, is given in Figure 13.35.

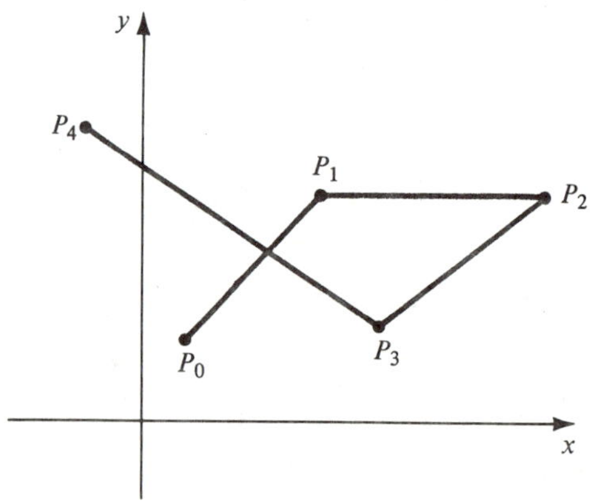

FIGURE 13.35

We are obligated, however, to show there actually is a *function* that defines such a curve. To see that such a function does exist, first consider two distinct points A and B in the plane. If **v** is the vector $\overrightarrow{AB}$, and if **A** is the position vector of A, then from 13.7 it is easily seen that the curve ρ_1 defined by the position vector

$$\rho_1(t) = \mathbf{A} + t\mathbf{v}, \qquad \text{for } 0 \le t \le 1,$$

is a 1–1 curve that has the segment AB as its trace.

It is possible to modify the function ρ_1 so as to obtain another 1–1 curve ρ_2 whose trace is still AB but whose domain is any prescribed interval $[c,d]$. Such a function ρ_2 may be defined by

$$\rho_2(t) = \rho_1\left(\frac{t-c}{d-c}\right), \qquad \text{for } c \le t \le d.$$

Now returning to the problem stated above, suppose that we are given a polygonal curve with vertices $P_0, P_1, P_2, \ldots, P_n$, and suppose that $[a,b]$ is any closed interval. Choose points $t_0, t_1, t_2, \ldots, t_n$ in $[a,b]$ with $a = t_0 < t_1 < t_2 < \cdots < t_n = b$. By the paragraph above, for each $i = 1, 2, \ldots, n$, there is a curve ρ_i with domain $[t_{i-1},t_i]$ and trace $P_{i-1}P_i$. Then the curve ρ defined by

13.20 $$\rho(t) = \rho_i(t) \text{ if } t \text{ is in } [t_{i-1},t_i], \qquad i = 1, 2, \ldots, n,$$

has domain $[a,b]$ and trace $P_0P_1 \cup P_1P_2 \cup \cdots \cup P_{n-1}P_n$.

For such a polygonal curve ρ we can naturally define its length $L(\rho)$ by the formula

$$L(\rho) = \sum_{i=1}^{n} d(P_{i-1},P_i).$$

Now let us consider the general case of a nonpolygonal curve λ with domain $[a,b]$. Suppose that $P_0, P_1, P_2, \ldots, P_n$ are any $n + 1$ points on the trace of λ, with $P_0 = \lambda(a)$, $P_n = \lambda(b)$, and no two consecutive points identical. Then, using 13.20, there are points $t_0, t_1, t_2, \ldots, t_n$ in $[a,b]$ with $a = t_0 < t_1 < t_2 < \cdots < t_n = b$, and a polygonal curve ρ with domain $[a,b]$, trace $P_0P_1 \cup P_1P_2 \cup \cdots \cup P_{n-1}P_n$, and $\rho(t_i) = P_i$ for all $i = 0, 1, 2, \ldots, n$. Such a polygonal curve ρ is called an *inscripture* of λ. An example of an inscripture, in which $n = 5$, is given in Figure 13.36. It is reasonable to expect that as we increase n, the number of segments of the inscripture ρ, then ρ approximates more and more closely the curve λ. Also, it is geometrically evident that as n increases, so does the length $L(\rho)$. These considerations provide the motivation for the following definition.

13.21 Definition

The *arc length* $L(\lambda)$ of the curve λ is the least upper bound, if it exists, of the set of numbers $\{L(\rho) \mid \rho \text{ is an inscripture of } \lambda\}$. If $L(\lambda)$ exists, we say that λ is *rectifiable*.

The notation $L_a^b(\lambda)$ will frequently be used to denote the arc length of the curve λ defined on the interval $[a,b]$.

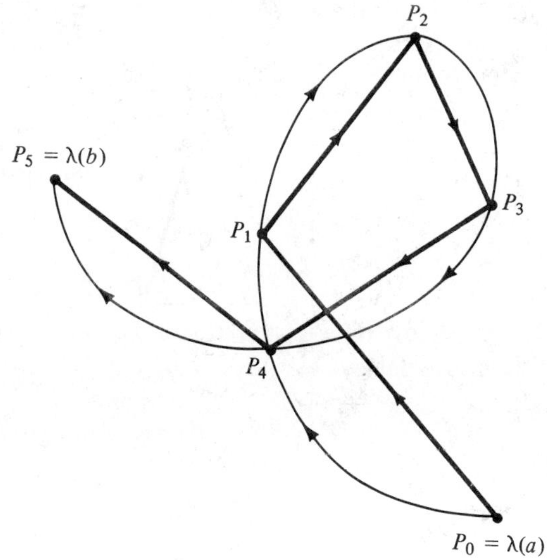

FIGURE 13.36

An "intuitively obvious" and desirable property of the above definition of arc length is that it is *additive*, in the sense that, if a curve λ is cut into two "pieces" λ_1 and λ_2, then

13.22 $$L(\lambda_1) + L(\lambda_2) = L(\lambda).$$

However, just as for other intuitively obvious properties, this requires proof. We shall give such a proof in an appendix (Section 12). In the meantime, we shall assume the validity of this result and use it without explicit comment.

It is important to realize that there exist continuous curves that are not rectifiable. An example of such a curve is given below.

Example 1 Let $P_0, P_1, P_2, \ldots, P_n, \ldots$ be an infinite sequence of points in the plane, with $P_0 = (1,0)$, $P_1 = (\frac{1}{2},\frac{1}{2})$, $P_2 = (\frac{1}{3},0)$, $P_3 = (\frac{1}{4},\frac{1}{4})$, and in general,

$$P_n = \left(\frac{1}{n+1}, \frac{1}{n+1}\right) \qquad \text{if } n \text{ is odd,}$$

$$P_n = \left(\frac{1}{n+1}, 0\right) \qquad \text{if } n \text{ is even.}$$

Following the procedure given at the beginning of this section, we may construct a curve λ whose trace is $P_0P_1 \cup P_1P_2 \cup P_2P_3 \cup \cdots \cup P_nP_{n+1} \cup \cdots$, and such that $\lambda(t)$ is defined for $0 < t \leq 1$. We then complete the definition of $\lambda(t)$ for all t in $[0,1]$ by defining $\lambda(0) = 0$. It should be geometrically obvious that λ is a continuous curve (see Figure 13.37). However, λ is not a polygonal curve, since the

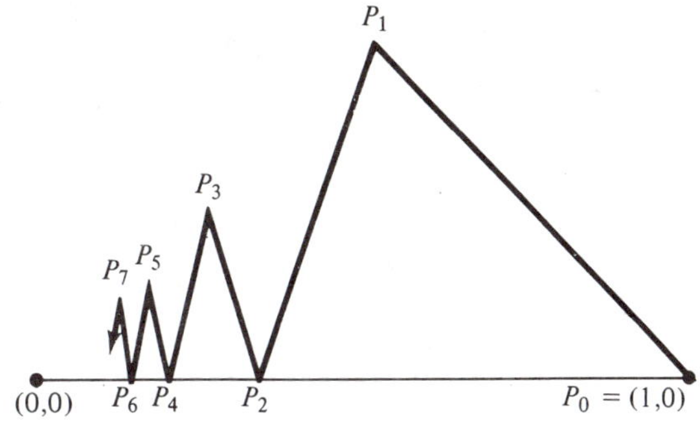

FIGURE 13.37

number of line segments in the trace of λ is infinite. Now from Figure 13.36 we see that

$$d(P_0,P_1) > \tfrac{1}{2},\quad d(P_1,P_2) > \tfrac{1}{2},$$
$$d(P_2,P_3) > \tfrac{1}{4},\quad d(P_3,P_4) > \tfrac{1}{4},$$
$$d(P_4,P_5) > \tfrac{1}{6},\quad d(P_5,P_6) > \tfrac{1}{6},$$
$$\cdots\cdots\cdots\cdots\cdots\cdots$$

Hence, if the arc length $L(\lambda)$ exists, this number must be greater than the sum of the infinite series

$$\tfrac{1}{2} + \tfrac{1}{2} + \tfrac{1}{4} + \tfrac{1}{4} + \tfrac{1}{6} + \tfrac{1}{6} + \cdots,$$

or the series

$$1 + \frac{1}{2} + \frac{1}{3} + \cdots + \frac{1}{n} + \cdots.$$

But this is simply the *harmonic series*, whose partial sums were shown to be *unbounded* (in Chapter 12, Section 2). Hence $L(\lambda)$ cannot exist.

However, *smooth* curves are rectifiable, according to the following theorem.

13.23 Theorem

Let λ be a smooth curve given by

$$\lambda(t) = (x(t), y(t)), \qquad \text{for } t \text{ in } [a,b].$$

Then λ is rectifiable, and $L(\lambda)$ is given by

$$L(\lambda) = \int_a^b \sqrt{[x'(t)]^2 + [y'(t)]^2}\, dt = \int_a^b |\lambda'(t)|\, dt.$$

We shall prove Theorem 13.23 in the next section. For the moment, we illustrate the use of this theorem with some examples and applications.

Example 2 Find the length of one arch of the cycloid λ defined by

$$\lambda(t) = (r(t - \sin t), r(1 - \cos t)).$$

Solution: One arch of the cycloid is traced out as t varies from 0 to 2π (see Figure 13.5). Clearly,

$$\lambda'(t) = \langle r(1 - \cos t), r \sin t \rangle$$

and $$|\lambda'(t)| = \sqrt{r^2(1 - \cos t)^2 + r^2 \sin^2 t} = r\sqrt{2(1 - \cos t)}.$$

Hence $$L_0^{2\pi}(\lambda) = r \int_0^{2\pi} \sqrt{2(1 - \cos t)}\, dt.$$

Using the identity $1 - \cos t = 2 \sin^2 (t/2)$, we have

$$L_0^{2\pi}(\lambda) = 2r \int_0^{2\pi} \sin \frac{t}{2}\, dt = -4r \cos \frac{t}{2}\bigg|_0^{2\pi} = 8r.$$

Example 3 The motion of a particle is given by

$$\lambda(t) = (t, 2t\sqrt{t}).$$

Find the distance traveled by the particle from $t = 0$ to $t = 4$.

Solution: Since

$$\lambda'(t) = \langle 1, 3\sqrt{t} \rangle,$$

the speed of the particle is given by

$$|\lambda'(t)| = \sqrt{1 + 9t}.$$

Hence

$$L_0^4(\lambda) = \int_0^4 \sqrt{1 + 9t}\, dt = \tfrac{2}{27}(1 + 9t)^{3/2}\bigg|_0^4 = \tfrac{2}{27}(37\sqrt{37} - 1) \doteq 16.6.$$

If f is a smooth real-valued function in the interval $[a,b]$, then the length L of its graph from a to b was defined in 6.8 to be

$$L = \int_a^b \sqrt{1 + f'^2(x)}\, dx.$$

We may obtain this formula from 13.23 by expressing the graph of f as the trace of the curve λ defined by

$$\lambda(t) = (t, f(t)).$$

Then $\lambda'(t) = \langle 1, f'(t) \rangle$ and $|\lambda'(t)| = \sqrt{1 + f'^2(t)}$. Thus 13.23 becomes 6.8 if we replace t by x in the integrand.

If the smooth real-valued function f is graphed in a polar coordinate plane, then, by 13.17, the graph of f is the same as the graph of the curve

$$\lambda(\theta) = (f(\theta) \cos \theta, f(\theta) \sin \theta)$$

in the associated rectangular coordinate plane. We may easily verify that

$$\lambda'(\theta) = \langle -f(\theta) \sin \theta + f'(\theta) \cos \theta, f(\theta) \cos \theta + f'(\theta) \sin \theta \rangle$$

and $|\lambda'(\theta)| = \sqrt{f^2(\theta) + f'^2(\theta)}$.

Hence the length L of the graph of f from $\theta = a$ to $\theta = b$ in polar coordinates is given by

13.24 $$L = \int_a^b \sqrt{f^2(\theta) + f'^2(\theta)}\, d\theta.$$

EXERCISES

In each of Exercises 1 to 5 find the length of the given curve.

1. $\lambda(t) = (t - \sin t, 1 - \cos t)$, domain $\lambda = [0, \pi/3]$
2. $\lambda(t) = (t^2, t - 1)$, domain $\lambda = [-1,1]$
3. $\lambda(t) = (\ln \sin t, t + 1)$, domain $\lambda = [\pi/6, \pi/2]$
4. $\lambda(t) = (t, \ln (t^2 - 1))$, domain $\lambda = [-3,-2]$
5. $\lambda(t) = (\cos t, \cos^2 t)$, domain $\lambda = [0,\pi]$

In each of Exercises 6 to 10, find the distance traveled by a particle with the given motion.

6. $\lambda(t) = (3 \cos 2t, 3 \sin 2t)$, between $t = 0$ and $t = 2$
7. $\lambda(t) = (2t + 1, t^2)$, between $t = 0$ and $t = 2$
8. $\lambda(t) = (3t^2, 2t^3)$, between $t = 0$ and $t = 3$
9. $\lambda(t) = (e^t \cos t, e^t \sin t)$, between $t = 0$ and $t = 2$
10. $\lambda(t) = (3t, t^3)$, between $t = 0$ and $t = 2$. (Approximate by Simpson's rule.)

In each of Exercises 11 to 15 find the length of the arc given in polar coordinates.

11. $r = e^\theta$, $\theta = 0$ to $\theta = \ln 4$
12. $r = a(1 - \cos \theta)$, complete curve
13. $r = \sin^2 \dfrac{\theta}{2}$, $\theta - 0$ to $\theta = \pi$
14. $r = a\theta^2$, $\theta = 0$ to $\theta = \pi$
15. $r = \cos^3 \dfrac{\theta}{3}$, complete curve
16. Show that the length of the graph of the equation $y = \cosh x$ between the points $(0,1)$ and $(x, \cosh x)$, $x > 0$, is $\sinh x$.

12 APPENDIX: PROOF OF THE ARC LENGTH FORMULA

We first prove 13.22. Let λ be a rectifiable curve with domain $[a,b]$. If $a < c < b$, we shall prove that

$$L_a^b(\lambda) = L_a^c(\lambda) + L_c^b(\lambda).$$

Proof: Let λ_1 and λ_2 denote the restrictions of the curve λ to the domains $[a,c]$ and $[c,b]$, respectively. (So λ is cut into the two "pieces" λ_1 and λ_2.) Let ρ_1 and ρ_2 be any inscriptures of λ_1 and λ_2, respectively. Then ρ_1 and ρ_2 together form an inscripture ρ of λ, and $L(\rho_1) + L(\rho_2) = L(\rho)$. Since $L(\rho) \leq L(\lambda)$ by Definition 13.21, we have

$$L(\rho_1) \leq L(\lambda) - L(\rho_2)$$

for every inscripture ρ_1 of λ_1. Thus $L(\lambda) - L(\rho_2)$ is an upper bound of the set of numbers $\{L(\rho_1) \mid \rho_1 \text{ is an inscripture of } \lambda_1\}$, and since $L(\lambda_1)$ is the l.u.b. of this set, we must have

$$L(\lambda_1) \leq L(\lambda) - L(\rho_2),$$

or

$$L(\rho_2) \leq L(\lambda) - L(\lambda_1).$$

Repeating the same argument, we then conclude that

(1) $$L(\lambda_2) \leq L(\lambda) - L(\lambda_1),$$

or

$$L(\lambda_1) + L(\lambda_2) \leq L(\lambda).$$

We now show that the reverse of inequality (1) also holds. Let ρ be any inscripture of λ, with vertices $P_0, P_1, P_2, \ldots, P_n$. The point $\lambda(c)$ must lie between two consecutive vertices P_{i-1} and P_i of ρ (or be identical with one of them). Form the inscripture ρ' with consecutive vertices $P_0, P_1, \ldots, P_{i-1}$, $\lambda(c), P_i, \ldots, P_n$ (if $\lambda(c)$ is identical with a vertex of ρ, then we simply have $\rho' = \rho$). It is geometrically evident that $L(\rho) \leq L(\rho')$. Now the vertices $P_0, P_1, \ldots, P_{i-1}, \lambda(c)$ define an inscripture ρ_1 of λ_1, and the vertices $\lambda(c)$, $P_i, \ldots, P_n$ define an inscripture ρ_2 of λ_2. Hence

$$L(\rho) \leq L(\rho') = L(\rho_1) + L(\rho_2) \leq L(\lambda_1) + L(\lambda_2).$$

Since this is true for any inscripture ρ of λ, by Definition 13.21 we have

$$L(\lambda) \leq L(\lambda_1) + L(\lambda_2).$$

Together with inequality (1) this means that

$$L(\lambda) = L(\lambda_1) + L(\lambda_2),$$

or

$$L_a^b(\lambda) = L_a^c(\lambda) + L_c^b(\lambda).$$

We now prove Theorem 13.23. By our hypothesis that λ is a smooth curve, the functions $x(t)$ and $y(t)$, of the parametric representation of λ, have continuous derivatives $x'(t)$ and $y'(t)$ for t in $[a,b]$. Hence the functions $[x'(t)]^2$ and $[y'(t)]^2$ are also continuous in $[a,b]$ and, by Theorem 4.2, these functions have maximum values $[x'(u)]^2$ and $[y'(v)]^2$, respectively, in $[a,b]$.

We first prove that λ is rectifiable. Let ρ be any inscripture of λ, with consecutive vertices $P_0, P_1, \ldots, P_i, \ldots, P_n$, and $P_0 = \lambda(a)$, $P_n = \lambda(b)$. According to 13.20, there are points $t_0, t_1, \ldots, t_i, \ldots, t_n$ in $[a,b]$ with $a = t_0 < t_1 < \cdots < t_n = b$ and $\rho(t_i) = P_i = (x(t_i), y(t_i))$ for each $i = 0, 1, \ldots, n$. Then

$$L(\rho) = \sum_{i=1}^{n} d(P_{i-1}, P_i)$$

$$= \sum_{i=1}^{n} \sqrt{[x(t_i) - x(t_{i-1})]^2 + [y(t_i) - y(t_{i-1})]^2}.$$

By the mean value theorem (4.8), there exist numbers w_i and z_i in each interval $[t_{i-1}, t_i]$ such that

$$x(t_i) - x(t_{i-1}) = x'(w_i)(t_i - t_{i-1}),$$
$$y(t_i) - y(t_{i-1}) = y'(z_i)(t_i - t_{i-1}).$$

Then we have

13.25
$$L(\rho) = \sum_{i=1}^{n} \sqrt{[x'(w_i)]^2 + [y'(z_i)]^2}(t_i - t_{i-1}).$$

But for each i we have $[x'(w_i)]^2 \leq [x'(u)]^2$ and $[y'(z_i)]^2 \leq [y'(v)]^2$. Hence

$$L(\rho) \leq \sqrt{[x'(u)]^2 + [y'(v)]^2} \sum_{i=1}^{n} (t_i - t_{i-1}).$$

The sum of the terms $(t_i - t_{i-1})$ is simply $(b - a)$, and so

13.26
$$L(\rho) \leq (b - a)\sqrt{[x'(u)]^2 + [y'(v)]^2}.$$

We have therefore exhibited an upper bound for the set of numbers $\{L(\rho) \mid \rho$ is an inscripture of $\lambda\}$, and hence λ is rectifiable according to Definition 13.21.

We may now define an *arc length function* s for the curve λ in the following way:

$$s(t) = L_a^t(\lambda) \qquad \text{for } t \text{ in } (a,b],$$

$$s(a) = 0.$$

Thus $s(t)$ is the length of the curve λ from a to t. We shall prove that the function s has a derivative given by

13.27
$$s'(t) = \sqrt{[x'(t)]^2 + [y'(t)]^2}.$$

Proof: If $h > 0$, then

$$s(t + h) - s(t) = L_a^{t+h}(\lambda) - L_a^t(\lambda) = L_t^{t+h}(\lambda),$$

by 13.22. If we apply 13.26 to the part of λ from t to $t + h$, we have (from 13.26 with $a = t$, $b = t + h$)

$$(1) \qquad s(t + h) - s(t) = L_t^{t+h}(\lambda) \leq h\sqrt{[x'(u)]^2 + [y'(v)]^2},$$

for some numbers u and v in the interval $[t, t + h]$.

Also, let us consider the curve λ from t to $t + h$, and take the particular inscripture ρ of this curve which contains just the single line segment from $\lambda(t)$ to $\lambda(t + h)$. Then from 13.25,

$$L(\rho) = h\sqrt{[x'(w)]^2 + [y'(z)]^2}$$

for some numbers w and z in the interval $[t, t + h]$.

Since $L(\rho) \leq L_t^{t+h}(\lambda)$, we have

$$(2) \qquad h\sqrt{[x'(w)]^2 + [y'(z)]^2} \leq s(t + h) - s(t).$$

We may combine (1) and (2) into one inequality

$$h\sqrt{[x'(w)]^2 + [y'(z)]^2} \leq s(t + h) - s(t) \leq h\sqrt{[x'(u)]^2 + [y'(v)]^2}.$$

Dividing through this inequality by the positive number h, we get

$$(3) \qquad \sqrt{[x'(w)]^2 + [y'(z)]^2} \leq \frac{s(t + h) - s(t)}{h} \leq \sqrt{[x'(u)]^2 + [y'(v)]^2}.$$

It may be verified that (3) holds even if $h < 0$.

Since the functions $x'(t)$ and $y'(t)$ are continuous, and the numbers u, v, w, and z are all between t and $t + h$, we have

$$\operatorname*{limit}_{h \to 0} [x'(w)]^2 = \operatorname*{limit}_{h \to 0} [x'(u)]^2 = [x'(t)]^2,$$

$$\operatorname*{limit}_{h \to 0} [y'(z)]^2 = \operatorname*{limit}_{h \to 0} [y'(v)]^2 = [y'(t)]^2,$$

and the two extremes of (3) have the same limit as h approaches 0. Thus

$$s'(t) = \operatorname*{limit}_{h \to 0} \frac{s(t + h) - s(t)}{h} = \sqrt{[x'(t)]^2 + [y'(t)]^2},$$

which proves 13.27.

Theorem 13.23 now follows directly, since

$$L_a^b(\lambda) = s(b) = s(b) - s(a) = \int_a^b s'(t)\, dt$$

$$= \int_a^b \sqrt{[x'(t)]^2 + [y'(t)]^2}\, dt.$$

13 AREA OF A SURFACE OF REVOLUTION

If a curve is rotated about a line in its plane, it sweeps out a surface of revolution. The theory of the preceding sections may be used to assign an area to such a surface.

Let λ be a smooth curve with domain $[a,b]$, with $\lambda(t) = (x(t),y(t))$ for every t in $[a,b]$. We shall also assume that the curve λ is above the x axis; i.e., that $y(t) \geq 0$ for every t in $[a,b]$. Let us find the area of the surface swept out by rotating λ about the x axis.

If $p = \{t_0, t_1, \ldots, t_n\}$ is a partition of $[a,b]$, let L_p designate the length of the polygonal curve ρ_p associated with λ, as in the preceding section. It seems reasonable that if ρ_p is rotated about the x axis, it will sweep out a surface with area approximating that of the given surface. The surface generated by ρ_p is made up of n frustums of cones. A frustum of a cone of slant height h and radii of bases r_1 and r_2 (Figure 13.38) has lateral surface area

$$\pi(r_1 + r_2)h,$$

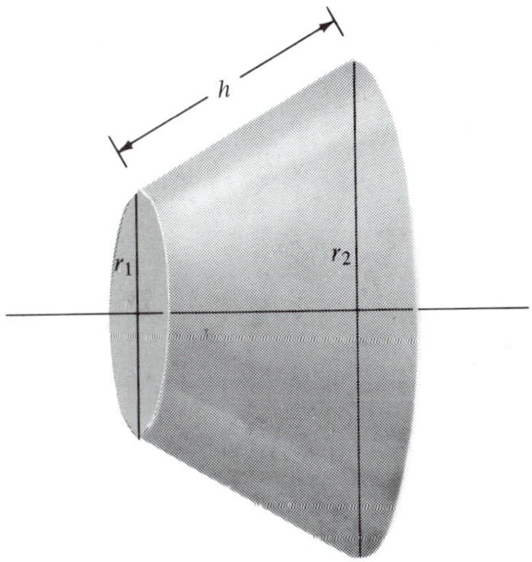

FIGURE 13.38

according to a formula of geometry. Hence the ith segment $\lambda(t_{i-1})\lambda(t_i)$ of ρ_p (Figure 13.39) sweeps out a frustum of a cone of lateral area

$$\pi[y(t_{i-1}) + y(t_i)]d(\lambda(t_{i-1}), \lambda(t_i)).$$

Using the value of $d(\lambda(t_{i-1}), \lambda(t_i))$ given in 13.25, we find that the total area S_p of the surface swept out by ρ_p is given by

$$S_p = \pi \sum_{i=1}^{n} [y(t_{i-1}) + y(t_i)]\sqrt{x'^2(w_i) + y'^2(z_i)}\, \Delta t_i,$$

where w_i and z_i are numbers in the interval $[t_{i-1},t_i]$.

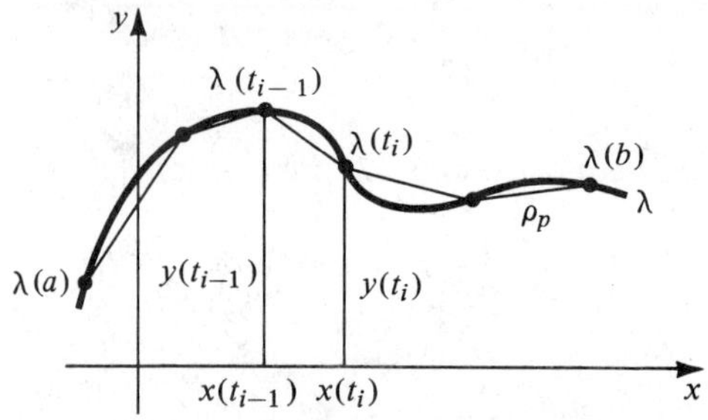

FIGURE 13.39

In the limit as the norm of p approaches zero, we can imagine that the above formula yields

13.28 $$S = 2\pi \int_a^b y(t)|\lambda'(t)|\, dt$$

as the area of the surface swept out by rotating curve λ about the x axis. Although we shall not give a formal proof of 13.28, we shall nevertheless accept its validity.

If the curve λ is rotated about the y axis, we get the same formula with the factor $y(t)$ replaced by $x(t)$ in the integrand.

If a curve λ is the graph of an equation

$$y = f(x),$$

then 13.28 becomes [letting $x = t$, $y = f(t)$]

13.29 $$S = 2\pi \int_a^b f(x)\sqrt{1 + f'^2(x)}\, dx.$$

Example 1 Find the area of the surface generated by rotating the curve

$$\lambda(t) = (t^2, 2t), \qquad \text{domain } \lambda = [0,4],$$

about the x axis.

Solution: By 13.28,

$$S = 2\pi \int_0^4 2t\sqrt{(2t)^2 + (2)^2}\, dt$$

$$= \frac{8\pi}{3}(t^2 + 1)^{3/2}\Big|_0^4 = \frac{8\pi}{3}(17\sqrt{17} - 1).$$

Example 2 Find the area of the surface generated by rotating one arch of the sine curve about its axis.

Solution: If we rotate the graph of the curve $y = \sin x$ between $x = 0$ and $x = \pi$ about the x axis, we obtain a surface with area (using 13.29)

$$S = 2\pi \int_0^\pi \sin x\sqrt{1 + \cos^2 x}\, dx.$$

This may be integrated by letting $u = \cos x$; thus we obtain

$$S = -2\pi \int_1^{-1} \sqrt{1 + u^2}\, du$$

$$= -2\pi \left[\frac{u}{2}\sqrt{1 + u^2} + \frac{1}{2}\ln\left(u + \sqrt{1 + u^2}\right)\right]\Bigg|_1^{-1}$$

$$= 2\pi\,[\sqrt{2} + \ln(2 + 1)].$$

EXERCISES

In each of Exercises 1 to 8 find the area of the surface obtained by rotating the given curve about the x axis.

1. $\lambda(t) = (t^2/2, t)$, domain $\lambda = [1,3]$
2. $\lambda(t) = (2t^3, 3t^2)$, domain $\lambda = [0,2]$
3. $\lambda(t) = (\cos^2 t, \sin t \cos t)$, domain $\lambda = [0, \pi/2]$
4. $\lambda(t) = (2 \ln t, t^2)$, domain $\lambda = [1,3]$
5. $y = x^3$, between $x = 0$ and $x = 2$
6. $y = \cosh x$, between $x = -1$ and $x = 1$
7. $y = e^x$, between $x = -2$ and $x = 2$
8. $y = 2\sqrt{x}$, between $x = 1$ and $x = 4$
9. Find the surface area of a sphere.
10. The cardioid $r = a(1 + \cos\theta)$ is rotated about the polar axis. Find the area of the surface generated.
11. One loop of the lemniscate $r^2 = a^2 \cos 2\theta$ is rotated about the polar axis. Find the area of the surface generated.
12. Find the area of the surface of an ellipsoid of revolution.

CHAPTER REVIEW

In each of Exercises 1 to 4 discuss the curve λ, describe its direction, and sketch its trace. Also, sketch the velocity and acceleration vectors at the given time t.

1. $\lambda(t) = (\cosh t, \sinh t)$, domain $\lambda = [0,2]$, $t = 1$
2. $\lambda(t) = (\sec t, \tan t)$, domain $\lambda = [0,\frac{1}{4}\pi]$, $t = \frac{1}{6}\pi$
3. $\lambda(t) = (t - 3, 3t + 4)$, domain $\lambda = [-4,2]$, $t = 0$
4. $\lambda(t) = (\cos t^2, \sin t^2)$, domain $\lambda = [0,\sqrt{2\pi}]$, $t = \sqrt{\pi/2}$

In Exercises 5 and 6, eliminate the parameter from the given parametric equations and then sketch the graph (either from the parametric equations or the rectangular equation).

5. $x = 2 + 5 \sin t$, $y = \cos t - 3$

6. $x = 4 \cot t$, $y = 4 \sin^2 t$

In Exercises 7–13, sketch the graph of the given equation in polar coordinates.

7. $r = a \tan^2 \theta \sec \theta$ (semi-cubical parabola)

8. $r = 2a \tan \theta \sin \theta$ (cissoid)

9. $r^2 = \dfrac{a}{\theta}$ (lituus)

10. $r = \dfrac{9}{4 + 5 \sin \theta}$

11. $r = \dfrac{9}{5 + 4 \cos \theta}$

12. $r = \dfrac{1}{\sqrt{2} - \cos \theta}$

13. $r = \dfrac{1}{1 + \sqrt{2} \sin \theta}$

In Exercises 14–17, a particle moves in the plane so that its x and y coordinates satisfy the given differential equations and initial conditions. Find parametric equations for the path of the particle and sketch the curve.

14. $dx/dt = 3t^2$, $dy/dt = x$, $x = 1$ and $y = 1$ when $t = 0$.

15. $dx/dt = x$, $dy/dt = 1/x^2$, $x = 1$ and $y = -2$ when $t = 0$.

16. $dx/dt = 1 + x^2$, $dy/dt = -x$, $x = 0$ and $y = 0$ when $t = 0$.

17. $dx/dt = -y$, $dy/dx = 2ty^2$, $x = 0$ and $y = -1$ when $t = 0$.

18. Find the length of the curve with parametric equations $x = a \cos^3 t$, $y = a \sin^3 t$ $(0 \le t \le 2\pi)$.

In each of Exercises 19–21, sketch the graph, find the area of the region bounded by the graph, and find the length of the graph.

19. $r = 2 + 2 \sin \theta$

20. $r = \cos \theta - 1$

21. $r = \cos^2 \theta/2$

In each of Exercises 22–24, find the area of the surface of revolution generated by revolving the graph of the given equation about the x-axis.

22. $x^{2/3} + y^{2/3} = a^{2/3}$

23. $x^3 = k^2 y$, between $x = 0$ and $x = k$.

24. $y = e^{-x}$ in $[0, \infty]$.

14

Three-dimensional Analytic Geometry

Coordinate systems may be introduced into three-dimensional space much as they are in the plane. They enable us to study graphs of equations in three variables. We also introduce vectors in three-dimensional space, and discuss their application to the study of lines, planes, and space curves.

1 THREE-DIMENSIONAL SPACE

Three mutually perpendicular coordinate lines in space allow us to introduce coordinates in space. The three coordinate lines, called the x axis, the y axis, and the z axis, are assumed to have the same scale and to meet at their origins. If the axes are oriented as in Figure 14.1, the coordinate system is said to be right-handed. If the x and y axes were interchanged, it would be left-handed. For the most part, we shall use a right-handed coordinate system. Three coordinate planes are determined by the axes, namely the xy plane containing the x and y axes, the xz plane, and the yz plane.

Each point P in space may be projected onto the coordinate axes. If the projections of P on the x, y, and z axes, respectively, have coordinates x, y, and z, then P itself is said to have coordinates (x,y,z). The three planes through P parallel to the coordinate planes form a parallelepiped with the coordinate planes. The coordinates of the vertices of this parallelepiped are shown in Figure 14.2.

Just as each point in space has a unique triple of coordinates, so each triple of numbers determines a unique point in space having the triple of num-

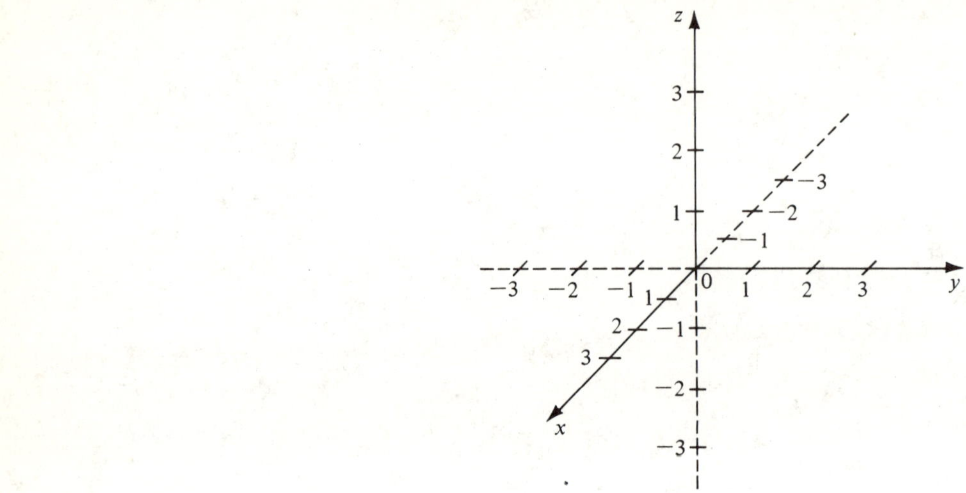

FIGURE 14.1

bers as coordinates. This association of triples of numbers with points in space is called a *rectangular coordinate system* in space.

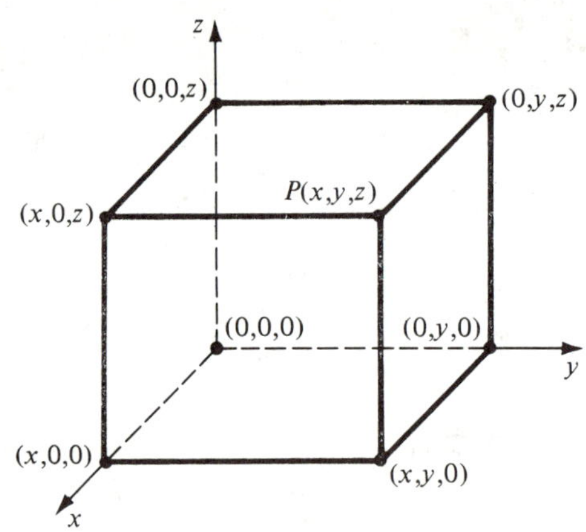

FIGURE 14.2

The three coordinate planes separate space into eight parts, called *octants*. We shall need to refer explicitly only to the *first octant*, consisting of all points $P(x,y,z)$ such that $x > 0$, $y > 0$, and $z > 0$.

Distances may be found between points in space by the following analog of the formula for the plane.

14.1 Distance Formula

The distance between the points $P(x_1,y_1,z_1)$ and $Q(x_2,y_2,z_2)$ is given by

$$|PQ| = \sqrt{(x_2 - x_1)^2 + (y_2 - y_1)^2 + (z_2 - z_1)^2}.$$

Proof: Construct a parallelepiped having its faces parallel to the coordinate planes and having P and Q as opposite vertices. If $A(x_2,y_1,z_1)$ and $B(x_2,y_2,z_1)$ are chosen as in Figure 14.3, then

$$|PA| = |x_2 - x_1|, \qquad |AB| = |y_2 - y_1|, \qquad |BQ| = |z_2 - z_1|.$$

Triangle PAB has a right angle at A, and triangle PBQ has a right angle at B. Hence $|PA|^2 + |AB|^2 = |PB|^2$, $|PB|^2 + |BQ|^2 = |PQ|^2$, and therefore

$$\begin{aligned} |PQ|^2 &= |PA|^2 + |AB|^2 + |BQ|^2 \\ &= (x_2 - x_1)^2 + (y_2 - y_1)^2 + (z_2 - z_1)^2. \end{aligned}$$

This establishes the distance formula.

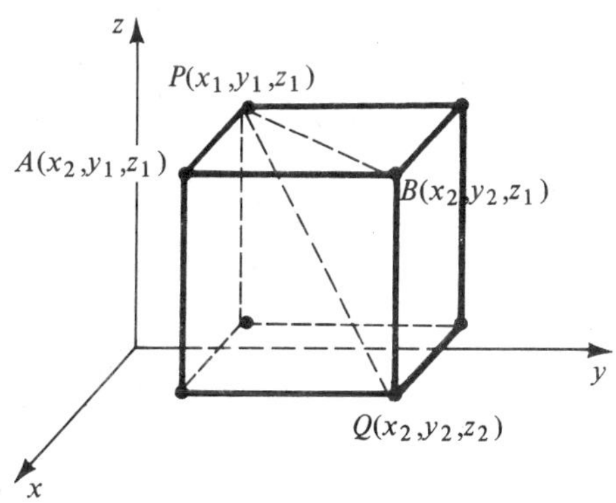

FIGURE 14.3

The parallelepiped of Figure 14.3 reduces to a rectangle (or a line segment) if P and Q lie in a plane parallel to a coordinate plane. In this case, either $|PA|$, $|AB|$, or $|BQ|$ is zero. However, the distance formula may easily be shown still to hold.

As a particular case of 14.1, the distance from the origin O to the point $P(x,y,z)$ is given by

$$|OP| = \sqrt{x^2 + y^2 + z^2}.$$

Thus the point $P(x,y,z)$ is on the sphere having radius r and center at the origin O if and only if

$$x^2 + y^2 + z^2 = r^2.$$

Therefore this is an equation of the sphere. A more general consequence of the distance formula is stated without proof below.

14.2 Theorem

The sphere of radius r having its center at the point (x_0,y_0,z_0) *has equation*

$$(x - x_0)^2 + (y - y_0)^2 + (z - z_0)^2 = r^2.$$

Since every equation of the form

$$x^2 + y^2 + z^2 + ax + by + cz + d = 0$$

can be put in the form of the equation of 14.2 by completing squares, its graph, if it exists, is a sphere.

Example 1 Discuss the graph of the equation

$$x^2 + y^2 + z^2 - 6x + 2y - z - \tfrac{23}{4} = 0.$$

Solution: We first complete squares as follows:

$$(x^2 - 6x + 9) + (y^2 + 2y + 1) + (z^2 - z + \tfrac{1}{4}) = \tfrac{23}{4} + 9 + 1 + \tfrac{1}{4},$$
$$(x - 3)^2 + (y + 1)^2 + (z - \tfrac{1}{2})^2 = 16.$$

Hence the graph is a sphere of radius 4 with center at $(3,-1,\frac{1}{2})$.

In subsequent chapters we shall have occasion to use the set of all points inside a sphere. Such a set is called an *open ball.* If $B(P,r)$ designates the open ball with center $P(x_0,y_0,z_0)$ and radius r, then, by 14.2,

$$B(P,r) = \{(x,y,z) \mid (x - x_0)^2 + (y - y_0)^2 + (z - z_0)^2 < r^2\}.$$

The *closed ball* $B[P,r]$ consists of the open ball together with the sphere,

$$B[P,r] = \{(x,y,z) \mid (x - x_0)^2 + (y - y_0)^2 + (z - z_0)^2 \le r^2\}.$$

If in Figure 14.3 we imagine a plane parallel to the yz plane cutting the parallelepiped into two equal parts, then this plane will bisect the line segments PA, PB, and PQ. Since the midpoint of PA has x coordinate $(x_1 + x_2)/2$, the midpoint of PQ will also have this x coordinate. A continuation of this argument yields the following result.

14.3 Midpoint Formula

The segment with endpoints (x_1,y_1,z_1) *and* (x_2,y_2,z_2) *has midpoint*

$$\left(\frac{x_1 + x_2}{2}, \frac{y_1 + y_2}{2}, \frac{z_1 + z_2}{2}\right).$$

Example 2 Show that $A(2,-1,1)$, $B(5,2,1)$, and $C(1,6,5)$ are the vertices of a right triangle with hypotenuse AC. Find an equation of the sphere having AC as a diameter and passing through the point B.

Solution: We have

$$|AB|^2 = (5 - 2)^2 + (2 + 1)^2 = 18,$$
$$|AC|^2 = (1 - 2)^2 + (6 + 1)^2 + (5 - 1)^2 = 66,$$
$$|BC|^2 = (1 - 5)^2 + (6 - 2)^2 + (5 - 1)^2 = 48.$$

Since $|AB|^2 + |BC|^2 = |AC|^2$, then ABC is a right triangle with hypotenuse AC. The point M with coordinates

$$\left(\frac{2 + 1}{2}, \frac{-1 + 6}{2}, \frac{1 + 5}{2}\right).$$

or $(\frac{3}{2},\frac{5}{2},3)$, is the midpoint of AC. Thus

$$(x - \tfrac{3}{2})^2 + (y - \tfrac{5}{2})^2 + (z - 3)^2 = \tfrac{33}{2},$$

or

$$x^2 + y^2 + z^2 - 3x - 5y - 6z + 1 = 0,$$

is an equation of the sphere with center M and radius $|MB| = \frac{1}{2}|AC| = \sqrt{66}/2$.

The reader will recall that we use R to denote the set of all real numbers, and R^2 to denote the set of all ordered pairs of real numbers. We identify R^2 of course with the set of all points in the plane. In the same spirit, we shall write R^3 for the set of all points in three-dimensional space; that is,

$$\mathsf{R}^3 = \{(x,y,z) \mid x,y,z \text{ in } \mathsf{R}\},$$

which is the set of all *ordered triples* of real numbers.

EXERCISES

In each of Exercises 1 to 6, A and B are the opposite vertices of a parallelepiped having its faces parallel to the coordinate planes. Sketch the parallelepiped, and find the coordinates of its other vertices.

1. $A(0,0,0)$, $B(7,2,3)$

2. $A(1,1,1)$, $B(3,4,2)$

3. $A(-1,1,2)$, $B(2,3,5)$

4. $A(0,-2,2)$, $B(2,0,-2)$

5. $A(0,-2,-1)$, $B(3,1,0)$

6. $A(2,-1,-3)$, $B(4,0,-1)$

In each of Exercises 7 to 10 show that the three given points are the vertices of a right triangle, and find the equation of the sphere passing through the three points and having its center on the hypotenuse.

7. $(4,4,1)$, $(1,1,1)$, $(0,8,5)$

8. $(2,1,3)$, $(0,1,2)$, $(1,3,0)$

9. $(-3,6,0)$, $(-2,-5,-1)$, $(1,4,2)$

10. $(2,5,-2)$, $(1,3,0)$, $(4,5,-1)$

In each of Exercises 11 to 16 discuss the graph of the equation.

11. $x^2 + y^2 + z^2 - 2x - 24 = 0$

12. $x^2 + y^2 + z^2 + 6x - 2y + 4z + 13 = 0$

13. $4x^2 + 4y^2 + 4z^2 + 12y - 4z + 1 = 0$

14. $x^2 + y^2 + z^2 - 2x + 4y - 8z + 21 = 0$

15. $x^2 + y^2 + z^2 - 6x + 2y + 11 = 0$

16. $x^2 + y^2 + z^2 - 3x + 4y - z = 0$

17. Show that (2,4,2), (2,1,5), (5,1,2), and (1,0,1) are the vertices of a regular tetrahedron, and sketch the tetrahedron.

18. Find an equation of the graph of all points equidistant from the points (3,1,2) and (7,5,6). What is the graph?

19. Find an equation of the graph of all points equidistant from the points $(-1,2,1)$ and $(1,-2,-1)$. What is the graph?

20. Let A, B, C, and D be four points in space, with no three of the points collinear, and let P, Q, R, and S be the midpoints of AB, BC, CD, and DA, respectively. Prove that $PQRS$ is a parallelogram.

In each of Exercises 21–28, determine the nature of the graph of the given equation in three-dimensional space.

21. $z = 1$

22. $x = 2$

23. $x = y$

24. $x^2 + y^2 = 4$

25. $y^2 + z^2 = 4$

26. $y^2 + z^2 = 0$

27. $|x| + |y| = 1$

28. $z = y^2$

2 THREE-DIMENSIONAL VECTORS

We define vectors in three-dimensional space just as we did in the plane. Thus, a *three-dimensional vector* is a directed line segment $\overrightarrow{PQ}$ in three-dimensional space. The point P is the *initial point* of the vector and Q is its *terminal point*.

A three-dimensional vector is determined by its *components*. If $\mathbf{v}$ is the vector $\overrightarrow{PQ}$, $P = (a,b,c)$, and $Q = (a + h, b + k, c + l)$, then the numbers h, k, l are the *components* of $\mathbf{v}$; and we write

$$\mathbf{v} = \langle h,k,l \rangle.$$

Two vectors are considered equal if and only if they have the same components.

A vector $\mathbf{v}$ is a *position vector* if its initial point is the origin $O = (0,0,0)$. Any vector can be considered as equal to a position vector $\mathbf{v}$. Thus, in Figure 14.4, the vector $\overrightarrow{PQ}$ is equal to the position vector $\mathbf{v}$ whose terminal point is the point (h,k,l).

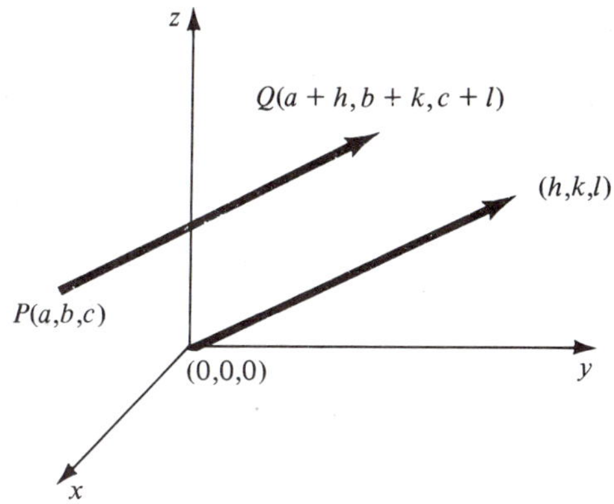

FIGURE 14.4

The set of all three-dimensional vectors is in one-to-one correspondence with the set R^3 of all points in three-dimensional space. For with each point (h,k,l) in R^3 we may associate the position vector $\mathbf{v}$ from $(0,0,0)$ to (h,k,l); and with each vector $\mathbf{v} = \langle h,k,l\rangle$ we may associate the point (h,k,l).

All the properties of two-dimensional vector algebra (Chapter 13, Section 3) carry over to the three-dimensional case. Thus the operations of addition, scalar multiplication and inner product are defined by

$$\langle h_1,k_1,l_1\rangle + \langle h_2,k_2,l_2\rangle = \langle h_1 + h_2, k_1 + k_2, l_1 + l_2\rangle$$
$$c\langle h,k,l\rangle = \langle ch,ck,cl\rangle$$
$$\langle h_1,k_1,l_1\rangle \cdot \langle h_2,k_2,l_2\rangle = h_1h_2 + k_1k_2 + l_1l_2$$

and satisfy the associative, commutative, and distributive properties listed in Chapter 13. We denote the zero vector by $\mathbf{0}$,

$$\mathbf{0} = \langle 0,0,0\rangle,$$

and the negative of a vector $\mathbf{v} = \langle h,k,l\rangle$ by $-\mathbf{v}$,

$$-\langle h,k,l\rangle = \langle -h,-k,-l\rangle.$$

The proofs of the various properties of addition, scalar multiplication, and inner product are almost identical with those given in Chapter 13, and hence are omitted.

Geometrically, vector addition is again the familiar triangular addition shown in Figure 14.5, where $\mathbf{u} = \langle h_1,k_1,l_1\rangle$ and $\mathbf{v} = \langle h_2,k_2,l_2\rangle$.

The *length* of a vector $\mathbf{v}$ is designated by $|\mathbf{v}|$ and defined to be

$$|\mathbf{v}| = \sqrt{\mathbf{v}\cdot\mathbf{v}}.$$

Thus if $\mathbf{v} = \langle h,k,l\rangle$, then

$$|\mathbf{v}| = \sqrt{h^2 + k^2 + l^2}.$$

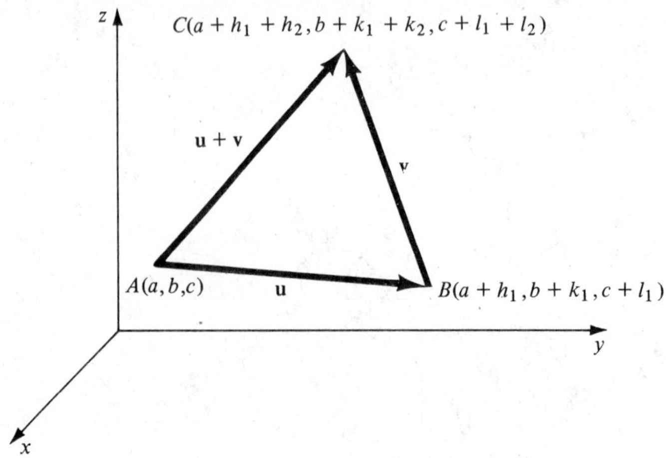

FIGURE 14.5

Clearly, $|\mathbf{v}| \geq 0$, with $|\mathbf{v}| = 0$ if and only if $\mathbf{v} = \mathbf{0}$. The length function has the usual properties stated in Chapter 13:

$$|c\mathbf{v}| = |c||\mathbf{v}|,$$

$$|\mathbf{u} \cdot \mathbf{v}| \leq |\mathbf{u}||\mathbf{v}|, \qquad \textit{Cauchy inequality},$$

$$|\mathbf{u} + \mathbf{v}| \leq |\mathbf{u}| + |\mathbf{v}|, \qquad \textit{triangle inequality}.$$

These properties hold for all three-dimensional vectors **u** and **v** and for every scalar c.

A vector **v** is called a *unit vector* if $|\mathbf{v}| = 1$. The unit vectors along the x axis, the y axis, and the z axis are designated by **i**, **j**, and **k**, respectively, and are defined by

$$\mathbf{i} = \langle 1,0,0 \rangle, \qquad \mathbf{j} = \langle 0,1,0 \rangle, \qquad \mathbf{k} = \langle 0,0,1 \rangle.$$

Another example of a unit vector is

$$\mathbf{u} = \langle \tfrac{2}{7}, -\tfrac{3}{7}, \tfrac{6}{7} \rangle,$$

since $(\frac{2}{7})^2 + (-\frac{3}{7})^2 + (\frac{6}{7})^2 = 1$.

The *angle* θ between two nonzero three-dimensional vectors **u** and **v** is defined precisely as it was in the two-dimensional case. Thus if we let **u** and **v** have the same initial point, θ is the angle of least nonnegative measure having **u** and **v** as its sides (Figure 14.6), $0 \leq \theta \leq \pi$.

By the same argument that was used in Chapter 13, we can show that if θ is the angle between the nonzero vectors **u** and **v**, then

14.4 $$\mathbf{u} \cdot \mathbf{v} = |\mathbf{u}||\mathbf{v}| \cos \theta.$$

Hence **u** and **v** are perpendicular if and only if $\mathbf{u} \cdot \mathbf{v} = 0$.

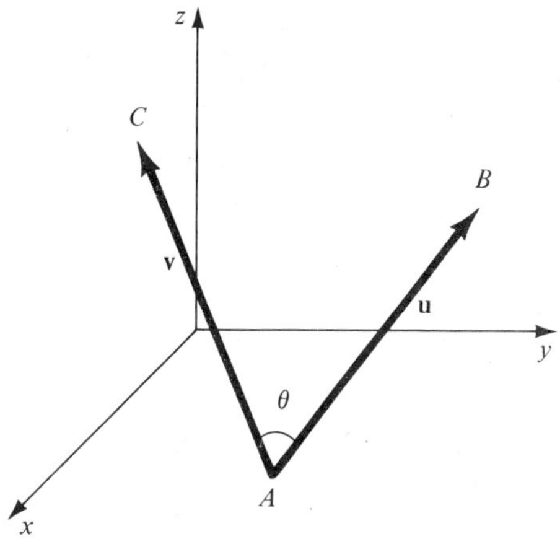

FIGURE 14.6

Each vector $\mathbf{v} = \langle r,s,t \rangle$ may be expressed as a linear combination of the basic vectors **i**, **j**, and **k**,

$$\mathbf{v} = r\mathbf{i} + s\mathbf{j} + t\mathbf{k}.$$

For example,

$$\langle 2,-7,3 \rangle = 2\mathbf{i} - 7\mathbf{j} + 3\mathbf{k}.$$

This fact will be used in subsequent sections of this chapter.

Example Find a unit vector perpendicular to both $\mathbf{v}_1 = \langle 3,4,-1 \rangle$ and $\mathbf{v}_2 = \langle 1,-1,2 \rangle$.

Solution: If the vector $\mathbf{u} = \langle h_1,h_2,h_3 \rangle$ is perpendicular to $\mathbf{v}_1$ and $\mathbf{v}_2$, then $\mathbf{u} \cdot \mathbf{v}_1 = 0$ and $\mathbf{u} \cdot \mathbf{v}_2 = 0$. Hence

$$\begin{aligned} 3h_1 + 4h_2 - h_3 &= 0, \\ h_1 - h_2 + 2h_3 &= 0. \end{aligned}$$

There are an infinite number of solutions of these equations for the unknowns h_1, h_2, and h_3. One solution (which can be found by setting $h_1 = 1$) is $h_1 = 1$, $h_2 = -1$, $h_3 = -1$. The vector $\mathbf{u} = \langle 1,-1,-1 \rangle$ has length $\sqrt{3}$, and so a unit vector perpendicular to $\mathbf{v}_1$ and $\mathbf{v}_2$ is $\langle 1/\sqrt{3}, -1/\sqrt{3}, -1/\sqrt{3} \rangle$.

Another method for solving this type of problem will be given in Section 5 of this chapter.

EXERCISES

1. Prove the Cauchy inequality in three dimensions.

2. Prove the triangle inequality in three dimensions.

3. Prove that the points $A(1,-1,2)$, $B(-1,3,0)$, and $C(3,-5,4)$ lie on a straight line by showing that the vectors $\overrightarrow{AB}$ and $\overrightarrow{BC}$ are parallel.
4. If $\mathbf{u} = \mathbf{i} + \mathbf{j} + \mathbf{k}$, $\mathbf{v} = 2\mathbf{i} - 3\mathbf{j} + 4\mathbf{k}$, and $\mathbf{w} = \mathbf{j} - 2\mathbf{k}$, then find:
   a. $|\mathbf{u}|$, $|\mathbf{v}|$, $|\mathbf{w}|$
   b. $\mathbf{u}\cdot\mathbf{v}$, $\mathbf{u}\cdot(\mathbf{u}+\mathbf{v}+\mathbf{w})$
   c. $\mathbf{u}+\mathbf{v}$, $\mathbf{u}-\mathbf{v}+\mathbf{w}$
   d. $|\mathbf{u}+\mathbf{v}+2\mathbf{w}|$
5. If $\mathbf{u}$ and $\mathbf{v}$ are unit vectors and α is the angle between them, then express $|\mathbf{u}-\mathbf{v}|$ in terms of α.
6. a. Show that the vectors $\mathbf{v}_1 = \langle -1,2,1\rangle$ and $\mathbf{v}_2 = \langle -1,1,-3\rangle$ are perpendicular.
   b. Show that the vector $\mathbf{u} = \langle -9,16,1\rangle$ is a linear combination of $\mathbf{v}_1$ and $\mathbf{v}_2$.
   c. Show that the vector $\mathbf{w} = \langle 3,4,-1\rangle$ can *not* be written as a linear combination of $\mathbf{v}_1$ and $\mathbf{v}_2$. (Also see Exercise 14.)
7. Show that $\left(\dfrac{\mathbf{u}}{|\mathbf{u}|^2} - \dfrac{\mathbf{v}}{|\mathbf{v}|^2}\right)^2 = \dfrac{(\mathbf{u}-\mathbf{v})^2}{|\mathbf{u}|^2|\mathbf{v}|^2}$. (By definition, $\mathbf{w}^2 = \mathbf{w}\cdot\mathbf{w}$.)
8. Prove that the vectors $\mathbf{u} = 2\mathbf{i} - \mathbf{j} + \mathbf{k}$, $\mathbf{v} = \mathbf{i} - 3\mathbf{j} - 5\mathbf{k}$, $\mathbf{w} = 3\mathbf{i} - 4\mathbf{j} - 4\mathbf{k}$ form the sides of a right triangle.
9. If $\mathbf{u}$, $\mathbf{v}$, and $\mathbf{w}$ are mutually perpendicular unit vectors, prove that
$$|\mathbf{u}+\mathbf{v}+\mathbf{w}|^2 = 3.$$
10. Prove that the vector $(\mathbf{u}\cdot\mathbf{u})\mathbf{v} - (\mathbf{u}\cdot\mathbf{v})\mathbf{u}$ is perpendicular to $\mathbf{u}$.

The projection of a vector $\mathbf{u}$ on a vector $\mathbf{v}$ in three dimensions is defined exactly as it was in Chapter 13, Section 3, Exercise 15. In Exercises 11 and 12, find the projection of the first vector on the second.

11. $\mathbf{i} + 3\mathbf{j} - 4\mathbf{k}$ on $4\mathbf{i} - \mathbf{j} + \mathbf{k}$
12. $2\mathbf{i} - 2\mathbf{j} + 3\mathbf{k}$ on $-\mathbf{i} + \mathbf{j} + 2\mathbf{k}$
13. a. Show that the vectors $\mathbf{u} = \frac{3}{7}\mathbf{i} + \frac{6}{7}\mathbf{j} + \frac{2}{7}\mathbf{k}$, $\mathbf{v} = \frac{2}{7}\mathbf{i} - \frac{3}{7}\mathbf{j} + \frac{6}{7}\mathbf{k}$, and $\mathbf{w} = \frac{6}{7}\mathbf{i} - \frac{2}{7}\mathbf{j} - \frac{3}{7}\mathbf{k}$ form a mutually perpendicular set of unit vectors.
    b. Express the vector $\mathbf{r} = 2\mathbf{i} + 3\mathbf{j} - \mathbf{k}$ as a linear combination of $\mathbf{u}$, $\mathbf{v}$, and $\mathbf{w}$.
14. If $\mathbf{v}$ is a given nonzero vector, show that any vector $\mathbf{u}$ can be expressed as a sum of a vector parallel to $\mathbf{v}$ and a vector perpendicular to $\mathbf{v}$.
15. Find a vector perpendicular to both $\mathbf{v}_1 = \langle 2,0,-1\rangle$ and $\mathbf{v}_2 = \langle 3,-4,0\rangle$.
16. Find a unit vector perpendicular to both $\mathbf{v}_1 = \langle 1,1,-1\rangle$ and $\mathbf{v}_2 = \langle 2,-3,1\rangle$.

3 LINES IN SPACE

Suppose that λ is a *function* that associates with each real number t in some interval I a position vector $\lambda(t)$ in three-dimensional space. The vector $\lambda(t)$ may be designated by
$$\lambda(t) = \langle x(t), y(t), z(t)\rangle,$$
where $x(t)$, $y(t)$, and $z(t)$ are real-valued functions that specify the components of $\lambda(t)$. If the vector-valued function λ is continuous in some appropriate sense, then the tip of the position vector $\lambda(t)$ can be considered to trace out a "space curve." We defer the general discussion of space curves until Section 8 of this

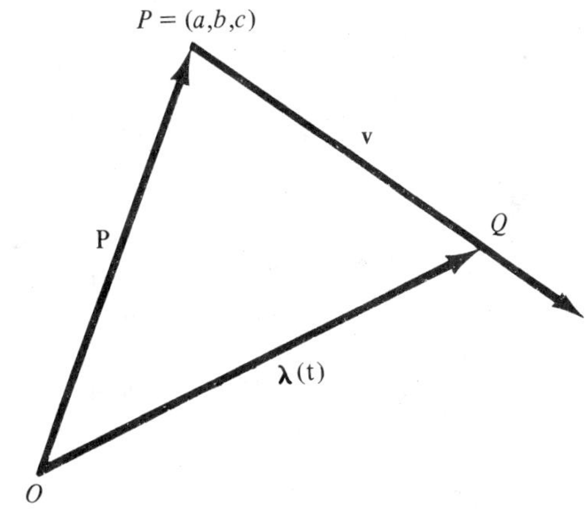

FIGURE 14.7

chapter, and investigate at this time only one special type of space curve, namely a straight line in space.

To find a vector equation for the straight line L through the point $P = (a,b,c)$ in the direction of a given nonzero vector $\mathbf{v}$, we proceed as in the two-dimensional case (Chapter 13, Section 5). Let $\mathbf{P}$ denote the position vector of the point P, and let $\mathbf{v}$ have its initial point at P (Figure 14.7). Then a point Q is on the line L if and only if the vector $\overrightarrow{PQ}$ is a scalar multiple of $\mathbf{v}$, say $\overrightarrow{PQ} = t\mathbf{v}$. Denoting the position vector $\overrightarrow{OQ}$ by $\lambda(t)$, the required vector equation of L is

14.5 $$\lambda(t) = \mathbf{P} + t\mathbf{v}, \qquad t \text{ in } R.$$

To obtain parametric equations for L, let $\mathbf{v} = \langle h,k,l \rangle$. Then

$$\begin{aligned}\lambda(t) &= \langle x(t), y(t), z(t) \rangle = \mathbf{P} + t\mathbf{v} \\ &= \langle a,b,c \rangle + t\langle h,k,l \rangle \\ &= \langle a + th,\ b + tk,\ c + tl \rangle.\end{aligned}$$

Hence the parametric equations for L are

14.5′ $$x(t) = a + th, \qquad y(t) = b + tk, \qquad z(t) = c + tl.$$

When a line L is given by an equation of the form 14.5 (or, equivalently, by 14.5′), there is a natural direction assigned to L, namely the direction of $\mathbf{v}$ (as in the two-dimensional case). The points on L corresponding to $t > 0$ in 14.5 constitute the *positive half* of L, while those corresponding to $t < 0$ form the *negative half* of L.

Suppose that $\mathbf{v}$ and $\mathbf{v}'$ are nonzero vectors with $\mathbf{v}' = c\mathbf{v}$ for some scalar $c \neq 0$. Then the lines L and L' given respectively by the equations

$$\lambda(t) = \mathbf{P} + t\mathbf{v}, \qquad \gamma(s) = \mathbf{P} + s\mathbf{v}', \qquad s,t \text{ in R},$$

are obviously the same, since $s\mathbf{v}' = t\mathbf{v}$ if $t = sc$. Thus, in particular, the vector equation 14.5 of any straight line L may be put in the form

14.5″ $$\lambda(t) = \mathbf{P} + t\mathbf{u}, \qquad t \text{ in R},$$

where $\mathbf{u}$ is a *unit* vector.

Now let us assume that the line L through the point P in the direction of the unit vector $\mathbf{u}$ is represented by an equation in the form 14.5″, and let $\mathbf{u} = \langle l,m,n \rangle$. Let the unit vectors $\mathbf{i}$, $\mathbf{j}$, $\mathbf{k}$ be drawn so that each has the point P as its initial point (Figure 14.8). Then the angles α, β, and γ between $\mathbf{u}$ and the vectors $\mathbf{i}$, $\mathbf{j}$, and $\mathbf{k}$, respectively, are called the *direction angles* of the directed line L. Since $\mathbf{u} \cdot \mathbf{i} = l$, $\mathbf{u} \cdot \mathbf{j} = m$, and $\mathbf{u} \cdot \mathbf{k} = n$, and also $|\mathbf{u}| = |\mathbf{i}| = |\mathbf{j}| = |\mathbf{k}| = 1$, we have, by 14.4, that

$$\cos\alpha = l, \qquad \cos\beta = m, \qquad \cos\gamma = n.$$

For this reason the coordinates l, m, and n of the unit vector $\mathbf{u}$ are called the *direction cosines* of the directed line L. Note that

$$\cos^2\alpha + \cos^2\beta + \cos^2\gamma = l^2 + m^2 + n^2 = |\mathbf{u}|^2 = 1.$$

If l, m, and n are direction cosines of a directed line, then

$$l^2 + m^2 + n^2 = 1.$$

Conversely, if l, m, and n are numbers, such that $l^2 + m^2 + n^2 = 1$, then there exists a directed line having l, m, and n as its direction cosines.

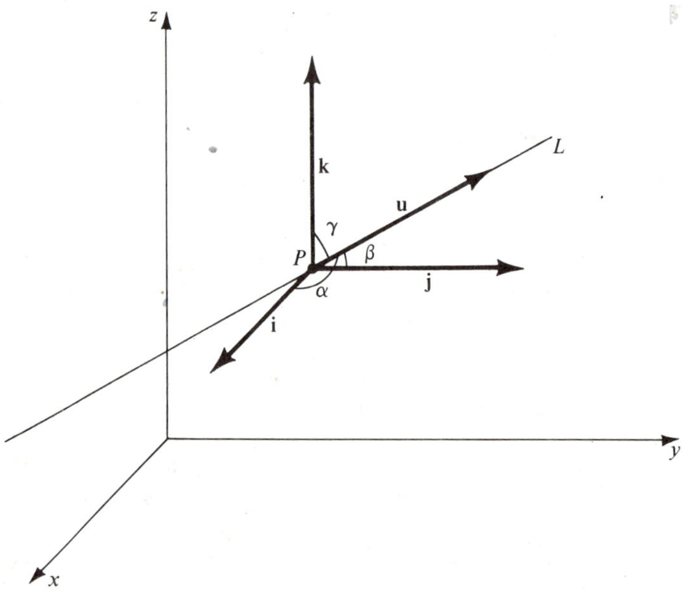

FIGURE 14.8

We have already proved the first half of 14.6. If l, m, and n are numbers such that $l^2 + m^2 + n^2 = 1$, then let $\mathbf{u} = \langle l,m,n \rangle$ and P be any point in R^3. The line L given by the equation $\lambda(t) = \mathbf{P} + t\mathbf{u}$, t in R, has direction $\mathbf{u}$, and hence has direction cosines l, m, and n. This proves 14.6.

For example, the directed line L of Figure 14.9 has direction cosines $-\frac{1}{2}, \frac{1}{2}, \sqrt{2}/2$ and direction angles 120°, 60°, and 45°.

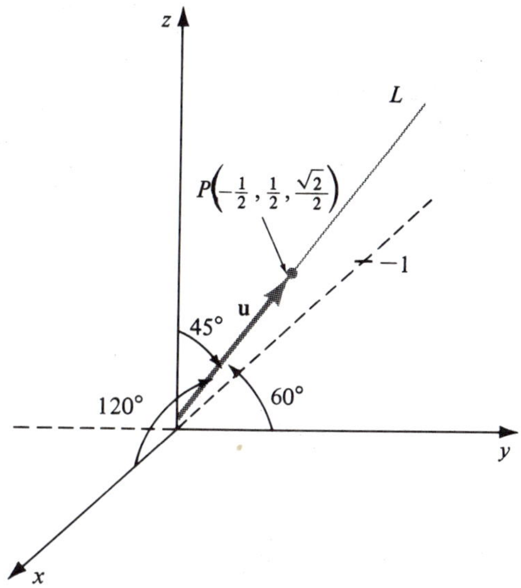

FIGURE 14.9

If we are given two points on a line L and a direction from one point toward the other, the following example shows how we can easily express L in one of the forms given above.

Example 1 If L is the line through $P = (-1,2,3)$ and $Q = (2,0,1)$, and L has the direction from P toward Q, find (i) a unit vector $\mathbf{u}$ in the direction of L, and (ii) parametric equations for L.

Solution: The line L has the direction of the vector

$$\mathbf{v} = \overrightarrow{PQ} = \langle 2-(-1), 0-2, 1-3 \rangle = \langle 3,-2,-2 \rangle.$$

Since $|\mathbf{v}| = \sqrt{3^2 + (-2)^2 + (-2)^2} = \sqrt{17}$, a unit vector $\mathbf{u}$ in the direction of $\mathbf{v}$ is given by

$$\mathbf{u} = \frac{1}{|\mathbf{v}|} \cdot \mathbf{v} = \left\langle \frac{3}{\sqrt{17}}, \frac{-2}{\sqrt{17}}, \frac{-2}{\sqrt{17}} \right\rangle.$$

A vector equation of L could be written in the form

$$\lambda(t) = \mathbf{P} + t\mathbf{v}, \qquad t \text{ in } \mathsf{R},$$

where **P** is the position vector of the point P. Corresponding to this vector equation, parametric equations of L are

$$x = -1 + 3t, \qquad y = 2 - 2t, \qquad z = 3 - 2t.$$

Example 2 Let L_1 and L_2 be straight lines given by the following sets of parametric equations:

$$L_1: \quad x = 3 - t, \quad y = 7 + 2t, \quad z = 1 + 3t,$$
$$L_2: \quad x = 9 + 4s, \quad y = 4 + s, \quad z = 3 + 8s.$$

Determine whether the lines L_1 and L_2 intersect.

Solution: If (x,y,z) is a point lying on both lines, then the values of x, y, z given by the two sets of parametric equations must be equal. In other words, for a point of intersection, t and s must satisfy the three simultaneous equations

$$\begin{aligned} 3 - t &= 9 + 4s, \\ 7 + 2t &= 4 + s, \\ 1 + 3t &= 3 + 8s. \end{aligned}$$

Solving the first two equations simultaneously, we get $t = -2$, $s = -1$. But the values $t = -2$, $s = -1$ also satisfy the third equation. Corresponding to these values of t and s, we see that the point $(5,3,-5)$ lies on both L_1 and L_2.

In connection with Example 2, it should be emphasized that *three* simultaneous equations in *two* unknowns t and s frequently fail to have a simultaneous solution. In this case the lines L_1 and L_2 fail to intersect.

Example 3 Let A be the point $(1,1,2)$ and suppose that the line L has parametric equations

$$x = 4 - t, \qquad y = 5 + 3t, \qquad z = 7 + t.$$

Find a point B on L so that the vector $\overrightarrow{AB}$ and the line L are perpendicular.

Solution: If $Q = (x,y,z)$ is any point on L, then the vector $\overrightarrow{AQ}$ is given by

$$\begin{aligned} \overrightarrow{AQ} &= \langle 4 - t - 1, 5 + 3t - 1, 7 + t - 2 \rangle \\ &= \langle 3 - t, 4 + 3t, 5 + t \rangle. \end{aligned}$$

The vector $\mathbf{v} = \langle -1,3,1 \rangle$ lies in the direction of the line L. If $\mathbf{v}$ and $\overrightarrow{AQ}$ are to be perpendicular, then their inner product must be 0; that is,

$$-1(3 - t) + 3(4 + 3t) + 1(5 + t) = 0.$$

Solving this equation for t, we get $t = -\frac{14}{11}$. Thus the point B on L corresponding to $t = -\frac{14}{11}$ has the property that $\overrightarrow{AB}$ and L are perpendicular. Substituting $t = -\frac{14}{11}$ in the parametric equations for L, we get $B = (\frac{58}{11}, \frac{13}{11}, \frac{63}{11})$.

Just as slopes may be used to determine when two lines in a plane are parallel, so may direction cosines be used to determine when two lines are parallel in space. Thus, suppose that L and L' are given by vector equations

$$\lambda(t) = \mathbf{P} + t\mathbf{u}, \qquad \gamma(t) = Q + t\mathbf{u}',$$

where $\mathbf{u}$ and $\mathbf{u}'$ are unit vectors. Then L and L' are parallel and have the same direction if and only if $\mathbf{u} = \mathbf{u}'$; they are parallel and have the opposite direction if and only if $\mathbf{u}' = -\mathbf{u}$.

Suppose that L and L' are lines intersecting in a point P, with vector equations

$$\lambda(t) = \mathbf{P} + t\mathbf{v}, \qquad \gamma(t) = \mathbf{P} + t\mathbf{v}'.$$

Then, as we saw in Section 2, L and L' are perpendicular if and only if $\mathbf{v} \cdot \mathbf{v}' = 0$.

The student undoubtedly recalls that in two-dimensional space an equation in the variables x and y of the form

$$ax + by + c = 0$$

has a straight line as its graph in the xy plane. He may therefore be tempted to ask about the nature of the graph of the "linear" equation

$$ax + by + cz + d = 0$$

in three-dimensional space. We shall see in the next section that the graph of this last equation is a *plane* in space, and not a straight line. Indeed, it is not possible to obtain a straight line in space as the graph of a *single* equation in the variables x, y, and z.

EXERCISES

I

1. Find parametric equations of the line passing through the point (2,3,1) and parallel to the vector $3\mathbf{i} - 7\mathbf{k} + 4\mathbf{j}$.
2. Find a unit vector parallel to the line $x = 3t + 1$, $y = 4t - 2$, $z = 9$.

In each of Exercises 3 to 5 find the direction cosines and direction angles of the half-line emanating from the origin and passing through the given point.

3. $(1,1,\sqrt{2})$
4. $(2,2,2)$
5. $(-3\sqrt{2},-3,-3)$

6. Find the direction cosines and direction angles of the positive coordinate axes and of the negative coordinate axes.
7. Under what conditions are l, m, and n direction cosines of a line parallel to a coordinate plane?
8. If $\pi/3$ and $3\pi/4$ are two of the direction angles of a half-line, find the third direction angle.
9. Prove that only one of the direction angles of a half-line can be less than 45°.

In Exercises 10 and 11, find parametric equations for the line passing through the given pair of points.

10. (5,1,2) and (7,2,4)   **11.** (2, −2, −2) and (0,0,1)

In each of Exercises 12 to 14 determine whether or not the given pair of lines intersect.

12. $x = 2 - t, y = -1 + 3t, z = t; x = -1 + 7s, y = 8 - 3s, z = 3 + s.$

13. $x = -3 + 3t, y = -2t, z = 7 + 6t; x = -6 + s, y = -5 - 3s, z = 1 + 2s_2$

14. $x = 2 + t, y = 1 - 3t, z = t - 1; x = s + 1, y = -s, z = 5 - 3s.$

15. Find the perpendicular distance from the point $A = (-2,2,4)$ to the line L whose parametric equations are

$$x = -1 + 2t, \qquad y = 6 + t, \qquad z = -5 - 4t.$$

(First find the foot of the perpendicular from A to L.)

16. Find the perpendicular distance from the origin to the line with parametric equations

$$x = 1 + t, \qquad y = 2t, \qquad z = 3 - t.$$

17. Find a general formula for the perpendicular distance from the point $A = (h,k,l)$ to the line L with parametric equations $x = x_0 + at, y = y_0 + bt, z = z_0 + bt.$

18. If the lines L_1 and L_2 are respectively parallel to the vectors $\mathbf{u} = \langle 1,0,1 \rangle$ and $\mathbf{v} = \langle -1,1,2 \rangle$, find parametric equations for the line through (2,3,0) that is perpendicular to both L_1 and L_2. (Find a vector perpendicular to both $\mathbf{u}$ and $\mathbf{v}$.)

II

1. Suppose the lines L_1 and L_2 have respective parametric equations $x = t, y = -t, z = 2t$, and $x = s + 2, y = s, y = -s + 2$.

a. Find points A on L_1 and B on L_2 such that $\overrightarrow{AB}$ is perpendicular to both L_1 and L_2.

b. Write parametric equations for the line through A and B.

2. Given lines L_1 and L_2 with respective parametric equations $x = 1 + 2t, y = 3 + t, z = -2 + t$ and $x = 1 + s, y = -2 - 4s, z = 9 + 2s$, find equations of the line L intersecting both L_1 and L_2 at right angles.

4 PLANES IN SPACE

Given a line L with vector equation

$$\lambda(t) = \mathbf{P} + t\mathbf{u}, \qquad t \text{ in } \mathsf{R},$$

passing through the point P in the direction of the vector $\mathbf{u}$, there is a unique plane p perpendicular to L at the point P (Figure 14.10). We shall now obtain an equation whose graph is the plane p.

Let $P = (x_0,y_0,z_0)$ and let $\mathbf{u} = \langle a,b,c \rangle$. If $Q = (x,y,z)$ is any point

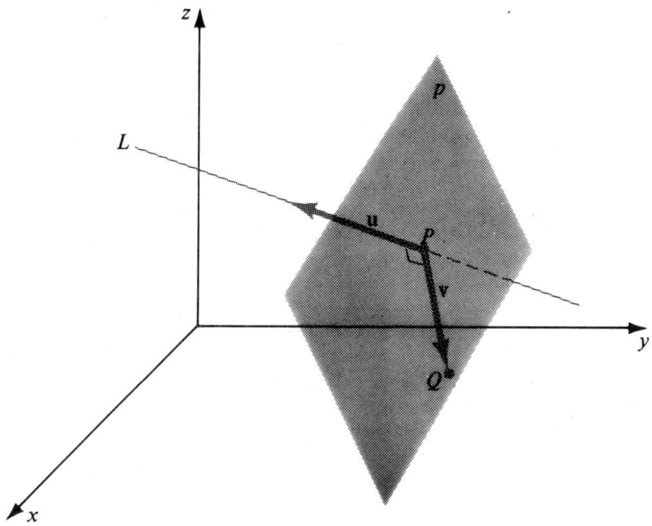

FIGURE 14.10

in the plane p, then the vector $\mathbf{v} = \overrightarrow{PQ}$ must be perpendicular to $\mathbf{u}$, and this is true if and only if $\mathbf{u} \cdot \mathbf{v} = 0$. Since

$$\mathbf{v} = \overrightarrow{PQ} = \langle x - x_0, y - y_0, z - z_0 \rangle,$$

the condition $\mathbf{u} \cdot \mathbf{v} = 0$ becomes

14.7 $$a(x - x_0) + b(y - y_0) + c(z - z_0) = 0.$$

This is an *equation of the plane through the point* (x_0,y_0,z_0) and *perpendicular to the vector* $\langle a,b,c \rangle$. The nonzero vector $\langle a,b,c \rangle$ perpendicular to p is called a *normal vector* of plane p.

Actually, the graph in R^3 of every linear equation of the form

14.8 $$ax + by + cz + d = 0$$

is a plane. We assume, of course, that a, b, and c are not all zero. To see that 14.8 is an equation of a plane, we need only select any point (x_0,y_0,z_0) such that

$$ax_0 + by_0 + cz_0 + d = 0.$$

Then $d = -(ax_0 + by_0 + cz_0)$ and 14.8 can be put in the form

$$a(x - x_0) + b(y - y_0) + c(z - z_0) = 0.$$

It follows from 14.7 that 14.8 is an equation of a plane having $\langle a,b,c \rangle$ as a normal vector.

Example 1 Find an equation of the plane passing through the point $(-2,1,3)$ and having normal vector $\langle 3,1,5\rangle$.

Solution: By 14.7,

$$3(x+2)+(y-1)+5(z-3)=0$$

or

$$3x+y+5z-10=0$$

is an equation of the plane.

Example 2 Describe the plane with equation

$$3x+y+z-6=0.$$

Solution: Points in the plane p may be found by giving arbitrary values to two of the variables in the equation and solving for the third variable. If we let $y=0$ and $z=0$, we get $x=2$; thus the point $(2,0,0)$ is in the plane. The other intercepts are found to be $(0,6,0)$ and $(0,0,6)$. These three intercepts completely determine p, as sketched in Figure 14.11. Other points in p are $(1,1,2)$, $(0,3,3)$, and $(2,1,-1)$.

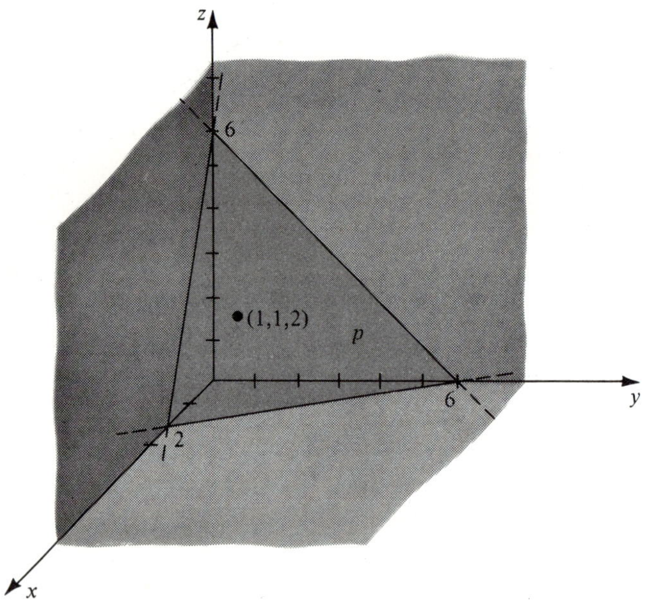

FIGURE 14.11

The points $(2,0,0)$ and $(0,6,0)$ determine a line in the xy plane called the *trace* of the plane p in the xy plane. Thus the trace of p in the xy plane consists of those points in p for which $z=0$. Substituting $z=0$ into the equation of p, we get

$$3x+y-6=0$$

as an equation of the trace of p in the xy plane. Similarly, an equation of the trace of p in the yz plane is

$$y + z - 6 = 0,$$

found by substituting $x = 0$ in the equation of p; and an equation of the trace of p in the xz plane is

$$3x + z - 6 = 0,$$

found by letting $y = 0$ in the equation of p.

The xy plane has equation $z = 0$, since the point $P(x,y,z)$ is in the xy plane if and only if $z = 0$. Each plane parallel to the xy plane has an equation of the form $z = k$. Similarly, the xz and yz planes have equations $y = 0$ and $x = 0$, respectively, and planes parallel to the xz and yz planes have equations of the form $y = k$ and $x = k$.

Example 3 Find an equation of the plane passing through the three points $(-1,1,2)$, $(2,0,-3)$, and $(5,1,-2)$.

Solution: The equation of the required plane has the form

$$ax + by + cz + d = 0, \tag{1}$$

where the constants a, b, c, and d are to be determined. Substituting the coordinates of the three given points in turn into equation (1), we obtain the three simultaneous equations

$$\begin{aligned} -a + b + 2c + d &= 0, \\ 2a \qquad - 3c + d &= 0, \\ 5a + b - 2c + d &= 0, \end{aligned} \tag{2}$$

in the four unknowns a, b, c, and d. However, these four unknowns are not all *independent*. For example, if in equation (1) we assume that $a \neq 0$, then we may divide through by a so as to make the x coefficient equal to 1. Hence in equations (2) we could set $a = 1$ to obtain

$$\begin{aligned} b + 2c + d &= 1 \\ -3c + d &= -2 \\ b - 2c + d &= -5. \end{aligned}$$

Solving these equations for b, c, and d, we find that $b = -\frac{9}{2}$, $c = \frac{3}{2}$, $d = \frac{5}{2}$. Hence the required equation of the plane is

$$x - \tfrac{9}{2}y + \tfrac{3}{2}z + \tfrac{5}{2} = 0,$$

or

$$2x - 9y + 3z + 5 = 0.$$

The use of determinants makes it simple to find an equation of the plane passing through three noncollinear points (x_1,y_1,z_1), (x_2,y_2,z_2), and (x_3,y_3,z_3). Such an equation is

14.9
$$\begin{vmatrix} x & y & z & 1 \\ x_1 & y_1 & z_1 & 1 \\ x_2 & y_2 & z_2 & 1 \\ x_3 & y_3 & z_3 & 1 \end{vmatrix} = 0.$$

It is evident that 14.9 is a linear equation. If in 14.9 we replace (x,y,z) by (x_i,y_i,z_i), $i = 1, 2, 3$, then two rows of the matrix are equal and therefore its determinant is zero. Thus each of the given points is on the graph of 14.9.

Another method for solving this type of problem will be given in Section 5 of this chapter.

Two planes with equations

$$a_1x + b_1y + c_1z + d_1 = 0$$
$$a_2x + b_2y + c_2z + d_2 = 0$$

are parallel if and only if their normal vectors have the same or opposite direction. Taking $\mathbf{v}_1 = \langle a_1,b_1,c_1\rangle$ and $\mathbf{v}_2 = \langle a_2,b_2,c_2\rangle$ as normal vectors, the planes are parallel if and only if

$$\mathbf{v}_2 = r\mathbf{v}_1 \qquad \text{for some } r \text{ in } \mathsf{R},$$

i.e., if and only if $a_2 = ra_1$, $b_2 = rb_1$, $c_2 = rc_1$ for some r in R.

The two planes with equations as above are perpendicular if and only if their normal vectors are perpendicular, i.e., if and only if $\mathbf{v}_1 \cdot \mathbf{v}_2 = 0$ or

$$a_1a_2 + b_1b_2 + c_1c_2 = 0.$$

For example, the plane with equation

$$ax + by + d = 0$$

is perpendicular to the xy plane with equation $z = 0$, since $0 \cdot a + 0 \cdot b + 1 \cdot 0 = 0$. In other words, the plane having an equation with no z term is parallel to the z axis. Analogously, the plane with equation

$$ax + cz + d = 0$$

is parallel to the y axis, and so on.

Two nonparallel planes intersect in a line that can be easily determined, as illustrated in the following example.

Example 4 Determine the line of intersection L of the two planes

$$2x - y + z - 4 = 0, \qquad x + 3y - z - 2 = 0.$$

Solution: Points on the line L may be determined by assigning a value to one coordinate arbitrarily, and then solving the two given equations simultaneously for the other two coordinates. Thus, if we set $x = 0$, the equations of the two planes become

$$-y + z - 4 = 0,$$
$$3y - z - 2 = 0.$$

Solving these simultaneously for y and z, we obtain $y = 3$, $z = 7$. Hence the point $P = (0,3,7)$ is on the line L. Setting $y = 0$ in the equations of the two planes, we obtain

$$2x + z - 4 = 0,$$
$$x - z - 2 = 0,$$

from which we get $x = 2$, $z = 0$. Hence the point $Q = (2,0,0)$ is on the line L.

The direction of L is determined by the vector $\mathbf{v} = \overrightarrow{PQ} = \langle 2,-3,-7\rangle$ (or by the vector $-\mathbf{v}$). A vector equation for L is

$$\lambda(t) = \mathbf{P} + t\mathbf{v},$$

where $P = (0,3,7)$ and $\mathbf{v} = \langle 2,-3,-7\rangle$. Parametric equations for L are given by

$$x = 2t, \qquad y = 3 - 3t, \qquad z = 7 - 7t.$$

We conclude this section by deriving a useful formula for the perpendicular distance from a point to a plane.

14.10 Theorem

Let p be the plane whose equation is $ax + by + cz + d = 0$, and let A be the point (x_0,y_0,z_0). Then the perpendicular distance from A to the plane p is given by

$$\frac{|ax_0 + by_0 + cz_0 + d|}{\sqrt{a^2 + b^2 + c^2}}.$$

Proof: The line L perpendicular to p through the point $A = (x_0,y_0,z_0)$ has parametric equations

$$(1) \qquad x = x_0 + at, \qquad y = y_0 + bt, \qquad z = z_0 + ct.$$

To find the point B in which L intersects the plane p, we substitute the expressions for x, y, and z given by (1) in the equation for p and then solve for t. We obtain

$$a(x_0 + at) + b(y_0 + bt) + c(z_0 + ct) + d = 0,$$
$$ax_0 + by_0 + cz_0 + d + (a^2 + b^2 + c^2)t = 0,$$

$$(2) \qquad t = \frac{-(ax_0 + by_0 + cz_0 + d)}{a^2 + b^2 + c^2}.$$

If $B = (x,y,z)$, then

$$[d(A,B)]^2 = (x - x_0)^2 + (y - y_0)^2 + (z - z_0)^2,$$

$$[d(A,B)]^2 = a^2t^2 + b^2t^2 + c^2t^2 = (a^2 + b^2 + c^2)t^2, \tag{3}$$

using equations (1). But the point B on L corresponds to the value of the parameter t given by (2). Substituting (2) into equation (3), we get

$$[d(A,B)]^2 = \frac{(ax_0 + by_0 + cz_0 + d)^2}{a^2 + b^2 + c^2},$$

from which the theorem follows.

EXERCISES

I

In each of Exercises 1 to 4, find an equation of the plane passing through the given point and perpendicular to the given vector.

1. $(3,1,3)$, $\langle 1,1,-1 \rangle$
2. $(0,2,-2)$, $\langle -1,2,-3 \rangle$
3. (a,b,c), $\langle 1,1,0 \rangle$
4. $(1,2,3)$, $\langle 0,5,0 \rangle$

In each of Exercises 5 to 8, find the traces in the coordinate planes of the given plane and sketch.

5. $2x + y + z = 4$
6. $3x + 4y + 6z = 12$
7. $4x + y = 6$
8. $2y - 3z = 4$

In each of Exercises 9 to 12, find an equation of the plane passing through the three given points.

9. $(0,0,0)$, $(1,1,1)$, $(-1,1,0)$
10. $(2,1,3)$, $(5,2,-1)$, $(3,0,1)$
11. $(2,1,-1)$ $(-1,3,1)$, $(4,0,1)$
12. $(a,0,0)$, $(0,b,0)$, $(0,0,c)$
13. Find an equation of the plane that passes through the points $(1,0,-1)$ and $(2,1,3)$ and is perpendicular to the plane $x + y - z + 2 = 0$.
14. Find an equation of the plane that passes through the point $(1,-1,4)$ and is perpendicular to each of the planes $2x + y - z + 2 = 0$ and $x - y + 3z - 1 = 0$.

In Exercises 15 and 16, find parametric equations for the line of intersection of the given pair of planes.

15. $x + 2y - z = 7$, $x - 2y + z = 3$
16. $x + z = 5$, $y - z = 2$
17. Find an equation of the plane containing the two lines in Exercise I-12, Section 3.

18. Find an equation of the plane containing the line $x = 1 + 2t$, $y = 2 - t$, $z = 4t$ and the point (1,0,3).

19. Find an equation of the plane containing the point (0,0,1) and the line of intersection of the planes $x + y + z = 2$, $x - 2y - 3z = 5$.

20. Find the angle of intersection of the normals to the planes $x - 2y + 2z = 7$, $3x - 2y + 2z = 1$.

21. Show that the following lines L_1 and L_2 intersect, and find an equation of the plane containing them.

$$L_1: \quad x = 3 + 2t, \quad y = 2 + t, \quad z = -4 - 3t,$$
$$L_2: \quad x = -3 + 4s, \quad y = 5 - 4s, \quad z = -6 + 5s.$$

22. Find an equation of the plane that contains the line $x = 1 + t$, $y = 5 - t$, $z = 4 - 2t$ and is parallel to the line $x = 2t$, $y = 1 - t$, $z = 7 + t$.

In Exercises 23 and 24, find the perpendicular distance from the given point to the given plane.

23. $(3,4,-5)$, $2x + 2y + z = 10$

24. $(1,0,0)$, $3x - y + 2z = 5$

25. Find the perpendicular distance between the parallel planes $ax + by + cz + d = 0$, $ax + by + cz + d' = 0$. (*Suggestion:* Find the distance from the origin to each plane.)

26. Find an equation of a plane whose perpendicular distance to the origin is $\frac{1}{3}$ and which passes through the points (1,0,0) and (0,0,2).

27. Find equations of the two planes whose perpendicular distances to the origin are 3 units, and which are perpendicular to the line through the points (7,3,1) and $(6,4,-1)$.

28. If a, b, c are the intercepts of a plane on the coordinate axes, and r is the perpendicular distance from the plane to the origin, prove that

$$\frac{1}{r^2} = \frac{1}{a^2} + \frac{1}{b^2} + \frac{1}{c^2}.$$

29. Use 14.9 above to work Example 3 of this section.

5 THE CROSS PRODUCT

Given three-dimensional vectors

$$\mathbf{u} = \langle l_1, m_1, n_1 \rangle \quad \text{and} \quad \mathbf{v} = \langle l_2, m_2, n_2 \rangle,$$

we find that the associated linear equation

$$\begin{vmatrix} x & y & z \\ l_1 & m_1 & n_1 \\ l_2 & m_2 & n_2 \end{vmatrix} = 0 \tag{1}$$

has $x = l_1, y = m_1, z = n_1$ as a solution, since the determinant of a matrix with two equal rows is zero. Similarly, $x = l_2$, $y = m_2$, $z = n_2$ is also a solution. If we expand (1) by minors of the first row, we obtain the equivalent equation

(2) $$\begin{vmatrix} m_1 & n_1 \\ m_2 & n_2 \end{vmatrix} x - \begin{vmatrix} l_1 & n_1 \\ l_2 & n_2 \end{vmatrix} y + \begin{vmatrix} l_1 & m_1 \\ l_2 & m_2 \end{vmatrix} z = 0$$

or

(3) $$\begin{vmatrix} m_1 & n_1 \\ m_2 & n_2 \end{vmatrix} x + \begin{vmatrix} n_1 & l_1 \\ n_2 & l_2 \end{vmatrix} y + \begin{vmatrix} l_1 & m_1 \\ l_2 & m_2 \end{vmatrix} z = 0.$$

Now if we define the vector **w** to be

$$\mathbf{w} = \left\langle \begin{vmatrix} m_1 & n_1 \\ m_2 & n_2 \end{vmatrix}, \begin{vmatrix} n_1 & l_1 \\ n_2 & l_2 \end{vmatrix}, \begin{vmatrix} l_1 & m_1 \\ l_2 & m_2 \end{vmatrix} \right\rangle,$$

then $\mathbf{w} \cdot \langle x,y,z \rangle = 0$ for every solution x, y, z of (3). Hence, by our remarks above, $\mathbf{w} \cdot \mathbf{u} = 0$ and $\mathbf{w} \cdot \mathbf{v} = 0$. Thus **w** is a vector perpendicular to both **u** and **v**.

Let us define the vector **w** above to be the product of **u** and **v**. Thus the *cross product* of vectors $\mathbf{u} = \langle l_1,m_1,n_1 \rangle$ and $\mathbf{v} = \langle l_2,m_2,n_2 \rangle$ is designated by $\mathbf{u} \times \mathbf{v}$ and is defined as follows:

14.11 $$\mathbf{u} \times \mathbf{v} = \langle m_1n_2 - m_2n_1, n_1l_2 - n_2l_1, l_1m_2 - l_2m_1 \rangle.$$

In particular, the cross products of the unit vectors **i**, **j**, and **k** are given by

$$\mathbf{i} \times \mathbf{j} = \mathbf{k},$$
$$\mathbf{j} \times \mathbf{i} = -\mathbf{k},$$

$$\mathbf{j} \times \mathbf{k} = \mathbf{i},$$
$$\mathbf{k} \times \mathbf{j} = -\mathbf{i},$$

$$\mathbf{k} \times \mathbf{i} = \mathbf{j},$$
$$\mathbf{i} \times \mathbf{k} = -\mathbf{j}.$$

Since $\mathbf{i} \times \mathbf{j} = -\mathbf{j} \times \mathbf{i}$, evidently the cross product is a noncommutative operation. This operation is also nonassociative, as we shall soon see. The important properties of the cross product are listed below. They hold for all vectors **u**, **v**, and **w** and for each scalar c.

P1 $\mathbf{u} \times \mathbf{v} = -\mathbf{v} \times \mathbf{u}$ (*anticommutative law*).

P2 $(c\mathbf{u}) \times \mathbf{v} = \mathbf{u} \times (c\mathbf{v}) = c(\mathbf{u} \times \mathbf{v})$.

P3 $(\mathbf{u} + \mathbf{v}) \times \mathbf{w} = \mathbf{u} \times \mathbf{w} + \mathbf{v} \times \mathbf{w}$.
$\mathbf{u} \times (\mathbf{v} + \mathbf{w}) = \mathbf{u} \times \mathbf{v} + \mathbf{u} \times \mathbf{w}$ (*distributive laws*).

P4 $\mathbf{v} \times \mathbf{0} = \mathbf{0} \times \mathbf{v} = \mathbf{0}$.

P5 $\mathbf{u} \cdot (\mathbf{v} \times \mathbf{w}) = (\mathbf{u} \times \mathbf{v}) \cdot \mathbf{w}$.

P6 $\mathbf{u} \times (\mathbf{v} \times \mathbf{w}) = (\mathbf{u} \cdot \mathbf{w})\mathbf{v} - (\mathbf{u} \cdot \mathbf{v})\mathbf{w}$.

P7 $|\mathbf{u} \times \mathbf{v}| = \sqrt{|\mathbf{u}|^2|\mathbf{v}|^2 - (\mathbf{u} \cdot \mathbf{v})^2} = |\mathbf{u}||\mathbf{v}| \sin \theta$, θ the angle between $\mathbf{u}$ and $\mathbf{v}$.

These properties may be proved by using 14.11 and the definitions of the other vector operations. We shall prove some of them below and leave the rest as exercises. Let

$$\mathbf{u} = \langle l_1, m_1, n_1 \rangle, \qquad \mathbf{v} = \langle l_2, m_2, n_2 \rangle, \qquad \mathbf{w} = \langle l_3, m_3, n_3 \rangle.$$

Proof of P1: $\mathbf{u} \times \mathbf{v}$ is given by 14.11. We may obtain $\mathbf{v} \times \mathbf{u}$ from 14.11 by interchanging subscripts 1 and 2. Thus

$$\mathbf{v} \times \mathbf{u} = \langle m_2 n_1 - m_1 n_2, n_2 l_1 - n_1 l_2, l_2 m_1 - l_1 m_2 \rangle.$$

Clearly, $\mathbf{v} \times \mathbf{u} = -\mathbf{u} \times \mathbf{v}$.

Proof of P5: Evidently,

$$\begin{aligned}\mathbf{u} \cdot (\mathbf{v} \times \mathbf{w}) &= l_1(m_2 n_3 - m_3 n_2) + m_1(n_2 l_3 - n_3 l_2) + n_1(l_2 m_3 - l_3 m_2),\\ (\mathbf{u} \times \mathbf{v}) \cdot \mathbf{w} &= (m_1 n_2 - m_2 n_1) l_3 + (n_1 l_2 - n_2 l_1) m_3 + (l_1 m_2 - l_2 m_1) n_3.\end{aligned}$$

It is easily verified that these two numbers are equal.

Proof of P6: Since

$$\mathbf{v} \times \mathbf{w} = \langle m_2 n_3 - m_3 n_2, n_2 l_3 - n_3 l_2, l_2 m_3 - l_3 m_2 \rangle,$$

we have

$$\begin{aligned}\mathbf{u} \times (\mathbf{v} \times \mathbf{w}) &= \langle m_1(l_2 m_3 - l_3 m_2) - (n_2 l_3 - n_3 l_2) n_1, n_1(m_2 n_3 - m_3 n_2)\\ &\qquad - (l_2 m_3 - l_3 m_2) l_1, l_1(n_2 l_3 - n_3 l_2) - (m_2 n_3 - m_3 n_2) m_1 \rangle\\ &= \langle (\mathbf{u} \cdot \mathbf{w}) l_2 - (\mathbf{u} \cdot \mathbf{v}) l_3, (\mathbf{u} \cdot \mathbf{w}) m_2 - (\mathbf{u} \cdot \mathbf{v}) m_3, (\mathbf{u} \cdot \mathbf{w}) n_2 - (\mathbf{u} \cdot \mathbf{w}) n_3 \rangle\\ &= (\mathbf{u} \cdot \mathbf{w})\langle l_2, m_2, n_2 \rangle - (\mathbf{u} \cdot \mathbf{v})\langle l_3, m_3, n_3 \rangle.\end{aligned}$$

This proves (P6).

Proof of P7: A straightforward calculation shows that

$$\begin{aligned}(l_1^2 + m_1^2 + n_1^2)(l_2^2 + m_2^2 + n_2^2)&\\ = (l_1 l_2 + m_1 m_2 + n_1 n_2)^2&\\ + (m_1 n_2 - m_2 n_1)^2 + (n_1 l_2 - n_2 l_1)^2 + (l_1 m_2 - l_2 m_1)^2&,\end{aligned}$$

and therefore that $|\mathbf{u} \times \mathbf{v}|^2 = |\mathbf{u}|^2|\mathbf{v}|^2 - (\mathbf{u} \cdot \mathbf{v})^2$. We recall from 14.4 that $\mathbf{u} \cdot \mathbf{v} = |\mathbf{u}||\mathbf{v}| \cos \theta$. Hence

$$|\mathbf{u} \times \mathbf{v}|^2 = |\mathbf{u}|^2|\mathbf{v}|^2(1 - \cos^2 \theta) = |\mathbf{u}|^2|\mathbf{v}|^2 \sin^2 \theta.$$

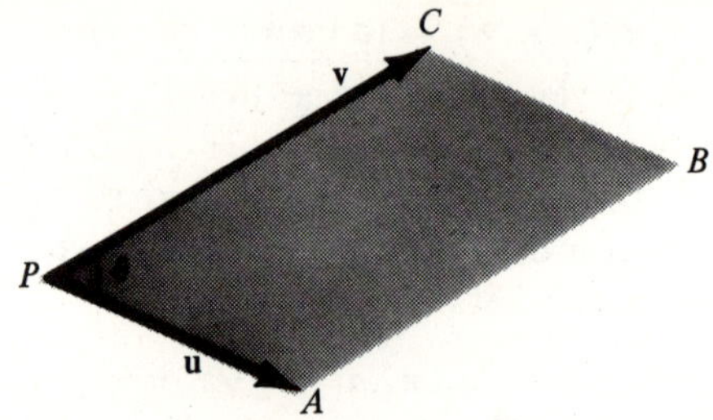

FIGURE 14.12

Since $\sin\theta \geq 0$ for θ in $[0,\pi]$, we may take square roots of the quantities above to obtain P7.

Geometrically, $|\mathbf{u} \times \mathbf{v}|$ is the area of the parallelogram $PABC$ determined by vectors $\mathbf{u}$ and $\mathbf{v}$ at any point P in R^3 (Figure 14.12.) This fact follows directly from P7, $|\mathbf{u} \times \mathbf{v}| = |\mathbf{u}||\mathbf{v}| \sin\theta$. If $\mathbf{u}$ and $\mathbf{v}$ are nonzero vectors, then $\mathbf{u} \times \mathbf{v} = \mathbf{0}$ if and only if the points P, A, B, and C of Figure 14.12 are collinear, i.e., if and only if $\mathbf{u} = c\mathbf{v}$ for some scalar c. We state this result in the following theorem.

14.12 Theorem

If $\mathbf{u}$ *and* $\mathbf{v}$ *are nonzero vectors, then*

$$\mathbf{u} \times \mathbf{v} = \mathbf{0} \text{ if and only if } \mathbf{u} = c\mathbf{v}$$

for some scalar c.

That cross product is a nonassociative operation is seen by the following example:

$$\mathbf{i} \times (\mathbf{i} \times \mathbf{j}) = \mathbf{i} \times \mathbf{k} = -\mathbf{j}, \qquad (\mathbf{i} \times \mathbf{i}) \times \mathbf{j} = \mathbf{0} \times \mathbf{j} = \mathbf{0}.$$

Therefore $\mathbf{i} \times (\mathbf{i} \times \mathbf{j}) \neq (\mathbf{i} \times \mathbf{i}) \times \mathbf{j}$.

Given three-dimensional vectors $\mathbf{u}$, $\mathbf{v}$, and $\mathbf{w}$, the scalar quantity of P5 above, namely,

$$\mathbf{u} \cdot (\mathbf{v} \times \mathbf{w}) = (\mathbf{u} \times \mathbf{v}) \cdot \mathbf{w},$$

is called the *scalar triple product* of $\mathbf{u}$, $\mathbf{v}$, and $\mathbf{w}$ (in that order). Its geometrical significance is seen in the following theorem.

14.13 Theorem

Let $\mathbf{u}$, $\mathbf{v}$, *and* $\mathbf{w}$ *be three noncoplanar vectors with the same initial point* P. *Then the volume of the parallelepiped with* $\mathbf{u}$, $\mathbf{v}$, *and* $\mathbf{w}$ *as three of its edges is given by* $|\mathbf{u} \cdot (\mathbf{v} \times \mathbf{w})|$.

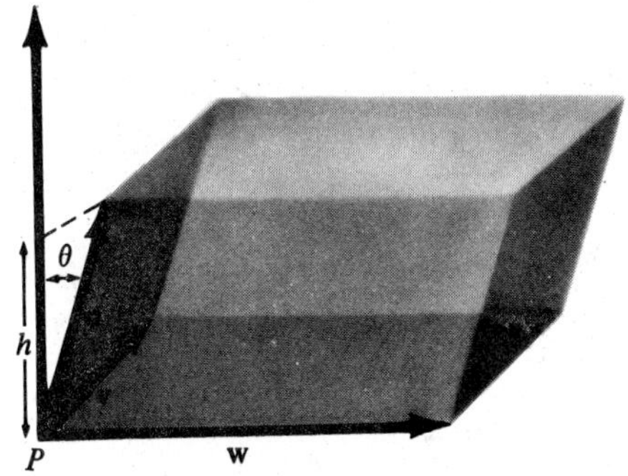

FIGURE 14.13

Proof: Referring to Figure 14.13, we see from our previous remarks that the area of the base of the parallelepiped is $|\mathbf{v} \times \mathbf{w}|$. Furthermore, $\mathbf{v} \times \mathbf{w}$ is a vector perpendicular to this base. The altitude h of the parallelepiped is the component of the vector $\mathbf{u}$ on $\mathbf{v} \times \mathbf{w}$; that is,

$$h = |\mathbf{u}| \cos \theta,$$

where θ is the angle between $\mathbf{u}$ and $\mathbf{v} \times \mathbf{w}$. But by 14.4,

$$h = |\mathbf{u}| \cos \theta = \frac{\mathbf{u} \cdot (\mathbf{v} \times \mathbf{w})}{|\mathbf{v} \times \mathbf{w}|}.$$

Since the volume of the parallelepiped is $h|\mathbf{v} \times \mathbf{w}|$, the theorem now follows.

It is left as an exercise for the student (Exercise 10, this section) to show that the scalar triple product of $\mathbf{u}$, $\mathbf{v}$, and $\mathbf{w}$ may be evaluated by the determinant whose rows are the components of the vectors $\mathbf{u}$, $\mathbf{v}$, and $\mathbf{w}$. This provides us with a simple method for computing volumes of parallelepipeds.

Cross products are often useful in problems in which it is required to compute a vector perpendicular to two given vectors. This is illustrated in the following examples.

Example 1 Find an equation of the plane that contains the points $A = (-1,2,2)$ and $B = (2,0,8)$ and is parallel to the line L through $C = (0,1,0)$ and $D = (4,1,-1)$.

Solution: It will be sufficient to find a vector $\mathbf{u}$ in the direction of the normal to the required plane. Such a vector $\mathbf{u}$ must be perpendicular to the vectors

$$\mathbf{v} = \overrightarrow{AB} = \langle 3,-2,6 \rangle$$

and

$$\mathbf{w} = \overrightarrow{CD} = \langle 4,0,-1 \rangle.$$

Thus we may take $\mathbf{u}$ to be the vector

$$\mathbf{v} \times \mathbf{w} = \langle 2,27,8 \rangle.$$

Hence an equation for the required plane is

$$2(x - 2) + 27y + 8(z - 8) = 0,$$

or

$$2x + 27y + 8z - 68 = 0.$$

Example 2 Find the perpendicular distance between the lines L_1 and L_2, if L_1 passes through the points $A(-1,1,1)$ and $B(2,1,5)$ and L_2 passes through $C(-2,-1,4)$ and $D(-1,3,0)$.

Solution: We first find a vector **w** perpendicular to both L_1 and L_2, or equivalently, to the vectors

$$\overrightarrow{AB} = \langle 3,0,4 \rangle,$$

and

$$\overrightarrow{CD} = \langle 1,4,-4 \rangle.$$

(See Figure 14.14.) We take **w** as the vector

$$\overrightarrow{AB} \times \overrightarrow{CD} = \langle -16,16,12 \rangle = 4\langle -4,4,3 \rangle.$$

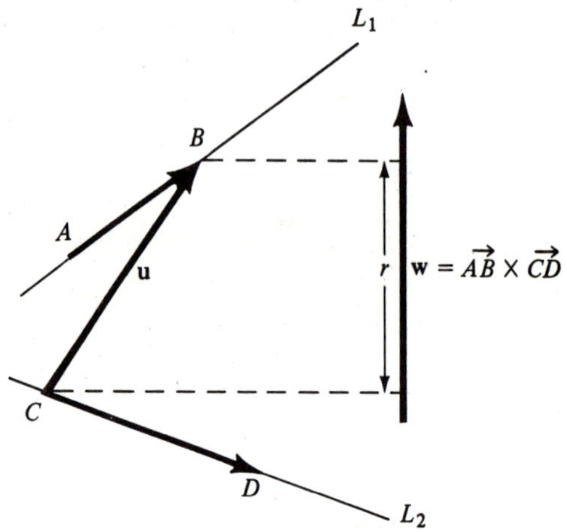

FIGURE 14.14

The perpendicular distance r from L_1 to L_2 may be found by taking any vector **u** from one line to the other, and projecting it onto the vector **w**. For example, let us take $\mathbf{u} = \overrightarrow{CB} = \langle 4,2,1 \rangle$. Then the component of **u** on **w** is $|\mathbf{u}||\cos\theta|$, where θ is the angle between **u** and **w**. Hence, by 14.4,

$$r = |\mathbf{u}||\cos\theta| = \frac{|\mathbf{u}\cdot\mathbf{w}|}{|\mathbf{w}|},$$

$$r = \frac{4|4(-4) + 2\cdot 4 + 1\cdot 3|}{4\sqrt{16 + 16 + 9}} = \frac{5}{\sqrt{41}}.$$

The reader may check the computation by taking **u** as any other vector between L_1 and L_2, say $\mathbf{u} = \overrightarrow{AD}$.

EXERCISES

I

In Exercises 1 to 6, evaluate each of the given expressions if $\mathbf{u} = \mathbf{i} + 2\mathbf{j} - 4\mathbf{k}$, $\mathbf{v} = 2\mathbf{j} - 3\mathbf{k}$, and $\mathbf{w} = -\mathbf{i} + 2\mathbf{k}$.

1. $\mathbf{u}\cdot\mathbf{v}$, $\mathbf{w}\cdot\mathbf{u}$
2. $\mathbf{u}\times\mathbf{v}$, $\mathbf{v}\times\mathbf{u}$
3. $\mathbf{u}\times\mathbf{u} + \mathbf{v}\times\mathbf{v}$
4. $\mathbf{u}\cdot\mathbf{v}\times\mathbf{w}$, $\mathbf{u}\times\mathbf{v}\cdot\mathbf{w}$
5. $\mathbf{u}\times(\mathbf{v}\times\mathbf{w})$, $(\mathbf{u}\times\mathbf{v})\times\mathbf{w}$
6. $(\mathbf{u}\times\mathbf{v})\cdot(\mathbf{u}\times\mathbf{w})$
7. Show that if $\mathbf{u} = \langle a_1,a_2,a_3\rangle$ and $\mathbf{v} = \langle b_1,b_2,b_3\rangle$ then

$$\mathbf{u}\times\mathbf{v} = \begin{vmatrix} \mathbf{i} & \mathbf{j} & \mathbf{k} \\ a_1 & a_2 & a_3 \\ b_1 & b_2 & b_3 \end{vmatrix},$$

where the determinant is defined as usual (although it contains vectors).

8. Prove that $(\mathbf{u} + \mathbf{v}) \times (\mathbf{u} - \mathbf{v}) = 2\mathbf{v} \times \mathbf{u}$.
9. If $\mathbf{u} + \mathbf{v} + \mathbf{w} = \mathbf{0}$, show that $\mathbf{u}\times\mathbf{v} = \mathbf{v}\times\mathbf{w} = \mathbf{w}\times\mathbf{u}$.
10. If $\mathbf{u} = \langle a_1,a_2,a_3\rangle$, $\mathbf{v} = \langle b_1,b_2,b_3\rangle$, and $\mathbf{w} = \langle c_1,c_2,c_3\rangle$, then show that

$$\mathbf{u}\cdot(\mathbf{v}\times\mathbf{w}) = (\mathbf{u}\times\mathbf{v})\cdot\mathbf{w} = \begin{vmatrix} a_1 & a_2 & a_3 \\ b_1 & b_2 & b_3 \\ c_1 & c_2 & c_3 \end{vmatrix}.$$

11. Find the volume of the parallelepiped three of whose edges are the position vectors $\mathbf{i} + \mathbf{j}$, $\mathbf{j} + \mathbf{k}$, and $\mathbf{k} + \mathbf{i}$.
12. Given the points $A = (-1,1,2)$, $B = (0,2,3)$, $C = (1,1,1)$, and $D = (-1,3,3)$, find the volume of the parallelepiped with $\overrightarrow{AB}$, $\overrightarrow{AC}$, and $\overrightarrow{AD}$ as three of its edges.
13. Find the area of the parallelogram that has the points $(-1,3,2)$, $(3,4,-1)$, and $(3,3,1)$ as three of its vertices.
14. Find the area of the triangle with vertices $(1,-2,4)$, $(6,4,-2)$, and $(5,3,-1)$.

Use cross products to do Exercises 15 to 20.

15. Find an equation of the plane that passes through the point $(1,3,-2)$ and is perpendicular to each of the planes $2x + y - z + 2 = 0$ and $x - y + 3z - 1 = 0$.
16. Find an equation of the plane containing the points $(1,1,0)$, $(5,3,-1)$, and $(-2,2,4)$.
17. Find an equation of the plane that contains the points $(2,0,-2)$ and $(3,4,1)$ and is perpendicular to the plane $x + y - z + 2 = 0$.

In Exercises 18 to 20, find the perpendicular distance between the given lines L_1 and L_2.

18. L_1 through $(-1,5,4)$ and $(3,1,1)$; L_2 through $(2,2,0)$ and $(2,-2,1)$.

19. L_1 is the line of intersection of the planes $x + 2z = 3$, $y - z = 1$; L_2 passes through (0,1,1) and (−3,2,2).

20. L_1 and L_2 have the respective parametric equations

$$x = 2 - t, \quad y = 3t, \quad z = 5 + 2t,$$
$$x = 7 + s, \quad y = 2 - 4s, \quad z = 1 - s.$$

21. Let **u** and **v** be vectors, neither of which is a scalar multiple of the other, with the same initial point A. Let p be the plane containing the point A and the vectors **u** and **v**. If **w** is any vector of the form $\mathbf{w} = r\mathbf{u} + s\mathbf{v}$ (for scalars r and s) with initial point at A, prove that **w** lies in the plane p. [*Suggestion:* Prove that **w** is perpendicular to the normal to p.]

II

1. Prove that for all vectors **u**, **v**, **w**:

$$\mathbf{u} \times (\mathbf{v} \times \mathbf{w}) + \mathbf{v} \times (\mathbf{w} \times \mathbf{u}) + \mathbf{w} \times (\mathbf{u} \times \mathbf{v}) = \mathbf{0}.$$

Prove that the following identities hold for all vectors **A**, **B**, **C**, and **D**.

2. $(\mathbf{A} \times \mathbf{B}) \times (\mathbf{C} \times \mathbf{D}) = [\mathbf{A} \cdot (\mathbf{B} \times \mathbf{D})]\mathbf{C} - [\mathbf{A} \cdot (\mathbf{B} \times \mathbf{C})]\mathbf{D}$

3. $(\mathbf{A} \times \mathbf{B}) \cdot (\mathbf{C} \times \mathbf{D}) = (\mathbf{A} \cdot \mathbf{C})(\mathbf{B} \cdot \mathbf{D}) - (\mathbf{A} \cdot \mathbf{D})(\mathbf{B} \cdot \mathbf{C})$

6 CYLINDERS AND SURFACES OF REVOLUTION

The graph in space of an equation in x, y, and z is called a *surface*. Two examples of surfaces have been studied thus far in this chapter, namely the sphere (14.2) and the plane (14.7).

Although it is easy to visualize a plane or a sphere, and even easy to sketch these surfaces on a plane, the general problem of visualizing and sketching a surface is much more difficult than the corresponding problem of sketching a curve in the plane. Of course, this is due to the fact that a surface in space can be sketched in perspectivity only on a piece of paper. Three-dimensional models of the common surfaces are available and are of great help in visualizing these surfaces.

One of the easiest surfaces to visualize is a cylinder. A *cylinder* is a surface that may be thought of as being generated by a line (called the *generator*) moving along a given curve in such a way as always to remain parallel to its original position. The curve along which the generating line moves is called a *directrix** of the cylinder. Each of the lines on the cylinder parallel to the generator is called a *ruling* of the cylinder.

If the directrix is a straight line, the cylinder is simply a plane. Other than a plane, the most common cylinder is the right circular cylinder, generated

* A directrix of a cylinder is in no way related to the directrix of a parabola.

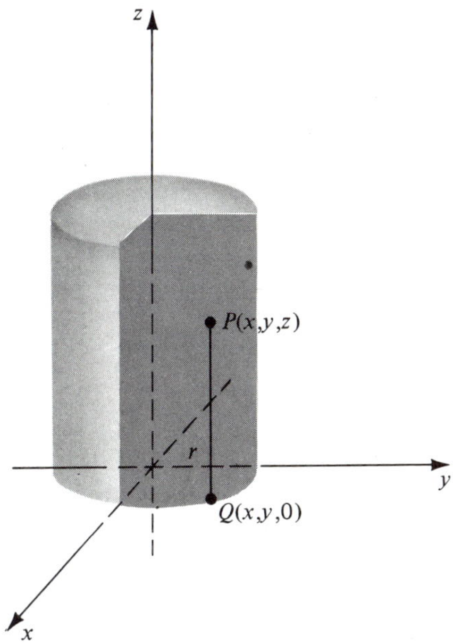

FIGURE 14.15

by a line moving along a circle so as always to be perpendicular to the plane of the circle.

The right circular cylinder sketched in Figure 14.15 has as its directrix a circle of radius r in the xy plane with its center at the origin. A point $P(x,y,z)$ is on this cylinder if and only if the projection $Q(x,y,0)$ of P on the xy plane is on the directrix of the cylinder. Thus $P(x,y,z)$ is a point on the cylinder if and only if

$$x^2 + y^2 = r^2,$$

and we conclude that this equation of the cylinder is just an equation of its directrix in the xy plane.

It is clear from the preceding example that an equation in x and y has as its graph in space a cylinder with generator parallel to the z axis. Similarly, an equation in x and z has its graph in space a cylinder with generator parallel to the y axis, and so on. In each case the directrix is the graph in a coordinate plane of the given equation in two variables.

Example 1 Discuss and sketch the graph of the equation

$$z = 4 - x^2.$$

Solution: The graph of this equation in the xz plane is a parabola symmetric to the z axis with vertex at (0,0,4). The graph in space is a parabolic cylinder with directrix

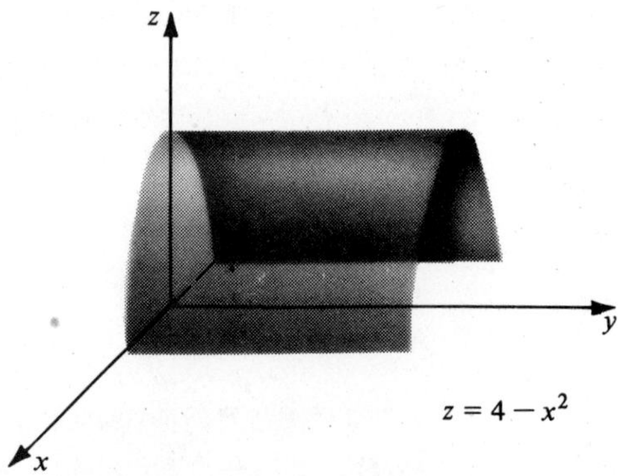

FIGURE 14.16

the parabola in the xz plane and generator parallel to the y axis. It is sketched in Figure 14.16.

Example 2 Discuss and sketch the graph of the equation

$$z = |y|.$$

Solution: The graph of this equation in the yz plane consists of two half-lines emanating from the origin. Thus its graph in space is a cylinder with generator parallel to the x axis and directrix the two half-lines. It looks like a trough made up of two half-planes meeting on the x axis at an angle of 90°, as sketched in Figure 14.17.

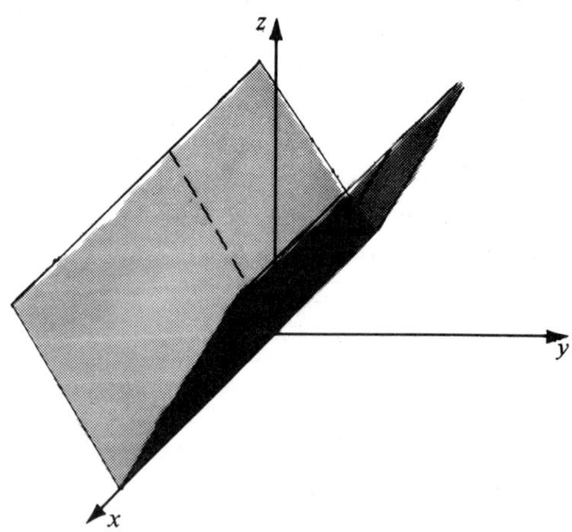

FIGURE 14.17

Example 3 Discuss and sketch the graph of the equation

$$y = e^x.$$

Solution: The cylinder has as its directrix the exponential curve $y = e^x$ in the xy plane and as its generator a line parallel to the z axis, as sketched in Figure 14.18.

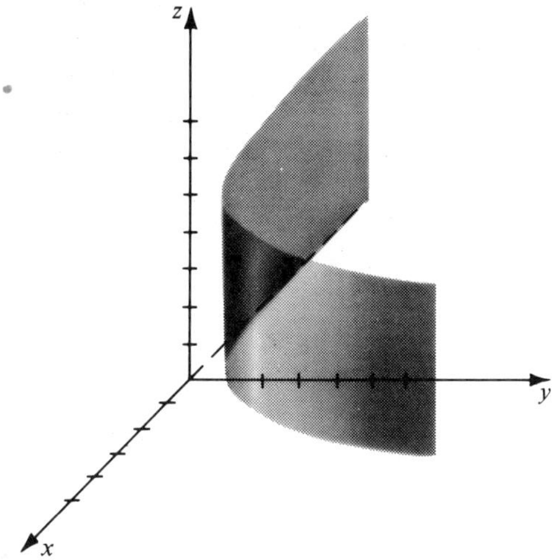

FIGURE 14.18

A fact that makes the sketching of a cylinder comparatively easy is that the cross section of the cylinder in each plane parallel to the plane of the directrix is the same as the directrix. By the *cross section* of a surface in a plane, we mean the set of all points of the surface that are on the given plane. For example, the cross section of a right circular cylinder in each plane parallel to the plane of the directrix is a circle.

Another surface that is easy to visualize is the surface of revolution discussed in previous chapters. We recall that a surface of revolution is generated by rotating a plane curve about some fixed axis in its plane. If a surface of revolution is generated by rotating a curve in a coordinate plane about a coordinate axis, then an equation of the surface may be easily found, as we shall now show.

Let the surface of revolution be formed by rotating the graph of an equation

$$f(x,y) = 0$$

about the x axis. We shall assume that $y \geq 0$ for each point (x,y) on the generating curve. A point $P(x,y,z)$ will be on this surface if and only if $Q(x,y_0,0)$ is on the generating curve, where

$$y_0 = |PA| = \sqrt{y^2 + z^2},$$

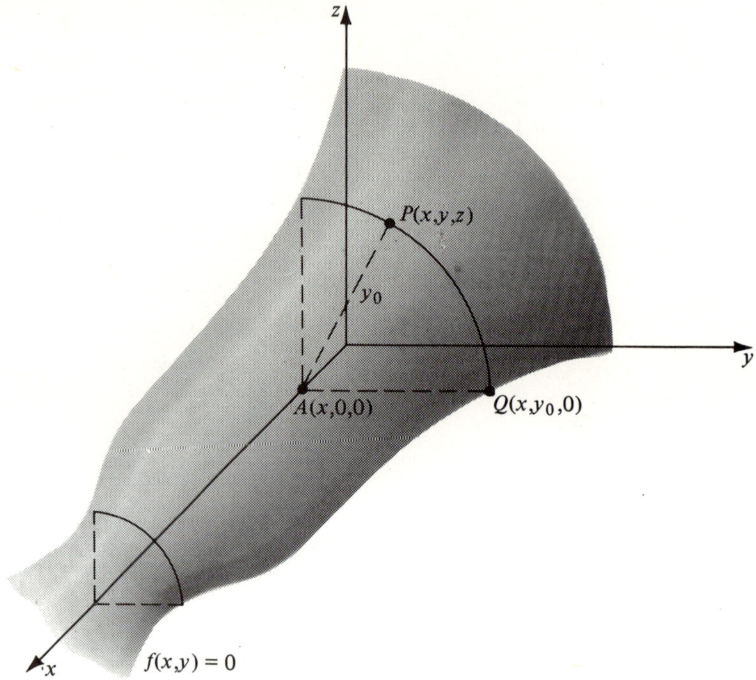

FIGURE 14.19

as indicated in Figure 14.19. Since $Q(x, y_0, 0)$ is on the generating curve if and only if $f(x, y_0) = 0$, the point $P(x,y,z)$ is on the surface of revolution if and only if

$$f(x, \sqrt{y^2 + z^2}) = 0.$$

Hence this equation in x, y, and z is an equation of the surface of revolution.

It is clear that similar remarks can be made if the generating curve is in any one of the coordinate planes and the axis of revolution is a coordinate axis in that plane. If, for example, the surface is generated by rotating a curve in the yz plane about the y axis, then the equation of the surface is obtained by replacing each z by $\sqrt{x^2 + z^2}$ in an equation of the generating curve.

We might summarize our remarks above by saying that the graph of an equation in x, y, and z is a surface of revolution if and only if two of the variables occur together in the form $x^2 + y^2$, $x^2 + z^2$, or $y^2 + z^2$.

Example 4 Find an equation of the surface generated by rotating the curve

$$x^2 + 4y^2 = 4$$

in the xy plane about the x axis.

Solution: We replace each y by $\sqrt{y^2 + z^2}$ in the equation of the generating curve to obtain

$$x^2 + 4(y^2 + z^2) = 4$$

as an equation of the surface of revolution. This surface, formed by rotating an ellipse about an axis, is called an *ellipsoid of revolution.*

In the above example the generating curve should be taken to be the semi-ellipse with equation

$$2y = \sqrt{4 - x^2}$$

in order to meet the assumption that $y \geq 0$ for each point (x,y) on the generating curve. When we replace y by $\sqrt{y^2 + z^2}$ in this equation, we get

$$2\sqrt{y^2 + z^2} = \sqrt{4 - x^2}$$

as an equation of the surface. It is evident that the graph of this equation is the same as that of the equation obtained in Example 4.

Example 5 Find an equation of the surface generated by rotating the curve

$$x^2 = 4z$$

in the xz plane about the z axis.

Solution: We replace each x^2 by $x^2 + y^2$ to obtain

$$x^2 + y^2 = 4z$$

as an equation of this surface. Since the surface is formed by rotating a parabola about its axis, it is called a *paraboloid of revolution.*

Example 6 Describe the graph of the equation

$$x^2 - 9y^2 + z^2 = 36.$$

Solution: Since this equation has the form

$$(x^2 + z^2) - 9y^2 = 36,$$

its graph is a surface of revolution obtained by rotating the curve

$$x^2 - 9y^2 = 36 \tag{1}$$

in the xy plane about the y axis. This surface, obtained by rotating the hyperbola (1) about an axis, is called a *hyperboloid of revolution*. Incidentally, the surface may also be generated by rotating the hyperbola

$$z^2 - 9y^2 = 36$$

in the yz plane about the y axis.

EXERCISES

In each of Exercises 1 to 16, discuss and sketch the graph in space of the equation.

1. $x^2 + y^2 = 9$
2. $y^2 + z^2 = 4$
3. $x^2 = 8z$
4. $x^2 - y^2 = 1$
5. $x^2 + y^2 + 9z^2 = 9$
6. $y^2 + z^2 = \sin x$
7. $y^2 = x^2 + z^2$
8. $xz = 1$
9. $x^2 = y^2$
10. $x^2 + y^2 = 4z$
11. $x^2(y^2 + z^2) = 1$
12. $4x^2 + y^2 + 4z^2 = 16$
13. $y^2 + z^2 = e^x$
14. $y = 9 - x^2$
15. $y^2 = 4 + z^2$
16. $x^2 - y^2 - z^2 = 1$
17. Find an equation of the ellipsoid of revolution obtained by rotating the ellipse $x^2/a^2 + y^2/b^2 = 1$ in the xy plane about (a) the x axis, (b) the y axis. Sketch.
18. Find an equation of the hyperboloid of revolution obtained by rotating the hyperbola $x^2/a^2 - y^2/b^2 = 1$ in the xy plane about (a) the x axis, (b) the y axis. Sketch.
19. Find an equation of the paraboloid of revolution obtained by rotating the parabola $y^2 = 4px$ in the xy plane about the x axis. Sketch.
20. Find an equation of the cone obtained by rotating the curve $|y| = mx$ in the xy plane about the y axis. Sketch.

In Exercises 21 to 24, find an equation of the surface of revolution obtained by rotating the given curve in the xy plane about the specified axis. Sketch the surface.

21. $x^3 = y^2$, y axis
22. $x = \sqrt{y}$, x axis
23. $y = \cos x$, x axis
24. $|x| + |y| = 1$, y axis
25. Find an equation of the surface obtained by rotating the circle $(y - k)^2 + z^2 = r^2$ in the yz plane about the z-axis (assume $k > r$). Describe the surface.

7 QUADRIC SURFACES

The graph of a second-degree equation in x, y, and z is called a *quadric surface*. We shall give some standard forms of equations of quadric surfaces in this section.

The graph of an equation of the form

14.14 $$\frac{x^2}{a^2} + \frac{y^2}{b^2} + \frac{z^2}{c^2} = 1$$

is called an *ellipsoid*. If $a^2 = b^2 = c^2$, 14.14 is an equation of a sphere. If

$a^2 = b^2$ (or $b^2 = c^2$, or $a^2 = c^2$), 14.14 is an equation of an ellipsoid of revolution, as discussed in the previous section.

If we let $z = 0$ in 14.14, we get

$$\frac{x^2}{a^2} + \frac{y^2}{b^2} = 1$$

as the equation of the cross section of the ellipsoid in the xy plane. Clearly, this cross section is an ellipse. The cross sections of 14.14 in the other coordinate planes are easily seen to be ellipses also.

Letting $z = k$ in 14.14, we get the equation

$$\frac{x^2}{a^2} + \frac{y^2}{b^2} = 1 - \frac{k^2}{c^2}$$

of the cross section of the ellipsoid in the plane $z = k$, a plane parallel to the xy plane. This cross section again is an ellipse if $k^2 < c^2$. Similar statements may be made for cross sections in planes parallel to the other coordinate planes. The graph of 14.14 is sketched in Figure 14.20.

The graph of an equation of the form

14.15 $$\frac{x^2}{a^2} + \frac{y^2}{b^2} = cz$$

is called an *elliptic paraboloid*, so named because each cross section of the surface in a plane $z = k$ is an ellipse (if $ck > 0$) whereas each cross section in a plane $x = k$ or $y = k$ is a parabola. If $a^2 = b^2$, 14.15 is an equation of a

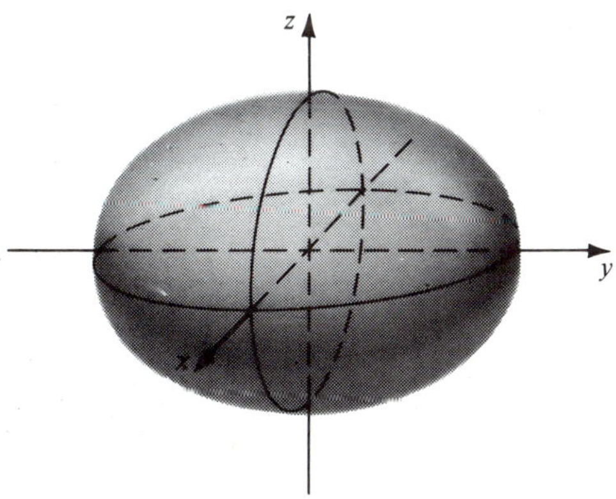

FIGURE 14.20

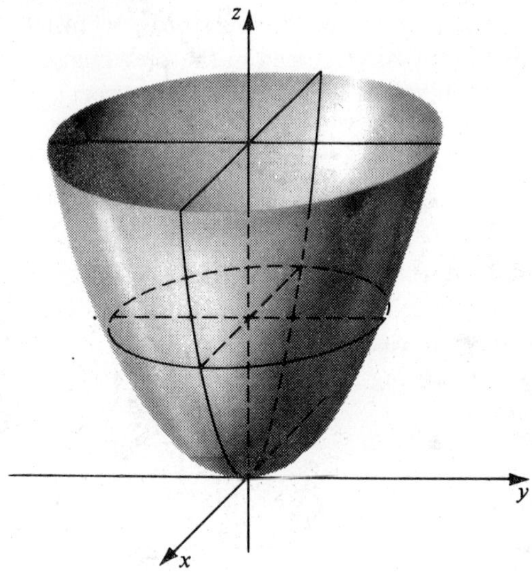

FIGURE 14.21

paraboloid of revolution with the z axis as axis of revolution. If $c > 0$, the graph of 14.15 is as shown in Figure 14.21.

The quadric surface with equation of the form

14.16 $$\frac{y^2}{b^2} - \frac{x^2}{a^2} = cz$$

is called a *hyperbolic paraboloid.* If $c > 0$, it is the saddle-shaped surface shown in Figure 14.22. We note that if $z = 0$, the cross section consists of the lines

$$\frac{x}{a} = \pm\frac{y}{b},$$

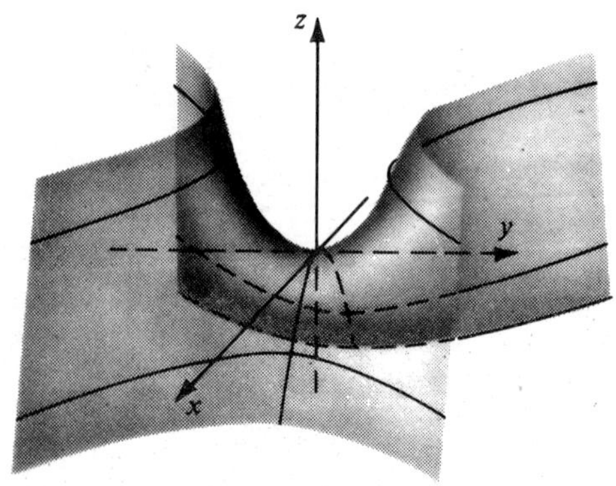

FIGURE 14.22

whereas if $z = k$, $k \neq 0$, the cross section is a hyperbola. The cross sections in planes parallel to the xz and yz planes are parabolas.

Quadric surfaces with equations of the form

14.17
$$\frac{x^2}{a^2} + \frac{y^2}{b^2} - \frac{z^2}{c^2} = 1,$$

or

14.18
$$\frac{x^2}{a^2} + \frac{y^2}{b^2} - \frac{z^2}{c^2} = -1$$

are called *hyperboloids*. For each surface the cross section in a plane $z = k$ is an ellipse, and the cross sections in the planes $x = k$ and $y = k$ are hyperbolas. If $a^2 = b^2$, these surfaces are hyperboloids of revolution. The graph of 14.17, shown in Figure 14.23, is called a *hyperboloid of one sheet*. The graph of 14.18, shown in Figure 14.24, is called a *hyperboloid of two sheets*.

Closely related to the hyperboloids is the quadric surface with equation

14.19
$$\frac{x^2}{a^2} + \frac{y^2}{b^2} - \frac{z^2}{c^2} = 0.$$

This surface is related to each of the hyperboloids as the asymptotes are to a

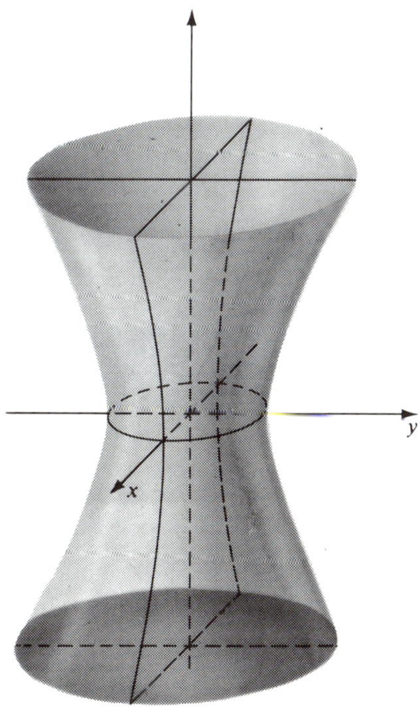

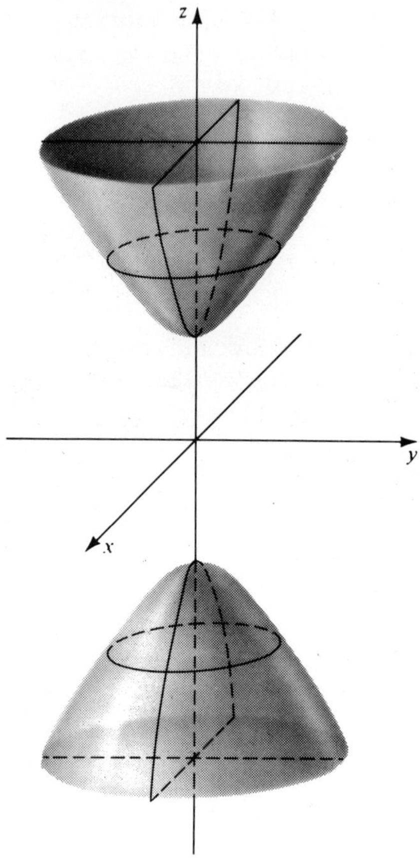

FIGURE 14.24

hyperbola. It is an *elliptic cone*, frequently called the asymptotic cone of each of the hyperboloids 14.17 and 14.18. For $z = k$ the cross section of 14.19 is an ellipse; for $x = k$ or $y = k$ the cross section is a hyperbola if $k \neq 0$ and a pair of lines intersecting at the origin if $k = 0$.

A cone may be thought of as a surface generated by a line moving along a given plane curve and passing through a fixed point, called the *vertex* of the cone. Thus, if V is the vertex of a cone and P is any point on the cone, every point on the line through V and P is also on the cone. The most common cone is the right circular cone generated by a line moving along a circle (14.19 if $a^2 = b^2$).

We may prove that 14.19 is a cone as follows. If $P(x_0, y_0, z_0)$ is any point on the graph of 14.19 other than the origin O, then the line L on P and O has parametric equations

$$x = x_0 t, \qquad y = y_0 t, \qquad z = z_0 t.$$

Each point on L is also on the graph of 14.19, since

$$\frac{(x_0t)^2}{a^2} + \frac{(y_0t)^2}{b^2} - \frac{(z_0t)^2}{c^2} = t^2\left(\frac{x_0^2}{a^2} + \frac{y_0^2}{b^2} - \frac{z_0^2}{c^2}\right) = 0.$$

Thus the graph of 14.19 is a cone.

EXERCISES

Name and sketch the graph of each of the following equations. (It is sometimes helpful to relabel the axes.)

1. $x^2 + 4y^2 + 9z^2 = 36$

2. $x^2 + 4y^2 = 4z$

3. $x^2 - y^2 + 4z^2 = 4$

4. $x^2 - y^2 - 4z^2 = 4$

5. $x^2 = z^2 + 4y$

6. $16x^2 + y^2 = 64 - 4z^2$

7. $4x^2 + 8y + z^2 = 0$

8. $x^2 + 9y^2 = z^2$

9. $16x^2 - 9y^2 - z^2 - 144 = 0$

10. $36 + 4y^2 = x^2 + 9z^2$

11. $x^2 + 25z^2 = 9y^2$

12. $y^2 = 4x + 4z^2$

13. $x^2 + 25y^2 - 50z = 0$

14. $x^2 + 4y^2 = 4z^2 - 4z + 1$

15. $x^2 + y^2 = 4x$

16. $x^2 + y^2 = 1 + z$

8 SPACE CURVES

A function λ that maps an interval I of R into three-dimensional space R^3 is called a *space curve*. The range of the function λ is called the *trace* of the curve. When we speak of a *curve* in this chapter, it is understood to be a space curve. (A plane curve may obviously be regarded as a special kind of space curve.)

If the function λ is specified by

$$\lambda(t) = (x(t),y(t),z(t)), \qquad \text{for } t \text{ in } I,$$

then the curve may also be represented by the parametric equations

$$x = x(t), \quad y = y(t), \quad z = z(t), \qquad \text{for } t \text{ in } I.$$

The curve λ has a *direction* determined by the order in I; if $t_1 < t_2$ in I, then we say that $\lambda(t_1)$ precedes $\lambda(t_2)$.

Continuity of a curve is defined just as it was in the two-dimensional case discussed in Chapter 13. The student should recall that $B(P,r)$ denotes the *open ball* of radius r centered at the point P. We then say that a curve λ is *continuous* at c in the domain I of λ if for every open ball $B(\lambda(c),\varepsilon)$ centered at the point $\lambda(c)$ with radius $\varepsilon > 0$, there exists a neighborhood N of c such that

$$\lambda(t) \text{ is in } B(\lambda(c),\varepsilon) \qquad \text{for every } t \text{ in } N \cap I.$$

By the obvious extension of 13.3 to space, it can be proved that λ is continuous at c if and only if the functions x, y, and z are continuous at c.

It will be convenient, just as it was in Chapter 13, to regard a curve as being defined by a *vector-valued* function. Corresponding to a curve λ with domain I, for each t in I we can consider the position vector $\boldsymbol{\lambda}(t)$ from the origin to the point $\lambda(t)$. The curve can then be represented by the vector-valued function

14.20 $$\boldsymbol{\lambda}(t) = \langle x(t),y(t),z(t)\rangle, \qquad \text{for } t \text{ in } I.$$

In this way we again regard the curve as being traced out by the tip of the position vector $\boldsymbol{\lambda}(t)$.

The definition of limit may be extended in the obvious way from the two-dimensional case (Definition 13.9). If $\mathbf{F}$ is a vector-valued function defined by

$$\mathbf{F}(t) = \langle g_1(t),g_2(t),g_3(t)\rangle, \qquad \text{for } t \text{ in } I,$$

and c is a number in I, then we define

$$\operatorname*{limit}_{t\to c} \mathbf{F}(t) = \langle \operatorname*{limit}_{t\to c} g_1(t), \operatorname*{limit}_{t\to c} g_2(t), \operatorname*{limit}_{t\to c} g_3(t)\rangle,$$

provided that the limits of the component functions exist.

Now if a space curve is given by the vector-valued function $\boldsymbol{\lambda}(t)$ in 14.20, and c is a number in I, we can define

14.21 $$\boldsymbol{\lambda}'(c) = \operatorname*{limit}_{h\to 0} \frac{1}{h} [\boldsymbol{\lambda}(c+h) - \boldsymbol{\lambda}(c)],$$

which is identical with definition 13.10. The vector $\boldsymbol{\lambda}'(c)$ is called the *tangent vector* to the curve λ at the point $\lambda(c)$. The analog of 13.11 also holds; in other words, if

$$\boldsymbol{\lambda}(t) = \langle x(t),y(t),z(t)\rangle,$$

then

$$\boldsymbol{\lambda}'(t) = \langle x'(t),y'(t),z'(t)\rangle$$

for each t in I. The *tangent line* to the curve at the point $\lambda(c)$ is the line through $\lambda(c)$ in the direction of the tangent vector $\boldsymbol{\lambda}'(c)$.

We say that the curve given by 14.20 is *differentiable* if the vector $\boldsymbol{\lambda}'(t)$ is defined for every t in I. The curve is *smooth* if, in addition, the curve defined by the vector function $\boldsymbol{\lambda}'(t)$ is continuous for t in I. Thus, by our previous remarks, λ is smooth if and only if the functions $x'(t)$, $y'(t)$, and $z'(t)$ are continuous for each t in I.

The higher derivatives of a curve λ can be defined in the usual way. Thus, if $\boldsymbol{\lambda}(t) = \langle x(t),y(t),z(t)\rangle$, the nth derivative of the vector-valued function $\boldsymbol{\lambda}$ is denoted by $\boldsymbol{\lambda}^{[n]}$ and is defined by

$$\boldsymbol{\lambda}^{[n]}(t) = \langle x^{[n]}(t),y^{[n]}(t),z^{[n]}(t)\rangle.$$

Example 1 Describe the curve defined by

$$\lambda(t) = \langle r\cos t, r\sin t, at\rangle, \qquad \text{for } t \geq 0,$$

where r and a are positive constants.

Solution: Parametric equations of this curve are given by

$$x = r\cos t, \qquad y = r\sin t, \qquad z = at.$$

If we eliminate t from the first two equations, we see that the x and y coordinates of any point (x,y,z) on the curve satisfy the equation $x^2 + y^2 = r^2$. Thus the curve lies on the circular cylinder whose equation is $x^2 + y^2 = r^2$. (See Figure 14.25.)

Referring to Figure 14.25, we can imagine the curve starting from the point $(r,0,0)$ when $t = 0$ and winding about the given cylinder in such a way that the z coordinate is proportional to the value of t (resembling the screw thread on a bolt).

This curve is called a *circular helix*. Its tangent vector is given by

$$\lambda'(t) = \langle -r\sin t, r\cos t, a\rangle.$$

It is interesting to note that the length of this tangent vector is constant:

$$|\lambda'(t)| = \sqrt{(-r\sin t)^2 + (r\cos t)^2 + a^2} = \sqrt{r^2 + a^2}.$$

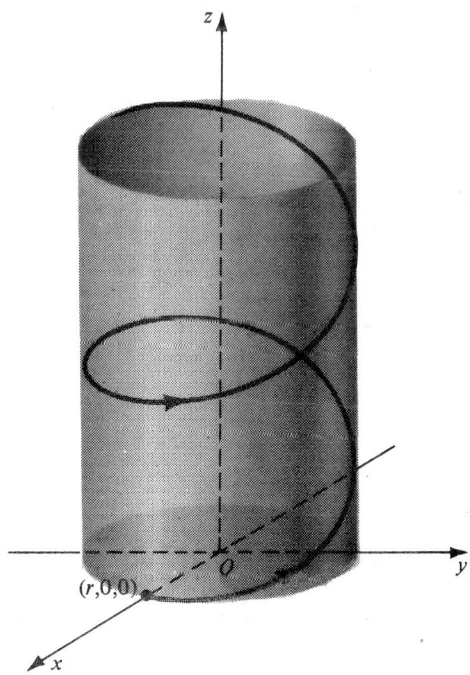

FIGURE 14.25

Space curves frequently arise in a natural way as the intersection of two surfaces. Suppose, for example, that the surfaces are represented by the equations $F(x,y,z) = 0$ and $G(x,y,z) = 0$. We can then often obtain a parametric representation for their curve of intersection by selecting one of the variables x, y, or z as the parameter t, and then expressing the other two variables in terms of t. This is illustrated in the following example.

Example 2 Describe the curve in the first octant formed by the intersection of the cone $z = 3 - \sqrt{x^2 + y^2}$ and the circular cylinder $x^2 + (y - 1)^2 = 1$. Obtain a parametric representation for this curve.

Solution: The surfaces and their curve of intersection are shown in Figure 14.26. The form of the equations suggests that we take the variable y as the parameter t. Solving the equation of the cylinder for x, we obtain

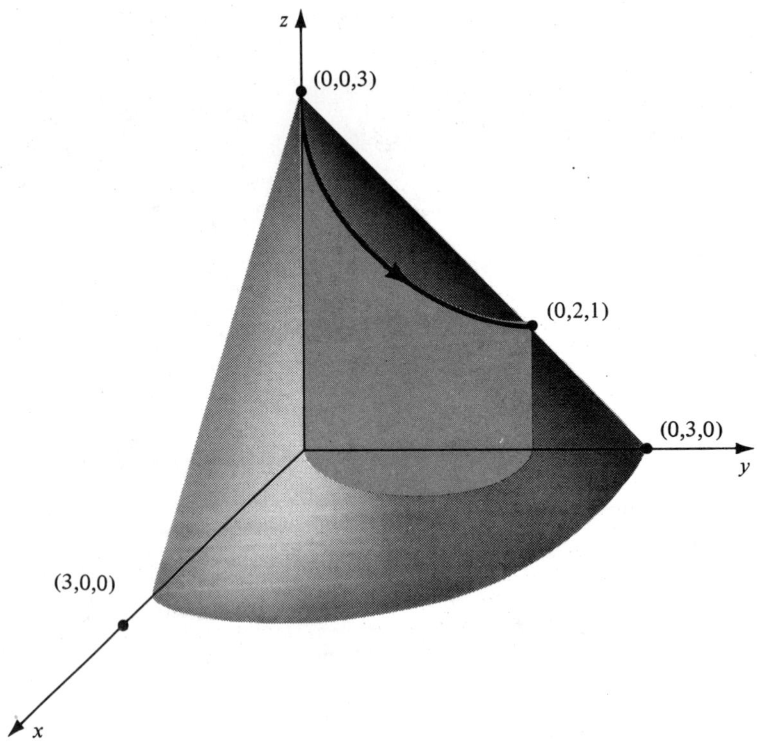

FIGURE 14.26

$$x^2 = 2y - y^2,$$
$$x = \sqrt{2y - y^2} = \sqrt{2t - t^2}.$$

Substituting $y = t$ and the above expression for x in the equation of the cone,

$$z = 3 - \sqrt{2t - t^2 + t^2} = 3 - \sqrt{2t}.$$

Hence the given curve is represented by the parametric equations

$$x = \sqrt{2t - t^2}, \quad y = t, \quad z = 3 - \sqrt{2t}, \qquad \text{for } t \text{ in } [0,2].$$

The direction thus determined on the curve is that of increasing t, from the point (0,0,3) toward the point (0,2,1).

If λ is a differentiable curve with domain I and c is a number in I such that the tangent vector $\boldsymbol{\lambda}'(c) \neq \mathbf{0}$, then the plane passing through the point $\lambda(c)$ and perpendicular to $\boldsymbol{\lambda}'(c)$ is called the *normal plane* to λ at c.

Example 3 Find an equation of the normal plane to the helix

$$\lambda(t) = \langle r \cos t, r \sin t, at \rangle, \qquad t \geq 0,$$

at any point on it.

Solution: Select a point on the curve corresponding to an arbitrary value t_0 of the parameter. Then

$$\boldsymbol{\lambda}'(t_0) = \langle -r \sin t_0, r \cos t_0, a \rangle.$$

An equation of the plane through the point $\lambda(t_0)$ with $\boldsymbol{\lambda}'(t_0)$ as its normal is

$$-r \sin t_0(x - r \cos t_0) + r \cos t_0(y - r \sin t_0) + a(z - at_0) = 0,$$

or

$$(-r \sin t_0)x + (r \cos t_0)y + az = a^2 t_0.$$

In the special case when $t_0 = 0$, this equation reduces to

$$ry + az = 0.$$

The reader will recall that a smooth curve λ in the plane, with domain $[a,b]$, is rectifiable and has arc length $L_a^b(\lambda)$ given by the formula 13.23. The discussion of arc length in Chapter 13 can be extended to space curves to yield the following result.

14.22 Theorem

Let λ be a smooth curve in space with position vector

$$\lambda(t) = \langle x(t), y(z), z(t) \rangle, \qquad \text{for } t \text{ in } [a,b].$$

Then λ is rectifiable, and its arc length $L_a^b(\lambda)$ is given by

$$L_a^b(\lambda) = \int_a^b |\boldsymbol{\lambda}'(t)|\, dt = \int_a^b \sqrt{[x'(t)]^2 + [y'(t)]^2 + [z'(t)]^2}\, dt.$$

Example 4 Find the length of the helix of Example 1 between the points where $t = 0$ and $t = k$ (k any positive number).

Solution: We saw previously that $|\lambda'(t)| = \sqrt{r^2 + a^2}$ for every number t. Hence

$$L_0^k(\lambda) = \int_0^k \sqrt{r^2 + a^2}\, dt = k\sqrt{r^2 + a^2}.$$

Thus, as might be anticipated, the length of a circular helix is a constant $\sqrt{r^2 + a^2}$ times the central angle k swept out in tracing the curve. For example, in one revolution around the cylinder the helix has length $2\pi\sqrt{r^2 + a^2}$.

EXERCISES

In each of Exercises 1 to 3, find equations of the tangent line and normal plane to the given space curve at the indicated point.

1. $\lambda(t) = \langle t - 3, t^2 + 1, t^2 \rangle$, $t = 3$ 2. $\lambda(t) = \langle t^2, t^{-1}, \sin \pi t \rangle$, $t = 1$
3. $\lambda(t) = \langle a \sin kt, b \cos kt, ct^2 \rangle$, $t = t_0$
4. Express in parametric form the curve of intersection of the two surfaces $x^2 + y^2 = 25z$ and $y^2 + z^2 = 17$. Find an equation of the normal plane to this curve at the point (3,4,1).
5. Find a parametric representation for the curve in the first octant formed by the intersection of the surfaces $z = x^2 + y^2$ and $y^2 + z^2 = 1$. Sketch the curve.
6. Prove that for the circular helix of Example 1, the tangent vector makes a constant angle with the z axis.
7. Show that the two curves defined by $\lambda_1(t) = \langle t, 2t^2, -t^{-1} \rangle$ and $\lambda_2(s) = \langle 1 - s, 2 \cos s, \sin s - 1 \rangle$ intersect at right angles.
8. **a.** Find all points of intersection of the two curves defined by $\lambda_1(t) = \langle t, t^2 - t, 1 - t/2 \rangle$ and $\lambda_2(s) = \langle 2 \sin s, 2 \sin s, \cos s \rangle$.
   b. Find the angle of intersection of λ_1 and λ_2 at each of their points of intersection.
9. Find the length of the space curve defined by $\lambda(t) = \langle t^2 + 1, t^2 - 1, 8t \rangle$, domain $\lambda = [1,3]$.
10. Find the length of the space curve defined by $\lambda(t) = \langle t \cos t, t \sin t, t^2 \rangle$, domain $\lambda = [0,a]$.
11. Find the length of the curve defined by $\lambda(t) = \langle t^2, 2t, \ln t \rangle$, between the points where $t = 1$ and $t = 3$.
12. Find the length of the curve in the first octant formed by the intersection of the elliptical cylinders $2y^2 + x^2 = 1$ and $2z^2 + x^2 = 1$.
13. Prove that the curve formed by the intersection of the surfaces $x^2 + y^2 = 2x$ and $z = x + 4$ lies on the cylinder $(z - 5)^2 + y^2 = 1$.
14. Find an equation for the cylinder that passes through the space curve

$$x = 2t, \qquad y = t^2, \qquad z = 3t^3$$

and has its generating lines parallel to the y axis.

15. Let **u**, **v**, and **w** be vector-valued functions of the real variable t, and let D_t denote differentiation with respect to t. Prove that

a. $D_t(\mathbf{u} \times \mathbf{v}) = (D_t\mathbf{u}) \times \mathbf{v} + \mathbf{u} \times (D_t\mathbf{v})$

b. $D_t(\mathbf{u} \cdot \mathbf{v} \times \mathbf{w}) = (D_t\mathbf{u}) \cdot \mathbf{v} \times \mathbf{w} + \mathbf{u} \cdot D_t\mathbf{v} \times \mathbf{w} + \mathbf{u} \cdot \mathbf{v} \times D_t\mathbf{w}$

9 CYLINDRICAL AND SPHERICAL COORDINATE SYSTEMS

There are two common generalizations in space of the polar coordinate system in a plane.

The basis of a *cylindrical coordinate system* is a plane p with a polar coordinate system on it and a z axis perpendicular to p with the origin of the z axis at the pole of p. Each point P in space then has coordinates (r,θ,z), where (r,θ) are the polar coordinates of the projection Q of P on p, and z is the coordinate of the projection R of P on the z axis (Figure 14.27).

The graph of an equation of the form

$$r = c, \qquad c > 0,$$

is a right circular cylinder of radius c having the z axis as its axis. This is the reason for the name "cylindrical" coordinate system. The graph of

$$\theta = c$$

is a half-plane emanating from the z axis; the graph of

$$z = c$$

is a plane parallel to the given polar coordinate plane p.

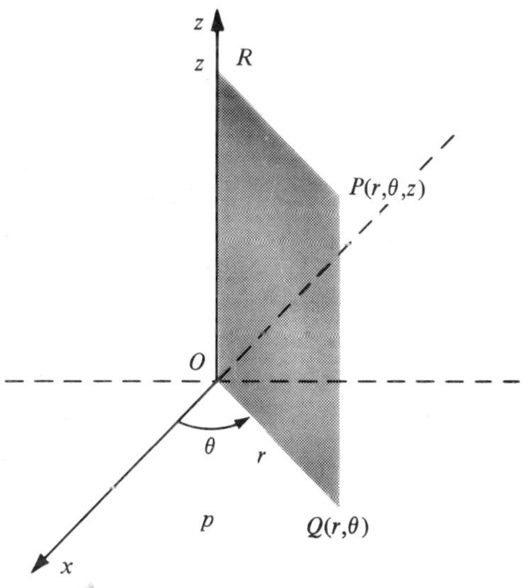

FIGURE 14.27

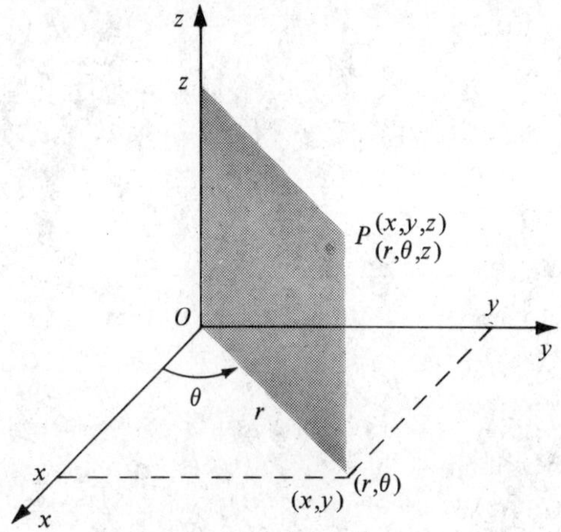

FIGURE 14.28

If a rectangular coordinate system and a cylindrical coordinate system are placed in space as in Figure 14.28, with the cylindrical coordinate system having the xy plane as its polar plane and the positive x axis as its polar axis, then each point P has two sets of coordinates, (x,y,z) and (r,θ,z), related by the equations

14.23 $$x = r\cos\theta, \qquad y = r\sin\theta, \qquad z = z.$$

This is evident from 13.14.

For a *spherical coordinate system*, we start off with a plane p having a polar coordinate system on it and a z axis perpendicular to p with the origin of the z axis meeting the plane p at the pole. Each point P in space has coordinates (ρ,θ,ϕ), where

$$\rho = |OP|,$$

θ is the polar angle associated with the projection Q of P on plane p, and ϕ is the direction angle of the half-line from the pole through P relative to the z axis (Figure 14.29). The origin has coordinates $(0,\theta,\phi)$ for any θ and ϕ. If $P(\rho,\theta,\phi)$ is a point different from the origin, then necessarily $\rho > 0$ and $0 \le \phi \le \pi$, with $\phi = 0$ if P is on the positive z axis and $\phi = \pi$ if P is on the negative z axis. There are no restrictions on the angle θ.

The graph of

$$\rho = c, \qquad c > 0,$$

is a sphere of radius c with center at the pole; hence the name "spherical" coordinate system. The graph of

$$\theta = c$$

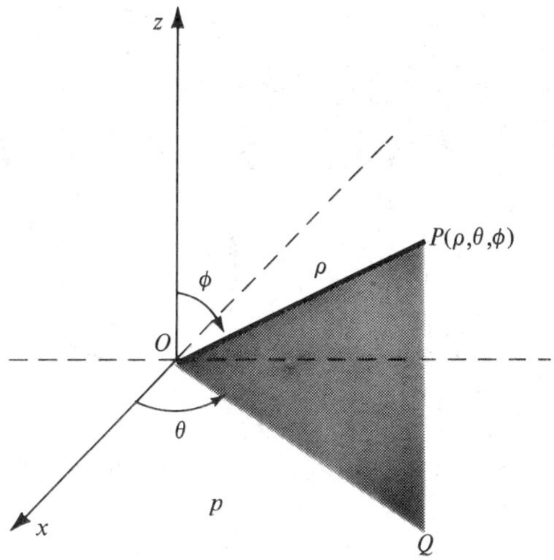

FIGURE 14.29

is a half-plane emanating from the z axis, and the graph of

$$\phi = c, \qquad 0 < c < \pi,$$

is a right circular cone having the z axis as its axis and the pole as its vertex.

If we place a rectangular coordinate system and a spherical coordinate system together, as indicated in Figure 14.30, then each point P in space has two sets of coordinates, (x,y,z) and (ρ,θ,ϕ). By 14.23, with $r = |OQ|$, we have

$$x = |OQ| \cos \theta, \qquad y = |OQ| \sin \theta.$$

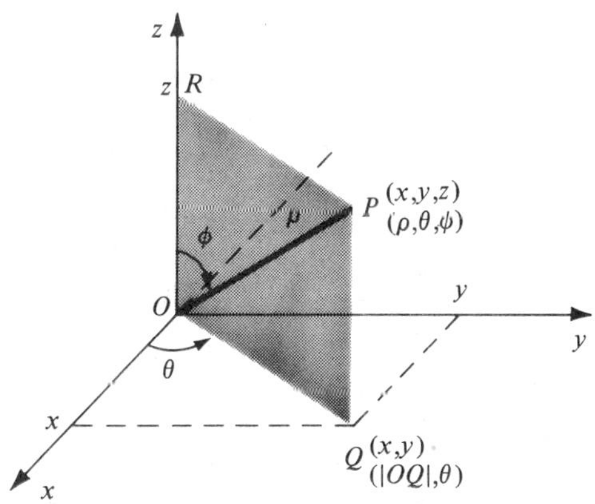

FIGURE 14.30

Since $|OQ| = |PR| = \rho \sin \phi$ and $z = \rho \cos \phi$, we obtain

14.24 $x = \rho \sin \phi \cos \theta, \quad y = \rho \sin \phi \sin \theta, \quad z = \rho \cos \phi$

as the equations relating x, y, and z to ρ, θ, and ϕ.

EXERCISES

1. Find spherical coordinates for the following points given in rectangular coordinates:
   a. $(4,2,-4)$ b. $(1,-\sqrt{3},4)$
2. Find cylindrical coordinates for the points in Exercise 1.
3. Find rectangular coordinates for the following points given in cylindrical coordinates:
   a. $(2, \arccos \frac{3}{5}, 6)$ b. $(10,-\frac{1}{2}\pi,4)$
4. Find rectangular coordinates for the following points given in spherical coordinates:
   a. $(2,\frac{1}{4}\pi,\frac{1}{3}\pi)$ b. $(3,\frac{1}{3}\pi,\frac{1}{6}\pi)$
5. Find equations in cylindrical coordinates of the graphs of the following equations given in rectangular coordinates:
   a. $(x + y)^2 = z - 5$ b. $x^2z^2 = 25 - y^2z^2$
   c. $\dfrac{x^2}{a^2} + \dfrac{y^2}{b^2} = 1$ d. $ax + by + cz = x^2 + y^2 + z^2$
6. The following surfaces are given in spherical coordinates. Find their equations in rectangular coordinates and describe the surfaces:
   a. $\cot \phi = \sin \theta + \cos \theta$ b. $\rho^2 \cos 2\phi = a^2$
   c. $\rho = a \sin \phi \sin \theta$ d. $\rho^2 \sin^2 \phi \sin 2\theta = a^2$
7. Find an equation in spherical coordinates for the sphere of radius 1 with center at (0,0,1).
8. Find an equation in cylindrical coordinates for the sphere of Exercise 7.

In Exercises 9 and 10, find an equation in (a) cylindrical and (b) spherical coordinates for the given surface.

9. The paraboloid $x^2 + y^2 = 2z$.
10. The hyperboloid $xy = z$.
11. Describe the surface $z = 2r$ (cylindrical coordinates), and obtain an equation for it in rectangular coordinates.
12. Obtain an equation in rectangular coordinates for the surface $z^2 = 1 - (r - 2)^2$.
13. Let C be the curve of intersection of the surfaces $\rho = 1$ and $z = \cos \theta$ (spherical coordinates). Find parametric equations for C in the form
$$x = f(\theta), \quad y = g(\theta), \quad z = h(\theta),$$
using θ as the parameter. Is the curve smooth? Consider, in particular, the points corresponding to $\theta = 0$ and $\theta = \pi$.

14. Derive the following formulas expressing ϕ and θ in terms of x, y, and z:

a. $\sin\phi = \sqrt{x^2 + y^2}/\sqrt{x^2 + y^2 + z^2}$

b. $\cos\phi = z/\sqrt{x^2 + y^2 + z^2}$

c. $\sin\theta = y/\sqrt{x^2 + y^2}$

d. $\cos\theta = x/\sqrt{x^2 + y^2}$

CHAPTER REVIEW

In Exercises 1–12, discuss and sketch the graph of the given equation in three-dimensional space.

1. $x^2 + y^2 + z^2 + 2x - 4y - 6z = 2$

2. $x^2 + 4y^2 + 9z^2 + 2x - 4y - 6z = -4$

3. $z = x^2 - y^2$

4. $xy = 0$

5. $xy = 1$

6. $z = x^3 + y^3$

7. $x^2 + y^2 - z^2 = 1$

8. $x^2 - y^2 - z^2 = 1$

9. $x^2 - y^2 = 0$

10. $z = y^3 + x$

11. $z = (x + 1)^3$

12. $z = (y - 3)^2$

13. Find the perpendicular distance from a vertex of a cube of side s to a diagonal which does not pass through that vertex.

14. Find the distance from the point (4,1,1) to the line of intersection of the planes $x + y + z = 4$, $x - 2y - z = 4$.

15. Let L_1 be the line of intersection of the planes $x - 2y + 4 = 0$, $x - 3z - 2 = 0$; and L_2 the line of intersection of $4y - 3z = 13$, $4x - 3z = 11$.

a. Prove that L_1 and L_2 intersect.

b. Find an equation of the plane containing L_1 and L_2.

16. Let L_1 and L_2 be the lines of intersection, respectively, of the planes

$$\begin{cases} 4x + z = 0 \\ 3y + z = 0, \end{cases} \qquad \begin{cases} 3x + y = 1 \\ 4z - y = 4. \end{cases}$$

Find the shortest distance between L_1 and L_2.

17. Find the center of the sphere that passes through the four points (1,2,3), (3,2,−1), (−1,1,2), and (1,−1,−2).

18. Prove that the three planes $2x + 5y + 3z = 0$, $x - y + 4z = 2$, and $7y - 5z + 4 = 0$ intersect in a straight line.

19. Find an equation of the plane through (1,4,3) perpendicular to the line of intersection of the planes $3x + 4y + 7z + 4 = 0$ and $x - y + 2z + 3 = 0$.

20. Find the volume of the rectangular parallelepiped three of whose edges consist of the line segments AB, AC, and AD, where $A = (-2,0,1)$, $B = (3,-2,-1)$, $C = (4,3,2)$, and $D = (2,2,5)$.

21. Find the area of the triangle whose vertices are $(-1,3,-6)$, $(-3,-4,2)$, and $(2,-1,0)$.

22. Find two points on the line

$$x = 3 + 6t, \qquad y = -1 + 2t, \qquad z = 2 - 3t$$

which are at a distance of 3 units from the point $(-3,-3,5)$.

23. For each of the following space curves, find equations of the tangent line and normal plane at the indicated point.

a. $\lambda(t) = (t^2 + 1, t^2 - 1, t)$, $t = 1$

b. $\lambda(t) = (2 \sin t, 3 \cos t, t^3)$, $t = \pi/2$

In Exercises 24 to 26, find the length of the given space curve.

24. $x = \sin^2 t$, $y = t - \sin t \cos t$, $z = 2 \sin t$, t in $[0, \pi/2]$

25. $x = \dfrac{t^2}{2} + t$, $y = \dfrac{t^2}{2} - t$, $z = \dfrac{\sqrt{2}}{2} \ln t$, t in $[1,2]$

26. $x = t$, $y = \ln \sec t$, $z = 0$, t in $[0, \pi/4]$

15

Differential Calculus of Functions of Several Variables

Our purpose in this chapter is to extend the concepts of limit, continuity, and differentiability to functions of two or more variables. The results obtained are then applied to the study of directional derivatives, tangent planes to surfaces, and extrema of functions of several variables.

1 LIMITS AND CONTINUITY

A mapping f having its domain in R^2 and its range in R is called a (real-valued) *function of two variables*. Similarly, a mapping g having its domain in R^3 and its range in R is called a (real-valued) *function of three variables*. If f is a function of two variables, then we shall denote its value at each point $P = (x,y)$ in its domain either by $f(P)$ or by $f(x,y)$, whichever is more convenient. A similar remark holds for functions of three variables.

For the most part we shall confine our remarks in this section to functions of two variables, leaving it to the reader to check that they also apply to functions of three (or more) variables.

An example of a function f of two variables is

$$f(x,y) = \sqrt{4 - x^2 - y^2}.$$

The domain S of f is taken to be $\{(x,y) \mid 4 - x^2 - y^2 \geq 0\}$; that is,

$$S = \{(x,y) \mid x^2 + y^2 \leq 4\},$$

the closed disk $B[O,2]$.

A function f of two variables with domain S has a graph in space defined by

$$\text{Graph } f = \{(x,y,f(x,y)) \mid (x,y) \text{ in } S\}.$$

For example, the graph of the function f above is the graph of the equation

$$z = \sqrt{4 - x^2 - y^2}.$$

This is the hemisphere on and above the xy plane with radius 2 and center the origin.

If

$$g(x,y) = x^2 + y^2,$$

then the graph of g is the graph of the equation

$$z = x^2 + y^2,$$

a paraboloid of revolution about the z axis.

Our first task is to extend the concept of limit to functions of two or more variables. The reader will observe that in Definition 2.1 of limit for a function of one variable the fundamental notion is that of a *neighborhood.* All that is required, therefore, is an appropriate definition of a *neighborhood of a point* in R^2 or R^3.

We have in essence already made use of such a definition in R^2 when we discussed the continuity of a curve in Chapter 13, Section 2. Let us denote the distance between any two points P and Q in either R^2 or R^3 by $d(P,Q)$ [in R^3, $d(P,Q)$ is given by formula 14.1]. The student will recall that in R^2 the *open disk* $B(A,r)$ with center A and radius r was defined by

$$B(A,r) = \{P \text{ in } \mathsf{R}^2 \mid d(A,P) < r\},$$

and in R^3 the *open ball* $B(A,r)$ with center A and radius r was defined by

$$B(A,r) = \{P \text{ in } \mathsf{R}^3 \mid d(A,P) < r\}.$$

15.1 Definition

By a *neighborhood of a point A in* R^2 we mean an open disk centered at A. A *neighborhood of a point A in* R^3 is an open ball centered at A. An *r neighborhood* of a point A (in R^2 or R^3) is a neighborhood of A of radius r.

It should be noted that for a point A in R, we have $\{x \text{ in } \mathsf{R} \mid d(A,x) < r\} = \{x \text{ in } \mathsf{R} \mid |x - A| < r\} = (A - r, A + r)$, which is a symmetric neighborhood of A. Definition 15.1 therefore does in fact generalize the one-dimensional case.

We also need the concept of a *deleted neighborhood.*

15.2 Definition

If A is a point in R^2 (or R^3) and r is a positive real number, then the *deleted r neighborhood* of A is the set

$$B'(A,r) = \{P \text{ in } \mathsf{R}^2 \text{ (or } \mathsf{R}^3) \mid d(A,P) < r, P \neq A\}.$$

Thus in R^2 or R^3, $B'(A,r)$ is the open disk or open ball, respectively, centered at A and of radius r, *but with its center removed.*

Now we are ready for the definition of limit. Let us assume that f is a real-valued function whose domain contains some deleted neighborhood of the point A in either R^2 or R^3.

15.3 Definition

The *limit* of the function f at A is b, and we write

$$\lim_{P \to A} f(P) = b,$$

if for every neighborhood N of b in R there exists a deleted neighborhood $B'(A,r)$ of A contained in the domain of f such that $f(P)$ is in N for every P in $B'(A,r)$.

The reader will observe that, except for a slight change in notation, Definition 15.3 is identical with Definition 2.1 of limit for a function of one variable.

If f is a function of two variables and $A = (c,d)$, then the limit in 15.3 is often written as

$$\lim_{(x,y) \to (c,d)} f(x,y) = b.$$

Example 1 Let f be the function of two variables defined by $f(x,y) = x$ for all (x,y) in R^2. If $A = (c,d)$ is any point in R^2, prove that

$$\lim_{(x,y) \to (c,d)} f(x,y) = c.$$

Solution: Let N be any neighborhood of c in R, say $N = (c - \varepsilon, c + \varepsilon)$. Note that the function f maps each point P in R^2 onto its projection on the x axis. So if we take the deleted neighborhood $B'(A,\varepsilon)$ of $A = (c,d)$, then whenever P is in $B'(A,\varepsilon)$ we have $f(P)$ in N (see Figure 15.1). Hence 15.3 is satisfied.

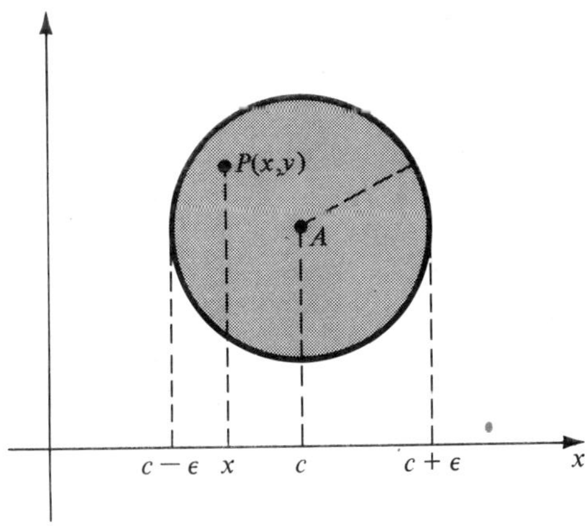

FIGURE 15.1

If g is the function defined by $g(x,y) = y$ for all (x,y) in R^2, then in a similar way we can show that

$$\lim_{(x,y)\to(c,d)} g(x,y) = d\,.$$

Furthermore, if h is a function with the constant value k, then $\lim_{P\to A} h(P) = k$ for any point A.

Since our definition of limit for functions of several variables is formally identical to that of the one-variable case, there is no difficulty in proving the analogs of Theorems 2.8 to 2.10 for several variables. We shall not go into these details, but shall use these results concerning limits of sums, products, and quotients without further comment.

Example 2

Find $\lim_{(x,y)\to(1,2)} \dfrac{x + y^2 + 2}{xy}$.

Solution: By Example 1,

$$\lim_{(x,y)\to(1,2)} x = 1, \qquad \lim_{(x,y)\to(1,2)} y = 2.$$

Using the appropriate theorems on limits of sums, products, and quotients, we have

$$\lim_{(x,y)\to(1,2)} \frac{x + y^2 + 2}{xy} = \frac{1 + 2\cdot 2 + 2}{1\cdot 2} = \frac{7}{2}.$$

The student should not assume, however, that the theory of limits for functions of several variables is in all respects similar to the one-variable theory. The existence of a limit for a function of two or more variables is a much *stronger* condition than in the one-variable case. We can see intuitively why this is so merely by observing that in R there are only two directions from which we may approach a point (from the left or from the right), while in R^2 we may approach a point from many different directions. The following example will make this clearer.

Example 3

Let f be the function defined by

$$f(x,y) = \frac{xy}{x^2 + y^2}$$

for all (x,y) in R^2 with $(x,y) \neq (0,0)$. Determine whether $\lim_{(x,y)\to(0,0)} f(x,y)$ exists.

Solution: If the point (x,y) lies on either of the coordinate axes, but $(x,y) \neq (0,0)$, we see that $f(x,y)$ has the value 0. Hence every deleted neighborhood of the origin contains points at which f takes on the value 0. From this we might be tempted to conclude that $\lim_{(x,y)\to(0,0)} f(x,y) = 0$, but this is not the case. For suppose that (x,y) lies on the line $y = x$, say $(x,y) = (a,a)$, with $a \neq 0$. Then $f(a,a) = a^2/2a^2 = \frac{1}{2}$, so that every deleted neighborhood of the origin also contains a point at which f takes on the value $\frac{1}{2}$. This means that we cannot have $\lim_{(x,y)\to(0,0)} f(x,y) = 0$.

For if $(-\varepsilon,\varepsilon)$ is any neighborhood of 0 in R with $0 < \varepsilon < \frac{1}{2}$, then *no* deleted neighborhood of the origin in R^2 can be mapped by f into $(-\varepsilon,\varepsilon)$. Also, $\underset{(x,y)\to(0,0)}{\text{limit}} f(x,y)$ cannot equal b for any real number $b \neq 0$. To see this, let $N = (b - \varepsilon, b + \varepsilon)$ be any neighborhood of b with $0 < \varepsilon < |b|$. Then no deleted neighborhood of the origin in R^2 is mapped by f into N [because every such neighborhood contains a point (x,y) with $f(x,y) = 0$]. Thus $\underset{(x,y)\to(0,0)}{\text{limit}} f(x,y)$ does not exist.

Example 4 If $f(x,y) = \dfrac{xy}{|x| + |y|}$ for $(x,y) \neq (0,0)$, prove that $\underset{(x,y)\to(0,0)}{\text{limit}} f(x,y) = 0$.

Solution: If r denotes the length of the position vector from the origin to the point (x,y), then we have $|x| \leq r$, $|y| \leq r$, and $|x| + |y| \geq r$ (by the triangle inequality). Hence

$$(1) \qquad |f(x,y)| = \frac{|xy|}{|x| + |y|} \leq \frac{r^2}{r} = r.$$

Thus, given any neighborhood $(-\varepsilon,\varepsilon)$ of 0 in R, consider the corresponding deleted neighborhood $B'(O,\varepsilon)$ of O in R^2. From the inequality (1) we see that $|f(x,y)| < \varepsilon$ whenever $r < \varepsilon$. This means that $f(x,y)$ is in $(-\varepsilon,\varepsilon)$ whenever (x,y) is in $B'(O,\varepsilon)$, which establishes the desired limit.

The concept of continuity may also be formulated exactly as in the one-variable case.

15.4 Definition

Let f be a function defined in some neighborhood of the point A in R^2 or R^3. We say that f is *continuous* at A if $\underset{P\to A}{\text{limit}} f(P)$ exists and has the value $f(A)$.

Definition 15.4 can be combined with 15.3 to give the following equivalent formulation.

15.4′ Definition

Let f be defined in some neighborhood of the point A. Then f is *continuous* at A if for every neighborhood N of $f(A)$ in R, there exists a neighborhood $B(A,r)$ of A contained in the domain of f such that $f(P)$ is in N for every P in $B(A,r)$.

We shall have to modify the above definition in subsequent work because we shall need to consider functions that might be defined at a point A in R^2, but might not be defined at all points in any neighborhood of A. In other words, the domain of f might fail to be an *open* set, as defined in Section 2 of Chapter 13. For example, consider the function f defined by

$$f(x,y) = \sqrt{x} + \sqrt{y}.$$

The domain of f is the set of all points (x,y) such that $x \geq 0$ and $y \geq 0$ (the "closed" first quadrant of the plane). How shall we formulate a condition of "continuity" for this function at a point such as (0,1)? Observe that Definition 15.4′ does not apply because *there is no neighborhood of* (0,1) *contained in the domain of f.* This difficulty can be overcome by the following definition.

15.5 Definition

Let f be a real-valued function whose domain is a subset S of R^2 or R^3. The function f is *continuous* at a point A in S if, given any neighborhood N of $f(A)$ in R, there exists some neighborhood $B(A,r)$ of A such that $f(P)$ is in N whenever P is in $S \cap B(A,r)$. The function f is *continuous in* S if it is continuous at each point of S.

Thus, according to 15.5, the function

$$f(x,y) = \sqrt{x} + \sqrt{y}$$

is continuous at all points in the first quadrant of the plane, *including those points on the coordinate axes.*

Corresponding to Theorems 2.8 to 2.10 on limits of sums, products, and quotients, we have the following theorem on continuity of functions of several variables.

15.6 Theorem

If functions f and g are continuous at point P in R^2 *or* R^3, *then*

(1) $f + g$ *is continuous at P.*
(2) fg *is continuous at P.*
(3) f/g *is continuous at P provided* $g(P) \neq 0$.

The proof of 15.6 is analogous to that of the corresponding theorem for functions of one variable, and hence it is omitted.

A function f having domain R^2 is called a *polynomial function* of two variables if $f(x,y)$ consists of a sum of terms of the form ax^my^n, where a is in R and m and n are nonnegative integers. For example, f defined by

$$f(x,y) = x^2 - 3xy + y^2 + 7$$

is a polynomial function. A function h such that $h = g/f$, where g and f are polynomial functions, is called a *rational function* of two variables.

Since every polynomial function can be represented as a sum of products of the simple functions

$$f(x,y) = x, \quad g(x,y) = y, \quad h(x,y) = c, \qquad c \text{ a constant},$$

and these three functions are continuous, every polynomial function is continuous by 15.6. In turn, each rational function is continuous since it is a quotient of continuous functions.

We can also form composite functions of several variables. In particular, we can form the composite $h \circ f$ of a function h of one variable and a function f of two variables, thereby obtaining a function $h \circ f$ of two variables.

For example, if

$$h(t) = \sqrt{t} \quad \text{and} \quad f(x,y) = 4 - x^2 - y^2,$$

then $$(h \circ f)(x,y) = h(f(x,y)) = h(4 - x^2 - y^2) = \sqrt{4 - x^2 - y^2}.$$

As another example, if

$$F(t) = \sin t \quad \text{and} \quad G(x,y) = xy - y^2,$$

then $$(F \circ G)(x,y) = F(xy - y^2) = \sin (xy - y^2).$$

In view of the following theorem, both functions $h \circ f$ and $F \circ G$ in the examples above are continuous.

15.7 Theorem

If h is a function of one variable and f a function of two variables such that f is continuous at point P and h is continuous at $f(P)$, then $h \circ f$ is also continuous at P.

We shall not prove Theorem 15.7, since its proof is essentially the same as that of Theorem 2.28 on functions of one variable.

Another important type of composite function is the composite of a function of two variables with two functions of one variable. For example, let

$$f(x,y) = x^2 - 3xy, \qquad g(t) = t^2, \qquad h(t) = 2t^3.$$

Then the composite function H defined by

$$H(t) = f(g(t),h(t)) = f(t^2,2t^3) = t^4 - 6t^5,$$

is a function of the single variable t. This composite function is continuous according to the following theorem, whose proof is again omitted.

15.8 Theorem

If g and h are functions of one variable that are continuous at a, and f is a function of two variables that is continuous at the point $(g(a),h(a))$, then the composite function H defined by

$$H(t) = f(g(t),h(t))$$

is continuous at a.

A special case of Theorem 15.8 arises in the following remarks.

Let f be a function of two variables, P a point in the domain of f, and L a line in R^2 passing through P. If we consider R^2 as the xy plane in R^3, then we

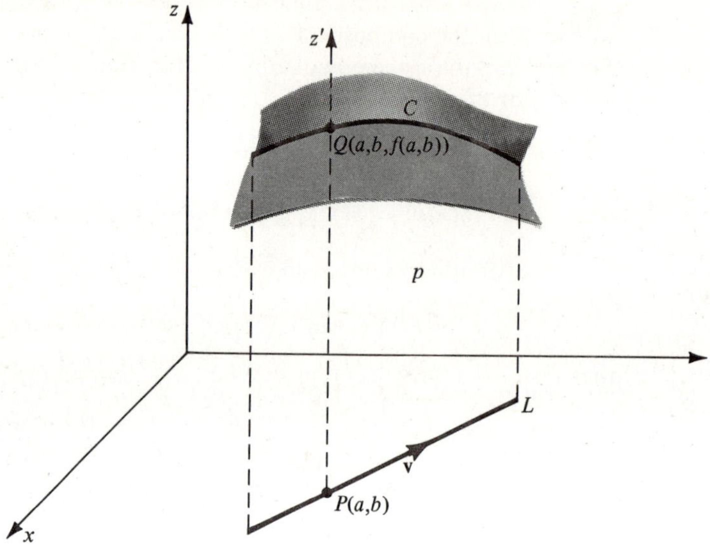

FIGURE 15.2

can construct a plane p in R^3 passing through line L and perpendicular to the xy plane. Plane p will intersect the graph of f in a curve C, called the cross section of the graph of f in p (Figure 15.2).

If L is the line through $P = (a,b)$ in the direction of a unit vector $\mathbf{v} = \langle r,s \rangle$, then parametric equations of L are

$$x(t) = a + rt, \qquad y(t) = b + st.$$

Let us define a function H of the one variable t by

$$H(t) = f(x(t),y(t)) = f(a + rt,\ b + st).$$

Then it is clear from Figure 15.2 that C is the graph of H in the plane p, if we let L be one coordinate axis, P be the origin, and the other axis have the same direction as the z axis. Since $\mathbf{v}$ is a unit vector, the unit of length is the same in the plane p as it is in the underlying coordinate system. Furthermore, if the function f is continuous at the point $P = (a,b)$, then by Theorem 15.8 the function H is continuous at $t = 0$.

The converse of the last statement is not true. Thus, it is possible that the function $H(t) = f(a + rt,\ b + st)$ could be continuous at $t = 0$ for *every* vector $\mathbf{v} = \langle r,s \rangle$, and yet f could fail to be continuous at $P = (a,b)$. An example to illustrate this unusual behavior is given in Exercise 24 below.

EXERCISES

In Exercises 1 to 6, (i) specify the set of all points (x,y) in $\mathbb{R}^2$ for which the given function is defined, and (ii) determine the range of the function.

1. $f(x,y) = \sqrt{x} + \sqrt{xy}$

2. $f(x,y) = \sqrt{\dfrac{x}{y}}$

3. $f(x,y) = \dfrac{1}{x^2 - y^2}$

4. $f(x,y) = \dfrac{1}{xy}$

5. $f(x,y) = \dfrac{1}{\sqrt{1 - x^2 - y^2}}$

6. $f(x,y) = e^{x/y}$

7. Let $f(x,y) = \dfrac{xy}{x+y}$, $g(t) = t^2$, $h(t) = 1 - t$. Find:

a. $g(f(x,y))$

b. $h(f(x,y))$

c. $f(g(t),h(t))$

d. $f(g(1),h(0))$

In Exercises 8 to 10, prove that $\lim_{(x,y)\to(0,0)} f(x,y)$ does not exist. [*Suggestion:* Consider the behavior of $f(x,y)$ on various straight lines through the origin.]

8. $f(x,y) = \dfrac{y^2}{x^2 + y^2}$

9. $f(x,y) = \dfrac{x + y}{x - y}$

10. $f(x,y) = \dfrac{x - y^2}{x^2 + y^2}$

In Exercises 11 to 12, find the indicated limits. Justify your assertions.

11. $\lim_{(x,y)\to(0,2)} \dfrac{y \sin x}{x}$

12. $\lim_{(x,y)\to(2,0)} \dfrac{x + \ln(1 + xy)}{1 + x + y}$

Use the methods of Example 4 to establish the indicated limits in Exercises 13 and 14.

13. $\lim_{(x,y)\to(0,0)} \dfrac{x^3 + y^3}{x^2 + y^2} = 0$

14. $\lim_{(x,y)\to(0,0)} \dfrac{x^2 + y^2}{|x| + |y|} = 0$

In each of Exercises 15 to 22, determine where the given function is continuous.

15. $F(x,y) = ax + by$

16. $G(x,y) = \dfrac{x + y}{1 + x^2}$

17. $f(x,y) = \sin(x^2 + y)$

18. $g(x,y) = \ln(1 + x^2 + y^4) + \sin xy$

19. $H(x,y) = \dfrac{x^2 y}{1 + x}$ if $x \neq -1$, $H(-1,y) = y$

20. $F(x,y) = \dfrac{x^2 y}{x^3 + y^3}$ if $(x,y) \neq (0,0)$; $F(0,0) = 0$

21. $G(x,y) = \dfrac{x^2 + y}{|x| + |y|}$ if $(x,y) \neq (0,0)$, $G(0,0) = 0$

22. $h(x,y) = \dfrac{x^3 + y^2}{x^2 + y}$ if $(x,y) \neq (0,0)$, $h(0,0) = 0$

23. Does $\displaystyle\lim_{(x,y)\to(0,0)} \frac{x^2y^2}{x^4 + y^4}$ exist?

24. Let $f(x,y) = \dfrac{xy^2}{x^2 + y^4}$ if $(x,y) \neq (0,0)$, $f(0,0) = 0$.

a. Prove that $\displaystyle\lim_{t\to 0} f(rt,st) = 0$ for every straight line $x = rt$, $y = st$ through the origin.

b. Despite the result of (a), prove that $\displaystyle\lim_{(x,y)\to(0,0)} f(x,y)$ does not exist, by considering values of the function along the parabola $x = y^2$.

If f is a function of x and y, then the graph in R^2 of the equation $f(x,y) = c$, c a constant, is called a *level curve* (or *contour line*) of the function f. By drawing a number of level curves of f, for various values of the constant c, one can often visualize the nature of the surface $z = f(x,y)$ in R^3, in exactly the same way as we visualize a land surface by examining the contour lines on a topographical map. Draw a number of level curves of each of the following functions.

25. $f(x,y) = xy$

26. $f(x,y) = x^2 + y^2$

27. $f(x,y) = 3x^2 + 4y^2$

28. $f(x,y) = \sin(x + y)$

29. $f(x,y) = e^{xy}$

2 PARTIAL DERIVATIVES

Let f be a function of two variables defined in a neighborhood of the point $P = (a,b)$. We define the *partial derivative of f with respect to x at the point P* by

15.9 $$D_1 f(P) = \lim_{h\to 0} \frac{f(a + h, b) - f(a,b)}{h}.$$

Similarly, the *partial derivative of f with respect to y at the point P* is

15.10 $$D_2 f(P) = \lim_{k\to 0} \frac{f(a, b + k) - f(a,b)}{k}.$$

Thus $D_1 f(P)$ is simply the ordinary derivative of f with respect to x with y held constant at the value $y = b$; and $D_2 f(P)$ is the ordinary derivative of f with respect to y with x held constant at the value $x = a$.

Actually, $D_1 f$ and $D_2 f$ can be regarded as defining new functions of x and y, since they may be obtained simply by differentiating $f(x,y)$ with respect to x and y, respectively.

Other common notations for the partial derivatives of f are

$$D_1 f = f_1 = \frac{\partial f}{\partial x} = f_x,$$

$$D_2 f = f_2 = \frac{\partial f}{\partial y} = f_y.$$

Example 1 Find the first partial derivatives of the function f defined by

$$f(x,y) = x^3 - 3x^2y + y^2 + x - 7.$$

Solution: We have

$$D_1 f(x,y) = \frac{\partial f}{\partial x} = 3x^2 - 6xy + 1,$$

$$D_2 f(x,y) = \frac{\partial f}{\partial y} = -3x^2 + 2y.$$

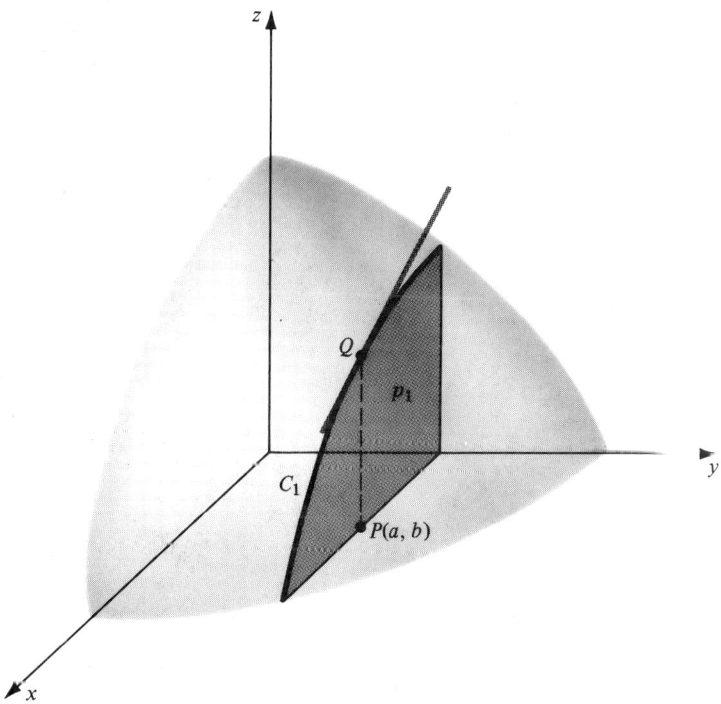

FIGURE 15.3

The partial derivatives of f at a point $P = (a,b)$ have an obvious geometrical interpretation. Let us again consider R^2 as the xy plane in R^3. Through the point $P = (a,b)$, pass a plane p_1 parallel to the xz plane (p has the equation $y = b$). Let the curve C_1 be the intersection of the graph of f with the plane p_1; that is, C_1 is the cross section of the graph of f in p_1 (see Figure 15.3). If Q is the point $(a,b,f(a,b))$ on the curve C_1 then $D_1f(a,b)$ represents the slope of the tangent line to the curve C_1 in the plane p_1 at the point Q.

Similarly, we may pass a plane p_2 through $P(a,b)$ parallel to the yz plane and obtain a cross section C_2 of the graph of f in p_2. Then $D_2f(a,b)$ represents the slope of the tangent line to the curve C_2 in the plane p_2 at the point Q (Figure 15.4).

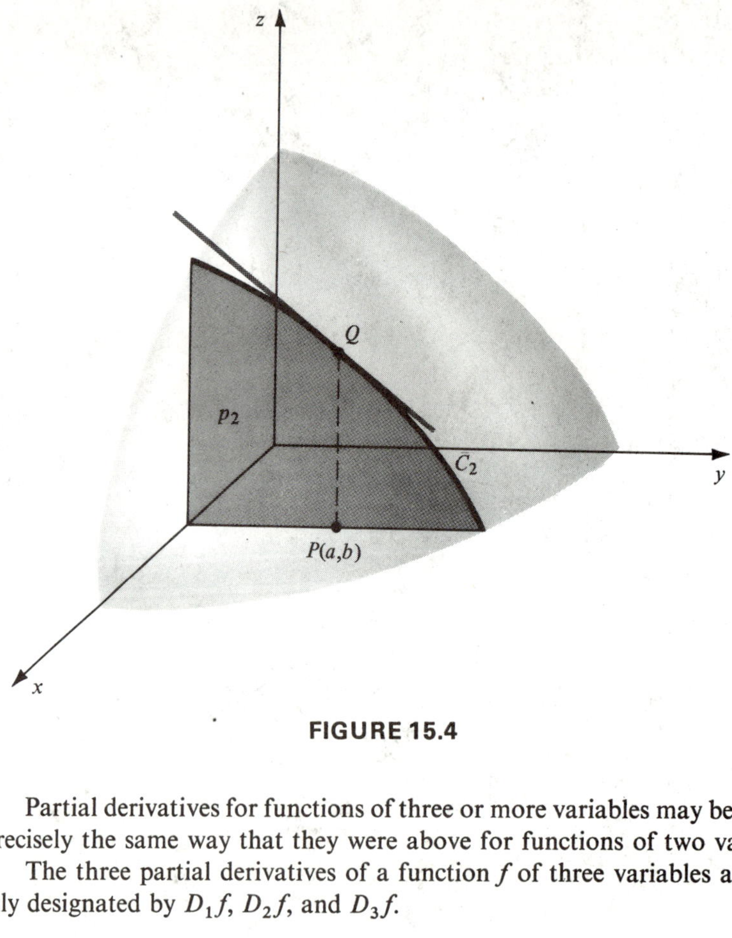

FIGURE 15.4

Partial derivatives for functions of three or more variables may be defined in precisely the same way that they were above for functions of two variables.

The three partial derivatives of a function f of three variables are commonly designated by D_1f, D_2f, and D_3f.

Example 2 Find the partial derivatives of the function f defined by

$$f(x,y,z) = \ln (x + y^2 + z^3).$$

Solution: We have

$$D_1f(x,y,z) = \frac{\partial f}{\partial x} = \frac{1}{x + y^2 + z^3},$$

$$D_2f(x,y,z) = \frac{\partial f}{\partial y} = \frac{2y}{x + y^2 + z^3},$$

$$D_3f(x,y,z) = \frac{\partial f}{\partial z} = \frac{3z^2}{x + y^2 + z^3}.$$

If f is a function of two variables, then D_1f and D_2f are also functions of two variables. As such, it might happen that D_1f and D_2f also possess first partial derivatives $D_1(D_1f)$, $D_2(D_1f)$, $D_1(D_2f)$, and $D_2(D_2f)$. If these four functions exist, they are called the *second partial derivatives* of f. Common notations for the second partial derivatives of f are

$$f_{11} = D_{11}f = D_1(D_1f) = \frac{\partial^2 f}{\partial x^2} = f_{xx},$$

$$f_{12} = D_{12}f = D_2(D_1f) = \frac{\partial^2 f}{\partial y\, \partial x} = f_{xy},$$

$$f_{21} = D_{21}f = D_1(D_2f) = \frac{\partial^2 f}{\partial x\, \partial y} = f_{yx},$$

$$f_{22} = D_{22}f = D_2(D_2f) = \frac{\partial^2 f}{\partial y^2} = f_{yy}.$$

A function of three variables has nine possible second partial derivatives (of which usually only six are distinct).

Higher partial derivatives of a function of several variables are defined in the obvious way. For example, if f is a function of two variables, then

$$f_{122} = D_{122}f = D_2(D_2(D_1f)) = \frac{\partial^3 f}{\partial y\, \partial y\, \partial x} = f_{xyy},$$

$$f_{2122} = D_{2122}f = D_2(D_2(D_1(D_2f))) = \frac{\partial^4 f}{\partial y\, \partial y\, \partial x\, \partial y} = f_{yxyy}.$$

Example 3 If $f(x,y,z) = xe^{yz} + yze^x$, find f_{213}, $\partial^3 f/\partial x^2\, \partial y$, and f_{12}.

Solution: We have

$$f_2(x,y,z) = \frac{\partial f}{\partial y} = xze^{yz} + ze^x,$$

$$f_{21}(x,y,z) = \frac{\partial^2 f}{\partial x\, \partial y} = \frac{\partial f_2}{\partial x} = ze^{yz} + ze^x,$$

$$f_{213}(x,y,z) = \frac{\partial^3 f}{\partial z\, \partial x\, \partial y} = \frac{\partial f_{21}}{\partial z} = e^{yz} + zye^{yz} + e^x,$$

$$\frac{\partial^3 f}{\partial x^2\, \partial y} = \frac{\partial f_{21}}{\partial x} = ze^x.$$

Also,

$$f_1(x,y,z) = e^{yz} + yze^x,$$
$$f_{12}(x,y,z) = ze^{yz} + ze^x.$$

Undoubtedly the reader observed in Example 3 that the *mixed* second partial derivatives f_{12} and f_{21} are identical. This is always true for well-behaved functions, and in fact we have the following theorem, which we state for functions of two variables.

15.11 Theorem

If f is a function of two variables for which $D_{12}f$ and $D_{21}f$ are continuous in some open set S of R^2, then $D_{12}f(P) = D_{21}f(P)$ for every P in S.

We shall prove 15.11 later, as Theorem 15.34 in Section 9. In the meantime, however, we shall make frequent use of it.

A consequent of 15.11 is that the order of differentiation may be interchanged for a function of several variables. For example, if f is a function of two variables having continuous higher partial derivatives in some open set of R^2, then

$$f_{211} = f_{121} = f_{112}, \qquad f_{2211} = f_{2121} = f_{1221},$$

and so on.

In Exercise II-1 at the end of this section we give an example to show that the equality $D_{12}f = D_{21}f$ may fail if these mixed second partial derivatives are not continuous.

EXERCISES

I

In Exercises 1 to 4, find f_1, f_2, f_{12}, and f_{21}. Verify in each case that $f_{12} = f_{21}$ (wherever these functions are defined).

1. $f(x,y) = \sin xy$.

2. $f(x,y) = y \ln \dfrac{y^2}{x}$

3. $f(x,y) = \dfrac{x - y}{x + y}$

4. $f(x,y) = xe^{xy}$

5. If $f(x,y) = x \ln \dfrac{y^2}{x}$, find the value of $f_1(4,2) + f_2(4,1)$.

6. If $f(x,y,z) = \sqrt{x^2 + y^2 + z^2}$, find f_{32} and f_{23}.

In each of Exercises 7 to 10 find the indicated partial derivatives of the given function.

7. F_{12} and F_{22}, where $F(x,y) = \ln (x + y)$

8. G_{123} and G_{312}, where $G(x,y,z) = x^3 + 3yz + \sin xyz$

9. F_{1223} and F_{1222}, where $F(x,y,z) = x^4 + y^4 - 2z^4 + 2x^2yz$

10. $\dfrac{\partial^2 F}{\partial x^2}$ and $\dfrac{\partial^2 F}{\partial y^2}$, where $F(x,y) = \ln(x^2 + y^2)$

11. If $f(x,y,z,w) = \dfrac{xy}{z + w}$, find f_{xyzw}.

12. If $F(x,y) = (y + ax)^2 e^{y+ax}$, show that $F_{xx} = a^2 F_{yy}$.

A function $F(x,y)$ with continuous second partial derivatives in an open set S and satisfying

$$\frac{\partial^2 F}{\partial x^2} + \frac{\partial^2 F}{\partial y^2} = 0, \qquad \textit{Laplace's equation,}$$

at all points of S is said to be *harmonic* in S. Show that the following functions are harmonic and specify the set where this is the case.

13. $F(x,y) = \ln \sqrt{x^2 + y^2}$

14. $F(x,y) = \dfrac{1}{\sqrt{x^2 + y^2}}$

15. $F(x,y) = e^x \cos y$

16. $F(x,y) = \tan^{-1} \dfrac{y}{x}$

17. $F(x,y) = \sin x \cosh y$

II

1. Let $F(x,y) = \dfrac{xy(x^2 - y^2)}{x^2 + y^2}$, $(x,y) \neq (0,0)$, $F(0,0) = 0$. Show that $F_{12}(0,0)$ and $F_{21}(0,0)$ both exist, but they are not equal.
2. Suppose that f is a function of two variables defined at all points inside an open disk centered at (a,b). If $f_1(x,y)$ and $f_2(x,y)$ exist and equal 0 for all (x,y) in S, prove that $f(x,y)$ is identically constant in S. *Suggestion:* Write

$$f(x,y) - f(a,b) = [f(x,y) - f(a,y)] + [f(a,y) - f(a,b)].$$

Apply the mean value theorem, for functions of one variable, to each bracket on the right; and conclude that $f(x,y) - f(a,b) = 0$ for all (x,y) in S.

3 DIFFERENTIALS

The reader will recall that a function of one variable is usually called "differentiable" at a certain point if its derivative exists at that point. However, for functions of several variables the situation is more complicated, since the concept of "differentiability" turns out to be a stronger condition than the simple requirement that the partial derivatives exist.

In order to motivate these somewhat subtle ideas, it will be advisable at this time to review some of the properties of the differential of a function of one variable. Suppose that f is a real-valued function of one variable defined in some neighborhood of the number a, and suppose that $f'(a)$ exists. Since, by definition,

$$f'(a) = \underset{h \to 0}{\text{limit}} \frac{f(a + h) - f(a)}{h},$$

it follows that

$$\underset{h \to 0}{\text{limit}} \left[\frac{f(a + h) - f(a)}{h} - f'(a)\right] = 0.$$

The quantity occurring on the left-hand side of the above formula is a function of h, which we shall denote by $\varepsilon(h)$. That is,

$$\varepsilon(h) = \frac{f(a + h) - f(a) - f'(a) \cdot h}{h},$$

or, rearranging slightly,

15.12 $$f(a + h) - f(a) = f'(a) \cdot h + \varepsilon(h) \cdot h.$$

The quantity on the left-hand side of 15.12 is the change, or *increment*, in the value of the function f corresponding to the change h in the value of the independent variable. This increment depends on the numbers a and h, and we shall denote it by $\Delta f(a,h)$. The quantity $f'(a) \cdot h$ occurring in 15.12, which also depends on both a and h, is called the *differential* of f and is denoted by $df(a,h)$. Thus,

$$\Delta f(a,h) = f(a + h) - f(a),$$

$$df(a,h) = f'(a) \cdot h.$$

The geometrical significance of these quantities is illustrated in Figure 15.5, in which the line L is the tangent line to the graph of f. The reader will of course recall that $f'(a)$ represents the slope of the line L.

The length $|\overrightarrow{TQ}|$ in Figure 15.5 represents the difference $\Delta f(a,h) - df(a,h)$, which according to 15.12 is the product $\varepsilon(h) \cdot h$ of two quantities, each of which approaches 0 as h does. Speaking intuitively, this means that $\Delta f(a,h) - df(a,h)$ approaches 0 more "rapidly" than h does, and consequently $df(a,h)$ should be a "good" approximation to $\Delta f(a,h)$ when h is "small."

In the following example we compute these quantities for a particular function.

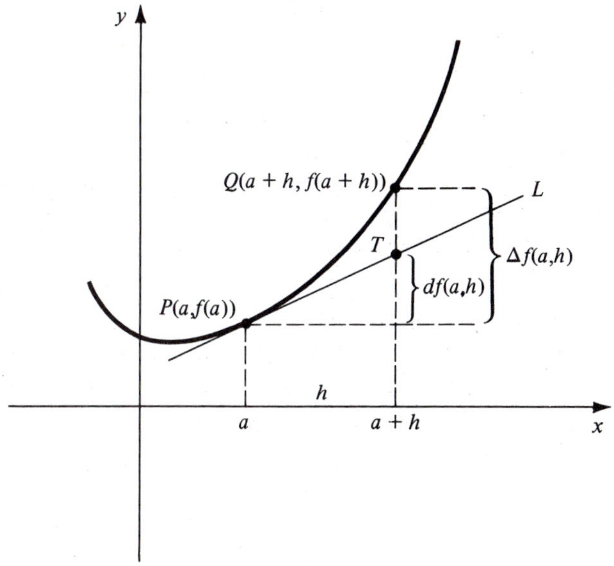

FIGURE 15.5

Example 1 Let $f(x) = x^3 + x$. Then

$$\Delta f(a,h) = f(a + h) - f(a)$$

$$= 3a^2h + 3ah^2 + h^3 + h$$

$$df(a,h) = (3a^2 + 1)h = 3a^2h + h.$$

Hence,

$$\varepsilon(h) = \frac{\Delta f(a,h) - df(a,h)}{h} = 3ah + h^2.$$

Note that in this case $\varepsilon(h)\, h = 3ah^2 + h^3$, which approaches 0 more rapidly than h.

We now turn our attention to the problem of extending these ideas to functions of two variables. Suppose that f is a function of two variables x and y, which is defined in some neighborhood of the point $P = (a,b)$. If the x and y coordinates of P are changed by amounts h and k, respectively, then P is moved to the new point $Q = (a + h, b + k)$. If $\mathbf{v}$ is the vector $\overrightarrow{PQ} = \langle h,k \rangle$, then we may think of the change in P as being specified by the vector $\mathbf{v}$. Corresponding to this change in P, there is an increment in f, namely,

$$\Delta f(P,\mathbf{v}) = f(a + h, b + k) - f(a,b).$$

This increment in f depends on the *four* real variables a, b, h, and k.

Let us calculate $\Delta f(P,\mathbf{v})$ for a specific function f, say $f(x,y) = x^3 + 3xy + y^2$.

$$\begin{aligned} f(a + h, b + k) - f(a,b) &= a^3 + 3a^2h + 3ah^2 + h^3 \\ &\quad + 3(ab + ak + bh + hk) - a^3 - 3ab - b^2 \\ &= (3a^2 + 3b)h + (3a + 2b)k + (3ah + h^2)h \\ &\quad + (3h + k)k. \end{aligned}$$

The reader will observe that the last equation may be put in the form

15.13 $$\Delta f(P,\mathbf{v}) = M \cdot h + N \cdot k + \varepsilon_1 \cdot h + \varepsilon_2 \cdot k,$$

where M and N are constants (depending only on a and b), but ε_1 and ε_2 are functions of h and k (for brevity we have omitted writing "$\varepsilon_1(h,k)$" and "$\varepsilon_2(h,k)$"). Furthermore, ε_1 and ε_2 both approach 0 as $(h,k) \to (0,0)$.

The form of 15.13 is clearly analogous to 15.12. [In fact, the reader undoubtedly noticed in 15.13 that $M = D_1f(a,b)$, $N = D_2f(a,b)$]. This suggests the following definition.

15.14 Definition

Let f be a function of x and y defined in some neighborhood D of $P = (a,b)$. The function f is *differentiable* at P if there are constants M, N, and functions ε_1 and ε_2 of h and k such that for any vector $\mathbf{v} = \langle h,k \rangle$ with $(a + h, b + k)$ in D,

(i) $\Delta f(P,\mathbf{v}) = M \cdot h + N \cdot k + \varepsilon_1 \cdot h + \varepsilon_2 \cdot k$.
(ii) limit ε_1 = limit $\varepsilon_2 = 0$ when $(h,k) \to (0,0)$.

The reader should note that if f is differentiable at P, then $D_1f(P)$ and $D_2f(P)$ exist and equal the M and N, respectively, of Definition 15.14. To see this, take $k = 0$ in 15.14(i); that is, consider the special case when $\mathbf{v} = \langle h,0 \rangle$. Then

$$D_1f(P) = \underset{h \to 0}{\text{limit}} \frac{f(a + h, b) - f(a,b)}{h} = \underset{h \to 0}{\text{limit}} \frac{\Delta f(P,\mathbf{v})}{h},$$

and so, from 15.14,

$$D_1f(P) = \underset{h \to 0}{\text{limit}} \frac{Mh + \varepsilon_1 h}{h}.$$

But $\underset{h \to 0}{\text{limit}}\ \varepsilon_1 = 0$, and so $D_1f(P) = M$. In a similar way it can be shown that $D_2f(P) = N$.

The converse of the above result does *not* hold. More precisely, the partial derivatives $D_1f(P)$ and $D_2f(P)$ may exist, but the function f may fail to be differentiable at P. An example to illustrate this will be given shortly. This behavior is in sharp contrast to the one-variable case, where the existence of $f'(a)$ is equivalent to the "differentiability" of f at a.

By analogy with the one-variable case, we now define the *differential* of f at P, denoted by $df(P,\mathbf{v})$, to be the quantity

$$M \cdot h + N \cdot k = D_1 f(P) \cdot h + D_2 f(P) \cdot k,$$

which occurs in 15.14(i).

For a function f of one variable we noticed that

$$\operatorname*{limit}_{h \to 0} \frac{\Delta f(a,h) - df(a,h)}{h} = 0.$$

(From this it followed that $df(a,h)$ is a good approximation to $\Delta f(a,h)$ for small values of h). A similar result holds for functions of two variables.

15.15 Theorem

If f is differentiable at P, and $\mathbf{v} = \langle h,k \rangle$, then

$$\operatorname*{limit}_{(h,k) \to (0,0)} \frac{\Delta f(P,\mathbf{v}) - df(P,\mathbf{v})}{|\mathbf{v}|} = 0.$$

Proof: By 15.14,

$$\Delta f(P,\mathbf{v}) - df(P,\mathbf{v}) = \varepsilon_1 h + \varepsilon_2 k,$$

where ε_1 and ε_2 approach 0 as (h,k) approaches $(0,0)$. Also,

15.16
$$\frac{\Delta f(P,\mathbf{v}) - df(P,\mathbf{v})}{|\mathbf{v}|} = \frac{\varepsilon_1 h}{\sqrt{h^2 + k^2}} + \frac{\varepsilon_2 k}{\sqrt{h^2 + k^2}}.$$

But

$$0 \le \left| \frac{\varepsilon_1 h}{\sqrt{h^2 + k^2}} \right| \le \varepsilon_1,$$

$$0 \le \left| \frac{\varepsilon_2 k}{\sqrt{h^2 + k^2}} \right| \le \varepsilon_2.$$

Hence the right-hand side of 15.16 approaches 0 as $(h,k) \to (0,0)$, which proves the theorem.

The following examples illustrate the use of the differential as an approximation to the actual increment in a function.

Example 2 An isosceles triangle T has dimensions as shown in Figure 15.6. Approximately what change occurs in the area of T if the lengths of the equal sides are increased by 1 in. and the vertex angle is increased by .04 radian?

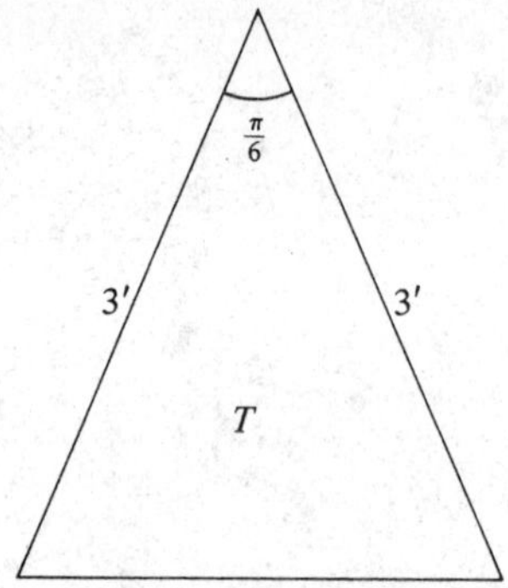

FIGURE 15.6

Solution: An isosceles triangle with equal sides of length x and vertex angle α has area

$$m(x,\alpha) = \tfrac{1}{2}x^2 \sin \alpha.$$

Thus the given triangle has area

$$m\left(3, \frac{\pi}{6}\right) = \frac{9}{4}.$$

Assuming that the function m of two variables is differentiable (which is true by 15.18 below), we have

$$D_1 m(x,\alpha) = x \sin \alpha, \qquad D_2 m(x,\alpha) = \tfrac{1}{2}x^2 \cos \alpha,$$
$$dm((x,\alpha),\mathbf{v}) = (x \sin \alpha)h + (\tfrac{1}{2}x^2 \cos \alpha)k,$$

where $\mathbf{v} = \langle h,k \rangle$. In particular, taking $P = (x,\alpha) = (3, \pi/6)$ and $\mathbf{v} = \langle h,k \rangle = \langle \frac{1}{12}, 0.04 \rangle$, we have

$$D_1 m\left(3, \frac{\pi}{6}\right) = \frac{3}{2}, \qquad D_2 m\left(3, \frac{\pi}{6}\right) = \frac{9\sqrt{3}}{4},$$

$$dm(P,\mathbf{v}) = \frac{3}{2} \cdot \frac{1}{12} + \frac{9\sqrt{3}}{4}(.04) \doteq .28.$$

Using $dm(P,\mathbf{v})$ as an approximation to $\Delta m(P,\mathbf{v})$ we conclude that the area of the triangle is increased by approximately 0.28 ft^2.

Example 3 Find the approximate value of $\sqrt{(2.98)^2 + (4.03)^2}$.

Solution: Let $f(x,y) = \sqrt{x^2 + y^2}$, and observe that $f(3,4) = 5$. Let us compute $df(P,\mathbf{v})$, where $P = (3,4)$ and $\mathbf{v} = \langle -.02, .03 \rangle$. Since

$$D_1 f(x,y) = \frac{x}{\sqrt{x^2 + y^2}}, \qquad D_2 f(x,y) = \frac{y}{\sqrt{x^2 + y^2}},$$

we have

$$df(P,\mathbf{v}) = \tfrac{3}{5}(-.02) + \tfrac{4}{5}(.03) = .012.$$

Using $df(P,\mathbf{v})$ as an approximation to $\Delta f(P,\mathbf{v})$, we have

$$f(2.98,4.03) \doteq f(3,4) + df(P,\mathbf{v}) = 5.012.$$

We now give an example to show that the existence of the partial derivatives $D_1f(P)$ and $D_2f(P)$ is not a sufficient condition for f to be differentiable at P.

Example 4 Let $f(x,y) = \sqrt{|xy|}$. Both $D_1f(0,0)$ and $D_2f(0,0)$ exist, since

$$D_1f(0,0) = \lim_{t\to 0}\frac{f(0+t,0) - f(0,0)}{t} = 0,$$

and likewise $D_2f(0,0) = 0$. Thus also $df((0,0),\mathbf{v}) = 0$ for every $\mathbf{v} = \langle h,k\rangle$. But

$$\frac{\Delta f((0,0),\mathbf{v}) - df((0,0),\mathbf{v})}{|\mathbf{v}|} = \frac{f(h,k) - f(0,0)}{\sqrt{h^2+k^2}} = \frac{\sqrt{|hk|}}{\sqrt{h^2+k^2}} = \sqrt{\frac{|hk|}{h^2+k^2}},$$

and, according to Example 3 of Section 1, this limit does not exist as $(h,k) \to (0,0)$. Hence the condition of Theorem 15.15 is not satisfied, and so f cannot be differentiable at (0,0).

The existence of $D_1f(P)$ and $D_2f(P)$ does not even imply that f is continuous at P, as the following example shows.

Example 5 Let $f(x,y) = \dfrac{xy}{x^2+y^2}$ for $(x,y) \neq (0,0)$ and $f(0,0) = 0$. By Example 3 of Section 1, f is not continuous at (0,0). But $D_1f(0,0)$ and $D_2f(0,0)$ exist, since

$$D_1f(0,0) = \lim_{t\to 0}\frac{f(0+t,0) - f(0,0)}{t} = 0$$

$$D_2f(0,0) = \lim_{t\to 0}\frac{f(0,0+t) - f(0,0)}{t} = 0.$$

However, *differentiability* implies continuity, as we now show.

15.17 Theorem

If f is differentiable at $P = (a,b)$, then f is continuous at P.

Proof: It is sufficient to prove that

$$\lim_{(h,k)\to(0,0)} f(a+h, b+k) = f(a,b)$$

or, equivalently, that

$$\lim_{(h,k)\to(0,0)} \Delta f(P,\mathbf{v}) = 0,$$

where $\mathbf{v} = \langle h,k\rangle$. But this follows immediately from 15.14.

Clearly it is of great importance to know when a function is differentiable. The following result provides a sufficient condition for this purpose.

15.18 Theorem

Let f be a function of two variables defined in some neighborhood D of the point $P = (a,b)$. If $D_1 f$ and $D_2 f$ exist in D and are continuous at P, then f is differentiable at P.

Proof: The idea of the proof is to exhibit functions ε_1 and ε_2 that satisfy the conditions of Definition 15.14. This may be done as follows. If $\mathbf{v} = \langle h,k \rangle$, then

$$\Delta f(P,\mathbf{v}) = f(a + h, b + k) - f(a,b).$$

Adding and subtracting $f(a + h, b)$,

$$\text{(1)} \qquad \begin{aligned} \Delta f(P,\mathbf{v}) = {} & [f(a + h, b) - f(a,b)] \\ & + [f(a + h, b + k) - f(a + h, b)]. \end{aligned}$$

In the terms of the first bracket in (1) the variable y is held constant at the value $y = b$; and so in this bracket f can be regarded as a function of x alone, whose derivative with respect to x is the function $D_1 f$. Hence the mean value theorem (4.8) may be applied to conclude that there is some number x_1, between a and $a + h$, such that

$$\text{(2)} \qquad f(a + h, b) - f(a,b) = D_1 f(x_1,b) \cdot h.$$

The number x_1 is actually a function of h, which approaches a as h approaches 0. If we write

$$\varepsilon_1 = D_1 f(x_1,b) - D_1 f(a,b),$$

then ε_1 is also a function of h. Furthermore, since $D_1 f$ is continuous at (a,b), ε_1 approaches 0 as x_1 approaches a. Thus, limit $\varepsilon_1 = 0$ as $h \to 0$. Also, (2) may be written as

$$\text{(3)} \qquad f(a + h, b) - f(a,b) = D_1 f(a,b) \cdot h + \varepsilon_1 h.$$

In the second bracket of equation (1) the variable x is held constant at the value $x = a + h$, and so in these terms f can be regarded as a function of y alone. Applying the mean value theorem again, we may assert that there is some number y_1, between b and $b + k$, such that

$$\text{(4)} \qquad f(a + h, b + k) - f(a + h, b) = D_2 f(a + h, y_1) \cdot k.$$

Here the number y_1 depends on *both* h and k. If we write

$$\varepsilon_2 = D_2 f(a + h, y_1) - D_2 f(a,b),$$

then ε_2 is a function of both h and k; and because of the continuity of

$D_2 f$ at (a,b), it follows that $\varepsilon_2 \to 0$ as $(h,k) \to (0,0)$. Now let us write (4) in the form

$$f(a + h, b + k) - f(a + h, b) = D_2 f(a,b) + \varepsilon_2 k. \tag{5}$$

Using (3) and (5), equation (1) may therefore be written in the form

$$\Delta f(P,\mathbf{v}) = D_1 f(a,b) \cdot h + D_2 f(a,b) \cdot k + \varepsilon_1 h + \varepsilon_2 k,$$

which satisfies all the conditions of 15.14. Hence f is differentiable at P.

The ideas of this section may also be extended to functions of three or more variables. For a function F of three variables, for example, a definition of differentiability may be given that is exactly analogous to 15.14. The differential $dF(P,\mathbf{v})$ is defined to be

$$dF(P,\mathbf{v}) = rD_1 F(P) + sD_2 F(P) + tD_3 F(P),$$

where P is a point in R^3 and $\mathbf{v}$ is the three-dimensional vector $\langle r,s,t \rangle$.

EXERCISES

I

For each of the functions in Exercises 1 to 4, write an expression for $\Delta f(P,\mathbf{v})$, where $P = (a,b)$ and $\mathbf{v} = \langle h,k \rangle$. Then determine the functions ε_1 and ε_2 of formula 15.13, and verify that ε_1 and ε_2 approach 0 as $(h,k) \to (0,0)$.

1. $f(x,y) = x^2 y + y^2 - 3x$

2. $f(x,y) = x^3 + y^3$

3. $f(x,y) = \dfrac{x}{y}$

4. $f(x,y) = \dfrac{y^2}{x + y}$

In Exercises 5 to 9, find the differential of the given function for the indicated point P and vector $\mathbf{v}$.

5. $f(x,y) = \tan^{-1} \dfrac{y}{x}$, $P = (2,1)$, $\mathbf{v} = \langle 1,1 \rangle$

6. $f(x,y) = \sqrt{x^2 + y^2}$, $P = (3,4)$, $\mathbf{v} = \langle -1,-1 \rangle$

7. $F(x,y,z) = \ln \sqrt{x^2 + y^2 + z^2}$, $P = (1,1,1)$, $\mathbf{v} = \langle r,s,t \rangle$

8. $g(x,y,z) = \sin(x + y)\sin(y + z)$, $P = (a,b,c)$, $\mathbf{v} = \langle h,k,l \rangle$

9. $F(a,y,z) = \sqrt{e^{x^2} + e^{y^2}} + \sin(x + y + z)$, $P = (0,0,0)$, $\mathbf{v} = \langle h,k,l \rangle$

In Exercises 10 to 14, approximate the given number by means of differentials.

10. $\sqrt{299^2 + 399^2}$ (note that $\sqrt{300^2 + 400^2} = 500$)

11. $\sqrt{100^2 + 199^2 + 201^2}$

12. $\sqrt{102^2 + 99^2 + 2503}$

13. $\sin 44° \cos 31°$

14. $\sin 28° \cos 29° \tan 44°$

15. The length, width, and height of a rectangular box are measured and found to be 10.2 in., 20.4 in., and 9.9 in., respectively. The volume is then found to be $10.2 \times 20.4 \times 9.9$, or 2059.992 in.^{3}. Are we justified in rounding off the volume to 2060.0 in.^{3}? (When we say that $L = 10.2$, we mean that $10.15 \leq L \leq 10.25$, etc.)

16. The radius of the base of a right circular cone is 15.3 in. and its slant height is 36.7. If the uncertainties in the radius of the base and slant height are $\pm.1$ and $\pm.2$, respectively, what is the uncertainty in the volume of the cone?

II

1. Let $f(x,y) = \dfrac{x^2y^2}{x^4 + y^4}$, $(x,y) \neq (0,0)$, $f(0,0) = 0$. Show that $D_1 f(0,0)$ and $D_2 f(0,0)$ both exist, but f is not differentiable at $(0,0)$. Reason as in Example 4 of this section.

4 THE CHAIN RULE

The reader will recall that, for differentiable functions of one variable, the so-called "chain rule" gives us a method for differentiating the composite of two functions,

$$D_x(f \circ g)(x) = f'(g(x))g'(x).$$

This rule may be extended to functions of several variables.

15.19 Chain Rule

Let F be a function of two variables, and let f and g be functions of one variable defined in some neighborhood N of the number t_0. Suppose that f and g are differentiable at t_0, and that F is differentiable at $P_0 = (f(t_0),g(t_0))$. Define the function G by

$$G(t) = F(f(t),g(t)), \qquad \text{for } t \text{ in } N.$$

Then $G'(t_0)$ exists and is given by

$$G'(t_0) = D_1F(P_0)f'(t_0) + D_2F(P_0)g'(t_0).$$

Proof: Corresponding to an increment h in the independent variable t, the functions f and g have increments

$$\Delta f = \Delta f(t_0,h) = f(t_0 + h) - f(t_0),$$
$$\Delta g = \Delta g(t_0,h) = g(t_0 + h) - g(t_0).$$

Then G has an increment

$$\Delta G(t_0,h) = F(f(t_0 + h), g(t_0 + h)) - F(f(t_0),g(t_0)),$$

which may also be written

$$(1) \qquad \Delta G(t_0,h) = \Delta F(P_0,\langle \Delta f,\Delta g\rangle).$$

Since by hypothesis F is differentiable at P_0, we may use 15.14 to write (1) in the form

$$(2) \qquad \Delta G(t_0,h) = D_1F(P_0)\,\Delta f + D_2F(P_0)\,\Delta g + \varepsilon_1\,\Delta f + \varepsilon_2\,\Delta g,$$

where ε_1 and ε_2 are functions of Δf and Δg which approach 0 with those quantities. Also note that Δf and Δg approach 0 as h approaches 0, since f and g are differentiable (and hence continuous) at t_0.

Now from (2) we have

$$(3) \qquad \frac{\Delta G(t_0,h)}{h} = D_1F(P_0)\frac{\Delta f}{h} + D_2F(P_0)\frac{\Delta g}{h} + \varepsilon_1\frac{\Delta f}{h} + \varepsilon_2\frac{\Delta g}{h}.$$

Since

$$\operatorname*{limit}_{h\to 0}\frac{\Delta G(t_0,h)}{h} = G'(t_0),$$

$$\operatorname*{limit}_{h\to 0}\frac{\Delta f}{h} = f'(t_0),$$

$$\operatorname*{limit}_{h\to 0}\frac{\Delta g}{h} = g'(t_0),$$

and

$$\operatorname*{limit}_{h\to 0}\varepsilon_1 = \operatorname*{limit}_{h\to 0}\varepsilon_2 = 0,$$

it follows immediately from (3) that

$$G'(t_0) = D_1F(P_0)f'(t_0) + D_2F(P_0)g'(t_0).$$

The chain rule may be written in a variety of different forms. For example, suppose we regard F as a function of the "intermediate" variables x and y, where $x = f(t)$, $y = g(t)$. Then we may write 15.19 in the suggestive form

$$\frac{dG}{dt} = \frac{\partial F}{\partial x}\frac{dx}{dt} + \frac{\partial F}{\partial y}\frac{dy}{dt},$$

in which it must be understood that dx/dt and dy/dt are to be evaluated at the number t_0, while $\partial F/\partial x$ and $\partial F/\partial y$ are to be evaluated at P_0.

Example 1 If $F(x,y) = xy^2 + x^3 + y$ and

$$f(t) = t^2 - 1, \qquad g(t) = 2t - t^3, \qquad G(t) = F(f(t),g(t)),$$

find $G'(t)$.

Solution: We have

$$F_1(x,y) = y^2 + 3x^2, \quad F_2(x,y) = 2xy + 1, \quad f'(t) = 2t, \quad g'(t) = 2 - 3t^2.$$

If we let $P(t) = (f(t),g(t)) = (t^2 - 1, 2t - t^3)$, then, by the chain rule,

$$\begin{aligned} G'(t) &= F_1(P(t))f'(t) + F_2(P(t))g'(t) \\ &= [(2t - t^3)^2 + 3(t^2 - 1)^2]2t + [2(t^2 - 1)(2t - t^3) + 1](2 - 3t^2). \end{aligned}$$

Similar to 15.19, there are chain rules for functions of three or more variables. For example, if F is a function of three variables and $x = x(s,t)$, $y = y(s,t)$, $z = z(s,t)$, and

$$G(s,t) = F(x(s,t),y(s,t),z(s,t)),$$

then

$$\frac{\partial G}{\partial s} = \frac{\partial F}{\partial x}\frac{\partial x}{\partial s} + \frac{\partial F}{\partial y}\frac{\partial y}{\partial s} + \frac{\partial F}{\partial z}\frac{\partial z}{\partial s}$$

$$\frac{\partial G}{\partial t} = \frac{\partial F}{\partial x}\frac{\partial x}{\partial t} + \frac{\partial F}{\partial y}\frac{\partial y}{\partial t} + \frac{\partial F}{\partial z}\frac{\partial z}{\partial t},$$

assuming that all functions involved are differentiable in appropriate domains. The proof that $\partial G/\partial s$ is as given above is exactly the same as the proof of 15.19, except that three variables are involved, and similarly for $\partial G/\partial t$.

More generally, if F is a function of n variables and $x_1, x_2, \ldots, x_n$ are functions of m variables, and if

$$G(t_1, t_2, \ldots, t_m) = F(x_1(t_1, \ldots, t_m), x_2(t_1, \ldots, t_m), \ldots, x_n(t_1, \ldots, t_m)),$$

then

15.20 $$\frac{\partial G}{\partial t_j} = \sum_{i=1}^{n} \frac{\partial F}{\partial x_i}\frac{\partial x_i}{\partial t_j}, \qquad j = 1, 2, \ldots, m.$$

Thus 15.20 is a set of m equations, one for each variable t_j, and each equation is a sum of n terms, one for each variable x_i. We are of course assuming appropriate differentiability conditions.

Example 2 The area of a rectangle is given by the formula

$$A = xy,$$

where x and y are the lengths of its sides. If u and θ are as indicated in Figure 15.7, then

$$x = u \cos \theta, \qquad y = u \sin \theta,$$

and A can be expressed in terms of u and θ. Find $\partial A/\partial u$ and $\partial A/\partial \theta$.

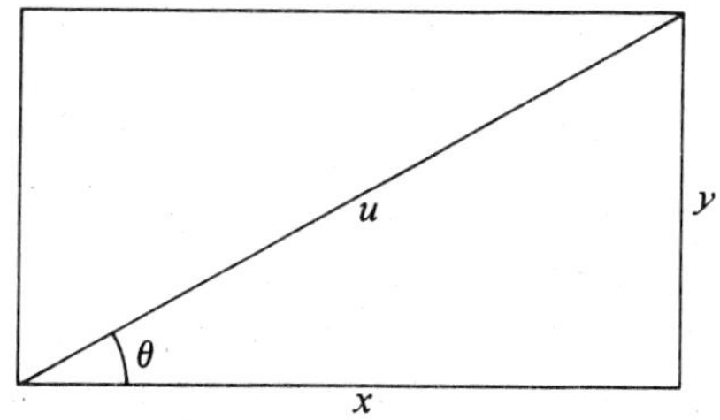

FIGURE 15.7

Solution: We are given $A = F(x,y) = xy$ and in turn

$$x = f(u,\theta) = u\cos\theta, \qquad y = g(u,\theta) = u\sin\theta.$$

Hence $A = F(f(u,\theta),g(u,\theta))$ and, according to 15.20,

$$\frac{\partial A}{\partial u} = F_1(u\cos\theta, u\sin\theta)f_1(u,\theta) + F_2(u\cos\theta, u\sin\theta)g_1(u,\theta).$$

Since $F_1(x,y) = y$ and $F_2(x,y) = x$, and $f_1(u,\theta) = \cos\theta$ and $g_1(u,\theta) = \sin\theta$, we have

$$\frac{\partial A}{\partial u} = (u\sin\theta)\cos\theta + (u\cos\theta)\sin\theta$$

$$= 2u\sin\theta\cos\theta = u\sin 2\theta.$$

Similarly,

$$\frac{\partial A}{\partial \theta} = F_1(u\cos\theta, u\sin\theta)f_2(u,\theta) + F_2(u\cos\theta, u\sin\theta)g_2(u,\theta)$$

$$= (u\sin\theta)(-u\sin\theta) + (u\cos\theta)(u\cos\theta)$$
$$= u^2(\cos^2\theta - \sin^2\theta) = u^2\cos 2\theta.$$

In this example, of course, we could have expressed A in the form

$$A = (u\cos\theta)(u\sin\theta) = \tfrac{1}{2}u^2\sin 2\theta,$$

and then found $\partial A/\partial u$ and $\partial A/\partial\theta$ directly.

EXERCISES

In each of Exercises 1 to 4 find dw/dt.

1. $w = x^2 + y^2$, $x = \dfrac{t-1}{t}$, $y = \dfrac{t}{t+1}$

2. $w = x\sin y + y\sin z$, $x = t^2$, $y = e^{2t}$, $z = \dfrac{1}{t}$

3. $w = t \sin xy$, $x = \ln t$, $y = t^3$
4. $w = x \tan y + y \tan x$, $x = te^t$, $y = te^{-t}$

In each of Exercises 5 to 10 find $\partial z/\partial u$ and $\partial z/\partial v$.

5. $z = x^2 + y^2$, $x = u \cos v$, $y = u \sin v$
6. $z = xe^y + ye^x$, $x = u \ln v$, $y = v \ln u$
7. $z = \sin^{-1} xy$, $x = u + v$, $y = u - v$
8. $z = \sqrt{x^2 + y^2}$, $x = u \cos v$, $y = u \sin v$
9. $z = e^{x/y}$, $x = 2u - v$, $y = u + 2v$
10. $z = \tan^{-1} \dfrac{y}{x}$, $x = u + \sin v$, $y = u - \cos v$

In Exercises 11 and 12, find $\partial w/\partial r$, $\partial w/\partial s$, and $\partial w/\partial t$.

11. $w = \dfrac{x + y}{z}$, $x = r - 2s + t$, $y = 2r + s - 3t$, $z = r^2 + s^2 + t^2$
12. $w = xy + yz + zx$, $x = r \cos s$, $y = r \sin t$, $z = st$

In Exercises 13 and 14 find $\partial w/\partial u$ and $\partial w/\partial v$.

13. $w = \sqrt{x^2 + y^2 + z^2}$, $x = u \sin v$, $y = u \cos v$, $z = uv$
14. $w = \dfrac{x^2 + y^2}{y^2 + z^2}$, $x = ue^v$, $y = ve^u$, $z = \dfrac{1}{u}$
15. If $z = F(x,y)$, $x = f(u,v)$, and $y = g(u,v)$, show that

$$\frac{\partial^2 z}{\partial u^2} = \frac{\partial^2 z}{\partial x^2}\left(\frac{\partial x}{\partial u}\right)^2 + 2\frac{\partial^2 z}{\partial x\,\partial y}\frac{\partial x}{\partial u}\frac{\partial y}{\partial u} + \frac{\partial^2 z}{\partial y^2}\left(\frac{\partial y}{\partial u}\right)^2 + \frac{\partial z}{\partial x}\frac{\partial^2 x}{\partial u^2} + \frac{\partial z}{\partial y}\frac{\partial^2 y}{\partial u^2}.$$

16. If $z = F(x,y)$, $x = f(u,v)$, and $y = g(u,v)$, find a formula for $\partial^2 z/\partial v\,\partial u$ analogous to that of Exercise 15.
17. If $z = x + f(u)$ and $u = xy$, show that

$$x\frac{\partial z}{\partial x} - y\frac{\partial z}{\partial y} = x.$$

18. If $z = f(u/v)/v$, show that $v(\partial z/\partial v) + u(\partial z/\partial u) + z = 0$. [*Hint:* Let $x = u/v$ and $y = 1/v$, so that $z = yf(x)$.]
19. If $z = f(u^2 + v^2)$, show that $u(\partial z/\partial v) - v(\partial z/\partial u) = 0$. (*Hint:* Let $x = u^2 + v^2$.)
20. If $z = f(x,y)$, $x = r \cos \theta$, and $y = r \sin \theta$, show that

$$\left(\frac{\partial z}{\partial r}\right)^2 + \frac{1}{r^2}\left(\frac{\partial z}{\partial \theta}\right)^2 = \left(\frac{\partial z}{\partial x}\right)^2 + \left(\frac{\partial z}{\partial y}\right)^2.$$

21. If $F(x,y) = f(y + ax) + g(y - ax)$, show that

$$\frac{\partial^2 F}{\partial x^2} = a^2 \frac{\partial^2 F}{\partial y^2}.$$

22. If $z = F(u,v)$ and $u = g(x,v)$, find $(\partial z/\partial x)_v$ and $(\partial z/\partial v)_x$. [*Note:* $(\partial z/\partial x)_v$ means v is kept fixed.]

23. If $H = E + PV$ and $E = g(P,V)$, find $\left(\dfrac{\partial H}{\partial V}\right)_P$.

24. If $E = G(V,T)$ and $PV = k$, where k is a constant, find

$$\left(\frac{\partial E}{\partial P}\right)_T \quad \text{and} \quad \left(\frac{\partial E}{\partial T}\right)_P.$$

25. If $z = f(x,y)$ and $y = g(x)$, find formulas for dz/dx and d^2z/dx^2.

26. If $z = f(x,y)$, $x = g(y,t)$, and $y = h(t)$, find a formula for dz/dt.

27. If $w = f(x,y,z)$, $y = g(x,u)$, and $z = h(x,v)$, find $\partial w/\partial x$, $\partial w/\partial u$, and $\partial w/\partial v$.

5 DIRECTIONAL DERIVATIVES

Let f be a function of two variables and $P(a,b)$ a point in its domain. Let $\mathbf{v}$ be any *unit* vector $\langle r,s\rangle$ in the xy plane with its initial point at P, and consider the line L through P in the xy plane in the direction of $\mathbf{v}$. Then the plane p through L perpendicular to the xy plane intersects the graph of f in a cross-sectional curve C (refer to Figure 15.2). Denoting the point $(a,b,f(a,b))$ by Q, and assuming appropriate differentiability conditions, we may ask the following natural questions.

1. How do we find the tangent vector to the curve C at Q?
2. Considering C as a curve in the plane p, how do we find the slope of its tangent line at Q?

To answer these questions, we first note that parametric equations for L are given by

$$x(t) = a + rt, \qquad y(t) = b + st.$$

Then the height of the curve C above the line L at the point $(x(t),y(t))$ is given by the composite function

$$g(t) = f(x(t),y(t)).$$

Considering C as a *space curve*, its vector equation can be written

$$\lambda(t) = \langle a + rt, b + st, g(t)\rangle,$$

where $\lambda(t)$ is the position vector of the curve. Then the tangent vector to C for any value of t is

$$\lambda'(t) = \langle r,s,g'(t)\rangle.$$

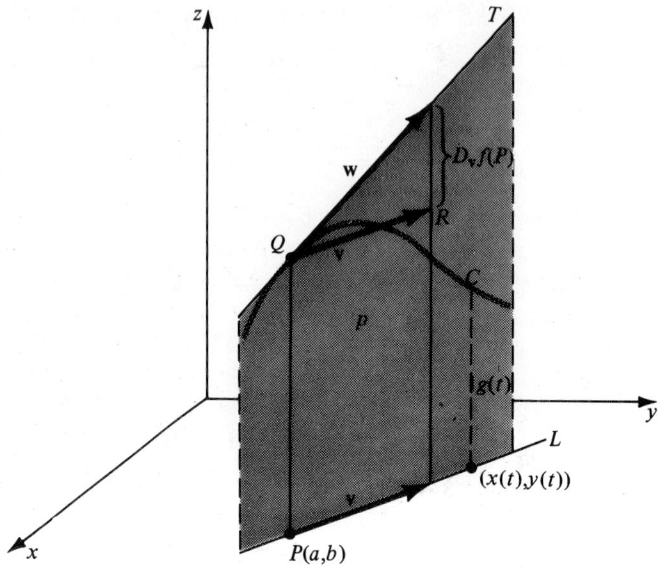

FIGURE 15.8

At the point Q we have $t = 0$, and so the tangent vector to C at Q is

$$\mathbf{w} = \lambda'(0) = \langle r,s,g'(0)\rangle.$$

The vector $\mathbf{w}$ is called the *tangent vector to the graph of f at Q in the direction of* $\mathbf{v}$ (Figure 15.8).

To compute $g'(t)$, we apply the chain rule (15.19) (assuming the function f is differentiable) to obtain

$$g'(t) = f_1(x(t),y(t))r + f_2(x(t),y(t))s.$$

Hence,

$$g'(0) = f_1(P)r + f_2(P)s.$$

To find the slope of the tangent line T to C at Q, think of $\mathbf{v}$ as emanating from Q as its initial point, and having its terminal point at R. Referring to Figure 15.8, and recalling that $\mathbf{v}$ is a *unit* vector, we see that the slope of T is simply the vertical distance from R to the line T in the plane p. But since $\mathbf{v} = \langle r,s\rangle$ and $\mathbf{w} = \langle r,s,g'(0)\rangle$, this vertical distance is $g'(0)$.

The quantity $g'(0)$ is called the *directional derivative of f at P in the direction of the unit vector* $\mathbf{v}$, and is denoted by $D_{\mathbf{v}}f(P)$. Thus

15.21 $$D_{\mathbf{v}}f(P) = f_1(P)r + f_2(P)s,$$

where $\mathbf{v} = \langle r,s\rangle$.

The reader has undoubtedly noticed that the above expression is simply the *differential* of f evaluated at P and $\mathbf{v}$; that is,

$$D_{\mathbf{v}}f(P) = df(P,\mathbf{v}).$$

Also, in those special cases when $\mathbf{v} = \mathbf{i}$ and $\mathbf{v} = \mathbf{j}$, the directional derivatives reduce to partial derivatives:

$$D_{\mathbf{i}}f(P) = f_1(P), \qquad D_{\mathbf{j}}f(P) = f_2(P).$$

Although the above formulas are meaningful for any nonzero vector $\mathbf{v}$, it should be emphasized that the geometrical interpretation of $D_{\mathbf{v}}f(P)$ as the slope of the curve C at Q is valid only if $\mathbf{v}$ is a *unit* vector.

Example 1 If $f(x,y) = \dfrac{x+y}{x-y}$, $P = (1,-1)$, and $\mathbf{v}$ is the unit vector $\left\langle \dfrac{1}{2}, \dfrac{\sqrt{3}}{2} \right\rangle$, find $D_{\mathbf{v}}f(P)$.

Solution: We have

$$f_1(x,y) = \frac{-2y}{(x-y)^2}, \qquad f_2(x,y) = \frac{2x}{(x-y)^2},$$

and hence $f_1(1,-1) = \frac{1}{2}$, $f_2(1,-1) = \frac{1}{2}$. Now using 15.21,

$$D_{\mathbf{v}}f(P) = \frac{1}{2}\cdot\frac{1}{2} + \frac{1}{2}\cdot\frac{\sqrt{3}}{2} = \frac{1+\sqrt{3}}{4}.$$

Example 2 If $f(x,y) = x^2 - xy + 5y$ and $P = (-1,2)$, find the directional derivative of f at P in the direction of the vector $\langle 3,-4 \rangle$.

Solution: The vector $\langle 3,-4 \rangle$ is not a unit vector, since its length is $\sqrt{3^2 + (-4)^2} = 5$. A unit vector in the direction of $\langle 3,-4 \rangle$ is the vector $\mathbf{v} = \langle \frac{3}{5}, -\frac{4}{5} \rangle$. Now we can apply 15.21. We have

$$f_1(x,y) = 2x - y, \qquad f_2(x,y) = -x + 5,$$

and so

$$D_{\mathbf{v}}f(P) = -4\cdot(\tfrac{3}{5}) + 6\cdot(-\tfrac{4}{5}) = -\tfrac{36}{5}.$$

It is also possible to define directional derivatives for functions of three variables. Let f be a function of x, y, and z, and $P(a,b,c)$ a point in R^3. Let $\mathbf{v}$ be a three-dimensional unit vector $\langle h,k,l \rangle$ with its initial point at P. Then the line L through P in the direction of $\mathbf{v}$ has parametric equations

$$x(t) = a + ht, \qquad y(t) = b + kt, \qquad z(t) = c + lt.$$

Now we may again consider the values of the function f along the line L. These values are given by the composite function

$$g(t) = f(x(t),y(t),z(t)).$$

The *directional derivative of* f *at* P *in the direction of* $\mathbf{v}$ is defined to be $g'(0)$, just as in the case of two variables. It is again denoted by $D_{\mathbf{v}}f(P)$, and using the chain rule it is easy to show that

15.22 $$D_{\mathbf{v}}f(P) = f_1(P)h + f_2(P)k + f_3(P)l,$$

where $\mathbf{v} = \langle h,k,l \rangle$.

Example 3 If $F(x,y,z) = x + xy - yz$, find $D_{\mathbf{v}}f(P)$ for $P = (-1,1,2)$ and $\mathbf{v} = \dfrac{1}{\sqrt{11}}\langle 3,1,1 \rangle$.

Solution: We have

$$F_1(x,y,z) = 1 + y, \quad F_2(x,y,z) = x - z, \quad F_3(x,y,z) = -y.$$

Using 15.22,

$$D_{\mathbf{v}}f(P) = 2\left(\frac{3}{\sqrt{11}}\right) + (-3)\cdot\left(\frac{1}{\sqrt{11}}\right) + (-1)\left(\frac{1}{\sqrt{11}}\right) = \frac{2}{\sqrt{11}}.$$

In connection with directional derivatives of a function f of two variables, at a point $P(a,b)$ in its domain, it is convenient to introduce a certain special vector called the *gradient* of f at P. It is denoted by $\nabla f(P)$ (read "del-f of P") and is defined by

$$\nabla f(P) = \langle f_1(P), f_2(P) \rangle.$$

In terms of the gradient $\nabla f(P)$ observe that formula 15.21 for the directional derivative may be written as an inner product:

15.23 $$D_{\mathbf{v}}f(P) = \nabla f(P) \cdot \mathbf{v}.$$

The significance of the gradient vector becomes apparent if we ask the question: In what direction from the point P does the directional derivative of f assume its *maximum value*? Or, in more geometrical language, if you were starting a walk on the graph of the function f from the point $(a,b,f(a,b))$, in what direction would you experience the steepest ascent?

To answer this question, note that if $\nabla f(P) = \langle 0,0 \rangle$, then $D_{\mathbf{v}}f(P) = 0$ for every $\mathbf{v}$ and 0 is the maximum (and minimum) value of $D_{\mathbf{v}}f(P)$. If $\nabla f(P) \neq \langle 0,0 \rangle$, then by 15.23 and the Cauchy inequality,

$$|D_{\mathbf{v}}f(P)| = |\nabla f(P) \cdot \mathbf{v}| \leq |\nabla f(P)||\mathbf{v}| = |\nabla f(P)|,$$

since $\mathbf{v}$ is a unit vector. Actually, $D_{\mathbf{u}}f(P)$ does assume the value $|\nabla f(P)|$ if $\mathbf{u}$ is a unit vector in the direction of $\nabla f(P)$; that is, if $\mathbf{u} = c\,\nabla f(P)$, where $c = |\nabla f(P)|^{-1}$. For then

$$\begin{aligned} D_{\mathbf{u}}f(P) &= \nabla f(P) \cdot c\,\nabla f(P) \\ &= c(\nabla f(P) \cdot \nabla f(P)) = c|\nabla f(P)|^2 = |\nabla f(P)|. \end{aligned}$$

This proves the following result.

15.24 Theorem

If function f is differentiable at point P, then $|\nabla f(P)|$ is the maximum value of the directional derivative $D_{\mathbf{v}}f(P)$. Furthermore, if $\nabla f(P) \neq \langle 0,0 \rangle$, the maximum value of $D_{\mathbf{v}}f(P)$ occurs when

$$\mathbf{v} = |\nabla f(P)|^{-1} \nabla f(P).$$

Example 4 If $f(x,y) = x^2 + xy$ and $P = (1,-1)$, find the maximum value of any directional derivative $D_{\mathbf{v}}f(P)$.

Solution: Since $D_1 f(x,y) = 2x + y$ and $D_2 f(x,y) = x$, we have

$$\nabla f(P) = \langle D_1 f(P), D_2 f(P) \rangle = \langle 1,1 \rangle.$$

Since $|\nabla f(P)| = \sqrt{2}$, the maximum value of $D_{\mathbf{v}}f(P)$ is $\sqrt{2}$ by 15.24. It occurs in the direction $\mathbf{v} = \langle 1/\sqrt{2}, 1/\sqrt{2} \rangle$.

Similar results hold for a function of three or more variables. For example, if f is a function of three variables x, y, and z, then the *gradient vector* $\nabla f(P)$ is defined by

$$\nabla f(P) = \langle f_1(P), f_2(P), f_3(P) \rangle.$$

For a differentiable function f, the maximum value of the directional derivative at P again occurs in the direction of $\nabla f(P)$ and is equal to $|\nabla f(P)|$.

There is another interesting fact about the direction of the gradient vector $\nabla f(P)$ at the point $P = (a,b)$. If $f(a,b) = c$, consider the curve in the xy plane defined by the equation $f(x,y) = c$. This is called the *level curve*, of height c, of the function f (see Exercises 23 to 26 of Section 1 of this chapter). The point $P = (a,b)$ clearly lies on this level curve. If we think of the vector $\nabla f(P)$ as emanating from the point $P = (a,b)$, then $\nabla f(P)$ *lies along the normal line to* $f(x,y) = c$ *at the point* P. To prove this, let us assume that the equation $f(x,y) = c$ defines y "implicitly" as a function $y = g(x)$ and that this function $g(x)$ is differentiable. (A fuller discussion of this matter is given in Section 7 of this chapter.) Applying the chain rule (15.19), we differentiate $f(x,y) = c$ with respect to x to obtain

$$f_1(x,y) \cdot 1 + f_2(x,y) \frac{dy}{dx} = 0.$$

At the point $P = (a,b)$,

$$f_1(a,b) + f_2(a,b)g'(a) = 0,$$

or

$$g'(a) = \frac{-f_1(a,b)}{f_2(a,b)}.$$

But $g'(a)$ is the slope of the tangent to $f(x,y) = c$ at P, while $f_2(a,b)/f_1(a,b)$ is the slope of the line through $\nabla f(P)$. The above equation shows that these quantities are negative reciprocals, which means that $\nabla f(P)$ lies along the normal to $f(x,y) = c$ at P.

EXERCISES

In each of Exercises 1 to 8, find the directional derivative of the given function at the indicated point and in the indicated direction.

1. $f(x,y) = ax^2 + 2byxy + cy^2$, $P = (1,1)$, $\mathbf{v} = \langle \frac{4}{5}, -\frac{3}{5} \rangle$
2. $g(x,y) = x^2 - y^2$, $P = (4,4)$, $\mathbf{v} = \langle \frac{5}{13}, -\frac{12}{13} \rangle$
3. $F(x,y,z) = x^2 + y^2 + xyz$, $P = (1,1,1)$, $\mathbf{v} = \langle \frac{1}{3}, \frac{2}{3}, \frac{2}{3} \rangle$
4. $G(x,y,z) = x + y + z$, $P = (a,b,c)$, $\mathbf{v} = \langle -\frac{2}{3}, -\frac{2}{3}, \frac{1}{3} \rangle$
5. $f(x,y) = xy^2$, $P = (1,2)$, in the direction making an angle of 45° with the positive x axis
6. $f(x,y) = \sin^2 xy$, $P = (\sqrt{3}, -2)$, in the direction making an angle of 60° with the positive x axis
7. $f(x,y) = x^3 - x^2y + y^2$, $P = (1,-1)$, in the direction from P to the point $Q = (4,3)$
8. $F(x,y,z) = x^3 - xyz + yz^2$, $P = (1,1,-1)$, in the direction from P to the point $Q = (3,-1,2)$

In Exercises 9 to 13, find a unit vector $\mathbf{v}$ at the given point P such that $D_{\mathbf{v}}f(P)$ takes on its maximum value.

9. $f(x,y) = x^2 + xy + y^3$, $P = (1,2)$
10. $f(x,y) = e^x \tan^{-1} y$, $P = (0,2)$
11. $f(x,y,z) = \sqrt{x^2 + y^2 + z^2}$, $P = (1,1,1)$
12. $f(x,y,z) = x^3 - xyz + y^2z$, $P = (1,-1,2)$
13. $f(x,y,z) = \dfrac{x + \ln y}{z}$, $P = (1,2,4)$

14. For which directions does the directional derivative of the function f at point (a,b) vanish if $f(x,y) = x^2 + y^2$?
15. Find the maximum and minimum directional derivatives of the function f defined by:
    a. $f(x,y) = x^2 + 2xy + y^2$ at point (a,b);
    b. $f(x,y) = x^2 + y^2 + x - y + 1$ at point $(1,-1)$.
16. What is the maximum directional derivative of the function f defined by $f(x,y,z) = x^2 + y^2 + z^2$ at point (a,b,c)?
17. Let f be a function of two variables that is differentiable at $P = (a,b)$. If $\mathbf{u}$ and $\mathbf{v}$ are any perpendicular two-dimensional unit vectors, prove that

$$[D_{\mathbf{u}}f(P)]^2 + [D_{\mathbf{v}}f(P)]^2$$

has a constant value, which is equal to $|\nabla f(P)|^2$.

18. Any two-dimensional unit vector $\mathbf{v}$ can be written as $\mathbf{v} = \langle \cos \theta, \sin \theta \rangle$, where θ is the angle between $\mathbf{v}$ and $\mathbf{i}$. Formula 15.21 therefore defines $D_{\mathbf{v}}f(P)$ as a function $h(\theta)$ of the single variable θ. Prove Theorem 15.24 by applying the one-variable theory of extrema to the function $h(\theta)$.

19. If $\rho(x,y,z) = \sqrt{x^2 + y^2 + z^2}$ and $\phi(x,y,z) = \cos^{-1} \dfrac{z}{\sqrt{x^2 + y^2 + z^2}}$, show that $\nabla\rho(P)$ and $\nabla\phi(P)$ are perpendicular at every point P (except the origin).

6 TANGENT PLANES

Let F be a differentiable function of three variables x, y, and z, and consider the equation

$$F(x,y,z) = 0. \tag{1}$$

The graph of this equation is some surface S in $\mathbf{R}^3$. Let $P = (x_0,y_0,z_0)$ be a fixed point on S, and suppose C is any curve lying on the surface S and passing through the point P. Let the position vector for the curve C be given by

$$\lambda(t) = \langle x(t),y(t),z(t)\rangle, \qquad \text{for } t \text{ in } [a,b],$$

and let $\lambda(t_0)$ be the position vector of the point P. Since every point of C is on S, the functions $x(t)$, $y(t)$, and $z(t)$ must satisfy equation (1); that is,

$$F(x(t),y(t),z(t)) = 0 \tag{2}$$

for all t in $[a,b]$. If we differentiate both sides of equation (2) with respect to t by the chain rule, and then set $t = t_0$, we obtain

$$F_1(P)x'(t_0) + F_2(P)y'(t_0) + F_3(P)z'(t_0) = 0. \tag{3}$$

What does equation (3) tell us? Recall from Section 5 that the gradient vector $\nabla F(P)$ is defined as

$$\nabla F(P) = \langle F_1(P),F_2(P),F_3(P)\rangle.$$

Also, the tangent vector $\mathbf{v}$ to the curve C at the point P is given by

$$\mathbf{v} = \lambda'(t_0) = \langle x'(t_0),y'(t_0),z'(t_0)\rangle.$$

So equation (3) states that the dot product $\nabla F(P) \cdot \mathbf{v}$ is equal to 0; that is, *the gradient vector $\nabla F(P)$ and the tangent vector $\mathbf{v}$ to C at P are perpendicular* (Figure 15.9). But since C denotes *any* curve on the surface S through P, this tells us that all tangent vectors $\mathbf{v}$ to such curves on S through P are all perpendicular to $\nabla F(P)$. Hence, all these tangent vectors $\mathbf{v}$ *lie in the same plane*, namely, the plane through P perpendicular to $\nabla F(P)$. This plane is called the *tangent plane to the surface S at P.*

It is easy to write an equation for this tangent plane. By 14.7, an equation of the plane through $P = (x_0,y_0,z_0)$ perpendicular to the vector $\nabla F(P) = \langle F_1(P),F_2(P),F_3(P)\rangle$ is given by

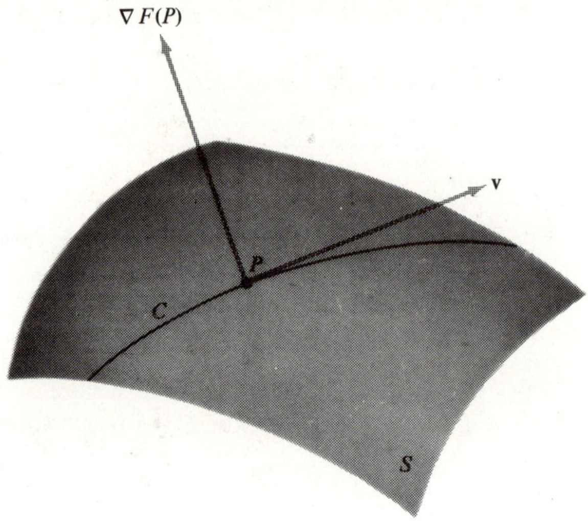

FIGURE 15.9

15.25 $$F_1(P)(x - x_0) + F_2(P)(y - y_0) + F_3(P)(z - z_0) = 0.$$

Frequently the equation of a surface S is given to us in the form

(4) $$z = f(x,y),$$

where z is expressed as a function $f(x,y)$ of two variables. We may reduce equation (4) to the form (1) by setting $F(x,y,z) = f(x,y) - z$, so that the equation of S now has the form $F(x,y,z) = 0$. To find an equation of the tangent plane to S at a point $P = (x_0,y_0,z_0)$ on S, we note that in this case

$$\nabla F = \langle f_1, f_2, -1\rangle,$$

so that

$$\nabla F(P) = \langle f_1(x_0,y_0), f_2(x_0,y_0), -1\rangle.$$

Equation 15.25 then takes the form

15.25′ $$f_1(x_0,y_0)(x - x_0) + f_2(x_0,y_0)(y - y_0) - (z - z_0) = 0.$$

Example 1 Find an equation of the tangent plane to the graph of $x^2 + y^2 + z^2 - 6z = 0$ at the point (2,2,4).

Solution: Let $F(x,y,z) = x^2 + y^2 + z^2 - 6z$. Then

$$F_1(x,y,z) = 2x, \qquad F_2(x,y,z) = 2y, \qquad F_3(x,y,z) = 2z - 6.$$

For $P = (2,2,4)$,

$$F_1(P) = 4, \qquad F_2(P) = 4, \qquad F_3(P) = 2.$$

By 15.25, an equation for the required tangent plane is

$$4(x - 2) + 4(y - 2) + 2(z - 4) = 0,$$

or

$$2x + 2y + z + 12 = 0.$$

Example 2 Find an equation of the tangent plane to the graph of $z = x^2 + 4y^2$ at the point where $x = -2$, $y = 1$.

Solution: Let $f(x,y) = x^2 + 4y^2$, $P = (-2,1,8)$. Then

$$f_1(x,y) = 2x, \qquad f_2(x,y) = 8y,$$

and

$$f_1(-2,1) = -4, \qquad f_2(-2,1) = 8.$$

By 15.25′, an equation for the required tangent plane is

$$-4(x + 2) + 8(y - 1) - (z - 8) = 0,$$

or

$$4x - 8y + z + 8 = 0.$$

EXERCISES

I

In each of Exercises 1 to 4 find an equation of the tangent plane to the graph of the given function at the point with the given x and y coordinates.

1. $f(x,y) = 4x^2 + y^2$, $(-1,2)$

2. $g(x,y) = x^2 + 3y^2$, $(3,-1)$

3. $f(x,y) = \sqrt{9 - x^2 - y^2}$, $(1,2)$

4. $G(x,y) = \dfrac{x + 2y}{2x + y}$, $(2,3)$

In each of Exercises 5 to 10 find the tangent plane to the graph of the given equation at the given point.

5. $z = x^2 - 4y^2$, $(2,1,0)$

6. $z^2 = x^2 + y^2$, $(3,3,3\sqrt{2})$

7. $x = y^2 + 9z^2$, $(13,-2,1)$

8. $xy + yz + zx + 1 = 0$, $(-1,1,-1)$

9. $x^2 + y^2 - 4z^2 = 4$, $(2,-2,1)$

10. $9x^2 - y^2 - 4z^2 = 1$, $(1,-2,1)$

A line perpendicular to the tangent plane at a given point on a surface is called a *normal line* of the surface.

11. If f is a differentiable function of two variables at the point $P = (a,b)$, then prove that the normal line to the graph of f at P has parametric equations $x = a + tf_1(a,b)$, $y = b + tf_2(a,b)$, $z = f(a,b) - t$.

12. **a.** Find an equation of the tangent plane to the sphere $x^2 + y^2 + z^2 = a^2$ at the point (h,k,l) on the surface.

b. Prove that every normal line of the sphere passes through the center of the sphere.

If a curve is the intersection of two surfaces, then the tangent line to the curve at a point is the line of intersection of the two tangent planes of the surfaces at that point. Use this fact in Exercises 13–15 to find parametric equations of the tangent line to the curve of intersection of the given surfaces at the indicated point.

13. $x^2 + y^2 + z^2 = 9$, $(x - 2)^2 + y^2 = 5$; $(1,2,2)$.
14. $x^2 + 4y^2 + z^2 = 33$, $z = x^2 + y^2$; $(-2,1,5)$.
15. $x + y + z + 2 = 0$, $x^2 + y^2 + z^2 = 9$; $(-2,1,-1)$.

II

1. Find an equation of the tangent plane at any point (a,b,c) of the surface $\sqrt{x} + \sqrt{y} + \sqrt{z} = \sqrt{k}$, and then show that the sum of the intercepts of this tangent plane on the coordinate axes is a constant.
2. Find an equation of the tangent plane at any point (a,b,c) of the surface $x^{2/3} + y^{2/3} + z^{2/3} = k^{2/3}$. Then show that the sum of the squares of the intercepts of this tangent plane on the coordinate axes is constant.
3. Prove that the tetrahedron formed by the coordinate planes and any tangent plane to the graph of $xyz = c^3$ has a constant volume.

7 THE IMPLICIT FUNCTION THEOREM

In Chapter 3 the student encountered the procedure known as "implicit differentiation." Suppose, for example, we are given an equation in x and h such as

$$F(x,y) = y^3 - y^2 - x^3 + x + 24 = 0, \tag{1}$$

and we consider the following questions.

(a) Does the above equation in some sense define y as a function of x, for some interval of values of x?
(b) If such a function $y = f(x)$ exists, is it differentiable and, if so, how does one compute its derivative?

In Chapter 3 it was assumed that for an equation such as (1) above the "implicit" function $y = f(x)$ exists [although certainly it is not at all obvious how equation (1) could be solved for y in terms of x]. Furthermore, the function $f(x)$ was also *assumed* to be differentiable and we learned how to obtain its derivative $f'(x)$ "implicitly." For example, under these assumptions it is possible to write (1) in the form

$$f^3(x) - f^2(x) - x^3 + x + 24 = 0. \tag{2}$$

Differentiating (2) with respect to x, we obtain

$$3f^2(x)f'(x) - 2f(x)f'(x) - 3x^2 + 1 = 0,$$

$$f'(x) = \frac{3x^2 - 1}{3f^2(x) - 2f(x)}. \tag{3}$$

Since formula (3) gives $f'(x)$ in terms of both x and $f(x)$, it is not very useful unless some pairs of values $(x,f(x))$ are known. If, for example, it is known that (3,1) is a point on the graph of (1), then we can compute $f'(3)$ from (3) and obtain the value $f'(3) = 26$.

Now it is certainly possible to give simple examples of equations $F(x,y) = 0$, in which our assumptions about affirmative answers to questions (a) and (b) above are quite unjustified. For example, consider

$$F(x,y) = x^2 + y^2 + 1 = 0. \tag{4}$$

There are no pairs of real numbers whatsoever that satisfy this equation. Hence it cannot define any function $y = f(x)$ implicitly.

For the equation

$$F(x,y) = x^2 + y^2 = 0,$$

there is precisely one pair of numbers, namely (0,0), that satisfies the equation. But there is *no interval of values of* x in which we can regard this equation as defining an implicit function $y = f(x)$.

We shall now state a general theorem that gives us sufficient conditions under which affirmative answers to questions (a) and (b) can be justified. It should be remarked that this is an example of what is called an "existence theorem," in that it allows us only to assert the *existence* of a certain function, but does not provide an explicit method for calculating its values.

15.26 Implicit Function Theorem

Let F be a function of x and y defined in some open set S in R^2, and suppose that the partial derivatives F_1 and F_2 exist and are continuous in S. Suppose, further, that $P = (a,b)$ is a point in S such that $F(a,b) = 0$, but $F_2(a,b) \neq 0$. Then there exists a rectangle $R^* = \{(x,y) \mid x_1 < x < x_2, y_1 < y < y_2\}$ containing (a,b), such that for every x with $x_1 < x < x_2$, the equation $F(x,y) = 0$ determines exactly one value $y = f(x)$ with $y_1 < y < y_2$. Furthermore,

(i) *$F(x,f(x)) = 0$ whenever $x_1 < x < x_2$.*
(ii) *The function $f(x)$ is continuous, differentiable, and*

$$f'(x) = \frac{-F_1(x,f(x))}{F_2(x,f(x))}$$

for all x with $x_1 < x < x_2$.

The proof of the implicit function theorem is usually deferred until a course in advanced calculus, and hence we shall not give it here. However, we can easily show how to obtain the formula for $f'(x)$, under the assumptions that the function f exists, is differentiable, and satisfies condition (i) of 15.26. Starting with (i),

$$F(x,f(x)) = 0,$$

we apply the chain rule (15.19) to obtain

$$D_xF(x,f(x)) = \frac{\partial F}{\partial x}\frac{dx}{dx} + \frac{\partial F}{\partial y}\frac{dy}{dx} = 0.$$

Solving for $dy/dx = f'(x)$, we get

$$\frac{dy}{dx} = -\frac{\partial F/\partial x}{\partial F/\partial y},$$

which is the formula of 15.26(ii).

Example 1 Consider the circle whose equation is

$$F(x,y) = x^2 + y^2 - 1 = 0.$$

Each point (a,b) of this circle, where $b \neq 0$, has a neighborhood in which the equation $F(x,y) = 0$ defines a unique function $y = f(x)$. Moreover,

$$f'(x) = \frac{-2x}{2y} = \frac{-x}{y}, \qquad \text{if } y \neq 0.$$

At the points $(-1,0)$ and $(1,0)$ the hypothesis of Theorem 15.26 fails, since $F_2(-1,0) = F_2(1,0) = 0$. Neither of these points possesses a neighborhood in which the equation $F(x,y) = 0$ defines y as a function of x. However, since $F_1(-1,0)$ and $F_1(1,0)$ are different from 0, we could interchange the roles of x and y for these points in Theorem 15.26; and in appropriate neighborhoods of these points the equation defines x as a function of y.

Example 2 Consider the equation

$$F(x,y) = x^3 + y^3 - 6xy = 0.$$

The graph of this equation is sketched in Figure 15.10. At the point (0,0) (where the curve "intersects itself"), F_1 and F_2 are both 0, so that 15.26 does not apply. However, near any point (x,y) where $y^2 \neq 2x$, the equation defines a function $y = f(x)$. For this function,

$$f'(x) = \frac{-F_1(x,y)}{F_2(x,y)} = -\frac{x^2 - 2y}{y^2 - 2x}.$$

The derivative $f'(x)$ gives us information about the curve. In particular, we see that $f'(x) = 0$ when $x^2 = 2y$. Substituting this for y in $F(x,y) = 0$, we find that the corresponding point on the curve is $(2\sqrt[3]{2}, 2\sqrt[3]{4})$. This is a relative maximum point on the graph.

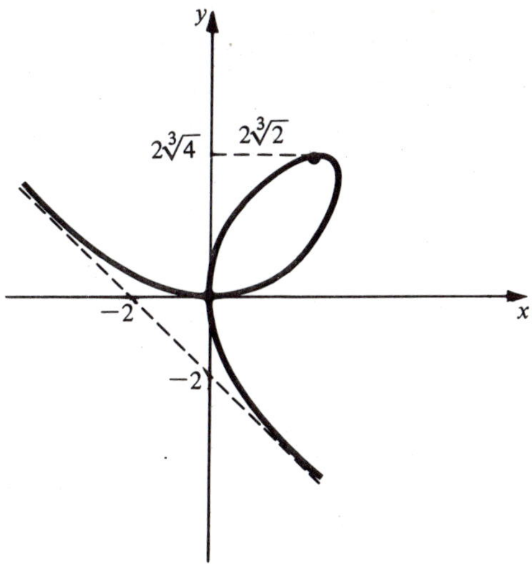

FIGURE 15.10

The implicit function theorem may be extended to functions of more than two variables. We shall state a version of Theorem 15.26 for functions of three variables, leaving the corresponding formulation for higher dimensions to the student.

15.27 Theorem

Let F be a function of x, y, and z defined in some open set S in $\mathbb{R}^3$, and suppose that the partial derivatives F_1, F_2, and F_3 exist and are continuous in S. Suppose that (a,b,c) is a point in S such that $F(a,b,c) = 0$, but $F_3(a,b,c) \neq 0$. Then there exists a rectangular parallelepiped

$$R^* = \{(x,y,z) \mid x_1 < x < x_2, y_1 < y < y_2, z_1 < z < z_2\}$$

containing (a,b,c), such that whenever x and y satisfy $x_1 < x < x_2$ and $y_1 < y < y_2$, the equation $F(x,y,z) = 0$ determines exactly one value $z = f(x,y)$ with $z_1 < z < z_2$. Furthermore,

(i) *$F(x,y,f(x,y)) = 0$ whenever $x_1 < x < x_2$ and $y_1 < y < y_2$.*
(ii) *The function $f(x,y)$ has continuous partial derivatives given by the formulas*

$$f_1(x,y) = \frac{-F_1(x,y,f(x,y))}{F_3(x,y,f(x,y))},$$

$$f_2(x,y) = \frac{-F_2(x,y,f(x,y))}{F_3(x,y,f(x,y))},$$

for $x_1 < x < x_2$ and $y_1 < y < y_2$.

The above differentiation formulas could in fact be obtained from Theorem 15.26(ii). For example, if we consider y to be held constant for the function $F(x,y,z)$, we obtain the formula for $f_1(x,y)$ directly from 15.26(ii), be treating F as a function of the two variables x and z.

Example 3 Find $\partial z/\partial x$ and $\partial z/\partial y$ if $z = f(x,y)$ satisfies the equation

$$xy^2 + yz^2 + z^3 + x^3 - 4 = 0.$$

Solution: If $F(x,y,z) = xy^2 + yz^2 + z^3 + x^3 - 4$, then

$$\frac{\partial F}{\partial x} = y^2 + 3x^2, \quad \frac{\partial F}{\partial y} = 2xy + z^2, \quad \frac{\partial F}{\partial z} = 2yz + 3z^2,$$

and

$$\frac{\partial z}{\partial x} = -\frac{y^2 + 3x^2}{2yz + 3z^2}, \quad \frac{\partial z}{\partial y} = -\frac{2xy + z^2}{2yz + 3z^2}.$$

EXERCISES

In each of Exercises 1 to 6 find the derivative of each differentiable function f such that $y = f(x)$ satisfies the equation.

1. $x + y^2 + \sin xy = 0$
2. $x \ln x - ye^y + 3 = 0$
3. $\tan^{-1} \dfrac{x}{y} + y^3 - 1 = 0$
4. $x \sec y + y \sec x - 2 = 0$
5. $x^3 - 2x^2y^2 + y^3 - 3xy + 1 = 0$
6. $x^2 \ln y - y^3 + e^x = 0$
7. Use the implicit function $y = f(x)$ defined by the equation $x^2 + xy + y^2 = 27$ to find the maximum and minimum points on the graph of this equation.
8. Show that the equation $x + y + z = \sin xyz$ defines an implicit function $z = f(x,y)$ in a neighborhood of (0,0,0). Find the partial derivatives of $f(x,y)$.

In each of Exercises 9 to 12, f is a function of two variables having continuous first partial derivatives such that $z = f(x,y)$ satisfies the equation. Find $\partial z/\partial x$ and $\partial z/\partial y$.

9. $x^2z + yz^2 + xy^2 - z^3 = 0$
10. $xe^{yz} + ye^{xz} - y^2 + 3 = 0$
11. $x^3 + y^3 + z^2(x + y) = 2z^3$
12. $\tan^{-1} x + \tan^{-1} y + \tan^{-1} z - 3 = 0$
13. The equations

$$xu + yv + x^2u^2 + y^2v^2 - 3 = 0,$$
$$xu^2 + x^2u - yv^2 - y^2v - 2 = 0$$

can be solved simultaneously to yield $x = f(u,v)$ and $y = g(u,v)$. Find $\partial x/\partial u$, $\partial x/\partial v$, $\partial y/\partial u$, and $\partial y/\partial v$ without actually solving the equations, by use of implicit differentiation.

14. If the equations for changing from spherical coordinates to rectangular coordinates

$$x = \rho \sin \phi \cos \theta, \qquad y = \rho \sin \phi \sin \theta, \qquad z = \rho \cos \phi,$$

were solved for ρ, ϕ, and θ in terms of x, y, and z, then the first-order partial derivatives of ρ, ϕ, and θ with respect to x, y, and z could be found directly. Show how to find these partial derivatives without actually solving the above equations.

15. If a differentiable function $y = f(x)$ is defined implicitly by the equation $F(x,y) = 0$, find a formula for $f''(x)$ in terms of partial derivatives of F.

16. Suppose $y = f(x)$ and $z = g(x)$ are differentiable functions that satisfy the equations

$$F(x,y,z) = 0, \qquad G(x,y,z) = 0$$

for all x in some interval I. Express $f'(x)$ and $g'(x)$ in terms of partial derivatives of F and G.

8 EXTREMA OF A FUNCTION OF TWO VARIABLES

The concept of an extremum of a function of one variable easily extends to a function of several variables, as we shall indicate in this section for functions of two variables.

If f is a function of two variables and S is a subset of the domain of f, and if P is a point of S such that $f(P) \geq f(Q)$ for every point Q in S, then $f(P)$ is called the *maximum value* of f in S. We call $f(P)$ a *relative maximum value* of f if there exists a disk $B(P,r)$ such that $f(P) > f(Q)$ for every Q in $B(P,r)$, $Q \neq P$. Relative minimum values of a function are similarly defined. If $f(P)$ is either a relative maximum or a relative minimum value of f, then $f(P)$ is called a *relative extremum* of the function f.

The reader will recall that a function of one variable that is continuous in a closed interval I assumes maximum and minimum values over I, according to Theorem 4.2. A similar theorem holds for continuous functions of two or more variables. We shall state a special case of this theorem that will be adequate for our purposes. By a *closed rectangle* in R^2 we mean a set S of the form

$$S = \{(x,y) \text{ in } \mathsf{R}^2 \mid a_1 \leq x < a_2, b_1 \leq y \leq b_2\},$$

where (a_1,b_1) and (a_2,b_2) are points in R^2.

15.28 Theorem

Let S be a closed rectangle or a closed disk. If f is a function of two variables that is continuous in S, then f assumes maximum and minimum values over S.

The proof of Theorem 15.28, which is omitted, is similar to that of 4.2.

The relative extrema of a function f of two variables occur at the "mountain peaks" and "valley bottoms" of the graph of f. Let us now show

how the relative extrema of a function of two variables are found. Henceforth, in discussing relative extrema of a function, we shall usually omit the word *relative*, it being implicitly understood that *extrema* means *relative* extrema.

In the first place, if $f(a,b)$ is a maximum value of f, then $(a,b,f(a,b))$ must be a maximum point on the cross section of the graph of f in each of the planes $x = a$ and $y = b$, and similarly for minimum values of f. Thus, if $f(a,b)$ is a relative extremum of f and if f has first partial derivatives at (a,b), then necessarily

$$f_1(a,b) = 0, \qquad f_2(a,b) = 0.$$

Hence, as is intuitively evident, the tangent plane at a relative extremum is parallel to the xy plane. The procedure to be followed in finding the extrema of a function f of two variables is clear from our remarks above. We find the simultaneous solutions of the equations

$$f_1(x,y) = 0, \qquad f_2(x,y) = 0,$$

and then test each of these solutions to see if the function has a maximum or minimum value there. An extremum might occur at a point (a,b) at which the partial derivatives do not exist, but we shall not consider such a possibility here.

Example 1 If $f(x,y) = 4 - x^2 - y^2$, find the extrema of f.

Solution: We have $f_1(x,y) = -2x$ and $f_2(x,y) = -2y$. The only solution of the equations

$$-2x = 0, \qquad -2y = 0$$

is $x = 0$ and $y = 0$. Thus $f(0,0)$ is the only possible extremum of f. Since $4 - x^2 - y^2 < 4$ if $(x,y) \neq (0,0)$, $f(0,0) = 4$ is a maximum value of f.

Example 2 If $g(x,y) = 4 + x^2 - y^2$, find the extrema of g.

Solution: We have $g_1(x,y) = 2x$, $g_2(x,y) = -2y$, so that again $g(0,0)$ is the only possible extremum of g. The cross section

$$z = 4 + x^2$$

of the surface $z = 4 + x^2 - y^2$ in the xz plane has a minimum point at $(0,0,4)$, whereas the cross section

$$z = 4 - y^2$$

of this surface in the yz plane has a maximum point at $(0,0,4)$. Therefore it is evident that $g(0,0) = 4$ is neither a maximum nor a minimum value of g. The point $(0,0,4)$ is called a *saddle point* of the surface, and it resembles the origin on the surface in Figure 14.22. The function g has no extrema.

For a more complicated example than those given above, it might be very difficult to decide whether or not a given point at which the first partial derivatives of f are zero leads to an extremum of f. Therefore we shall give a test, corresponding to the second derivative test for functions of one variable, for extrema of a function of two variables.

If $f_1(a,b) = 0$ and $f_2(a,b) = 0$, and if the second partial derivatives $f_{11}(a,b)$ and $f_{22}(a,b)$ exist and are nonzero, then the cross sections of the graph of f in the planes $x = a$ and $y = b$ must be concave downward and we must have

$$f_{11}(a,b) < 0, \qquad f_{22}(a,b) < 0,$$

if $f(a,b)$ is a maximum value of f. Similarly, if $f(a,b)$ is a minimum value of f,

$$f_{11}(a,b) > 0, \qquad f_{22}(a,b) > 0.$$

If $f_{11}(a,b)$ and $f_{22}(a,b)$ differ in sign, then $f(a,b)$ cannot be an extremum of f.

15.29 Test for Extrema

Let f be a function of two variables defined in an open set S of R^2, and possessing continuous first and second partial derivatives in S. Define a function F by

$$F(Q) = f_{11}(Q)f_{22}(Q) - f_{12}^2(Q), \qquad \textit{domain } F = S.$$

If P is a point in S such that

$$f_1(P) = 0 \qquad \textit{and} \qquad f_2(P) = 0,$$

then

(1) *$f(P)$ is a maximum value of f if $F(P) > 0$ and $f_{11}(P) < 0$.*
(2) *$f(P)$ is a minimum value of f if $F(P) > 0$ and $f_{11}(P) > 0$.*
(3) *$f(P)$ is not an extremum of f if $F(P) < 0$.*

If $F(P) > 0$, then evidently the first term of $F(P)$ is positive, $f_{11}(P)f_{22}(P) > 0$. Hence $f_{11}(P)$ and $f_{22}(P)$ must agree in sign, and $f_{11}(P)$ can be replaced by $f_{22}(P)$ in either part (1) or part (2) of the test.

The theorem gives no information about $f(P)$ if $F(P) = 0$. In such a case a direct analysis (as in Examples 1 and 2) may be made to determine if $f(P)$ is an extremum of f.

Example 3 Find the extrema of the function f defined by

$$f(x,y) = x^3 - 12xy + 8y^3.$$

Solution: Evidently,

$$f_1(x,y) = 3x^2 - 12y, \qquad f_2(x,y) = -12x + 24y^2.$$

The two equations

$$3x^2 - 12y = 0, \qquad -12x + 24y^2 = 0$$

have as their simultaneous solutions (0,0) and (2,1). The second partial derivatives of f are

$$f_{11}(x,y) = 6x, \qquad f_{12}(x,y) = -12, \qquad f_{22}(x,y) = 48y.$$

We easily compute

$$F(0,0) = -144, \qquad F(2,1) = 432.$$

Since $F(0,0) < 0$, $f(0,0)$ is not an extremum of f, by 15.29(3). And since $F(2,1) > 0$ and $f_{11}(2,1) > 0$, $f(2,1) = -8$ is a minimum value of f, by 15.29(2). This is the only extremum of f.

Proof of 15.29(1): By assumption, $f_{11}(P) < 0$, $F(P) > 0$, and the second partial derivatives of f are continuous in S. It follows immediately that $f_{11}(Q) < 0$ and $F(Q) > 0$ for every point Q in some disk $B(P,\varepsilon)$. We shall prove the theorem by showing that $D_{\mathbf{v}}^2 f(Q) < 0$ for every Q in $B(P,\varepsilon)$ and every two-dimensional unit vector $\mathbf{v}$. This will mean that $D_{\mathbf{v}} f$ is a strictly decreasing function in $B(P,\varepsilon)$ along every vector $\mathbf{v}$, which in turn will mean that $f(P)$ is the maximum value of f in $B(P,\varepsilon)$ along every $\mathbf{v}$. It will follow that $f(P)$ is the maximum value of f in $B(P,\varepsilon)$, and hence that $f(P)$ is a relative maximum value of f.

If $\mathbf{v} = \langle h,k \rangle$, then $D_{\mathbf{v}} f(Q) = \nabla f(Q) \cdot \mathbf{v} = hf_1(Q) + kf_2(Q)$ and

$$\begin{aligned} D_{\mathbf{v}}^2 f(Q) &= D_{\mathbf{v}}(D_{\mathbf{v}} f)(Q) = \nabla D_{\mathbf{v}} f(Q) \cdot \mathbf{v} \\ &= \langle hf_{11} + kf_{21}, hf_{12} + kf_{22} \rangle \cdot \mathbf{v}, \end{aligned}$$

where each second partial derivative of f is evaluated at Q. Since $f_{12} = f_{21}$, we have

$$D_{\mathbf{v}}^2 f(Q) = h^2 f_{11} + 2hkf_{12} + k^2 f_{22}.$$

It is a simple exercise in algebra to change this into the form

$$D_{\mathbf{v}}^2 f(Q) = f_{11}\left[h + \frac{f_{12}}{f_{11}} k\right]^2 + \frac{k^2}{f_{11}} F(Q).$$

Since $f_{11}(Q) < 0$ and $F(Q) > 0$, and h and k are not both zero, evidently $D_{\mathbf{v}}^2 f(Q) < 0$ for every Q in $B(P,\varepsilon)$. This proves 15.29(1).

The proof of 15.29(2) is similar to that for 15.29(1) and hence is omitted. We shall leave the proof of 15.29(3) as an exercise for the reader.

Example 4 A rectangular box without a top is to have a given volume. How should the box be made so as to use the least amount of material?

Solution: Let k designate the volume of the box. Hence, if the base of the box is x by y units and the altitude is z units, then

$$k = xyz.$$

The surface area S of the box is given by

$$S = xy + 2xz + 2yz.$$

Since $z = k/xy$, we may express S in the form

$$S = xy + \frac{2k}{y} + \frac{2k}{x}.$$

The problem is to find the minimum value of S. Hence we solve simultaneously the equations

$$S_1 = y - \frac{2k}{x^2} = 0, \qquad S_2 = x - \frac{2k}{y^2} = 0.$$

Since $x^2y = 2k$ and $xy^2 = 2k$, clearly $x = y$ and $x^3 = y^3 = 2k$. It is easily established that $F(\sqrt[3]{2k},\sqrt[3]{2k}) > 0$ and $S_{11} > 0$, so that $(\sqrt[3]{2k},\sqrt[3]{2k})$ gives a minimum value for S. We have

$$z = \frac{k}{xy} = \frac{kx}{x^3} = \frac{x}{2},$$

and therefore we conclude that the box should have a square base and an altitude half the length of the base.

EXERCISES

In each of Exercises 1 to 10, find the extrema of the function.

1. $f(x,y) = x^2 + xy + y^2$
2. $g(x,y) = x^2 - xy + y^2 + 6x$
3. $F(x,y) = x^3 - 12x + y^2$
4. $G(x,y) = x^3 + x^2 - y^3 + y^2$
5. $g(x,y) = x^2 - 2xy + y^2 - y^3$
6. $f(x,y) = (2x - y)^2 - x^3$
7. $f(x,y) = \dfrac{1}{x} + xy - \dfrac{8}{y}$
8. $F(x,y) = xy(4 - x - y)$
9. $F(x,y) = x^2 + \dfrac{2}{xy^2} + y^2$
10. $g(x,y) = \sin x + \sin y + \sin (x + y)$
11. Find the minimum distance between the point $(1,-1,2)$ and the plane $3x + y - 2z = 4$. [*Hint:* Express in terms of x and y the distance between the point $(1,-1,2)$ and the point (x,y,z) on the plane.]
12. Find the minimum distance between the point $(0,-2,-4)$ and the plane $x + y - 4z = 5$.
13. Show that the volume of the largest rectangular parallelepiped (with sides parallel to the coordinate planes) that can be inscribed in the ellipsoid

$$\frac{x^2}{a^2} + \frac{y^2}{b^2} + \frac{z^2}{c^2} = 1$$

is $8abc/3\sqrt{3}$.

14. Find the minimum distance between the lines with parametric equations $x = t$, $y = 3 - 2t$, $z = 1 + 2t$, and $x = -1 - s$, $y = s$, $z = 4 - 3s$.
15. A rectangular box without top has a surface area of S ft^2. Determine the dimensions of the box if the volume is to be a maximum. (*Note:* This is the dual of Example 4, Section 8).

16. A rectangular parallelepiped has three of its faces in the coordinate planes and its vertex opposite the origin in the first octant and on the plane $2x + y + 3z = 6$. Find the maximum volume this box can have.
17. Indicate the changes that must be made in the proof of 15.29(1) to give the proof of 15.29(2).
18. Prove 15.29(3).
19. Find the points on the surface $z^2 = xy + 1$ at least distance from (0,0,0).
20. Let C be the curve of intersection of the surfaces

$$x^2 + z^2 = 2y, \qquad x - y + z + 3 = 0.$$

Find the point on C at the greatest distance from the xz plane.
21. A storage tank is designed in the form of a right circular cylinder surmounted by a cone, and is required to have a given volume V. Show that the surface area is a minimum if the altitude of the cone is twice that of the cylinder.
22. Find three numbers whose sum is s and whose product is a maximum.

9 APPENDIX: FUNCTIONS DEFINED BY INTEGRALS

In this section we discuss some important theorems in whose proofs the notion of *uniform continuity* for functions of two variables plays a crucial role. This notion was discussed for functions of one variable in the appendix to Chapter 5. We repeat the definition here for a function of two variables, in a form that is equivalent to 5.46.

15.30 Definition

Let f be a function of two variables defined in a subset S of R^2. The function f is *uniformly continuous* in S if for every number $\varepsilon > 0$ there exists a number $\delta > 0$ such that $|f(P) - f(Q)| < \varepsilon$ for all pairs of points P, Q, in S satisfying $d(P,Q) < \delta$.

The proof of the following theorem, which is similar to that of the corresponding theorem in the one-variable case, will be omitted.

15.31 Theorem

If f is a function of two variables that is continuous in a closed rectangle S, then f is uniformly continuous in S.

The following notation for closed rectangles will be useful. If $A = (a_1,a_2)$ and $B = (b_1,b_2)$ are points in R^2 with $a_1 < b_1$ and $a_2 < b_2$, then the closed rectangle with vertices A, (b_1,a_2), B, and (a_1,b_2) will be denoted by $[A,B]$. Thus (Figure 15.11),

$$[A,B] = \{(x,y) \text{ in } \mathrm{R}^2 \mid a_1 \le x \le b_1,\ a_2 \le y \le b_2\}.$$

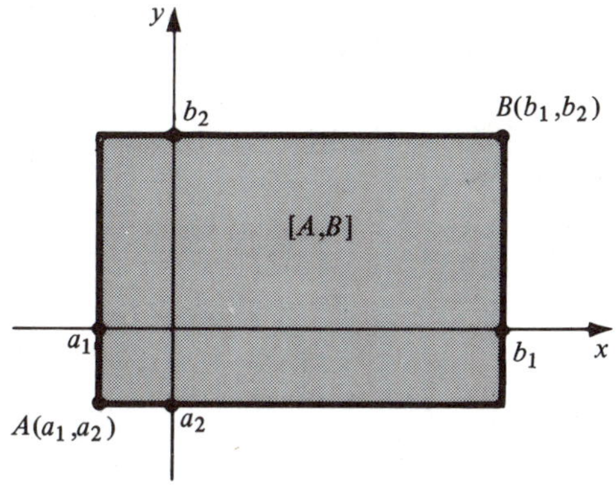

FIGURE 15.11

We now turn our attention to certain functions definable in terms of integrals. Suppose f is a continuous function of two variables in the closed rectangle $[A,B]$. For each x in $[a_1,b_1]$ the function $f(x,t)$ is a continuous function of t, and hence the integral

$$\int_{a_2}^{b_2} f(x,t)\,dt$$

exists for each x in $[a_1,b_1]$. Its value depends on the particular choice of x, and hence this integral defines a function G of the variable x over the interval $[a_1,b_1]$.

A basic property of this integral is given below.

15.32 Theorem

If the function f of two variables is continuous in $[A,B]$, then the function G defined by

$$G(x) = \int_{a_2}^{b_2} f(x,t)\,dt, \qquad \textit{domain of } G = [a_1,b_1],$$

is continuous in $[a_1,b_1]$.

Proof: If $a_2 = b_2$, $G(x) = 0$ for each x and G is obviously continuous. So let us assume that $a_2 \neq b_2$. For all x and c in $[a_1,b_1]$, we have

$$|G(x) - G(c)| = \left| \int_{a_2}^{b_2} [f(x,t) - f(c,t)]\,dt \right| \leq (b_2 - a_2)M,$$

where M is the maximum value of $|f(x,t) - f(c,t)|$ for t in $[a_2,b_2]$. By the uniform continuity of f, for each number $\varepsilon > 0$ there exists a number $\delta > 0$ such that

$$|f(P) - f(Q)| < \frac{\varepsilon}{b_2 - a_2} \quad \text{if } d(P,Q) < \delta, \qquad P \text{ and } Q \text{ in } [A,B].$$

If we select x and c so that $|x - c| < \delta$, then $d((x,t), (c,t)) < \delta$ for every t in $[a_2,b_2]$ and hence $|f(x,t) - f(c,t)| < \varepsilon/(b_2 - a_2)$. Therefore

$$|G(x) - G(c)| < (b_2 - a_2)\cdot\frac{\varepsilon}{b_2 - a_2} = \varepsilon \qquad \text{if } |x - c| < \delta.$$

This proves that G is continuous at c. Thus 15.32 is proved.

If

$$H(y) = \int_{a_1}^{b_1} f(t,y)\,dt, \qquad \text{domain of } H = [a_2,b_2],$$

then H is continuous by a similar proof.

The question sometimes arises as to how a function defined by an integral such as that in 15.32 may be differentiated. The answer is the expected one given below. (From now on we shall frequently use the notation f_x for the partial derivative of the function f with respect to x.)

15.33 Theorem

Let f be a function of two variables x and t, and suppose that f and f_x are both continuous in the closed rectangle $S = [A,B]$. Suppose that the function F is defined by

$$F(x) = \int_{a_2}^{b_2} f(x,t)\,dt, \qquad \text{for } a_1 \le x \le b_1.$$

Then for each x,

$$F'(x) = \int_{a_2}^{b_2} f_x(x,t)\,dt.$$

Proof: To prove the theorem, it will be sufficient to show that, for any $\varepsilon > 0$, we have

$$(1) \qquad \left|\frac{F(x + h) - F(x)}{h} - \int_{a_2}^{b_2} f_x(x,t)\,dt\right| < \varepsilon$$

for all values of h sufficiently close to 0: since from this it follows that

$$\frac{dF}{dx} = \lim_{h\to 0}\frac{F(x + h) - F(x)}{h} = \int_{a_2}^{b_2} f_x(x,t)\,dt.$$

Now to prove (1) first note that

$$F(x + h) - F(x) = \int_{a_2}^{b_2} [f(x + h, t) - f(x,t)]\,dt.$$

(Here x is considered as *fixed*, and h is a variable.) By the mean value theorem (4.8), there is a number z, between x and $x + h$, such that

$$f(x + h, t) - f(x,t) = h\cdot f_x(z,t).$$

Hence

$$\left|\frac{F(x+h)-F(x)}{h}-\int_{a_2}^{b_2} f_x(x,t)\,dt\right| = \left|\int_{a_2}^{b_2} f_x(z,t)\,dt-\int_{a_2}^{b_2} f_x(x,t)\,dt\right|$$

$$= \left|\int_{a_2}^{b_2} [f_x(z,t)-f_x(x,t)]\,dt\right|.$$

Since f_x is continuous, it is also uniformly continuous in the closed rectangle S. From this it follows that, given $\varepsilon > 0$, there is some $\delta > 0$ such that

$$|f_x(z,t)-f_x(x,t)| < \frac{\varepsilon}{b_2-a_2}$$

for all h with $|h| < \delta$ (recall that z is between x and $x + h$). Hence we have

$$\left|\frac{F(x+h)-F(x)}{h}-\int_{a_2}^{b_2} f_x(x,t)\,dt\right| < \frac{\varepsilon}{b_2-a_2}\cdot(b_2-a_2)=\varepsilon,$$

whenever $|h| < \delta$. This proves (1).

Evidently, a result analogous to 15.33 holds for $\int_{a_1}^{b_1} f(t,y)\,dt$.

The rule given by Theorem 15.33 is usually known as "differentiation under the integral sign." It may sometimes be used to evaluate definite integrals. The following example illustrates this technique.

Example 1 Compute $\int_0^1 \frac{x^\alpha-1}{\log x}\,dx$, for $\alpha \geq 0$.

Solution: The presence of the constant α under the integral sign suggests that we consider the value of the integral as a function of α. Thus, set

$$F(\alpha)=\int_0^1 \frac{x^\alpha-1}{\log x}\,dx.$$

Then, by 15.33,

$$F'(\alpha)=\int_0^1 \frac{x^\alpha \log x}{\log x}\,dx=\int_0^1 x^\alpha\,dx$$

$$=\left.\frac{x^{\alpha+1}}{\alpha+1}\right|_0^1=\frac{1}{\alpha+1}.$$

Hence, $F(\alpha) = \log(\alpha + 1) + C$, where the constant C must be determined. But by the definition of F,

$$F(0)=\int_0^1 0\,dx=0.$$

Hence $0 = \log 1 + C$, and $C = 0$. Thus we obtain

$$F(\alpha)=\int_0^1 \frac{x^\alpha-1}{\log x}\,dx=\log(\alpha+1).$$

Example 2 Suppose a function F is defined by

$$F(x) = \int_a^{u(x)} f(x,t)\, dt,$$

where the function u is differentiable, and f and f_x are continuous. Find a formula for $F'(x)$.

Solution: Define a function H of the two variables x and u by the formula

$$H(x,u) = \int_a^u f(x,t)\, dt.$$

Then $F(x) = H(x,u(x))$. By the chain rule,

$$F'(x) = \frac{dH}{dx} = H_x(x,u)\frac{dx}{dx} + H_u(x,u)\frac{du}{dx}.$$

But by 15.33,

$$H_x(x,u) = \int_a^u f_x(x,t)\, dt,$$

and by the fundamental theorem of calculus (5.20),

$$H_u(x,u) = f(x,u).$$

Hence, replacing u by $u(x)$, we have

$$F'(x) = \int_a^{u(x)} f_x(x,t)\, dt + f(x,u(x))\frac{du}{dx},$$

which is the required formula.

As another application of Theorem 15.33, we may now prove Theorem 15.11 on the equality of mixed second partial derivatives.

15.34 Theorem

If f is a function of two variables x and y, and if f_{xy} and f_{yx} are continuous in some open set S of R^2, then $f_{xy}(x,y) = f_{yx}(x,y)$ for every (x,y) in S.

Proof: Let us write $h(x,y) = f_y(x,y)$, for (x,y) in S. Then, if a is any constant such that (x,a) is in S, we have

$$\begin{aligned} f(x,y) &= f(x,a) + f(x,y) - f(x,a) \\ &= f(x,a) + \int_a^y h(x,t)\, dt. \end{aligned}$$

Then by 15.33 we have

$$f_x(x,y) = f_x(x,a) + \int_a^y h_x(x,t)\, dt.$$

Now differentiating with respect to y, and using the fundamental theorem of calculus (5.20), we obtain

$$f_{xy}(x,y) = h_x(x,y) = f_{yx}(x,y),$$

which completes the proof.

EXERCISES

In Exercises 1 to 5, find $F'(x)$.

1. $F(x) = \int_0^{\pi} (1 - x\cos t)^2\,dt$
2. $F(x) = \int_0^{x^2} \tan^{-1}\frac{t}{x^2}\,dt$
3. $F(x) = \int_0^{x^3} \sqrt{x}\,dx$
4. $F(x) = \int_{x^2}^{x^3} e^{-xy}\,dy$
5. $F(x) = \int_0^{\pi} \sin\frac{x}{2}\,dx$
6. Starting with integration formulas for $\int \sin ax\,dx$ and $\int \cos ax\,dx$, use Theorem 15.33 to obtain formulas for $\int x^2 \sin ax\,dx$ and $\int x^2\cos ax\,dx$.
7. Apply Theorem 15.33 to each of the following integration formulas to obtain a new formula. Check your answer by use of the fundamental theorem of the calculus.

   a. $\int \frac{1}{t^2 + x^2}\,dt = \frac{1}{x}\tan^{-1}\frac{t}{x}$

   b. $\int (x^2 - t^2)^{-1/2}\,dt = \sin^{-1}\frac{t}{x}$

   c. $\int \frac{1}{t\sqrt{t^2 - x^2}}\,dt = \frac{1}{x}\sec^{-1}\frac{t}{x}$

   d. $\int t^x\,dt = \frac{t^{x+1}}{x+1}$

8. Starting from $\int e^{xt}\,dt = \frac{e^{xt}}{x} + c$, use Theorem 15.33 to obtain $\int t^n e^{xt}\,dt$ (n integer > 0).

CHAPTER REVIEW

In each of Exercises 1 to 3 verify that $F_{12} = F_{21}$, $F_{112} = F_{211}$, $F_{221} = F_{122}$.

1. $F(x,y) = xe^{x+y}$
2. $F(x,y) = \frac{x}{1 + y^2}$
3. $F(x,y) = e^x \sin(x + y)$
4. Use a differential to approximate (99.99)(49.98)(9.97).
5. Suppose it is known that ln 5 is approximately 1.61. Use a differential to approximate $\ln[(1.02)^2 + (2.03)^2]$.
6. If $z = x/(x + y)$, $x = 2 + t$, and $y = g(t)$, find a formula for dz/dt.
7. Suppose $z = f(x,y)$, $x = r\cos\theta$, $y = r\sin\theta$. Express $\partial z/\partial r$ in terms of r, θ, and partial derivatives of f with respect to x and y. Then differentiate your result to find a formula for $\partial^2 z/\partial r^2$.

8. Suppose $z = f(x,y)$, $x = ue^v$, $y = ue^{-v}$. Find a formula for $\partial^2 z/\partial u^2$.

9. Find $\partial z/\partial y$ if it is assumed that z is defined as an implicit function of x and y by the equation

$$\frac{x^2 + y^2}{z^3} = \frac{1}{z} = 2.$$

10. Find the maximum directional derivative of the function $F(x,y) = 100 - 2x^2 - y^2$ at the point (5,5).

11. Find the directional derivative of $f(x,y,z) = e^{-x} \sin y \cos z$ in the direction of the vector $\langle \sqrt{2}/2, \frac{1}{2}, \frac{1}{2} \rangle$ at the point $P = (0,0,0)$.

12. Let $f(x,y) = x^2 + xy^2 - y^3$, and let P be the point (3,1).

a. Find the directional derivative $D_{\mathbf{u}}f(P)$ if $\mathbf{u}$ is a unit vector in the direction of the line through P and the point (4,3).

b. Find a unit vector $\mathbf{u}$ such that $D_{\mathbf{u}}f(P)$ has its maximum value.

c. Find a unit vector $\mathbf{u}$ such that $D_{\mathbf{u}}f(P)$ has its minimum value.

d. Find a unit vector $\mathbf{u}$ such that $D_{\mathbf{u}}f(P) = 0$.

In each of Exercises 13–16, sketch the surface and find the tangent plane to the surface at the given point.

13. $z = 10 - x^2 - y^2$, (1,3,0)

14. $z = \dfrac{9}{1 + x^2 + y^2}$, (2,2,1)

15. $y = \sqrt{x + z}$, (1,2,3)

16. $x^2 - y^2 + z^2 = 6$, (1,2,−3)

In Exercises 17 and 18, find parametric equations of the tangent line of the curve of intersection of the given surfaces at the indicated point.

17. $z = 1 - y^2$, $x^2 + 4y^2 = 4$; $(\sqrt{3}, \frac{1}{2}, \frac{3}{4})$

18. $z = 4x^2 + y^2$, $x^2 - 4y^2 + z^2 = 49$; $(-1,-2,8)$

Find the extrema of the function in each of Exercises 19–22.

19. $f(x,y) = x^2e^{x+y}$

20. $F(x,y) = 4xy - 2x^2 - y^4$

21. $g(x,y) = (x - 1) \ln y - x^2$

22. $z = x/\sqrt{x^2 + y^2 + 1}$

23. A rectangular tank with an open top is required to have a volume of 81 cu. ft. The cost of the material for the sides is $1 per sq ft, and for the base $3 per sq ft. Find the dimensions of the most economical tank.

24. Prove that if a triangle has fixed perimeter, then its area is a maximum when all its sides are equal.

25. It is required that the sum of the lengths of the edges of a rectangular box be a constant. Find its dimensions if the surface area is to be a maximum.

26. The equations

$$xv + uy + 1 = 0$$
$$yv - ux - 1 = 0$$

can be solved simultaneously for x and y in terms of u and v. Find $\dfrac{\partial x}{\partial u}$, $\dfrac{\partial x}{\partial v}$,

$\frac{\partial y}{\partial u}, \frac{\partial y}{\partial v}$ by implicit differentiation.

27. Assuming that the equations

$$2x + 3y - u + v + 2 = 0$$
$$x^2 - y^2 + u^2 - v^2 = 0$$

define x and y as differentiable functions of u and v, find $\frac{\partial x}{\partial u}, \frac{\partial x}{\partial v}, \frac{\partial y}{\partial u}$, and $\frac{\partial y}{\partial v}$ by implicit differentiation. What restrictions must be placed on x and y?

28. Assume that the equations

$$x^2 + yu + v^3 = 1$$
$$yv^2 + u = 2$$

define u and v as implicit functions of x and y. Find $\partial v/\partial x$ as a function of x, y, u, and v.

16

Multiple Integration

The concept of the integral of a function of one variable, given in Chapter 5, is extended in this chapter to functions of several variables. We then discuss a number of geometrical and physical applications.

1 DOUBLE INTEGRALS

In this section we begin the study of integration for functions of two variables. Our discussion will parallel the definition of the integral for a function of one variable, which was presented in Chapter 5. However, in order to obtain an adequate theory in the two-variable case, it will be necessary to enlarge the class of functions that we consider. The reader will recall that in Chapter 5 the integral was defined, and shown to exist, for a function continuous in a closed interval I. For functions of two variables we shall no longer be able to restrict ourselves to the continuous case (as will be evident in the discussion of integrals over non-rectangular regions in Section 5 of this chapter).

It is for this reason that we consider *bounded* functions in this section. A function f with domain S is said to be *bounded in* S if there exist numbers r and s such that $r \leq f(P) \leq s$ for all points P in S. In other words, f is bounded if the range of f is a bounded set of real numbers. As shown in Chapter 5, any bounded set of real numbers has a least upper bound (l.u.b.) and a greatest lower bound (g.l.b.). Hence if f is bounded in S, its range has a l.u.b. M and a g.l.b. m. We refer to these numbers M and m as the l.u.b. and g.l.b., respectively, *of the function* f *over* S.

Our discussion in this section will refer to an arbitrary bounded function defined in a closed rectangle in R^2 (which is the analog of a closed interval in R). If $A = (a_1,a_2)$ and $B = (b_1,b_2)$ are points in R^2, then, as in Chapter 15, we shall let $Q = [A,B]$ denote the closed rectangle with vertices A, (b_1,a_2), B, and (a_1,b_2). Let f be a given real-valued function of the two variables x and y, which is defined and bounded in Q. Let m and M denote the g.l.b. and l.u.b., respectively, of f over Q; that is,

$$m = \text{g.l.b. } \{f(x,y) \mid (x,y) \text{ in } Q\},$$
$$M = \text{l.u.b. } \{f(x,y) \mid (x,y) \text{ in } Q\}.$$

Our purpose is now to define the concept of an integral of the function f over Q. In order to do this, we shall start by defining what is meant by a *partition* of the rectangle $Q = [A,B]$. Let us assume that $a_1 < b_1$ and $a_2 < b_2$. If $P_1 = \{x_0, x_1, \ldots, x_r\}$ is a partition of $[a_1,b_1]$ and $P_2 = \{y_0, y_1, \ldots, y_s\}$ is a partition of $[a_2,b_2]$, and if Q_{ij} is the closed rectangle

$$Q_{ij} = [(x_{i-1},y_{j-1}),(x_i,y_j)],$$

then

$$p = \{Q_{ij} \mid i = 1, \ldots, r, j = 1, \ldots, s\}$$

is called a *partition* of rectangle $[A,B]$. Evidently, p consists of rs subrectangles of $[A,B]$, as shown in Figure 16.1, where $r = 4$ and $s = 3$. If we let

$$\Delta x_i = x_i - x_{i-1}, \qquad i = 1, \ldots, r,$$
$$\Delta y_j = y_j - y_{j-1}, \qquad j = 1, \ldots, s,$$

then evidently $\Delta x_i \, \Delta y_j = A(Q_{ij})$, the area of the rectangle Q_{ij}. The length of the longest diagonal of any Q_{ij} of partition p is called the *norm* of p and is denoted by $\|p\|$. Thus $\|p\|$ is the largest of the numbers

$$\sqrt{\Delta x_i^2 + \Delta y_j^2}, \qquad i = 1, \ldots, r, \quad j = 1, \ldots, s.$$

Clearly, no two points of any subrectangle Q_{ij} of $[A,B]$ are farther apart than $\|p\|$. The norm of a partition of a rectangle will play the same role here as the norm of a partition of an interval in Chapter 5.

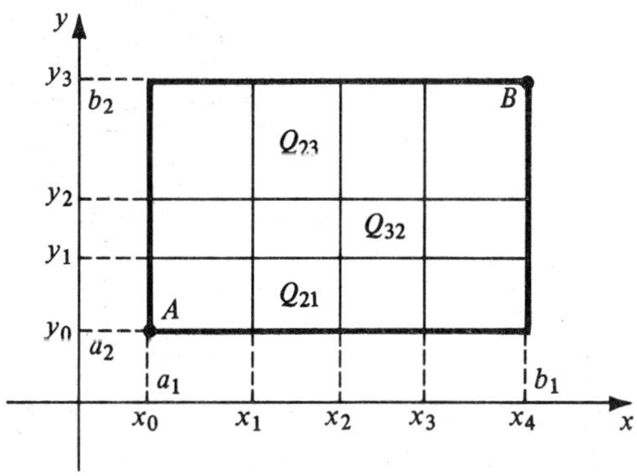

FIGURE 16.1

If p and p^* are both partitions of Q, then, corresponding to each of these, there are associated partitions P_1 and P_1^* of $[a_1,b_1]$, and partitions P_2 and P_2^* of $[a_2,b_2]$. We say that p^* is a *refinement* of p if P_1^* is a refinement of P_1 and P_2^* is a refinement of P_2.

Now, given a partition p of Q, say

$$p \doteq \{Q_{ij} \mid i = 1, \ldots, r, j = 1, \ldots, s\},$$

let us write

$$\begin{aligned} m_{ij} &= \text{g.l.b. } \{f(x,y) \mid (x,y) \text{ in } Q_{ij}\}, \\ M_{ij} &= \text{l.u.b. } \{f(x,y) \mid (x,y) \text{ in } Q_{ij}\}. \end{aligned}$$

In the event the function f should happen to be continuous in Q, the numbers m_{ij} and M_{ij} would simply be the minimum and maximum values, respectively, of the function f in the set Q_{ij} [and they correspond to the numbers $f(u_i)$ and $f(v_i)$ of Chapter 5, Section 4].

Corresponding to any given partition p of Q, we form the sums

$$l(p) = \sum_{i=1}^{r} \sum_{j=1}^{s} m_{ij} A(Q_{ij}),$$

$$u(p) = \sum_{i=1}^{r} \sum_{j=1}^{s} M_{ij} A(Q_{ij}).$$

The numbers $l(p)$ and $u(p)$ are called the *lower* and *upper sums*, respectively, of the function f relative to the given partition p.

The numbers $l(p)$ and $u(p)$ have properties that are exactly analogous to those of the lower and upper sums for functions of one variables. We state the following lemmas without proof, since the proofs are nearly identical with those of 5.11, 5.15, and 5.16.

16.1 Lemma

For any partition p of Q,

$$mA(Q) \leq l(p) \leq u(p) \leq MA(Q).$$

16.2 Lemma

If p^* is a refinement of p, then

$$l(p^*) \geq l(p), \; u(p^*) \leq u(p).$$

16.3 Lemma

If p and p' are any partitions of Q, then

$$l(p) \leq u(p').$$

From the above results it follows that the sets L and U of all lower and upper sums, respectively, of f over Q are bounded sets of real numbers. Hence the numbers l.u.b. L and g.l.b. U both exist for any function f that is bounded over Q. Furthermore,

16.4 $$\text{l.u.b. } L \leq \text{g.l.b. } U,$$

for any f bounded over Q. The case when the two numbers of 16.4 happen to be equal is of fundamental importance.

16.5 Definition

A bounded function f is *integrable* on the rectangle Q if and only if the numbers l.u.b. L and g.l.b. U are equal. If f is integrable on Q, then the common value of l.u.b. L and g.l.b. U is called the *double integral of f over Q*, and is denoted by $\int_Q f$.

We have the following useful criterion for the integrability of a function. It holds also in the one-variable case, in which the same proof would apply.

16.6 Theorem

The function f is integrable over the rectangle Q if and only if, for every $\varepsilon > 0$, there exists a partition p of Q such that $u(p) - l(p) < \varepsilon$.

Proof: Let us write $\alpha =$ g.l.b. U, $\beta =$ l.u.b. L. That is,

$$\alpha = \text{g.l.b. } \{u(p) \mid p \text{ is a partition of } Q\},$$
$$\beta = \text{l.u.b. } \{l(p) \mid p \text{ is a partition of } Q\}.$$

Now suppose f is integrable over Q, and an arbitrary $\varepsilon > 0$ is given. By the definition of l.u.b., there must be some partition p_1 of Q such that

$$0 \leq \beta - l(p_1) < \frac{\varepsilon}{2}.$$

Likewise there must be some partition p_2 of Q such that

$$0 \leq u(p_2) - \alpha < \frac{\varepsilon}{2}.$$

If p is the common refinement of p_1 and p_2, then by 16.2 we also have

$$0 \leq \beta - l(p) < \frac{\varepsilon}{2},$$

$$0 \leq u(p) - \alpha < \frac{\varepsilon}{2}.$$

But $\alpha = \beta$, since f is integrable. Hence, by adding the above inequalities it follows that $u(p) - l(p) < \varepsilon$.

To prove the converse, simply note that for every partition p of Q,

$$l(p) \leq \beta \leq \alpha \leq u(p),$$

and hence

$$0 \leq \alpha - \beta \leq u(p) - l(p)$$

for every p. Therefore, if the difference $u(p) - l(p)$ can be made arbitrarily close to 0, it follows that α and β must be equal, and hence f is integrable.

We state the following four theorems without proof. The proof of 16.8 is similar to that of 5.19.

16.7 Theorem

If f is integrable on the rectangle $Q = [A,B]$, and if R is a closed subrectangle of Q with sides parallel to the coordinate axes, then f is integrable on R.

16.8 Theorem

Suppose that the rectangle $Q = [A,B]$ is the union of two closed subrectangles R_1 and R_2, whose sides are parallel to the coordinate axes. If f is integrable on R_1 and R_2, then f is also integrable on Q and

$$\int_Q f = \int_{R_1} f + \int_{R_2} f.$$

16.9 Theorem

If f and g are both integrable on $Q = [A,B]$, then so is the function $f + g$, and

$$\int_Q (f + g) = \int_Q f + \int_Q g.$$

16.10 Theorem

If f and g are integrable functions on $Q = [A,B]$, and if $f(x,y) \leq g(x,y)$ for all (x,y) in Q, then $\int_Q f \leq \int_Q g$.

The double integral of a function f over a rectangle Q may also be expressed as the limit of a sequence of Riemann sums. Thus, suppose that

$$p = \{Q_{ij} \mid i = 1, \ldots, r, j = 1, \ldots, s\}$$

is a partition of Q. Then a sum of the form

$$R(p) = \sum_{i=1}^{r} \sum_{j=1}^{s} f(x_i', y_j')A(Q_{ij}),$$

where (x_i', y_j') is any point in Q_{ij}, is called a *Riemann sum* of the function f over Q. Now suppose that f is integrable over Q. If $p_1, p_2, \ldots, p_n, \ldots$ is any sequence of partitions of Q with

$$\operatorname*{limit}_{n \to \infty} \|p_n\| = 0,$$

then, the double integral analog of Theorem 5.39 states, whenever $R(p_1)$, $R(p_2), \ldots, R(p_n), \ldots$ is any associated sequence of Riemann sums for f over Q,

$$\operatorname*{limit}_{n \to \infty} R(p_n) = \int_Q f.$$

The proof of this result is similar to 5.39.

If f is a continuous function on the closed rectangle Q, it is possible to prove that f is integrable over Q. (The proof is similar to the one-variable case discussed in Chapter 5.) In the next section, however, we shall state a somewhat more general theorem concerning the integrability of functions of two variables, which will be needed in our subsequent work. This theorem tells us that a function is integrable under certain conditions even if it possesses some points of discontinuity.

We now give an example to show how a double integral may be computed from some Riemann sums.

Example Compute $\int_Q f$, where $f(x,y) = x^2y$ and $Q = \{(x,y) \mid 0 \le x \le 1, 0 \le y \le 1\}$.

Solution: Since f is continuous, we know that it is integrable over Q. Let p_n be the partition of Q that subdivides Q into n^2 equal squares. In each subsquare Q_{ij}, choose the upper right-hand vertex, namely, the point $(i/n, j/n)$; and form the corresponding Riemann sum,

$$\begin{aligned} R(p_n) &= \sum_{i=1}^{n}\sum_{j=1}^{n} f\left(\frac{i}{n},\frac{j}{n}\right) A(Q_{ij}) \\ &= \sum_{i=1}^{n}\sum_{j=1}^{n} \left(\frac{i}{n}\right)^2 \left(\frac{j}{n}\right) \frac{1}{n^2} \\ &= \frac{1}{n^5}\left(\sum_{i=1}^{n} i^2\right)\left(\sum_{j=1}^{n} j\right). \end{aligned}$$

But by 5.6 and 5.7,

$$\sum_{j=1}^{n} j = \frac{n(n+1)}{2}, \qquad \sum_{i=1}^{n} i^2 = \frac{n(n+1)(2n+1)}{6}.$$

Hence,

$$\begin{aligned} R(p_n) &= \frac{1}{n^5}\left[\frac{n(n+1)(2n+1)}{6}\right]\left[\frac{n(n+1)}{2}\right] \\ &= \frac{2n^5 + 5n^4 + 3n^3 + n^2}{12n^5}. \end{aligned}$$

It follows that

$$\int_Q f = \lim_{n\to\infty} R(p_n) = \tfrac{1}{6}.$$

EXERCISES

In each of Exercises 1 to 4, Q is the rectangle $[(0,0),(4,2)]$ and p_n is the partition of Q into n^2 equivalent rectangles. The numbers l_n and u_n are the lower and upper sums, respectively, of f relative to p_n.

1. If $f(x,y) = 2x + y$, find l_2, l_4, u_2, and u_4. Also find l_n in terms of n, and then find $\int_Q f$.

2. If $f(x,y) = 4 - x - y$, find l_3 and u_3. Also, find u_n in terms of n, and then find $\int_Q f$.

3. If $f(x,y) = x^2 + y$, find l_4 and u_4. Also, find l_n in terms of n, and then find $\int_Q f$.

4. If $f(x,y) = xy$, find l_3 and u_3. Also, find u_n in terms of n, and then find $\int_Q f$.

5. Compute $\int_Q f$, if $f(x,y) = xy^3$, and Q is the rectangle $[(0,0),(4,2)]$.

2 A CONDITION FOR INTEGRABILITY

In our subsequent work (particularly in connection with integration over non-rectangular regions) we shall need to know that a function is integrable even if it possesses certain points of discontinuity, and if, in a certain sense, its set of points of discontinuity is not too "large." In order to formulate such a theorem in a precise way, we introduce the following definition.

16.11 Definition

A subset E of R^2 has *content zero* if and only if for every number $\varepsilon > 0$, there is a finite set $R_1, \ldots, R_n$ of rectangles whose union contains E, and whose total area is less than ε.

An important example of a set of content zero is provided by the following theorem, whose proof we shall omit.

16.12 Theorem

If f is a real-valued continuous function defined in the closed interval $[a,b]$, then the graph of f is a subset of R^2 with content zero.

We now state the desired condition for the integrability of a function of two variables. The proof of this theorem is rather difficult and will be omitted. The reader will find this proof, and also that of 16.12, in most texts on advanced calculus.

16.13 Theorem

Let f be a bounded function in a closed rectangle $Q = [A,B]$. If the set E of all points in Q at which f is discontinuous is a set of content zero, then f is integrable on Q.

Corresponding to Theorem 16.13, there is an analogous theorem concerning the integrability of a function of one variable. Without going into the details, it is possible to define the concept of "length zero" for a subset of R, in a manner analogous to Definition 16.11. It is then possible to prove that *a function defined in a closed interval I is integrable on I if its set of discontinuities in I has length zero.* We shall not actually need this theorem in the sequel, but only the special case of it which asserts that *a function defined in the closed interval I is integrable in I if it has only a finite number of points of discontinuity in I.* This result will be used in Section 5 of this chapter.

3 REPEATED INTEGRALS

If Q is the rectangle $[A,B]$, where $A = (a_1,a_2)$, $B = (b_1,b_2)$, and if f is a real-valued function defined in Q, then we may define a function G by the formula

$$G(x) = \int_{a_2}^{b_2} f(x,y)\,dy, \qquad a_1 \le x \le b_1,$$

provided that this integral exists. Here, of course, the integration is performed with respect to the variable y, while x is held constant during the integration. If G is also an integrable function, for x in the interval $[a_1,b_1]$, then we write

$$(1) \qquad \int_{a_1}^{b_1}\left(\int_{a_2}^{b_2} f(x,y)\,dy\right)dx \qquad \text{or} \qquad \int_{a_1}^{b_1} dx \int_{a_2}^{b_2} f(x,y)\,dy$$

for $\int_{a_1}^{b_1} G(x)\,dx$.

Likewise, assuming that the integral exists, we may define a function H of y by the formula

$$H(y) = \int_{a_1}^{b_1} f(x,y)\,dx, \qquad a_2 \le y \le b_2.$$

If H is integrable for y in the interval $[a_2,b_2]$, then we write

$$(2) \qquad \int_{a_2}^{b_2}\left(\int_{a_1}^{b_1} f(x,y)\,dx\right)dy \qquad \text{or} \qquad \int_{a_2}^{b_2} dy \int_{a_1}^{b_1} f(x,y)\,dx$$

for $\int_{a_2}^{b_2} H(y)\,dy$.

Each of the integrals (1) and (2) is called a *repeated* (or *iterated*) integral of f over $[A,B]$.

Example If $f(x,y) = x^2 - 2xy$, find $\int_{-1}^{2} dx \int_{1}^{4} f(x,y)\,dy$.

Solution: In evaluating $\int_1^4 (x^2 - 2xy)\,dy$, we are to assume that x is a constant. Hence

$$\int_1^4 (x^2 - 2xy)\,dy = (x^2y - xy^2)\Big|_1^4 = (4x^2 - 16x) - (x^2 - x)$$
$$= 3x^2 - 15x,$$

and

$$\int_{-1}^{2} dx \int_1^4 (x^2 - 2xy)\,dy = \int_{-1}^{2} (3x^2 - 15x)\,dx = x^3 - \tfrac{15}{2}x^2\Big|_{-1}^{2} = \tfrac{27}{2}.$$

Under very general conditions it is possible to prove that a *double* integral may be evaluated by computing a *repeated* integral. We now state an important theorem of this type whose proof will be found in an appendix to this chapter (Section 12). Observe that no hypothesis of continuity is involved.

16.14 Theorem

Suppose that:

(i) *The function f is integrable in $Q = [A,B]$.*
(ii) $\int_{a_2}^{b_2} f(x,y)\,dy$ *exists for each x in $[a_1,b_1]$.*

Then the function $G(x) = \int_{a_2}^{b_2} f(x,y)\,dy$ *is integrable in the interval $[a_1,b_1]$ and*

$$\int_Q f = \int_{a_1}^{b_1} G(x)\,dx = \int_{a_1}^{b_1} dx \int_{a_2}^{b_2} f(x,y)\,dy.$$

It should be clear that a proof exactly similar to that of 16.14 could be given to show that, with similar hypotheses,

$$\int_Q f = \int_{a_2}^{b_2} dy \int_{a_1}^{b_1} f(x,y)\,dx.$$

Thus we have the following result.

16.15 Theorem

If the function f is integrable in $Q = [A,B]$, and if the repeated integrals

$$\int_{a_1}^{b_1} dx \int_{a_2}^{b_2} f(x,y)\,dy, \qquad \int_{a_2}^{b_2} dy \int_{a_1}^{b_1} f(x,y)\,dx$$

both exist, then the two repeated integrals of f over Q are equal, and their common value is $\int_Q f$.

If the function f is continuous in $Q = [A,B]$, then the hypotheses of 16.14 and 16.15 are all satisfied. For if f is continuous in Q, then the functions

$$G(x) = \int_{a_2}^{b_2} f(x,y)\,dy,$$
$$(Hy) = \int_{a_1}^{b_1} f(x,y)\,dx,$$

are each continuous, in their respective intervals, by Theorem 15.32. Thus the functions f, G, and H are all integrable, and we have the following theorem.

16.16 Theorem

If f is continuous in $Q = [A,B]$, then the repeated integrals of f over Q both exist and are equal; and their common value is $\int_Q f$.

EXERCISES

In each of Exercises 1 to 4, evaluate the double integral of f over the given rectangle Q by computing a repeated integral.

1. $f(x,y) = x^2y(x - y)$, $Q = [(0,0),(1,1)]$
2. $f(x,y) = \cos(x + y)$, $Q = [(0,0), (\pi/2, \pi/2)]$
3. $f(x,y) = \cos^2 x \cos^2 y$, $Q = [(0,0),(\pi,\pi)]$
4. $f(x,y) = y \sin x - xe^y$, $Q = [(0,-1), (\pi/2, 1)]$

In Exercises 5 to 7, verify Theorem 16.15 by direct integration of each side of the equation.

5. $\int_a^b dx \int_0^1 (3x^2 + xy + y^2)\, dy = \int_0^1 dy \int_a^b (3x^2 + xy + y^2)\, dx$

6. $\int_1^3 dx \int_1^2 \ln(x + 2y)\, dy = \int_1^2 dy \int_1^3 \ln(x + 2y)\, dx$

7. $\int_0^1 dx \int_0^1 x\sqrt{x^2 + y}\, dy = \int_0^1 dy \int_0^1 x\sqrt{x^2 + y}\, dx$

8. Suppose that the function f is defined in a rectangle $Q = [A,B]$, and that $f(x,y) = g(x)h(y)$ for all (x,y) in Q. Assuming appropriate integrability hypotheses, prove that the double integral $\int_Q f$ is equal to the product

$$\left[\int_{a_1}^{b_1} g(x)\, dx\right] \cdot \left[\int_{a_2}^{b_2} h(y)\, dy\right]$$

of two single integrals. To which of Exercises 1 to 7 does this result apply?

4 VOLUME

If the function f of two variables is continuous and nonnegative in a closed rectangle $Q = [(a_1,a_2), (b_1,b_2)]$, then the graph of f and the planes $z = 0$, $x = a_1$, $x = b_1$, $y = a_2$, and $y = b_2$ bound a solid (Figure 16.2):

$$S(f,Q) = \{(x,y,z) \mid (x,y) \text{ in } Q, 0 \leq z \leq f(x,y)\}.$$

The volume of this solid can be found by integration, as we shall show below.

A parallelepiped P in R^3 having its edges parallel to the coordinate axes has volume equal to the product abc of the lengths of the three edges meeting at each vertex (Figure 16.3). In turn, the volume of a solid made up of non-overlapping parallelepipeds is the sum of the volumes of the individual parallelepipeds.

Each lower sum

$$l = \sum_{i=1}^{r} \sum_{j=1}^{s} f(T_{ij})A(Q_{ij})$$

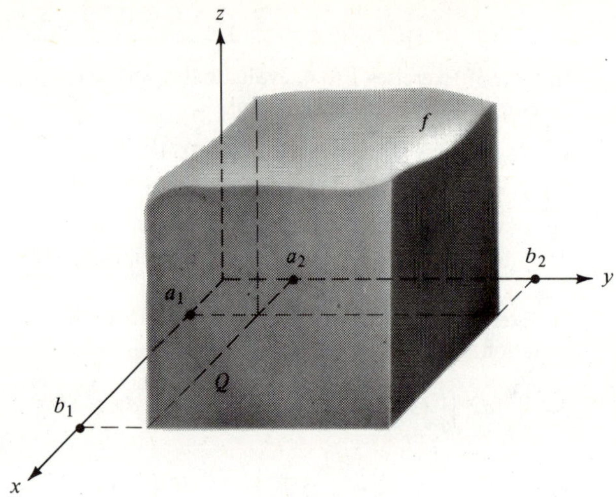

FIGURE 16.2

of f over Q is the volume of a solid made up of rs parallelepipeds and inscribed in $S(f,Q)$ and, similarly, each upper sum u of f over Q is the volume of a solid circumscribing $S(f,Q)$. The volume of solid $S(f,Q)$ is defined to be the l.u.b. of the set L of all lower sums of f over Q, or, equivalently, the g.l.b. of the set U of all upper sums of f over Q. Hence, the solid $S = S(f,Q)$ has volume $V(S)$ defined by

$$V(S) = \int_Q f.$$

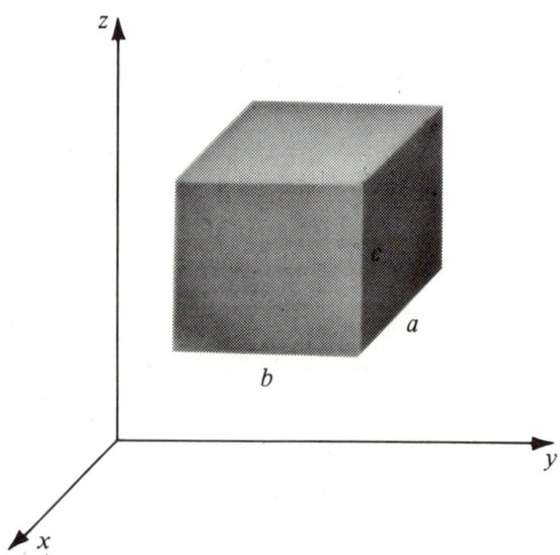

FIGURE 16.3

In actually computing the volume of a solid, we use the fact that the double integral equals a repeated integral, as shown in the following example.

Example If $f(x,y) = 4 - \frac{1}{100}(25x^2 + 16y^2)$ and $Q = [(0,0),(2,3)]$, find the volume of solid $S = S(f,Q)$ (Figure 16.4).

Solution: We have

$$V(S) = \int_Q f = \int_0^2 dx \int_0^3 [4 - \tfrac{1}{100}(25x^2 + 16y^2)]\, dy$$

$$= \int_0^2 [4y - \tfrac{1}{4}x^2y - \tfrac{4}{75}y^3]\Big|_{y=0}^{y=3} dx$$

$$= \int_0^2 (\tfrac{264}{25} - \tfrac{3}{4}x^2)\, dx = 19.12.$$

Hence the volume of S is 19.12 cubic units.

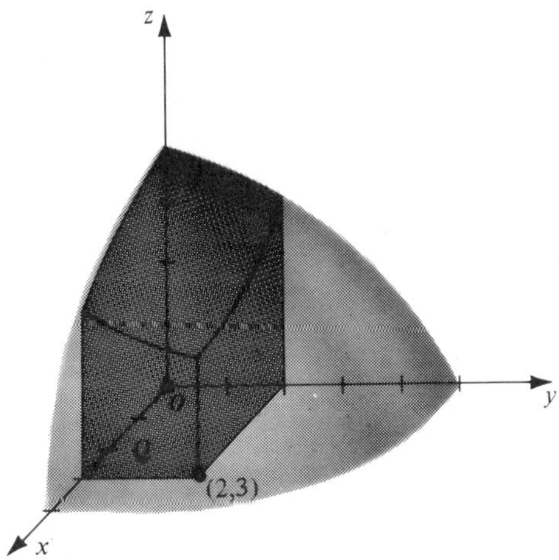

FIGURE 16.4

5 INTEGRALS OVER NONRECTANGULAR REGIONS

So far we have discussed the double integral only for functions defined in a rectangle. However, in many applications it is necessary to integrate functions defined over more general kinds of subsets of the plane. In this section we shall consider two such general kinds of plane regions, which are bounded by line segments and graphs of smooth functions. These regions are illustrated in Figures 16.5 and 16.6, and we shall call them, respectively, regions of type I and type II.

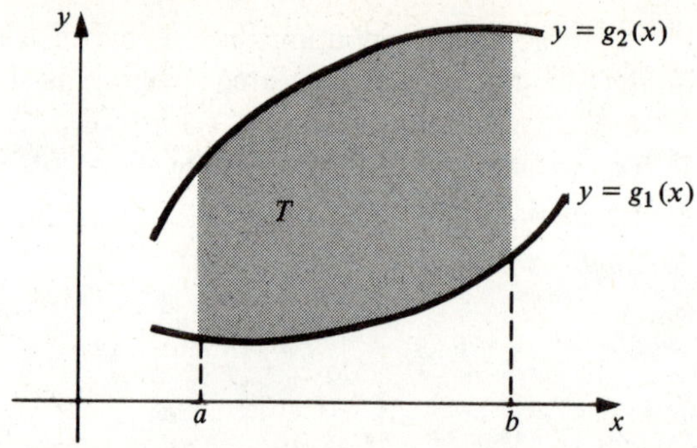

FIGURE 16.5

A region T of type I is a plane set bounded by the vertical lines $x = a$, $x = b$, and the graphs of the smooth functions $y = g_1(x)$, $y = g_2(x)$, where it is assumed that $g_1(x) \leq g_2(x)$ for all x in $[a,b]$. Such a region T may be expressed in the form

$$T = \{(x,y) \mid a \leq x \leq b \text{ and } g_1(x) \leq y \leq g_2(x)\}.$$

Similarly, a region T of type II is a plane set bounded by the horizontal lines $y = c$, $y = d$, and the graphs of smooth functions $x = g_1(y)$, $x = g_2(y)$, where $g_1(y) \leq g_2(y)$ for all y in $[c,d]$. A region of type II is expressible in the form

$$T = \{(x,y) \mid c \leq y \leq d \text{ and } g_1(y) \leq x \leq g_2(y)\}.$$

Now let f be a bounded function of the two variables x and y defined in a "standard" region T of type I or type II. Enclose T in a rectangle Q with

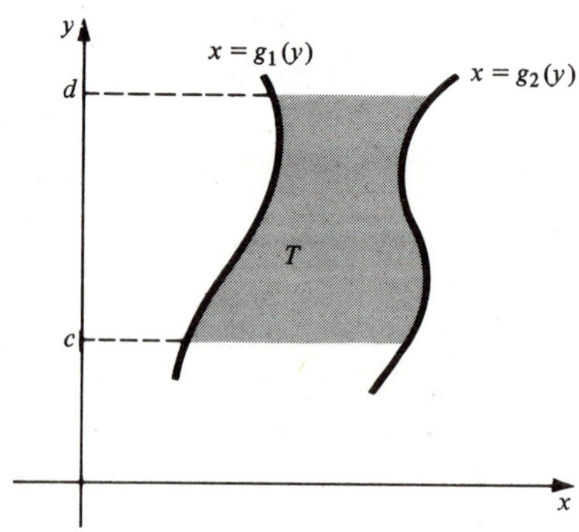

FIGURE 16.6

sides parallel to the axes. Let us define a new function F, with domain Q, as follows:

$$\begin{aligned} F(x,y) &= f(x,y), && \text{if } (x,y) \text{ is in } T, \\ F(x,y) &= 0, && \text{if } (x,y) \text{ is in } Q \text{ but not in } T. \end{aligned}$$

If the new function F is integrable on Q, then we say that the function f is integrable on T, and *by definition*,

16.17
$$\int_T f = \int_Q F.$$

It is easy to show that this definition is independent of the choice of the containing rectangle Q.

We shall prove that if f is continuous in T, then the double integral $\int_T f$ exists and may be evaluated by repeated integration. Specifically, we have the following fundamental theorem.

16.18 Theorem

Let T be a region of type I defined by

$$T = \{(x,y) \mid a \le x \le b \text{ and } g_1(x) \le y \le g_2(x)\}.$$

Suppose that the function f is defined and bounded in T, and continuous in the interior of T. Then:

(i) *The double integral $\int_T f$ exists.*

(ii) *The double integral $\int_T f$ may be evaluated by*

$$\int_T f = \int_a^b dx \int_{g_1(x)}^{g_2(x)} f(x,y)\, dy.$$

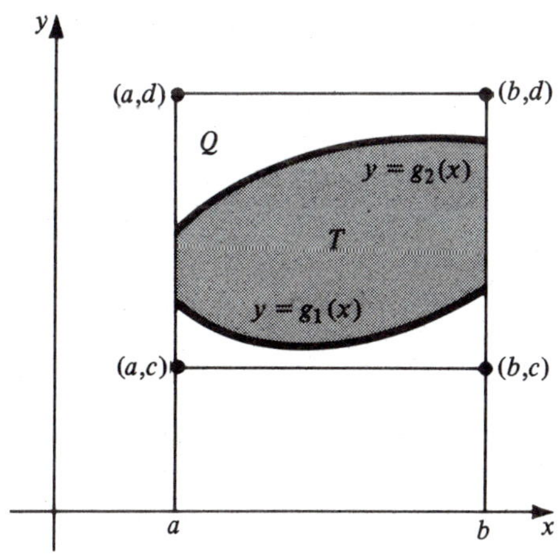

FIGURE 16.7

Proof: Enclose T in a rectangle $Q = [(a,c),(b,d)]$ (Figure 16.7). Define the function F as in the preceding paragraph. Then *the only possible discontinuities of the function F lie along the boundary curves of the region T.* But these curves constitute sets of content zero in R^2, by Theorem 16.12. Hence, by Theorem 16.13, the function F is integrable on Q. By definition 16.17, f is integrable on T and $\int_T f = \int_Q F$.

To prove (ii) we apply Theorem 16.14 to the function F. Observe that for each fixed x in $[a,b]$, the integral $\int_c^d F(x,y)\,dy$ exists, because there are only two possible discontinuities of $F(x,y)$ for $c \le y \le d$. These discontinuities can occur of course only when $y = g_1(x)$ or $y = g_2(x)$. By our remarks at the end of Section 2, the integral $\int_c^d F(x,y)\,dy$ exists for each x in $[a,b]$. Hence the hypotheses of Theorem 16.14 are satisfied, and it follows that

$$\int_Q F = \int_a^b dx \int_c^d F(x,y)\,dy.$$

But $F(x,y) = 0$ for $y < g_1(x)$ and for $y > g_2(x)$, and so

$$\int_c^d F(x,y)\,dy = \int_{g_1(x)}^{g_2(x)} f(x,y)\,dy$$

for all x in $[a,b]$. This completes the proof of (ii).

The same proof obviously applies to regions of type II and yields the following result.

16.19 Theorem

Let T be a region of type II defined by

$$T = \{(x,y) \mid c \le y \le d \text{ and } g_1(y) \le x \le g_2(y)\}.$$

Suppose that the function f is defined and bounded in T, and continuous in its interior. Then $\int_T f$ exists and may be evaluated by

$$\int_T f = \int_c^d dy \int_{g_1(y)}^{g_2(y)} f(x,y)\,dx.$$

Example 1 If

$$f(x,y) = \frac{2y - 1}{x + 1},$$

and if the region T is bounded by $x = 0$, $y = 0$, and $2x - y - 4 = 0$, find the double integral of f over T.

Solution: The region T (Figure 16.8) may be thought of as being bounded by the curves $y = 0$ and $y = 2x - 4$ between $x = 0$ and $x = 2$. Hence by 16.18,

$$\int_T f = \int_0^2 dx \int_{2x-4}^0 \frac{2y-1}{x+1}\, dy = \int_0^2 \frac{y^2 - y}{x+1}\Bigg|_{y=2x-4}^{y=0} dx$$

$$= -\int_0^2 \frac{4x^2 - 16x + 16 - 2x + 4}{x+1}\, dx$$

$$= -2\int_0^2 \left(2x - 11 + \frac{21}{x+1}\right) dx$$

$$= -2(x^2 - 11x + 21 \ln (x+1))\Big|_0^2 = -6(7 \ln 3 - 6).$$

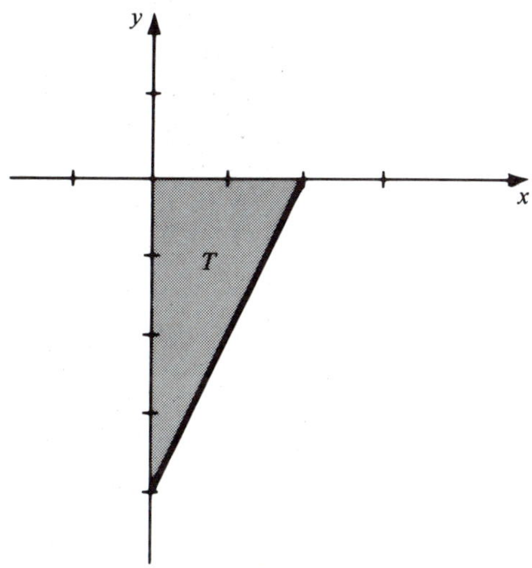

FIGURE 16.8

We could think of the region T as being bounded by the curves $x = 0$ and $x = (y + 4)/2$ between $y = -4$ and $y = 0$, in which case (by 16.19)

$$\int_T f = \int_{-4}^0 dy \int_0^{(y+4)/2} \frac{2y-1}{x+1}\, dx$$

$$= \int_{-4}^0 (2y-1) \ln (x+1)\Big|_{x=0}^{x=(y+4)/2} dy$$

$$= \int_{-4}^0 (2y-1) \ln \frac{y+6}{2}\, dy,$$

and so on. It is clear that the integration is easier in the first case.

If the continuous function f is nonnegative in a region T, then the double integral $\int_T f$ can be interpreted as the volume of a cylindrical solid S having region T as its base, generators parallel to the z axis, and the graph of f as its top just as before when T was a rectangle. In the particular case that $f(x,y) = 1$ for all x and y, $\int_T f$ is the area of region T. Some examples are given below.

Example 2 Find the area of the region T bounded by the parabolas $y = x^2$ and $y = 4 - x^2$ (Figure 16.9).

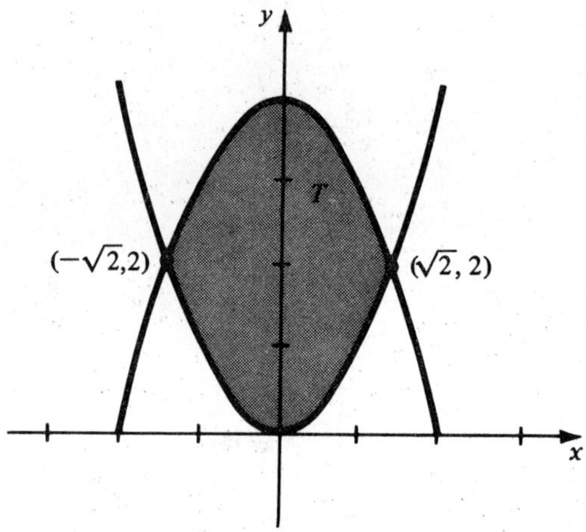

FIGURE 16.9

Solution: Using 16.18, we have the area given by

$$\int_T 1 = \int_{-\sqrt{2}}^{\sqrt{2}} dx \int_{x^2}^{4-x^2} dy = \int_{-\sqrt{2}}^{\sqrt{2}} (4 - x^2 - x^2)\, dx = \tfrac{16}{3}\sqrt{2}.$$

Example 3 Find the volume of the solid S under the surface

$$z = 4 - x^2 - 4y^2$$

and above the region T of the xy plane bounded by $x = 0$, $y = 0$, and $x + 2y - 2 = 0$ (Figure 16.10).

Solution: The volume $V(S)$ of this solid is given by

$$V(S) = \int_T z.$$

Using 16.19, we evaluate this double integral by a repeated integral

$$V(S) = \int_0^1 dy \int_0^{2-2y} (4 - x^2 - 4y^2)\, dx.$$

For a given y the inner integral

$$\int_0^{2-2y} (4 - x^2 - 4y^2)\, dx = 8(1 - y) - \tfrac{8}{3}(1 - y)^3 - 8y^2 + 8y^3$$

is the area of a section of the solid S parallel to the xz plane, as indicated in Figure 16.10. Then the volume can be interpreted as the sum of the volumes of all slices between $y = 0$ and $y = 1$. Continuing the integration, we obtain

$$V(S) = \int_0^1 [8(1 - y) - \tfrac{8}{3}(1 - y)^3 - 8y^2 + 8y^3]\, dy = \tfrac{8}{3}.$$

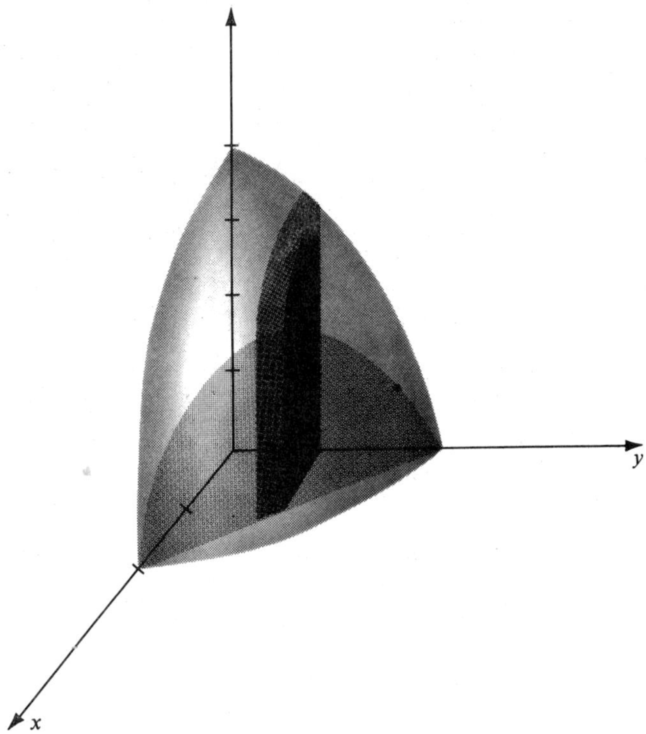

FIGURE 16.10

Example 4 Find the volume of the solid S under the plane $z = x + 2y$ and over the quarter-circle of radius 2 in the first quadrant of the xy plane (Figure 16.11).

Solution: The quarter-circle may be described as the region T bounded by the curves $y = 0$ and $y = \sqrt{4 - x^2}$ between $x = 0$ and $x = 2$. Hence the volume of solid S under the plane $z = x + 2y$ and above T is given by 16.18,

$$V(S) = \int_0^2 dx \int_0^{\sqrt{4-x^2}} (x + 2y)\, dy$$

$$= \int_0^2 (x\sqrt{4 - x^2} + 4 - x^2)\, dx = \left[-\tfrac{1}{3}(4 - x^2)^{3/2} + 4x - \tfrac{1}{3}x^3\right]\Big|_0^2 = 8.$$

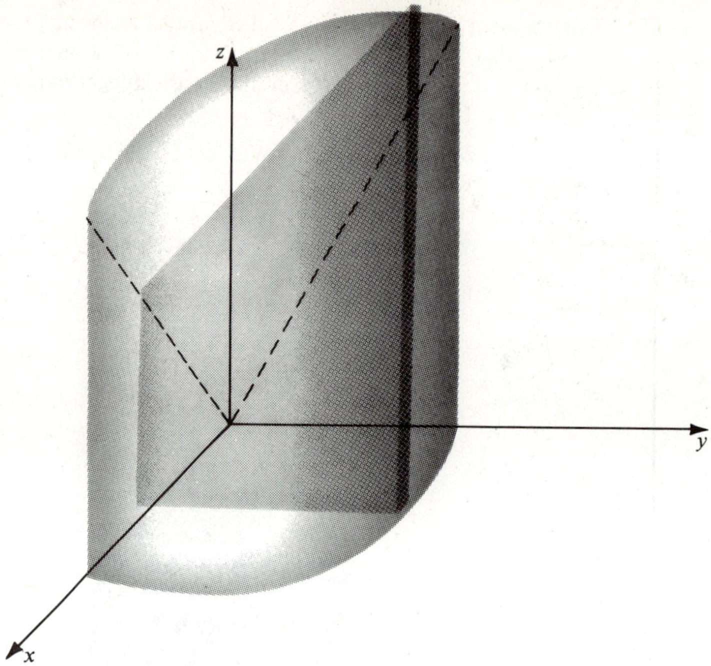

FIGURE 16.11

Again the inner integral is the area of a section of S, this time parallel to the yz plane, as indicated in Figure 16.11, and the outer integral sums up the volumes of the slices from $x = 0$ to $x = 2$.

Example 5 Compute $\int_0^4 dy \int_{y/2}^2 e^{x^2}\, dx$.

Solution: Since we cannot find an antiderivative of e^{x^2} with respect to x, the first integration does not seem feasible. Observing the limits on x, we see that the first integration runs from $x = y/2$ to $x = 2$. Hence the repeated integral extends over the region T shown in Figure 16.12.

Now let us set up the repeated integral with the *order of integration reversed.* We obtain

$$\int_0^2 dx \int_0^{2x} e^{x^2}\, dy = \int_0^2 [ye^{x^2}]\Big|_0^{2x} dx$$

$$= \int_0^2 2xe^{x^2}\, dx = e^{x^2}\Big|_0^2 = e^4 - 1.$$

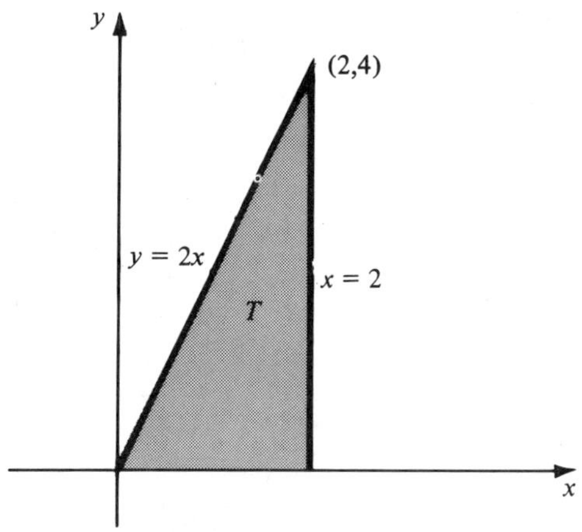

FIGURE 16.12

Example 6 Compute the double integral $\int_T f$, where $f(x,y) = x$ and T is the shaded region in Figure 16.13 bounded by the curves $x = -1$, $x = (y + 1)^2$, $y = -2$ and $y = \sin \pi x$.

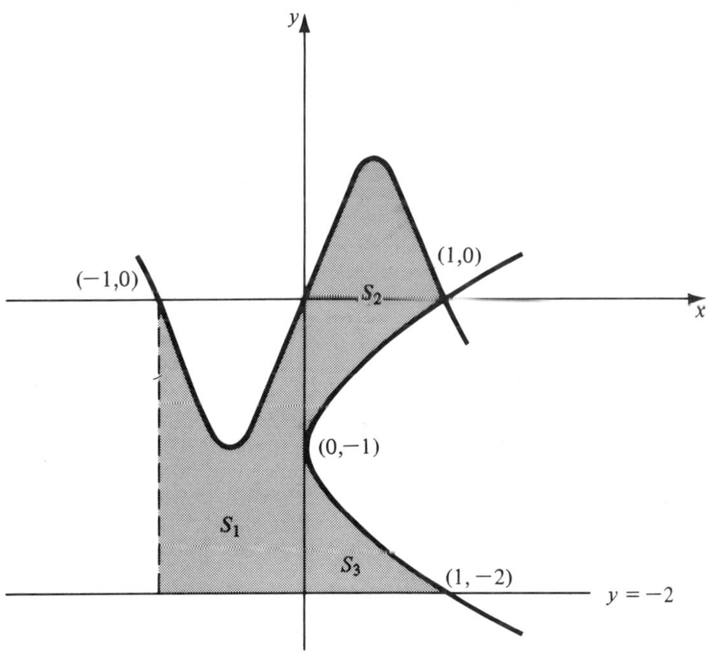

FIGURE 16.13

Solution: The region T is not a region of either of the standard types I or II. However, T may be decomposed into three regions S_1, S_2, and S_3, indicated in Figure 16.13, each of which is of type I. Specifically,

$$S_1 = \{(x,y) \mid -1 \le x \le 0, \ -2 \le y \le \sin \pi x\},$$
$$S_2 = \{(x,y) \mid 0 \le x \le 1, \ \sqrt{x} - 1 \le y \le \sin \pi x\},$$
$$S_3 = \{(x,y) \mid 0 \le x \le 1, \ -2 \le y \le -\sqrt{x} - 1\}.$$

Hence,

$$\int_T f = \int_{S_1} f + \int_{S_2} f + \int_{S_3} f$$
$$= \int_{-1}^{0} dx \int_{-2}^{\sin \pi x} x\, dy + \int_0^1 dx \int_{\sqrt{x}-1}^{\sin \pi x} x\, dy + \int_0^1 dx \int_{-2}^{-\sqrt{x}-1} x\, dy.$$

Each of the above repeated integrals may be readily evaluated. The details are left for the student.

EXERCISES

I

In each of Exercises 1 to 10, by the use of repeated integrals, find the double integral of the function over the given region S of the plane, and sketch the region S.

1. $f(x,y) = 2x - y + 4$; S bounded by $x = 1, x = 4, y = -1, y = 2$
2. $g(x,y) = 3x - y + 2xy$; S bounded by $x = -2, x = 0, y = 0, y = 3$
3. $F(x,y) = y \sin x - xe^y$; S bounded by $x = \pi/2, x = 0, y = -1, y = 1$
4. $G(x,y) = \dfrac{y^2}{1 + x^2}$; S bounded by $x = -1, x = 1, y = 0, y = 2$
5. $g(x,y) = 3x - y + 1$; S bounded by $x = 0, y = 0, x + 3y - 3 = 0$
6. $f(x,y) = 2x + 2y - 1$; S bounded by $x = y, x + y - 4 = 0, x = 0$
7. $G(x,y) = xy - x^2 + 1$; S bounded by $x - 2y + 2 = 0, x + 3y - 3 = 0, y = 0$
8. $F(x,y) = 2x^2y - x + 3$; S bounded by $3y = x, x + y - 4 = 0, y = -1$
9. $f(x,y) = \dfrac{2y - 1}{x^2 - 1}$; S bounded by $y = 4 - x^2, y = 0$
10. $g(x,y) = 2xy - 3x^2$; S bounded by $y = \ln |x|, y = 0, y = -2$

In each of Exercises 11 to 16, by the use of repeated integrals find the area of the region bounded by the given curves.

11. $x = y^2, x - y - 2 = 0$
12. $y = x^3 - x, x = -1, 2x + y - 2 = 0$
13. $y = \sin \pi x, y = x^2 - x$

14. $y = \ln x,\ 2x + y - 2 = 0,\ y = 2$
15. $y = |x|,\ 4y = 4x^2 + 1$
16. $x^2 + 4y^2 = 16,\ x^2 = 12y$ (smaller region)

17. Find the volume of the solid under the plane $z = 2x + y + 1$ and above the region bounded by $x = 0$, $y = 0$, and $x + 3y - 3 = 0$. Sketch the solid, and describe an alternate way of finding its volume.
18. Find the volume of the solid under the plane $x - z + y + 2 = 0$ and above the region bounded by $y = x$, $x + 2y = 2$, and $y = 1$. Sketch.
19. Find the volume of the solid under the plane $z = 3x + y$ and above the part of the ellipse $4x^2 + 9y^2 \leq 36$ in the first quadrant. Sketch.
20. Find the volume of the solid under the plane $z = 2x + y + 10$ and above the circle $x^2 + y^2 \leq 16$. Sketch.
21. Find the volume of the solid under the plane $z = 2y$ and above the region bounded by $y = x^2$, $y = 0$, and $x = 2$. Sketch.
22. Find the volume of the solid in the first octant bounded by the cylinder $y^2 + z^2 = 4$, the plane $y = x$, and the xy plane. Sketch.
23. Find the volume of the solid in the first octant bounded by the paraboloid $z = 16 - x^2 - 4y^2$. Sketch.
24. Find the volume of the solid in the first octant bounded by the cylinder $z = x^2$ and the planes $x = 2y$ and $x = 2$. Sketch.
25. Find the volume of the solid in the first octant bounded by the cylinder $x^2 + y^2 = 9$ and the planes $y = 0$, $z = 0$, $z = x$. Sketch.
26. Find the volume of the solid in the first octant bound by the two cylinders $z = 4 - x^2$ and $y = 4 - x^2$. Sketch.

In each of Exercises 27 to 32 describe and sketch a solid whose volume is given by the repeated integral, and find each volume.

27. $\int_0^2 dx \int_0^3 x\, dy$

28. $\int_1^3 dy \int_1^4 (2 + x + y)\, dx$

29. $\int_0^r dy \int_0^y \sqrt{r^2 - y^2}\, dx$

30. $\int_0^1 dx \int_{-3x}^{\sqrt{1-x^2}} (3x + y)\, dy$

31. $\int_0^2 dx \int_0^{\sqrt{4-x^2}} (x^2 + 4y^2)\, dy$

32. $\int_0^2 dy \int_{2-y}^{2y+2} (x^2 + 4y^2)\, dx$

33. Express each of the repeated integrals in Exercises 29 to 32 in terms of repeated integrals with the order of integration reversed.

II

1. Set up a repeated integral for the volume of the ellipsoid $x^2/a^2 + y^2/b^2 + z^2/c^2 = 1$.
2. Show that the volume V of the solid bounded by the graph of the equation $\sqrt{x/a} + \sqrt{y/b} + \sqrt{z/c} = 1$ and the coordinate planes is given by $V = abc/90$.

3. Show that the volume V of the solid bounded by the graph of the equation $x^{2/3} + y^{2/3} + z^{2/3} = a^{2/3}$ is given by $V = 4\pi a^3/35$.
4. Use repeated integrals to find the volume of the solid common to two right circular cylinders of radius r whose axes intersect orthogonally.

6 POLAR COORDINATES

Suppose that f is a real-valued function defined for each point P in some region S of R^2, and suppose that its value $f(P)$ at each point is given in terms of the polar coordinates (r,θ) of P. It is then natural to ask how the integral $\int_S f$ may be computed, and in particular, how it may be represented as a repeated integral. We shall provide an answer to this question in this section, although our discussion will be only on a very intuitive level. A more precise treatment of a more general "change of variable" theorem, including polar coordinates as a special case, will be given in Section 4 of Chapter 17.

We start with the simplest case when S is the region in the polar coordinate plane bounded by the lines $\theta = \alpha$ and $\theta = \beta$ and by the circles $r = a$ and $r = b$. (This is the analog of a rectangle in the xy plane, with sides parallel to the axes.) The region S may be divided into mn subregions by a partition

$$p = \{S_{ij} \mid i = 1, 2, \ldots, n, j = 1, 2, \ldots, m\},$$

where the i,jth subregion S_{ij} of p is bounded by the lines $\theta = \theta_{i-1}$ and $\theta = \theta_i$ and by the circles $r = r_{j-1}$ and $r = r_j$ (Figure 16.14). The area $A(S_{ij})$ of S_{ij} is given by

$$A(S_{ij}) = \tfrac{1}{2}\,\Delta\theta_i(r_j^2 - r_{j-1}^2) = \bar{r}_j\,\Delta r_j\,\Delta\theta_i,$$

where $\Delta\theta_i = \theta_i - \theta_{i-1}$, $\Delta r_j = r_j - r_{j-1}$, and $\bar{r}_j = (r_{j-1} + r_j)/2$.

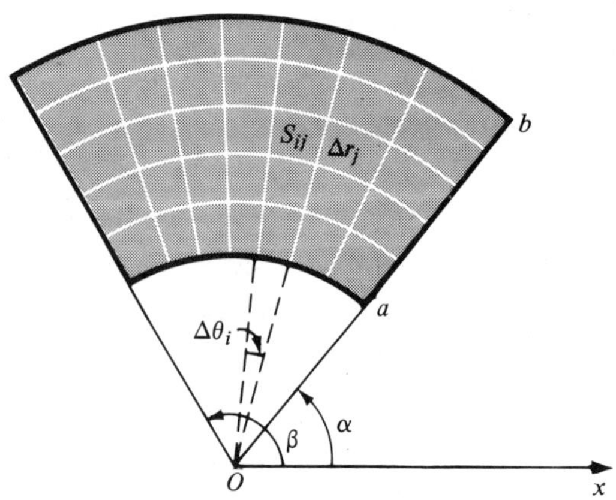

FIGURE 16.14

If f is a continuous function of two variables in S, then relative to a partition p of S we can define a Riemann sum s_p of f as follows:

$$S_p = \sum_{i=1}^{n} \sum_{j=1}^{m} f(r'_j,\theta'_i)\bar{r}_j\ \Delta r_j\ \Delta\theta_i,$$

where (r'_j,θ'_i) is any point of S_{ij}. By taking a sequence of partitions of S with norms approaching 0, the limit of the corresponding sequence of Riemann sums of f will be the double integral of f over S, just as for a rectangular coordinate plane. In turn, this double integral of f over S may be shown to equal a repeated integral having one of two possible forms:

16.20 $$\int_S f = \int_\alpha^\beta d\theta \int_a^b f(r,\theta)r\ dr = \int_a^b dr \int_\alpha^\beta f(r,\theta)r\ d\theta.$$

Note that the integrand in each repeated integral is not just $f(r,\theta)$, but r times $f(r,\theta)$.

The double integral of a continuous function f of two variables can be defined over regions of the polar coordinate plane other than the ones described above. If, for example, the region S is bounded by the graphs of the smooth curves $r = g_1(\theta)$ and $r = g_2(\theta)$ and by the lines $\theta = \alpha$ and $\theta = \beta$ (Figure 16.15), where $g_1(\theta) \leq g_2(\theta)$ in $[\alpha,\beta]$, then it may be shown that the double integral of f over S exists and equals a repeated integral as follows:

16.21 $$\int_S f = \int_\alpha^\beta d\theta \int_{g_1(\theta)}^{g_2(\theta)} f(r,\theta)r\ dr.$$

If the region S is bounded by the smooth curves $\theta = h_1(r)$ and $\theta = h_2(r)$ and by the circles $r = a$ and $r = b$, as in Figure 16.16, where $h_1(r) \leq h_2(r)$ in $[a,b]$, then

16.22 $$\int_S f = \int_a^b dr \int_{h_1(r)}^{h_2(r)} f(r,\theta)r\ d\theta.$$

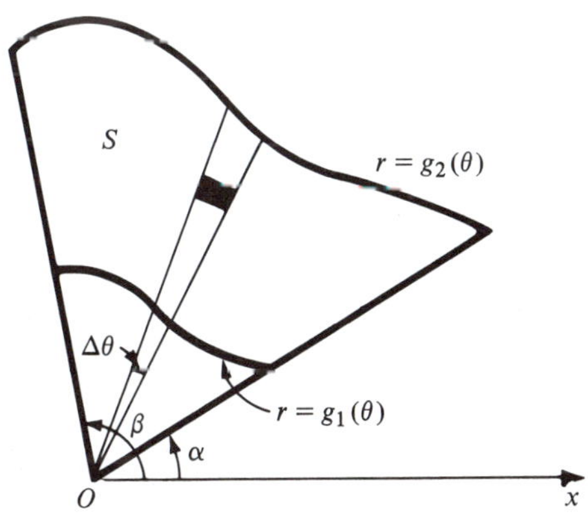

FIGURE 16.15

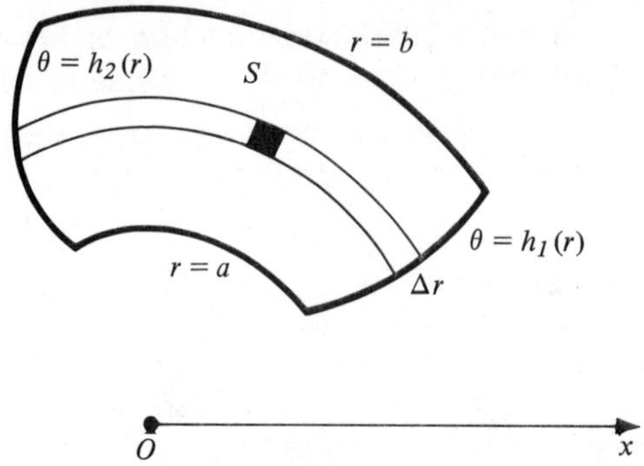

FIGURE 16.16

We may figure out which of the above evaluations of a double integral of f over a region S to use by a device similar to that used for a rectangular coordinate plane. Thus, if we can cut S into strips by lines drawn through the pole, then

$$\Delta\theta \int_{g_1(\theta)}^{g_2(\theta)} f(r,\theta)r\, dr$$

is approximately equal to the double integral of f over the strip at θ (shown in Figure 16.15), where $g_1(\theta)$ and $g_2(\theta)$ are the values of r at each end of the strip. On summing up the double integrals over all the strips and taking a limit, we obtain 16.21.

If the region S can be cut into strips by circles with centers at the pole, as in Figure 16.16, then the double integral of f over S will be evaluated by 16.22. In many examples either 16.21 or 16.22 may be used.

The double integral of a function f over a region S of a polar coordinate plane may be interpreted as a volume of a solid in a cylindrical coordinate space. Thus it may be shown that the volume of the solid under the surface

$$z = f(r,\theta)$$

and above the region S of the polar coordinate plane is given by

$$V = \int_S f.$$

Example 1 Find the volume of the solid under the surface $z = 4 - r^2$ and above the region S bounded by the circle $r = 1$.

Solution: The surface $z = 4 - r^2$ is a paraboloid of revolution about the z axis, so that the solid is a circular cylinder with a parabolic top. Its volume is given by

$$\begin{aligned} V &= \int_S z \\ &= \int_0^1 dr \int_0^{2\pi} (4 - r^2)r\, d\theta \\ &= 2\pi \int_0^1 (4r - r^3)\, dr = \tfrac{7}{2}\pi. \end{aligned}$$

Example 2 Find the volume of the solid bounded by the cylinder $r = 2\cos\theta$, the cone $z = r$, $r \geq 0$, and the plane $z = 0$.

Solution: The solid in question is under the surface $z = r$, $r \geq 0$, and above the circular region S bounded by $r = 2\cos\theta$. If we imagine the region S cut into strips by lines through the pole, then we may evaluate the volume of this solid as follows:

$$V = \int_S z = \int_{-\pi/2}^{\pi/2} d\theta \int_0^{2\cos\theta} r^2\,dr = \tfrac{8}{3}\int_{-\pi/2}^{\pi/2} \cos^3\theta\,d\theta = \tfrac{32}{9}.$$

Example 3 Use polar coordinates to compute $\int_S f$, if $f(x,y) = \dfrac{1}{1 + x^2 + y^2}$ and S is the closed disk of radius a centered at the origin.

Solution: In terms of polar coordinates r and θ,

$$S = \{(r,\theta) \mid 0 \leq r \leq a,\ 0 \leq \theta \leq 2\pi\}.$$

Hence, by 16.20,

$$\begin{aligned}\int_S f &= \int_0^{2\pi} d\theta \int_0^a \frac{r}{1 + r^2}\,dr \\ &= \tfrac{1}{2}\int_0^{2\pi} \ln(1 + r^2)\Big|_0^a\,d\theta \\ &= \tfrac{1}{2}\int_0^{2\pi} \ln(1 + a^2)\,d\theta = \pi\ln(1 + a^2).\end{aligned}$$

EXERCISES

I

Find the volume of the solid bounded by the following surfaces.

1. The cylinder $r = 2$ and the sphere $z^2 + r^2 = 9$.
2. The cylinder $r = 2\sin\theta$, the paraboloid $z = 4 - r^2$, and the plane $z = 0$.
3. The ellipsoid $z^2 + 4r^2 = 4$.
4. The ellipsoid $z^2 + 4r^2 = 4$ and the cylinder $r = \cos\theta$.
5. The cone $z = 2r$, $r \geq 0$, the cylinder $r = 1 - \cos\theta$, and the plane $z = 0$.
6. The cylinder $r = 1 + \cos\theta$, and the planes $z = r\cos\theta$ and $z = 0$, in the first octant.

In Exercises 7 to 11, use polar coordinates to compute $\int_S f$, for the given function f and region S.

7. $f(x,y) = \dfrac{1}{x^2 + y^2}$; S is the region in the first quadrant between the circles $x^2 + y^2 = a^2$ and $x^2 + y^2 = b^2$, $0 < a < b$.

8. $f(x,y) = \dfrac{x + y}{\sqrt{x^2 + y^2}}$; S is the region of Exercise 7.

9. $f(x,y) = x^2 + y^2$; S is the closed disk of radius a centered at (0,0).

10. $f(x,y) = xy$; S is the region of Exercise 9.

11. $f(x,y) = \dfrac{x^2}{(x^2 + y^2)^2}$; S is the region of Exercise 7.

II

1. Find the volume of the solid drilled out of a sphere of radius R by a bit of radius a if the axis of the bit passes through the center of the sphere.

7 CENTER OF MASS OF A LAMINA

Single integrals were used in Chapter 10 to find the center of mass of a homogeneous lamina. We shall now employ double integrals to find the center of mass of any lamina, whether or not it is homogeneous.

Let a given lamina have the shape of region S in a rectangular coordinate plane, and let ρ be the density function of the lamina. Thus ρ is a continuous function of two variables in S, and the mass of any piece of the lamina of area ΔS lies between $\rho_m \, \Delta S$ and $\rho_M \, \Delta S$, where ρ_m is the minimum value and ρ_M is the maximum value of ρ in the piece. It is clear by the usual arguments that the mass W of the lamina is given by

$$W = \int_S \rho.$$

The moment of a piece of the lamina about the x axis is between $y_m\rho_m \, \Delta S$ and $y_M\rho_M \, \Delta S$, where y_m is the minimum and y_M the maximum y coordinate of any point in the piece, and ρ and ΔS are as described above. Hence the moment about the x axis of the whole lamina is given by

$$M_x = \int_S y\rho,$$

and, similarly, its moment about the y axis is

$$M_y = \int_S x\rho.$$

The center of mass of the lamina is the point $(\bar{x},\bar{y})$, where

$$\bar{x} = \frac{M_y}{W}, \qquad \bar{y} = \frac{M_x}{W}.$$

Each of the double integrals above may be evaluated by a repeated integral. We illustrate the method in the following examples.

Example 1 Find the center of mass of a homogeneous lamina (of constant density ρ) having the shape of the region bounded by the parabola $y = 2 - 3x^2$ and the line $3x + 2y - 1 = 0$ (Figure 16.17).

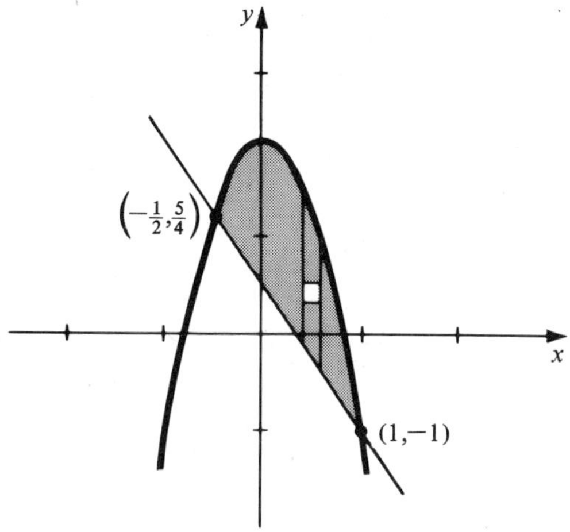

FIGURE 16.17

Solution: It is easily verified that the parabola and the line meet at the points $(-\frac{1}{2},\frac{5}{4})$ and $(1,-1)$. Integrating first with respect to y, we have

$$W = \int_{-1/2}^{1} dx \int_{(1-3x)/2}^{2-3x^2} \rho\, dy = \rho \int_{-1/2}^{1} [2 - 3x^2 - \tfrac{1}{2}(1 - 3x)]\, dx = \tfrac{27}{16}\rho,$$

$$M_x = \int_{-1/2}^{1} dx \int_{(1-3x)/2}^{2-3x^2} \rho y\, dy = \frac{\rho}{2} \int_{-1/2}^{1} [(2 - 3x^2)^2 - \tfrac{1}{4}(1 - 3x)^2]\, dx = \tfrac{27}{20}\rho,$$

$$M_y = \int_{-1/2}^{1} dx \int_{(1-3x)/2}^{2-3x^2} \rho x\, dy = \rho \int_{-1/2}^{1} [2x - 3x^3 - \tfrac{1}{2}(x - 3x^2)]\, dx = \tfrac{27}{64}\rho.$$

Hence $\bar{x} = M_y/W = \frac{1}{4}$, $\bar{y} = M_x/W = \frac{4}{5}$, and the center of mass of the lamina is the point $(\frac{1}{4},\frac{4}{5})$.

Example 2 Find the center of mass of a rectangular lamina $ABCD$ if the density of the lamina at any point P is the product of the distances of P from AB and BC.

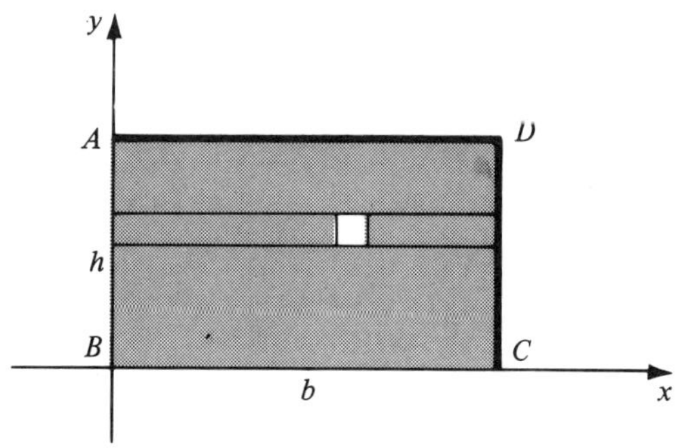

FIGURE 16.18

Solution: Let us choose coordinate axes as in Figure 16.18. By assumption, the density $\rho(x,y)$ at point (x,y) is given by

$$\rho(x,y) = xy.$$

Hence

$$W = \int_0^h dy \int_0^b xy\, dx = \frac{b^2}{2} \int_0^h y\, dy = \frac{b^2h^2}{4},$$

$$M_x = \int_0^h dy \int_0^b xy^2\, dx = \frac{b^2}{2} \int_0^h y^2\, dy = \frac{b^2h^3}{6},$$

$$M_y = \int_0^h dy \int_0^b x^2y\, dx = \frac{b^3}{3} \int_0^h y\, dy = \frac{b^3h^2}{6}.$$

Hence $(\bar{x},\bar{y}) = (\frac{2}{3}b,\frac{2}{3}h)$ is the center of mass.

The formulas for W, M_x, and M_y hold for a polar coordinate system, M_x being the moment of the lamina about the polar axis and M_y being the moment of the lamina about the line perpendicular to the polar axis at the pole. Thus, for a lamina having the shape of a region S of the polar coordinate plane and density function ρ,

$$W = \int_S \rho, \qquad M_x = \int_S \rho r \sin\theta, \qquad M_y = \int_S \rho r \cos\theta.$$

These integrals are equal to certain repeated integrals, as shown in the previous section.

Example 3 Find the center of mass of a semicircular lamina if the density of the lamina at any point P is proportional to the distance between P and the center of the circle.

Solution: Let us choose a polar coordinate system as in Figure 16.19, with the semicircle having equation

$$r = a, \qquad 0 \le \theta \le \pi.$$

The density $\rho(r,\theta)$ at point (r,θ) is given by

$$\rho(r,\theta) = kr$$

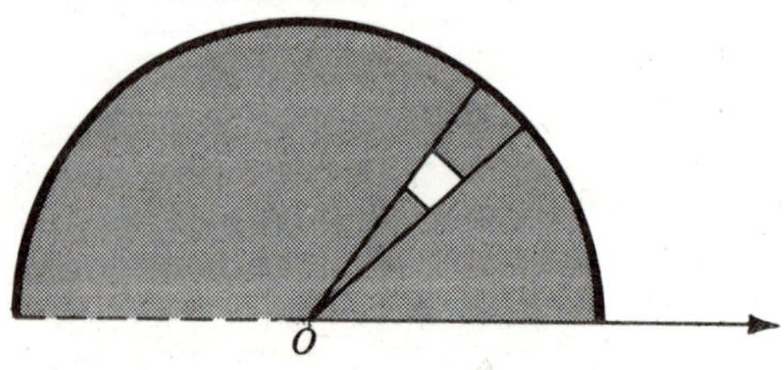

FIGURE 16.19

for some constant k. Hence (by 16.21)

$$W = \int_0^\pi d\theta \int_0^a (kr)r\,dr = \frac{ka^3}{3}\int_0^\pi d\theta = \frac{\pi k a^3}{3},$$

$$M_x = \int_0^\pi d\theta \int_0^a (kr)(r\sin\theta)r\,dr = \frac{ka^4}{4}\int_0^\pi \sin\theta\,d\theta = \frac{ka^4}{2},$$

$$M_y = \int_0^\pi d\theta \int_0^a (kr)(r\cos\theta)r\,dr = \frac{ka^4}{4}\int_0^\pi \cos\theta\,d\theta = 0,$$

and $(0, 3a/2\pi)$ in rectangular coordinates, or $(3a/2\pi, \pi/2)$ in polar coordinates, is the center of mass.

EXERCISES

I

In each of Exercises 1 to 10, find the center of mass of a lamina having density function ρ and the shape of a region bounded by the given curves. Sketch each region.

1. $y = \sqrt{x}$, $y = 0$, $x = 4$; ρ a positive constant
2. $y = \sqrt{x}$, $y = 0$, $x = 4$; $\rho(x,y) = x + y$
3. $y = 0$, $y = h$, $x = 0$, $x = b$; $\rho(x,y) = kx$, h, b, k, positive constants
4. $y = \ln x$, $y = 0$, $x = e$; $\rho(x,y) = y$
5. $r = a$, $\theta = 0$, $\theta = \frac{1}{2}\pi$; $\rho(r,\theta) = \theta$, a a positive constant
6. $r = 1 + \cos\theta$, $\theta = 0$, $\theta = \frac{1}{2}\pi$; $\rho(r,\theta) = r\sin\theta$
7. $x^2 - y^2 = 1$, $x = 3$; $\rho(x,y) = x$
8. $y = \sin x$, $y = 0$, $x = 0$, $x = \pi$; $\rho(x,y) = ky$, k a positive constant
9. $r = \cos 2\theta$, $\theta = -\frac{1}{4}\pi$, $\theta = \frac{1}{4}\pi$; $\rho(r,\theta) = r$
10. $r = \cos 2\theta$, $\theta = 0$, $\theta = \frac{1}{4}\pi$; $\rho(r,\theta) = r\theta$
11. The density at a point P of a triangular lamina with base of length b and altitude h is proportional to the distance of P from the base. Find the distance from the base to the center of mass of the lamina.
12. A lamina has the shape of the region cut off from a parabola by its latus rectum. If the density at point P of the lamina is proportional to the distance of P from the latus rectum, find the center of mass of the lamina.
13. The density of a semicircular lamina at any point P is proportional to the square of the distance of P from the center of the circle. Find the center of mass of the lamina.
14. The density of a lamina in the shape of a quarter of an ellipse (bounded by the semimajor and semiminor axes) at any point P equals the sum of its distances from the axes. Find the center of mass of the lamina.

II

In Exercises 1 and 2, find the center of mass of a lamina having density function ρ and the shape of a region bounded by the given curves. Sketch each region.

1. $\sqrt{x} + \sqrt{y} = \sqrt{a}$, $x = 0$, $y = 0$; $\rho(x,y) = xy$
2. $x^{2/3} + y^{2/3} = a^{2/3}$, $x = 0$, $y = 0$; $\rho(x,y) = x^2y^2$

8 MOMENTS OF INERTIA

If a particle of mass m is d units from a line L (Figure 16.20), then the number md^2 is called the *moment of inertia* of the particle about L. The moment of a particle studied in Section 7 is frequently called the *first moment*, and the moment of inertia the *second moment*, of the particle about L.

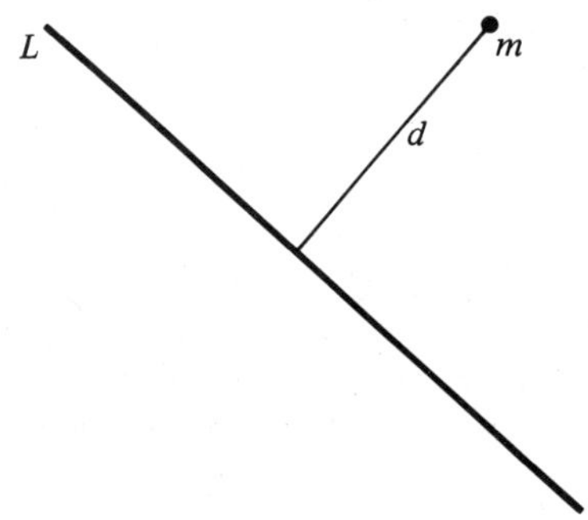

FIGURE 16.20

A system of n particles of masses $m_1, m_2, \ldots, m_n$ and at distances $d_1, d_2, \ldots, d_n$ units, respectively, from a line L has a moment of inertia I defined as the sum of the moments of the individual particles:

$$I = \sum_{i=1}^{n} m_i d_i^2.$$

It is clear that by our usual limiting process the moment of inertia of a lamina having the shape of a plane region S and density function ρ can be found about any line L. In particular, it is clear that the moments of inertia I_x and I_y of the lamina about the x and y axes, respectively, are given by

$$I_x = \int_S \rho y^2, \qquad I_y = \int_S \rho x^2.$$

Example 1 Find I_x and I_y for the homogeneous lamina having the shape of the region S bounded by the curve $y = \sqrt{x}$ and by the lines $y = 0$ and $x = 4$ (Figure 16.21).

Solution: Since ρ is a constant by assumption, we have

$$I_x = \int_S \rho y^2 = \int_0^4 dx \int_0^{\sqrt{x}} \rho y^2\, dy = \frac{\rho}{3}\int_0^4 x^{3/2}\, dx = \tfrac{64}{15}\rho,$$

$$I_y = \int_S \rho x^2 = \int_0^4 dx \int_0^{\sqrt{x}} \rho x^2\, dy = \rho \int_0^4 x^{5/2}\, dx = \tfrac{256}{7}\rho.$$

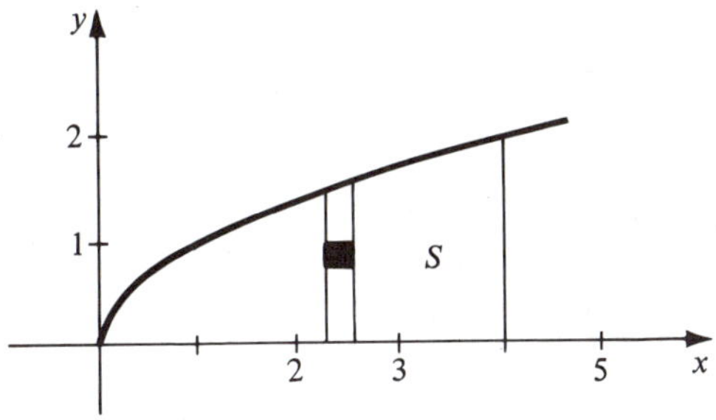

FIGURE 16.21

The physical meaning of the moment of inertia is to be found in the study of the *kinetic energy* of a moving particle. A particle of mass m moving at a speed v has kinetic energy

$$K = \tfrac{1}{2}mv^2.$$

If a particle of mass m at a distance of d units from a line L is rotating about L with an angular velocity of ω radians per unit of time, then $|\omega d|$ is the speed of the particle and

$$K = \tfrac{1}{2}m(\omega d)^2 = \tfrac{1}{2}I\omega^2$$

is its kinetic energy, where I is the moment of inertia of the particle. Noting the similarity between $\frac{1}{2}I\omega^2$ and $\frac{1}{2}mv^2$, we can say that I is the "rotation mass" of the particle. Similarly, if I is the moment of inertia of a system of particles about L, then $\frac{1}{2}I\omega^2$ is again the kinetic energy of the system when rotated about L with an angular velocity of ω.

A lamina (or system of particles) of total mass m and moment of inertia I about a line L has *radius of gyration* d defined by the equation

$$I = md^2.$$

Accordingly, a particle of mass m located d units from L has the same moment of inertia as the given lamina; i.e., in computing the moment of inertia of the lamina, the mass of the lamina may be considered to be concentrated at a point d units from L.

Example 2 Find the radius of gyration of a semicircular lamina about its diameter, if the density of the lamina at a point is proportional to the distance of the point from the diameter.

Solution: If we select the coordinate axes as in Figure 16.22, the density of the lamina at point (x,y) is given by $\rho(x,y) = ky$ (k a positive constant). Evidently,

$$I_x = \int_S \rho y^2 = \int_{-a}^{a} dx \int_0^{\sqrt{a^2-x^2}} ky^3\, dy = \frac{k}{4}\int_{-a}^{a} (a^2 - x^2)^2\, dx = \tfrac{4}{15}ka^5.$$

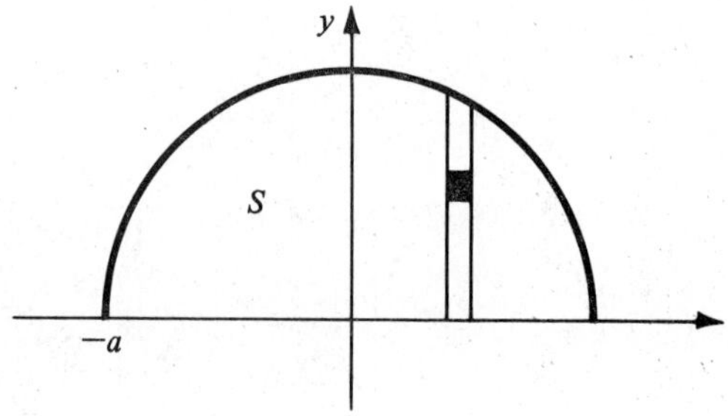

FIGURE 16.22

The mass m of the lamina is given by

$$m = \int_S \rho = \int_{-a}^{a} dx \int_0^{\sqrt{a^2-x^2}} ky\, dy = \frac{k}{2}\int_{-a}^{a} (a^2 - x^2)\, dx = \tfrac{2}{3}ka^3.$$

Hence the radius of gyration d of the lamina satisfies the equation

$$\tfrac{4}{15}ka^5 = (\tfrac{2}{3}ka^3)d^2,$$

from which we conclude that $d = (\sqrt{10}/5)a \doteq .63a$.

If a lamina has the shape of a region S of the polar coordinate plane, then its moment of inertia about the polar axis is given by

$$I_x = \int_S \rho r^2 \sin^2\theta,$$

where ρ is the density function of the lamina, and its moment of inertia about the axis perpendicular to the polar axis at the pole (the "y axis") is given by

$$I_y = \int_S \rho r^2 \cos^2\theta.$$

Since the distance of a point (x,y) in the xy plane from the z axis is $\sqrt{x^2 + y^2}$, the moment of inertia of a particle of mass m at (x,y) about the z axis is $m(x^2 + y^2)$. Similarly, a lamina having the shape of a region S of the xy plane and density function ρ will have

$$I_z = \int_S \rho(x^2 + y^2)$$

as its moment of inertia about the z axis. Clearly,

$$I_z = I_x + I_y.$$

If the region S is in the polar coordinate plane of a cylindrical coordinate system then

$$I_z = \int_S \rho r^2.$$

Example 3 Find I_z for the lamina of Example 2.

Solution: We are assuming that the z axis is perpendicular to the plane of the semicircle at the origin. In polar coordinates the semicircle has equation $r = a$, $0 \le \theta \le \pi$, and $\rho = kr \sin\theta$. Hence

$$I_z = \int_S \rho r^2 = k \int_0^{\pi} d\theta \int_0^a r^4 \sin\theta \, dr = \tfrac{2}{5}ka^5.$$

Example 4 Find the radius of gyration d of a homogeneous lamina in the shape of a right triangle about an axis perpendicular to the plane of the triangle at the vertex of the right angle.

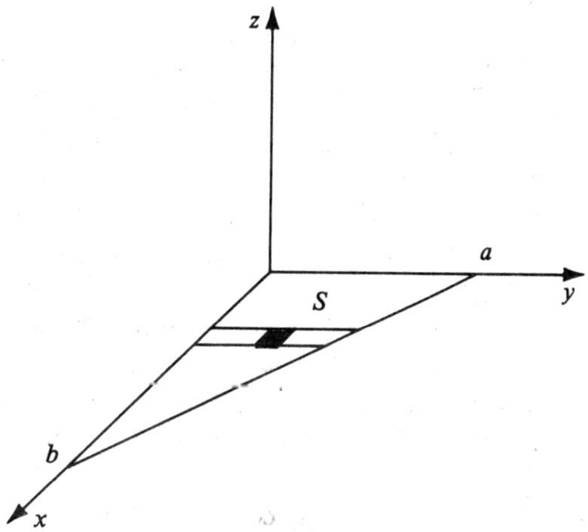

FIGURE 16.23

Solution: If the coordinate axes are chosen as in Figure 16.23, the hypotenuse has equation

$$y = -\frac{a}{b}x + a$$

in the xy plane. Hence (with ρ the constant density)

$$I_z = \int_S \rho(x^2 + y^2) = \rho \int_0^b dx \int_0^{-ax/b+a} (x^2 + y^2)\, dy$$

$$= \rho \int_0^b \left[x^2\left(-\frac{a}{b}x + a\right) + \frac{1}{3}\left(-\frac{a}{b}x + a\right)^3\right] dx = \tfrac{1}{12}\rho ab(a^2 + b^2).$$

The mass of the lamina is $\rho ab/2$, and therefore

$$\tfrac{1}{12}\rho ab(a^2 + b^2) = \tfrac{1}{2}\rho ab\, d^2,$$

and $d = \sqrt{a^2 + b^2}/\sqrt{6}$.

EXERCISES

I

A rectangular lamina $ABCD$ has density function ρ. Find the radius of gyration of the lamina about the following.

1. AB if ρ is a constant
2. AB if the density at a point P is the sum of the distances of P from AB and BC
3. The line perpendicular to the lamina at B if ρ is a constant
4. The line perpendicular to the lamina at B if the density at a point P is the sum of the distances of P from AB and BC
5. The line perpendicular to the lamina at its center of mass if ρ is a constant
6. The line perpendicular to the lamina at its geometric center O if the density at a point P is proportional to the distance $|OP|$

A circular lamina with center O and radius a has density function ρ. Find the radius of gyration of the lamina about the following.

7. A diameter if ρ is a constant
8. A line perpendicular to the lamina at O if ρ is a constant
9. A tangent line if ρ is a constant
10. A diameter if the density at P is proportional to the distance of point P from the diameter
11. A line perpendicular to the lamina at O if the density at P is proportional to the distance of point P from O
12. A tangent line if the density at P is proportional to the distance of P from the point of tangency
13. A lamina has the shape of a triangle with sides of lengths a, b, and c. Assuming ρ is a constant, find the moment of inertia of the lamina about the side of length c.
14. Find the moment of inertia in Exercise 13 when the density at point P is proportional to the distance of P from the side of length c.
15. A homogeneous lamina is bounded by one loop of the curve $r^2 = \cos 2\theta$ in the polar plane. Find its radius of gyration about an axis perpendicular to the polar plane at the pole.
16. A homogeneous lamina is bounded by the curve $r = 1 + \cos\theta$ in the polar plane. Find the radius of gyration of the lamina about an axis perpendicular to the polar plane at the pole.

II

1. Show that the moment of inertia of any plane lamina about an axis in its plane is equal to its moment of inertia about a parallel axis passing through the center of mass plus the product of the mass times the square of the distance between the two axes (theorem of parallel axes). This shows that of all parallel axes the moment of inertia about the one passing through the center of mass is the least.

9 TRIPLE INTEGRALS

A region S of R^3 of the form

$$S = \{(x,y,z) \mid a_1 \leq x \leq b_1, a_2 \leq y \leq b_2, a_3 \leq z \leq b_3\}$$

is a rectangular parallelepiped with edges parallel to the coordinate axes. Let f be a bounded real-valued function defined for all points in S. The methods of Section 1 of this chapter may be used to define an integral, called a *triple integral*, of the function f over S. Since these ideas are probably now familiar to the student, we give only a brief indication of the procedure.

The parallelepiped S above may be partitioned into a set

$$p = \{S_{ijk} \mid i = 1, \ldots, r, j = 1, \ldots, s, k = 1, \ldots, t\}$$

of smaller mutually disjoint parallelepipeds with edges parallel to the axes. Corresponding to each S_{ijk}, we define

$$\begin{aligned} m_{ijk} &= \text{g.l.b.}\ \{f(x,y,z) \mid (x,y,z) \text{ in } S_{ijk}\}, \\ M_{ijk} &= \text{l.u.b.}\ \{f(x,y,z) \mid (x,y,z) \text{ in } S_{ijk}\}. \end{aligned}$$

Corresponding to any given partition p of S, form the sums

$$l(p) = \sum_{i=1}^{r} \sum_{j=1}^{s} \sum_{k=1}^{t} m_{ijk} V(S_{ijk}),$$

$$u(p) = \sum_{i=1}^{r} \sum_{j=1}^{s} \sum_{k=1}^{t} M_{ijk} V(S_{ijk}),$$

where $V(S_{ijk})$ denotes the volume of S_{ijk}. The numbers $l(p)$ and $u(p)$, as usual, are called the *lower* and *upper sums*, respectively, of the function f relative to the given partition p.

Let α denote the g.l.b. of the set of all upper sums of f, and let β denote the l.u.b. of the set of all lower sums of f, taken over all partitions of S. If the numbers α and β are equal, then their common value is called the *triple integral* of f over S, and is denoted by

$$\int_S f.$$

If $\int_S f$ exists, then as usual we say that f is integrable over S.

If f is continuous on S, then it can be proved that f is integrable on S. Actually, using the ideas of Section 2 of this chapter, it is possible to formulate and prove a criterion for the integrability of f that is similar to Theorem 16.13, but we shall not go into these details.

We may also consider *repeated integrals* of f over S. For a function of three variables there are six possible choices of the order of integration, corresponding to the six possible ways in which we may order the variables x, y, and z. One such ordering is

$$\int_{a_1}^{b_1} dx \int_{a_2}^{b_2} dy \int_{a_3}^{b_3} f(x,y,z)\, dz,$$

where we integrate f first with respect to z (holding x and y constant), then with respect to y (holding x constant), and finally with respect to x. It is possible to prove (using methods similar to Theorem 16.16) that, if f is continuous on S, then the six different repeated integrals of f over S all have the same value, which coincides with the triple integral $\int_S f$. Thus, in particular,

$$\int_S f = \int_{a_1}^{b_1} dx \int_{a_2}^{b_2} dy \int_{a_3}^{b_3} f(x,y,z)\, dz.$$

Example 1 If $f(x,y,z) = 3(x^2y + y^2z)$, find the triple integral of f over the rectangular parallelepiped S bounded by the planes $x = 1$, $x = 3$, $y = -1$, $y = 1$, $z = 2$, and $z = 4$.

Solution: We have

$$\begin{aligned}\int_S f &= \int_2^4 dz \int_{-1}^1 dy \int_1^3 3(x^2y + y^2z)\, dx \\ &= \int_2^4 dz \int_{-1}^1 (x^3y + 3xy^2z)\Big|_{x=1}^{x=3} dy \\ &= \int_2^4 dz \int_{-1}^1 (26y + 6y^2z)\, dy = \int_2^4 (13y^2 + 2y^3z)\Big|_{y=-1}^{y=1} dz \\ &= \int_2^4 4z\, dz = 24.\end{aligned}$$

We could equally well have started with

$$\int_S f = \int_{-1}^1 dy \int_2^4 dz \int_1^3 3(x^2y + y^2z)\, dx,$$

or any other one of the six possible threefold repeated integrals.

The triple integral of a continuous function f of three variables can be defined over a region of space other than a rectangular parallelepiped. For example, it can be defined over the region S of Figure 16.24, a region bounded below by the surface $z = h_1(x,y)$, above by $z = h_2(x,y)$, and laterally by the cylinders $y = g_1(x)$ and $y = g_2(x)$ and by the planes $x = a_1$ and $x = a_2$, where the functions involved are smooth. It may be shown that

$$\int_S f = \int_{a_1}^{a_2} dx \int_{g_1(x)}^{g_2(x)} dy \int_{h_1(x,y)}^{h_2(x,y)} f(x,y,z)\, dz.$$

The region S might be oriented differently with respect to the axes, in which case the repeated integral might have to be taken in a different order. If $f(x,y,z) = 1$ throughout the region S, then the triple integral of f over S is simply the volume V of the region S:

$$V = \int_S 1.$$

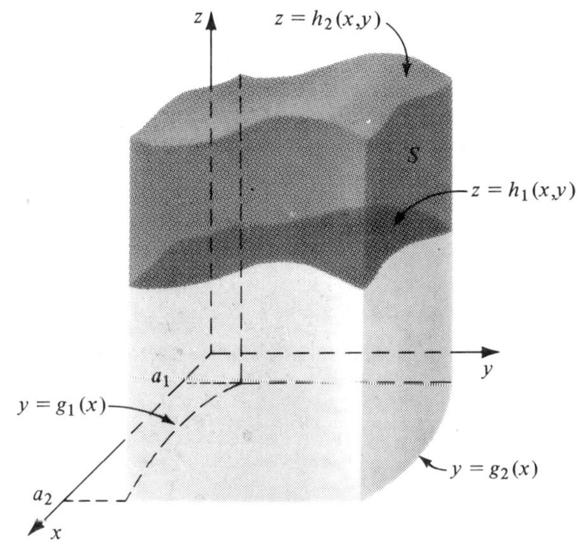

FIGURE 16.24

Example 2 Find the volume of the solid bounded above by the paraboloid $z = 4 - x^2 - y^2$ and below by the plane $z = 4 - 2x$.

Solution: The solid is sketched in Figure 16.25. If we eliminate z between the two given equations, we obtain $4 - 2x = 4 - x^2 - y^2$, or

$$y^2 = 2x - x^2,$$

as the equation of a cylinder containing the curve of intersection of the given paraboloid and plane. Thus $y = -\sqrt{2x - x^2}$ and $y = \sqrt{2x - x^2}$ are the y limits of integration, and the volume of the solid is given by

$$\begin{aligned} V &= \int_0^2 dx \int_{-\sqrt{2x-x^2}}^{\sqrt{2x-x^2}} dy \int_{4-2x}^{4-x^2-y^2} dz \\ &= \int_0^2 dx \int_{-\sqrt{2x-x^2}}^{\sqrt{2x-x^2}} [(4 - x^2 - y^2) - (4 - 2x)]\, dy \\ &= \int_0^2 (-x^2y - \tfrac{1}{3}y^3 + 2xy)\Big|_{y=-\sqrt{2x-x^2}}^{y=\sqrt{2x-x^2}} dx \\ &= \tfrac{4}{3}\int_0^2 (2x - x^2)^{3/2}\, dx = \frac{\pi}{2}. \end{aligned}$$

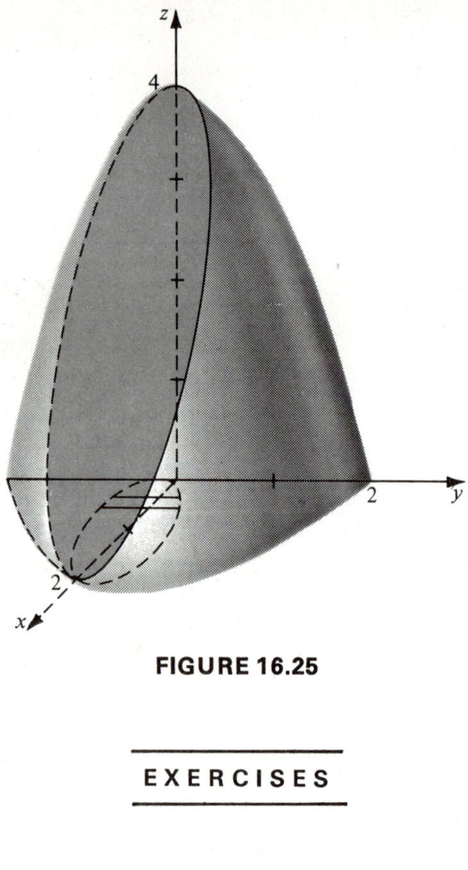

FIGURE 16.25

EXERCISES

I

Find the volume of each of the following regions of space, and also find the value of the triple integral of the given function f over each region. Sketch each region.

1. Region S bounded by the planes $x = -1$, $x = 2$, $y = 0$, $y = 3$, $z = 1$, and $z = 4$; $f(x,y,z) = x - 2y + z$

2. Region S bounded by the planes $x = 0$, $x = 1$, $y = -1$, $y = 2$, $z = 0$, and $z = 5$; $f(x,y,z) = 3xyz$
3. Region S bounded by the cylinder $x^2 + y^2 = 16$ and the planes $z = 0$ and $z = 3$; $f(x,y,z) = xz + yz$
4. Region S bounded by the cylinder $y^2 + z^2 = 9$ and the planes $x = 0$ and $x + z = 3$; $f(x,y,z) = 3y + yz$
5. Region S bounded by the cylinders $x^2 = z$ and $x^2 = 4 - z$, and the planes $y = 0$ and $z + 2y = 4$; $f(x,y,z) = 2x - z$
6. Region S bounded by the cylinder $x = \sqrt{4 + y^2}$ and the planes $z = 0$ and $x + 2z = 4$; $f(x,y,z) = xy$
7. Region S bounded by the surface $z = y/(1 + x^2)$ and the planes $x = 0$, $y = 0$, $z = 0$, and $x + y = 1$; $f(x,y,z) = y + x^2y$
8. Region S bounded by the surface $y^2 + z^2 = 2x$ and the plane $x + y = 1$; $f(x,y,z) = 3z$
9. Region S in the first octant bounded by the cylinders $x^2 + y^2 = a^2$ and $y^2 + z^2 = a^2$; $f(x,y,z) = xyz$
10. Region S bounded by the ellipsoid $x^2/a^2 + y^2/b^2 + z^2/c^2 = 1$; $f(x,y,z) = xz$

II

Follow the same directions as in Part I.

1. Region S bounded by the graphs of the equations $x = 0$, $y = 0$, $z = 0$, and $\sqrt{x/a} + \sqrt{y/b} + \sqrt{z/c} = 1$; $f(x,y,z) = xyz$
2. Region S bounded by the graphs of the equation $(x/a)^{2/3} + (y/b)^{2/3} + (z/c)^{2/3} = 1$; $f(x,y,z) = x^2 + y^2 + z^2$

10 PHYSICAL APPLICATIONS OF TRIPLE INTEGRALS

If a material object has the shape of a region S of space and has a constant density ρ, then the mass W of the object is given by

$$W = \int_S \rho.$$

It could be argued that W is the mass of the object even if ρ is variable, but we shall not consider such a possibility here.

A particle of mass m located at the point (x,y,z) has moments mx, my, and mz with respect to the yz, xz, and xy planes, respectively. Using familiar arguments, we find that the moments with respect to the coordinate planes of a homogeneous material object of density ρ and having the shape of a region S of space are given by

$$M_{xy} = \int_S \rho z, \qquad M_{xz} = \int_S \rho y, \qquad M_{yz} = \int_S \rho x.$$

The center of mass of the object is the point $(\bar{x},\bar{y},\bar{z})$, where

$$\bar{x} = \frac{M_{yz}}{W}, \qquad \bar{y} = \frac{M_{xz}}{W}, \qquad \bar{z} = \frac{M_{xy}}{W}.$$

Example 1 Find the center of mass of a homogeneous material object bounded by the coordinate planes, the plane $x + y = 1$, and the paraboloid

$$z = 4 - x^2 - 4y^2.$$

Solution: The object is sketched in Figure 16.26. Evidently, its mass is given by

$$W = \int_0^1 dx \int_0^{1-x} dy \int_0^{4-x^2-4y^2} \rho \, dz,$$

and we easily show that

$$W = \tfrac{19}{12}\rho.$$

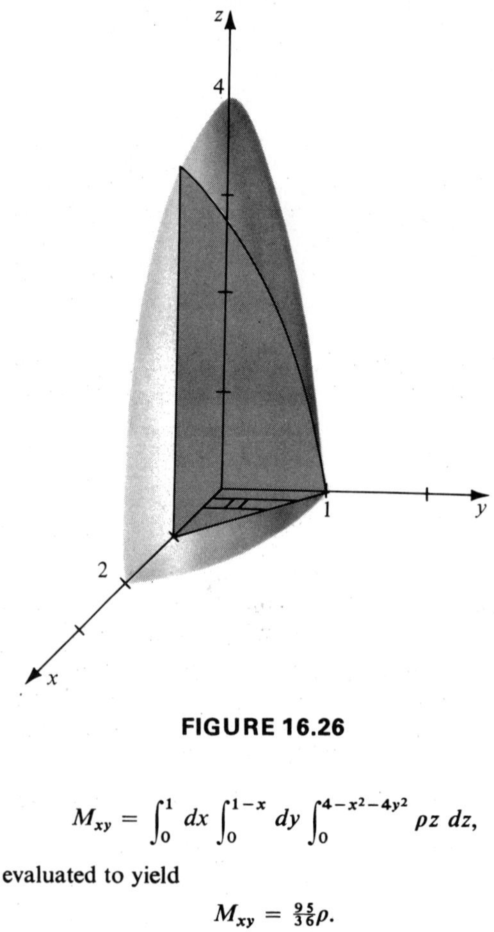

FIGURE 16.26

Also, $$M_{xy} = \int_0^1 dx \int_0^{1-x} dy \int_0^{4-x^2-4y^2} \rho z \, dz,$$

which may be evaluated to yield

$$M_{xy} = \tfrac{95}{36}\rho.$$

Similarly, $$M_{xz} = \int_0^1 dx \int_0^{1-x} dy \int_0^{4-x^2-4y^2} \rho y \, dz = \tfrac{9}{20}\rho,$$

$$M_{yz} = \int_0^1 dx \int_0^{1-x} dy \int_0^{4-x^2-4y^2} \rho x \, dz = \tfrac{11}{20}\rho.$$

Thus $(\frac{33}{95}, \frac{27}{95}, \frac{5}{3})$ is the center of mass of the object.

We may use triple integrals to find the moment of inertia of a material object about some line. Since a particle of mass m located at the point (x,y,z) in space has $m(y^2 + z^2)$ as its moment of inertia about the x axis, it seems reasonable that the moment of inertia about the x axis of a material object of constant density ρ having the shape of region S of space is given by

$$I_x = \int_S \rho(y^2 + z^2).$$

Similarly, $$I_y = \int_S \rho(x^2 + z^2), \qquad I_z = \int_S \rho(x^2 + y^2).$$

Example 2 Find the moment of inertia and radius of gyration about the z axis of the homogeneous solid of density ρ bounded by the coordinate planes and the plane

$$\frac{x}{a} + \frac{y}{b} + \frac{z}{c} = 1, \qquad a, b, c \text{ positive.}$$

Solution: The solid is the tetrahedron shown in Figure 16.27. Since the trace of the given plane in the xy plane has equation

$$y = b - \frac{b}{a}x,$$

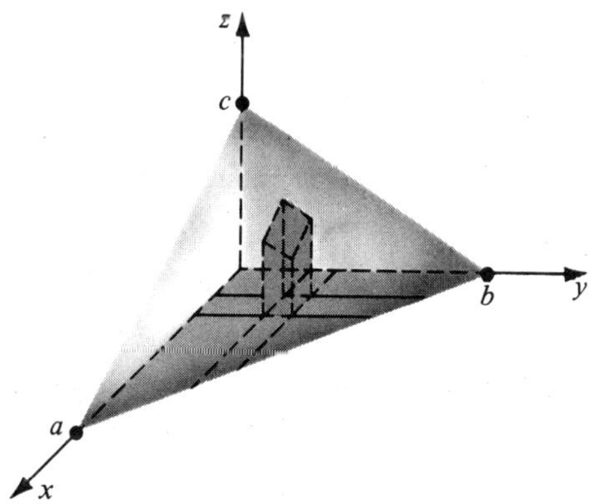

FIGURE 16.27

we have

$$
\begin{aligned}
I_z &= \int_0^a dx \int_0^{b-bx/a} dy \int_0^{c-cx/a-cy/b} \rho(x^2+y^2)\,dz \\
&= \rho c \int_0^a dx \int_0^{b-bx/a} (x^2+y^2)\left(1-\frac{1}{a}x-\frac{1}{b}y\right) dy \\
&= \rho bc \int_0^a \left[\frac{1}{2}x^2 - \frac{1}{a}x^3 + \frac{1}{2a^2}x^4 + \frac{b^2}{12}\left(1-\frac{1}{a}x\right)^4\right] dx \\
&= \frac{\rho abc}{60}(a^2+b^2).
\end{aligned}
$$

The volume of the given solid is $abc/6$, and its mass is therefore $\rho abc/6$. Hence the radius r of gyration is given by

$$\frac{\rho abc}{60}(a^2+b^2) = \frac{\rho abc}{6} r^2$$

and $r = \sqrt{a^2+b^2}/\sqrt{10}$.

EXERCISES

In each of Exercises 1 to 10, find the center of mass of the homogeneous solid having the given shape.

1. The tetrahedron with vertices (0,0,0), $(a,0,0)$, $(0,b,0)$, $(0,0,c)$
2. The first octant of the sphere $x^2 + y^2 + z^2 = a^2$ (*Hint:* $\bar{x} = \bar{y} = \bar{z}$ by symmetry.)
3. The region bounded by the xy plane and the paraboloid $z = 1 - x^2/a^2 - y^2/b^2$
4. The region in the first octant bounded by the cylinder $x = y^2$ and the planes $x = 4$, $z = 0$, and $z = 2$
5. Region S of Exercise 4, Section 9, I
6. Region S of Exercise 5, Section 9, I
7. Region S of Exercise 6, Section 9, I
8. Region S of Exercise 8, Section 9, I
9. Region S of Exercise 9, Section 9, I
10. The first octant of the ellipsoid $x^2/a^2 + y^2/b^2 + z^2/c^2 = 1$

Set up an integral for I_z for each of the following homogeneous solids of density ρ.

11. A hemisphere, z the axis of symmetry
12. The solid bounded by the plane $z = h$ and the paraboloid $z = x^2/a^2 + y^2/b^2$
13. The solid bounded by the plane $z = h$ and the paraboloid $z = x^2 + y^2$
14. A right circular cylinder of radius r and altitude h, z the axis of symmetry

15. The solid having the shape of region S in Exercise 7

16. The solid having the shape of region S in Exercise 9

17. A rectangular parallelepiped of length a, width b, and height c, the z axis along an edge of length c (compute)

18. The rectangular parallelepiped of Exercise 17, the z axis along a diagonal

11 CYLINDRICAL AND SPHERICAL COORDINATES

Triple integrals may be defined for continuous functions of three variables over regions of a cylindrical or spherical coordinate space much as they are over regions of a rectangular coordinate space. We shall indicate in this section how such triple integrals may be evaluated by repeated integrals.

A region S of cylindrical coordinate space bounded by the curves $r = a_1$, $r = a_2$, $\theta = b_1$, $\theta = b_2$, $z = c_1$, and $z = c_2$ may be partitioned into subregions of the same type. One such subregion is shown in Figure 16.28. This subregion

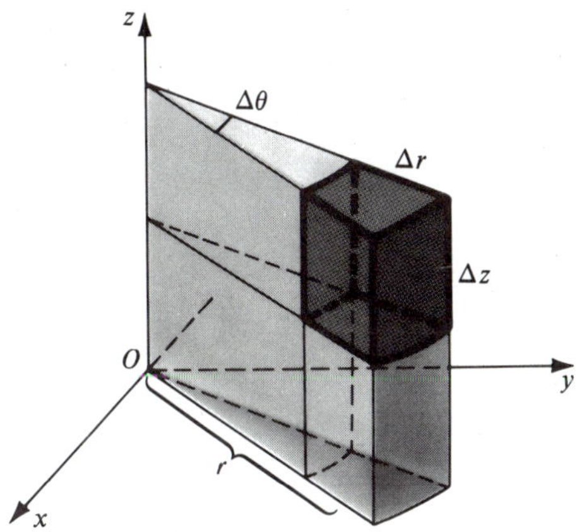

is a cylinder having its generator parallel to the z axis. Hence its volume, $V(S)$, is the product of the area of its base, $r\,\Delta r\,\Delta\theta$, and its altitude Δz,

$$V(S) = r\,\Delta r\,\Delta\theta\,\Delta z,$$

where r is the average radius of its base.

If f is a continuous function of three variables over S, then the triple integral of f over S may be defined as usual in terms of the l.u.b. of the set of all lower sums of f over partitions of S. Then it may be shown that the triple integral of f over S can be expressed in the form

$$\int_S f = \int_{a_1}^{a_2} dr \int_{b_1}^{b_2} d\theta \int_{c_1}^{c_2} f(r,\theta,z)r\,dz.$$

Again there are six possible permutations of the single integrals on the right side of the above equation.

The usual modifications of the limits of integration must be made if the triple integral of f is taken over a region S of space not bounded by coordinate surfaces like the region described above. Centers of mass and moments of inertia of an object may be found if the formulas of Section 10 are modified in the obvious way.

Example 1 Find the center of mass and moment of inertia about the z axis of the homogeneous solid bounded by the cylinder $r = a$, the cone $z = r$, and the plane $z = 0$.

Solution: This solid S is a right circular cylinder with a cone hollowed out of it, as shown in Figure 16.29. If ρ designates its constant density, then its mass is given by

$$W = \rho(\pi a^3 - \tfrac{1}{3}\pi a^3) = \tfrac{2}{3}\rho\pi a^3.$$

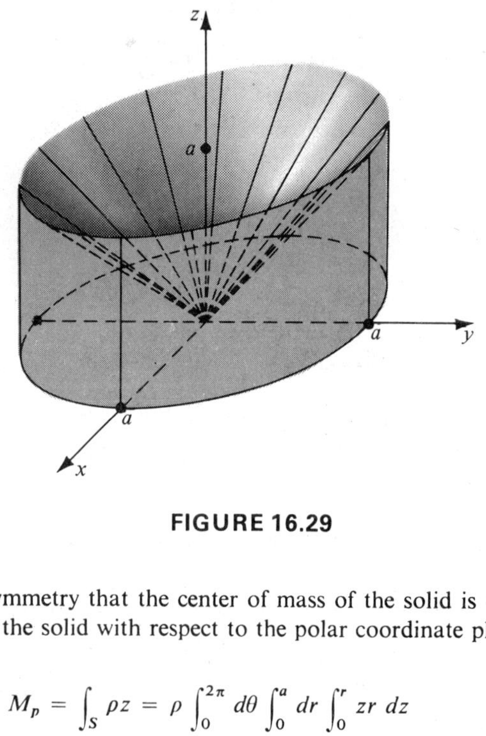

FIGURE 16.29

It is clear by symmetry that the center of mass of the solid is on the z axis. The moment M_p of the solid with respect to the polar coordinate plane p is given by

$$M_p = \int_S \rho z = \rho \int_0^{2\pi} d\theta \int_0^a dr \int_0^r zr\,dz$$

$$= \frac{\rho}{2}\int_0^{2\pi} d\theta \int_0^a r^3\,dr = \frac{\rho}{8}\int_0^{2\pi} a^4\,d\theta = \frac{\rho}{4}\pi a^4.$$

Hence $\bar{z} = M_p/W = 3a/8$, and the point $(0, 0, 3a/8)$ is the center of mass of the solid.

A particle of mass m located at the point (r,θ,z) is r units from the z axis. Hence mr^2 is its moment of inertia about the z axis. Therefore it is clear that the moment of inertia about the z axis of the solid of Figure 16.29 is given by

$$I_z = \int_S \rho r^2 = \rho \int_0^{2\pi} d\theta \int_0^a dr \int_0^r r^3\,dz$$

$$= \rho \int_0^{2\pi} d\theta \int_0^a r^4\,dr = \frac{\rho}{5}\int_0^{2\pi} a^5\,d\theta = \tfrac{2}{5}\rho\pi a^5.$$

The basic region of spherical coordinate space is the region S bounded by the spheres $\rho = a_1$ and $\rho = a_2$, the planes $\theta = b_1$ and $\theta = b_2$, and the cones $\phi = c_1$ and $\phi = c_2$. If such a region is partitioned into subregions in the usual way, then one such subregion is shown in Figure 16.30. The volume $V(S)$ of this subregion may be shown by geometry to be approximately equal to

$$V(S) = \rho^2 \sin\phi\,\Delta\rho\,\Delta\theta\,\Delta\phi.$$

Although we shall not prove it, we can imagine that the triple integral of a continuous function f over S is given by

16.23 $$\int_S f = \int_{a_1}^{a_2} d\rho \int_{b_1}^{b_2} d\theta \int_{c_1}^{c_2} f(\rho,\theta,\phi)\rho^2 \sin\phi\,d\phi.$$

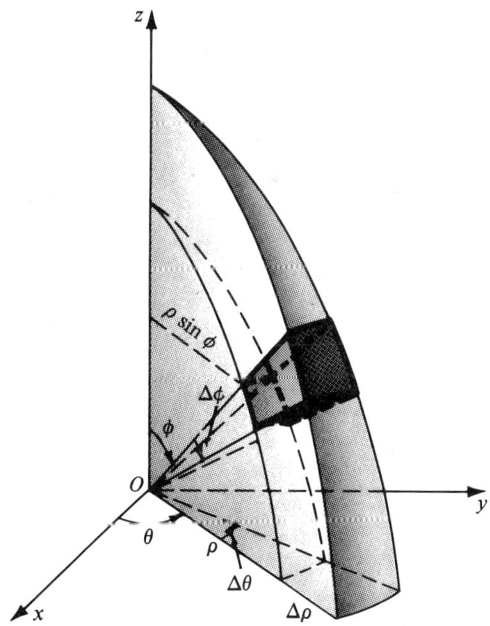

FIGURE 16.30

There are six possible permutations of the single integrals in the above repeated integral.

An application of spherical coordinates to the problem of finding the center of gravity and moment of inertia of a solid is indicated in the following example.

Example 2 Find the volume, center of mass, and moment of inertia about the axis of symmetry of the solid (of constant density 1) bounded above by the sphere $\rho = a$ and below by the cone $\phi = k$.

Solution: The solid is a cone with a spherical top; a quarter of it is shown in Figure 16.31. The volume of the solid is given by

$$V = \int_0^{2\pi} d\theta \int_0^k d\phi \int_0^a \rho^2 \sin\phi \, d\rho = \tfrac{2}{3}\pi a^3(1 - \cos k),$$

and its moment about the polar coordinate plane p (recalling that $z = \rho\cos\phi$) by

$$\begin{aligned} M_p &= \int_0^{2\pi} d\theta \int_0^k d\phi \int_0^a (\rho\cos\phi)\rho^2 \sin\phi \, d\rho \\ &= \frac{a^4}{4}\int_0^{2\pi} d\theta \int_0^k \sin\phi\cos\phi \, d\phi \\ &= \frac{a^4}{8}\sin^2 k \int_0^{2\pi} d\theta = \tfrac{1}{4}\pi a^4 \sin^2 k. \end{aligned}$$

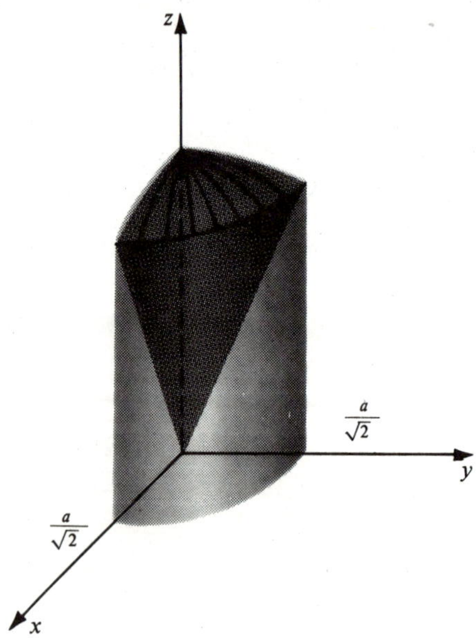

FIGURE 16.31

By symmetry, the center of mass is on the vertical z axis above the pole at the distance of

$$\bar{z} = \frac{M_p}{V} = \tfrac{3}{8}a(1 + \cos k).$$

Thus $(\frac{3}{8}a(1 + \cos k),0,0)$ is the center of mass of the solid. Note that if $k = \pi/2$ we obtain $(\frac{3}{8}a,0,0)$ as the center of mass of a hemisphere.

The distance r of a point (ρ,θ,ϕ) from the vertical z axis is given by $r = \rho \sin \phi$. Hence the moment of inertia about the z axis of a particle of mass m located at the point (ρ,θ,ϕ) is $m\rho^2 \sin^2 \phi$. With this in mind, we find that evidently the moment of inertia about the z axis of the given solid (assuming a constant density of 1) is

$$\begin{aligned} I_z &= \int_0^{2\pi} d\theta \int_0^k d\phi \int_0^a (\rho^2 \sin^2 \phi)\rho^2 \sin \phi \, d\rho \\ &= \frac{a^5}{5} \int_0^{2\pi} d\theta \int_0^k \sin^3 \phi \, d\phi \\ &= \tfrac{2}{15}\pi a^5(\cos^3 k - 3 \cos k + 2). \end{aligned}$$

Example 3 For the solid of Example 2, assume that the cone has the equation $\phi = \pi/4$. Set up the integrals for the moment of inertia about the z axis in both rectangular and cylindrical coordinates.

Solution: In rectangular coordinates, the sphere $\rho = a$ has the equation $x^2 + y^2 + z^2 = a^2$, and the cone $\phi = \pi/4$ has the equation $x^2 + y^2 = z^2$. In cylindrical coordinates, the sphere has the equation $z^2 = a^2 - r^2$, and the cone has the equation $z = r$.

To set up the required integrals, we shall need to find the curve of intersection of the sphere and the cone. For a point (x,y,z) on this curve,

$$x^2 + y^2 = a^2 - z^2 \quad \text{and} \quad x^2 + y^2 = z^2.$$

Hence $z^2 = a^2 - z^2$ and $z = a/\sqrt{2}$. From Figure 16.31 we now see that the projection of the region of integration onto the xy plane is the interior of the circle $x^2 + y^2 = a^2/2$. Thus the required integrals are

$$I_z = \int_{-(a/\sqrt{2})}^{a/\sqrt{2}} dx \int_{-\sqrt{(a^2/2)-x^2}}^{\sqrt{(a^2/2)-x^2}} dy \int_{\sqrt{x^2+y^2}}^{\sqrt{a^2-x^2-y^2}} (x^2 + y^2) \, dz,$$

$$I_z = \int_0^{2\pi} d\theta \int_0^{a/\sqrt{2}} dr \int_r^{\sqrt{a^2-r^2}} r^3 \, dz.$$

Example 4 Use spherical coordinates to compute the triple integral $\int_S f$, if

$$f(x,y,z) = \frac{z^2}{\sqrt{x^2 + y^2 + z^2}}$$

and S is the region between the spheres of radius a and b centered at the origin $(0 < a < b)$ and above the xy plane.

Solution: In terms of spherical coordinates, $\sqrt{x^2 + y^2 + z^2} = \rho$ and $z = \rho \cos \phi$ (from 14.24). Hence, by 16.23,

$$\int_S f = \int_a^b d\rho \int_0^{2\pi} d\theta \int_0^{\pi/2} \frac{\rho^2 \cos^2 \phi}{\rho^3} \rho^2 \sin \phi \, d\phi$$

$$= \int_a^b d\rho \int_0^{2\pi} d\theta \int_0^{\pi/2} \rho \cos^2 \phi \sin \phi \, d\phi$$

$$= \int_a^b d\rho \int_0^{2\pi} \rho \frac{(-\cos^3 \phi)}{3}\bigg|_0^{\pi/2} d\theta$$

$$= \frac{1}{3} \int_a^b d\rho \int_0^{2\pi} \rho \, d\theta = \frac{2\pi}{3} \int_a^b \rho \, d\rho = \frac{\pi}{3}(b^2 - a^2).$$

EXERCISES

1. Given a right circular cone having radius of base a and altitude h, find:
   a. Its center of mass
   b. Its moment of inertia about the axis of symmetry
   c. Its moment of inertia about a diameter of the base
2. Given a right circular cylinder of diameter a and altitude h, find:
   a. Its moment of inertia about the axis of symmetry
   b. Its moment of inertia about a generator
   c. Its moment of inertia about a diameter of the base
3. Given a hemispherical shell having inner radius a and outer radius b, find:
   a. Its center of mass
   b. Its moment of inertia about the axis of symmetry
   c. Its moment of inertia about a diameter of the base
4. Given the cone $\phi = k$ (spherical coordinates) cut out of the solid hemisphere $\rho = a, 0 \leq \phi \leq \pi/2$, find:
   a. The center of mass of the solid remaining
   b. Its moment of inertia about the axis of symmetry
   c. Its moment of inertia about a diameter of the base
5. Given a solid bounded by the cylinder $r = a \cos \theta$, the paraboloid $z = br^2$, and the plane $z = 0$ (cylindrical coordinates), find:
   a. Its volume
   b. Its center of mass

In Exercises 6 to 9, use spherical or cylindrical coordinates to compute the triple integral $\int_S f$, for the given function f and region S.

6. $f(x,y,z) = \dfrac{1}{x(9 - x^2 - y^2 - z^2)}$; S is the set of all points (x,y,z) with $x^2 + y^2 + z^2 \leq 4$ and $z^2 \geq x^2 + y^2$.

7. $f(x,y,z) = x^2 + y^2$; S is the set of all (x,y,z) with $x^2 + y^2 + z^2 \le a^2$, $x \ge 0$, $y \ge 0$, $z \ge 0$.
8. $f(x,y,z) = x^2$; S is the set of all (x,y,z) with $x^2 + y^2 \le a^2$, $0 \le z \le x^2 + y^2$.
9. $f(x,y,z) = \dfrac{z}{\sqrt{x^2 + y^2 + z^2}}$; S is the set of all (x,y,z) with $a^2 \le x^2 + y^2 \le b^2$, $0 \le z^2 \le x^2 + y^2$.

In Exercises 10 to 14, compute the volume of the given region.

10. Inside the cylinder $r = 3$ and outside the cone $z = r$
11. Outside the paraboloids $z = \pm r^2$ and inside the cylinder $r^2 = a^2 \cos 2\theta$
12. Inside both of the spheres $\rho = 1$ and $\rho = 2 \cos \phi$
13. Outside the cone $\phi = \pi/6$ and inside the sphere $\rho = 2a \cos \phi$
14. Above the plane $z = a$ and inside the sphere $\rho = 2a$

12 APPENDIX: EQUALITY OF THE DOUBLE AND REPEATED INTEGRALS

In this section we present the proof of Theorem 16.14. As in the statement of the theorem in Section 3, let G be the function defined by

$$G(x) = \int_{a_2}^{b_2} f(x,y)\, dy, \qquad \text{for } x \text{ in } [a_1,b_1].$$

We shall show that, given any $\varepsilon > 0$, there is a partition P_1 of $[a_1,b_1]$ such that $T(P_1) - S(P_1) < \varepsilon$, where $T(P_1)$ and $S(P_1)$ denote the upper and lower sums of G relative to the partition P_1. The integrability of G will then follow by the one-variable analog of Theorem 16.6.

Since $f(x,y)$ is integrable in Q, corresponding to any given $\varepsilon > 0$, we may choose a partition p of Q such that $u(p) - l(p) < \varepsilon$, where $u(p)$ and $l(p)$ are the upper and lower sums of $f(x,y)$ relative to p. If we write

$$p = \{Q_{ij} \mid i = 1, \ldots, r, j = 1, \ldots, s\},$$

then p is determined by two partitions P_1 and P_2 of $[a_1,b_1]$ and $[a_2,b_2]$, respectively, say

$$P_1 = \{a_1 = x_0, x_1, \ldots, x_r = b_1\},$$
$$P_2 = \{a_2 = y_0, y_1, \ldots, y_s = b_2\},$$

and

$$Q_{ij} = [(x_{i-1},y_{j-1}), (x_i,y_j)],$$

for each $i = 1, \ldots, r, j = 1, \ldots, s$.

Let us use the following notation:

$$m_i = \text{g.l.b. } G(x), \qquad \text{for } x \text{ in } [x_{i-1},x_i],$$
$$M_i = \text{l.u.b. } G(x), \qquad \text{for } x \text{ in } [x_{i-1},x_i],$$
$$m'_{ij} = \text{g.l.b. } f(x,y), \qquad \text{for } (x,y) \text{ in } Q_{ij},$$
$$M'_{ij} = \text{l.u.b. } f(x,y), \qquad \text{for } (x,y) \text{ in } Q_{ij},$$

$$\Delta x_i = x_i - x_{i-1},$$
$$\Delta y_j = y_j - y_{j-1}.$$

Now, with respect to the partition P_1:

$$m_i = \text{g.l.b.} \int_{a_2}^{b_2} f(x,y)\,dy, \qquad \text{for } x \text{ in } [x_{i-1},x_i]$$

$$= \text{g.l.b.} \left[\sum_{j=1}^{s} \int_{y_{j-1}}^{y_j} f(x,y)\,dy \right], \qquad \text{for } x \text{ in } [x_{i-1},x_i].$$

But clearly we must have

$$\int_{y_{j-1}}^{y_j} f(x,y)\,dy \geq m'_{ij}\,\Delta y_j, \qquad \text{for each } x \text{ in } [x_{i-1},x_i],$$

and so
$$m_i \geq \sum_{j=1}^{s} m'_{ij}\,\Delta y_j.$$

Similarly, we have

$$M_i = \text{l.u.b.} \int_{a_2}^{b_2} f(x,y)\,dy, \qquad \text{for } x \text{ in } [x_{i-1},x_i],$$

$$= \text{l.u.b.} \left[\sum_{j=1}^{s} \int_{y_{j-1}}^{y_j} f(x,y)\,dy \right], \qquad \text{for } x \text{ in } [x_{i-1},x_i],$$

$$\leq \sum_{j=1}^{s} M'_{ij}\,\Delta y_j.$$

Hence

$$S(P_1) = \sum_{i=1}^{r} m_i\,\Delta x_i \geq \sum_{i=1}^{r} \sum_{j=1}^{s} m'_{ij}\,\Delta x_i\,\Delta y_j = l(p),$$

$$T(P_1) = \sum_{i=1}^{r} M_i\,\Delta x_i \leq \sum_{i=1}^{r} \sum_{j=1}^{s} M'_{ij}\,\Delta x_i\,\Delta y_j = u(p).$$

It follows that

(*)
$$l(p) \leq S(P_1) \leq T(P_1) \leq u(p).$$

But $u(p) - l(p) < \varepsilon$, by our assumption; and so the above inequalities imply that also $T(P_1) - S(P_1) < \varepsilon$. Hence G is integrable in $[a_1,b_1]$ by 16.6. Furthermore, we see from (*) that $\int_{a_1}^{b_1} G(x)\,dx$ lies between $l(p)$ and $u(p)$ for every partition p of Q. Since the only number with this property is $\int_Q f$, it follows that

$$\int_Q f = \int_{a_1}^{b_1} G(x)\,dx,$$

completing the proof of Theorem 16.14.

CHAPTER REVIEW

I

In each of Exercises 1 to 6, compute the integral.

1. $\int_0^1 dx \int_0^2 (x + y)\, dy$

2. $\int_0^2 dx \int_0^1 (x + y)\, dy$

3. $\int_{-1}^0 dx \int_1^2 dz \int_0^3 (x^2 - y + z)\, dy$

4. $\int_1^2 dz \int_0^3 dy \int_{-1}^0 (x^2 - y + z)\, dx$

5. $\int_0^2 dx \int_0^x (x^2 + y^2)\, dy$

6. $\int_1^2 dy \int_0^{y^2} (x + 2y)\, dx$

7. Interpret each of the integrals in Exercises 1, 2, 5, and 6 as the volume of a solid.

In Exercises 8 to 10, express the integrals in terms of repeated integrals with the order of integration reversed.

8. $\int_0^2 dy \int_{y/2}^{3-y} f(x,y)\, dx$

9. $\int_0^3 dy \int_0^{y/3} f(x,y)\, dx + \int_3^4 dy \int_0^{\sqrt{4-y}} f(x,y)\, dx$

10. $\int_{-2}^4 dx \int_{(x^2-4)/2}^{x+2} f(x,y)\, dy$

In Exercises 11 to 13, compute the integral by first reversing the order of integration.

11. $\int_0^1 dx \int_{2x}^2 e^{y^2}\, dy$

12. $\int_0^2 dx \int_{x/2}^1 \cos y^2\, dy$

13. $\int_0^1 dx \int_0^{\tan^{-1} x} x\, dy$

In each of Exercises 14 to 18 set up integrals for the volume and the center of mass of the solid with the given boundaries.

14. $x + y + z = 2, (x = 0, y = 0, z = 0)$

15. $z = \dfrac{1}{x^2 + y^2}, (z = 0; x^2 + y^2 = 1)$

16. $z = \cos (x^2 + y^2), \left(z = 0; x^2 + y^2 \leq \dfrac{\pi}{2}\right)$

17. $z = \cos (x^2 + y^2), \left(z = 0; \dfrac{3\pi}{2} \leq x^2 + y^2 \leq \dfrac{5\pi}{2}\right)$

18. $\dfrac{x^2}{a^2} + \dfrac{y^2}{b^2} + \dfrac{z^2}{c^2} = 1$

In Exercises 19 and 20 set up the triple integral for the moment of inertia of the given uniform solid about the indicated axis.

19. $\dfrac{x^2}{a^2} + \dfrac{y^2}{b^2} + \dfrac{z^2}{c^2} \leq 1$; z axis

20. $x^2 + y^2 + z^2 \leq a^2, z \geq x^2 + y^2$; y axis

II

1. Show by evaluating the repeated integrals that

$$\int_0^1 dx \int_0^1 \frac{x^2 - y^2}{(x^2 + y^2)^2}\, dy \neq \int_0^1 dy \int_0^1 \frac{x^2 - y^2}{(x^2 + y^2)^2}\, dx.$$

Explain.

2. Compute the triple integral $\int_S f$, where

$$f(x,y,z) = \frac{1}{x^2 + y^2 + (z - \frac{1}{2})^2},$$

and S is the sphere $x^2 + y^2 + z^2 \leq 1$. Use spherical coordinates.

17

Further Topics in Integration

The definition of line integral is another extension of the concept of integral of a function of one variable. The line integral is related to the double integral by means of a fundamental result known as Green's theorem. We discuss the general notion of a transformation and apply it to the problem of changing variables in multiple integrals.

1 LINE INTEGRALS

Just as we used the intuitive concept of *area* to motivate our definition of the integral of a real-valued function defined in an interval $[a,b]$, so we can also make use of the physical concept of *work* to define an integral of a *vector-valued* function defined over a smooth curve. This type of integral is known as a *line integral*. Our discussion will be presented in terms of curves in the plane, but the reader will easily see that there is no essential difference in the three-dimensional case.

Suppose that $\mathbf{F}$ is a constant force acting on a particle that it displaces along a vector $\mathbf{v}$. It will be recalled that the *work* done by this force is then defined to be $\mathbf{F} \cdot \mathbf{v}$. This is simply the component of $\mathbf{F}$ in the direction of $\mathbf{v}$ multiplied by the length of $\mathbf{v}$. This may be seen by using 13.4, since $\mathbf{F} \cdot \mathbf{v} = |\mathbf{F}| \cdot |\mathbf{v}| \cos \theta$, where θ is the angle between $\mathbf{F}$ and $\mathbf{v}$.

Now suppose that λ is a smooth curve in R^2 with parametric representation $\lambda(t) = (x(t), y(t))$, for t in $[a,b]$. For each point (x,y) on λ, suppose there is a *variable* force $\mathbf{F}(x,y)$ defined, with components $A(x,y)$ and $B(x,y)$. Thus $\mathbf{F}(x,y) = \langle A(x,y), B(x,y)\rangle$ is a *vector-valued* function of x and y, which is often called a *force field.* (We assume, furthermore, that the functions $A(x,y)$ and $B(x,y)$ are continuous.) Now we ask the question: how can we define the *work* done by the variable force $\mathbf{F}(x,y)$ in displacing a particle along the curve λ?

In order to handle this problem, consider a partition $p = \{t_0, t_1, \ldots, t_n\}$ of $[a,b]$ into n subintervals. Then the points $\lambda(t_i) = (x(t_i),y(t_i))$, for $i = 0, 1, 2, \ldots, n$, partition the curve λ into n subarcs. For $i = 1, 2, \ldots, n$, let us write $\Delta t_i = t_i - t_{i-1}$, $\Delta x_i = x(t_i) - x(t_{i-1})$, $\Delta y_i = y(t_i) - y(t_{i-1})$, and let the vector from $\lambda(t_{i-1})$ to $\lambda(t_i)$ be denoted by $\Delta\mathbf{r}_i$. See Figure 17.1.

By the mean value theorem (4.8), for each $i = 1, 2, \ldots, n$, there is a number c_i in (t_{i-1},t_i) such that $\Delta x_i = x'(c_i)\,\Delta t_i$. Also, there is a number d_i in (t_{i-1},t_i) such that $\Delta y_i = y'(d_i)\,\Delta t_i$. Let $P_i = \lambda(c_i)$, $Q_i = \lambda(d_i)$, for $i = 1, 2, \ldots, n$. Then for each i the vector $\mathbf{F}_i = \langle A(P_i),B(Q_i)\rangle$ is a constant that can be used as an approximation to $\mathbf{F}(x,y)$ along the subarc from $\lambda(t_{i-1})$ to $\lambda(t_i)$ (for even though P_i and Q_i might be different points, the vector $\mathbf{F}_i$ will nevertheless be close to the values assumed by $\mathbf{F}(x,y)$). Also, the subarc from $\lambda(t_{i-1})$ to $\lambda(t_i)$ can be approximated by the vector $\Delta\mathbf{r}_i$. Then the work done by the constant vector $\mathbf{F}_i$ in moving a particle along $\Delta\mathbf{r}_i$ is

$$\begin{aligned}\mathbf{F}_i \cdot \Delta\mathbf{r}_i &= A(P_i)\,\Delta x_i + B(Q_i)\,\Delta y_i\\ &= A(P_i)x'(c_i)\,\Delta t_i + B(Q_i)y'(d_i)\,\Delta t_i.\end{aligned}$$

Therefore we can approximate the work done by $\mathbf{F}(x,y)$ along λ (which we are trying to *define*) by the sums

17.1 $$\sum_{i=1}^{n} A(P_i)x'(c_i)\,\Delta t_i + \sum_{i=1}^{n} B(Q_i)y'(d_i)\,\Delta t_i.$$

Now the ith term of the first sum is

$$A(x(c_i),y(c_i))x'(c_i)\,\Delta t_i,$$

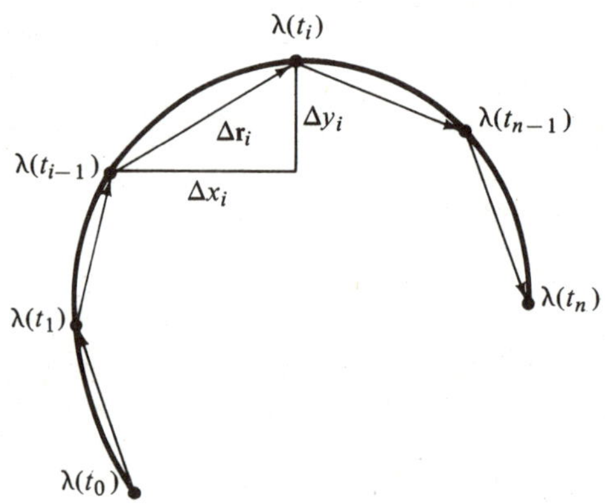

FIGURE 17.1

while the ith term of the second sum is

$$B(x(d_i),y(d_i))y'(d_i)\,\Delta t_i.$$

We recognize, therefore, that the first sum in 17.1 is simply a Riemann sum for the function $A(x(t),y(t))x'(t)$, while the second is a Riemann sum for $B(x(t),y(t))y'(t)$, in the interval $a \le t \le b$. Hence, as $n \to \infty$, the sums 17.1 approach the integral

17.2 $$\int_a^b [A(x(t),y(t))x'(t) + B(x(t),y(t))y'(t)]\,dt,$$

which can be written in the briefer form

$$\int_a^b [\mathbf{F}(x(t),y(t)) \cdot \lambda'(t)]\,dt.$$

This justifies the following definition.

17.3 Definition

If λ is a smooth curve with domain $[a,b]$ and $\mathbf{F}(x,y)$ is a force acting at each point (x,y) of λ, then the *work* done by $\mathbf{F}(x,y)$ in displacing a particle along λ is

$$\int_a^b [\mathbf{F}(x(t),y(t)) \cdot \lambda'(t)]\,dt.$$

If $\mathbf{F}(x,y)$ is any continuous vector-valued function (not necessarily to be interpreted physically as a force) that is defined for (x,y) on λ, then the integral 17.3 is meaningful and we call it the *line integral* of $\mathbf{F}$ over λ.

Many notations for line integrals are common. Let the components of $\mathbf{F}(x,y)$ be $A(x,y)$ and $B(x,y)$, and let us write $dx = x'(t)\,dt$, $dy = y'(t)\,dt$. Then 17.2 suggests as a convenient notation for a line integral,

$$\int_\lambda (A\,dx + B\,dy),$$

where of course A and B are to be considered as functions of t in the interval of integration $\lfloor a,b \rfloor$.

We may also consider a smooth curve λ in space, and a three-dimensional vector-valued function $\mathbf{F}(x,y,z)$ defined for (x,y,z) on λ. If the components of $\mathbf{F}(x,y,z)$ are $A(x,y,z)$, $B(x,y,z)$, and $C(x,y,z)$, then the corresponding line integral may be written in the form

$$\int_\lambda (A\,dx + B\,dy + C\,dz).$$

In most applications of line integrals, it is necessary to integrate over a somewhat more general type of curve λ, which is called *piecewise smooth.* A curve λ is *piecewise smooth* if the function $\lambda(t)$ has a bounded derivative $\lambda'(t)$ continuous everywhere in its domain except possibly at a finite number of points. (In geometrical terms, this means that λ has a tangent line at all but a finite number of points.) For example, a continuous curve made up of a finite number of line segments is piecewise smooth.

Example 1 Compute $\int_\lambda (x^2y\,dx + y^3\,dy)$, where λ is the arc of the parabola $y = x^2$ from (0,0) to (1,1).

Solution: λ may be represented parametrically by $x = t$, $y = t^2$, for $0 \le t \le 1$. Then $dx = dt$, $dy = 2t\,dt$, and

$$\int_\lambda (x^2y\,dx + y^3\,dy) = \int_0^1 [t^2 \cdot t^2 + t^6(2t)]\,dt$$
$$= \int_0^1 (t^4 + 2t^7)\,dt = \tfrac{9}{20}.$$

Example 2 Compute $\int_\lambda y\,dx + z\,dy + x\,dz$, where λ is the arc of the helix $\lambda(t) = (\cos t, \sin t, t)$ from $t = 0$ to $t = 2\pi$.

Solution: From the parametric representation for λ, we have $x = \cos t$, $y = \sin t$, $z = t$, $dx = -\sin t\,dt$, $dy = \cos t\,dt$, $dz = dt$. Hence

$$\int_\lambda y\,dx + z\,dy + x\,dz = \int_0^{2\pi} (-\sin^2 t + t\cos t + \cos t)\,dt.$$

Using Formulas 50 and 54 of the Table of Integrals, the above integral is found to be

$$\left(\tfrac{1}{2}\sin t\cos t - \frac{t}{2}\right) + (\cos t + t\sin t) + (\sin t)\Bigg|_0^{2\pi} = -\pi.$$

EXERCISES

1. Compute the integral of Example 1 if λ is the arc of the curve $y^3 = x^2$ from (0,0) to (1,1).

2. Evaluate $\int_\lambda \left(\frac{y}{x}\,dx + dy\right)$ along the curve $y = \ln x$ from (1,0) to $(e,1)$.

3. Evaluate $\int_\lambda [(x + y)\,dx + (x - y)\,dy]$ if λ is the ellipse $b^2x^2 + a^2y^2 = a^2b^2$ traversed in a counterclockwise direction.

4. Compute $\int_\lambda y^2\,dx + x^2\,dy$, where λ is the arc of the semicircle $y = \sqrt{1 - x^2}$ from (1,0) to (−1,0).

5. Compute $\int_\lambda y^2\,dx - z\,dy + xy\,dz$, where λ is the straight line joining (0,0,0) and (1,1,1).

6. Evaluate the line integral $\int x\,dy - y\,dx$ around (a) the circle $(x - h)^2 + (y - k)^2 = a^2$; (b) the ellipse $\lambda(t) = (a\cos t, b\sin t)$, domain $\lambda = [0,2\pi]$.

7. Evaluate the line integral $\int_\lambda \frac{x\,dy - y\,dx}{x^2 + y^2}$ around (a) the circle $(x - 2a)^2 + (y - 2a)^2 = a^2$; (b) the circle $x^2 + y^2 = a^2$.

8. Evaluate the line integral $\int y^2\,dx + 2xy\,dy$ around the two curves in Exercise 6.

9. Evaluate the line integral $\int_\lambda (A\,dx + B\,dy + C\,dz)$ for (a) $\lambda(t) = (a\cos t, b\sin t, t^2)$, domain $\lambda = [0,2\pi]$, $A = x + y$, $B = y + z$, $C = z + x$; (b) $\lambda(t) = (at,bt^2,ct^3)$ domain $\lambda = [0,1]$, $A = x + y + z$, $B = xyz$, $C = x^2y^2z^2$.

10. Evaluate $\int_\lambda (y^2\,dx + x\,dy)$, where (a) λ is the square with vertices (0,0), (2,0), (2,2), (0,2); (b) λ is the square with vertices $(\pm 1, \pm 1)$. (Each curve is to be traversed in the counterclockwise direction.)

11. Compute $\int_\lambda [(2x - y)\,dx + (x + y)\,dy]$ over each of the following curves.

$$\lambda_1(t) = (t, t + 2), \qquad t \text{ in } [0,2],$$

$$\lambda_2(t) = (t,t^2), \qquad t \text{ in } [0,2].$$

Note that the results are different although the endpoints of λ_1 and λ_2 are the same.

2 PATH-INDEPENDENT LINE INTEGRALS

From the discussion in Section 1, it is clear that in general the value of a line integral $\int_\lambda (A\,dx + B\,dy)$ depends on the functions A and B, the endpoints of the curve λ, and the curve λ itself. Under certain special conditions, however, the value of the integral will not depend on the curve λ, but only on the *endpoints* of the curve. When these conditions hold, we say that the line integral $\int_\lambda (A\,dx + B\,dy)$ is *independent of the path.*

It turns out that the property of independence of the path for a line integral $\int (A\,dx + B\,dy)$ is directly related to the question as to whether there exists a scalar function $f(x,y)$ whose gradient vector $\nabla f(x,y)$ is equal to the vector function $\langle A(x,y),B(x,y)\rangle$. If this condition holds, that is, if there exists a function $f(x,y)$ such that

$$\nabla f(x,y) = \langle f_x(x,y),f_y(x,y)\rangle = \langle A(x,y),B(x,y)\rangle$$

for all (x,y) in some set S in R^2, then we say that $f(x,y)$ is a *potential function* of $\langle A(x,y),B(x,y)\rangle$ in S. The following theorem gives a necessary and sufficient condition for $\langle A(x,y),B(x,y)\rangle$ to possess a potential function.

17.4 Theorem

Let A and B be functions of two variables having continuous partial derivatives in a rectangle Q in $\mathbb{R}^2$. *Then there exists a function f such that*

$$\nabla f(x,y) = \langle A(x,y), B(x,y)\rangle$$

for all (x,y) in Q if and only if the condition

$$A_y(x,y) = B_x(x,y)$$

holds for all (x, y) in Q.

Proof: (i) Suppose $\langle A(x,y),B(x,y)\rangle$ has a potential function $f(x,y)$ in Q. Then $f_x(x,y) = A(x,y)$ and $f_y(x,y) = B(x,y)$ for all (x,y) in Q. But then we also have

$$f_{xy}(x,y) = A_y(x,y), \qquad f_{yx}(x,y) = B_x(x,y)$$

for (x,y) in Q. Since by hypothesis these partial derivatives are continuous, we must have $f_{xy}(x,y) = f_{yx}(x,y)$ by 15.11. Hence it follows that $A_y(x,y) = B_x(x,y)$ for all (x,y) in Q.

(ii) Now suppose that the condition $A_y(x,y) = B_x(x,y)$ holds for all (x,y) in Q. We must prove that $\langle A(x,y),B(x,y)\rangle$ possesses a potential function in Q.

Let us choose any fixed point (a,b) in Q, and let us define the function f by

$$f(x,y) = \int_a^x A(t,y)\,dt + \int_b^y B(a,t)\,dt,$$

for all (x,y) in Q. Using the fundamental theorem of calculus (5.20), we have

$$\begin{aligned} f_x(x,y) &= \frac{\partial}{\partial x}\int_a^x A(t,y)\,dt + \frac{\partial}{\partial x}\int_b^y B(a,t)\,dt \\ &= A(x,y) + 0. \end{aligned}$$

Also, using 15.33,

$$\begin{aligned} f_y(x,y) &= \frac{\partial}{\partial y}\int_a^x A(t,y)\,dt + \frac{\partial}{\partial y}\int_a^y B(a,t)\,dt \\ &= \int_a^x A_y(t,y)\,dt + B(a,y) = \int_a^x B_x(t,y)\,dt + B(a,y) \\ &= B(t,y)\Big|_a^x + B(a,y) = B(x,y) - B(a,y) + B(a,y) = B(x,y). \end{aligned}$$

So $\nabla f(x,y) = \langle A(x,y),B(x,y)\rangle$ and the proof is complete.

The following example illustrates a convenient method for finding a potential function (rather than applying 17.4 directly).

Example 1 Find a function f such that

$$\nabla f(x,y) = \langle 2xy + 1, x^2 + 3y^2 \rangle.$$

Solution: Let $A(x,y) = 2xy + 1$, $B(x,y) = x^2 + 3y^2$. Then clearly $A_y = B_x$, so that the condition of Theorem 17.4 holds. Since we want to find f such that $f_x = A$ and $f_y = B$, it is reasonable to expect that

$$f(x,y) = \int (2xy + 1)\,dx + g(y),$$

where $g(y)$ is a function of y alone. Hence

$$f(x,y) = x^2y + x + g(y).$$

Now differentiating with respect to y,

$$f_y(x,y) = x^2 + g'(y).$$

But, by assumption, $f_y(x,y) = x^2 + 3y^2$. Hence

$$x^2 + g'(y) = x^2 + 3y^2$$
$$g'(y) = 3y^2$$
$$g(y) = y^3 + C.$$

Thus we have $f(x,y) = x^2y + x + y^3 + C$.

We now state our basic theorem on independence of the path for line integrals.

17.5 Theorem

Suppose A and B are functions of two variables having continuous partial derivatives in a rectangle Q, and satisfying $A_y(x,y) = B_x(x,y)$ for all (x,y) in Q. Let λ be any smooth curve lying in Q, defined by

$$\lambda(t) = (x(t),y(t)), \qquad t \text{ in } [t_1,t_2],$$

and let $P_1 = \lambda(t_1)$, $P_2 = \lambda(t_2)$. If f is any function with $\nabla f(x,y) = \langle A(x,y),B(x,y)\rangle$ for all (x,y) in Q, then

$$\int_\lambda (A\,dx + B\,dy) = f(P_2) - f(P_1).$$

Thus, the line integral $\int_\lambda (A\,dx + B\,dy)$ does not depend on the curve λ but only on the endpoints P_1 and P_2.

Proof: By definition of line integral (17.3),

$$(1) \quad \int_\lambda (A\,dx + B\,dy) = \int_{t_1}^{t_2} [\langle A(x(t),y(t)),B(x(t),y(t))\rangle \cdot \boldsymbol{\lambda}'(t)]\,dt.$$

But if f is any potential function of $\langle A(x,y),B(x,y)\rangle$, then by the chain rule,

$$\begin{aligned} D_t f(x(t), y(t)) &= f_x(x(t), y(t))x'(t) + f_y(x(t), y(t))y'(t) \\ &= A(x(t), y(t))x'(t) + B(x(t), y(t))y'(t). \end{aligned}$$

Hence $f(x(t), y(t))$ is an antiderivative of the integrand in (1) above. Thus

$$\begin{aligned} \int_\lambda (A\,dx + B\,dy) &= \int_{t_1}^{t_2} D_t f(x(t), y(t))\,dt \\ &= f(x(t_2), y(t_2)) - f(x(t_1), y(t_1)) = f(P_2) - f(P_1). \end{aligned}$$

The reader should note the similarity between Theorem 17.5 and the Fundamental Theorem of Calculus (5.20). It should also be observed that in the proof of 17.5 an arbitrary constant could be added to the potential function f, but this would not affect the resulting value $f(P_2) - f(P_1)$.

Example 2 Compute $\int_\lambda [(x + y)\,dx + (x - y)\,dy]$, where λ is any smooth curve joining the points (0,1) and (2,3).

Solution: Set $A(x,y) = x + y$ and $B(x,y) = x - y$. Then $A_y(x,y) = 1 = B_x(x,y)$. Hence, by Theorem 17.4, there exists a function f with $f_x(x,y) = x + y$ and $f_y(x,y) = x - y$. Using the method of Example 1, we find that such a function f is given by

$$f(x,y) = \frac{x^2}{2} + xy - \frac{y^2}{2}.$$

By Theorem 17.5 we have

$$\int_\lambda [(x + y)\,dx + (x - y)\,dy] = f(2,3) - f(0,1) = 4.$$

EXERCISES

I

In each of Exercises 1–5, verify that $A_y = B_x$, and use the method of Example 1 to find a function f such that $f_x = A$ and $f_y = B$.

1. $A = \sqrt{1 - y^2}$, $B = \dfrac{1 - xy}{\sqrt{1 - y^2}}$
2. $A = y \sec x \tan x$, $B = \sec x$
3. $A = 4x^3 + 6xy^2 + 7y$, $B = 6x^2y + 7x + 3y^2$
4. $A = \ln y$, $B = x/y$.
5. $A = ye^{xy} + y/(x^2 + y^2)$, $B = xe^{xy} - x/(x^2 + y^2)$

In Exercises 6–9, show that the given line integral is independent of the path, and then evaluate it over any smooth curve λ joining the given points.

6. $\int_\lambda [2xy\,dx + (x^2 - y^2)\,dy]$; (0,0) to (4,4).

7. $\int_\lambda [(x^2 + y)\,dx + (y^2 + x)\,dy]$; $(-1,2)$ to $(3,4)$.

8. $\int_\lambda (y \cos x\,dx + \sin x\,dy)$; $(0,1)$ to $(\pi,2)$.

9. $\int_\lambda (\ln y\,dx + (x/y)\,dy)$; $(1,1)$ to $(1,e)$.

10. Compute $\int_\lambda [x/(x^2 - y^2)\,dx + y/(y^2 - x^2)\,dy]$, where λ is any smooth curve joining $(1,0)$ to $(3,2)$ and not touching the lines $y = \pm x$.

II

1. Let A, B, and C be functions of three variables having continuous partial derivatives in a rectangular parellalepiped Q in $\mathbf{R}^3$. Prove that there exists a function f such that $\nabla f = \langle A,B,C\rangle$ at each point of Q if and only if the conditions

$$A_y = B_x, \qquad A_z = C_x, \qquad B_z = C_y$$

hold at each point of Q. *Suggestion:* Let (a,b,c) be any fixed point in Q and define

$$f(x,y,z) = \int_a^x A(t,y,z)\,dt + \int_b^y B(a,t,z)\,dt + \int_c^z C(a,b,t)\,dt.$$

Then look at the proof of 17.4.

2. Find a function f of three variables such that $\nabla f = \langle A,B,C\rangle$, where

$$A(x,y,z) = 2xy + z^2, \qquad B(x,y,z) = x^2 + 6yz^2, \qquad C(x,y,z) = 2xz + 6y^2z.$$

3. State conditions under which a line integral $\int_\lambda (A\,dx + B\,dy + C\,dz)$ is independent of the path, and prove that these conditions are sufficient (see 17.5).

In Exercises 4 and 5, show that the given line integral is independent of the path, and then evaluate it over any smooth curve λ joining the given points.

4. $\int_\lambda [(y + z)\,dx + (x + z)\,dy + (x + y)\,dz]$; $(1,1,0)$ to $(2,1,1)$.

5. $\int_\lambda [\sin yz\,dx + xz \cos yz\,dy + xy \cos yz\,dz]$; $(0,0,0)$ to $(1, 1, \pi/2)$.

3 GREEN'S THEOREM

There is a very important relationship between double integrals and line integrals, which we shall now discuss. This concerns line integrals over piecewise smooth simple closed curves. (See Section 1 of this chapter and Section 1 of Chapter 13 for these definitions.)

Before stating this important theorem, we shall introduce one more bit of notation. If λ is a piecewise smooth simple closed curve in the plane, and if we wish to integrate the vector-valued function $\langle A(x,y),B(x,y)\rangle$ over λ in the *counterclockwise* direction, then we denote this line integral by

$$\oint_\lambda (A\,dx + B\,dy).$$

17.6 Green's Theorem

Let S be an open set in the plane, and let λ be a piecewise smooth simple closed curve in S whose interior also lies in S. If $A(x,y)$ and $B(x,y)$ are real-valued functions that are continuous and that have continuous first partial derivatives in S, then

$$\oint_\lambda (A\,dx + B\,dy) = \int_R \left(\frac{\partial B}{\partial x} - \frac{\partial A}{\partial y}\right),$$

where R is the region in the plane formed by λ and its interior.

We shall not prove Green's theorem in the general form stated above, but only in the case when the region R is of a simple elementary type. The special type of region R that we shall consider is that in which each horizontal line and each vertical line in the plane intersects λ in at most two points. This corresponds to the situation in which R is representable in both of the following forms:

(1) $$R = \{(x,y) \mid a \le x \le b, f(x) \le y \le g(x)\},$$

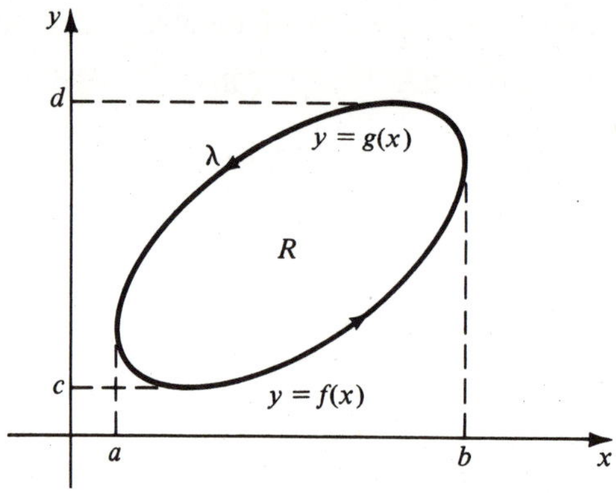

FIGURE 17.2

(2) $$R = \{(x,y) \mid c \le y \le d,\ u(y) \le x \le v(y)\},$$

for appropriate functions f, g, u, and v. See Figure 17.2, which provides an illustration of a representation of the form (1).

We shall prove Green's theorem in this case by showing that each of the following formulas holds:

17.7 $$\oint_\lambda A\,dx = -\int_R \frac{\partial A}{\partial y},$$

17.8 $$\oint_\lambda B\,dy = \int_R \frac{\partial B}{\partial x},$$

and the theorem then follows by addition.

To prove 17.7, let us write

$$\int_R \frac{\partial A}{\partial y} = \int_a^b dx \int_{f(x)}^{g(x)} \frac{\partial A}{\partial y}\,dy,$$

making use of the representation (1) for R. Then

$$\int_R \frac{\partial A}{\partial y} = \int_a^b A(x,y)\Big|_{y=f(x)}^{y=g(x)} dx$$

$$= \int_a^b [A(x,g(x)) - A(x,f(x))]\,dx.$$

Now we also compute $\oint_\lambda A\,dx$. This can be done by writing

$$\oint_\lambda A\,dx = \int_{\lambda_1} A\,dx + \int_{\lambda_2} A\,dx,$$

where λ_1 is the graph of $y = f(x)$ from $x = a$ to $x = b$, and λ_2 is the graph of $y = g(x)$ from $x = b$ to $x = a$ (note the order). Then

$$\oint_\lambda A\,dx = \int_a^b A(x,f(x))\,dx + \int_b^a A(x,g(x))\,dx$$

$$= \int_a^b [A(x,f(x)) - A(x,g(x))]\,dx$$

$$= -\int_R \frac{\partial A}{\partial y}.$$

The reader may prove 17.8 in a similar way, using the representation (2) for R.

For the proof of Green's theorem in its general form, we refer the reader to T. M. Apostol, *Mathematical Analysis*, Addison-Wesley, 1957.

Example 1 Use Green's theorem to find the value I of $\oint_\lambda (y\,dx + x^2y\,dy)$, where λ is the closed curve formed by $y^2 = x$ and $y = x$ between (0,0) and (1,1).

Solution: Here $B_x = 2xy$, $A_y = 1$, so that by Green's theorem,

$$I = \int_0^1 dy \cdot \int_{y^2}^{y} (2xy - 1)\,dx$$

$$= \int_0^1 (x^2y - x)\Big|_{y^2}^{y} dx = -\tfrac{1}{12}.$$

Example 2 Various line integrals can be used to compute the area of a region R in the plane. For example, if R is bounded by the piecewise smooth simple closed curve λ, then by Green's theorem each of the integrals

$$\oint_\lambda x\,dy, \qquad \oint_\lambda (-y\,dx), \qquad \tfrac{1}{2}\oint_\lambda (-y\,dx + x\,dy)$$

is equal to $\int_R (1) = A(R)$.

EXERCISES

In each of Exercises 1 to 4, use Green's theorem to evaluate the given line integral.

1. $\oint_\lambda (x^2y\,dx + y^3\,dy)$, where λ is the closed curve formed by $y = x$ and $y^3 = x^2$ from (0,0) to (1,1).
2. $\oint_\lambda [(xy - x^2)\,dx + x^2y\,dy]$, where λ is the closed curve formed by $y = 0$, $x = 1$, and $y = x$.
3. $\oint_\lambda [(2x^3 - y^3)\,dx + (x^3 + y^3)\,dy]$, where λ is the circle $x^2 + y^2 = 1$. (*Hint:* Convert the resulting double integral to polar coordinates.)
4. $\oint_\lambda ((1/y)\,dx + (1/x)\,dy)$, where λ is the closed curve formed by $y = 1$, $x = 4$, and $y = \sqrt{x}$.
5. What is the value of $\oint_\lambda (ay\,dx + bx\,dy)$ around any simple closed curve λ?
6. Suppose that $A(x,y)$ and $B(x,y)$ are real-valued functions with continuous first partial derivatives, and for which $A_y = B_x$ throughout some open set S in the plane. Use Green's theorem to prove that

$$\oint_\lambda (A\,dx + B\,dy) = 0,$$

   if λ is any piecewise smooth simple closed curve in S whose interior also lies in S.
7. Prove the result of Exercise 6 above by applying Theorems 17.4 and 17.5.
8. Verify by direct computation that

$$\oint_\lambda \left(\frac{y\,dx}{x^2 + y^2} - \frac{x\,dy}{x^2 + y^2}\right) \neq 0,$$

   where λ is the circle $x^2 + y^2 = 1$. But, since $A_y = B_x$, the application of Green's theorem would give the value 0 for this integral. Explain why this is not a contradiction.
9. **a.** Use one of the line integrals of Example 2 to compute the area of the circle $x = a\cos t$, $y = a\sin t$, $0 \le t \le 2\pi$.
   b. Compute the area of the ellipse $x = a\cos t$, $y = b\sin t$, $0 \le t \le 2\pi$.

4 TRANSFORMATION OF COORDINATES

In our study of line integrals we have already encountered functions $\mathbf{F}(x,y)$ of two variables x and y whose values are ordered pairs of real numbers. In that connection we interpreted the values of such a function as vectors. However, we could just as well consider the values of such a function as *points* in the plane, and it is to this interpretation that we now give our attention.

Suppose we have a function F that associates with each point (x,y) in R^2 another point (u,v) in R^2. Then u and v are real-valued functions of x and y, which we can write as

17.9
$$\begin{aligned} u &= u(x,y) \\ v &= v(x,y), \end{aligned}$$

where, for convenience, we have used the same letters u and v for the corresponding functions. The *coordinate functions* $u(x,y)$ and $v(x,y)$ thus determine the function F, which we can think of as *mapping* or *transforming* each point (x,y) in the plane into another point (u,v) in the plane. We shall therefore refer to such a function F as a *transformation*. This is a "dynamic" interpretation of the mapping F, since we think of each point in the plane as being *moved* to a new point.

However, we can give another interpretation of F in "static" terms. We simply think of each point (x,y) in the plane as remaining *fixed*, but being *relabeled with new coordinates u and v*. Thus, in this interpretation, F is thought of as a *change of coordinates*.

The "dynamic" interpretation of F as a transformation is easier to visualize geometrically, and our pictures can be considerably improved if we employ *two planes*. We label one plane with x and y axes (the "xy plane") and another plane with u and v axes (the "uv plane"). We then think of the transformation F, determined by equations 17.9, as mapping each point (x,y) in the xy plane into some point (u,v) in the uv plane (Figure 17.3).

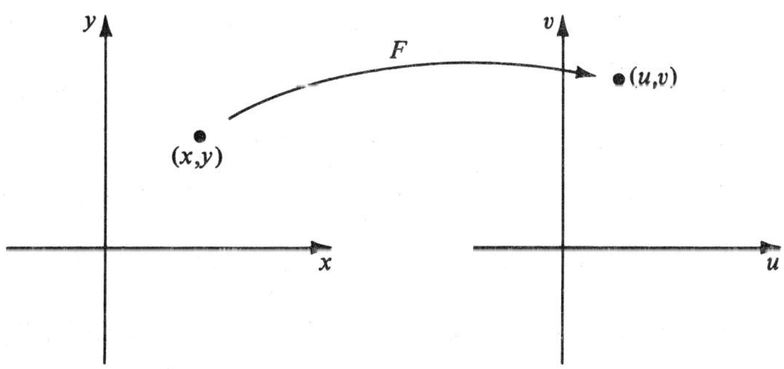

FIGURE 17.3

In the *uv* plane, the vertical and horizontal lines u = constant and v = constant form a rectangular coordinate grid. Thus the *curves* $u(x,y)$ = constant and $v(x,y)$ = constant can be thought of as forming a "curvilinear" coordinate grid in the *xy* plane (Figure 17.4). We shall refer to these curves $u(x,y)$ = constant and $v(x,y)$ = constant as the *u curves* and *v curves*, respectively, of the transformation F.

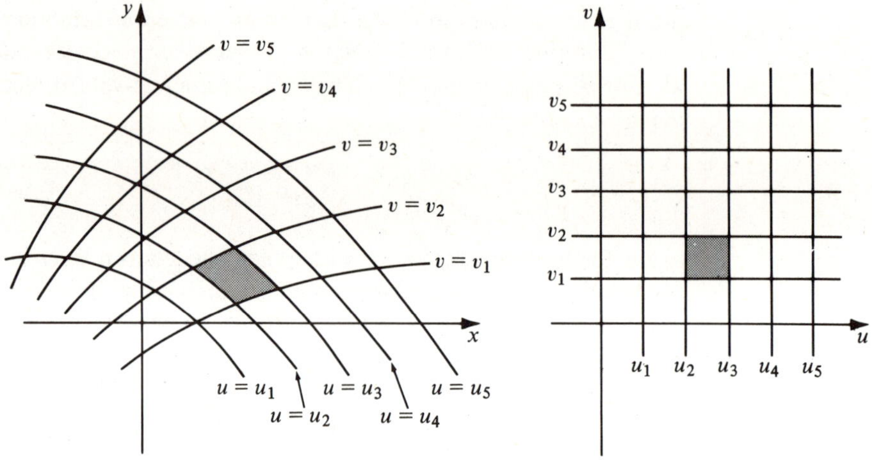

FIGURE 17.4

Example 1

Let F be the transformation defined by $u = x^2 - y^2$, $v = 2xy$.

(a) Sketch the u curves and v curves of F.
(b) Find the images in the *uv* plane of a vertical line $x = c$ and of a horizontal line $y = k$, in the *xy* plane.

Solution:

(a) The u curves are rectangular hyperbolas of the form $x^2 - y^2 = c$, for c constant. These hyperbolas have as asymptotes the lines $y = x$ and $y = -x$. The v curves are of the form $2xy = k$ (k a constant), and are rectangular hyperbolas with the x and y axes as asymptotes. See Figure 17.5.
(b) To find the curve in the *uv* plane into which F maps the line $x = c$, we substitute $x = c$ in the equations for the coordinate functions and obtain $u = c^2 - y^2$, $v = 2cy$. These may be regarded as *parametric equations* for the desired curve in the *uv* plane, where y plays the role of the *parameter*. Or, alternatively, we can eliminate y between these two equations. Thus, $y = v/2c$, and so

$$u = c^2 - \frac{v^2}{4c^2},$$

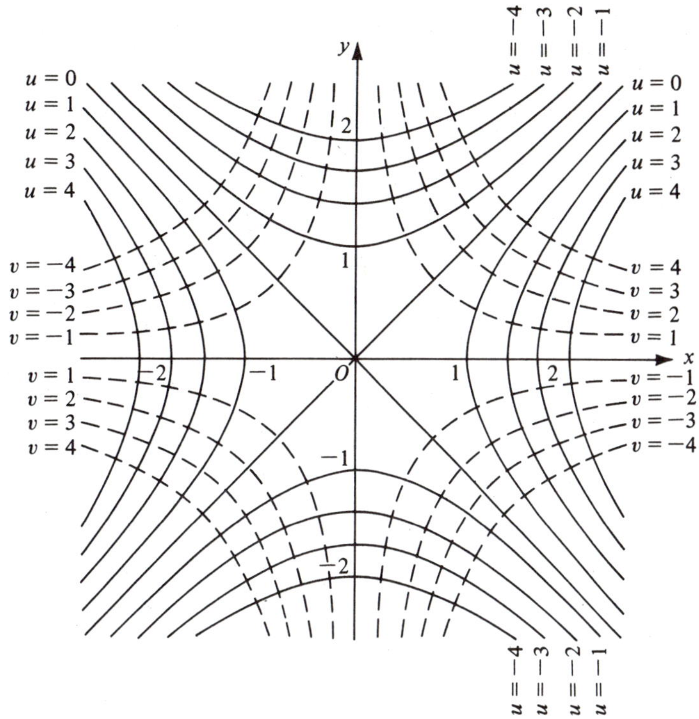

FIGURE 17.5

or
$$v^2 = 4c^2(c^2 - u),$$

which is an equation of a parabola in the uv plane.

The image of the line $y = k$ is also a parabola, whose equation is $v^2 = 4k^2(k^2 + u)$.

We shall say that a transformation F is 1–1 if distinct points in the xy plane are mapped by F into distinct points in the uv plane. If F is 1–1, then F has an *inverse transformation* F^{-1}, whose coordinate functions would be given by equations of the form

17.10
$$x = x(u,v),$$
$$y = y(u,v).$$

Example 2 (*Polar Coordinates*) The reader will recall that every point $(x,y) \neq (0,0)$ in the xy plane can be assigned polar coordinates (r,θ) in a unique way if we restrict θ so that $0 \leq \theta < 2\pi$. As opposed to this "static" interpretation of a change in coordinates, we can also give a "dynamic" interpretation. Let us write u for r and v for θ, and think of a uv plane where the u and v axes are horizontal and vertical, respectively.

Then define a transformation F from the xy plane into the uv plane by

$$u = \sqrt{x^2 + y^2},$$
$v =$ the nonnegative angle $< 2\pi$ from the positive end of the x axis to the radius vector from (0,0) to (x,y).

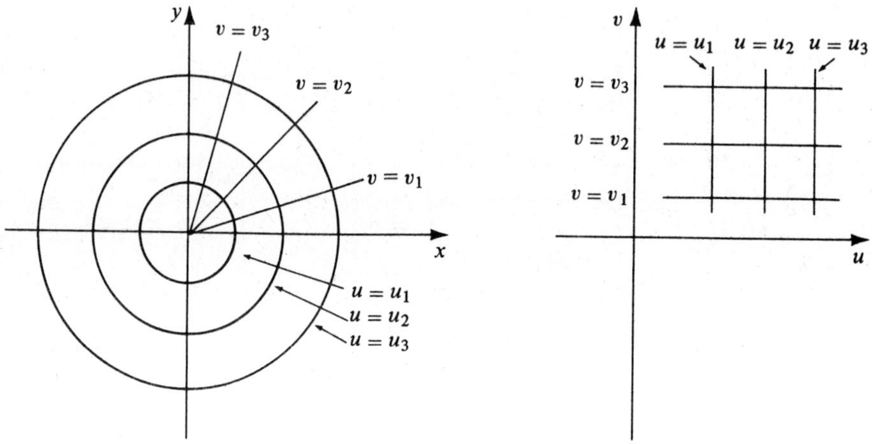

FIGURE 17.6

Note that we are now treating u and v as *rectangular coordinates* in the uv plane. Also note that F maps all of the xy plane (with the origin removed) onto the "strip"

$$S = \{(u,v) \mid 0 \le v < 2\pi \text{ and } u > 0\}$$

in the uv plane.

Since F is 1–1, it has an inverse mapping F^{-1} defined for all (u,v) in S. The coordinate functions for F^{-1} are given by the equations $x = u \cos v$, $y = u \sin v$. The u curves in the xy plane are circles centered at (0,0), while the v curves are rays emanating from (0,0). These give us the familiar polar coordinate grid in the xy plane. See Figure 17.6.

EXERCISES

I

1. Let F be the transformation defined by

$$u = \frac{x}{x^2 + y^2}, \quad v = \frac{y}{x^2 + y^2}; \qquad (x,y) \neq (0,0).$$

a. Determine the u curves and v curves.

b. Show that F is 1–1 and find the coordinate functions for F^{-1}.

2. Find the u curves and v curves for the transformation defined by

$$u = -x + \sqrt{x^2 + y^2}, \qquad v = -x - \sqrt{x^2 + y^2}.$$

3. Let F be the transformation defined by $u = e^x \cos y$, $v = e^x \sin y$. Find the region in the uv plane onto which F maps the "strip"

$$\{(x,y) \mid x \le 0 \text{ and } 0 \le y \le \pi\}.$$

4. Let F be as in Exercise 3. Find the region in the uv plane onto which F maps the rectangle with vertices $(0,0)$, $(1,0)$, $(1,\pi)$, $(0,\pi)$.
5. If F is the transformation of Exercise 1, find the image under F of the hyperbola $x^2 - y^2 = 1$. (*Hint:* Polar coordinates in the uv plane may be useful.)
6. Let F be the transformation $u = x^2 - y^2$, $v = 2xy$, of Example 1. Find the region in the xy plane that F maps onto the rectangle in the uv plane bounded by the lines $u = 1$, $u = 2$, $v = 1$, and $v = 2$.
7. Two curves in the xy plane are said to be *orthogonal* if their tangent lines are perpendicular at any point of intersection of the curves. Let F be a transformation $u = u(x,y)$, $v = v(x,y)$, where u and v have continuous first partial derivatives. Find a necessary and sufficient condition on $u(x,y)$ and $v(x,y)$ for each u curve to be orthogonal to each v curve.
8. Use the result of Exercise 7 to show that for every transformation discussed in this section and these exercises, the u curves and v curves are orthogonal.

II

1. If λ and γ are curves in the xy plane that intersect at a point P, we can define their angle of intersection at P to be the angle between their corresponding tangent vectors. Let F be the transformation $u = u(x,y)$, $v = v(x,y)$. We say that F is *conformal* if, for any curves λ and γ in the xy plane, the angle between the images of λ and γ (in the uv plane) is the same as the angle between λ and γ. Find a necessary and sufficient condition on $u(x,y)$ and $v(x,y)$ for F to be conformal.
2. Use Exercise II-1 to show that every transformation discussed in this section is conformal (except possibly at the origin).

5 CHANGE OF VARIABLES IN MULTIPLE INTEGRALS

The reader will recall the change of variable formula 5.25 for single integrals. If we interchange the roles of x and u in 5.25, we may rewrite that formula in the following form, more suitable to our present purposes.

17.11
$$\int_a^b f(g(u))g'(u)\,du = \int_c^d f(x)\,dx,$$

where x is expressed as a function $x = g(u)$, and $g(a) = c$, $g(b) = d$.

Now suppose that we have a double integral, say $\int dy \int f(x,y)\,dx$, which we wish to evaluate over some region R in the xy plane. It might happen that a change of variables, say $x = g(u,v)$, $y = h(u,v)$, could simplify the form of the integrand. How do we transform the given integral to a new integral with variables of integration u and v?

First note that we can consider the change of variables $x = g(u,v)$, $y = h(u,v)$ as defining a transformation G from the uv plane into the xy plane. We then ask if there is some integral over an appropriate region S in the uv plane whose value is the same as our given integral over R, and such that G maps the region S of the uv plane into the region R of the xy plane.

In order to handle this problem we need to make a number of assumptions about the functions involved. First we assume that the region R is the interior of a piecewise smooth simple closed curve λ in the xy plane, and that the function f has continuous first partial derivatives in some open set containing λ and its interior R. We assume that the region S is the interior of a piecewise smooth simple closed curve γ in the uv plane, that the transformation G maps S into R, and that G maps the curve γ in a 1–1 manner into the curve λ. (See Figure 17.7.) Finally, we assume that the coordinate functions of G, $x = g(u,v)$ and $y = h(u,v)$, have continuous second partial derivatives.

We shall introduce a function that plays a fundamental role in the theory of transformations. Corresponding to the transformation G, we define a function J_G of the variables u and v by

17.12
$$J_G(u,v) = g_u(u,v)h_v(u,v) - g_v(u,v)h_u(u,v) = \begin{vmatrix} \dfrac{\partial g}{\partial u} & \dfrac{\partial g}{\partial v} \\[2ex] \dfrac{\partial h}{\partial u} & \dfrac{\partial h}{\partial v} \end{vmatrix}.$$

J_G is a continuous real-valued function of u and v, which is called the *jacobian* of G.

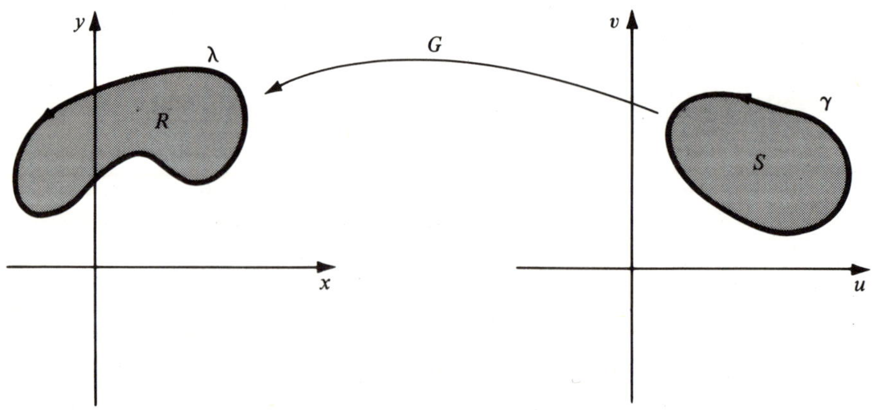

FIGURE 17.7

Now we can state the fundamental theorem on change of variables in a double integral.

17.13 Theorem

Under the assumptions of the above paragraphs,

$$\int_R dy \int f(x,y)\,dx = \pm \int_S dv \int f(g(u,v), h(u,v)) \cdot J_G(u,v)\,du,$$

where the + or − sign is to be chosen according as the point (x,y) traverses λ in the counterclockwise or clockwise direction, when the corresponding point (u,v) traverses γ in the clockwise direction.

(The notation $\int_R dy \int f(x,y)\,dx$ means, of course, that the limits of integration of this iterated integral are to be chosen so as to cover the region R.)

To prove Theorem 17.13, we first choose any function $B(x,y)$ such that $D_1B(x,y) = f(x,y)$, and such that B has continuous first partial derivatives. Then Green's theorem (17.6) tells us that

17.14
$$\int_R f = \int_R \frac{\partial B}{\partial x} = \oint_\lambda B\,dy.$$

We shall now make a change of variable in the line integral $\oint B\,dy$ in 17.14, in order to convert it to a line integral over the curve γ in the uv plane. Let us suppose that $u = u(t)$, $v = v(t)$, for $a \le t \le b$, is a parametric representation for the curve γ in the uv plane. Then $x = g(u(t),v(t))$, $y = h(u(t),v(t))$, for $a \le t \le b$, is a parametric representation for the curve λ in the xy plane. (However, as t increases from a to b, the direction in which the point (x,y) traverses λ might be the opposite of that in which the corresponding (u,v) traverses γ.) From the definition of a line integral (Section 1), we have

$$\oint_\lambda B\,dy = \int_a^b B[g(u(t),v(t)), h(u(t),v(t))] \cdot \frac{dy}{dt}\,dt,$$

where, by the chain rule,

$$\frac{dy}{dt} = \frac{\partial h}{\partial u} \cdot u'(t) + \frac{\partial h}{\partial v} \cdot v'(t).$$

If we introduce the notation

$$B^*(u,v) = B(g(u,v),h(u,v)),$$

then

$$\oint_\lambda B\,dy = \int_a^b B^*(u(t),v(t)) \left[\frac{\partial h}{\partial u} u'(t) + \frac{\partial h}{\partial v} v'(t)\right] dt$$

$$= \int_a^b \left[B^* \frac{\partial h}{\partial u} u'(t) + B^* \frac{\partial h}{\partial v} v'(t)\right] dt.$$

17.15 Or
$$\oint_\lambda B\,dy = \pm \int_\gamma \left(B^* \frac{\partial h}{\partial u}\,du + B^* \frac{\partial h}{\partial v}\,dv\right),$$

where the + or − sign is to be chosen according as (x,y) traverses λ in the same or opposite direction from that in which (u,v) traverses γ, as t goes from a to b.

The line integral on the right-hand side of 17.15 is of the form

$$\oint_\gamma (M\,du + N\,dv),$$

where $M = B^* \dfrac{\partial h}{\partial u}$, $N = B^* \dfrac{\partial h}{\partial v}$. Hence Green's theorem may be applied, and we obtain

$$\oint_\gamma (M\,du + N\,dv) = \int_S \left(\frac{\partial N}{\partial u} - \frac{\partial M}{\partial v}\right)$$

$$\int_S \left(\frac{\partial B^*}{\partial u}\frac{\partial h}{\partial v} + B^* \frac{\partial^2 h}{\partial u\,\partial v} - \frac{\partial B^*}{\partial v}\frac{\partial h}{\partial u} - B^* \frac{\partial^2 h}{\partial v\,\partial u}\right)$$

$$= \int_S \left[\left(\frac{\partial B}{\partial x}\frac{\partial g}{\partial u} + \frac{\partial B}{\partial y}\frac{\partial h}{\partial u}\right)\frac{\partial h}{\partial v} - \left(\frac{\partial B}{\partial x}\frac{\partial g}{\partial v} + \frac{\partial B}{\partial y}\frac{\partial h}{\partial v}\right)\frac{\partial h}{\partial u}\right]$$

$$= \int_S \frac{\partial B}{\partial x}\left(\frac{\partial g}{\partial u}\frac{\partial h}{\partial v} - \frac{\partial g}{\partial v}\frac{\partial h}{\partial u}\right)$$

$$= \int_S dv \int f(g(u,v), h(u,v)) \cdot J_G(u,v)\,du.$$

The theorem now follows from this last equation and equations 17.14 and 17.15.

The reader will note the analogy between the formulas 17.11 and 17.13. In Theorem 17.13 we see that the jacobian $J_G(u,v)$ plays the role corresponding to $g'(u)$ in 17.11.

For functions of three (or more) variables there is also a corresponding version of Theorem 17.13. Suppose that we have a transformation G defined by

$$\begin{aligned} x &= x(u,v,w) \\ y &= y(u,v,w) \\ z &= z(u,v,w). \end{aligned}$$

We may interpret G as a mapping from a uvw space into an xyz space. Define the jacobian $J_G(u,v,w)$ of G by

$$J_G(u,v,w) = \begin{vmatrix} \dfrac{\partial x}{\partial u} & \dfrac{\partial x}{\partial v} & \dfrac{\partial x}{\partial w} \\[2ex] \dfrac{\partial y}{\partial u} & \dfrac{\partial y}{\partial v} & \dfrac{\partial y}{\partial w} \\[2ex] \dfrac{\partial z}{\partial u} & \dfrac{\partial z}{\partial v} & \dfrac{\partial z}{\partial w} \end{vmatrix}.$$

Now suppose that R and S are appropriate regions of xyz and uvw space such that R is the image of S with respect to the mapping G. With appropriate assumptions on all the functions involved, we then have the formula

17.16

$$\int_R dz \int dy \int f(x,y,z)\, dx$$
$$= \pm \int_S dw \int dv \int f(x(u,v,w),y(u,v,w),z(u,v,w))J_G(u,v,w)\, du.$$

Returning to 17.13, we note that if the function f is chosen identically equal to 1, then the left-hand side of 17.13 is simply the area of R. Thus

17.17

$$A(R) = \pm \int_S dv \int J_G(u,v)\, du.$$

Therefore, *if the jacobian $J_G(u,v)$ is never negative for (u,v) in S, only the + sign can hold in* 17.13 (and a similar statement holds concerning 17.16). If $J_G(u,v)$ is always negative or 0, then the minus sign must be chosen in 17.13.

We shall illustrate the use of 17.13 and 17.16 with some examples.

Example 1 Let R be the region in the upper half of the xy plane bounded by the parabolas $y^2 = 4(1 - x)$, $y^2 = 4(1 + x)$, and the x axis. Compute $\int_R dy \int \sqrt{x^2 + y^2}\, dx$ by making the change of variables $x = u^2 - v^2$, $y = 2uv$.

Solution: We consider the change of variables as a mapping G of the uv plane into the xy plane. The reader may verify that G maps the lines $u = 1$ and $v = 1$ into the parabolas $y^2 = 4(1 - x)$ and $y^2 = 4(1 + x)$, respectively. Also G maps the positive half of the u axis into the positive half of the x axis, and the positive half of the v axis into the negative half of the x axis. Hence the rectangle S, with vertices (0,0), (1,0), (1,1), and (0,1) in the uv plane, is mapped by G into the region R. See Figure 17.8. Furthermore, (x,y) traces the boundary of R in the counterclockwise direction as the corresponding point (u,v) traces the boundary of S in a counterclockwise manner.

The jacobian of G is

$$J_G(u,v) = \begin{vmatrix} 2u & -2v \\ 2v & 2u \end{vmatrix} = 4(u^2 + v^2).$$

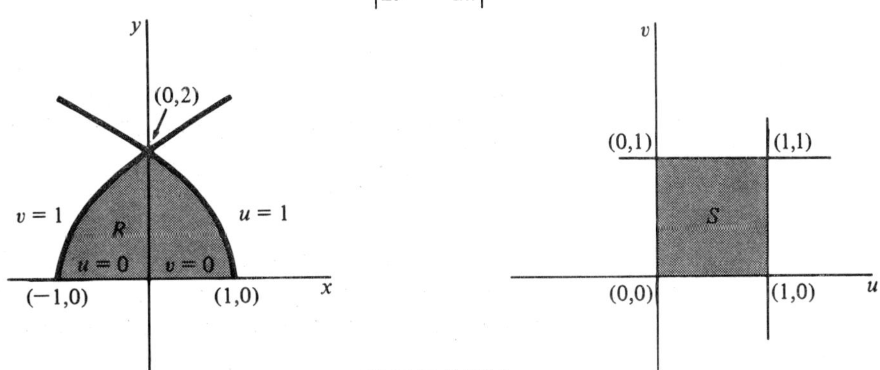

FIGURE 17.8

Hence,

$$\int_R dy \int \sqrt{x^2 + y^2}\, dx = \int_S dv \int \sqrt{(u^2 - v^2)^2 + (2uv)^2} \cdot 4(u^2 + v^2)\, du$$

$$= 4 \int_S dv \int (u^2 + v^2)(u^2 + v^2)\, du$$

$$= 4 \int_0^1 dv \int_0^1 (u^2 + v^2)^2\, du = \tfrac{112}{45}.$$

Example 2 The change from polar to rectangular coordinates is defined by a transformation G with coordinate functions $x = r\cos\theta$, $y = r\sin\theta$. G can be regarded as a mapping of an $r\theta$ plane into the xy plane (see Example 2, Section 4). The jacobian of G is

$$J_G(r,\theta) = \begin{vmatrix} \cos\theta & -r\sin\theta \\ \sin\theta & r\cos\theta \end{vmatrix} = r.$$

Suppose, for example, it is required to compute

$$\int_R dy \int e^{-x^2-y^2}\, dx,$$

where R is the interior of the circle $x^2 + y^2 = 1$. With respect to the above transformation G, the rectangle $S = \{(r,\theta) \mid 0 \le r \le 1, 0 \le \theta \le 2\pi\}$ in the $r\theta$ plane is mapped into the region R. Thus, from 17.13 we have

$$\int_R dy \int e^{-x^2-y^2}\, dx = \int_0^{2\pi} d\theta \int_0^1 e^{-r^2} \cdot r\, dr$$

$$= \int_0^{2\pi} \left(-\frac{e^{-r^2}}{2}\right)\Bigg|_0^1 d\theta$$

$$= -\tfrac{1}{2} \int_0^{2\pi} (e^{-1} - 1)\, d\theta = (1 - e^{-1})\pi.$$

Example 3 (*Spherical Coordinates*) The rectangular coordinates (x,y,z) of a point in space are related to its spherical coordinates (ρ,θ,ϕ) by the equations (14.24):

$$\begin{aligned} x &= \rho\sin\phi\cos\theta \\ y &= \rho\sin\phi\sin\theta \\ z &= \rho\cos\phi. \end{aligned}$$

These equations may be regarded as defining a transformation G from $\rho\phi\theta$ space to xyz space. The jacobian of G is

$$J_G(\rho,\phi,\theta) = \begin{vmatrix} \sin\phi\cos\theta & \rho\cos\phi\cos\theta & -\rho\sin\phi\sin\theta \\ \sin\phi\sin\theta & \rho\cos\phi\sin\theta & \rho\sin\phi\cos\theta \\ \cos\phi & -\rho\sin\phi & 0 \end{vmatrix}$$

$$= \rho^2\sin\phi.$$

Thus, in this case, the change of variable formula 17.16 for triple integrals becomes

$$\int_R dz \int dy \int f(x,y,z)\, dx = \int_S d\rho \int d\phi \int f^*(\rho,\phi,\theta)\rho^2 \sin\phi\, d\theta,$$

where $f^*(\rho,\phi,\theta) = f(\rho \sin\phi \cos\theta, \rho \sin\phi \sin\theta, \rho\cos\phi)$. This agrees with the result of our intuitive discussion in Section 11, Chapter 16.

EXERCISES

In each of Exercises 1 to 3, convert the given integral to polar coordinates and evaluate.

1. $\int_0^2 dx \int_0^{\sqrt{2x-x^2}} (x^2 + y^2)\, dy$

2. $\int_0^a dy \int_{-\sqrt{a^2-y^2}}^0 x\sqrt{x^2 + y^2}\, dx$

3. $\int_0^1 dx \int_{x^2}^x (x^2 + y^2)^{-1/2}\, dy$

4. Compute $\int_R dx \int xy\, dy$, where R is the interior of the circle $x^2 + y^2 \leq 1$, by making the change of variable $x = u^2 - v^2$, $y = 2uv$.
5. Let R be the region in the xy plane bounded by $x = 2$, $y = x$, and $y = -x^2$. Let G be the transformation defined by $x = u + v$, $y = v - u^2$.
   a. Find the region S in the uv plane that G maps into R.
   b. Use 17.17 to find the area $A(R)$ by evaluating an integral over S.
6. Let R be the same region as in Exercise 5. Using the change of variables in Exercise 5, compute $\int_R dx \int (x - y)\, dy$.
7. Let R be the triangular region in the xy plane bounded by $x = 0$, $y = 0$, and $x + y = 1$. Compute $\int_R dx \int \exp\,[(x - y)/(x + y)]\, dy$. [*Hint:* The form of the integrand suggests setting $u = x - y$, $v = x + y$. However, we must take for G the *inverse* of this transformation, namely $x = \frac{1}{2}(u + v)$, $y = \frac{1}{2}(-u + v)$.]
8. Use the method of Exercise 7 to compute $\int_R dx \int [(y - x)/(y + x)]^{1/2}\, dy$, where R is the trapezoid in the xy plane bounded by the lines $x = 0$, $y = x$, $x + y = 1$, and $x + y = 2$.
9. a. Compute the volume of the ellipsoid

$$\frac{x^2}{a^2} + \frac{y^2}{b^2} + \frac{z^2}{c^2} \leq 1.$$

   (*Hint:* The transformation $x = au$, $y = bv$, $z = cw$ maps the sphere $u^2 + v^2 + w^2 \leq 1$ in uvw space into the given ellipsoid.)

   b. Compute $\int dx \int dy \int xyz\, dz$ over the ellipsoid in part (a) above.

10. Let R be the region in the first quadrant of the xy plane bounded by $xy = 1$, $xy = 2$, $y = x$, and $y = 4x$. Compute $\iint_R xy\,dy\,dx$ by making the change of variables $x = u/v$, $y = v$.

11. Let R be the region in the xy plane bounded by $x = 0$, $y = 3x$, $2x + y = 2$, and $2x + y = 4$. Compute $\iint_R x\,dy\,dx$ by making the change of variables $u = 3x - y$, $v = 2x + y$.

CHAPTER REVIEW

I

1. Evaluate the line integral $\int x^2\,dy + y^2\,dz + z^2\,dx$ along the curve

$$\lambda(t) = (t^2, t + 1, t - 1), \qquad \text{domain } \lambda = [0,1].$$

2. Compute $\int_\lambda x\,dx + 2y\,dy + z\,dz$, where λ is the curve of intersection of the surfaces $y^2 = x$ and $z^2 = x$ from (0,0,0) to (1,1,1).

3. Compute $\int_\lambda y\,dx + z\,dy + x\,dz$, where λ is the curve of intersection of the surfaces $x^2 + y^2 = 1$ and $y^2 + z^2 = 1$ (taken in the counterclockwise direction when viewed from above).

4. Let R be the bounded region in the first quadrant of the xy plane enclosed by the curves $xy = 1$, $xy = 3$, $x(1 - y) = 1$, and $x(1 - y) = 2$. Compute $\int dx \int x\,dy$ over R by making the change of variables $x = u + v$, $y = v/(u + v)$.

5. Compute $\int_0^\infty dx \int_0^\infty e^{-x^2-y^2}\,dy$ by changing to polar coordinates.

6. Let R^4 denote the four-dimensional space consisting of the set of all ordered quadruples (x,y,z,w) of real numbers. Consider a change of coordinates in R^4 defined by

$$x = r\cos\theta, \qquad y = r\sin\theta, \qquad z = \rho\cos\phi, \qquad w = \rho\sin\phi.$$

   a. What should be the formula for making this change of variable in a fourfold repeated integral?

   b. Use (a) to show that the repeated integral $\int dx \int dy \int dz \int dw$ over the region $S = \{(x,y,z,w) \mid x^2 + y^2 + z^2 + w^2 \le a^2\}$ in R^4 has the value $\frac{1}{2}\pi^2 a^4$. This result can be interpreted as the "volume" of the "sphere" S in R^4.

II

1. Find the area of the loop of the folium of Descartes, given by $x^3 + y^3 - 3axy = 0$, using one of the line integrals of Example 3, Section 2. Obtain a parametric representation for this curve by setting $y = tx$.

2. Find the area inside the hypocycloid $x = a\cos^3 t$, $y = a\sin^3 t$.
3. By changing to polar coordinates, compute the integral

$$\int dx \int \frac{dy}{(1 + x^2 + y^2)^2}$$

over a loop of the lemniscate $(x^2 + y^2)^2 - (x^2 - y^2) = 0$.
4. Compute the integral in Exercise 3 over the triangle with vertices (0,0), (1,0), (0,1).

18

Differential Equations

The theory of differential equations is a large part of mathematics, and the application of the results of this theory constitutes a strong tool of science. Parts of the brief treatment of the subject in this chapter suggest the character of the general theory. A surprising number of applications may be made of the limited set of topics covered.

1 INTRODUCTION

If G is a function of $n + 2$ variables, the equation

$$G(x, y, y', y'', \ldots, y^{[n]}) = 0,$$

where $y', y'', \ldots, y^{[n]}$ formally designate the first, second, ..., nth derivative of y at x, is called an *ordinary differential equation of order n.* A function f is a *solution* of this equation if

$$G(x, f(x), f'(x), f''(x), \ldots, f^{[n]}(x)) = 0$$

for every x in the domain of f.

The separable differential equation

18.1 $$M(x) + N(y)y' = 0$$

studied in Chapter 9 is an example of an ordinary differential equation of order 1.

The equation

$$y'' - 4y = 0$$

is an example of an ordinary differential equation of order 2. It is easily verified that the function f defined by $f(x) = e^{2x}$ is a solution of this equation.

In contrast to ordinary differential equations, an equation such as

$$\frac{\partial^2 z}{\partial x^2} = \frac{\partial^2 z}{\partial y^2}$$

is called a *partial differential equation.* A function f of two variables is a solution of this equation if

$$\frac{\partial^2 f}{\partial x^2} = \frac{\partial^2 f}{\partial y^2}.$$

We shall focus our attention in this chapter on ordinary differential equations of the more elementary types and of order 1 or 2.

If M and N are continuous functions, then it was shown in 9.7 that the separable differential equation 18.1 has solution

18.2 $$\int M(x)\,dx + \int N(y)\,dy = C.$$

That is, every solution f (with continuous derivative) of 18.1 satisfies 18.2 for some constant C, and vice versa.

For example, the differential equation

(1) $$3x^2 + 1 + e^y y' = 0$$

has solution

(2) $$x^3 + x + e^y = C,$$

by 18.2. Thus each differentiable function f that is a solution of (1) satisfies (2) [with $y = f(x)$] for some constant C, and vice versa. Equation (2) is called an *implicit solution* of (1). We may solve (2) for y, thereby obtaining an explicit solution

$$y = \ln (C - x^3 - x)$$

of (1). The function f defined by $f(x) = \ln (C - x^3 - x)$ is a solution of (1) for every constant C.

Equation (2) is typical of the solution of a differential equation of order 1 in that it contains one arbitrary parameter C. Such a description of the solution of a differential equation in terms of one or more parameters is called the *general solution* of the equation.

For example, the differential equation

(3) $$y'' - 4y = 0$$

may be shown to have the general solution

(4) $$y = C_1 e^{2x} + C_2 e^{-2x}.$$

We note in this example that the differential equation is of order 2 and the general solution has two parameters C_1 and C_2.

On the other hand, the second-order partial differential equation

$$\frac{\partial^2 z}{\partial x^2} = \frac{\partial^2 z}{\partial y^2}$$

has solutions described in terms of two arbitrary *functions* f and g:

$$z = f(x - y) + g(x + y).$$

Example 1

Verify that, for any constants C_1 and C_2, equation (4) above is a solution of (3).

Solution: If y is as given in (4), then

$$y' = 2C_1e^{2x} - 2C_2e^{-2x}, \qquad y'' = 4C_1e^{2x} + 4C_2e^{-2x},$$

and $y'' = 4y$. Thus (4) is a solution of (3) for any constants C_1 and C_2.

In this example the solution of the differential equation is given explicitly, so that we might verify it by direct computation of y' and y'' and subsequent substitution of these functions into the given differential equation. If the solution of a differential equation is given implicitly, then the solution may be verified by implicit differentiation, as illustrated in the following example.

Example 2

Show that, for every constant C,

$$x^2 - 2xy + y^4 = C$$

is a solution of the differential equation

$$x - y + (2y^3 - x)y' = 0.$$

Solution: If f is a differentiable function such that $y = f(x)$ satisfies

$$x^2 - 2xy + y^4 = C$$

for some constant C, then we have by implicit differentiation that

$$2x - 2y + (-2x + 4y^3)y' = 0$$

or

$$x - y + (2y^3 - x)y' = 0.$$

Example 3

Find the general solution of the differential equation

$$\frac{1}{x} + \frac{y'}{y} = 0.$$

Solution: This separable differential equation has the general solution

$$\int \frac{1}{x}\,dx + \int \frac{1}{y}\,dy = C_1,$$

or

(1) $$\ln|x| + \ln|y| = C_1,$$

where C_1 is a parameter. That is, a function f is a solution of the given differential equation if and only if $y = f(x)$ is a solution of (1) for some constant C_1. It is evident that the solution (1) may be put in the form $\ln |xy| = C_1$, or

$$|xy| = e^{C_1} = C_2. \tag{2}$$

We can show, moreover, that

$$xy = C \tag{3}$$

(where $C = \pm C_2$) is also a solution for each nonzero C. Thus

$$y = \frac{C}{x}$$

is an explicit solution of the given equation.

That (2) implies (3) is a consequence of the following remarks. If the function G is continuous in an interval $[a,b]$ and if $G(x) \neq 0$ in this interval, then either $G(x) > 0$ or $G(x) < 0$ in $[a,b]$. Hence either $|G(x)| = G(x)$ or $|G(x)| = -G(x)$ in $[a,b]$.

EXERCISES

In each of Exercises 1 to 7 verify that if C, C_1, C_2 are any constants then the given relation between x and y satisfies the corresponding differential equation.

1. $y = Ce^{2x};\ y' = 2y$

2. $y = Ce^{x^2};\ y' = 2xy$

3. $y = C_1 + C_2x;\ y'' = 0$

4. $y = C_1 \sin x + C_2 \cos x;\ y'' + y = 0$

5. $xy + \cos x = C;\ xy' + y = \sin x$

6. $y = \dfrac{C + x}{x^2 + 1};\ y' = \dfrac{1 - 2xy}{x^2 + 1}$

7. $y = C_1(x^2 + C_2);\ y' = y''x$

In each of Exercises 8 to 11, find a differential equation that has the given relation as its general solution.

8. $x = y + C(1 - xy)$ (*Suggestion:* Solve for C and then differentiate both sides.)

9. $xy = C(x^3 + y^3)$

10. $y = C_1e^{3x} + C_2e^{-x}$

11. $y = e^{2x}(C_1 \cos x + C_2 \sin x)$

12. Show that if f and g are any functions possessing second derivatives and if $z = f(x - y) + g(x + y)$, then $\dfrac{\partial^2 z}{\partial x^2} = \dfrac{\partial^2 z}{\partial y^2}$.

Find the general solution of the following differential equations.

13. $x - yy' = 0$

14. $\sin x - (\sin y)y' = 0$

15. $dy/dx = y/x$

16. $x + y(1 + x^2)D_xy = 0$

17. $xy' = 2y$

18. $y' = x^{-2}$

19. $y'' = x^{-1}$

20. $y''' = x$

2 FAMILIES OF CURVES

The differential equation

$$y' = 2$$

has the general solution

$$y = 2x + C.$$

We may interpret this solution as the set, or *family*, of all straight lines in the plane having slope 2 (Figure 18.1).

Similarly, the equation

$$xy' + y = 0$$

has the general solution

$$xy = C,$$

which may be interpreted as the family of all hyperbolas in the plane, each having the coordinate axes as asymptotes. Note that in these examples the parameter C of the general solution of a differential equation is the parameter of the family of curves.

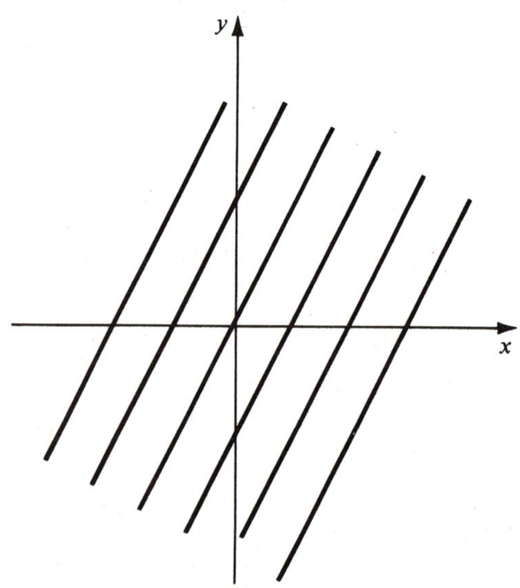

FIGURE 18.1

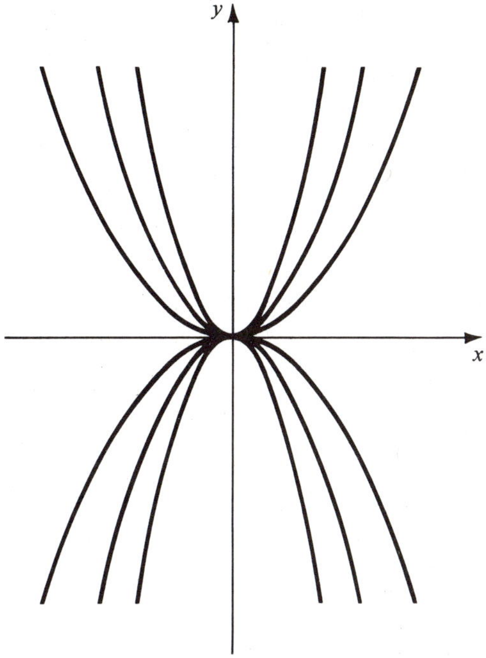

FIGURE 18.2

It is true, conversely, that a family of curves described with one parameter may often be shown to be the general solution of a first-order differential equation.

For example, the equation

$$y = Cx^2 \tag{1}$$

describes the family of parabolas, each of which has its vertex at the origin and axis along the y axis (Figure 18.2). If we differentiate (1), we obtain

$$y' = 2Cx. \tag{2}$$

We may eliminate C between (1) and (2), thereby getting the differential equation

$$xy' = 2y. \tag{3}$$

It is easily shown that (1) is the general solution of (3), so that (3) is completely descriptive of the given family of parabolas.

According to (3), $y' = 2y/x$; that is, at each point (x,y) (other than the origin) on each of the parabolas of the given family, the slope of the tangent line to that parabola is $2y/x$. Therefore this is a property of every member of the family. Equation (2), on the other hand, expresses a property of a particular member of the family.

Example 1 Find a differential equation describing the family of hyperbolas

(1) $$xy = Cx - 1.$$

Solution: We may differentiate (1), obtaining

(2) $$xy' + y = C.$$

On eliminating C between (1) and (2), we get

$$xy = (xy' + y)x - 1$$

or

$$x^2y' = 1$$

as the differential equation of the family of hyperbolas.

Example 2 Each member of a family of curves has the property that its slope is $-x/y$ at each point (x,y), $y \neq 0$, on the curve. Find an equation describing this family.

Solution: By assumption, the family is described by the differential equation $y' = -x/y$, or

$$x + yy' = 0.$$

This separable differential equation has the solution

$$x^2 + y^2 = C,$$

which is therefore an equation of the given family of curves. Clearly, this family consists of all circles in the plane, each with center at the origin.

3 BOUNDARY CONDITIONS

It was shown above that the differential equation $x + yy' = 0$ has a family of circles $x^2 + y^2 = C$ as its general solution. There is one and only one circle of this family passing through a given point of the plane.

For example, to find the circle passing through the point (2,1), we must determine the value of C such that $2^2 + 1^2 = C$. Clearly, $C = 5$ and

$$x^2 + y^2 = 5$$

is the particular circle of the family passing through the point (2,1).

The equation $x^2 + y^2 = 5$ is called a *particular solution* of $x + yy' = 0$, as distinguished from the general solution $x^2 + y^2 = C$. It is the unique solution satisfying the condition $y = 1$ when $x = 2$. Such a condition is called a *boundary condition* of the given differential equation.

A differential equation of order 1 has one parameter in its general solution, and one boundary condition suffices to determine a particular solution. For a differential equation of order 2, two boundary conditions are needed to determine a particular solution (for example, $y = b$ and $y' = c$ when $x = a$). Similar remarks may be made for higher order equations.

Example 1 Determine the member of the family of parabolas $y = Cx^2$ that passes through the point $(10, -3)$.

Solution: Substituting $x = 10$ and $y = -3$ in $y = Cx^2$, we get $-3 = 100C$ and hence $C = -.03$. Thus

$$y = -.03x^2$$

is the parabola desired.

Example 2 Find the particular solution of the differential equation

$$(1 - \tan^2 \theta) + 2r \tan \theta \frac{d\theta}{dr} = 0$$

satisfying the boundary condition $r = 1$ when $\theta = \pi/8$.

Solution: The given differential equation may be put in the form

$$\frac{1}{r} + \frac{2 \tan \theta}{1 - \tan^2 \theta} \frac{d\theta}{dr} = 0,$$

or

$$\frac{1}{r} + \tan 2\theta \frac{d\theta}{dr} = 0.$$

Its general solution is therefore

$$\ln |r| + \tfrac{1}{2} \ln |\sec 2\theta| = C_1,$$

or

$$r^2 = C \cos 2\theta.$$

If $\theta = \pi/8$ and $r = 1$, we find $C = \sqrt{2}$. Hence

$$r^2 = \sqrt{2} \cos 2\theta$$

is the desired particular solution.

Example 3 Find a function f that is a solution of the differential equation

$$x + e^x + 3y^2y' = 0$$

and is such that $f(0) = 3$.

Solution: We are asked to find an explicit solution $y = f(x)$ of the equation satisfying the boundary condition $y = 3$ when $x = 0$. The given differential equation has the general solution

$$\tfrac{1}{2}x^2 + e^x + y^3 = C,$$

or

$$y = \sqrt[3]{C - \tfrac{1}{2}x^2 - e^x}.$$

If $y = 3$ when $x = 0$, then $C = 28$. Hence

$$f(x) = \sqrt[3]{28 - \tfrac{1}{2}x^2 - e^x}$$

defines the solution for which $f(0) = 3$.

EXERCISES

Sketch four different members of each of the following families of curves, and find a differential equation of the family.

1. $y = Cx$
2. $y^2 = x - C$
3. $y = x^2 + C$
4. $\frac{x^2}{4} + \frac{y^2}{9} = C$
5. $x^2 - y^2 = C$
6. $x^2 + (y - C)^2 = 1$
7. $(x - C)^2 + (y - C)^2 = C^2$
8. $\frac{x^2}{C^2} + \frac{y^2}{9} = 1$

Solve each of the following differential equations, and sketch that member of the family of solutions passing through the given point.

9. $y' = 2x$; $(-2,0)$
10. $y' = -3$; $(1,-1)$
11. $xy' + y = 0$; $(-1,3)$
12. $y' = 2\sqrt{y}$; $(2,4)$
13. $yy' + x - 2 = 0$; $(0,0)$
14. $(y - 1)y' = x$; $(0,1)$

In each of Exercises 15 to 18 find the particular solution of each of the following differential equations that satisfies the given boundary condition.

15. $r^2 \frac{dr}{d\theta} = \sin \theta$; $r = 1$ when $\theta = \frac{\pi}{4}$
16. $\frac{dz}{du} = \ln u$; $z = 2$ when $u = e$
17. $yy' = x + 1$; $y = 3$ when $x = 0$
18. $e^{y-x}y' + x = 0$; $y = 2$ when $x = 2$
19. Show that $y'' = 0$ is the differential equation of the family of curves $y = C_1x + C_2$. Interpret this result geometrically.
20. Show that the particular solution of the differential equation

$$M(x) + N(y)y' = 0$$

satisfying the boundary condition $y = y_0$ when $x = x_0$ is given by

$$\int_{x_0}^{x} M(x)\,dx + \int_{y_0}^{y} N(y)\,dy = 0.$$

Solve Exercises 17 and 18 by this method.

21. If one family of curves has the differential equation $y' = G(x,y)$ and another family of curves has the differential equation $y' = -1/G(x,y)$, then each member of the first family is *orthogonal* to each member of the second family (i.e., they meet at right angles). The second family is called the set of *orthogonal trajectories* of the first family, and vice versa. Show that the two families with equations $y^2 = Cx^3$ and $2x^2 + 3y^2 = K$ are orthogonal trajectories of each other.

In each of Exercises 22 to 27 find the orthogonal trajectories of the family of curves. (*Hint:* Find the differential equation of the family, replace y' by $-1/y'$, and then solve the resulting differential equation.)

22. $y = Cx$ **23.** $xy = Cx - 1$

24. $y^2 = x^2 + C$ **25.** $y^2 = Cx$

26. $x^2 + y^2 = C^2$ **27.** $y = e^{Cx}$

28. Show that the family of curves $y^2 = 4C(x + C)$ is self-orthogonal; i.e., each two members of the family that meet necessarily meet at right angles.

29. Show that the family of circles $(y - 1)^2 + (x - C)^2 = 1$ has the differential equation $y^2 - 2y + (y - 1)^2(y')^2 = 0$. Show that this differential equation also has the solutions $y = 0$ and $y = 2$. Explain this phenomenon geometrically.

4 EXACT DIFFERENTIAL EQUATIONS

We shall study in this section methods of solving a large class of differential equations of the first order, among which are the separable differential equations.

18.3 Definition

The differential equation

$$M(x,y) + N(x,y)y' = 0$$

is called *exact* if

$$\frac{\partial M}{\partial y} = \frac{\partial N}{\partial x}.$$

For example,

$$2x - y + (y^2 - x)y' = 0$$

is an exact differential equation, since

$$\frac{\partial}{\partial y}(2x - y) = -1 = \frac{\partial}{\partial x}(y^2 - x).$$

We shall always assume in this chapter that the functions M and N have continuous first partial derivatives. The reader will recall from Theorem 17.4 that if $\partial M/\partial y = \partial N/\partial x$, then there exists a function F of the two variables x and y such that $\nabla F = \langle M,N \rangle$; that is, there is a function F such that

$$F_x(x,y) = M(x,y), \qquad F_y(x,y) = N(x,y).$$

This enables us to prove the following theorem.

18.4 Theorem

The exact differential equation

$$M(x,y) + N(x,y)y' = 0$$

has the general solution

$$F(x,y) = C$$

where F is any smooth function of two variables such that $\nabla F = \langle M,N\rangle$.

Proof: If f is a solution of the differential equation, then by the chain rule

$$\begin{aligned} D_xF(x,f(x)) &= D_1F(x,f(x)) + D_2F(x,f(x))f'(x) \\ &= M(x,f(x)) + N(x,f(x))f'(x) = 0. \end{aligned}$$

Hence $F(x,f(x)) = C$ for some constant C.

On the other hand, if $F(x,f(x)) = C$, then $D_xF(x,f(x)) = D_xC = 0$. Therefore, using the chain rule as above, we have

$$M(x,f(x)) + N(x,f(x))f'(x) = 0.$$

Hence f is a solution.

The method for finding a potential function given in Example 1, Section 2, Chapter 17, may also be used for solving an exact differential equation.

Example

Solve the differential equation

$$2x - y + (y^2 - x)y' = 0.$$

Solution: We verified above that this equation is exact. Its solution will be $F(x,y) = C$, where F is a function such that

$$\text{(1)} \qquad F_x(x,y) = 2x - y,$$

$$\text{(2)} \qquad F_y(x,y) = y^2 - x.$$

Integrating (1) with respect to x, we obtain

$$F(x,y) = x^2 - xy + g(y),$$

where g is a function only of the variable y. Hence, by (2),

$$F_y(x,y) = -x + g'(y) = y^2 - x,$$

so that $g'(y) = y^2$ and $g(y) = y^3/3$. Thus

$$F(x,y) = x^2 - xy + \tfrac{1}{3}y^3,$$

and the given differential equation has the general solution

$$x^2 - xy + \tfrac{1}{3}y^3 = C.$$

For the separable differential equation 18.1,

$$M(x) + N(y)y' = 0,$$

we clearly have $\partial M/\partial y = \partial N/\partial x = 0$. Thus this equation is exact. If we define the function F by

$$F(x,y) = \int M(x)\,dx + \int N(y)\,dy,$$

then evidently $F_x = M$ and $F_y = N$. Hence, by 18.4,

$$\int M(x)\,dx + \int N(y)\,dy = C$$

is the general solution of 18.1, and we have another proof of 9.7.

5 THE DIFFERENTIAL NOTATION

The first-order differential equation

$$R(x,y) + S(x,y)\frac{dy}{dx} = 0$$

is equivalent to the equation

$$R(x,y)\,dx + S(x,y)\,dy = 0$$

if we use the differential notation. That is, letting $dy = y'\,dx$, $y = f(x)$ is a solution of the first equation above if and only if it is a solution of the second one when $dx \neq 0$. This second equation is called the *symmetric form* of the given differential equation.

Example 1 Solve the equation $y' = \dfrac{2xy}{y^2 - x^2}$.

Solution: If we let $y' = dy/dx$, this equation has symmetric form

$$2xy\,dx + (x^2 - y^2)\,dy = 0.$$

We recognize the equation above as an exact differential equation. If

$$F(x,y) = x^2y - \tfrac{1}{3}y^3,$$

then $D_1F(x,y) = 2xy$ and $D_2F(x,y) = x^2 - y^2$. Hence

$$x^2y - \tfrac{1}{3}y^3 = C$$

is the general solution of the given equation.

Example 2 Find the particular solution of the equation

$$(\sin y + y \sin x)\,dx + (x \cos y - \cos x)\,dy = 0$$

that satisfies the boundary condition $y = \pi/2$ when $x = \pi$.

Solution: The given equation is exact. If

$$F(x,y) = x \sin y - y \cos x,$$

then $D_1F(x,y) = \sin y + y \sin x$ and $D_2F(x,y) = x \cos y - \cos x$. Hence

$$x \sin y - y \cos x = C$$

is the general solution of the given differential equation. If $y = \pi/2$ when $x = \pi$, then $C = 3\pi/2$, and

$$x \sin y - y \cos x = \frac{3\pi}{2}$$

is the desired particular solution.

EXERCISES

Solve the following differential equations.

1. $(x + y)\,dx + (x + 2y)\,dy = 0$
2. $1 + r \cos \theta + \sin \theta \dfrac{dr}{d\theta} = 0$
3. $ye^x - x + (e^x + 1)y' = 0$
4. $ye^x - y + (e^x + 1)y' = 0$
5. $(x \sin y - y)y' = \cos y$
6. $(x \sin y - x)y' = \cos y$
7. $(ye^{xy} + 2xy)\,dx + (xe^{xy} + x^2)\,dy = 0$
8. $(r + e^\theta)\,d\theta + (\theta + e^r)\,dr = 0$
9. $y \sec^2 x\,dx + \tan x\,dy = 0$
10. $y' = \dfrac{x(6xy + 2)}{3y - 2x^3}$
11. $(e^x \sin y + y)y' = e^x \cos y$
12. $\ln (y^2 + 1) + \dfrac{2xy}{y^2 + 1} y' = 0$

In Exercises 13 and 14 find the particular solution satisfying the given boundary condition for each equation.

13. $(x - y)\,dx + (2y^3 - x)\,dy = 0$; $y = 1$ when $x = 2$
14. $y \cos xy + (1 + x \cos xy)y' = 0$; $y = -1$ when $x = \pi/4$
15. **a.** Show that if $M(x,y)\,dx + N(x,y)\,dy = 0$ is an exact differential equation then the solution which satisfies the boundary condition $x = x_0$, $y = y_0$ can be given in the form

$$\int_{x_0}^{x} M(s,y)\,ds + \int_{y_0}^{y} N(x_0,t)\,dt = 0.$$

b. Solve Exercises 13 and 14 by this method.

If $I(x,y)M(x,y)\,dx + I(x,y)N(x,y)\,dy = 0$ is an exact differential equation, then $I(x,y)$ is said to be an *integrating factor* for the equation

$$M(x,y)\,dx + N(x,y)\,dy = 0.$$

Show in each of the following exercises that $I(x,y)$ is an integrating factor, and solve the equation.

16. $y\,dx - x\,dy + 0$; $I(x,y) = y^{-2}$
17. $y\,dx - x\,dy = 0$; $I(x,y) = x^{-2}$
18. $y\,dx - x\,dy = 0$; $I(x,y) = \dfrac{1}{xy}$
19. $x + y + y' = 0$; $I(x,y) = e^x$
20. $xy' = x - 3y$; $I(x,y) = x^2$

21. $(y + x)\,dx + (y - x)\,dy = 0;\ I(x,y) = (x^2 + y^2)^{-1}$

22. Show that $1/(x^2 + y^2)$ and $2/(x^2 - y^2)$ are also integrating factors for Exercises 16, 17, and 18.

23. Using Exercise 22, solve:

a. $x\,dy - y\,dx = (x^2 - y^2)\,dx$ **b.** $x\dfrac{dy}{dx} - y = x^3 + xy^2$

6 HOMOGENEOUS EQUATIONS

A function F of two variables is said to be *homogeneous of degree n* if

18.5
$$F(tx,ty) = t^nF(x,y)$$

for every number t and every number pair (x,y) such that both (x,y) and (tx,ty) are in the domain of F.

The polynomial functions defined by

$$\begin{aligned} f(x,y) &= ax + by, \\ g(x,y) &= ax^2 + bxy + cy^2, \\ h(x,y) &= ax^3 + bx^2y + cxy^2 + dy^3, \end{aligned}$$

and so on, are examples of homogeneous functions, f of degree 1, g of degree 2, h of degree 3, and so on. As another example, the function F defined by

$$F(x,y) = x^2 + \frac{x^3 + 2y^3}{y}$$

is homogeneous of degree 2, since

$$F(tx,ty) = (tx)^2 + \frac{(tx)^3 + 2(ty)^3}{ty} = t^2F(x,y).$$

Also, the function G defined by

$$G(x,y) = \frac{1}{x + y}\sin\frac{x - y}{x + y}$$

is homogeneous of degree -1, since

$$G(tx,ty) = \frac{1}{tx + ty}\sin\frac{tx - ty}{tx + ty} = t^{-1}G(x,y).$$

The equation

18.6
$$R(x,y) + S(x,y)y' = 0$$

is called a *homogeneous differential equation* if the functions R and S are homogeneous of the same degree. Let us show how such an equation may be solved.

We shall seek a solution of 18.6 of the form

$$y = xg(x), \qquad x \neq 0,$$

for some differentiable function g. If we let $v = g(x)$, then

$$y = xv, \qquad y' = v + xv',$$

and 18.6 takes on the form

$$R(x,xv) + S(x,xv)(v + xv') = 0.$$

If R and S are homogeneous of degree n, then, by 18.5,

$$R(x,xv) = x^n R(1,v), \qquad S(x,xv) = x^n S(1,v),$$

and the above differential equation becomes (on dividing out x^n)

$$R(1,v) + S(1,v)(v + xv') = 0,$$

or

18.7
$$\frac{1}{x} + \frac{S(1,v)}{R(1,v) + vS(1,v)} v' = 0.$$

Retracing our steps, we see that if $v = g(x)$ is a solution of 18.7, then $y = xg(x)$ is a solution of 18.6.

Thus we are always able to transform a homogeneous differential equation 18.6 into a separable differential equation 18.7. The transformation $y = xv$, $(v = g(x))$, is said to *reduce* 18.6 to the simpler form 18.7. Examples of reductions of other first- and second-order differential equations will be given in the exercises.

In the differential notation the substitution

$$y = xv, \qquad dy = x\,dv + v\,dx$$

transforms the homogeneous differential equation

$$R(x,y)\,dx + S(x,y)\,dy = 0$$

into the separable differential equation

$$\frac{1}{x}\,dx + \frac{S(1,v)}{R(1,v) + vS(1,v)}\,dv = 0.$$

Example 1 Solve the differential equation

$$(y - 4x)\,dx + (y + 2x)\,dy = 0.$$

Solution: In this example

$$R(x,y) = y - 4x, \qquad S(x,y) = y + 2x,$$

and R and S are homogeneous of degree 1. The substitution

$$y = xv, \qquad dy = x\,dv + v\,dx$$

changes the given equation into the form

$$(xv - 4x)\,dx + (xv + 2x)(x\,dv + v\,dx) = 0,$$

or, on dividing by $x \neq 0$,

$$(v - 4)\,dx + (v + 2)(x\,dv + v\,dx) = 0.$$

A separation of variables yields the equation

$$\frac{1}{x}\,dx + \frac{v + 2}{v^2 + 3v - 4}\,dv = 0,$$

whose solution is

$$\ln|x| + \tfrac{2}{5}\ln|v + 4| + \tfrac{3}{5}\ln|v - 1| = C_1.$$

This may be put into the form

$$\ln|x^5(v + 4)^2(v - 1)^3| = 5C_1,$$
$$x^5(v + 4)^2(v - 1)^3 = C,$$

or

$$(xv + 4x)^2(xv - x)^3 = C.$$

Since $xv = y$, we have

$$(y + 4x)^2(y - x)^3 = C$$

as the general solution of the given differential equation.

Example 2 Solve the differential equation

$$\left(x - y\tan\frac{y}{x}\right)dx + x\tan\frac{y}{x}\,dy = 0.$$

Solution: This equation again is homogeneous of degree 1. Substituting

$$y = xv, \qquad dy = x\,dv + v\,dx$$

in this equation, we get the separable differential equation

$$\frac{1}{x}\,dx + \tan v\,dv = 0,$$

whose solution is

$$\ln|x| + \ln|\sec v| = C_1,$$

or

$$x\sec v = C.$$

Hence the original equation has solution

$$x\sec\frac{y}{x} = C.$$

EXERCISES

Solve each of the following equations.

1. $(x - 2y)\,dx + x\,dy = 0$

2. $y' = \dfrac{x^2 + y^2}{x^2}$

3. $(2ye^{y/x} - x)\,dy + (2x + y)\,dx = 0$
4. $(y^2 - x^2 + 2xy)\,dx + (y^2 - x^2 - 2xy)\,dy = 0$
5. $x^2\,dy + (y\sqrt{x^2 - y^2} - xy)\,dx = 0$
6. $(y^2 - xy - x^2)\,dx + x^2\,dy = 0$
7. $y\,dx = \left(x + y\cos\frac{x}{y}\right)dy$
8. $x\,dy = y\left(1 + \frac{y}{\sqrt{x^2 + y^2}}\right)dx$

Show that each of the equations in Exercises 9 and 10 is both homogeneous and exact, and solve it by each of the corresponding methods.

9. $(x^2 + y^2)\,dx + 2xy\,dy = 0$
10. $(2x + y)\,dx + (x + 3y)\,dy = 0$
11. Let the point (h,k) be the point of intersection of the lines

$$4x + 3y + 1 = 0,$$
$$x + y + 1 = 0.$$

Show that the transformation

$$X = x - h, \quad dX = dx,$$
$$Y = y - k, \quad dY = dy,$$

reduces the equation

$$(4x + 3y + 1)\,dx + (x + y + 1)\,dy = 0$$

to a homogeneous equation, and solve the equation.

12. Show that the transformation

$$z = 2x + y, \qquad dz = 2\,dx + dy$$

reduces the equation

$$(2x + y)\,dx + (1 - 4x - 2y)\,dy = 0$$

to a separable equation, and solve the equation. Why does this problem require a solution different from that of Exercise 11?

Solve each of the following equations, using the methods of Exercises 11 and 21.

13. $(x + 2y + 1)\,dx + (x + y)\,dy = 0.$
14. $(2x + y + 1)\,dx + (2x + y)\,dy = 0.$
15. $(4x - y - 5)\,dx + (x - 4y - 5)\,dy = 0.$
16. $(x - y + 1)\,dx + (x + y)\,dy = 0.$

A second-order differential equation of the form $F(x,y',y'') = 0$ in which y is missing may be reduced to one of the first order by the transformation $p = y'$, $p' = y''$. Solve each of the following equations.

17. $y'' = y'$
18. $y'' = -2y'$

19. $y'' = \sin x$ **20.** $xy'' = y' + 1$

21. $y'' = \sqrt{(y')^2 + 1}$ **22.** $y'' - e^x y' = 0$

7 FIRST-ORDER LINEAR DIFFERENTIAL EQUATIONS

The equation

18.8 $$y' + P(x)y = Q(x),$$

where P and Q are continuous functions, is called a *linear differential equation of the first order*. We shall show below how this equation can be solved.

If the function Q is zero, the resulting equation $y' + P(x)y = 0$ may be put in the form

$$\frac{1}{y}y' + P(x) = 0.$$

This separable differential equation has solution

$$y = C \exp\left(-\int P(x)\,dx\right),$$

or

$$y \exp\left(\int P(x)\,dx\right) = C.$$

To return to the solution of 18.8, we first note that

$$D_x y \exp\left(\int P(x)\,dx\right) = y' \exp\left(\int P(x)\,dx\right) + yP(x) \exp\left(\int P(x)\,dx\right).$$

Hence, if we multiply each side of 18.8 by $I(x)$, where

$$I(x) = \exp\left(\int P(x)\,dx\right),$$

the left side becomes $D_x yI(x)$ and 18.8 has the form

$$D_x yI(x) = Q(x)I(x).$$

Integrating, we get

$$yI(x) = \int Q(x)I(x)\,dx + C,$$

or

18.9 $$y = \exp\left(-\int P(x)\,dx\right)\left[\int Q(x) \exp\left(\int P(x)\,dx\right)dx + C\right]$$

as the general solution of 18.8. We call $I(x)$ an *integrating factor* for the equation 18.8.

Example 1 Solve the differential equation

$$y' + y = x.$$

Solution: This equation has the form 18.8 with $P(x) = 1$ and $Q(x) = x$. If we multiply each side of the equation by the integrating factor

$$I(x) = e^{\int P(x)\,dx} = e^x,$$

we obtain the equation

$$e^x y' + e^x y = xe^x,$$

or

$$D_x(ye^x) = xe^x.$$

Hence $ye^x = \int xe^x\,dx + C = (x - 1)e^x + C$, and

$$y = x - 1 + Ce^{-x}$$

is the general solution of the given equation.

Example 2 Solve the differential equation

$$\cos y\,dx = (2 + 2x \sin y)\,dy.$$

Solution: This equation obviously does not fit the form of 18.8. However, dividing by $\cos y\,dy$, it may be rewritten as

$$\frac{dx}{dy} - 2x \tan y = \frac{2}{\cos y}.$$

This has the form

$$\frac{dx}{dy} + P(y)x = Q(y),$$

which is 18.8 with the roles of x and y interchanged. Hence the equation has an integrating factor

$$\begin{aligned} I(y) &= e^{-2\int \tan y\,dy} = e^{-2\ln|\sec y|} \\ &= e^{2\ln|\cos y|} = e^{\ln \cos^2 y} = \cos^2 y. \end{aligned}$$

Multiplying through by $I(y)$, we obtain

$$\cos^2 y \frac{dx}{dy} - 2x \sin y \cos y = 2 \cos y,$$

$$D_y(x \cos^2 y) = 2 \cos y.$$

Hence the general solution is

$$x \cos^2 y = 2 \sin y + C.$$

The differential equation

18.10 $$y' + R(x)y = S(x)y^k, \qquad k \neq 0, 1,$$

is called a *Bernoulli equation.** Since k is neither 0 nor 1, this equation is not linear. However, it may be reduced to a linear equation by a suitable transformation, as we now show.

Assuming that $y \neq 0$, let us multiply each side of 18.10 by $(1 - k)y^{-k}$ to obtain

$$(1 - k)y^{-k}y' + (1 - k)R(x)y^{1-k} = (1 - k)S(x).$$

If $v = y^{1-k}$, then $v' = (1 - k)y^{-k}y'$, and the above equation may be put in the form

$$v' + (1 - k)R(x)v = (1 - k)S(x),$$

which is a linear differential equation with $P(x) = (1 - k)R(x)$ and $Q(x) = (1 - k)S(x)$ (in 18.8). This equation may be solved, as in 18.9, for v. Then $y^{1-k} = v$ is a solution of 18.10.

Example 3 Solve the differential equation

$$y' + \frac{1}{x}y = x^5y^4.$$

Solution: This is a Bernoulli equation (18.10) with $k = 4$. If we let $v = y^{-3}$ as above, and multiply each side of the given equation by $-3y^{-4}$, we obtain the equation

$$-3y^{-4}y' - \frac{3}{x}y^{-3} = -3x^5,$$

or

(1) $$v' - \frac{3}{x}v = -3x^5, \qquad v = y^{-3}.$$

This differential equation may be solved by multiplying each side by the integrating factor

$$\begin{aligned} I(x) &= e^{\int(-3/x)\,dx} \\ &= e^{-3\ln|x|} \\ &= |x|^{-3}. \end{aligned}$$

Thus $I(x) = x^{-3}$ if $x > 0$, whereas $I(x) = -x^{-3}$ if $x < 0$. In either case (1) becomes

$$\begin{aligned} x^{-3}v' - 3x^{-4}v &= -3x^2, \\ D_x x^{-3}v &= -3x^2, \end{aligned}$$

and thus

$$x^{-3}v = -x^3 + C.$$

* Jacob Bernoulli (1654–1705) proposed this equation for solution in 1695. The solution given here is that of Leibnitz published in 1696. Jacob Bernoulli is credited with the first use of the word *integral*. He was a member of a very large and famous family of Swiss mathematicians.

Hence

$$v = -x^6 + Cx^3$$

is the solution of (1), and

$$y^{-3} = -x^6 + Cx^3,$$

or

$$(-x^6 + Cx^3)y^3 = 1,$$

is the solution of the given equation.

EXERCISES

Solve each of the following differential equations.

1. $y' + xy = x$
2. $y' - 2y = 3$
3. $y' + by = c$ (b, c constants)
4. $y' - y/x = x \sin x$
5. $y' + y \tan x = \sec x$
6. $y' + y \tan x = \sin x$
7. $y' = e^{2x} + 3y$
8. $y' = e^{ax} + ay$ (a constant)
9. $(x^2 + 1)y' = 2x(x^2 + 1)^2 + 2xy$
10. $y' \sin x + y \cos x = 1$
11. $y' + xy = xy^2$
12. $y' = y^3 e^{2x} + 3y$
13. $yy' - 2y^2 = e^x$
14. $xy' + y = x^2 e^x y^2$
15. $y\,dx + (2x - y^4)\,dy = 0$
16. $y^2\,dx + (x - 2xy - 2y^2)\,dy = 0$
17. $dx + (x - x^2 e^{2y})\,dy = 0$
18. $x^3\,dx + (x^4 y - ye^{-y^2})\,dy = 0$
19. $\dfrac{dy}{dx} + \left(\dfrac{y^2 - 1}{y}\right)x = 0$
20. $2x^2 y\,dx + (y^2 - x^3)\,dy = 0$
21. $s\,ds + (s^2 - t)\,dt = 0$
22. $(r^3 - 2\cos t)\,dt + 3r^2\,dr = 0$

8 APPLICATIONS

The solutions of many physical problems are naturally described by differential equations. For example, it was shown in Chapter 7 that the amount of a radioactive substance left after a period of time is given by the differential equation

$$\frac{dy}{dt} = ky.$$

Other examples are given in this section.

Example 1 A body of mass m falls from rest in a straight line toward the earth. Describe the motion of the body, assuming that the force due to air resistance on the body is proportional to its speed.

Solution: Let us choose an x axis directed downward, with its origin at the point from which the body is dropped. Let $x = x(t)$ be the position of the body at time t, and let us assume that $x = 0$ when $t = 0$. Then $v = x'(t)$ is the velocity of the body at time t. By assumption, $x = 0$ and $v = 0$ when $t = 0$.

The weight of the body is mg, where g is the acceleration due to gravity. Thus mg is the force due to gravity acting on the body in the direction of the positive x axis. The force due to air resistance on the body will be a vector directed upward (in direction opposite to that of the motion). By assumption, its magnitude is $-k_1 v$ for some number $k_1 > 0$. Hence the sum of the forces acting on the body is $mg - k_1 v$. By Newton's second law, this force equals ma where a is the acceleration of the body. Therefore

$$ma = mg - kv.$$

It is convenient to let $k_1 = mk$, in which case the equation above yields $a = g - kv$, or

$$v' = g - kv, \tag{1}$$

as the differential equation of the motion.

Equation (1) is separable and has the solution

$$-\frac{1}{k}\ln|g - kv| = +C_1,$$

or

$$g - kv = Ce^{-kt}.$$

Since $v = 0$ when $t = 0$, we have $C = g$ and

$$v = \frac{g}{k}(1 - e^{-kt}). \tag{2}$$

We note in passing that the body has a limiting velocity given by

$$\lim_{t \to \infty} v = \lim_{t \to \infty} \frac{g}{k}(1 - e^{-kt}) = \frac{g}{k}.$$

Since $v(t) = x'(t)$, (2) is the differential equation

$$x' = \frac{g}{k}(1 \quad e^{-kt}).$$

This equation has the solution

$$x = \frac{g}{k}\left(t + \frac{1}{k}e^{-kt}\right) + C_2.$$

If we let $x = 0$ when $t = 0$, we get $C_2 = -g/k^2$. Hence

$$x = \frac{g}{k}t + \frac{g}{k^2}(e^{-kt} - 1)$$

is the equation of motion of the body.

It is interesting to note that $x'' = ge^{-kt}$, and that

$$\lim_{t \to \infty} x'' = 0.$$

That is, the forces of gravity and air resistance tend to balance each other, giving rise to the limiting velocity mentioned above.

Example 2 (*Catenary Problem*) A flexible rope is suspended from two fixed points and hangs at rest under its own weight. Find the curve in which the rope hangs.

Partial solution: Let $Q(x_0, y_0)$ be any fixed point on the curve and let $P(x,y)$ be any other point on the curve (Figure 18.3). If H_0 and H are the horizontal components of the forces acting at Q and P, respectively, then

$$H_0 = H$$

since the rope is at rest. We interpret the term *flexible rope* to mean that y' is continuous. Hence the length L of the rope between P and Q is given by

$$L = \int_{x_0}^{x} \sqrt{1 + y'^2}\, dx.$$

If w is the weight of a unit length of the rope, then wL is the weight of the rope between P and Q. The sum $V_0 + V$ of the vertical components of the forces acting at Q and P must be the weight of the rope between P and Q; i.e.,

(1) $$V_0 + V = w \int_{x_0}^{x} \sqrt{1 + y'^2}\, dx.$$

The force vector at P lies along the tangent line to the curve, and therefore

$$y' = \frac{V}{H}.$$

Hence $V = Hy' = H_0 y'$, and (1) becomes

(2) $$\frac{V_0}{H_0} + y' = k \int_{x_0}^{x} \sqrt{1 + y'^2}\, dx,$$

where $k = w/H_0$. We may differentiate each side of (2), obtaining the differential equation

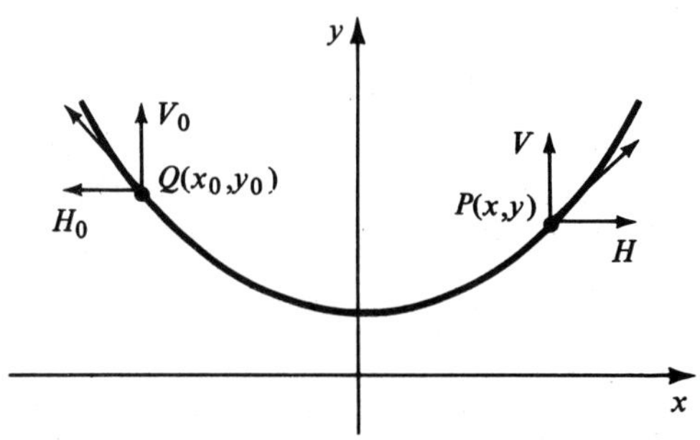

FIGURE 18.3

$$y'' = k\sqrt{1 + y'^2}. \tag{3}$$

One of the solutions of (3) has as its graph the curve in which the rope hangs.

We may start solving (3) by letting $p = y'$, thereby obtaining the equation

$$p' = k\sqrt{1 + p^2},$$

which may be solved for p. In turn, this solution is a first-order differential equation which may be solved for y. When $x = 0$, it is convenient to choose $y' = 0$ and $y = 1/k$ as boundary conditions.

Example 3 A mirror has the shape of a surface of revolution. If there exists a point F on the axis of the mirror such that light emitted from this point will be reflected from the mirror in rays parallel to the axis, prove that the mirror is parabolic and that F is the focus of the generating parabola.

Partial solution 1: Let the surface be generated by rotating the curve $r = f(\theta)$ in a polar coordinate plane about the polar axis, and let F be at the pole. If L is a light ray reflected from the surface at the point (r,θ) so that $\alpha = \beta$ in Figure 18.4, then $\theta = \alpha + \beta = 2\alpha$ if L is parallel to the polar axis. By 13.18, $r' = r\cot(\alpha - \theta) = -r\cot(\theta/2)$; that is,

$$\cot\frac{\theta}{2} = -\frac{r'}{r}.$$

Since $\cot(\theta/2) = \sin\theta/(1 - \cos\theta)$, we have

$$\frac{r'}{r} = -\frac{\sin\theta}{1 - \cos\theta}$$

as the differential equation satisfied by $r = f(\theta)$. Each solution of this equation may be shown to be a parabola with focus at F.

Partial solution 2: Let the surface be generated by rotating the curve $y = g(x)$ in a rectangular coordinate plane about the x axis, and let F be at the origin (Figure 18.5). If L is a light ray reflected from the surface at the point (x,y) so that $\alpha = \beta$ in the figure, then $\theta = \alpha + \beta = 2\alpha$ if L is parallel to the x axis. Evidently,

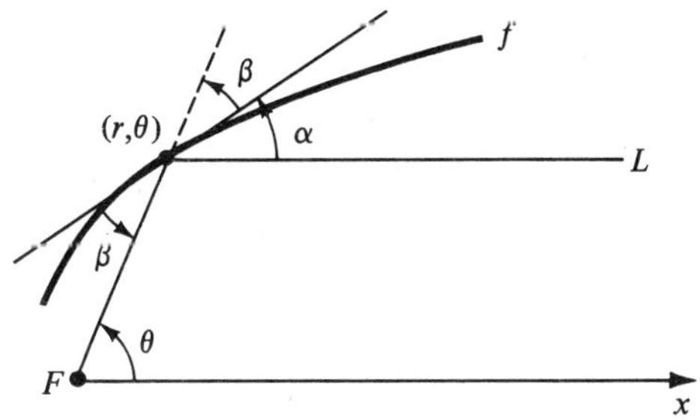

FIGURE 18.4

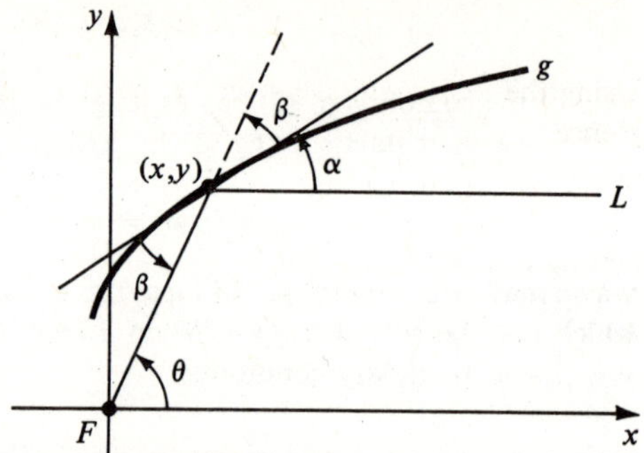

FIGURE 18.5

$$\tan \theta = \frac{y}{x}$$

and $\tan \alpha = y'$. In solving the resulting differential equation satisfied by $y = g(x)$, it is reasonable to assume that $y' > 0$.

Example 4 A large tank contains 100 gal of brine in which initially 50 lb of salt are dissolved. Brine containing 1 lb of salt per gallon then flows into the tank at the rate of 3 gal/min while the mixture runs out of the tank at the rate of 2 gal/min. How much salt is in the tank after 30 min?

Solution: Let $x = x(t)$ denote the amount of salt (in pounds) in the tank after t minutes. The fundamental principle involved in this and similar problems is that the rate of change of x with respect to t is given by

(1) $$\frac{dx}{dt} = (\text{incoming rate}) - (\text{outgoing rate}).$$

Clearly the incoming rate is 3 lb/min.. To compute the outgoing rate, note that the number of gallons in the tank at time t is $100 + t$, while the *concentration* of the salt in the tank at time t is $\frac{x}{100 + t}$ lb/gal. Hence the outgoing rate is $\frac{2x}{100 + t}$, and equation (1) becomes

$$\frac{dx}{dt} = 3 - \frac{2x}{100 + t},$$

or

$$\frac{dx}{dt} + \left(\frac{2}{100 + t}\right) x = 3.$$

We recognize this as a linear differential equation in the variable x. Using the methods of Section 7, we find its general solution to be

$$x(100 + t)^2 = (100 + t)^3 + C.$$

Using the initial condition that $x = 50$ when $t = 0$, we find that $C = -50(100)^2$. Hence

$$x = 100 + t - \frac{50(100)^2}{(100 + t)^2}.$$

Setting $t = 30$, we obtain

$$x = 130 - 50(\tfrac{100}{130})^2 \doteq 100.41 \text{ lb}.$$

EXERCISES

1. An object falls from rest in a straight line toward the earth. Assume the drag due to air resistance is proportional to the square of the velocity. Show that the differential equation of motion is $a = g - k^2v^2$, and solve the equation. It is convenient to let $r^2 = gk^2$ and to give the result in terms of r.
2. Complete the solution of Example 2.
3. Complete both solutions of Example 3.

In the circuit of Figure 18.6, R is the resistance, L is the inductance, and E is an impressed emf. If $I = I(t)$ is the current in the circuit at the time t, then $I(t)R$ is the voltage drop across the resistor and $I'(t)L$ is the voltage drop across the inductor. The sum of the voltage drops in the circuit must equal the impressed emf, that is,

$$LI' + RI = E.$$

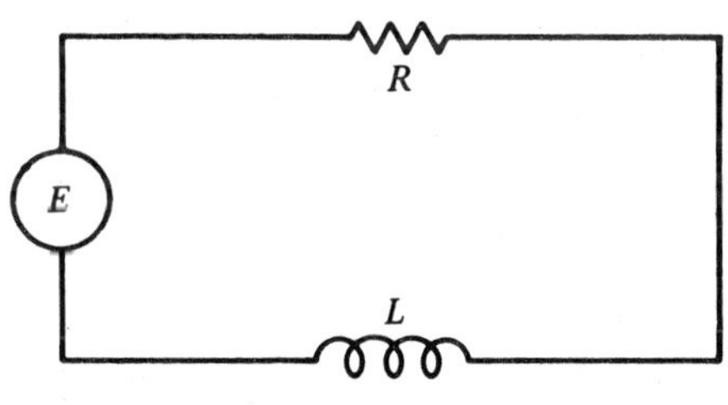

FIGURE 18.6

The numbers L and R are constants. If $I(0) = 0$, determine $I(t)$ in each of Exercises 4 to 6.

4. $L = 10, R = 5, E = 1$
5. $L = 5, R = 20, E = 1$
6. $L = 1, R = 2, E = \sin t$
7. Solve the equation

$$LI' + RI = E(t).$$

8. A tank contains 100 gal of brine in which 40 lb of salt are dissolved. Brine containing 2 lb of salt per gallon runs into the tank at the rate of 3 gal/min, while the mixture runs out of the tank at 3 gal/min. How much salt is in the tank at the end of 20 min?

9. In Exercise 8 suppose that the mixture runs out of the tank at the rate of only 1 gal/min, while the other conditions remain the same. How much salt is in the tank after 20 min?

10. A tank contains 100 gal of brine in which initially 50 lb of salt are dissolved. Fresh water then runs into the tank at the rate of 3 gal/min and the mixture runs out at the rate of 2 gal/min. How long does it take for the amount of salt in the tank to be reduced to 40 lb?

11. A tank initially contains 1000 gal of fresh water. Brine containing 3 lb/gal of salt flows into the tank at a rate of 2 gal/min, and the mixture, kept uniform by stirring, runs out at the same rate. How long will it take for the quantity of salt in the tank to reach 100 lb? As this process continues indefinitely, what is the maximum quantity of salt in the tank that can be attained?

12. A liter flask is filled with a gas A containing 10 percent of a gas B. A pure form of gas B then flows slowly through a tube into the flask, and the mixture escapes through another tube. If the process is slow enough so that the mixture in the flask may be considered uniform, what is the percentage of gas B in the flask after 10 liters have passed through?

Let A and B be chemical substances which combine to form a chemical substance C. If a and b are the amounts of A and B present at the time $t = 0$ and if $x = x(t)$ is the amount of C formed at the time t, then, under certain conditions, $a - x(t)$ and $b - x(t)$ will be the amounts of A and B, respectively, remaining at the time t. We assume $x(0) = 0$. If the rate of increase of $x(t)$ is proportional to the product of the amounts of A and B present at the time t, the chemical action is known as a *second-order process*, and the differential equation is

$$x' = k(a - x)(b - x).$$

Solve this equation in each of the cases given in Exercises 13 to 16.

13. $a = 10, b = 5$

14. $a = b = 10$

15. $a \neq b$

16. $a = b$

17. A radioactive substance A changes into a substance B, which in turn changes into a substance C. Let a be the amount of A present at the time $t = 0$. Let $x(t)$ be the amount of B which has been formed up to the time t, and let $y(t)$ be the amount of C which has been formed up to the time t. Then $x(t) - y(t)$ is the amount of B present at the time t. Assuming that the radioactivity is proportional to the amount present at a given time, we get the equations

$$x' = k_1(a - x), \qquad y' = k_2(x - y).$$

Solve for $x(t)$ and $y(t)$.

In each of Exercises 18 to 20, find the orthogonal trajectories of the given family of curves (see Exercise 21, Section 3).

18. $y = x + Ce^{-x}$ **19.** $x^2 + y^2 = Cy^3$

20. $3x^2 + y^2 = Cx$

21. Find the orthogonal trajectories of the family of all circles passing through the origin with centers on the y axis.

9 COMPLEX NUMBERS

Our purpose in this section is to show how the set R^2 of all points in the plane may be made into a *number system* in which addition, subtraction, multiplication, and division are defined, and in which these operations have many of the desirable properties enjoyed by real numbers. This number system will be called the system of *complex numbers.* It is of fundamental importance to many branches of mathematics, and in particular, will provide a useful tool in our study of differential equations.

A *complex number* is defined as an ordered pair (x,y) of real numbers. In other words, the set of all complex numbers is identical with the set of all points in R^2. Addition of two complex numbers (a,b) and (c,d) is defined by the rule

18.11 $$(a,b) + (c,d) = (a + c, b + d).$$

The reader will observe that this is identical with the definition of addition for two-dimensional vectors, which was discussed in Chapter 13.

A suitable definition of multiplication for complex numbers may be slightly less natural looking. We shall define multiplication of two complex numbers (a,b) and (c,d) by

18.12 $$(a,b) \cdot (c,d) = (ac - bd, ad + bc).$$

In particular, it should be observed that

18.13 $$\text{(i)} \qquad (a,b)(1,0) = (1,0)(a,b) - (a,b)$$

$$\text{(ii)} \qquad (a,b)(0,1) = (0,1)(a,b) = (-b,a).$$

The second of the above equations states, in geometrical language, that multiplying (a,b) by $(0,1)$ rotates the point (a,b) through a counterclockwise angle of 90°.

Complex numbers that lie on the x-axis behave just like real numbers. In particular,

$$(x_1,0) + (x_2,0) - (x_1 + x_2, 0),$$

$$(x_1,0) \cdot (x_2,0) = (x_1x_2,0).$$

It is therefore reasonable to denote the complex number $(x,0)$ simply by x, and to regard the set of all complex numbers on the x-axis as identical with the set of real numbers.

The special complex number (0,1) will be denoted by i. (The reader should not confuse this notation with that of Chapter 13, where the vectors $\langle 1,0\rangle$ and $\langle 0,1\rangle$ were denoted by **i** and **j**, respectively). Using this notation, we may write any complex number (x,y) as follows:

$$\begin{aligned} (x,y) &= (x,0) + (0,y), && \text{by 18.11,} \\ &= (x,0) + (y,0)\cdot(0,1), && \text{by 18.13(ii),} \\ &= x + yi. \end{aligned}$$

The notation $x + yi$ in place of (x,y) has many advantages, and we shall use it frequently.

The reader probably noticed from 18.13(ii) that

$$i^2 = ii = (0,1)\cdot(0,1) = (-1,0) = -1.$$

Thus, in the complex number system there is a number, namely i, whose square is -1.

Definition 18.12 now takes the form

$$(a + bi)\cdot(c + di) = ac - bd + (ad + bc)i.$$

The right-hand side of the above equation could be obtained simply by multiplying all the quantities involved as if they were real numbers, and then replacing i^2 by -1. It thus becomes unnecessary for the student to memorize 18.12. Simply perform multiplication using the notation $x + yi$, and replace i^2 by -1 whenever it occurs.

A complex number is frequently denoted by a single letter. If we let $z = (x,y) = x + yi$, then x is called the *real part* of z and y is called the *imaginary part* of z. (This terminology persists from times past, when complex numbers were considered "unreal" or "imaginary".)

For any complex number $z = x + yi$, the number $\bar{z} = x - yi$ is called the *conjugate* of z. The reader will notice that

18.14 $$z\bar{z} = (x + yi)(x - yi) = x^2 + y^2.$$

If we denote the non-negative real number $\sqrt{x^2 + y^2}$ by $|z|$, then equation 18.14 states that $z\bar{z} = |z|^2$. The number $|z|$ is called the *absolute value* or *modulus* of z.

Equation 18.14 also states that $z(\bar{z}/|z|^2) = 1$, provided $z \neq 0$. It follows that any non-zero complex number z has a multiplicative inverse z^{-1} given by

18.15 $$z^{-1} = \bar{z}/|z|^2, \qquad \text{for } z \neq 0.$$

It is therefore also possible to compute the quotient z_1/z_2 of two complex numbers, if $z_2 \neq 0$. The following example shows how this may be done by using conjugates.

Example 1 Compute $\dfrac{1+i}{3+4i}$.

Solution: Multiply numerator and denominator by the conjugate of the denominator:

$$\frac{1+i}{3+4i} = \frac{(1+i)(3-4i)}{(3+4i)(3-4i)} = \frac{3+4+(3-4)i}{3^2+4^2} = \frac{7}{25} - \frac{i}{25}.$$

The operation of taking conjugates satisfies the following rules, for all complex numbers z_1 and z_2.

18.16

$$\text{(i)} \quad \overline{z_1 + z_2} = \bar{z}_1 + \bar{z}_2,$$

$$\text{(ii)} \quad \overline{z_1 z_2} = z_1 \bar{z}_2,$$

$$\text{(iii)} \quad \overline{\left(\frac{z_1}{z_2}\right)} = \frac{\bar{z}_1}{\bar{z}_2}, \qquad \text{for } z_2 \neq 0,$$

The reader may prove these rules directly from the definition of conjugate.

Absolute values possess the following properties, for all z_1 and z_2.

18.17

$$\text{(i)} \quad |z_1 z_2| = |z_1| \cdot |z_2|,$$

$$\text{(ii)} \quad \left|\frac{z_1}{z_2}\right| = \frac{|z_1|}{|z_2|}, \qquad \text{for } z_2 \neq 0,$$

$$\text{(iii)} \quad |z_1 + z_2| \leq |z_1| + |z_2|.$$

The proofs of 18.17(i) and 18.17(ii) will be left as exercises. The inequality 18.17(iii) is simply the *triangle inequality*, which was proved in Chapter 13, Section 3.

The introduction of polar coordinates will give us further insight into the multiplication of complex numbers. If $z = (x,y)$, then the point (x,y) possesses polar coordinates (r,θ), where r is the non-negative real number $|z| = \sqrt{x^2 + y^2}$. The angle θ is called the *argument* of z, and is denoted by $\arg z$. The rectangular coordinates x and y are related to the polar coordinates r and θ by the formulas

$$x = r\cos\theta, \qquad y = r\sin\theta,$$

and so z may be written as

18.18 $$z = x + yi = r(\cos\theta + i\sin\theta).$$

The formula 18.18 is called the *polar form* of the complex number z.

Now suppose z_1 and z_2 are complex numbers with the polar forms

$$z_1 = r_1(\cos\theta_1 + i\sin\theta_1), \qquad z_2 = r_2(\cos\theta_2 + i\sin\theta_2).$$

Let us compute the product $z_1 z_2$:

$$\begin{aligned} z_1z_2 &= [r_1(\cos\theta_1 + i\sin\theta_1)][r_2(\cos\theta_2 + i\sin\theta_2)] \\ &= r_1r_2[\cos\theta_1\cos\theta_2 - \sin\theta_1\sin\theta_2 + i(\cos\theta_1\sin\theta_2 + \cos\theta_2\sin\theta_1)], \end{aligned}$$

and making use of the addition formulas for the sine and cosine (Appendix A),

18.19 $$z_1z_2 = r_1r_2[\cos(\theta_1 + \theta_2) + i\sin(\theta_1 + \theta_2)].$$

The last equation gives z_1z_2 in its polar form. Hence it tells us that the argument of z_1z_2 is equal to $\theta_1 + \theta_2$; that is,

$$\arg(z_1z_2) = \arg z_1 + \arg z_2.$$

If in 18.19 we take z_1 and z_2 to be the same number $z = r(\cos\theta + i\sin\theta)$, we obtain

$$z^2 = r^2(\cos 2\theta + i\sin 2\theta).$$

In general, if n is any positive integer, we can prove by induction that

18.20 $$z^n = r^n(\cos n\theta + i\sin n\theta).$$

In the special case when $r = 1$, this gives the formula

18.21 $$(\cos\theta + i\sin\theta)^n = \cos n\theta + i\sin n\theta,$$

which is known as DeMoivre's Theorem.

DeMoivre's Theorem also holds when n is a negative integer. To prove it for $n = -1$, let z be any complex number with $|z| = 1$, say $z = \cos\theta + i\sin\theta$. Then by 18.15,

$$z^{-1} = \bar{z} = \cos\theta - i\sin\theta = \cos(-\theta) + i\sin(-\theta).$$

The theorem may then be proved for all negative integers by induction.

The following examples illustrate some uses of DeMoivre's Theorem.

Example 2 Compute $(1 + i)^7$.

Solution: Instead of expanding by the binomial theorem, it is much simpler to use polar coordinates. In polar form,

$$1 + i = \sqrt{2}\left(\cos\frac{\pi}{4} + i\sin\frac{\pi}{4}\right).$$

Hence $$\begin{aligned} (1 + i)^7 &= 2^{7/2}\left(\cos\frac{7\pi}{4} + i\sin\frac{7\pi}{4}\right) \\ &= 8\sqrt{2}\left(\frac{1}{\sqrt{2}} - \frac{i}{\sqrt{2}}\right) = 8 - 8i. \end{aligned}$$

Example 3 Find the complex numbers z that satisfy the equation $z^3 = 1$.

Solution: Let $z = r(\cos\theta + i\sin\theta)$. Then

$$z^3 = r^3(\cos\theta + i\sin\theta)^3 = r^3(\cos 3\theta + i\sin 3\theta) = 1.$$

But the complex number 1 can be written in polar form three ways:

$$1 = 1(\cos 0 + i \sin 0) = 1(\cos 2\pi + i \sin 2\pi) = 1(\cos 4\pi + i \sin 4\pi).$$

Hence we must have $r = 1$, and $3\theta = 0, 2\pi$, or 4π. This gives $\theta = 0, \frac{2}{3}\pi$, or $\frac{4}{3}\pi$. There are then three solutions for z, namely

$$z_1 = 1(\cos 0 + i \sin 0) = 1$$

$$z_2 = 1\left(\cos \frac{2\pi}{3} + i \sin \frac{2\pi}{3}\right) = -\frac{1}{2} + \frac{\sqrt{3}}{2} i$$

$$z_3 = 1\left(\cos \frac{4\pi}{3} + i \sin \frac{4\pi}{3}\right) = -\frac{1}{2} - \frac{\sqrt{3}}{2} i.$$

These are the three *cube roots* of 1. (The student should note that in the above computation the use of additional values for the argument of 1, say 6π, 8π, and so on, leads only to the same solutions z_1, z_2, or z_3 over again.)

EXERCISES

Perform the indicated computations in Exercises 1–6, expressing your answers in the form $a + bi$.

1. $i(2 + i)(1 - 3i)$
2. $i^3 + i^4 + i^6$
3. $\dfrac{4 + 5i}{i}$
4. $\dfrac{2 + 3i}{1 - 2i}$
5. $\dfrac{2 + i}{i(1 - i)(1 + i)}$
6. $\dfrac{4 + 3i}{1 + 2i + i(2 + i)}$

In Exercises 7 and 8, express the roots of the given quadratic equations in the form $a + bi$.

7. $x^2 - 6x + 11 = 0.$
8. $36x^2 - 36x + 13 = 0.$
9. Solve the equation $z^2 = 8 + 6i$. *Suggestion:* Write $z = x + yi$ and equate real and imaginary parts, respectively, of the two sides of the equation.
10. Prove that the solutions of the equation $(x + yi)^2 = a + bi$, where a and b are arbitrary real numbers, are given by

$$x^2 = \tfrac{1}{2}(\sqrt{a^2 + b^2} + a) \qquad y^2 = \tfrac{1}{2}(\sqrt{a^2 + b^2} - a).$$

11. Compute $\left(\dfrac{1}{2} + \dfrac{\sqrt{3}}{2} i\right)^{15}$.
12. Compute $(1 + i)^{10}$.
13. Find the cube roots of $-i$.
14. Find the fourth roots of -16.
15. Find the sixth roots of 8.

16. Find the fourth roots of $-2 - 2\sqrt{3}i$.
17. Find all roots of the equation $x^6 + 7x^3 - 8 = 0$.
18. Prove 18.17(i) and 18.17(ii). *Suggestion:* For the proof of (i) write $|z_1z_2|^2 = (z_1z_2)(\overline{z_1z_2})$ and use 18.16(ii).
19. Let z_0 be a fixed complex number and k a positive real number. Show that the set of all complex numbers that satisfy the equation

$$z\bar{z} - \bar{z}_0 z - z_0\bar{z} + z_0\bar{z}_0 = k^2$$

consists of the circle of radius k centered at z_0.
20. What is the set of all complex numbers z that satisfy $z^2 + \bar{z}^2 = 2$?
21. Use DeMoivre's Theorem to prove the following trigonometric identities:
a. $\cos 3\theta = \cos^3\theta - 3\cos\theta\sin^2\theta$,
b. $\sin 3\theta = 3\cos^2\theta\sin\theta - \sin^3\theta$.
Suggestion: Expand $(\cos\theta + i\sin\theta)^3$ two ways.
22. Consider the polynomial equation

$$a_0 + a_1z + \cdots + a_nz^n = 0,$$

where $a_0,a_1,\ldots,a_n$ are real constants. Prove that if z_0 satisfies this equation, then so does $\bar{z}_0$. *Suggestion:* Take the conjugate of both sides, noting that the conjugate of a real number a is simply a itself.

10 EULER'S FORMULA

The number e^x is well-defined for any real number x. Is it possible to define e^z, where the exponent z is now a complex number? More precisely, can we define e^z for complex values of z so that certain desirable properties of e^x, such as the laws of exponents, carry over to the complex case?

Some insight into an appropriate definition will be obtained by recalling the MacLaurin series for e^x:

$$e^x = 1 + x + \frac{x^2}{2!} + \frac{x^3}{3!} + \frac{x^4}{4!} + \cdots.$$

Suppose that, in a mechanical way, we simply replace x by a complex number of the form $i\theta$, ignoring all questions about convergence. We obtain

$$e^{i\theta} = 1 + i\theta + \frac{i^2\theta^2}{2!} + \frac{i^3\theta^3}{3!} + \frac{i^4\theta^4}{4!} + \cdots.$$

But $i^2 = -1$, $i^3 = -i$, $i^4 = 1$, $i^5 = i$, and so on. Hence

$$e^{i\theta} = 1 + i\theta - \frac{\theta^2}{2!} - \frac{i\theta^3}{3!} + \frac{\theta^4}{4!} + \cdots$$

$$= \left(1 - \frac{\theta^2}{2!} + \frac{\theta^4}{4!} + \cdots\right) + i\left(\theta - \frac{\theta^3}{3!} + \frac{\theta^5}{5!} - \cdots\right).$$

We recognize the two series in parentheses as the MacLaurin series for $\cos\theta$ and $\sin\theta$, respectively. Thus, if we want a definition of $e^{i\theta}$ that extends the theory of infinite series to complex variables, it is clear that we must make this definition as follows:

18.22

$$e^{i\theta} = \cos\theta + i\sin\theta.$$

Formula 18.22 is known as *Euler's Formula.*

More generally, if z is the complex number $x + yi$, we define

18.23

$$e^z = e^x e^{iy} = e^x(\cos y + i\sin y).$$

It should be noted that 18.23 gives e^z in its *polar form*: Thus, if $z = x + iy$, then $|e^z| = e^x$ and $\arg e^z = y$. In particular, $e^{\pi i} = \cos\pi + i\sin\pi = -1$, $e^{2\pi i} = \cos 2\pi + i\sin 2\pi = 1$.

Example

Find (a) $e^{2+\pi i/4}$, (b) $e^{i(2+3i)}$.

Solution:

(a) $e^{2+\pi i/4} = e^2(\cos(\pi/4) + i\sin(\pi/4)) = e^2(\sqrt{2}/2 + (\sqrt{2}/2)i)$.

(b) $e^{i(2+3i)} = e^{-3+2i} = e^{-3}(\cos 2 + i\sin 2)$.

We conclude this section with some remarks on complex-valued functions. The reader has already studied vector-valued functions of a scalar variable in Chapter 13. It will be recalled that such a function $\mathbf{F}$ is defined by a formula of the type

(1) $$\mathbf{F}(x) = \langle u(x), v(x)\rangle,$$

where u and v are real-valued functions, and the domain of the variable x is some set of real numbers. But since a vector can also be written as a complex number, formula (1) can be rewritten

(2) $$F(x) = u(x) + iv(x),$$

where $F(x)$ is a complex number. Thus, formula (2) now defines a complex-valued function F of the real variable x. The functions u and v are called the *real* and *imaginary parts*, respectively, of the function F.

How should the derivative of such a complex-valued function F be defined? Recall that by 13.11 the derivative of the vector-valued function (1) is given by

$$\mathbf{F}'(x) = \langle u'(x), v'(x)\rangle.$$

Thus it is obvious that we must define the derivative of the complex-valued function (2) by

18.24
$$F'(x) = u'(x) + iv'(x).$$

This definition enables us to extend many of the familiar rules for differentiation to complex-valued functions (see Exercise 7 below).

The following theorem will be of considerable importance in the next section.

18.25 Theorem

Suppose that the complex-valued function

$$F(x) = u(x) + iv(x)$$

is a solution of the second-order differential equation

$$y'' + by' + cy = 0,$$

where b and c are real constants. Then the functions u(x) and v(x) are also solutions of this differential equation.

Proof: By hypothesis,

$$(u(x) + iv(x))'' + b(u(x) + iv(x))' + c(u(x) + iv(x)) = 0.$$

Hence

$$u''(x) + iv''(x) + bu'(x) + ibv'(x) + cu(x) + icv(x) = 0,$$
$$u''(x) + bu'(x) + cu(x) + i(v''(x) + bv'(x) + cv(x)) = 0.$$

But a complex number is 0 if and only if its real and imaginary parts are 0. Hence

$$u''(x) + bu'(x) + u(x) = 0$$
$$v''(x) + bv'(x) + v(x) = 0,$$

completing the proof.

EXERCISES

1. Evaluate the following:
   a. $e^{\pi i/3}$ **b.** $e^{1+\pi i/2}$ **c.** ie^{1-i}
   d. $e^{\ln 2+3i}$ **e.** $e^{\pi/(1+2i)}$
2. Prove:
   a. $e^{z+2\pi i} = e^z$ for all complex numbers z.
   b. $e^{z+\pi i} = -e^z$ for all complex numbers z.
3. Prove that the following laws of exponents hold for all complex numbers z_1 and z_2:
   a. $e^{z_1}e^{z_2} = e^{z_1+z_2}$, **b.** $e^{z_1}/e^{z_2} = e^{z_1-z_2}$.
4. If $z = x + iy$, prove that e^z is a real number if and only if $y = 0, \pm\pi, \pm 2\pi, \pm 3\pi, \ldots$.

In Exercises 5 and 6, find all values of z that satisfy the given equation.

5. $e^z = -2$ **6.** $e^z = i$

7. Let m be a complex constant $\alpha + \beta i$, and let f be the complex-valued function defined by $f(x) = e^{mx}$ for all real x. Prove that $f'(x) = me^{mx}$ for all x. *Suggestion:* Find the real and imaginary parts of f and differentiate using 18.24.

11 SECOND-ORDER LINEAR DIFFERENTIAL EQUATIONS

A differential equation of the form

$$y^{[n]} + a_1(x)y^{[n-1]} + \cdots + a_{n-1}(x)y' + a_n(x)y = G(x)$$

is called a *linear differential equation of order n*. We shall not study the general linear equation above, but rather shall limit our remarks to the second-order linear differential equation

18.26 $$y'' + by' + cy = G(x)$$

with constant coefficients b and c.

Let us first solve the so-called *homogeneous* linear differential equation

18.27 $$y'' + by' + cy = 0.$$

The solutions of 18.27 will be used in finding the solutions of 18.26.

18.28 Theorem

If $y = u(x)$ and $y = v(x)$ are solutions of 18.27, then so is $y = C_1u(x) + C_2v(x)$ for any numbers C_1 and C_2.

Proof: By assumption,

$$u''(x) + bu'(x) + cu(x) = 0,$$
$$v''(x) + bv'(x) + cv(x) = 0.$$

Hence

$$D_x^2[C_1u(x) + C_2v(x)] + bD_x[C_1u(x) + C_2v(x)] + c[C_1u(x) + C_2v(x)]$$
$$= C_1[u''(x) + bu'(x) + cu(x)] + C_2[v''(x) + bv'(x) + cv(x)] = 0,$$

and the theorem is proved.

This theorem will allow us to express the general solution of 18.27 in terms of two particular solutions.

We first inquire whether the exponential function defined by

$$y = e^{mx}$$

is a solution of 18.27. Since $D_xe^{mx} = me^{mx}$ and $D_x^2e^{mx} = m^2e^{mx}$, this function is a solution of 18.27 if and only if

$$m^2e^{mx} + bme^{mx} + ce^{mx} = 0,$$

or, on dividing by the nonzero number e^{mx}, if and only if

18.29
$$m^2 + bm + c = 0.$$

Equation 18.29 is called either the *characteristic equation* or the *auxiliary equation* of 18.27.

The roots of 18.29 are given by

$$m = \frac{-b \pm \sqrt{b^2 - 4c}}{2}.$$

Let us consider three cases as follows.

Case 1. $b^2 - 4c > 0$. Then 18.29 has distinct real roots m_1 and m_2, and $y = e^{m_1x}$ and $y = e^{m_2x}$ are particular solutions of 18.27. The general solution is given by

18.30
$$y = C_1e^{m_1x} + C_2e^{m_2x}.$$

Case 2. $b^2 - 4c = 0$. Then 18.29 has a double root $m = -b/2$, and $y = e^{mx}$ is a particular solution of 18.27. It is easily verified that $y = xe^{mx}$ also is a solution:

$$\begin{aligned} D_x^2xe^{mx} + b\,D_xxe^{mx} + cxe^{mx} &= (m^2xe^{mx} + 2me^{mx}) \\ &\quad + b(mxe^{mx} + e^{mx}) + cxe^{mx} \\ &= (m^2 + bm + c)xe^{mx} \\ &\quad + (2m + b)e^{mx} = 0. \end{aligned}$$

Hence

18.31
$$y = C_1e^{mx} + C_2xe^{mx}$$

is the general solution of 18.27 in this case.

Case 3. $b^2 - 4c < 0$. Then 18.29 has distinct complex roots $m_1 = \alpha + \beta i$ and $m_2 = \alpha - \beta i$, where

$$\alpha = -\frac{b}{2}, \qquad \beta i = \frac{\sqrt{b^2 - 4c}}{2}.$$

The functions f and g defined by

$$f(x) = e^{m_1x}, \qquad g(x) = e^{m_2x},$$

are again solutions of the differential equation 18.27, although in this case they are *complex-valued.* The student may verify directly that f and g satisfy 18.27, using the fact that $D_xe^{mx} = me^{mx}$ and $D_x^2e^{mx} = m^2e^{mx}$ for any complex number m (see Exercise 7, Section 10). By Euler's Formula (18.23), the function f is given by

$$f(x) = e^{m_1x} = e^{(\alpha+\beta i)x} = e^{\alpha x}(\cos\beta x + i\sin\beta x).$$

By 18.25, the real and imaginary parts of f, defined by

$$u(x) = e^{\alpha x} \cos \beta x, \qquad v(x) = e^{\alpha x} \sin \beta x,$$

are, therefore, also solutions of equation 18.27. Thus, in Case 3 we obtain

18.32 $$y = e^{\alpha x}(C_1 \cos \beta x + C_2 \sin \beta x)$$

for the general solution of 18.27.

The reader should observe that if we use the function g instead of f, in the above argument, we still arrive at the same general solution 18.32.

Example 1 Solve the differential equation

$$y'' - y' - 6y = 0.$$

Solution: The characteristic equation

$$m^2 - m - 6 = 0$$

has roots $m_1 = 3$ and $m_2 = -2$. Hence the given equation has solution (18.30),

$$y = C_1 e^{3x} + C_2 e^{-2x}.$$

Example 2 Solve the differential equation

$$y'' + 2\sqrt{3}\, y' + 3 = 0.$$

Solution: The characteristic equation

$$m^2 + 2\sqrt{3}\, m + 3 = 0$$

has a double root $m = -\sqrt{3}$. Therefore, by 18.31,

$$y = (C_1 + C_2 x)e^{-\sqrt{3}x}$$

is the solution of the given equation.

Example 3 Solve the differential equation

$$y'' - 6y' + 13y = 0.$$

Solution: The characteristic equation

$$m^2 - 6m + 13 = 0$$

has imaginary roots $3 \pm 2i$. Thus $\alpha = 3$ and $\beta = 2$ in 18.32, and

$$y = e^{3x}(C_1 \cos 2x + C_2 \sin 2x)$$

is the general solution of the given equation.

Example 4 Solve the differential equation

$$y''' - y = 0.$$

Solution: This is a homogeneous linear differential equation with constant coefficients, although it is of the *third order*. However, the theory developed in this section is easily extended to such higher order equations. In this case the characteristic equation is

$$m^3 - 1 = 0,$$

whose solutions are the cube roots of 1 found in Example 3, Section 10:

$$m_1 = 1, \qquad m_2 = -\frac{1}{2} + \frac{\sqrt{3}}{2} i, \qquad m_3 = -\frac{1}{2} - \frac{\sqrt{3}}{2} i.$$

Corresponding to the root $m_1 = 1$ of the characteristic equation, there is a solution e^x of the given differential equation: while corresponding to the root m_2, there are solutions

$$e^{-x/2} \cos \frac{\sqrt{3}}{2} x, \qquad e^{-x/2} \sin \frac{\sqrt{3}}{2} x.$$

Hence the general solution of the given equation is

$$y = C_1 e^x + e^{-x/2} \left(C_2 \cos \frac{\sqrt{3}}{2} x + C_3 \sin \frac{\sqrt{3}}{2} x \right).$$

18.33 Theorem

Let $y = u(x)$ and $y = v(x)$ be solutions of the differential equation

$$y'' + by' + cy = 0$$

such that

$$u(x)v'(x) - v(x)u'(x) \neq 0$$

for all x. Then for any given numbers x_0, y_0, and y_1 there exists a solution

$$f(x) = C_1 u(x) + C_2 v(x)$$

of the given equation such that

$$f(x_0) = y_0, \qquad f'(x_0) = y_1.$$

Proof: We wish to show that the constants C_1 and C_2 can be determined so that $f(x_0) = y_0$ and $f'(x_0) = y_1$, that is, so that

$$y_0 = C_1 u(x_0) + C_2 v(x_0), \qquad y_1 = C_1 u'(x_0) + C_2 v'(x_0).$$

This pair of linear equations in C_1 and C_2 has the unique solution

$$C_1 = \frac{y_0 v'(x_0) - y_1 v(x_0)}{w}, \qquad C_2 = \frac{y_1 u(x_0) - y_0 u'(x_0)}{w}$$

where, by assumption, $w = u(x_0)v'(x_0) - v(x_0)u'(x_0) \neq 0$. That $y = f(x)$ is a solution of the differential equation for this choice of C_1 and C_2 follows from 18.28.

In each of the three cases considered above, we have chosen the two solutions u and v so that the hypotheses of 18.33 are satisfied. For example, in Case 1, $u(x) = e^{m_1 x}$ and $v(x) = e^{m_2 x}$ with $m_1 \neq m_2$. Hence

$$u(x)v'(x) - v(x)u'(x) = (m_2 - m_1)e^{(m_1+m_2)x} \neq 0.$$

That the hypotheses of 18.33 are satisfied in the other two cases will be left for the reader to verify.

We shall prove in a later section that if $y = f(x)$ and $y = g(x)$ are solutions of 18.27 such that

$$f(x_0) = g(x_0), \qquad f'(x_0) = g'(x_0)$$

for some number x_0, then $f(x) = g(x)$ for every x. Now if $y = g(x)$ is any solution of 18.27, then, by 18.33, there exists a solution of 18.27 of the form

$$f(x) = C_1 u(x) + C_2 v(x)$$

such that $f(x_0) = g(x_0)$, $f'(x_0) = g'(x_0)$. Hence it will follow that $g(x) = C_1 u(x) + C_2 v(x)$, that is, that $C_1 u(x) + C_2 v(x)$ is the general solution of 18.27, as we contended in each of the cases above.

EXERCISES

1. Prove that in each of the following cases $u(x)v'(x) - v(x)u'(x) \neq 0$.

a. $u(x) = e^{mx}$, $v(x) = xe^{mx}$

b. $u(x) = e^{ax} \cos \beta x$, $v(x) = e^{ax} \sin \beta x$, $(\beta \neq 0)$

Solve each of the following differential equations.

2. $y'' + 9y = 0$

3. $y'' - 9y = 0$

4. $y'' + 2y' + y = 0$

5. $y'' + 2y' - y = 0$

6. $y'' - y' - 6y = 0$

7. $y'' - 3y' - 10y = 0$

8. $y'' + y' = 0$

9. $y'' + y' + 3y = 0$

10. $y'' - 4y' + 29y = 0$

11. $y'' - 2y' + 3y = 0$

12. $y'' + 2y' + 3y = 0$

13. $y'' + \sqrt{2}y' + 7y = 0$

14. $6y'' + y' - 2y = 0$

15. $3y'' - 2y' + 5y = 0$

16. $y'' - 2y' + (1 - \pi)y = 0$

17. $y'' - 2\sqrt{3}y' + (3 + \pi^2)y = 0$

In each of Exercises 18 to 22 find the particular solution of the equation satisfying the given boundary conditions.

18. $y'' + y = 0$; $y = 0$, $y' = 1$ when $x = 0$

19. $y'' - 3y' + 5y = 0$; $y = 0$, $y' = 0$ when $x = 0$

20. $y'' - 2y' - 8y = 0$; $y = 10$, $y' = 1$ when $x = 0$

21. $y'' - 5y' + 6y = 0$; $y = 1$, $y' = 1$ when $x = 1$

22. $y'' - 4y' + 4y = 0$; $y = 3$, $y' = 0$ when $x = 2$

23. An object is suspended from a fixed standard by a spring. There is a point at which this system is in equilibrium. The resultant of the force of the spring and the weight of the object is proportional to the vertical displacement of the object from the point of equilibrium. Describe the motion of the object.

Solve the differential equations in Exercises 24–27.

24. $y''' + 5y'' + 6y' = 0.$ **25.** $y''' + 2y'' - y' - 2y = 0.$

26. $y^{(4)} - y = 0.$ **27.** $y^{(4)} + y = 0.$

12 NONHOMOGENEOUS LINEAR EQUATIONS

We shall consider in this section the problem of solving the nonhomogeneous second-order linear differential equation

18.34 $$y'' + by' + cy = G(x),$$

where b and c are constants and G is a continuous function.

In discussing the solutions of 18.34, it is convenient to introduce the notation $L(y)$ for the left side of 18.34:

$$L(y) = y'' + by' + cy.$$

The "operator" L is linear in the sense that

$$L(y_1 \pm y_2) = L(y_1) \pm L(y_2).$$

Using the L notation, the function g is a solution of 18.34 if and only if $L(g(x)) = G(x)$.

Let g_p designate a particular solution of 18.34 and let g be any other solution. If $f(x) = g(x) - g_p(x)$, then

$$L(f(x)) = L(g(x)) - L(g_p(x)) = 0;$$

that is, f is a solution of the *complementary equation*

18.35 $$y'' + by' + cy = 0.$$

Conversely, if g_p is a particular solution of 18.34, and f is any solution of 18.35, then $g(x) = g_p(x) + f(x)$ also is a solution of 18.34. Consequently, the general solution of 18.34 has the form

$$y = g_p(x) + f_c(x),$$

where g_p is a particular solution of 18.34 and f_c is the general solution of 18.35. Since we can find the general solution of the homogeneous linear differential equation 18.35 by the methods of Section 11, we can find the general solution of 18.34 provided we can find a particular solution of it. Various ways of finding particular solutions of 18.34 are given in the remainder of this section.

Example 1 Solve the differential equation

$$y'' + y = 2x.$$

Solution: We see by inspection that $y_p = 2x$ is a solution. The complementary equation $y'' + y = 0$ has general solution

$$y_c = C_1 \cos x + C_2 \sin x.$$

Hence $$y = C_1 \cos x + C_2 \sin x + 2x$$

is the general solution of the given equation.

The following procedure, called the method of *variation of parameters*, will yield a particular solution of 18.34. Let

$$y_c = C_1 u(x) + C_2 v(x)$$

be the general solution of 18.35; $L(y_c) = 0$. We shall find functions u_1 and v_1 which, when put in place of C_1 and C_2 in y_c, will give a particular solution

$$y_p = u_1(x)u(x) + v_1(x)v(x)$$

of 18.34.

We first establish the condition that

$$u_1'u + v_1'v = 0.$$

This condition simplifies subsequent computations. Now if $y = u_1 u + v_1 v$, then $y' = u_1 u' + u_1' u + v_1' v + v_1 v'$, and, using the condition above,

$$y' = u_1 u' + v_1 v'.$$

Hence

$$y'' = u_1 u'' + v_1 v'' + u_1' u' + v_1' v',$$

and $$L(u_1 u + v_1 v) = u_1 L(u) + v_1 L(v) + u_1' u' + v_1' v' = u_1' u' + v_1' v'.$$

Therefore $L(u_1 u + v_1 v) = G(x)$ if and only if $u_1' u' + v_1' v' = G(x)$. Consequently, if the functions u_1 and v_1 are chosen so that

18.36 $$u_1' u + v_1' v = 0, \qquad u_1' u' + v_1' v' = G(x),$$

then $y_p = u_1 u + v_1 v$ is a particular solution of 18.34.

Equations 18.36 are two linear equations in the unknowns u_1' and v_1'. Since $uv' - vu' \neq 0$ by results of the preceding section, these equations may be solved for u_1' and v_1', yielding

$$u_1' = -\frac{vG(x)}{uv' - vu'}, \qquad v_1' = \frac{uG(x)}{uv' - vu'}.$$

Hence $$u_1 = -\int \frac{vG(x)}{uv' - vu'}\,dx, \qquad v_1 = \int \frac{uG(x)}{uv' - vu'}\,dx$$

always exist.

Example 2 Solve the differential equation

$$y'' + y = \sec x.$$

Solution: The complementary equation $y'' + y = 0$ has general solution

$$y_c = C_1 \sin x + C_2 \cos x.$$

We wish to determine functions u_1 and v_1 such that

$$y_p = u_1(x) \sin x + v_1(x) \cos x$$

is a particular solution of the given equation. For this example $G(x) = \sec x$, $u = \sin x$, $v = \cos x$, $u' = \cos x$, $v' = -\sin x$, and 18.36 becomes

$$u_1' \sin x + v_1' \cos x = 0, \qquad u_1' \cos x - v_1' \sin x = \sec x.$$

Solving for u_1' and v_1', we obtain

$$u_1' = 1, \qquad v_1' = -\sin x \sec x = -\tan x.$$

Hence
$$u_1 = x, \qquad v_1 = \ln |\cos x|,$$

and
$$y_p = x \sin x + (\ln |\cos x|) \cos x$$

is a particular solution. The general solution of the given equation is therefore $y_p + y_c$, or

$$y = (x + C_1) \sin x + (\ln |\cos x| + C_2) \cos x.$$

Another technique for finding a particular solution of 18.34, called the *method of undetermined coefficients*, is illustrated below.

Example 3 Determine A_1 and A_2 so that

$$y_p = A_1 \sin x + A_2 \cos x$$

is a particular solution of the differential equation

$$y'' - y' - 6y = \sin x.$$

Solution: Evidently,

$$y_p' = A_1 \cos x - A_2 \sin x, \qquad y_p'' = -A_1 \sin x - A_2 \cos x,$$

so that

$$y_p'' - y_p' - 6y_p = (A_2 - 7A_1) \sin x + (-A_1 - 7A_2) \cos x.$$

Hence y_p is a solution if

$$(A_2 - 7A_1) \sin x + (-A_1 - 7A_2) \cos x = \sin x,$$

that is, if

$$A_2 - 7A_1 = 1, \qquad -A_1 - 7A_2 = 0.$$

Solving these equations for A_1 and A_2, we obtain

$$A_1 = -\tfrac{7}{50}, \qquad A_2 = \tfrac{1}{50}.$$

Thus

$$y_p = -\tfrac{7}{50} \sin x + \tfrac{1}{50} \cos x$$

is a particular solution of the given equation. The general solution is

$$y = C_1 e^{3x} + C_2 e^{-2x} - \tfrac{7}{50} \sin x + \tfrac{1}{50} \cos x.$$

EXERCISES

1. If $L(y) = y'' - 3y'$, find $L(\sin x)$, $L(e^x)$, $L(e^{3x})$, $L(xe^{3x})$, $L(x)$.
2. If $L(y) = y'' + y$, find $L(e^x)$, $L(\sin x)$, $L(\cos x)$, $L(A_1x^2 + A_2x + A_3)$.
3. Determine A_1, A_2, A_3 so that
$$y_p = A_1x^2 + A_2x + A_3$$
is a solution of the equation
$$y'' - 2y' + 2y = x^2 - 1,$$
and find the general solution.
4. Find a particular solution $y_p = Ax + B$, and find the general solution of the equation
$$y'' - 2y' - 8y = x + 3.$$
5. If $L(y) = y'' + 2y' - 2y$, find $L(A \sin x + B \cos x)$ and hence solve the equation
$$y'' + 2y' - 2y = 2 \sin x - \cos x.$$

In each of Exercises 6 to 11 use the method of variation of parameters to solve the equation.

6. $y'' + y = \csc x$
7. $y'' + y = \tan x$
8. $y'' - 4y' + 4y = x^2e^{2x}$
9. $y'' - y = e^x \sin x$
10. $y'' - y = \cos x$
11. $y'' + 2y' = xe^{-2x}$
12. If $L(y) = y'' - 3y' + 2y$, find $L(e^x)$ and $L(xe^x)$. Hence solve the equation
$$y'' - 3y' + 2y = 3e^x.$$
13. If $L(y) = y'' + 4y$, find $L(x \sin 2x)$ and $L(x \cos 2x)$, and solve the equation
$$y'' + 4y = \cos 2x.$$

Solve by any method each of the following equations.

14. $y'' - 3y = x$
15. $y'' - 2y = 5x$
16. $y'' - y' = 3x$
17. $y'' + y' = e^x$
18. $y'' + y' + y = \sin x$
19. $y'' - 3y' - y = e^x$

Use the method of undetermined coefficients to do Exercises 20 and 21, assuming the existence of a y_p of the indicated form.

20. $y'' + 3y' + 2y = x^2$, $y_p = Ax^2 + Bx + C$
21. $y'' + y = xe^x$, $y_p = Axe^x + Be^x$

In the circuit of Figure 18.7 the current at the time t satisfies the equation
$$L\frac{dI}{dt} + RI + \frac{1}{C}\int I\,dt = E(t),$$
or (1)
$$L\frac{d^2I}{dt^2} + R\frac{dI}{dt} + \frac{1}{C}I = E'(t),$$

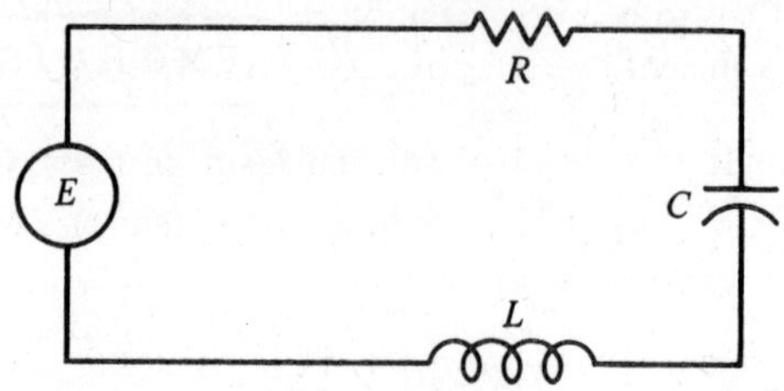

FIGURE 18.7

where $E(t)$ is the impressed emf, L, R, and C are the constants of inductance, resistance, and capacitance, respectively. Solve equation (1) in each of the following cases:

22. $R = 0$, $C = 1$, $L = 10$, $E = 5$

23. $R = 10$, $C = \frac{1}{5}$, $L = 10$, $E = \sin t$

24. $R = 1$, $C = 1$, $L = 1$, $E = 10 \sin 2t$

25. Given that $m^3 - 6m^2 + 11m - 6 = (m - 1)(m - 2)(m - 3)$, solve the equation $y''' - 6y'' + 11y' - 6y = 0$.

26. Solve the equation $y''' - 3y'' - y' + 3y = 0$.

27. Solve the equation $y''' - 3y'' + y' - 3y = 0$.

13 SOLUTIONS IN SERIES

The infinite series

$$f(x) = C_0 + C_1(x - x_0) + C_2(x - x_0)^2 + \cdots + C_n(x - x_0)^n + \cdots$$

defines a function f in the interval of convergence of the series. It is frequently possible to find solutions of differential equations in the form of infinite series, as is illustrated in the following example.

Example Find an infinite series solution of the differential equation

$$y'' + xy' + y = 0.$$

Solution: If

$$y = \sum_{k=0}^{\infty} C_k x^k$$

is a solution, then (by 12.21),

$$y' = \sum_{k=1}^{\infty} kC_k x^{k-1}, \qquad y'' = \sum_{k=2}^{\infty} k(k - 1)C_k x^{k-2} = \sum_{k=0}^{\infty} (k + 2)(k + 1)C_{k+2}x^k.$$

Hence $\quad y'' + xy' + y = \sum_{k=0}^{\infty} [C_k + kC_k + (k + 2)(k + 1)C_{k+2}]x^k.$

If y is to be a solution, the right side of the above equation must equal zero, and the coefficient of each power of x must be zero:

$$(1 + k)C_k + (k + 2)(k + 1)C_{k+2} = 0$$

or $C_k + (k + 2)C_{k+2} = 0$. Thus

$$C_{k+2} = -\frac{1}{k+2}C_k, \qquad k = 0, 1, 2, \ldots, n, \ldots.$$

In particular,

$$C_2 = -\frac{1}{2}C_0, \qquad C_3 = -\frac{1}{3}C_1,$$

$$C_4 = -\frac{1}{4}C_2 = \frac{1}{2\cdot 4}C_0, \qquad C_5 = -\frac{1}{5}C_3 = \frac{1}{3\cdot 5}C_1,$$

$$C_6 = -\frac{1}{6}C_4 = -\frac{1}{2\cdot 4\cdot 6}C_0, \qquad C_7 = -\frac{1}{7}C_5 = -\frac{1}{3\cdot 5\cdot 7}C_1,$$

and so on, with

$$C_{2n} = \frac{(-1)^n}{2\cdot 4\cdot \cdots \cdot 2n}C_0, \qquad C_{2n+1} = \frac{(-1)^n}{1\cdot 3\cdot \cdots \cdot (2n+1)}C_1.$$

Thus y may be expressed as a sum of two series, one containing the even powers of x and the other containing the odd powers:

$$y = C_0\sum_{k=0}^{\infty}\frac{(-1)^k}{2\cdot 4\cdot \cdots \cdot 2k}x^{2k} + C_1\sum_{k=0}^{\infty}\frac{(-1)^k}{1\cdot 3\cdot \cdots \cdot (2k+1)}x^{2k+1}.$$

The ratio test establishes that each of the series is convergent everywhere. This is the general solution of the given differential equation.

A function f which can be expressed as a power series in some interval is said to be *analytic* in that interval. Thus we are seeking analytic solutions of differential equations in this section.

The theory of Taylor's series (Chapter 12) gives us a criterion for determining whether or not a function is analytic. Let us use Taylor's theory to establish the following result.

18.37 Theorem

If f is a solution of the differential equation

$$y'' = ky$$

in an interval $[x_1,x_2]$, then f is analytic in this interval.

Proof: Since f is a solution of $y'' = ky$, then f' and f'' exist in $[x_1,x_2]$. Furthermore, $f''(x) = kf(x)$, $f'''(x) = kf'(x)$, $f^{iv}(x) = kf''(x) = k^2f(x)$, and so on. That is, f possesses derivatives of every order in $[x_1,x_2]$.

By Taylor's formula (11.9),

$$f(x) = P_n(x) + R_n(x),$$

where

$$P_n(x) = \sum_{k=0}^{n} \frac{f^{[k]}(x_0)}{k!}(x - x_0)^k, \qquad R_n(x) = \frac{f^{[n+1]}(z_n)}{(n + 1)!}(x - x_0)^{n+1},$$

for x_0 in $[x_1,x_2]$ and some z_n between x and x_0. If

$$\operatorname*{limit}_{n\to\infty} R_n(x) = 0$$

for every x in $[x_1,x_2]$, then f is represented by its Taylor series (that is, f is analytic) in $[x_1,x_2]$,

$$f(x) = \sum_{k=0}^{\infty} \frac{f^{[k]}(x_0)}{k!}(x - x_0)^k,$$

according to 12.25.

Since f and f' are continuous in $[x_1,x_2]$, there exists a number M such that

$$|f(x)| \le M, \qquad |f'(x)| \le M$$

in $[x_1,x_2]$. Let $m = |k|$ if $|k| > 1$ and $m = 1$ if $|k| \le 1$. Then

$$\begin{aligned} |f''(x)| &= |kf(x)| \le Mm \le Mm^2, \\ |f'''(x)| &= |kf'(x)| \le Mm \le Mm^3, \\ |f^{iv}(x)| &= |kf''(x)| \le m|f''(x)| \le Mm^4, \end{aligned}$$

and, in general,

$$|f^{[n]}(x)| \le Mm^n$$

for every x in $[x_1,x_2]$. Hence

$$|R_n(x)| = \frac{|f^{[n+1]}(z_n)|}{(n + 1)!}|x - x_0|^{n+1} \le \frac{Mm^{n+1}}{(n + 1)!}|x - x_0|^{n+1},$$

or $$|R_n(x)| \le \frac{M}{(n + 1)!}|m(x - x_0)|^{n+1}.$$

Now

$$\operatorname*{limit}_{n\to\infty} \frac{M}{(n + 1)!}|m(x - x_0)|^{n+1} = 0,$$

and therefore $\operatorname*{limit}_{n\to\infty} R_n = 0$. This proves the theorem.

Let us consider the possibility that the equation

$$y'' = ky$$

has two solutions g_1 and g_2 in a given interval such that

$$g_1(x_0) = g_2(x_0), \qquad g_1'(x_0) = g_2'(x_0)$$

at some number x_0 in the interval. If we let

$$f(x) = g_1(x) - g_2(x),$$

then f is a solution of the given homogeneous linear differential equation such that

$$f(x_0) = 0, \qquad f'(x_0) = 0.$$

Since $f''(x_0) = kf(x_0)$, $f'''(x_0) = kf'(x_0)$, $f^{iv}(x_0) = k^2 f(x_0)$, and so on, it is clear that

$$f^{[n]}(x_0) = 0$$

for every n.

By the proof of 18.37, f is represented by its Taylor series

$$f(x) = \sum_{k=0}^{\infty} \frac{f^{[k]}(x_0)}{k!}(x - x_0)^k$$

in the given interval. Since each coefficient of this power series is zero, evidently $f(x) = 0$. Hence

$$g_1(x) = g_2(x)$$

at every x in the interval. In other words, the boundary conditions

$$y = y_0, \qquad y' = y_1,$$

when $x = x_0$, determine a unique solution (see 18.33) of the differential equation $y'' = ky$.

Since the differential equation

$$w'' + bw' + cw = 0$$

can be reduced to the equation

$$y'' = ky$$

by the transformation

$$y = we^{bx/2},$$

we have established the uniqueness of the solution of any second-order linear differential equation with constant coefficients.

EXERCISES

1. Prove that every solution of the equation $y''' = ky$ is analytic.

In each of Exercises 2 to 6 find series solutions for the equation, and determine the interval of convergence of the series.

2. $y' = y$

3. $y' = xy$

4. $y'' = xy$

5. $y'' + x^2y' + xy = 0$

6. $(1 - x^2)y'' + 2xy' + 4y = c$

7. Solve the equation $y'' = y$ by infinite series to get the solution
$$y = A_1 \cosh x + A_2 \sinh x.$$
Show that this solution is equivalent to $y = C_1e^x + C_2e^{-x}$.

8. Solve the equation
$$xy'' - (x + 2)y' + 2y = 0$$
by infinite series, and then show that the solution has the equivalent form
$$y = C_1\left(1 + x + \frac{x^2}{2}\right) + C_2e^x.$$

9. Find a series solution of the equation
$$(1 + x^2)y'' - 4xy' + 6y = 0,$$
and show that it converges for all x.

CHAPTER REVIEW

Solve the differential equations in Exercises 1 to 16.

1. $\dfrac{dy}{dx} = \dfrac{xy}{1 + x^2}$

2. $\dfrac{dy}{dx} + \dfrac{1 + y^2}{1 + x^2} = 0$ and $y = 1$ at $x = 0$

3. $x\dfrac{dy}{dx} + y = 2\sqrt{xy}$

4. $\dfrac{dy}{dx} = \dfrac{xe^{y/x} + y}{x}$

5. $(x + y + 1)\,dx + (6x + 10y + 14)\,dy = 0$

6. $(x + y + 4)\,dx - (2x + 2y - 1)\,dy = 0$

7. $(x + 2y)\,dx + (2x - 3y)\,dy = 0$

8. $(x^2 - \sin y)\,dy + (2xy - x^2)\,dx = 0$

9. $y\,dx + (3x - y^3)\,dy = 0$

10. $\dfrac{dy}{dx} - \dfrac{y}{x} = \dfrac{x + y}{x - 1}$

11. $3x\,dy + (y + x^2y^4)\,dx = 0$

12. $x\dfrac{dy}{dx} - y = \sqrt{x^2 + y^2}$

13. $\dfrac{dy}{dx} + \dfrac{3x^2 + 6xy^2}{6x^2y + 4y^2} = 0$

14. Solve the equation $\dfrac{dx}{dy} = \cos(x + y)$ by making the change of variable $z = x + y$.

15. $r \ln r\,ds + (s - r)\,dr = 0$

16. $dx + (y + x \cot y)\,dy = 0$

In each of Exercises 17 to 20 solve the differential equation by two methods, one method being by series expansion. Compare the results.

17. $y' = 3x^2y$

18. $y' = y(3x^2 + 2x)$

19. $y'' + 2xy' = 2x$

20. $x^2y'' - 6xy' + 10y = 0$ (*Hint:* Let $x = e^z$.)

Solve the following linear equations.

21. $y'' - 5y' - 6y = 0$

22. $y'' - 5y' - 6y = e^{-x}$

23. $y''' - 5y'' - 6y' = 0$

24. $y''' - 5y'' - 6y' = \sin x$

25. $y'' + 4y = x^2$

26. $y''' + 4y' = x^2$

27. $y''' - y'' + 4y' = \sin x$

28. $y''' - y'' + 4y' - 4y = e^x$

29. $y''' - 7y' + 6y = e^{-x}$

30. Determine the curves such that the lengths of the tangent and normal at any point of the curve, intercepted by the coordinate axes, are equal.

APPENDIXES

A

Facts and Formulas from Trigonometry

If angle θ is placed in a rectangular coordinate plane as in Figure A.1, with its initial side on the positive x axis, then the coordinates of the point on the terminal side of θ one unit from the origin are $\cos\theta$ and $\sin\theta$. The other trigonometric functions are defined as follows:

$$\tan\theta = \frac{\sin\theta}{\cos\theta}, \qquad \cot\theta = \frac{\cos\theta}{\sin\theta},$$

$$\sec\theta = \frac{1}{\cos\theta}, \qquad \csc\theta = \frac{1}{\sin\theta}.$$

The basic trigonometric identities are as follows:

$$\sin^2\theta + \cos^2\theta = 1, \qquad \tan^2\theta + 1 = \sec^2\theta,$$

$$\sin(\theta + \phi) = \sin\theta\cos\phi \pm \cos\theta\sin\phi,$$

$$\cos(\theta \pm \phi) = \cos\theta\cos\phi \mp \sin\theta\sin\phi,$$

$$\tan(\theta \pm \phi) \frac{\tan\theta \pm \tan\phi}{1 \mp \tan\theta\tan\phi},$$

$$\sin\theta\sin\phi = -\tfrac{1}{2}[\cos(\theta + \phi) - \cos(\theta - \phi)],$$

$$\sin\theta\cos\phi = \tfrac{1}{2}[\sin(\theta + \phi) + \sin(\theta - \phi)],$$

$$\cos\theta\cos\phi = \tfrac{1}{2}[\cos(\theta + \phi) + \cos(\theta - \phi)],$$

$$\sin 2\theta = 2\sin\theta\cos\theta, \qquad \cos 2\theta = \cos^2\theta - \sin^2\theta,$$

$$\sin^2\frac{\theta}{2} = \frac{1 - \cos\theta}{2}, \qquad \cos^2\frac{\theta}{2} = \frac{1 + \cos\theta}{2}.$$

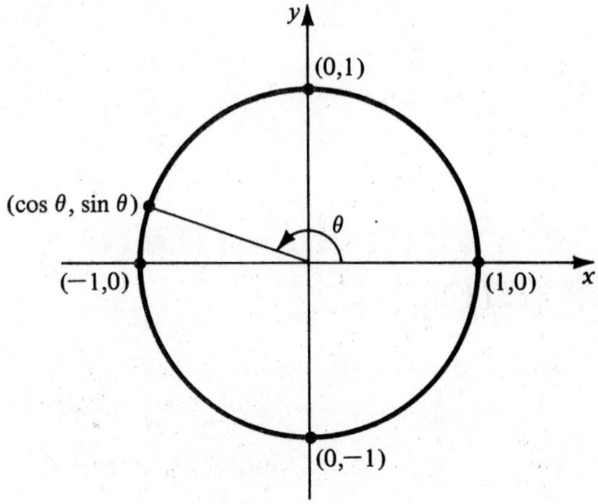

FIGURE A.1

If A, B, C are the angles of a triangle and a, b, c are the lengths of the respective opposite sides, then the following formulas hold:

Law of sines: $$\frac{\sin A}{a} = \frac{\sin B}{b}.$$

Law of cosines: $$a^2 = b^2 + c^2 - 2bc \cos A.$$

B

Table of Integrals

1. $\int [f(x) \pm g(x)]\, dx = \int f(x)\, dx \pm \int g(x)\, dx.$

2. $\int cf(x)\, dx = c \int f(x)\, dx.$

3. $\int f(g(x))g'(x)\, dx = \int f(u)\, du \Big|_{u=g(x)}.$

4. $\int x^n\, dx = \dfrac{1}{n+1} x^{n+1} + C,\ n \neq -1.$

5. $\int \dfrac{1}{x}\, dx = \ln |x| + C.$

6. $\int \sin x\, dx = -\cos x + C.$

7. $\int \cos x\, dx = \sin x + C.$

8. $\int \sec^2 x\, dx = \tan x + C.$

9. $\int \csc^2 x\, dx = -\cot x + C.$

10. $\int \sec x \tan x\, dx = \sec x + C.$

11. $\int \csc x \cot x\, dx = -\csc x + C.$

12. $\int e^x\, dx = e^x + C.$

13. $\int \dfrac{1}{\sqrt{a^2 - x^2}}\, dx = \sin^{-1} \dfrac{x}{a} + C.$

14. $$\int \frac{1}{a^2 + x^2}\,dx = \frac{1}{a}\tan^{-1}\frac{x}{a} + C.$$

15. $$\int \frac{1}{x\sqrt{x^2 - a^2}}\,dx = \frac{1}{a}\sec^{-1}\frac{x}{a} + C.$$

16. $$\int f(x)g'(x)\,dx = f(x)g(x) - \int g(x)f'(x)\,dx.$$

17. $$\int \frac{1}{x\sqrt{ax + b}}\,dx = \frac{1}{\sqrt{b}}\ln\left|\frac{\sqrt{ax + b} - \sqrt{b}}{\sqrt{ax + b} + \sqrt{b}}\right| + C,\ b > 0.$$

18. $$\int \frac{1}{x\sqrt{ax + b}}\,dx = \frac{2}{\sqrt{-b}}\tan^{-1}\sqrt{\frac{ax + b}{-b}} + C,\ b < 0.$$

19. $$\int \frac{1}{x^n\sqrt{ax + b}}\,dx = -\frac{1}{b(n - 1)}\frac{\sqrt{ax + b}}{x^{n-1}} - \frac{(2n - 3)a}{(2n - 2)b}\int \frac{1}{x^{n-1}\sqrt{ax + b}}\,dx,$$

$$n \neq 1.$$

20. $$\int \frac{\sqrt{ax + b}}{x}\,dx = 2\sqrt{ax + b} + b\int \frac{1}{x\sqrt{ax + b}}\,dx.$$

21. $$\int \frac{1}{x^2 - a^2}\,dx = \frac{1}{2a}\ln\left|\frac{x - a}{x + a}\right| + C.$$

22. $$\int \frac{1}{(ax + b)(cx + d)}\,dx = \frac{1}{bc - ad}\ln\left|\frac{cx + d}{ax + b}\right| + C,\ bc - ad \neq 0.$$

23. $$\int \frac{x}{(ax + b)(cx + d)}\,dx = \frac{1}{bc - ad}\left\{\frac{b}{a}\ln|ax + b| - \frac{d}{c}\ln|cx + d|\right\} + C,$$

$$bc - ad \neq 0.$$

24. $$\int \frac{1}{(ax + b)^2(cx + d)}\,dx = \frac{1}{bc - ad}\left\{\frac{1}{ax + b} + \frac{c}{bc - ad}\ln\left|\frac{cx + d}{ax + b}\right|\right\} + C,$$

$$bc - ad \neq 0.$$

25. $$\int \frac{x}{(ax + b)^2(cx + d)}\,dx$$

$$= -\frac{1}{bc - ad}\left\{\frac{b}{a(ax + b)} + \frac{d}{bc - ad}\ln\left|\frac{cx + d}{ax + b}\right|\right\} + C,\ bc - ad \neq 0.$$

26. $$\int \sqrt{x^2 \pm a^2}\,dx = \frac{x}{2}\sqrt{x^2 \pm a^2} \pm \frac{a^2}{2}\ln|x + \sqrt{x^2 \pm a^2}| + C.$$

27. $$\int \frac{1}{\sqrt{x^2 \pm a^2}}\,dx = \ln|x + \sqrt{x^2 \pm a^2}| + C.$$

28. $\int x^2\sqrt{x^2 \pm a^2}\,dx = \frac{x}{8}(2x^2 \pm a^2)\sqrt{x^2 \pm a^2} - \frac{a^4}{8}\ln|x + \sqrt{x^2 \pm a^2}| + C.$

29. $\int \frac{x^2}{\sqrt{x^2 \pm a^2}}\,dx = \frac{x}{2}\sqrt{x^2 \pm a^2} \mp \frac{a^2}{2}\ln|x + \sqrt{x^2 \pm a^2}| + C.$

30. $\int (x^2 \pm a^2)^{3/2}\,dx = x(x^2 \pm a^2)^{3/2} - 3\int x^2\sqrt{x^2 \pm a^2}\,dx.$

31. $\int \frac{1}{(x^2 \pm a^2)^{3/2}}\,dx = \frac{\pm x}{a^2\sqrt{x^2 \pm a^2}} + C.$

32. $\int \frac{x^2}{(x^2 \pm a^2)^{3/2}}\,dx = \frac{-x}{\sqrt{x^2 \pm a^2}} + \ln|x + \sqrt{x^2 \pm a^2}| + C.$

33. $\int \frac{1}{x^2\sqrt{x^2 \pm a^2}}\,dx = \mp\frac{\sqrt{x^2 \pm a^2}}{a^2 x} + C.$

34. $\int \frac{\sqrt{x^2 \pm a^2}}{x^2}\,dx = -\frac{\sqrt{x^2 \pm a^2}}{x} + \ln|x + \sqrt{x^2 \pm a^2}| + C.$

35. $\int \frac{\sqrt{x^2 \pm a^2}}{x}\,dx = \sqrt{x^2 \pm a^2} \pm a^2\int \frac{1}{x\sqrt{x^2 \pm a^2}}\,dx.$

36. $\int \frac{1}{x\sqrt{x^2 + a^2}}\,dx = -\frac{1}{a}\ln\left|\frac{a + \sqrt{x^2 + a^2}}{x}\right| + C.$

37. $\int \sqrt{a^2 - x^2}\,dx = \frac{x}{2}\sqrt{a^2 - x^2} + \frac{a^2}{2}\sin^{-1}\frac{x}{a} + C.$

38. $\int x^2\sqrt{a^2 - x^2}\,dx = -\frac{x}{4}(a^2 - x^2)^{3/2} + \frac{a^2}{4}\int \sqrt{a^2 - x^2}\,dx.$

39. $\int \frac{x^2}{\sqrt{a^2 - x^2}}\,dx = -\frac{x}{2}\sqrt{a^2 - x^2} + \frac{a^2}{2}\sin^{-1}\frac{x}{a} + C.$

40. $\int (a^2 - x^2)^{3/2}\,dx = \frac{x}{4}(a^2 - x^2)^{3/2} + \frac{3a^2}{4}\int \sqrt{a^2 - x^2}\,dx.$

41. $\int \frac{1}{(a^2 - x^2)^{3/2}}\,dx = \frac{x}{a^2\sqrt{a^2 - x^2}} + C.$

42. $\int \frac{x^2}{(a^2 - x^2)^{3/2}}\,dx = \frac{x}{\sqrt{a^2 - x^2}} - \sin^{-1}\frac{x}{a} + C.$

43. $\int \frac{1}{x\sqrt{a^2 - x^2}}\,dx = -\frac{1}{a}\ln\left|\frac{a + \sqrt{a^2 - x^2}}{x}\right| + C.$

44. $\int \frac{1}{x^2\sqrt{a^2 - x^2}}\,dx = -\frac{\sqrt{a^2 - x^2}}{a^2x} + C.$

45. $\int \frac{\sqrt{a^2 - x^2}}{x}\,dx = \sqrt{a^2 - x^2} - a \ln \left|\frac{a + \sqrt{a^2 - x^2}}{x}\right| + C.$

46. $\int \frac{\sqrt{a^2 - x^2}}{x^2}\,dx = -\frac{\sqrt{a^2 - x^2}}{x} - \sin^{-1}\frac{x}{a} + C.$

47. $\int \frac{1}{(x^2 + a^2)^n}\,dx$

$$= \frac{1}{2(n-1)a^2}\left\{\frac{x}{(x^2 + a^2)^{n-1}} + (2n - 3)\int \frac{1}{(x^2 + a^2)^{n-1}}\,dx\right\},\ n \neq 1.$$

48. $\int x \sin x\,dx = \sin x - x \cos x + C.$

49. $\int x^n \sin x\,dx = -x^n \cos x + nx^{n-1} \sin x - n(n-1)\int x^{n-2} \sin x\,dx.$

50. $\int x \cos x\,dx = \cos x + x \sin x + C.$

51. $\int x^n \cos x\,dx = x^n \sin x + nx^{n-1} \cos x - n(n-1)\int x^{n-2} \cos x\,dx.$

52. $\int \sin^m \cos^n x\,dx$

$$= \begin{cases} \dfrac{1}{m+n}\left[-\sin^{m-1} x \cos^{n+1} x + (m-1)\displaystyle\int \sin^{m-2} x \cos^n x\,dx\right] \\ \dfrac{1}{m+n}\left[\sin^{m+1} x \cos^{n-1} x + (n-1)\displaystyle\int \sin^m x \cos^{n-2} x\,dx\right], \end{cases}$$

$m + n \neq 0.$

53. $\int \sin^n x\,dx = -\frac{1}{n}\sin^{n-1} x \cos x + \frac{n-1}{n}\int \sin^{n-2} x\,dx,\ n \geq 2.$

54. $\int \sin^2 x\,dx = -\frac{1}{2}\sin x \cos x + \frac{x}{2} + C.$

55. $\int \cos^n x\,dx = \frac{1}{n}\sin x \cos^{n-1} x + \frac{n-1}{n}\int \cos^{n-2} x\,dx,\ n \geq 2.$

56. $\int \cos^2 x\,dx = \frac{1}{2}\sin x \cos x + \frac{x}{2} + C.$

57. $\int \sin^2 x \cos^2 x\,dx = -\frac{1}{4}\sin x \cos^3 x + \frac{1}{8}\sin x \cos x + \frac{x}{8} + C.$

58. $\int \tan x\,dx = \ln|\sec x| + C.$

59. $\int \tan^2 x\,dx = \tan x - x + C.$

60. $\int \tan^n x\,dx = \frac{1}{n-1}\tan^{n-1} x - \int \tan^{n-2} x\,dx,\ n \geq 2.$

61. $\int \cot x\,dx = \ln|\sin x| + C.$

62. $\displaystyle\int \cot^2 x\,dx = -\cot x - x + C.$

63. $\displaystyle\int \cot^n x\,dx = -\frac{1}{n-1}\cot^{n-1} x - \int \cot^{n-2} x\,dx,\ n \geq 2.$

64. $\displaystyle\int \sec x\,dx = \ln|\sec x + \tan x| + C.$

65. $\displaystyle\int \sec^n x\,dx = \frac{1}{n-1}\left\{\sec^{n-2} x\tan x + (n-2)\int \sec^{n-2} x\,dx\right\},\ n \geq 2.$

66. $\displaystyle\int \csc x\,dx = \ln|\csc x - \cot x| + C.$

67. $\displaystyle\int \csc^n x\,dx = \frac{1}{n-1}\left\{-\csc^{n-2} x\cot x + (n-2)\int \csc^{n-2} x\,dx\right\},\ n \leq 2.$

68. $\displaystyle\int xe^{ax}\,dx = \frac{1}{a^2}(ax-1)e^{ax} + C.$

69. $\displaystyle\int x^n e^{ax}\,dx = \frac{x^n}{a}e^{ax} - \frac{n}{a}\int x^{n-1}e^{ax}\,dx.$

70. $\displaystyle\int e^{ax}\sin bx\,dx = \frac{1}{a^2+b^2}(a\sin bx - b\cos bx)e^{ax} + C.$

71. $\displaystyle\int e^{ax}\cos bx\,dx = \frac{1}{a^2+b^2}(a\cos bx + b\sin bx)e^{ax} + C.$

72. $\displaystyle\int \ln|x|\,dx = x\ln|x| - x + C.$

73. $\displaystyle\int x^m \ln^n|x|\,dx = \frac{1}{m+1}\left\{x^{m+1}\ln^n|x| - n\int x^m \ln^{n-1}|x|\,dx\right\},\ m \neq -1.$

74. $\displaystyle\int \ln^n|x|\,dx = x\ln^n|x| \quad n\int \ln^{n-1}|x|\,dx.$

75. $\displaystyle\int x^n \ln|x|\,dx = \frac{x^{n+1}}{n+1}\left(\ln|x| - \frac{1}{n+1}\right) + C,\ n \neq -1.$

76. $\displaystyle\int \frac{\ln^n|x|}{x}\,dx = \frac{1}{n+1}\ln^{n+1}|x| + C,\ n \neq -1.$

77. $\displaystyle\int \frac{1}{x\ln|x|}\,dx - \ln|\ln|x|| + C.$

78. $\displaystyle\int \sin^{-1} x\,dx = x\sin^{-1} x + \sqrt{1-x^2} + C.$

79. $\displaystyle\int x^n \sin^{-1} x\,dx = \frac{1}{n+1}\left\{x^{n+1}\sin^{-1} x - \int \frac{x^{n+1}}{\sqrt{1-x^2}}\,dx\right\},\ n \neq -1.$

80. $\displaystyle\int \tan^{-1} x\,dx = x\tan^{-1} x - \tfrac{1}{2}\ln(x^2+1) + C.$

81. $\displaystyle\int x^n \tan^{-1} x\,dx = \frac{1}{n+1}\left\{x^{n+1}\tan^{-1} x - \int \frac{x^{n+1}}{x^2+1}\,dx\right\},\ n \neq -1.$

82. $\displaystyle\int \sec^{-1} x\,dx = x\sec^{-1} x - \ln|x + \sqrt{x^2-1}| + C.$

C

Numerical Tables

TABLE I. *Logarithms to Base 10*

N	0	1	2	3	4	5	6	7	8	9
1.0	.0000	.0043	.0086	.0128	.0170	.0212	.0253	.0294	.0334	.0374
1.1	.0414	.0453	.0492	.0531	.0569	.0607	.0645	.0682	.0719	.0755
1.2	.0792	.0828	.0864	.0899	.0934	.0969	.1004	.1038	.1072	.1106
1.3	.1139	.1173	.1206	.1239	.1271	.1303	.1335	.1367	.1399	.1430
1.4	.1461	.1492	.1523	.1553	.1584	.1614	.1644	.1673	.1703	.1732
1.5	.1761	.1790	.1818	.1847	.1875	.1903	.1931	.1959	.1987	.2014
1.6	.2041	.2068	.2095	.2122	.2148	.2175	.2201	.2227	.2253	.2279
1.7	.2304	.2330	.2355	.2380	.2405	.2430	.2455	.2480	.2504	.2529
1.8	.2553	.2577	.2601	.2625	.2648	.2672	.2695	.2718	.2742	.2765
1.9	.2788	.2810	.2833	.2856	.2878	.2900	.2923	.2945	.2967	.2989
2.0	.3010	.3032	.3054	.3075	.3096	.3118	.3139	.3160	.3181	.3201
2.1	.3222	.3243	.3263	.3284	.3304	.3324	.3345	.3365	.3385	.3404
2.2	.3424	.3444	.3464	.3483	.3502	.3522	.3541	.3560	.3579	.3598
2.3	.3617	.3636	.3655	.3674	.3692	.3711	.3729	.3747	.3766	.3784
2.4	.3802	.3820	.3838	.3856	.3874	.3892	.3909	.3927	.3945	.3962
2.5	.3979	.3997	.4014	.4031	.4048	.4065	.4082	.4099	.4116	.4133
2.6	.4150	.4166	.4183	.4200	.4216	.4232	.4249	.4265	.4281	.4298
2.7	.4314	.4330	.4346	.4362	.4378	.4393	.4409	.4425	.4440	.4456
2.8	.4472	.4487	.4502	.4518	.4533	.4548	.4564	.4579	.4594	.4609
2.9	.4624	.4639	.4654	.4669	.4683	.4698	.4713	.4728	.4742	.4757
3.0	.4771	.4786	.4800	.4814	.4829	.4843	.4857	.4871	.4886	.4900
3.1	.4914	.4928	.4942	.4955	.4969	.4983	.4997	.5011	.5024	.5038
3.2	.5051	.5065	.5079	.5092	.5105	.5119	.5132	.5145	.5159	.5172
3.3	.5185	.5198	.5211	.5224	.5237	.5250	.5263	.5276	.5289	.5302
3.4	.5315	.5328	.5340	.5353	.5366	.5378	.5391	.5403	.5416	.5428
3.5	.5441	.5453	.5465	.5478	.5490	.5502	.5514	.5527	.5539	.5551
3.6	.5563	.5575	.5587	.5599	.5611	.5623	.5635	.5647	.5658	.5670
3.7	.5682	.5694	.5705	.5717	.5729	.5740	.5752	.5763	.5775	.5786
3.8	.5798	.5809	.5821	.5832	.5843	.5855	.5866	.5877	.5888	.5899
3.9	.5911	.5922	.5933	.5944	.5955	.5966	.5977	.5988	.5999	.6010
4.0	.6021	.6031	.6042	.6053	.6064	.6075	.6085	.6096	.6107	.6117
4.1	.6128	.6138	.6149	.6160	.6170	.6180	.6191	.6201	.6212	.6222
4.2	.6232	.6243	.6253	.6263	.6274	.6284	.6294	.6304	.6314	.6325
4.3	.6335	.6345	.6355	.6365	.6375	.6385	.6395	.6405	.6415	.6425
4.4	.6435	.6444	.6454	.6464	.6474	.6484	.6493	.6503	.6513	.6522
4.5	.6532	.6542	.6551	.6561	.6571	.6580	.6590	.6599	.6609	.6618
4.6	.6628	.6637	.6646	.6656	.6665	.6675	.6684	.6693	.6702	.6712
4.7	.6721	.6730	.6739	.6749	.6758	.6767	.6776	.6785	.6794	.6803
4.8	.6812	.6821	.6830	.6839	.6848	.6857	.6866	.6875	.6884	.6893
4.9	.6902	.6911	.6920	.6928	.6937	.6946	.6955	.6964	.6972	.6981
5.0	.6990	.6998	.7007	.7016	.7024	.7033	.7042	.7050	.7059	.7067
5.1	.7076	.7084	.7093	.7101	.7110	.7118	.7126	.7135	.7143	.7152
5.2	.7160	.7168	.7177	.7185	.7193	.7202	.7210	.7218	.7226	.7235
5.3	.7243	.7251	.7259	.7267	.7275	.7284	.7292	.7300	.7308	.7316
5.4	.7324	.7332	.7340	.7348	.7356	.7364	.7372	.7380	.7388	.7396
N	0	1	2	3	4	5	6	7	8	9

TABLE I. *Logarithms to Base 10 (continued)*

N	0	1	2	3	4	5	6	7	8	9
5.5	.7404	.7412	.7419	.7427	.7435	.7443	.7451	.7459	.7466	.7474
5.6	.7482	.7490	.7497	.7505	.7513	.7520	.7528	.7536	.7543	.7551
5.7	.7559	.7566	.7574	.7582	.7589	.7597	.7604	.7612	.7619	.7627
5.8	.7634	.7642	.7649	.7657	.7664	.7672	.7679	.7689	.7694	.7701
5.9	.7709	.7716	.7723	.7731	.7738	.7745	.7752	.7760	.7767	.7774
6.0	.7782	.7789	.7796	.7803	.7810	.7818	.7825	.7832	.7839	.7846
6.1	.7853	.7860	.7868	.7875	.7882	.7889	.7896	.7903	.7910	.7917
6.2	.7924	.7931	.7938	.7945	.7952	.7959	.7966	.7973	.7980	.7987
6.3	.7993	.8000	.8007	.8014	.8021	.8028	.8035	.8041	.8048	.8055
6.4	.8062	.8069	.8075	.8082	.8089	.8096	.8102	.8109	.8116	.8122
6.5	.8129	.8136	.8142	.8149	.8156	.8162	.8169	.8176	.8182	.8189
6.6	.8195	.8202	.8209	.8215	.8222	.8228	.8235	.8241	.8248	.8254
6.7	.8261	.8267	.8274	.8280	.8287	.8293	.8299	.8306	.8312	.8319
6.8	.8325	.8331	.8338	.8344	.8351	.8357	.8363	.8370	.8376	.8328
6.9	.8388	.8395	.8401	.8407	.8414	.8420	.8426	.8432	.8439	.8445
7.0	.8451	.8457	.8463	.8470	.8476	.8482	.8488	.8494	.8500	.8506
7.1	.8513	.8519	.8525	.8531	.8537	.8543	.8549	.8555	.8561	.8567
7.2	.8573	.8579	.8585	.8591	.8597	.8603	.8609	.8615	.8621	.8627
7.3	.8633	.8639	.8645	.8651	.8657	.8663	.8669	.8675	.8681	.8686
7.4	.8692	.8698	.8704	.8710	.8716	.8722	.8727	.8733	.8739	.8745
7.5	.8751	.8756	.8762	.8768	.8774	.8779	.8785	.8791	.8797	.8802
7.6	.8808	.8814	.8820	.8825	.8831	.8837	.8842	.8848	.8854	.8859
7.7	.8865	.8871	.8876	.8882	.8887	.8893	.8899	.8904	.8910	.8915
7.8	.8921	.8927	.8932	.8938	.8943	.8949	.8954	.8960	.8965	.8971
7.9	.8976	.8982	.8987	.8993	.8998	.9004	.9009	.9015	.9020	.9025
8.0	.9031	.9036	.9042	.9047	.9053	.9058	.9063	.9069	.9074	.9079
8.1	.9085	.9090	.9096	.9101	.9106	.9112	.9117	.9122	.9128	.9133
8.2	.9138	.9143	.9149	.9154	.9159	.9165	.9170	.9175	.9180	.9186
8.3	.9191	.9196	.9201	.9206	.9212	.9217	.9222	.9227	.9232	.9238
8.4	.9243	.9248	.9253	.9258	.9263	.9269	.9274	.9279	.9284	.9289
8.5	.9294	.9299	.9304	.9309	.9315	.9320	.9325	.9330	.9335	.9340
8.6	.9345	.9350	.9355	.9360	.9365	.9370	.9375	.9380	.9385	.9390
8.7	.9395	.9400	.9405	.9410	.9415	.9420	.9425	.9430	.9435	.9440
8.8	.9445	.9450	.9455	.9460	.9465	.9469	.9474	.9479	.9484	.9489
8.9	.9494	.9499	.9504	.9509	.9513	.9518	.9523	.9528	.9533	.9538
9.0	.9542	.9547	.9552	.9557	.9562	.9566	.9571	.9576	.9581	.9586
9.1	.9590	.9595	.9600	.9605	.9609	.9614	.9619	.9624	.9628	.9633
9.2	.9638	.9643	.9647	.9652	.9657	.9661	.9666	.9671	.9675	.9680
9.3	.9685	.9689	.9694	.9699	.9703	.9708	.9713	.9717	.9722	.9727
9.4	.9731	.9736	.9741	.9745	.9750	.9754	.9759	.9763	.9768	.9773
9.5	.9777	.9782	.9786	.9791	.9795	.9800	.9805	.9809	.9814	.9818
9.6	.9823	.9827	.9832	.9836	.9841	.9845	.9850	.9854	.9859	.9863
9.7	.9868	.9872	.9877	.9881	.9886	.9890	.9894	.9899	.9903	.9908
9.8	.9912	.9917	.9921	.9926	.9930	.9934	.9939	.9943	.9948	.9952
9.9	.9956	.9961	.9965	.9969	.9974	.9978	.9983	.9987	.9991	.9996
N	0	1	2	3	4	5	6	7	8	9

TABLE II. *Logarithms to Base e*

N	.0	.1	.2	.3	.4	.5	.6	.7	.8	.9
1	0.000	0.095	0.182	0.262	0.336	0.405	0.470	0.531	0.588	0.642
2	0.693	0.742	0.788	0.833	0.875	0.916	0.956	0.993	1.030	1.065
3	1.099	1.131	1.163	1.194	1.224	1.253	1.281	1.308	1.335	1.361
4	1.386	1.411	1.435	1.459	1.482	1.504	1.526	1.548	1.569	1.589
5	1.609	1.629	1.649	1.668	1.686	1.705	1.723	1.740	1.758	1.775
6	1.792	1.808	1.825	1.841	1.856	1.872	1.887	1.902	1.917	1.932
7	1.946	1.960	1.974	1.988	2.001	2.015	2.028	2.041	2.054	2.067
8	2.079	2.092	2.104	2.116	2.128	2.140	2.152	2.163	2.175	2.186
9	2.197	2.208	2.219	2.230	2.241	2.251	2.262	2.272	2.282	2.293
10	2.303	2.313	2.322	2.332	2.342	2.351	2.361	2.370	2.380	2.389

TABLE III. e^x *and* e^{-x}

x	e^x	e^{-x}	x	e^x	e^{-x}
0.0	1.00	1.00	3.1	22.2	.045
0.1	1.11	.905	3.2	24.5	.041
0.2	1.22	.819	3.3	27.1	.037
0.3	1.35	.741	3.4	30.0	.033
0.4	1.49	.670	3.5	33.1	.030
0.5	1.65	.607	3.6	36.6	.027
0.6	1.82	.549	3.7	40.4	.025
0.7	2.01	.497	3.8	44.7	.022
0.8	2.23	.449	3.9	49.4	.020
0.9	2.46	.407	4.0	54.6	.018
1.0	2.72	.368	4.1	60.3	.017
1.1	3.00	.333	4.2	66.7	.015
1.2	3.32	.301	4.3	73.7	.014
1.3	3.67	.273	4.4	81.5	.012
1.4	4.06	.247	4.5	90.0	.011
1.5	4.48	.223	4.6	99.5	.010
1.6	4.95	.202	4.7	110	.0091
1.7	5.47	.183	4.8	122	.0082
1.8	6.05	.165	4.9	134	.0074
1.9	6.69	.150	5.0	148	.0067
2.0	7.39	.135	5.1	164	.0061
2.1	8.17	.122	5.2	181	.0055
2.2	9.02	.111	5.3	200	.0050
2.3	9.97	.100	5.4	221	.0045
2.4	11.0	.091	5.5	245	.0041
2.5	12.2	.082	5.6	270	.0037
2.6	13.5	.074	5.7	299	.0033
2.7	14.9	.067	5.8	330	.0030
2.8	16.4	.061	5.9	365	.0027
2.9	18.2	.055	6.0	403	.0025
3.0	20.1	.050			

TABLE IV. *Trigonometric Functions*

Degrees	*Radians*	*Sin*	*Tan*	*Cot*	*Cos*		*Degrees*
0°	**.000**	**.000**	**.000**		**1.000**	**1.571**	**90°**
1°	.017	.017	.017	57.29	1.000	1.553	89°
2°	.035	.035	.035	28.64	.999	1.536	88°
3°	.052	.052	.052	19.081	.999	1.518	87°
4°	.070	.070	.070	14.301	.998	1.501	86°
5°	**.087**	**.087**	**.087**	**11.430**	**.996**	**1.484**	**85°**
6°	.105	.105	.105	9.514	.995	1.466	84°
7°	.122	.122	.123	8.144	.993	1.449	83°
8°	.140	.139	.141	7.115	.990	1.431	82°
9°	.157	.156	.158	6.314	.988	1.414	81°
10°	**.175**	**.174**	**.176**	**5.671**	**.985**	**1.396**	**80°**
11°	.192	.191	.194	5.145	.982	1.379	79°
12°	.209	.208	.213	4.705	.978	1.361	78°
13°	.227	.225	.231	4.331	.974	1.344	77°
14°	.244	.242	.249	4.011	.970	1.326	76°
15°	**.262**	**.259**	**.268**	**3.732**	**.996**	**1.309**	**75°**
16°	.279	.276	.287	3.487	.961	1.292	74°
17°	.297	.292	.306	3.271	.956	1.274	73°
18°	.314	.309	.325	3.078	.951	1.257	72°
19°	.332	.326	.344	2.904	.946	1.239	71°
20°	**.349**	**.342**	**.364**	**2.747**	**.940**	**1.222**	**70°**
21°	.367	.358	.384	2.605	.934	1.204	69°
22°	.384	.375	.404	2.475	.927	1.187	68°
23°	.401	.391	.424	2.356	.921	1.169	67°
24°	.419	.407	.445	2.246	.914	1.152	66°
25°	**.436**	**.423**	**.466**	**2.144**	**.906**	**1.134**	**65°**
26°	.454	.438	.488	2.050	.899	1.117	64°
27°	.471	.454	.510	1.963	.891	1.100	63°
28°	.489	.469	.532	1.881	.883	1.082	62°
29°	.506	.485	.554	1.804	.875	1.065	61°
30°	**.524**	**.500**	**.577**	**1.732**	**.866**	**1.047**	**60°**
31°	.541	.515	.601	1.664	.857	1.030	59°
32°	.559	.530	.625	1.600	.848	1.012	58°
33°	.576	.545	.649	1.540	.839	.995	57°
34°	.593	.559	.675	1.483	.829	.977	56°
35°	**.611**	**.574**	**.700**	**1.428**	**.819**	**.960**	**55°**
36°	.628	.588	.727	1.376	.809	.942	54°
37°	.646	.602	.754	1.327	.799	.925	53°
38°	.663	.616	.781	1.280	.788	.908	52°
39°	.681	.629	.810	1.235	.777	.890	51°
40°	**.698**	**.643**	**.839**	**1.192**	**.766**	**.873**	**50°**
41°	.716	.656	.869	1.150	.755	.855	49°
42°	.733	.669	.900	1.111	.743	.838	48°
43°	.750	.682	.933	1.072	.731	.820	47°
44°	.768	.695	.966	1.036	.719	.803	46°
45°	**.785**	**.707**	**1.000**	**1.000**	**.707**	**.785**	**45°**
Degrees		*Cos*	*Cot*	*Tan*	*Sin*	*Radians*	*Degrees*

Answers to Odd-numbered Exercises

D

CHAPTER 0

Section 1, Part I

1. $(-\infty,2)$
3. $(-\infty,\frac{3}{2})$
5. $(-\infty,\frac{5}{3}]$
7. $(4.9,5.1)$
9. $[-1.575,-1.425]$
11. $(-\infty,\infty)$
13. $(0,\frac{8}{5})$
15. $[4,\infty)$
17. $(-\infty,b^2/4a]$ if $a > 0$
19. $(\frac{3}{2},\infty)$
21. $(0,6)$
23. $(-1,3)$
25. $(-\infty,-3)$
27. ϕ
29. $(-\infty,\infty)$
31. ϕ
33. $\{-2\}$
41. $(-7,3)$
43. $[1.65,1.68\bar{3}]$
45. $[0,1]$
47. $(-2-\pi, 2-\pi)$
49. $(-\infty,-2) \cup (2,\infty)$
51. $(-\infty,-7] \cup [3,\infty)$
53. $\left(\frac{-\delta-7}{3},\frac{\delta-7}{3}\right)$
55. $\left(7-\frac{\delta}{3},7+\frac{\delta}{3}\right)$

Section 2, Part I

11. $x^2 + y^2 - 9x + y + 8 = 0$
13. $x^2 + y^2 - 2x = 24$
15. $x^2 + y^2 - 6y + 5 = 0$
17. $x^2 + y^2 + 2x + 6y + 1 = 0$
19. $x^2 + y^2 - 4x - 8y = 0$
21. $x^2 + y^2 + 8x - 8y + 16 = 0$
23. $\{(x,y) \mid (x+1)^2 + (y-2)^2 > 7\}$
25. $(0,0)$, $r = 4$
27. $(1,2)$, $r = 2$
29. $(-3,5)$, $r = 3$
31. $(-2,0)$, $r = \frac{1}{2}$

Section 2, Part II

1. $(8,0)$, $(\frac{8}{5},\frac{24}{5})$
(First subtract one equation from the other.)

Section 3, Part I

7. The area of a triangle whose vertices are (x_r, y_r), $r = 1, 2, 3$, is given by the determinant

$$A = \frac{1}{2}\begin{vmatrix} x_1 & y_1 & 1 \\ x_2 & y_2 & 1 \\ x_3 & y_3 & 1 \end{vmatrix}.$$

If $A = 0$, the points are collinear.

9. $x - 3y = 2$
11. $3x - y = 3$
13. $y - 4x = 2$
15. $y - 6x = 2$
17. $y = 1$
19. $mx - y = ma$
21. $m = 2, a = -\frac{3}{2}, b = 3$
23. $m = -\frac{1}{2}, a = -6, b = -3$
25. No slope, $a = -\frac{5}{3}$
27. $m = -5, a = -3, b = -15$
29. $2x + y = 9,\ 2y - x = 3$
31. $3x - 2y = 8,\ 2x + 3y = 1$
33. $4x + 7y = 0,\ 7x - 4y = 0$
35. $(2,10)$
37. $(5,-4)$
39. $(4,2), (-2,-6)$
41. $(2,4)$

Section 3, Part II

1.
$$\theta = \alpha_2 - \alpha_1$$
$$\tan\theta = \tan(\alpha_2 - \alpha_1)$$
$$\tan\theta = \frac{\tan\alpha_2 - \tan\alpha_1}{1 + \tan\alpha_2 \tan\alpha_1} = \frac{m_2 - m_1}{1 + m_1 m_2}$$

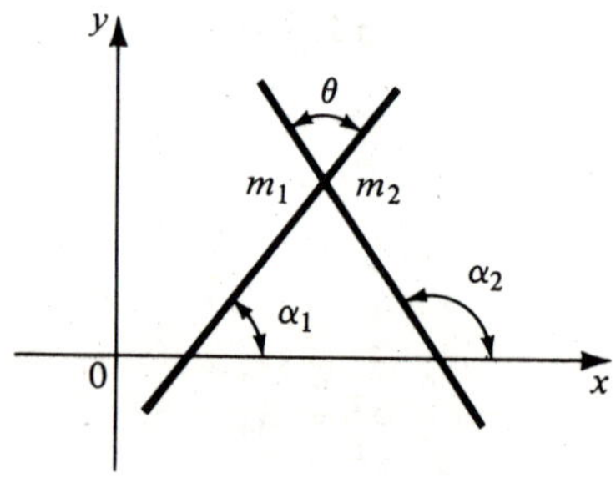

Section 4, Part I

13. $y^2 = 16x$
15. $(x + 3)^2 = -12y$
17. $(y - 2)^2 = 8(x - 2)$
19. $(x - 2)^2 = 16(y - 1)$

Section 4, Part II

1. Consider, separately, the regions $x \geq y^2$, $x < y^2$.

Review

1. $(-\frac{12}{17}, \infty)$
3. $(-\infty, \frac{3}{14}]$
5. $(\frac{11}{5}, \infty) \cup (-\infty, 2)$
7. $(4,10)$
9. $(-\infty,1) \cup (2,\infty)$
11. $\{-2\}$
13. $\{\pm\sqrt{2}, \pm 4\}$
15. ϕ
27. $(-\frac{4}{3}, \infty)$
29. $y - 5 = -\sqrt{3}(x - 1)$
31. $y + 3 = \frac{1}{4}(x - 9)$
33. $(x + 2)^2 = 4(y - 2)$
35. $(y - 6)^2 = 36(x - 6)$
37. $x = 5$
39. $24xy + 32y^2 - 60x + 364y = 1043$

41. (5,0), 5
(find the intersection of perpendicular bisectors of the sides)
43. $(\pm s/\sqrt{2},0)$, $(0, \pm s/\sqrt{2})$; $(\pm s/2, \pm s/2)$

CHAPTER 1

Section 2, Part I

1. $1; 5; 4 + 3\sqrt{3}$ **3.** $x + a - 3; 2x + h - 3$
5. $1 - h^{-1}; (x + y - 1)/(x + y + 1); 1 - 2x; 1/x$
7. $0, 1; 0, 3/(a + 1), a \neq -1; x = \pm\sqrt{2}/2$
9. Domain is set of all real numbers; range is $\{y \mid y \geq -2\}$.
11. Domain is $\{x \mid x \geq 0\}$; range is $\{y \mid y \geq 2\}$.
13. Domain is $\{x \mid x \leq -3 \text{ or } x > 4\}$; range is $\{y \mid 0 \leq y < 1 \text{ or } y > 1\}$.

Section 2, Part II

1. $81x^4 - 1; (x + 1)^3$

Section 4

1. $x^2 + 3x; x^2 - 3x - 2; 3x^3 + x^2 - 3x - 1; (x^2 - 1)/(3x + 1); 9x^2 + 6x;$
$3x^2 - 2$
3. $(x^2 + 2x - 1)/(x^2 - 1); (x^2 + 1)/(1 - x^2); x/(x^2 - 1); (x - 1)/x(x + 1);$
$(x - 1)/(2x - 1); -1/x$
5. $g(x) = \sqrt{x}, x \geq 0; g(f(x)) = x$ if $x \geq 0$, $g(f(x)) = -x$ if $x < 0$
7. $|x|$ **9.** $4x^2 - 8x + 8$ **13.** $x; x^{-1}$
15. $a + ab + b^2x; a + ab + ab^2 + b^3x$
17. $f(x) = a^x$ for any $a > 0$ works.
19. $f(x) = \sqrt{x}$, $g(x) = 1/x$ is one example.
21. 8

Section 5, Part I

1. $f^{-1}(y) = \frac{1}{2}(y + 1)$ **3.** $g^{-1}(y) = 2(y - 5)$
5. $f^{-1}(y) = \sqrt{y}$, domain $f^{-1} = [1,16]$
7. $h^{-1}(y) = y^3 - 8$
9. $g^{-1}(y) = \sqrt[3]{y} + 1$
11. $f^{-1}(y) = \sqrt{y^2 - 1}$, domain $f^{-1} = [1,\infty)$
13. **a.** $\{x \mid x \leq -3 \text{ or } x \geq 3\}$
b. No
c. Two possibilities are domain $f = \{x \mid x \geq 3\}$ and domain $f = \{x \mid x \leq -3\}$
d. For domain $f = \{x \mid x \geq 3\}$, $f^{-1}(x) = \sqrt{x^2 + 9}$, domain $f^{-1} = \{x \mid x \geq 0\}$

Review

1. Set of all real numbers
3. $\{x \mid x^2 \neq 1\}$
5. $\{h \mid h < 6\}$
7. $\{y \mid y \neq -2 \text{ and } y \neq -3\}$
9. $\{t \mid t \neq -3\}$
19. a. Even
    b. Odd
    c. Even
    d. Odd
    e. Neither
    f. Even
21. $f^{-1}(x) = \dfrac{x+5}{3}$
23. $h^{-1}(x) = \sqrt[3]{x} - 3$
25. $F^{-1}(t) = \dfrac{4}{t} - 1$
27. b. Two possibilities are domain $f = \{x \mid x \geq 0\}$ and domain $f = \{x \mid x \leq 0\}$.
    c. For domain $f = \{x \mid x \geq 0\}$, $f^{-1}(x) = \frac{3}{2}\sqrt{x^2 - 4}$, domain $f^{-1} = \{x \mid x \geq 2\}$.

CHAPTER 2

Section 1, Part I

1. -13
3. -6
5. -1
7. $\frac{1}{2}$
9. 12
11. 0
13. 0, 0
15. $1/\sqrt{15}$
17. $\dfrac{1}{4}$ $\left(\text{multiply by } \dfrac{\sqrt{x}+2}{\sqrt{x}+2}\right)$
19. $\dfrac{1}{2\sqrt{2}}$ $\left(\text{multiply by } \dfrac{\sqrt{2+h}+\sqrt{2}}{\sqrt{2+h}+\sqrt{2}}\right)$
21. $1/2\sqrt{7}$
23. $\frac{1}{10}$

Section 1, Part II

1. No limit
3. $\dfrac{1}{4}$ $\left(\text{multiply by } \dfrac{\sqrt{4+m+m^2}+2}{\sqrt{4+m+m^2}+2}\right)$
5. 3 [*Note:* $t^3 - 1 = (t-1)(t^2+t+1)$]

Section 2, Part I

19. $\dfrac{x+2}{x+1} = 1 + \dfrac{1}{x+1}$
21. $|y| - y = 0$ for $y > 0$
    $= -2y$ for $y < 0$
23. $x/|x| = 1$ for $x > 0$
    $= -1$ for $x < 0$

Section 2, Part II

1. $\dfrac{x^2+4}{x-2} = x + 2 + \dfrac{6}{x-2}$
3. See illustrated Example 3.
7. $x^3 + 1 = (x+1)(x^2 - x + 1)$
9. Consider $x = 5 + \varepsilon$ and $x = 5 - \varepsilon$.
11. See text.

Section 3, Part I

1. -7 **3.** 3 **5.** $\frac{1}{8}$
7. $\frac{1}{2}$ **9.** $\frac{431}{12}$ **11.** $\dfrac{a^2}{a^2+1}$
13. 6 **15.** 0 **17.** No

Section 4, Part I

1. -2 **3.** 0 **5.** -1
7. 5 **9.** 1 **11.** 2
13. 4, 3, no limit **15.** 9, 4, no limit **19.** Continuous

Section 4, Part II

1. 4
3. No limit

Section 5, Part I

1. ∞ **3.** $-\infty$ **5.** $-\infty$ **7.** 0
9. 7 **11.** ∞ **13.** 0 **15.** ∞
17. Horizontal asymptote $y = 0$, vertical asymptote $x = 2$

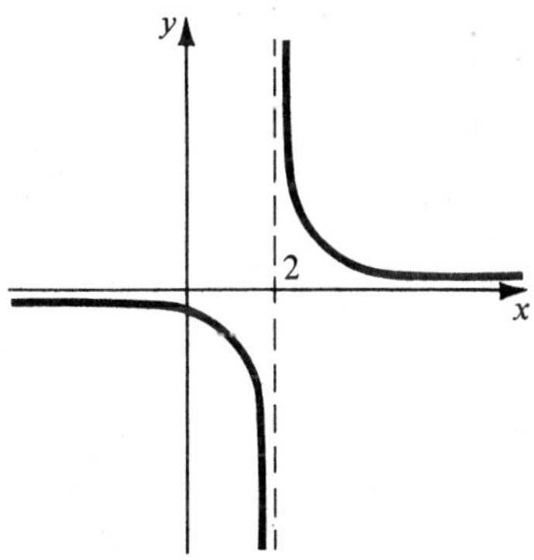

19. Horizontal asymptote $y = 2$, vertical asymptote $x = -2$
21. Horizontal asymptote $y = 2$, vertical asymptote $x = -2$

Section 5, Part II

1. $F(x) = \dfrac{(x-1)(x+1)}{(x-2)(x+2)}$ (symmetric about y axis)

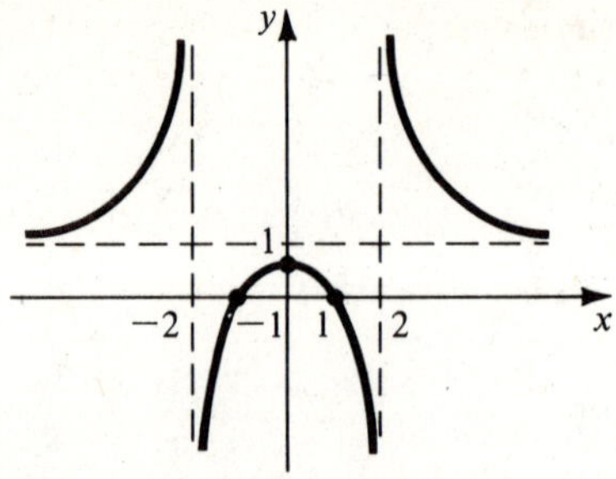

3. $F(x) = \dfrac{x^2 + 1}{(x - 2)(x + 2)}$ (symmetric about y axis)

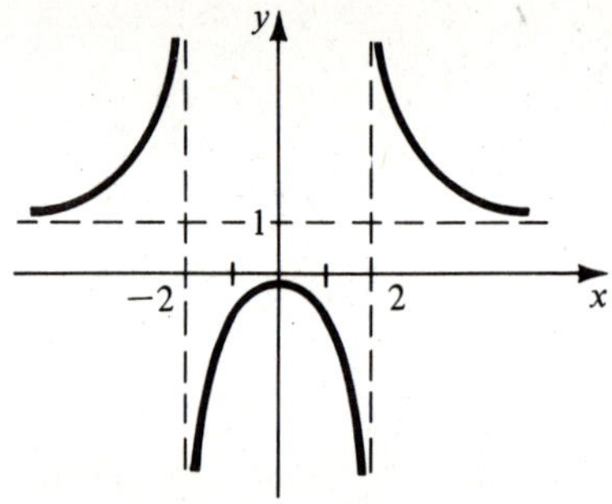

Section 6, Part I

1. $a^5 + 3a^2 + 2a + 1$ **3.** 0 **5.** $\frac{3}{2}$
7. -1 **9.** 0 **11.** 2
13. $\sqrt{3}\,|a|$ **15.** $\frac{11}{3}$ **17.** $-\dfrac{1}{a^2}$

Section 6, Part II

1. 4 $\left(\text{multiply by } \dfrac{2 + \sqrt{r^2 + 3}}{2 + \sqrt{r^2 + 3}}\right)$

3. 1 $\left(\text{multiply by } \dfrac{\sqrt{4 + m + m^2} + 2}{\sqrt{4 + m + m^2} + 2} \cdot \dfrac{\sqrt{4 + m - m^2} + 2}{\sqrt{4 + m - m^2} + 2}\right)$

Review, Part I

1. 5 **3.** $\frac{3}{2}$ **5.** $2\sqrt{2}$ **7.** $\frac{4}{7}$
9. 0 **11.** 0 **13.** 1 **15.** $\frac{3}{2}$
17. ∞ **19.** Continuous **21.** Discontinuous
23. Discontinuous at $t = 4, 6$, continuous at $t = 5$
25. Horizontal asymptote $y = 1$, vertical asymptote $x = 0$
27. Horizontal asymptote $y = 3$, vertical asymptote $x = -3$

29. Horizontal asymptote $y = 1$, vertical asymptote $x = \pm 2$

39. a. 1 b. 4 c. No limit d. 1
e. 1 f. 1 g. 4 h. -1

Review, Part II

1. 0
3. 0

CHAPTER 3

Section 1

1. 3
3. $2 - 2x$
5. $4x$
7. $9(3x + 1)^2$
9. $\dfrac{5}{2\sqrt{5x - 6}}$
11. $\dfrac{-1}{(x - 2)^2}$
13. $\frac{3}{2}\sqrt{x}$
15. 5
17. $\frac{1}{6}$
19. 60
21. 0
23. Yes; no

Section 2, Part I

1. $y = 2x - 3,\ -2y = x - 4$
3. $4y = x + 4,\ y = -4x + 18$
5. $y = -3x,\ 3y = x$
7. $2y = x,\ y = -2x + 5$
9. $3y = -2x + 5,\ 2y = 3x + 12$
11. $y - \frac{1}{4} = -\frac{1}{2}(x + \frac{63}{16})$
13. $y + 3x + 8 = 0$

Section 2, Part II

1. $y' = 6x = 6$; $y' = 6x^2 = 6$. Also, $3 = 3(1)^2$ and $3 = 2(1)^3 + 1$.

3. Slope $= D_x\sqrt{4px} = \dfrac{2p}{\sqrt{4px}} = \dfrac{2p}{y}$. Tangent line: $y - y_1 = \dfrac{2p}{y_1}(x - x_1)$.

Section 5, Part I

1. $4x^3 - 8x$
3. $30x^5 - 12x^3$
5. $12t^3 - 21t^2 + 10t$
7. $2/(y + 1)^2$
9. $2x + 2/x^3$
11. $\dfrac{3t^2 + 6t}{(t + 1)^2}$
13. $-\dfrac{2}{x^2}\left(2 + \dfrac{1}{x}\right) - \dfrac{1}{x^2}\left(1 + \dfrac{2}{x}\right)$
15. $\dfrac{1 - y}{2\sqrt{y}\,(y + 1)^2}$
17. $(2x^2 + 1)(3x^3 + 1) + 4x(x + 1)(3x^3 + 1) + 9x^2(x + 1)(2x^2 + 1)$
19. $\dfrac{4}{t^2} - \dfrac{14}{t^3}$
21. $\dfrac{4t^4 + 12t^2}{(t^2 + 1)^2}$

23. $9w^2 + 6w + 2$

25. Tangent line: $y = 12x - 9$
Normal line: $x + 12y = 37$

27. Tangent line: $y = \frac{11}{16}x + \frac{15}{4}$
Normal line: $y - \frac{13}{2} = -\frac{16}{11}(x - 4)$

Section 5, Part II

1. $f'(x) = 2x, |x| > 2$
$f'(x) = -2x, |x| < 2$
3. 0 (x nonintegral)

Section 7, Part I

1. $24x^2(1 + 2x^3)^3$

3. $\dfrac{3x^2 + 1}{2\sqrt{x^3 + x}}$

5. $-\frac{3}{4}(8x - 1)(4x^2 - x)^{-7/4}$

7. $(1 - 2x^2)/\sqrt{1 - x^2}$

9. $(3 + 3t)\sqrt{2t + t^2}$

11. $(3t + 1)^{-2/3}$

13. $-1/x^2\sqrt{x^2 + 1}$

15. $3/2\sqrt{2z}$

17. $4(y - 2) = x - 3;\ 2 - y = 4(x - 3)$

19. $y = x;\ y = -x$

21. $16(y - 2) = x - 4,\ 2 - y = 16(x - 4)$

Section 7, Part II

1. $\dfrac{1}{2\sqrt{x + \sqrt{x + \sqrt{x}}}}\left(1 + \dfrac{1}{2\sqrt{x + \sqrt{x}}}\left(1 + \dfrac{1}{2\sqrt{x}}\right)\right)$

3. $\frac{2}{3}(2x)^{-2/3}$ if $x > 0$; 0 if $x < 0$

Section 9

1. -4

3. $-\frac{6}{7}$

5. $\dfrac{x}{3y^2 - y}$

7. $\dfrac{10x^2 - 4xy - 2(2x - y)^{1/2}}{x^2 - 6(2x - y)^{1/2}}$

9. $-y^2/x^2$

11. $24(3x^2 + 1)^3 + 432x^2(3x^2 + 1)^2$

13. $-\frac{3}{8}(1 + x)^{-3/2}$

15. $18t - \dfrac{2}{t^3}$

17. $\dfrac{3v - 54}{8(9 - v)^{5/2}}$

19. $-2;\ \frac{3}{2}$

21. $-\frac{6}{5};\ -\frac{42}{125}$

23. $(3y^2 - 2y)^{-1},\ (2 - 6y)/(3y^2 - 2y)^3$

25. $4(x + 1)^{-2} - 8(x + 1)^{-3};\ -8(x + 1)^{-3} + 24(x + 1)^{-4}$

27. $y + 2 = -(x + 2)$

29. $y + 1 = -(x + 1)$

31. $f'(x) = \begin{cases} -3x^2 & \text{if } x < 0 \\ 3x^2 & \text{if } x > 0 \end{cases}$ $\qquad f''(x) = \begin{cases} -6x & \text{if } x < 0 \\ 6x & \text{if } x > 0 \end{cases}$

Domain of f', f'' is $\{x \mid x \neq 0\}$.

33. $f'(x) = \begin{cases} 2x \text{ if } |x| > 2 \\ -2x \text{ if } |x| < 2 \end{cases}$ $\qquad f''(x) = \begin{cases} 2 \text{ if } |x| > 2 \\ -2 \text{ if } |x| < 2 \end{cases}$

Domain of f', f'' is $\{x \mid |x| \neq 2\}$.

Section 10

1. $(3x^2 + 4x)\,\Delta x + (3x + 2)(\Delta x)^2 + (\Delta x)^3$

3. $\dfrac{(1 - x^2)\,\Delta x - x(\Delta x)^2}{(x^2 + 1)[(x + \Delta x)^2 + 1]}$

5. .0703; .07; .0003

7. $-.875$; -1.5; .625

9. .02076; .02083; $-.00007$

11. 3.008 $(= 3 + \frac{1}{120})$

13. .196 $(= \frac{1}{5} - \frac{1}{250})$

15. 2.0125 $(= 2 + \frac{1}{80})$

17. $\pm 3.6\pi$ in.^{2}

19. ± 2.4 in.^{2}

21. 157 ft^3 $(= 50\pi$ ft$^3)$

Review, Part I

1. -11

3. $-\frac{1}{2}$

5. $5x^4 - 5$

7. $6t^5 - 3t^2$

9. $-\dfrac{2}{(2x + 3)^2}$

11. $-\dfrac{1}{y^2} + \dfrac{6}{y^4}$

13. $5(2x - x^2)^4(2 - 2x)$

15. $-\dfrac{t}{\sqrt{1 - t^2}}$

17. $\dfrac{2y + 1}{\sqrt{y + 1}}$

19. $(t^3 + \sqrt{t^4 + 9})^{-2/3}[t^2 + \frac{2}{3}t^3(t^4 + 9)^{-1/2}]$

21. $y = 0,\ x = 1$

23. $y = -\frac{3}{100}x + \frac{13}{25},\ y = \frac{100}{3}x - \frac{1994}{15}$

25. $y = 6x - 18$

27. $\dfrac{3x^2 \quad 3y}{3x - 2y}$

29. $\sqrt{\dfrac{y + 1}{x + 1}}$

31. $\dfrac{2 - 24x(3x^2 + y) - (x + y)^{-1/2}}{4(3x^2 + y) + (x + y)^{-1/2}}$

33. $(2,8),\ (-2,-8)$

35. $(1,0),\ (-1,-4)$

37. .06757, .06667, .00090

39. $-.3095$, $-.4615$, .1520

41. $2x - 9y - 11 = 0$

43. Differentiate three times implicitly:

$$\begin{aligned} 1 &= 5y^4y' + y', \\ 0 &= 20y^3y'^2 + 5y^4y'' + y'', \\ 0 &= 60y^2y'^3 + 60y^3y'y'' + 5y^4y''' + y'''. \end{aligned}$$

Now solve for y' in the first equation, then y'' in the second, then y''' in the third.

45. **a.** $10 - \frac{1}{75}$ **b.** 9.95

Review, Part II

1. Intuitively, 2, 1, 0, depending whether the point is outside, on, or inside the

parabola, respectively. Analytically, let the parabola be $y^2 = 2ax$ and the given point be (h,k) and the corresponding tangent point be (x,y), then,

$$\frac{y-k}{x-h} = \text{slope of tangent} = \frac{a}{y}, \qquad \text{by implicit differentiation,}$$

or

$$y(y-k) = \frac{y^2}{2} - ah.$$

Since this is a quadratic equation, there are at most two real roots.

3. $y^2 = 4ax$, focus $(a,0)$. Point C is chosen such that $AB = BC$ [that is, $(x + AB, y)$]. Now show that the midpoint of AC lies on the normal line through B or, equivalently, that AC is parallel to the tangent line at B.

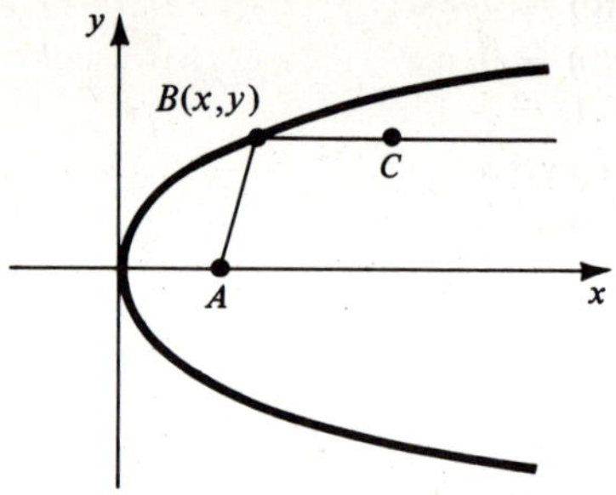

5. If the parabola be $y^2 = 4ax$, focus F is at $(a,0)$. Let A have coordinates (h,k) where $k^2 = 4ah$. Tangent line at A is $yk = 2a(x + h)$ which intercepts the y axis at point $B(0, k/2)$. Slope of $AB = 2a/k$, slope of $BF = -k/2a$. Product of slopes $= -1$.

7. 10.0033 **9.** .4534

CHAPTER 4

Section 1

1. $-\frac{5}{2}$ **3.** None **5.** $0, \pm\sqrt{2}$
7. $0, \{x \mid -3 \le x \le -2 \text{ or } 2 \le x \le 3\}$
9. 0, 1 **13.** $\sqrt{2}/2$ **15.** 1
17. $0, \pm\sqrt{3}; \pm 1$ **19.** $0, a; 0, na/(n+1)$ **21.** $\frac{1}{2}$
23. $\frac{5}{4}$ **25.** $\sqrt{3}$ **27.** $\frac{5}{4}$
29. 3 **31.** $c = \sqrt{3}$ **33.** F is discontinuous.
35. $f'(1)$ is not defined.

Section 2

1. $[1,\infty); (-\infty,1]$ **3.** $(-\infty,1], [3,\infty); [1,3]$
5. $(-\infty,\frac{3}{4}]; [\frac{3}{4},\infty)$ **7.** $(-\infty,-1), (-1,0); [0,1), (1,\infty)$

9. $[0,\infty)$
11. $(-\infty,-1]$, $[1,\infty)$; $[-1,0)$, $(0,1]$
13. $[-3,\infty)$; $(-\infty,-3]$

Section 3, Part I

1. $f(2) = 0$, min
3. $F(-\frac{5}{2}) = \frac{37}{4}$, max
5. No extrema
7. No extrema
9. $g(-2) = 0$, max; $g(0) = -108$, min
11. $f(\pm 2) = 0$, min; $f(0) = 16$, max
13. $F(-2) = -16$ and $F(1) = 38$, max; $F(-1) = -38$ and $F(2) = 16$, min
15. $f(1) = -4$, min; $f(-3) = 28$, max
17. No extrema
19. No extrema
21. $g(0) = 0$, max; $g(\pm 1) = -1$, min
23. $f(0) = 0$, max; $f(\frac{6}{5}) = -26{,}244/3125$, min
25. $g(0) = 0$, max; $g(2) = -3\sqrt[3]{4}$, min
27. $F(1) = -1$, min
29. $f(\sqrt{3}) = \sqrt[3]{-6\sqrt{3}}$, min; $f(-\sqrt{3}) = \sqrt[3]{6\sqrt{3}}$, max
31. $G(\pm 1) = 2$, min

Section 3, Part II

1. a. $F(1)$, min; $f(-1)$, min; $f((q-p)/(q+p))$, max
   b. $f(1)$, min; $f((q-p)/(q+p))$, max
   c. $f(-1)$, max; $f((q-p)/(q+p))$, min
   d. $f((q-p)/(q+p))$, min

Section 4

1. $f(\frac{1}{5}) = \frac{4}{5}$, min
3. $g(-3) = 37$, max; $g(1) = 5$, min
5. $F(2) = 3$, min
7. $f(9) = \frac{2}{3}$, min
9. $h(0) = 0$, min
11. $(-\infty,-1)$, $(\frac{1}{2},\infty)$; $(-1,\frac{1}{2})$; pts. of infl. at $(-1,-6)$, $(\frac{1}{2},-\frac{9}{16})$
13. Concave downward in $(-\infty,\infty)$, no pts. of infl.
15. $(-\infty,0)$; $(0,\infty)$; no pts. of infl.
17. $G(-1) = 4$, max
19. $F(-\frac{1}{3}) = \frac{59}{27}$, max; $F(1) = 1$, min; $(\frac{1}{3},\frac{43}{27})$, pt. of infl.
21. $f(-1) = -2$, min; $f(0) = 3$, max; $f(2) = -29$, min; pts. of infl. at $x = (1 \pm \sqrt{7})/3$
23. $f(5) = 7$, max; $f(17) = -13$, min
25. $f(1) = f(4) = \sqrt[3]{4}$, max; $f(-1) = -2\sqrt[3]{2}$, min
27. $f(-3) = f(3) = 4 + \sqrt[3]{9}$, max; $f(0) = 4$, min
29. $g(-\sqrt{5}) = 10\sqrt{5}$, max; $g(\sqrt{5}) = -10\sqrt{5}$, min
31. $G(\frac{5}{8}) = -15\sqrt[3]{25}/256$, min
33. $F(-3) = 0$, max; $F(-2) = -2$, min
35. $f(0) = 0$, min; $f(\frac{1}{2}) = 9\sqrt[3]{2}/8$, max; $f(2) = 0$, min
37. $g\left(\dfrac{1}{a}\right) = \dfrac{a}{|a|}\sqrt{a^2+1}$, max if $a > 0$; min if $a < 0$

39. $f(-1) = 0$, min
41. $F(0) = 0$, min
43. $g(0) = -\frac{1}{3}$, max
45. $(\pm 1, \frac{1}{4})$
47. $(1, \frac{5}{6})$; $15x - 6y = 10$
49. $a = 1$, $b = -5$, min

Section 5, Part I

1. 100×150
3. $(6 - 2\sqrt{3}) \times 4\sqrt{3} \times (12 + 4\sqrt{3})$
5. Radius = height
7. Length = 2(width)
9. If (x, y) is to be the closest point, we must then minimize $(x - 11)^2 + (x^3 - 3x - 1)^2$
11. $5\sqrt{5}$
15. 3 miles from P
17. 500
19. $(1,1)$; $(-1,-1)$

Section 5, Part II

1. $a^2 + \dfrac{h^2}{4} = r^2$

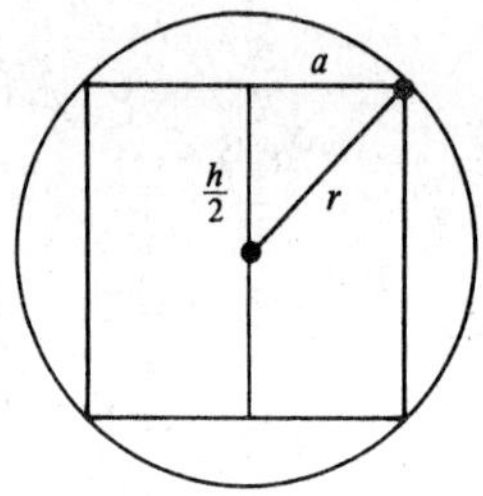

$$\text{Vol.} = \pi h a^2 = \pi\left(r^2 h - \frac{h^3}{4}\right)$$

$$\frac{dv}{dh} = 0 = r^2 - \frac{3h^2}{4}$$

$$h = \frac{2r}{\sqrt{3}},\ a = \frac{r\sqrt{6}}{3}$$

3. $\dfrac{H - h}{H} = \dfrac{r}{R}$ (by similar triangles)

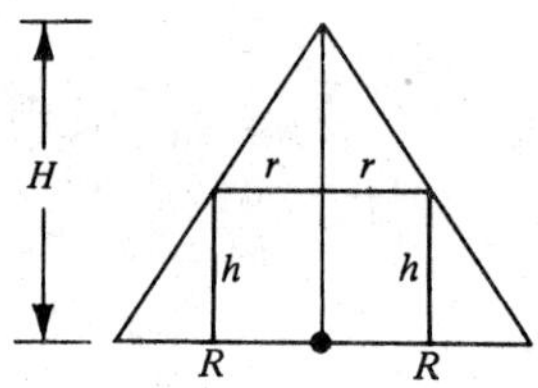

$$\text{Vol.} = \pi r^2 h = \pi H\left(r^2 - \frac{r^3}{R}\right)$$

$$\frac{dV}{dr} = 0 = 2r - \frac{3r^2}{R} \text{ or } r = \frac{2R}{3}$$

Then, $h = \dfrac{H}{3}$

5. Eq. of line: $\dfrac{x}{a} + \dfrac{y}{b} = 1$

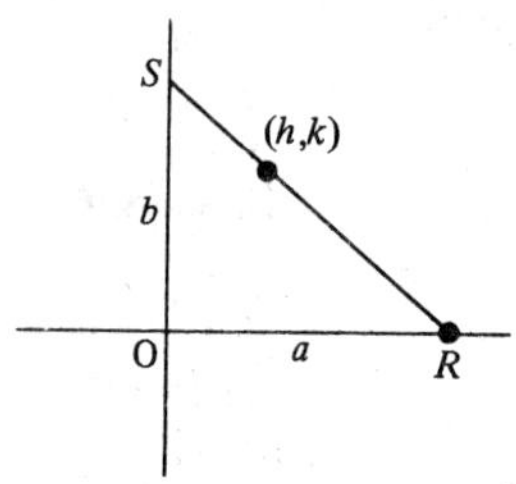

Thus $\dfrac{h}{a} + \dfrac{k}{b} = 1$

a. $RS = \sqrt{a^2 + b^2}$

$\dfrac{dRS^2}{da} = 2a + 2b\dfrac{db}{da}$; also $-\dfrac{h}{a^2} - \dfrac{k}{b^2}\dfrac{db}{da} = 0$

(implicit diff.)

$a - b\left(\dfrac{b^2 h}{k a^2}\right) = 0$, or $\dfrac{a}{b} = \sqrt[3]{\dfrac{h}{k}}$

Subst. back in $\dfrac{h}{a} + \dfrac{k}{b} = 1$.

Finally, $RS_{min} = (h^{2/3} + k^{2/3})^{3/2}$.

b. $OR + OS = a + b$;

$$\frac{d}{da}(OR + OS) = 1 + \frac{db}{da} = 1 - \frac{b^2h}{ka^2} = 0, \text{ or } \frac{a}{b} = \sqrt{\frac{h}{R}}$$

Thus, $(OR + OS)_{min} = (\sqrt{h} + \sqrt{k})^2$

c. $OR \cdot OS = ab$

$$\frac{dOR \cdot OS}{da} = b + a\frac{db}{da} = b - a\left(\frac{b^2h}{a^2k}\right) = 0$$

or $\frac{h}{a} = \frac{k}{b} = \frac{1}{2}$ and $(OR \cdot OS)_{min} = 4hk$.

Section 6

1. $v = -32t + 80$, $a = -32$; $s(\frac{5}{2}) = 100$, max; $s(0) = 0$, min

3. $v = 2t - 8$, $a = 2$; $s(4) = -12$, min, $s(-2) = 24$, max

5. $v = 3t^2 - 3$, $a = 6t$; $s(-1) = 2$, max, $s(1) = -2$, min

7. $v = 1 - 2t - 3t^2$, $a = -2 - 6t$; $s(-1) = 1$, min, $s(\frac{1}{3}) = \frac{59}{27}$, max

9. $v = 2t - 16t^{-2}$, $a = 2 + 32t^{-3}$; $s(2) = 12$, min

Section 7

1. $x - 2x^2 + 3x^3$

3. $4x - x^3 + \frac{1}{5}x^5$

5. $\frac{2}{5}x^{5/2} + \frac{2}{3}x^{3/2} - 5x$

7. $-\frac{1}{2x}$

9. $-x^{-1} + x^{-2} - \frac{1}{3}x^{-3}$

11. $3x^{4/3} - \frac{2}{3}x^{3/2}$

13. $4t^2 + 7t$

15. $3x + \frac{1}{3}x^3 - \frac{1}{4}x^4 - \frac{5}{3}$

17. 49 ft, 3.5 sec

19. 40 ft/sec

21. $s(t) = 4t^2$; $5\sqrt{2}$ sec; 34 ft/sec

23. 80 ft/sec

Section 8

1. $r' = (9\pi t^2)^{-1/3}$; .082 in/sec

3. $d(t) = 4\sqrt{41t^2 - 120t + 225}$; closest at 11:28 a.m.

5. $h(t) = \sqrt{2t/3}$; $\frac{1}{6}$ ft/min

7. Decreasing by $\frac{28}{3}$ in^2/sec

9. $\frac{40}{3}$ min

11. 550 mph

13. Increasing by 734 mi/h

15. 80.5 ft/min

Review

1. $(0,-4)$, min

3. $(\sqrt{3}/3, -2\sqrt{3}/9)$, min; $(-\sqrt{3}/3, 2\sqrt{3}/9)$, max; $(0,0)$, pt. of infl.

5. $(3,-756)$, min; $(-3,756)$, max; $(0,0)$, $(-2,-494)$, $(2,494)$, pts. of infl.

7. $(1,4)$, min; $(-\frac{1}{3},\frac{140}{27})$, max; $(\frac{1}{3},\frac{124}{27})$, pt. of infl.
9. $(\sqrt{3},\sqrt{3}/6)$, max; $(-\sqrt{3},-\sqrt{3}/6)$, max; $(0,0)$, $(3,\frac{1}{4})$, $(-3,-\frac{1}{4})$, pts. of infl.
11. No extrema or pts. of infl.
13. $\left(\frac{1-\sqrt{7}}{3}, \frac{14\sqrt{7}-47}{27}\right)$, max; $(0,-1)$, min; $(\frac{1}{3},-\frac{11}{27})$, pt. of infl.
15. $(\frac{3}{2}\sqrt{2},\frac{9}{2})$, max; $(-\frac{3}{2}\sqrt{2},\frac{9}{2})$, min; $(0,0)$, pt. of infl.
17. $(0,0)$, min; $(0,0)$, min; $(1/4m^2, -1/4m)$, max
19. $\frac{2-\sqrt{7}}{3}$
21. n odd; n even
23. $f(x) = x^3, x \geq 0;\ f'''(x) = 3, x > 0$
$f(x) = -x^3, x < 0;\ f'''(x) = -3, x < 0$
No pt. of infl.
25. 9 in. × 6 in.; 54 in.^{2}
27. 12.5 knots
29. $\frac{15}{8}$ sec, $\frac{625}{4}$ ft, 60 ft/sec, 5 sec
31.

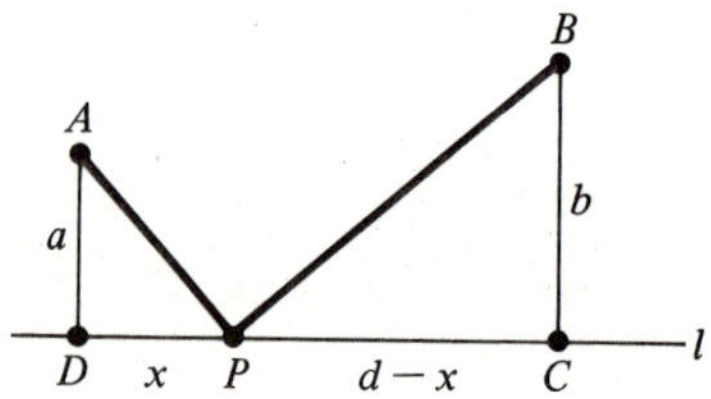

$$AP + BP = \sqrt{a^2 + x^2} + \sqrt{b^2 + (d-x)^2}$$

Differentiating,

$$\frac{x}{\sqrt{a^2 + x^2}} = \frac{d-x}{\sqrt{b^2 + (d-x)^2}},$$

Thus the two triangles are similar and $\angle APD = \angle BPC$. Alternatively, we can use the solution of Heron, the early Greek geometer. The shortest distance will be the length of the straight line segment BA' where A' is the image of A across DC.
33. $1, \frac{3}{5}$
35. min $(|a| + |b|)^2$, no max
37. $\frac{4}{5}x^5 + \frac{1}{4}x^4 - \frac{1}{3}x^3 - \frac{1}{2}x^2 + 3x$
39. $\frac{9}{4}x^{4/3} - \frac{14}{11}x^{11/7}$
41. $\frac{4}{3}(y-3)^{3/4}$
43. $v = -1 + 4t - 8t^3$, $a = 4 - 24t^2$, $s\left(-\frac{1+\sqrt{5}}{4}\right) = \frac{82 + 13\sqrt{5}}{32}$, max
45. $v = 3 - \frac{6}{t^2}$, $a = \frac{12}{t^3}$, $s(2) = 9$, min

CHAPTER 5

Section 1, Part I

1. 3, 0 **3.** $\sqrt{53}$, 3 **5.** 5, 5
7. $\sqrt{3}$, $-\sqrt{3}$ **9.** $\frac{1}{2}$, -1 **11.** No l.u.b.; g.l.b. = 7
13. No l.u.b. or g.l.b.

Section 1, Part II

3. Any finite set. $a_n = k^2(1 - 1/n)$. Any finite set. $a_n = k^2/n$.
5. l.u.b. $S = \max_n a_n$; g.l.b. $S = \min_n a_n$.

Section 2, Part I

1. a. $\frac{63}{4}$, $\frac{87}{4}$ **b.** $\frac{275}{16}$, $\frac{323}{16}$
3. a. $\frac{7}{16}$, $\frac{15}{16}$ **b.** $\frac{175}{256}$, $\frac{207}{256}$
5. a. $\frac{61}{144}$, $\frac{49}{36}$ **b.** .55, 1.02
7. 0, $\frac{3}{2}$
9. 16, 29

Section 2, Part II

1.

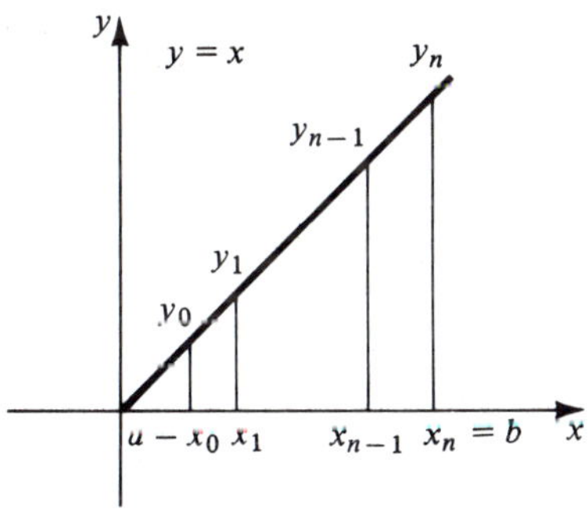

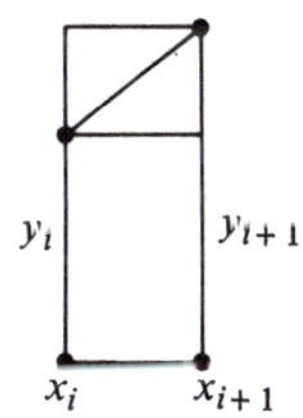

Consider the $(i + 1)$st interval $[x_i, x_{i+1}]$.

$$I_{i+1} = y_i(x_{i+1} - x_i),\ C_{i+1} = y_{i+1}(x_{i+1} - x_i).$$

Now

$$x_i(x_{i+1} - x_i) < \frac{x_i + x_{i+1}}{2}(x_{i+1} - x_i) < x_{i+1}(x_{i+1} + x_i)$$

or

$$I_{i+1} < \frac{x_{i+1}^2 - x_i^2}{2} < C_{i+1}.$$

Letting $i = 0, 1, 2, \ldots,$

$$I_1 < \frac{x_1^2 - a^2}{2} < C_1,$$

$$I_2 < \frac{x_2^2 - x_1^2}{2} < C_2,$$

$$\cdots\cdots\cdots\cdots\cdots$$

$$I_n < \frac{b^2 - x_{n-1}^2}{2} < C_n.$$

Now adding up,

$$I(P) = I_1 + I_2 + \cdots + I_n < \frac{b^2 - a^2}{2} < C_1 + C_2 + \cdots + C_n = C(P).$$

3. If $p < q$ then, $p^3 < \dfrac{p^3 + p^2q + pq^2 + q^3}{4} < q^3$. By multiplying by $q - p$,

$p^3(q - p) < \dfrac{q^4 - p^4}{4} < q^3(q - p)$. Now see solution of previous exercise.

Section 3

1. n^2
3. 42
5. $\dfrac{3n + 1}{2n}$
7. $\frac{7}{30}$
9. $217 - 7t$
11. 684
13. $a^2n(n + 1)(2n + 1)/6 + abn(n + 1) + b^2n$
15. $n(n + 1)(n + 2)/3$
17. $[F(1) - F(0)] + [F(2) - F(1)] + \cdots + [F(n) - F(n - 1)] = F(n) - F(0)$

Section 4, Part I

1. $-1.75, .25$
3. .5, 2.5
5. $-5.56, -2.78$
7. $-1.83, -1.08$
9. .55, 1.02
11. 20
13. 16
15. $\frac{2}{3}$

Section 5

1. 21
3. 8
5. 588
7. $-\frac{119}{3}$
9. 3
11. -7
13. $\frac{2}{9}$
15. $\frac{3}{2}\sqrt[3]{2} + \frac{4}{3}\sqrt[4]{8} - \frac{25}{12}$
17. $\frac{2}{3}x^{3/2} + \frac{1}{2}x^2$
19. $\frac{1}{4}[(b + 4)^4 - (a + 4)^4]$
21. 2
23. $\frac{39}{2}$
25. $\sqrt{3} - \sqrt{2}$
27. $-\dfrac{1}{1 + x^3}$
29. $-3x^2(1 + x^6)^{1/3}$

Section 6, Part I

1. 2
3. 15
5. $\frac{136}{3}$
7. 10.1
9. $\frac{32}{3}$
11. $\frac{3}{5}t^{5/3} - \frac{8}{7}t^{7/4} + C$

13. $\frac{1}{3}x^3 + \frac{2}{x} + \frac{2}{5}x^{5/2} + C$

15. $3a$

17. $\frac{1}{3}tx^3 + rx + C$

Section 6, Part II

1. 1

3. $\frac{75}{4}$; $x^3 - x = x(x - 1)(x + 1)$;

$$\int_{-2}^{3} |x^3 - x|\, dx = \int_{-2}^{-1} (x - x^3)\, dx + \int_{-1}^{0} (x^3 - x)\, dx + \int_{0}^{1} (x - x^3)\, dx + \int_{1}^{3} (x^3 - x)\, dx$$

Section 7, Part I

1. $\frac{14}{3}$

3. 0 (by odd symmetry) or by integration

5. -14

7. $\frac{1}{3}$

9. 0 (by odd symmetry) or by integration

11. $\frac{1}{2}\sqrt{x^4 + 2} + C$

13. $\frac{1}{2}(2 + \sqrt{x})^4 + C$

15. $\frac{1}{6}[1 + (2x - 1)^2]^{3/2} + C$

17. $\dfrac{2}{1 - \sqrt{x}} + C$

19. $\frac{3}{14}(10 + y^7)^{2/3} + C$

21. $b^2/6$

23. $\frac{16}{3025}$

25. 0

27. $\frac{4}{3}(1 + \sqrt{x})^{3/2} + C$ (let $y = \sqrt{x}$)

Section 7, Part II

1. $\frac{2}{3}(x + 1)^{3/2} - 2\sqrt{x + 1} + C$

3. $-\dfrac{1}{9}\left(2 + \dfrac{3}{x^2}\right)^{3/2} + C$

5. $\frac{1}{3}x^3 - \frac{1}{3}(x^2 - 1)^{3/2} + C$

Section 8, Part I

1. $3, \frac{3}{2}, 1, \frac{3}{4}, \frac{3}{3}; 0$

3. $\frac{1}{2}, \frac{4}{5}, 1, \frac{8}{7}, \frac{5}{4}; 2$

5. $-1, \frac{1}{4}, -\frac{1}{9}, \frac{1}{16}, -\frac{1}{25}; 0$

7. $-\frac{1}{3}, \frac{1}{3}, \frac{7}{11}, \frac{7}{9}, \frac{23}{27}; 1$

9. $0, \frac{7}{12}, \frac{5}{6}, \frac{39}{40}, \frac{16}{15}; \frac{3}{2}$

11. $\frac{4}{3}, -\frac{8}{3}, -\frac{108}{77}, -\frac{64}{63}, -\frac{500}{621}; 0$

13. $1/\sqrt{2}, 1/\sqrt{5}, 1/\sqrt{10}, 1/\sqrt{17}, 1/\sqrt{26}; 0$

15. $\frac{1}{3}, \frac{1}{9}, \frac{1}{27}, \frac{1}{81}, \frac{1}{243}; 0$

17. $\sqrt{3} - \sqrt{2}, 2 - \sqrt{3}, \sqrt{5} - 2, \sqrt{6} - \sqrt{5}, \sqrt{7} - \sqrt{6}; 0$

19. $\frac{1}{5}, \frac{5}{13}, \frac{13}{41}, \frac{41}{122}, \frac{61}{182}; \frac{1}{3}$

21. $2, \frac{3}{4}, 2, \frac{1}{2}, 2$; no limit

Section 8, Part II

1. 0

3. 1

5. 2

Section 9

1. $\frac{3}{2}$ 3. $\frac{5}{6}$ 5. 6
7. 0 9. $12s + 4r$

Review, Part I

1. $\dfrac{1}{2(3 - x^2)} + C$ 3. $\dfrac{1}{3}x^3 + \dfrac{4}{x} + C$ 5. $-\dfrac{1}{6(t^3 + 1)^2} + C$
7. $21\sqrt[3]{2} - \frac{3}{2}\sqrt[3]{4} - \frac{39}{2}$ 9. $\frac{1}{28}(4x - 3)^7 + C$
11. $\frac{25}{2}$ 13. $2 + 3/\sqrt{2}$
15. $-\frac{1}{25}[(1 + 5x - x^5)^5]_0^1 = 5^{-2} - 5^3$
17. $\dfrac{a^{2n+2} - (a^2 + 1)^{n+1}}{2n + 2}$ 19. 1
21. No limit 23. 2, $\frac{1}{3}$ 25. 1, $\frac{2}{3}$
27. π, 3.1 29. $\frac{3}{4}$
31. $\frac{4}{5}$ 33. $-\frac{5}{2}$ 35. $-\frac{1}{2}$
37. $2\sqrt{3} - \frac{4}{3}\sqrt{2} + \frac{9}{2}$

Review, Part II

1. $f(g(x))g'(x)$
3. The graph of x vs. u consists of two sections with a discontinuity at $x = 0$. To make the change of variables properly, we first break up the interval of integration, i.e.,

$$I = \int_{-1}^{0} \frac{1}{1 + x^2}\,dx + \int_{0}^{1} \frac{1}{1 + x^2}\,dx.$$

Now if we let

$$x = \frac{1}{u}$$

we obtain

$$I = \lim_{x \to 0^-} \int_{-1}^{1/x} \frac{1}{1 + u^2}\,du + \lim_{x \to 0^+} \int_{1/x}^{1} \frac{1}{1 + u^2}\,du.$$

The integrals

$$\int_{-1}^{-\infty} \frac{1}{1 + u^2}\,du \qquad \text{and} \qquad \int_{\infty}^{1} \frac{1}{1 + u^2}\,du$$

are called *improper integrals* and will be treated subsequently.

CHAPTER 6

Section 1, Part I

1. 20 3. $\frac{1}{3}$ 5. $4\sqrt{3} - \frac{20}{3}$
7. $\frac{703}{42}$ 9. $\frac{32}{3}$ 11. $\frac{64}{3}$
13. $\frac{4}{3}$ 15. $\frac{1}{2}$ 17. $\frac{8}{3}$ 19. $\frac{5}{6}$

Section 1, Part II

1. Tangent line at $(0, -3)$ is $y = 4x - 3$. By symmetry,

$$A = 2\int_0^2 [(4x - 3) - (-x^2 + 4x - 3)]\,dx = \tfrac{16}{3}.$$

3. $A = 4\int_0^a \sqrt{a^2x^6 - x^8}\,dx = 4\int_0^a x^3\sqrt{a^2 - x^2}\,dx.$ Let $a^2 - x^2 = y^2$, giving

$$A = \int_0^a 4y(a^2 - y^2)\,dy = a^4.$$

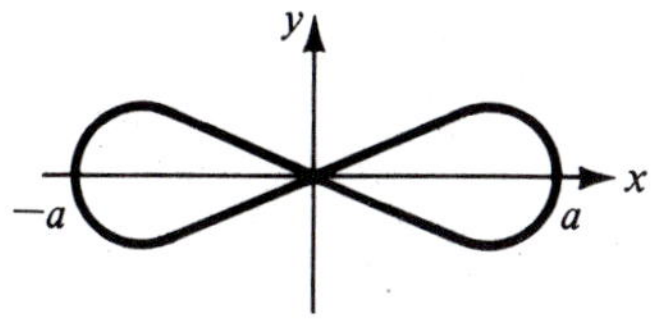

Symmetric with respect to both axes and tangent to x axis at the origin.

Section 2, Part I

1. $\dfrac{32\pi}{5}$
3. 8π
5. $\dfrac{2\pi}{3}$
7. $\dfrac{\pi}{30}$
9. 8π
11. $\dfrac{128\pi}{5}$
13. 16π
15. $\dfrac{16\pi}{3}$
17. $\dfrac{80\pi}{3}$
19. $\dfrac{\pi h(R^2 + Rr + r^2)}{3}$
21. $\dfrac{32\pi p^3}{15}$

Section 2, Part II

1. $\frac{4}{3}\pi(r^2 - R^2)^{3/2}$

Section 3

1. 72 in./lb
3. $84{,}375\pi/4$ ft/lb
5. $-25k/72$; 0
7. $1{,}125\pi$ ft/lb
9. 12,500 ft-lb

Section 4, Part I

1. $\frac{2}{27}(37^{3/2} - 1)$
3. $\frac{14}{3}$
9. $\frac{33}{8}$
5. $\frac{134}{27}$
7. $\frac{10}{3}$

Section 4, Part II

1.

$$3ay^2 = x(x - a)^2$$

$$6ayy' = (x - a)^2 + 2x(x - a) = (x - a)(3x - a)$$

$$(1 + y'^2)^{1/2} = \left\{\frac{36a^2y^2 + (x - a)^2(3x - a)^2}{36a^2y^2}\right\}^{1/2}$$

$$= \frac{\sqrt{3}}{6\sqrt{a}}\left(3\sqrt{x} + \frac{a}{\sqrt{x}}\right)$$

$$L = 2\int_0^a \sqrt{1 + y'^2}\, dx = \frac{4a}{\sqrt{3}}$$

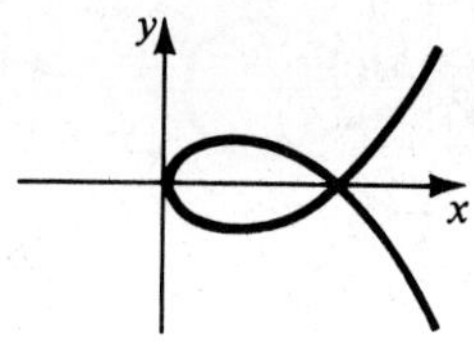

Section 5

1. 1.10 **3.** 3.00 **5.** 11.35 **7.** −.2723

Section 6, Part I

1. 1.10 **3.** 2.31 **5.** 3.15
7. 3.140 **9.** 1.03

Section 6, Part II

1. Let $F(x) = A + 2Bx + 3Cx^2 + 4Dx^3$. It then follows that

$$\int_a^b F(x)\, dx = (Ax + Bx^2 + Cx^3 + Dx^4)\Big|_a^b$$

$$= (b - a)(A + B(b + a) + C(b^2 + ba + a^2) + D(b^3 + b^2a + ba^2 + a^3))$$

$$= \frac{b - a}{6}\left\{(A + 2Ba + 3Ca^2 + 4Da^3) + (A + 2Bb + 3Cb^2 + 4Db^3) + 4\left(A + 2B\left(\frac{a + b}{2}\right) + 3C\left(\frac{a + b}{2}\right)^2 + 4D\left(\frac{a + b}{2}\right)^3\right)\right\}$$

Review

1. $\frac{125}{6}$ **3.** $4 - 2\sqrt{3}$ **5.** $16\pi\sqrt{pb^5}/5$
7. $\frac{7168}{15}\pi$ **9.** $\pi a^5/80p^2$

11. *Hint:* Show that the volume here is to the volume of Exercise 14 as the area of a regular pentagon is to the area of a square (of same length of edge).
13. $104\pi a^3/105$; same by symmetry.
15. 2.41, 2.43
17. $\frac{91}{27}$
19. 2.025×10^6 ft-lb

CHAPTER 7

Section 2

1. 1, 2, 3, 4, -1, -2, -3, $\frac{1}{2}$, $-\frac{1}{3}$
3. 1, e, e^{-1}, e^{-3}, $2 + e^3$, $\pm\sqrt{5 - e^{-2}}$
9. If $F(u) = \ln u + u$, then $F(1) > 0$, $F(1/e) < 0$. By continuity and monotonicity ($F'(u) > 0$), there exists a unique root.
11. $\ln(-x)$

Section 3

1. $D_x e^x = e^x > 0$, or $e^x - e^y = (e^{x-y} - 1)e^y > 0$ for $x > y$
3. $1 - e^{-2}$
5. $f(-1) = -e^{-1}$, min; $f(-2) = -e^{-2}$, pt. of infl.

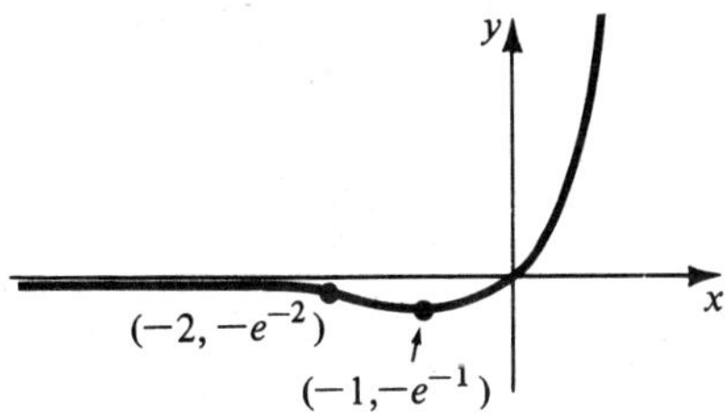

7. $f(1) \doteq e$, min

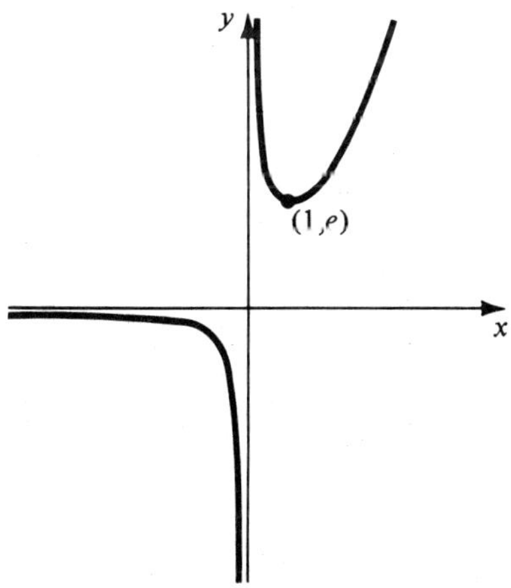

9. $F(x) - F(0) = xF'(\theta), 0 < \theta < x.\ \ e^x - 1 = xe^{\theta} \geq x$; similarly for $F = e^{-x}$.

13. Let $u = at$; $\int_a^{ax} \frac{1}{u}\,du = \ln(ax) - \ln a = \int_1^x \frac{1}{t}\,dt = \ln x$.

Section 4, Part I

1. $2x^2e^{2x} + 2xe^{2x}$

3. $-18xe^{-3x^2}(1 + e^{-3x^2})^2$

5. $\ln|x^2 - 1| + \dfrac{2x}{x - 1}, x^2 \neq 1$

7. $\dfrac{2x}{\ln x} - \dfrac{x}{\ln^2 x}$

9. $\dfrac{e^x}{e^x + 1}$

11. $\dfrac{3e^{3x} + 2x}{e^{3x} + x^2}$

13. $\dfrac{x - 2}{2(x - 1)^{3/2}} \exp \dfrac{x}{\sqrt{x - 1}}$

15. $\dfrac{(2x - 1)e^{2x} + 5}{xe^{2x} - 5x}$

17. $F(e^{-1}) = -e^{-1}$, min

19. $F(e) = e^{-1}$, max; $F(e^{3/2}) = \frac{3}{2}e^{-3/2}$, pt. of infl.

21. $\dfrac{1 - e^{-10}}{5} \doteq \dfrac{1}{5}$

23. $2e^{\sqrt{x}} + C$

25. $\dfrac{x^3 + a^3 - a^3}{x + a} = x^2 - ax + a^2 - \dfrac{a^3}{x + a}$;

$$\int = \frac{x^3}{3} - \frac{ax^2}{2} + a^2x - a^3 \ln(x + a) + C$$

27. $\dfrac{\ln^2 x}{2}$

29. $\frac{1}{2} \ln \frac{5}{4}$

31. $2 \ln 3$

33. $e^{\sqrt{2}}$

35. $\dfrac{2 \ln x\, e^{\ln^2 x}}{x}$

37. $(2x \ln x + x + 3x^2)x^{x^2}e^{x^3}$

39. $\left[3 \ln(x^2 + 1) + \dfrac{6x^2}{x^2 + 1}\right](x^2 + 1)^{3x}$

41. $\dfrac{(x - 1)y}{x(y + 2)}$

43. $\dfrac{2xy - ye^y}{xye^y + 1}$

Section 4, Part II

1. $e^x \exp(e^x), (e^x + e^{2x}) \exp(e^x)$

3. $-e^x \exp(-e^x), (e^{2x} - e^x) \exp(-e^x)$

5. $\dfrac{1}{x + 1) \ln(x + 1)}, -\dfrac{1}{(x + 1)^2 \ln(x + 1)} - \dfrac{1}{(x + 1)^2 \ln^2(x + 1)}$

7. $\ln \ln \ln x + C$

9. $\int e^x e^{e^x}\,dx = e^{e^x} + C$

Section 5, Part I

1. $\cosh x$, $\sinh x$

3. $\operatorname{sech}^2 x$, $-2\operatorname{sech}^2 x \tanh x$

5. $-\operatorname{csch} x \coth x$, $\dfrac{\cosh^2 x + 1}{\sinh^3 x}$ or $\operatorname{csch} x \coth^2 x + \operatorname{csch}^3 x$

13. $\dfrac{\sinh(\ln x + 1)}{x}$

15. $2x \operatorname{sech}^2 (x^2 + 1)e^{\tanh(x^2+1)}$

17. $6 \cosh 2x - 6 \sinh 3x$

19. $-\operatorname{sech} x + C$

21. $-(\operatorname{sech}^2 x)/2 + C$

23. $e^{\sinh x} + C$

25. $2 \sinh \sqrt{x} + C$

27. $\sinh (e - e^{-1})$

29. $\pi a^3(2 + \sinh 2)/4$

Section 5, Part II

1. Let $x = y$ in 7.26

3. From Exercises 1 and 2, $\tanh 2x = \dfrac{\sinh 2x}{\cosh 2x} = \dfrac{2 \sinh x \cosh x}{\cosh^2 x + \sinh^2 x} = \dfrac{2 \tanh x}{1 + \tanh^2 x}$

5. Set $y = 2x$ in 7.26 and then use Exercises 1 and 2.

7. Use Exercise 4.

9. $\tanh x$ (use Exercise I-3).

Section 6

1. $\dfrac{-2 \log_2 e}{5 - 2x}$

3. $2x \log_{10} (3 + 2x^2) + \dfrac{4x^3 \log_{10} e}{3 + 2x^2}$

5. $\pi (\ln x + 1)^{\pi-1}/x$

7. $(1 + e)(1 + \sqrt{x})^e/2\sqrt{x}$

9. $2^{2x} (2 \ln 2)$

11. $(\frac{1}{2} \ln x + 1)x^{\sqrt{x}-1/2}$

13. $\dfrac{\sqrt{\log_{10} e\, x}}{(x^2 + 1)\sqrt{\ln (x^2 + 1)}}$

15. $\log_{10} e \dfrac{\sinh (\log_{10} x)}{x}$

17. $3/\ln 2$

19. $990/\ln 10$

21. $\dfrac{1}{\ln 3} 3^{\ln x} + C$

23. $\dfrac{10e - 1}{\ln 10 + 1}$

25. $\dfrac{1}{3 \ln 7} 7^{3x} + C$

27. Let $m = \log_a b$, $n = \log_b x$. Then $b = a^m$, $x = b^n$, and $x = a^{mn}$. Thus $mn = \log_a x = (\log_a b)(\log_b x)$. Letting $x = a$, $1 = (\log_a b)(\log_b a)$.

Section 7

1. $50\sqrt{2}$, $\frac{25}{4}$; $1600(\ln .9)/\ln .5$

3. $P = P_0e^{kt}$, $625 \times 10^3/18$

5. $(4 \ln 20)/\ln 3$

7. $200 \times 10^6 \cdot 2^{2/5}$

9. $15\dfrac{\ln 4}{\ln \frac{5}{4}}$

11. e^{-2x}

13. $100 \cdot 2^{-x}$

15. $\frac{25}{8}$

Review

1. $3e^{3x}$

3. $2x \sinh (x^2 + 1)$

5. $2\pi x^{2\pi - 1}$

7. $\sqrt{2}(e^x + 1)(e^x + x)^{\sqrt{2}-1}$

9. $\log_{10} 8$

11. $\dfrac{1}{x} - 3$

13. $\dfrac{3x^2y + x}{y - 3xy \ln y}$

15. $2x(\sinh x^2 + x \cosh x^2)$

17. $x(1 + 2 \ln x)$

19. $(2 \ln x)/x$

21. $(1 + \ln \sqrt{x})/\sqrt{x}$

23. $1/\sqrt{x^2 - 1}$

25. $-2xy/(1 - x^4)$

27. $-xy/(y + 2)$

29. $x^{10^x}\left\{\dfrac{10^x}{x} + 10^x(\ln x)(\ln 10)\right\} - (10)^{x^{10}}\{10x^9 \ln 10\}$

31. $2a^2 \sinh 1$

33. $\pi a^2(e^{2bc} - 1)/2b$

35. $x - a \ln (x + a) + c$

37. $x - \frac{2}{3} \ln (1 + e^{3x}) + c$

39. $x - \ln (x + 3) + C$

41. $2 \cosh \sqrt{x} + C$

43. $\frac{1}{4} \ln (e + e^{4x}) + C$

45. $\frac{2}{3}$

47. $\dfrac{1}{\ln (e/2)} \dfrac{e^x}{2^x} + C$

CHAPTER 8

Section 2

1. $4 \cos 4x$

3. $6 \sin 3x \cos 3x$

5. $6 \cos 6x$

7. $\dfrac{1 + \sin x}{\cos^2 x}$

9. $-\dfrac{\cos x}{2\sqrt{1 - \sin x}}$

11. $\cot x$

13. $-e^x \tan e^x$

15. $\dfrac{\pi - 3 \cos^2 x \sin x}{2\sqrt{\cos^3 x + \pi x}}$

17. $\dfrac{y \sin x + \cos y}{\cos x - x \sin y}$

19. $\dfrac{\sin^2 y + 3x^2}{1 - 2x \sin y \cos y}$

21. 1

23. $-\frac{1}{2} \cos (2t + 1) + C$

25. 0

27. $-\dfrac{\sin 3x + \cos 3x}{3} + C$

29. $\sin e^x + C$

31. $\pi^5/5$

33. $e^{\sin x} + C$

35. 1

37. $\cosh(\sin x) + C$

39. $f\left(\dfrac{\pi}{4}\right) = \sqrt{2}$, max; $f\left(\dfrac{3\pi}{4}\right) = -\sqrt{2}$, min

41. $g(0) = g(2\pi) = 3$, max; $g\left(\dfrac{2\pi}{3}\right) = g\left(\dfrac{4\pi}{3}\right) = -\dfrac{3}{2}$, min

43. $\frac{1}{2}$

45. $\sqrt{2}$

Section 3, Part I

1. $4 \sec^2 4x$

3. $\dfrac{3 \sec 3x(\tan 3x - 1)}{(1 + \tan 3x)^2}$

5. $2 \sec^2 x \tan x\,(\tan^2 x + \sec^2 x)$

7. $\sec x$

9. $-\dfrac{2 \tan 2x \sec^2 2x}{\sqrt{1 - \tan^2 2x}}$

11. $e^{\tan x} \sec x\,(\sec^2 x + \tan x)$

13. $\dfrac{4 \sec \sqrt[3]{t} \tan \sqrt[3]{t}}{3\sqrt[3]{t^2}}$

15. $\csc x$

17. $y'' = 12 \csc^2 2x(3 \cot^2 2x + \csc^2 2x)$

19. $y'' = 4 \sec^2 x \tan x(3 \sec^2 x - 1)$

21. $\sec y$

23. $-\dfrac{y \cos x + \sin(x + y)}{\sin x + \sin(x + y)}$

25. $-\dfrac{y \sec xy \tan xy}{x \sec xy \tan xy + 1}$

27. No extrema

29. $g\left(\dfrac{\pi}{6}\right) = \sqrt{3}$ min; $g\left(\dfrac{5\pi}{6}\right) = -\sqrt{3}$ max

31. 1

33. $\ln \sin \theta + C$

35. $-\dfrac{4}{\pi} \csc \dfrac{\pi\theta}{4} + C$

37. $\frac{1}{3} \tan^3 x + C$

39. $\frac{1}{2} \ln 2$

41. $\ln 2$

Section 3, Part II

1. $\dfrac{\sin(x + y)}{\cos(x + y)} = \dfrac{\sin x \cos y + \cos x \sin y}{\cos x \cos y - \sin x \sin y} = \dfrac{\tan x + \tan y}{1 - \tan x \tan y}$

3. $\cos 2x = \cos^2 x - \sin^2 x = 2 \cos^2 x - 1$

5. $\cos 3x = \cos(x + 2x) = \cos x \cos 2x - \sin x \sin 2x$.
Now using Exercise 3,
$\cos 3x = \cos x(2\cos^2 x - 1) - \sin x(2 \sin x \cos x) = 4\cos^3 x - 3\cos x$.
7. $(2ax^2 + \sin 2ax^2)/8a + c$; let $ax^2 = y$, then use Exercise 3.
9. $(9 \sin 2x^3 + \sin 6x^3)/72$; let $2x^3 = y$, then use Exercise 5.
11. $(12ax + 8 \sin 2ax + \sin 4ax)/32a$;

$$\cos^4 ax = \left(\frac{1 + \cos 2ax}{2}\right)^2 = \tfrac{1}{4} + \tfrac{1}{2}\cos 2ax + \tfrac{1}{4}\cos^2 2ax.$$

Now use Exercise 3 again.
13. a. $s(t) = 5 \sin 2t$
b. $s(t) = 5 \cos 2t$
c. $s(t) = 6 \sin 2t + 5 \cos 2t$

Section 4

1. $\pi/6$ **3.** $\pi/4$ **5.** $-\pi/4$ **7.** $1/\sqrt{2}$

9. $\dfrac{-2}{\sqrt{3}}$ **11.** $\dfrac{\sqrt{3}}{2}$ **13.** $\frac{13}{5}$

15. If $\theta = \sin^{-1} x$, also $\theta = \tan^{-1} \dfrac{x}{\sqrt{1 - x^2}}$. Thus $\tan(\sin^{-1} x) = \dfrac{x}{\sqrt{1 - x^2}}$.

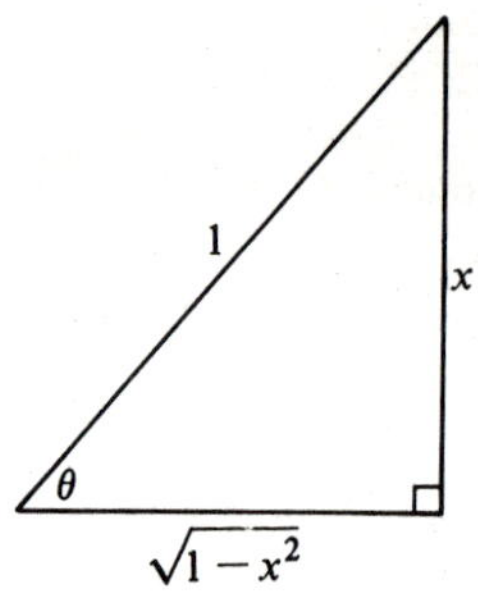

17. Since $\sin 3a = 3 \sin a - 4 \sin^3 a$ (Exercise II-4, Section 3), $\sin(3 \sin^{-1} x) = 3x - 4x^3$.

Section 5

1. $-1/\sqrt{1 - x^2}$
5. $2/\sqrt{1 - 4x^2}$
7. $1/x\sqrt{16x^2 - 1}$
9. $6(1 + \arcsin 3x)/\sqrt{1 - 9x^2}$
11. $2/(1 + e^{4x}) - (2 \arctan e^{2x})/e^{2x}$
13. $\sin^{-1} x$
15. $1/x\sqrt{x^2 - 1}$
17. $1/2x\sqrt{x - 1}$
19. $1/(1 + x^2) \arctan x$
21. $2x^2/\sqrt{1 - x^2}$
23. $6x^2 \tan^{-1} x$
25. $\dfrac{e^x}{\arcsin e^x \sqrt{1 - e^{2x}}}$

27. $\dfrac{\cosh x}{\sinh x\sqrt{\sinh^2 x - 1}}$

29. $\dfrac{3x^2}{x^6 + 1}$

31. $c = -\pi/4$

33. $\pi/3$

35. $\pi/18$

37. $\pi/12$

39. $\arctan \sin x + c$

41. $\dfrac{1}{\sqrt{7}} \sin^{-1} \sqrt{\dfrac{7}{3}}\, x + C$

43. $\sin^{-1} \frac{1}{3}$

45. $\dfrac{1}{12} \tan^{-1} \dfrac{x^3}{4} + C$

47. $\dfrac{x + y}{x - y}$

49. $\dfrac{1 + (x + y)^2 - \sqrt{1 - (e^y + x)^2}}{\sqrt{1 - (e^y + x)^2} - e^y[1 + (x + y^2)]}$

51. π

53. Note that $\tanh x = \dfrac{e^{2x} - 1}{e^{2x} + 1} = 1 - \dfrac{2}{1 + e^{2x}}$

Review, Part I

1. $\cot x$

3. $(2/x) \sec^2 \ln x^2$

5. $\dfrac{m \cos x}{\sqrt{1 - m^2 \sin^2 x}}$

7. $\dfrac{x \arcsin x}{\sqrt{1 - x^2}}$

9. O; note that $-|x| \le x \sin \dfrac{1}{x} \le |x|$, now use the squeeze theorem.

11. O; note that $\dfrac{\sin x^2}{x} = x \cdot \dfrac{\sin x^2}{x^2}$

13. 4; use Exercise II-2, Section 3

15. $2(\sec x^3 + \sec^{-1} x)\left(3x^2 \sec x^3 \tan x^3 + \dfrac{1}{x\sqrt{x^2 - 1}}\right)$

17. $\dfrac{[\tan^{-1}(x + 1)]^{-4/5}}{5[1 + (x + 1)^2]}$

19. $\dfrac{e^x \cos e^x}{2\sqrt{\sin e^x}}$

21. $4 \sec 4x \tan^2 4x + 4 \sec^3 4x$

23. $\dfrac{-\exp(\sec^{-1} x^{-1})}{\sqrt{1 - x^2}}$

25. $\dfrac{[e^x(x + y)^2 - 1](x^2 + y)\sqrt{(x^2 + y)^2 - 1} - 2x(x + y)^2}{(x + y)^2\sqrt{(x^2 + y)^2 - 1} + (x + y)^2}$

27. $\dfrac{(1 + y^2)[\sqrt{1 - x^2} \tan^{-1} y - y]}{\sqrt{1 - x^2}[(1 + y^2) \sin^{-1} x - x]}$

29. $2\pi/3$

31. Period of $\sin 2x$ is π, period of $2 \cos 4x$ is $\pi/2$; thus period of sum is π.

33. Period of $\sin x/3$ is 6π, period of $-\cos x/6$ is 12π, period of $\tan x/9$ is 18π; thus period of sum is 36π.

35. $f(x) = \sin x(1 + \cos x)$; f is an odd function with period 2π.
$f'(x) = (2\cos x - 1)(\cos x + 1)$. Slope $= 0$ at $x = \pi/3$ and f has a local maximum at $x = \pi/3$.

37. No extrema

39. $\frac{1}{2}\sec^{-1} x^2 + c$; let $y = x^2$

41. $3\tan\dfrac{x}{3} - x + c$; $\tan^2 y = \sec^2 y - 1$

43. $\dfrac{3x - \tan^{-1} 3x}{27} + c$; $\dfrac{x^2}{9x^2+1} = \dfrac{1}{9}\left(1 - \dfrac{1}{9x^2+1}\right)$.

45. $-\ln(\cos t - 3t) + C$

47. $2\sqrt{2} - 2$

49. $\frac{1}{2}\sec(x^2 + 4) + C$

51. $-\frac{1}{15}\tan^{-1}\dfrac{\cot 3x}{5} + C$

53. $\frac{1}{16}\tan^4 4x + C$

55. $9\sinh 1$

57. 8; $y' = \dfrac{1 \mp x}{\sqrt{1 - x^2}}$.

$$L = \int_{-1}^{1}\sqrt{1 + \left(\frac{1-x}{\sqrt{1-x^2}}\right)^2}\,dx + \int_{-1}^{1}\sqrt{1 + \left(\frac{1+x}{\sqrt{1-x^2}}\right)^2}\,dx = 8.$$

59. $V = \displaystyle\int_0^{\pi/a}\pi\sin^2 ax\,dx = \pi\int_0^{\pi/a}\frac{1 - \cos 2ax}{2}\,dx = \frac{\pi^2}{2a}$

61. $y - (\sqrt{2} + 1) = \dfrac{\sqrt{2} - 2}{2}\left(x - \dfrac{\pi}{4}\right)$

Review, Part II

1. $\pi^2/32$; let $y = \tan^{-1} x$

3. $\cos^{-1} e^{-x}$; $\dfrac{1}{\sqrt{e^{2x} - 1}} = \dfrac{e^{-x}}{\sqrt{1 - e^{-2x}}}$; now let $y = e^{-x}$

5. $2\tan^{-1} e^a - \dfrac{\pi}{2}$; $\operatorname{sech} x = \dfrac{2e^x}{e^{2x} + 1}$.

CHAPTER 9

Section 1

1. $(\frac{2}{7})(x - 2)^{7/2} + (\frac{4}{5})(x - 2)^{5/2} + C$

3. $\frac{4}{3}x\sqrt{x} - 8x + 64\sqrt{x} + 256\ln(4 + \sqrt{x}) + C$

5. $\dfrac{x^2 - 6}{5}(x^2 + 9)^{3/2} + C$

7. $(\cos^3 x - 3\cos x)/3 + C$

9. $\frac{1}{8}x - \frac{1}{32}\sin 4x + C$

11. $(x - 2)^2/2 + 4\ln(x + 2) + C$

13. $2x^{3/2}(60x^2 + 84x + 35)/105 + C$

15. $\frac{1}{5}\cos^5 x + \frac{2}{3}\cos^3 x - \cos x + C$

17. $\frac{1}{5}\tan^5 x + \frac{1}{3}\tan^3 x + C$

19. $\frac{1}{3}\arctan(x - 2)/3 + C$

21. $\arcsin(x - 1) + C$

23. $\dfrac{1}{2\sqrt{2}}\arctan\dfrac{3x^2 - 1}{\sqrt{2}}$

25. $\sin^{-1}\dfrac{x + 3}{\sqrt{3}} + C$

27. $\frac{1}{2}\cos^2 x - \ln|\cos x| + C$

29. $\sec x + \cos x + C$

31. $-\ln(1 + e^{-x}) + C$

33. $2\sqrt{1 + \cos x}\left(1 - \dfrac{1 + \cos x}{3}\right) + C$; let $y = \cos x$, then $t^2 = 1 + y$.

35. Since $\cos 2x = 1 - 2\sin^2 x$, the difference is merely one of form.

Section 2

1. $x^2(2\ln x - 1)/4 + C$

3. $4(3\sqrt{2}\ln 2 - 2\sqrt{2} + 1)/9$

5. $[(x^2 + 1)\tan^{-1} x - x]/2 + C$

7. $x\sin x + \cos x + C$

9. $(x^2/4) - (x\sin 2x)/4 - (\cos 2x)/8$

11. $e^x(x^2 - 2x + 2) + C$

13. $(x - 1)\sqrt{2x + 1}/3 + C$

15. $e^{2x}(2\sin 3x - 3\cos 3x)/13 + C$

17. $\frac{2}{15}$

19. $x\ln(x^2 + 1) - 2x + 2\tan^{-1} x + C$

21. $x^{r+1}[(r + 1)\ln x - 1]/(r + 1)^2 + C$

23. $-\pi^2/6 - 4/27$

25. $-\frac{1}{2}\csc x\cot x - \frac{1}{2}\ln|\csc x - \cot x| + C$

27. $x\sin^{-1} ax + \sqrt{1 - a^2x^2}/a + C$

29. $x\tan^{-1} ax - (1/2a)\ln(1 + a^2x^2) + C$

31. $[2ax + 2a^2x^2\tanh^{-1} ax + \ln|ax - 1|/|ax + 1|]/4a^2 + C$

33. $x(\sin^{-1} x)^2 + 2\sin^{-1} x\sqrt{1 - x^2} - 2x + C$

35. $x(\ln x)^2 - 2x\ln x + 2x + C$

37. $\dfrac{x}{2}[\sin(\ln x) - \cos(\ln x)] + C$

39. $x^n e^x - n\int x^{n-1}e^x\,dx$

41. $-\dfrac{\sin^{n-1} x\cos x}{n} + \dfrac{n - 1}{n}\int \sin^{n-2} x\,dx$

43. $\dfrac{1}{n + 1}\left(x^{n+1}\sin^{-1} x - \int \dfrac{x^{n+1}}{\sqrt{1 - x^2}}\,dx\right)$

Section 3

1. $[25\sin^{-1} x/5 + x\sqrt{25 - x^2}]/2 + C$

3. $\sin^{-1} x/5 + C$

5. $(x/2)\sqrt{9x^2 - 4} - (\frac{2}{3})\ln|3x + \sqrt{9x^2 - 4}| + C$

7. $-(\frac{1}{3}) \ln |3 + \sqrt{x^2 + 9}|/|x| + C$
9. $-[4x/(x^2 - 4) + \ln |x - 2|/|x + 2|]/32 + C$
11. $\sqrt{x^2 - a^2} - a \sec^{-1} x/a + C$
13. $-\sqrt{a^2 - x^2}/x - \sin^{-1} x/a + C$
15. $[x\sqrt{x^2 + a^2} + a^2 \ln (x + \sqrt{x^2 + a^2})]/2 + C$
17. $[x\sqrt{x^2 - a^2} + a^2 \ln |x + \sqrt{x^2 - a^2}|]/2 + C$
19. $-(1/a) \ln |a + \sqrt{a^2 - x^2}|/|x|$
21. $[\tan^{-1} (x - 2) + (x - 2)/(x^2 - 4x + 5)]/2 + C$

23. $-\dfrac{\sqrt{16 - x^2}}{16x} + C$

25. $\dfrac{x - 1}{\sqrt{x^2 - 2x + 5}} + C$

27. $\dfrac{1}{2}\left(\dfrac{\sqrt{e^4 - 1}}{e^4} - \dfrac{\sqrt{e^2 - 1}}{e^2} + \tan^{-1} \sqrt{e^4 - 1} - \tan^{-1} \sqrt{e^2 - 1}\right)$

Section 4

1. $\ln (x - 1)^2/|x| + C$
3. $[2x^2 + 8x + 27 \ln |x - 3| + \ln |x + 1|]/4 + C$
5. $[3 \ln |2x - 1| - (2x - 1)^{-1}]/4 + C$
7. $[5 \ln |x + 2| + 4 \ln |x - 1| - 3 \ln |x|]/6 + C$
9. $[\ln |x - 2|/|x + 2| + 2 \tan^{-1} x/2]/8 + C$
11. $[8x - 2 \ln |x| + 9 \ln |x - 2| - 7 \ln |x + 2|]/8 + C$
13. $[4(x - 2)^2 \ln |x - 2| - 8(x - 2) - 9]/2(x - 2)^2 + C$
15. $[2 \ln (x^2 + 4) + \tan^{-1} x/2 + 3/(x^2 + 4)]/2 + C$
17. $[\ln |x + 2|/|x - 2| - 14/(x - 2) + 10/(x + 2)]/32 + C$
19. $[\ln |cx + d|/|ax + b|]/(bc - ad) + C$, $bc - ad \neq 0$,

$$\frac{1}{(ax + b)(cx + d)} = \frac{1}{bc - ad}\left(\frac{c}{cx + d} - \frac{a}{ax + b}\right).$$

21. See table of integrals at end of book,

$$\frac{x}{(ax + b)^2(cx + d)} = \frac{1}{a}\left(\frac{ax + b}{(ax + b)^2(cx + d)} - \frac{b}{(ax + b)^2(cx + d)}\right).$$

The first term on the right-hand side corresponds to Exercise 19. To integrate the second term, note that from Exercise 19 above,

$$\frac{1}{(ax + b)^2(cx + d)} = \frac{1}{bc - ad}\left(\frac{c}{(ax + b)(cx + d)} - \frac{a}{(ax + b)^2}\right).$$

23. $[x^2 + 9 \ln |x^2 - 9|]/2 + C$

25. $\dfrac{x^4}{x^2 - 1} = \dfrac{x^4 - 1}{x^2 - 1} + \dfrac{1}{x^2 - 1} = x^2 + 1 + \dfrac{1}{2}\left(\dfrac{1}{x - 1} - \dfrac{1}{x + 1}\right)$; now integrate.

27. $\arctan x - 1/x - 1/3x^3 + C$; let $x = 1/y$ to give

$$-\int \frac{y^4}{y^2 + 1}\, dy = -\left\{\int (y^2 - 1)\, dy + \int \frac{1}{y^2 + 1}\, dy\right\}.$$

29. $\left(\frac{1}{6}\right)\frac{\ln(x+1)^2}{x^2-x+1}+\frac{1}{\sqrt{3}}\arctan\frac{2x-1}{3}+C;$

$$\frac{1}{x^3+1}=\frac{1}{(x+1)(x^2-x+1)}=\frac{a}{x+1}+\frac{bx+c}{x^2-x+1}.$$

Now determine a, b, and c. Then note that

$$\frac{bx+c}{x^2-x+1}=\frac{b}{2}\left\{\frac{2x-1}{x^2-x+1}+\frac{2c/b+1}{(x-\frac{1}{2})^2+\frac{3}{4}}\right\}.$$

31. $-\frac{1}{2(e^{2x}+4)}+C$

Section 5

1. $xy=C$
3. $x^2+2/y=C$
5. $y-x+\ln|y-1|/|x|=C$
7. $e^{-x}+e^{-y}=C$
9. $(\sec x+\tan x)\sec y=C$
11. $y-x+\ln(|y-1||x+1|)=C$
13. $(y-1)/xy=C$
15. $\frac{3}{2}x^2-2x-15$
17. a. $45°$ b. 4 hr, 19 min after immersion
19. a. $\frac{1}{100}$ b. $\frac{300}{13}$ grams

Review

1. $x+1-6\ln|x+1|-\frac{12}{x+1}+\frac{4}{(x+1)^2}+C;\ \left(\frac{x-1}{x+1}\right)^3=\left(1-\frac{2}{x+1}\right)^3,$ now expand.
3. $x(\ln^3|x|-3\ln^2|x|+6\ln|x|-6)+C$
5. $2e^{\sqrt{t}}(\sqrt{t}-1)+C$
7. $x\ln|x+\sqrt{x^2-a^2}|-\sqrt{x^2-a^2}+C$
9. $\tan x+C$
11. $(\arctan x)^3/3+C$
13. $-\cos e^x+C$
15. $\frac{\sin 5x}{10}+\frac{\sin x}{2}+C$
17. $\frac{5^x}{\ln^2 5}(x\ln 5-1)+C$
19. $\frac{(x^3+1)\ln|1+x|}{3}-\frac{x^3}{9}+\frac{x^2}{6}-\frac{x}{3}+C$
21. $2e^{\sqrt{x}}(x-2\sqrt{x}+2)+C$
23. Let $y^2=1+e^{2x}$, then $dx=\frac{y}{y^2-1}\,dy$ giving $\int\frac{y^2}{y^2-1}\,dy$.

But

$$\frac{y^2}{y^2-1}=1+\frac{1}{2}\left\{\frac{1}{y-1}-\frac{1}{y+1}\right\}.$$

Now integrate termwise.

25. $-e^{-x^2}(x^4+2x^2+2)/2+C$; let $-x^2=y$ and integrate by parts.
27. $\frac{16}{7}\ln 5+\frac{9}{7}\ln 3-\frac{103}{14}\ln 2$
29. $\frac{1}{24}$

31. $\frac{15}{2} - 8 \ln 2$

33. $\frac{1}{4} \cos 2\theta - \frac{1}{20} \cos 10\theta + C$

35. $\frac{1}{112} \ln |y + 2| - \frac{23}{784} \ln |y - 2| - \frac{1}{28(y - 2)} + \frac{1}{98} \ln (y^2 + 3) + \frac{2}{49\sqrt{3}} \tan^{-1} \frac{y}{\sqrt{3}} + C$

37. $x(\ln x)^n - n \int (\ln x)^{n-1}\, dx$

39. $\frac{1}{2(n + 1)} x^{n+1} + \frac{n}{16} x^{n-1} - \frac{1}{4} x^n \sin 2x - \frac{n}{8} x^{n-1} \cos 2x - \frac{n(n - 1)}{8} \int x^{n-2} \sin^2 x\, dx$

43. $2\pi(1 + 4 \ln 2 + 3 \ln 3)$

45. $9 \sinh (a/3)$

47. $61.8°$, 37.5 min

CHAPTER 10

Section 1, Part I

1. Vertices $(\pm 2,0)$, foci $(\pm \sqrt{3},0)$

3. Vertices $(0, \pm 5)$, foci $(0, \pm 4)$

5. Vertices $(\pm \frac{3}{4},0)$, foci $(\pm \frac{9}{20},0)$

7. Vertices $(0, \pm 1)$, foci $(0, \pm \sqrt{3}/2)$

9. $9x^2 + 25y^2 = 225$

11. $25x^2 + 4y^2 = 100$

13. $9x^2 + 4y^2 = 36$

17. $\sqrt{7}/5$ ft

Section 1, Part II

1. The coordinates of the square must have the form $(\pm m, \pm m)$. If the equation of the ellipse is

$$\frac{x^2}{a^2} + \frac{y^2}{b^2} = 1,$$

then

$$m^2 \left(\frac{1}{a^2} + \frac{1}{b^2}\right) = 1$$

and

$$\text{Area} = 4m^2 = \frac{4a^2b^2}{a^2 + b^2}.$$

3. $L = 4 \int_0^a \sqrt{1 + y'^2}\, dx$. Now choosing $x = a \cos \theta$, $y = b \sin \theta$ (which are parametric equations for the ellipse),

$$dx = -a \sin \theta\, d\theta, \qquad \frac{dy}{dx} = -\frac{b^2 x}{a^2 y} = -\frac{b^2}{a^2} \frac{a \cos \theta}{b \sin \theta}.$$

Thus

$$L = 4a \int_0^{\pi/2} \sqrt{1 - e^2 \cos^2 \theta}\, d\theta = 4a \int_0^{\pi/2} \sqrt{1 - e^2 \sin^2 \theta}\, d\theta.$$

Section 2

1. Vertices $(\pm 2,0)$, foci $(\pm \sqrt{5},0)$, asymptotes $x = \pm 2y$
3. Vertices $(0,\pm 3)$, foci $(0,\pm 5)$, asymptotes $3x = \pm 4y$
5. Vertices $(\pm 2,0)$, foci $(\pm 2\sqrt{2},0)$, asymptotes $x = \pm y$
7. Vertices $(0,\pm \frac{4}{3})$, foci $(0,\pm 2\sqrt{13}/3)$, asymptotes $2x = \pm 3y$
9. $3x^2 - y^2 = 12$
11. $16x^2 - 9y^2 = 576$
13. $4x^2 - y^2 = 16$
15. $3y^2 - 4x^2 = 12$

Section 3, Part I

1. Parabola, vertex $(3,-2)$, focus $(3,-1)$
3. Ellipse, center $(2,1)$, vertices $(-1,1)$, $(5,1)$, foci $(2 \pm 2\sqrt{2}, 1)$
5. Ellipse, center $(1,-2)$, vertices $(1 \pm \sqrt{3}, -2)$, foci $(0,-2)$, $(2,-2)$
7. Hyperbola, center $(2,-2)$, vertices $(2,-1)$, $(2,-3)$, foci $(2, -2 \pm \sqrt{2})$, asymptotes $y + 2 = \pm(x - 2)$
9. Parabola, vertex $(\frac{3}{4},0)$, focus $(-\frac{5}{4},0)$
11. Hyperbola, center $(3,0)$, vertices $(1,0)$, $(5,0)$, foci $(3 \pm \sqrt{13}, 0)$, asymptotes $2y = \pm 3(x - 3)$
13. $(y - k)^2 = 4p(x - h)$
15. $(x - h)^2/a^2 + (y - k)^2/(a^2 - c^2) = 1$
17. $(x - h)^2/a^2 - (y - k)^2/(c^2 - a^2) = 1$
19. Since, equivalently, $y + \dfrac{b^2 - 4ac}{4a} = a\left(x + \dfrac{b}{2a}\right)^2$ (by completing the square),

    the vertex is at $\left(-\dfrac{b}{2a}, -\dfrac{b^2 - 4ac}{4a}\right)$.

Section 3, Part II

7. $5\pi/4$
9. $\theta = \pi/4$; $x'^2 + 3y'^2 - 8\sqrt{2}\,x' = 40$
11. $\cot 2\theta = \frac{24}{7}$; $x'^2 - y'^2 - \frac{3}{25} = 0$

Section 4, Part I

1. $(0,\frac{2}{7})$
3. $(4,4)$
5. The figure (see top of next page) has a vertical axis of symmetry and its centroid must lie on it.

$$\begin{aligned} m &= \rho(12 \times 12 - 2(2 \times 2) - 8 \times 4) = 104\rho. \\ m\bar{y} &= \rho[(8 \times 6) \times 3 + 2(10 \times 2) \times 7 + (8 \times 2) \times 11]. \\ \bar{y} &= \tfrac{600}{104} = \tfrac{75}{13}. \end{aligned}$$

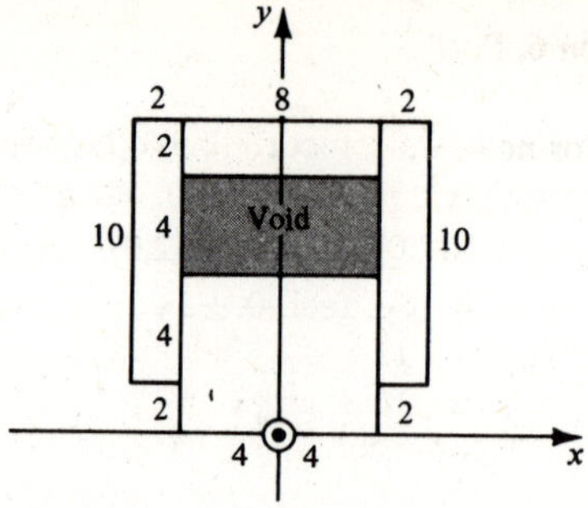

Section 4, Part II

1. We could divide the perforated circle into segments. But since we do not know the location of a centroid of a segment as yet, we use another method which is much simpler. We fill in the void and add its mass taken negatively at its centroid. This gives a mass of 100ρ acting at (0,0) and a mass of -48ρ acting at (4,3). Now taking moments,

$$(100\rho - 48\rho)\bar{x} = 100\rho \times 0 - 48\rho \times 4; \qquad \bar{x} = \tfrac{48}{13}.$$
$$(100\rho - 48\rho)\bar{y} = 100\rho \times 0 - 48\rho \times 3; \qquad \bar{y} = \tfrac{36}{13}.$$

Section 5, Part I

1. $(\frac{12}{5},\frac{3}{4})$
3. $(\frac{8}{5},\frac{16}{7})$
5. $(\pi/2 - 1, \pi/8)$
7. $((e^2 + 1)/4, e/2 - 1)$
9. $\bar{x} = (\sqrt{2} - 1)/\ln(1 + \sqrt{2}), \bar{y} = \pi/8 \ln(1 + \sqrt{2})$
11. $(3,\frac{72}{5})$
13. $(\frac{3}{8},\frac{9}{4})$
15. $\bar{x} = 0, \bar{y} = 4a/3\pi$
17. $\bar{x} = (a \cosh a - \sinh a)/(\cosh a - 1), \bar{y} = (\sinh 2a - 2a)/8(\cosh a - 1)$

Section 5, Part II

1. $\bar{x} = \bar{y} = 4r/3\pi$
3. $\bar{x} = (2r \sin\theta)/3\theta, \bar{y} = 0$, where (0,0) is at the center of the circle and the x axis is the axis of symmetry; $\bar{x} = 2r/3, \bar{y} = 0$

Section 6, Part I

1. $(\frac{8}{3},0)$
3. $((\pi^2 + 4)/4\pi, 0)$
5. $(\ln 5/2 \tan^{-1} 2, 0)$
7. Three-fourths the way down the altitude from the vertex
9. $\bar{y} = \dfrac{2p}{3}$ for parabola $y^2 = 4px$
11. $(0,\frac{81}{10})$
13. $(0,\frac{89}{42})$

Section 6, Part II

1. For now, we consider simple type regions as shown:

$$V = \int_{x_1}^{x_2} (y_2^2 - y_1^2)\, dx, \qquad A = \int_{x_1}^{x_2} (y_2 - y_1)\, dx,$$

$$\bar{y} = \int_{x_1}^{x_2} (y_2 - y_1)\left(\frac{y_1 + y_2}{2}\right) dx \div A.$$

Thus, $2\pi\bar{y}A = V$. The theorem follows more easily for more complicated regions by double integration which comes later.

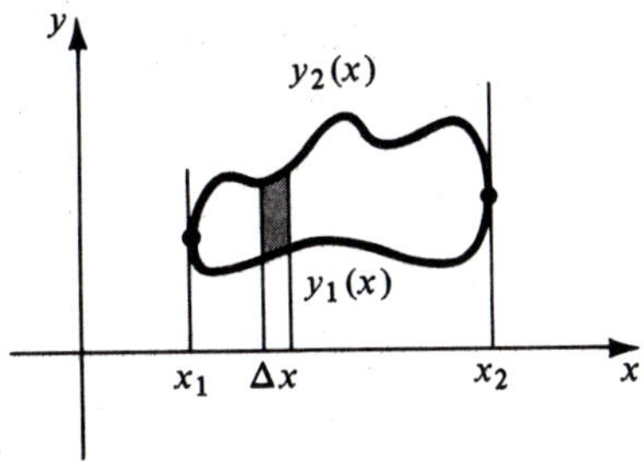

3. $\frac{128}{3} + 12\pi$

5. Since the volume of a sphere is $\frac{4}{3}\pi r^3$, we get $\frac{4}{3}\pi r^3 = 2\pi\bar{y}(\pi r^2/2)$. Thus $\bar{y} = 4r/3\pi$ (above the base).

7. Rotating semiellipse about the x axis gives

$$V = 2\pi\bar{y}A = 2\pi\left(\frac{4b}{3\pi}\right)\frac{\pi ab}{2} = \frac{4}{3}\pi ab^2.$$

(For $\bar{y}$ see Exercise I-16, Section 5.)

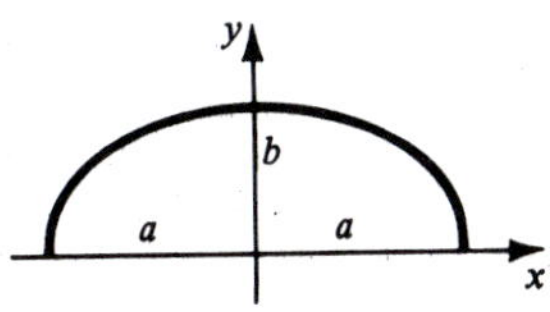

Section 7, Part I

1. 416,667 lb

3. 9750 lb

5. 3×10^6 lb

7. 216.5 lb

Section 7, Part II

1. a. For now, we consider simple types of regions as shown.

$$F = \int_{y_1}^{y_2} [x_2(y) - x_1(y)]\rho y\, dy$$

$$A = \int_{y_1}^{y_2} [x_2(y) - x_1(y)]\, dy$$

$$\bar{y} = \int_{y_1}^{y_2} [x_2(y) - x_1(y)]y\, dy \div A.$$

Thus $F = \rho\bar{y}A$.

b. $F = \rho 12(6 \times 4) = 288\rho \doteq 18{,}000$ lb

Review, Part I

1. $(1 - e^2)x^2 + y^2 - 6x = 9$
5. Ellipse, center $(2,-3)$, vertices $(-2,-3)$, $(6,-3)$, foci $(2 \pm \sqrt{7}, -3)$
7. Slope $= \pm e$
9. $\bar{x} = (\pi/8 + \frac{1}{4})a$, $\bar{y} = 0$
11. $\bar{x} = e - 2$, $\bar{y} = (e^2 - 1)/8$
13. $\bar{x} = \bar{y} = 0$ by centrosymmetry
15. $\bar{x} = 1$, $\bar{y} = 2$
17. $\bar{x} = 3\pi/16$, $\bar{y} = 0$
19. $(e^2 - 1)/4$
21. $(2 + \sinh 2)/2$
23. $\dfrac{19 - 3e^2}{2(e^2 - 5)}$ from origin on x axis

Review, Part II

1. By completing the square, the equation of the graph is

$$\frac{(x + 2)^2}{4^2} - \frac{(y + 1)^2}{3^2} = 1.$$

Since the point $(-1,-1)$ is on the axis of the hyperbola, the closest point will be the vertex $(2,-1)$ and the shortest distance 3.

3. Subtracting one equation from the other, we get a linear equation which corresponds to the straight line containing the points of intersection. The latter line can only intersect an ellipse in at most two points. In general, two quadratic graphs will intersect in 4 points (real or complex).

CHAPTER 11

Section 1, Part I

1. 1
3. 1
5. 1
7. 0
9. 2
11. -1
13. 0
15. $\frac{1}{6}$
17. $\frac{1}{9}$
19. $\frac{2}{3}$
21. $\ln 10 - 1$
23. 0
25. na^{n-1}
27. 0

Section 1, Part II

1. $\frac{1}{30}$
3. 0
5. $\frac{1}{20}$

Section 2, Part I

1. 0
3. 0
5. 0
7. $1/e$
9. 0
11. 1
13. e^3
15. 1
17. e^a
19. $e^{c(a+b)}$

Section 2, Part II

1. 1
3. 0
5. 1

Section 3, Part I

1. 2
3. Doesn't exist
5. 2
7. 2
9. $\frac{3}{2}$
11. $\pi/4$
13. Doesn't exist
15. $1/2e$
17. **b.** Let $f(x) = x^{2n-1}$, $n = 1, 2, 3, \ldots$
19. $\pi/2$
21. $3(1 + \sqrt[3]{2})$
25. 1; integrate by parts.
27. -1

Section 3, Part II

1. $\ln \frac{4}{3}$

Section 4

1. $f(x) = -\left(x - \frac{\pi}{2}\right) + \frac{1}{3!}\left(x - \frac{\pi}{2}\right)^3 - \frac{1}{5!}\left(x - \frac{\pi}{2}\right)^5 - \frac{1}{6!}\left(x - \frac{\pi}{2}\right)^6 \cos z$, z between $\frac{\pi}{2}$ and x

3. $F(x) = \frac{\pi}{4} + \frac{1}{2}(x-1) - \frac{1}{4}(x-1)^2 + \frac{1}{12}(x-1)^3 - \frac{(x-1)^4 z(z^2-1)}{(z^2+1)^4}$, z between 1 and x

5. $f(x) = \sqrt{2} + \sqrt{2}\left(x - \frac{\pi}{4}\right) + \frac{3\sqrt{2}}{2}\left(x - \frac{\pi}{4}\right)^2 + \frac{11\sqrt{2}}{6}\left(x - \frac{\pi}{4}\right)^3$
$+ \frac{1}{24}\left(x - \frac{\pi}{4}\right)^4 (24\sec^5 z - 20\sec^3 z + \sec z)$, z between $\frac{\pi}{4}$ and x

7. $f(x) = x - \frac{x^3}{3!} + \frac{x^5}{5!} - \frac{x^7}{7!}\cos z$, z between 0 and x

9. $g(x) = x - \frac{x^2}{2} + \frac{x^3}{3} - \frac{x^4}{4} + \frac{x^5}{5} - \frac{x^6}{6}(z + 1)^{-6}$, z between 0 and x

11. $F(x) = x + \frac{x^3}{3!} + \frac{3x^4}{4!}(1 - z^2)^{-7/2}(3z + 2z^3)$, z between 0 and x

13. $f(x) = 1 + \frac{x}{2} + \frac{1 \cdot 3}{2!\,2^2}x^2 + \frac{1 \cdot 3 \cdot 5}{3!\,2^3}x^3 + \frac{1 \cdot 3 \cdot 5 \cdot 7}{4!\,2^4}x^4$

$+ \frac{1 \cdot 3 \cdot 5 \cdot 7 \cdot 9}{5!\,2^4}x^5(1 - z)^{-11/2}$, z between 0 and x

15. $g(x) = x + \frac{x^3}{3} + \frac{x^5}{15}(15\sec^4 z - 15\sec^2 z + 2)\sec^2 z$, z between 0 and x

17. $F(x) = \frac{1}{2} - \frac{1}{4}x + \frac{1}{48}x^3 + \frac{x^4}{24}(e^{3z} - 11e^{2z} + 11e^z - 1)$, z between 0 and x

19. $f(x) = 1 + \frac{1}{2}x + \frac{3}{8}x^2 + \frac{5}{16}x^3 + \frac{35}{128}x^4 + \frac{63x^5}{256(1 - z)^{11/2}}$, z between 0 and x

21. $3y^2y' + xy' + y = 0$; hence $y'(0) = -\frac{1}{3}$.
$3y^2y'' + 6yy'^2 + 2y' + xy'' = 0$; hence $y''(0) = 0$.
$3y^2y''' + 18yy'y'' + 6y'^3 + 3y'' + xy''' = 0$; hence $y'''(0) = \frac{1}{81}$.
$y(x) = 1 - \frac{x}{3} + \frac{x^3}{81} + \cdots$.

Section 5, Part I

1. $\sin x = x - (x^3/3!) + (x^5/5!) \pm .000002$; .479427

3. $e^x = 1 + x/1! + x^2/2! + x^3/3! + x^4/4! + x^5/5! \pm .005$; 2.716

5. $\ln(1 + x) = x - \frac{x^2}{2} + \frac{x^3}{3} - \frac{x^4}{4} + \frac{x^5}{5} - \frac{x^6}{6} \pm .002$; .405

7. $\sinh x = x + \frac{x^3}{3!} + \frac{x^5}{5!} \pm .008$; 1.175

9. $\tan^{-1} x = x - \frac{x^3}{3} \pm .004$; .1973

11. Since $F^{[n+1]}(x) = 0$, $F(x)$ is the nth Taylor polynomial of F at any number c

13. .695
15. .933
17. .742
19. 1.649
21. 1.947

Section 5, Part II

1. $1 + \dfrac{x^2}{2} + \dfrac{5x^4}{24}$

3. $x^2 - \dfrac{x^4}{3}$ [Use $\sin^2 x = \frac{1}{2}(1 - \cos 2x)$]

5. $x = \dfrac{\pi}{2(1 + k)}$

Review

1. 1 | **3.** 1 | **5.** 0
7. 0 | **9.** Does not exist | **11.** 1

13. -2

15. $\dfrac{b}{a^2 + b^2}$ (integrate by parts)

17. Does not exist

19. $\dfrac{\pi}{4}$ (let $y = e^x$)

21. $\frac{20}{3}$

23. $\dfrac{\pi\sqrt{156}}{39}$

15. $\dfrac{1}{2 + x} = 1 - (x + 1) + (x + 1)^2 - (x + 1)^3 + (x + 1)^4 - \dfrac{(x + 1)^5}{(2 + z)^5}$

27. $(2 + x^2)^{-1} = \dfrac{1}{3} - \dfrac{2}{9}(x - 1) + \dfrac{1}{27}(x - 1)^2 + \dfrac{2}{81}(x - 1)^3 + \dfrac{10z^4 - 21z^2}{2(2 + z^2)^5}(x - 1)^4$, z between c and x

31. $\sin 2x = 2x - \dfrac{4}{3}x^3 + \dfrac{4}{15}x^5 - \dfrac{8x^7}{315}\cos 2z$, z between 0 and x

33. $\tan 3x = 3x + 6x^3 + \frac{54}{5}(2 \sec^2 3z \tan^4 3z + 11 \sec^4 3z \tan^2 3z + 2 \sec^6 3z)x^5$, z between 0 and x

35. 0.27633 (use the Maclaurin expansion for e^{-x^2})

CHAPTER 12

Section 1, Part I

1. $\frac{35}{6}$

3. $\pi\sqrt{2}/(\sqrt{2} - 1)$

5. $\frac{68}{111}$

7. 4

9. $\displaystyle\sum_{k=1}^{\infty} \frac{1}{4k^2 - 1}$; converges to $\frac{1}{2}$

11. $\displaystyle\sum_{k=1}^{\infty} \frac{k^2 + k - 1}{k(k + 1)}$; diverges

13. $\frac{1}{2} - \frac{1}{2^2} - \frac{1}{2^3} - \cdots - \frac{1}{2^n} - \cdots = 0$

15. $\sum_{k=1}^{\infty} \frac{2k - 1}{(k^2 + 1)(k^2 - 2k + 2)}$; converges to 1

17. $\lim_{n \to \infty} \frac{n}{n + 1} = 1$

19. $r = \frac{3}{2} > 1$

23. 140 ft

25. $\sum_{k=1}^{n} (b_{k+1} - b_k) = b_{n+1} - b_1$ and $\lim (b_{n+1} - b_1) = \infty$

$$\sum_{k=1}^{n} \left(\frac{1}{b_k} - \frac{1}{b_{k+1}}\right) = \frac{1}{b_1} - \frac{1}{b_{n+1}} \text{ and } \lim \left(\frac{1}{b_1} - \frac{1}{b_{n+1}}\right) = \frac{1}{b_1}$$

27. Geometric series with $a = x^2$ and $r = x^3$. Thus,

$$S = \frac{x^2}{1 - x^3} \text{ for } |x| < 1.$$

29. Geometric series with $a = 1$, $r = (-1)(2x - 1)^2$. Thus,

$$S = \frac{1}{1 + (2x - 1)^2} \text{ for } |2x - 1| < 1 \text{ or } 0 < x < 1.$$

31. 1

Section 2, Part I

1. Convergent
3. Convergent
5. Divergent
7. Divergent
9. Convergent
11. Convergent
13. Convergent

Section 2, Part II

3. $\int_2^{\infty} \frac{1}{n(\ln n)^r}\, dn = \int_{\ln 2}^{\infty} \frac{1}{t^r}\, dt$. Converges if $r > 1$, diverges if $r \leq 1$

Section 3, Part I

1. Convergent, $\frac{k + 1}{(k + 2)2^k} < \frac{1}{2^k}$

3. Divergent, $\frac{k + 1}{\ln (k + 2)} > \frac{1}{k}$

5. Convergent, $ke^{-k^2} < \frac{k}{1 + k^2 + \frac{k^4}{2}} < \frac{2}{k^3}$

7. Convergent, $\frac{1}{\sqrt{k(k + 1)(k + 2)}} < \frac{1}{\sqrt[2]{k^3}}$

9. Convergent, $\dfrac{\sqrt{k}}{k^2 - \sin^2 100k} \leq \dfrac{\sqrt{k}}{k^2 - 1} < \dfrac{2}{k^{3/2}}$

11. Divergent, $\dfrac{\sqrt{k^4 + 1}}{k^3 \ln k} > \dfrac{1}{k \ln k}$

13. Convergent. For n sufficiently large, $\dfrac{n^3}{n!} < \dfrac{n^3}{n^5}$

15. Divergent, $\dfrac{|\sec n|}{\sqrt{n}} > \dfrac{1}{\sqrt{n}}$

17. Divergent

19. Convergent, $\dfrac{n!}{n^n} = \dfrac{1 \cdot 2 \cdots n}{n \cdot n \cdots n} \leq \dfrac{2}{n^2}$.

21. Divergent: compare with $\sum \dfrac{1}{n}$.

Section 3, Part II

1. Divergent

3. Divergent

5. Could either converge or diverge. Consider the cases $a_k = k^2$, $a_k = k$.

7. $\left(a_n - \dfrac{1}{n}\right)^2 \geq 0$ or $a_n^2 + \dfrac{1}{n^2} \geq \dfrac{2a_n}{n}$. Thus

$$\sum_{n=1}^{\infty} \frac{2a_n}{n} < \sum_{n=1}^{\infty} \left(a_n^2 + \frac{1}{n^2}\right).$$

Result is also valid if a_n is not ≥ 0. This will follow from subsequent theorem 12.14.

Section 4

1. Converges

3. Converges

5. Converges

7. Diverges

9. Converges

11. Diverges

13. The first few terms of the series are

$$1 - 1 + \frac{1}{2} - \frac{1}{4} + \frac{1}{3} - \frac{1}{9} + \cdots + \frac{1}{k} - \frac{1}{k^2} + \cdots$$

which is *not* an alternating decreasing one. Here, $\sum_{k=1}^{\infty} a_{2k}$ converges while $\sum_{k-1}^{\infty} a_{2k-1}$ diverges. Thus, $\sum_{k=1}^{\infty} a_k$ diverges.

15. .800 (series can be summed exactly to this number)

17. .969

19. Converges

21. Converges

Section 5, Part I

1. Converges absolutely
3. Diverges
5. Converges, but not absolutely
7. Converges absolutely
9. Converges, but not absolutely
11. Converges, but not absolutely
13. Converges absolutely
15. Converges absolutely
17. Converges absolutely
19. Diverges
21. Converges, but not absolutely
23. Diverges
25. $a_k^2 + b_k^2 \geq 2|a_k b_k|$, since $(a_k - b_k)^2 \geq 0$.

Section 5, Part II

1. $a_n^2 = |a_n| \cdot |a_n| < |a_n|$ for n sufficiently large

Section 6, Part I

1. $(-\infty,\infty)$
3. $[-1,1)$
5. $(-\infty,\infty)$
7. $(-\frac{2}{3},\frac{2}{3})$
9. $(-1,1)$
11. $(-2,4)$
13. $[-\sqrt{5} - 1, \sqrt{5} - 1]$
15. $(-4, 4)$
17. $[-1,1]$
19. $[0,4)$
21. Converges for $x > 1$ or $x < -1$
23. Converges for $x \geq -\frac{1}{2}$
25. $\left(\dfrac{1}{e}, e\right)$
27. Converges for $|x| > 1$

Section 6, Part II

1. Pair consecutive terms of the series. Thus, let

$$u_k = a_{2k-1}x^{2k-1} + a_{2k}x^{2k} = 2^{-2k+2}x^{2k-1} + 2^{-2k}x^{2k} = 2^{-2k}x^{2k}\left(\frac{4}{x} + 1\right).$$

Then show that $\lim\limits_{k\to\infty} |u_{k+1}/u_k| = x^2/4$.

Section 7

1. $\ln 2 = .182$
3. $\ln 3 = 1.099$
5. $\sum\limits_{n=2}^{\infty} (-1)^{n+1} \dfrac{x^n}{n^2}$; using Exercise 1, integrate the expansion of $\dfrac{\ln(1+t)}{t}$ termwise.
7. $S'(x) = \dfrac{x}{1} + \dfrac{x^2}{2} + \dfrac{x^3}{3} + \cdots + \dfrac{x^{n-1}}{n-1} + \cdots = -\ln(1-x)$

$$S(x) = -\int_0^x \ln(1-t)\,dt = (1-x)\ln(1-x) + x$$

9. If $S(x) = \sum_{n=2}^{\infty} \frac{x^{3n}}{2n}$, then $2S(\sqrt[3]{x}) + x = \sum_{n=1}^{\infty} \frac{x^n}{n} = -\ln(1-x)$ (by Exercise 2).

Thus $S(x) = -\frac{1}{2}\{x^3 + \ln(1 - x^3)\}$.

11. $\frac{1}{(1-x)^3} = \frac{1}{2}[2 + 2 \cdot 3x + 3 \cdot 4x^2 + 4 \cdot 5x^3 + \cdots]$

13. $\frac{\cos x}{1+x} = 1 - x + \frac{x^2}{2} - \frac{x^3}{2} + \frac{13x^4}{24} - \frac{13x^5}{24} + \cdots$

15. $\sec x = 1 + \frac{x^2}{2} + \frac{5x^4}{24} + \cdots$

17. $\frac{1-x}{1-x+x^2} = 1 - x^2 - x^3 + x^5 + x^6 - x^8 - x^9 + \cdots$

19. $\frac{\ln(1+x)}{2+x} = \frac{1}{2}\left(x - x^2 + \frac{5}{6}x^3 - \frac{2}{3}x^4 + \frac{8}{15}x^5 - \cdots\right)$

21. .4940

Section 8, Part I

1. $\frac{\sqrt{2}}{2}\left[1 - \frac{(x-\pi/4)}{1!} - \frac{(x-\pi/4)^2}{2!} + \frac{(x-\pi/4)^3}{3!} + \frac{(x-\pi/4)^4}{4!} - \cdots\right]$, $r = \infty$

3. $-\sum_{n=1}^{\infty} (x+1)^n/n$, $r = 1$

5. $\sum_{n=1}^{\infty} \frac{x^{2n-1}}{(2n-1)!}$, $r = \infty$

7. $\sum_{n=1}^{\infty} \frac{(-1)^{n+1}x^{2n-1}}{(2n)!}$, $r = \infty$

9. $\sum_{n=1}^{\infty} \frac{(-1)^{n+1}2^{2n-1}x^{2n}}{(2n)!}$, $r = \infty$

11. $63 - 101(x+1) - (x+1)^2 - 31(x+1)^3 + 4(x+1)^4$

13. $\sum_{n=0}^{\infty} (-1)^n(n+1)x^n$, $r = 1$

15. $\sum_{n=0}^{\infty} \binom{2n}{n}\left(\frac{x}{2}\right)^{2n}$, $r = 1$

17. $8x - 3x^2 + \frac{3}{16}x^3 + \cdots + \frac{3 \cdot 1 \cdot 3 \cdot 5 \cdots (2n-5)}{2^{3(n-1)}n!}x^{2n} + \cdots$, $r = 4$

19. $\frac{1}{4}\left[1 - \frac{x^4}{2 \cdot 16} + \frac{1 \cdot 3x^8}{2!\,2^2 \cdot 16^2} - \frac{1 \cdot 3 \cdot 5x^{12}}{3!\,2^3 \cdot 16^3} + \cdots\right]$, $r = 2$

21. $2\sum_{n=0}^{\infty} \binom{2n}{n} \frac{(x/2)^{2n+1}}{2n+1}$, $r = 1$

23. $\sec^{-1}\frac{1}{x} = \cos^{-1} x = \frac{\pi}{2} - \sin^{-1} x$. Now use Exercise 9.

25. 1.3956

27. .021372

29. 1.11039

31. .4848 33. $\frac{1}{3}$ 35. $\frac{1}{12}$

Section 8, Part II

1. $f'(x) = \frac{2}{x^3}e^{-1/x^2}$ if $x \neq 0$

$$f'(0) = \lim_{x\to 0} \frac{e^{-1/x^2}}{x} = 0$$

$$f''(x) = \left(\frac{4}{x^6} - \frac{6}{x^4}\right)e^{-1/x^2} \quad \text{if } x \neq 0$$

$$f''(0) = \lim_{x\to 0} \frac{f'(x)}{x} = 0$$

Similarly, $f^{(n)}(0) = 0$. Here the Maclaurin series $= 0$. Also,

$$\lim_{n\to\infty} R_n(x) \neq 0.$$

Review

1. Converges
3. Diverges
5. Converges absolutely
7. Converges absolutely
9. Converges
11. Converges absolutely
13. Converges absolutely. Use the fact that $0 \leq \sin x < x$ for $0 \leq x \leq \pi/2$.
15. Converges absolutely
17. $(-\frac{5}{2}, -\frac{1}{2})$
19. Converges for $x \geq \frac{1}{2}$
21. $[-e,e]$
23. $\sum_{n=0}^{\infty} \frac{(\ln a)^n x^n}{n!}$
25. $e^5 \sum_{n=0}^{\infty} \frac{(x-2)^n}{n!}$
27. $x + x^2 + \frac{x^3}{3} - \frac{x^5}{30} + \cdots$
29. $\frac{1}{x} + \frac{x}{6} + \frac{7}{360}x^3 + \cdots$
31. $\frac{x^2}{2} + \frac{x^4}{12} + \frac{x^6}{45} + \cdots$
33. $\frac{1}{2}$
35. $\frac{1}{4}$
37. .23385

CHAPTER 13

Section 2, Part I

1. Straight line
3. Circle

5. Semiellipse

7. Ellipse

9. Hyperbola

21. $x = h + r\cos\theta,\ y = k + r\sin\theta$

Section 2, Part II

1. $x = t^2(3 - 2t),\ y = t^3(3 - 2t)$

3. $x = \dfrac{4t}{(1 + t^2)^2},\ y = \dfrac{4t^2}{(1 + t^2)^2}$

Section 3, Part I

1. $\langle 0,0\rangle,\ \langle 8,7\rangle,\ 119,\ \langle 0,0\rangle,\ -45$

3. $\langle \frac{4}{5},-\frac{3}{5}\rangle,\ \langle \frac{3}{5},-\frac{4}{5}\rangle,\ \langle 1/\sqrt{2},-1/\sqrt{2}\rangle,\ \langle \sin\theta, -\cos\theta\rangle,\ \langle b/\sqrt{a^2+b^2}, -a/\sqrt{a^2+b^2}\rangle$

7. **(a)** $(7,-8)$ **(b)** $(-2,13)$

9. $(4,-1)$

11. $(b_1 - c_1)\mathbf{v}_1 + (b_2 - c_2)\mathbf{v}_2 = 0$ implies $b_1 - c_1 = 0$ and $b_2 - c_2 = 0$.

13. $\overrightarrow{CE} = \mathbf{u} - \frac{1}{3}\mathbf{v},\ \overrightarrow{CF} = \frac{3}{4}\mathbf{u} - \frac{1}{4}\mathbf{v} = \frac{3}{4}\overrightarrow{CE}$

15. Note that $\cos\theta = \dfrac{\mathbf{u}\cdot\mathbf{v}}{|\mathbf{u}|\,|\mathbf{v}|}$ and $\mathbf{v}_1 = \dfrac{\mathbf{v}}{|\mathbf{v}|}$.

17. $\frac{4}{5}\mathbf{i} + \frac{8}{5}\mathbf{j}$

19. $\mathbf{r} = \left(1 + \dfrac{3\sqrt{3}}{2}\right)\mathbf{u} + \left(\sqrt{3} - \dfrac{3}{2}\right)\mathbf{v}$

21. 22.4 mph, 26.5° south of east

Section 3, Part II

1. $(a_1^2 + a_2^2)(b_1^2 + b_2^2) - (a_1b_1 + a_2b_2)^2 = (a_1b_2 - a_2b_1)^2 \geq 0$; with equality if and only if the two vectors are parallel.

Section 4, Part I

1. $\overrightarrow{AD} = \frac{3}{4}\mathbf{B} + \frac{1}{4}\mathbf{C} - \mathbf{A}$

3. $(\frac{1}{2},7)$

Section 4, Part II

3.

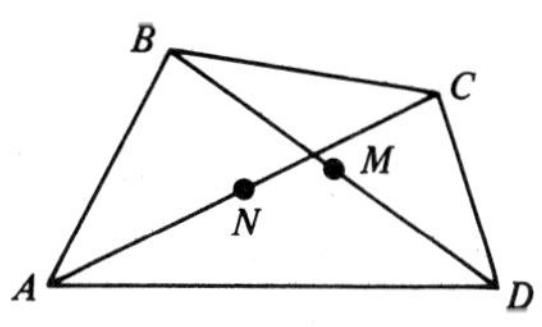

Let M denote midpoint of BD, let N denote midpoint of AC, and let $\overrightarrow{AB} = \mathbf{u}$, $\overrightarrow{AC} = \mathbf{v}$, and $\overrightarrow{AD} = \mathbf{w}$. Then

$$\overrightarrow{BC} = \mathbf{v} - \mathbf{u}, \qquad \overrightarrow{CD} = \mathbf{w} - \mathbf{v}, \qquad \overrightarrow{BD} = \mathbf{w} - \mathbf{u},$$

$$\overrightarrow{AN} = \frac{\mathbf{v}}{2}, \qquad \overrightarrow{AM} = \frac{\mathbf{u} + \mathbf{w}}{2}, \qquad \overrightarrow{NM} = \frac{\mathbf{u} + \mathbf{w} - \mathbf{v}}{2}$$

$$\overrightarrow{AB}^2 + \overrightarrow{BC}^2 + \overrightarrow{CD}^2 + \overrightarrow{DA}^2 = 2(\mathbf{u}\cdot\mathbf{u} + \mathbf{v}\cdot\mathbf{v} + \mathbf{w}\cdot\mathbf{w} - \mathbf{v}\cdot\mathbf{u} - \mathbf{v}\cdot\mathbf{w})$$

$$\overrightarrow{BD}^2 + \overrightarrow{AC}^2 + 4\overrightarrow{MN}^2 = 2(\mathbf{u}\cdot\mathbf{u} + \mathbf{v}\cdot\mathbf{v} + \mathbf{w}\cdot\mathbf{w} - \mathbf{v}\cdot\mathbf{u} - \mathbf{v}\cdot\mathbf{w})$$

Section 5, Part I

1. a. $x(t) = 2 + 3t$, $y(t) = 3 + 4t$
b. $x(t) = 1 + t$, $y(t) = 2$
c. $x(t) = -1 + 4t$, $y(t) = 5 - 9t$

3. $\langle 4,12 \rangle$

5. $\langle 2,-2 \rangle$

7. $t = \pi/2$

9. $r = 0, \pi/2, \pi, 3\pi/2$

17. $e^{-2t} - te^{-2t}$, $2te^{-3t} - 3e^{-3t}$

19. $-\tan\theta$, $\dfrac{1}{3a}\sec^4\theta\csc\theta$

Section 5, Part II

1. Follows immediately from the definition of tangent line.

3. The parametric equations of the tangent line are (see Exercise II-1):

$$x = a\cos^4 t_0 - 4at\cos^3 t_0 \sin t_0,$$
$$y = a\sin^4 t_0 + 4at\cos t_0 \sin^3 t_0.$$

For $x = 0$, $t = \dfrac{\cos t_0}{4\sin t_0}$ and $y = a\sin^4 t_0 + a\cos^2 t_0 \sin^2 t_0$.

For $y = 0$, $t = -\dfrac{\sin t_0}{4\cos t_0}$ and $x = a\cos^4 t_0 + a\cos^2 t_0 \sin^2 t_0$.

Thus sum of intercepts $= a[\cos^2 t_0 + \sin^2 t_0]^2 = a$.

5. The tangent line at $t = a$ is given by

$$x = r(a - \sin a) + rt(1 - \cos a),$$
$$y = r(1 - \cos a) + rt\sin a.$$

The highest point of the rolling circle at $t = a$ is given by $x = ra$, $y = 2r$. It is easily verified that the latter point is on the tangent line. Similarly for the normal line. Here, the lowest point of the rolling circle at $t = a$ is $x = ra$, $y = 0$.

Section 6

1. $\langle 2t,2 \rangle$, $\langle 2,0 \rangle$, $2\sqrt{5}$

3. $\langle 2\cos t, -2\sin t \rangle$, $\langle -2\sin t, -2\cos t \rangle$, 2

5. $\langle -\sin t, -2\sin 2t \rangle$, $\langle -\cos t, -4\cos 2t \rangle$, 1

7. $\langle e^t, -e^{-t} \rangle$, $\langle e^t, e^{-t} \rangle$, $\sqrt{2}$

9. $\langle 2\cos t, 2\sin t \rangle$, $\langle -2\sin t, 2\cos t \rangle$, 2

11. $\langle -ab\sin bt, ab\cos bt \rangle$, $\langle -ab^2\cos bt, -ab^2\sin bt \rangle$, ab

13. The particle must be moving on a circle.

15. $y = v_0 t\sin\alpha - 16t^2$, $x = v_0 t\cos\alpha$.

$y = 0$ for $t = 0$ and $t = \dfrac{v_0\sin\alpha}{16}$.

$$\text{Range} = v_0 \cos \alpha \cdot \frac{v_0 \sin \alpha}{16} = \frac{v_0^2 \sin 2\alpha}{32}.$$

At maximum height, $\dfrac{dy}{dt} = 0 = v_0 \sin \alpha - 32t$ or $t = \dfrac{v_0 \sin \alpha}{32}$.

$$\text{Maximum height} = v_0 \sin \alpha \left(\frac{v_0 \sin \alpha}{32}\right) - 16\left(\frac{v_0 \sin \alpha}{32}\right)^2 = \frac{v_0^2 \sin^2 \alpha}{64}.$$

$$\text{Total time of flight} = \frac{v_0 \sin \alpha}{16}.$$

Trajectory; $y = v_0 \sin \alpha \left(\dfrac{x}{v_0 \cos \alpha}\right) - 16\left(\dfrac{x}{v_0 \cos \alpha}\right)^2$, or $y = x \tan \alpha - \dfrac{16x^2}{v_0^2} \sec^2 \alpha$.

For maximum range, $\sin 2\alpha = 1$ or $\alpha = \pi/4$ and range (max) $= v_0^2/32$.

$$\sin 2\alpha = \sin 2\left(\frac{\pi}{2} - \alpha\right).$$

17. $\alpha = 19°54'$ or $70°4'$ approximately.

Section 7, Part I

21. $r = 3$

23. $r^2 = 2 \csc 2\theta$

25. $r = -4 \cos \theta$

27. $x^2 + y^2 = 2y$

29. $(x^2 + y^2 + y)^2 = x^2 + y^2$

31. $y = 2$

33. $|PQ|^2 = r_1^2 + r_2^2 - 2r_1r_2 \cos(\theta_2 - \theta_1)$

Section 7, Part II

1. $r = 4 \cos 3\theta$

11. $(0,0)$, $(\sqrt{3},\pi/3)$, $(\sqrt{3},2\pi/3)$

13. $r = 0$, $(\pm 2,\pi/2)$, $(.4384, \pm 141°20')$

15. $(6,\pi/6)$, $(6,5\pi/6)$, $(2,-\pi/6)$, $(2,-5\pi/6)$

Section 8, Part II

1. Note that $r = \dfrac{2}{1 - \sqrt{2}\cos(\theta - 45°)}$.

3. The asymptotes correspond to the denominator being zero.

5. Let $x = r \cos \theta$, $y = r \sin \theta$ in $\dfrac{x^2}{a^2} - \dfrac{y^2}{b^2} = 1$.

Section 9

3. At the origin, 0°; at two other points, $\arctan 3\sqrt{3}$.

5. The curves do not intersect.

7. $\dfrac{\pi}{4}$

9. $\dfrac{\pi}{2}$

11. $r = 2\theta$

13. $r = \csc \theta$

Section 10, Part I

1. $\pi^3/48$
3. 1
5. $(e^{2\pi} - 1)/4$
7. $(\pi + 4)/4$
9. 25π
11. $3\pi/2$
13. $\pi/2$
15. $\frac{1}{2}$
17. $\pi a^2/4n$
19. $(\pi - 2)/8$
21. $8\pi - 16$

Section 10, Part II

1. At intersection, $\cos 2\theta = \frac{1}{2}$.

$$\frac{A}{4} = \frac{1}{2}\int_0^{\pi/6} [2a^2 \cos 2\theta - a^2]\, d\theta,$$

$$A = a^2 \frac{3\sqrt{3} - \pi}{3}.$$

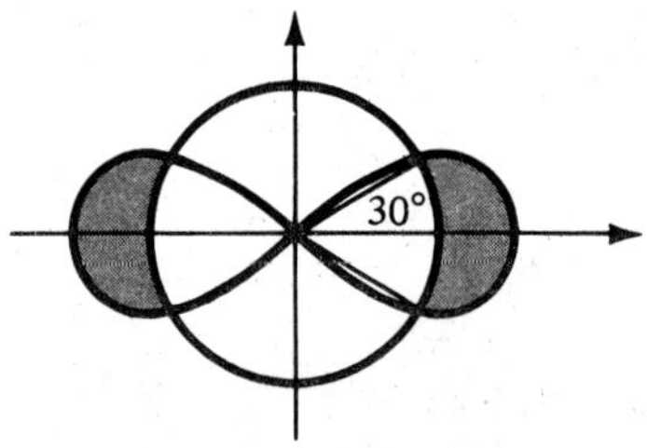

3. $A = a^2 \int_0^{\pi/4} \frac{\cos^2 2\theta}{\cos^2 \theta}\, d\theta = a^2 \int_0^{\pi/4} [\sec^2 \theta + 2\cos 2\theta - 2]\, d\theta = a^2 \frac{4 - \pi}{2}$

5. In rectangular coordinates, the equation is $x^3 + y^3 = 6xy$. It now follows that $x + y = -2$ is the asymptote. (Consider $x + y = \frac{6xy}{x^2 - xy + y^2}$ and let $y \to a - x$ in the right-hand side.)

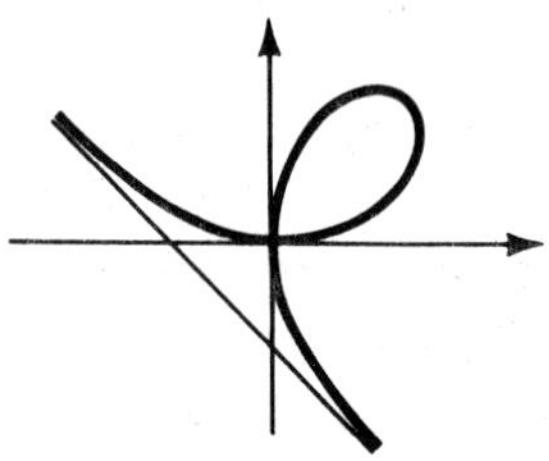

$$A = \int_{3\pi/4}^{\pi} - \left\{\frac{36 \sin^2 \theta \cos^2 \theta}{(\sin^3 \theta + \cos^3 \theta)^2} - \frac{4}{(\sin \theta + \cos \theta)^2}\right\} d\theta + 2.$$

Let $x = \tan \theta$; then

$$A - 2 = \int_{-1}^{0} \left\{\frac{4}{x^2 - x + 1} - \frac{6}{(x^2 - x + 1)^2} - \frac{12x - 6}{(x^2 - x + 1)^2}\right\} dx.$$

To integrate the first two terms in the integrand, let $x - \frac{1}{2} = \frac{\sqrt{3}}{2}\tan\phi$. This will lead to a zero contribution. Then

$$A = 2 + \frac{6}{x^2 - x + 1}\Big]_{-1}^{0} = 6.$$

Section 11

1. $2 - \sqrt{3}$ **3.** $\ln(2 - \sqrt{3})$ **5.** $\sqrt{5} + \frac{1}{2}\ln(2 + \sqrt{5})$
7. $2\sqrt{5} + \ln(2 + \sqrt{5})$ **9.** $\sqrt{2}(e^2 - 1)$
11. $3\sqrt{2}$ **13.** 2 **15.** $3\pi/2$

Section 13

1. $4\pi(5\sqrt{10} - \sqrt{2})/3$ **3.** π
5. $\pi(145\sqrt{145} - 1)/27$
7. $\pi\left[\frac{e^6 - 1}{e^4}\sqrt{e^4 + 1} + 2 + \ln\frac{e^2 + \sqrt{e^4 + 1}}{1 + \sqrt{e^4 + 1}}\right]$
9. $4\pi r^2$ **11.** $2\pi a^2$

Review

5. $\frac{(x - 2)^2}{25} + (y + 3)^2 = 1$
15. $x = e^t, y = -\frac{1}{2}(e^{-2t} + 3)$
17. $x = \tan^{-1} t, y = \frac{-1}{t^2 + 1}$
19. 6π; 16
21. $3\pi/8$; 4
23. $\pi k^2(10\sqrt{10} - 1)/27$

CHAPTER 14

Section 1

1. (0,0,3), (7,0,3), (0,2,3), (0,2,0), (7,0,0), (7,2,0)
3. (−1,1,5), (2,1,5), (−1,3,5), (2,1,2), (2,3,2), (−1,3,2)
5. (0,−2,0), (3,−2,0), (0,1,0), (3,−2,−1), (3,1,−1), (0,1,−1)
7. $x^2 + y^2 + z^2 - x - 9y - 6z + 13 = 0$
9. $x^2 + y^2 + z^2 + 5x - y + z = 24$
11. Sphere: center (1,0,0), radius 5 **13.** Sphere: center $(0,\frac{3}{2},-\frac{1}{2})$, radius $\frac{3}{2}$
15. No graph **19.** $x = 2y + z$; a plane

21. Plane parallel to the xy plane
23. Plane passing through the z axis
25. Right circular cylinder with generators parallel to the x axis
27. Square cylinder with generators parallel to the z axis

Section 2

1. Let $\mathbf{u} = \langle a,b,c \rangle$, $\mathbf{v} = \langle r,s,t \rangle$, then $|\mathbf{u}|^2|\mathbf{v}|^2 - |\mathbf{u}\cdot\mathbf{v}|^2 = (bt - cs)^2 + (cr - at)^2 + (as - br)^2 \geq 0$, with equality if and only if $\frac{a}{r} = \frac{b}{s} = \frac{t}{c}$ or $\mathbf{u} = k\mathbf{v}$.

5. $|\mathbf{u} - \mathbf{v}|^2 = \mathbf{u}^2 + \mathbf{v}^2 - 2\mathbf{u}\cdot\mathbf{v} = 2(1 - \cos\alpha) = 4\sin^2\frac{\alpha}{2}$, $|\mathbf{u} - \mathbf{v}| = 2\left|\sin\frac{\alpha}{2}\right|$

7. Square out each side and compare terms.
9. Expand $(\mathbf{u} + \mathbf{v} + \mathbf{w})\cdot(\mathbf{u} + \mathbf{v} + \mathbf{w})$.
11. $-\frac{1}{6}\langle 4,-1,1\rangle$
13. **b.** $\mathbf{r} = \frac{22}{7}\mathbf{u} - \frac{11}{7}\mathbf{v} + \frac{9}{7}\mathbf{w}$
15. $\langle 4,3,8\rangle$ is one such vector.

Section 3, Part I

1. $x = 2 + 3t$, $y = 3 - 7t$, $z = 1 + 4t$
3. $\frac{1}{2}, \frac{1}{2}, \sqrt{2}/2; \pi/3, \pi/3, \pi/4$
5. $-\frac{\sqrt{2}}{2}, -\frac{1}{2}, -\frac{1}{2}; \frac{3\pi}{4}, \frac{2\pi}{3}, \frac{2\pi}{3}$
7. $lmn = 0$
11. $P = (0,0,1)$, $\mathbf{v} = \langle 2,-2,-3\rangle$
13. Intersects
15. $(-5,4,3)$; distance is $\sqrt{14}$

17. If B is the foot of the perpendicular from A to L, then B corresponds to the value of t given by the orthogonality condition

$$a(x_0 + at - h) + b(y_0 + bt - k) + c(z_0 + ct - l) = 0,$$

or

$$t = \frac{a(h - x_0) + b(k - y_0) + c(l - z_0)}{a^2 + b^2 + c^2}.$$

Now use the formula for the distance from A to B.

Section 3, Part II

1. $A = \left(\frac{9}{7}, \frac{-9}{7}, \frac{18}{7}\right)$, $B = \left(\frac{8}{7}, \frac{-6}{7}, \frac{20}{7}\right)$

Section 4

1. $x + y - z = 1$
3. $x + y = a + b$
5. $2x + y = 4; y + z = 4, 2x + z = 4$
7. $4x + y = 6; 2x = 3; y = 6$
9. $x + y = 2z$
11. $6x + 10y - z = 23$
13. $x - y = 1$

15. $x = 5, y = 1 + t, z = 2t$

17. $3x + 4y - 9z = 2$

19. $3x - z + 1 = 0$

21. $7x + 22y + 12z = 17$

23. $\frac{1}{3}$

25. $\dfrac{|d - d'|}{\sqrt{a^2 + b^2 + c^2}}$

27. $x - y + 2z = \pm 3\sqrt{6}$

Section 5

1. 16, −9

3. 0

5. $\langle 16, -18, -5\rangle, \langle 6, -6, 3\rangle$

11. 1

13. 9

15. $2x - 7y - 3z + 13 = 0$

17. $7x - 4y + 3z = 8$

19. $\sqrt{2}/2$

Section 6,

17. $x^2/a^2 + y^2/b^2 + z^2/b^2 = 1; x^2/a^2 + y^2/b^2 + z^2/a^2 = 1$

19. $y^2 + z^2 = 4px$

21. $x^2 + z^2 = y^{4/3}$

23. $y^2 + z^2 = \cos^2 x$

25. $(x^2 + y^2 + z^2 + k^2 - r^2)^2 = 4k^2(x^2 + y^2)$. The surface is a doughnut-shaped solid (torus).

Section 7

1. Ellipsoid

3. Hyperboloid of one sheet

5. Hyperbolic paraboloid

7. Elliptic paraboloid

9. Hyperboloid of two sheets

11. Elliptic cone

13. Elliptic paraboloid

15. Right circular cylinder

Section 8

1. $x = t, y = 10 + 6t, z = 9 + 6t; x + 6y + 6z = 114$

3. $x = a \sin kt_0 + tak \cos kt_0, y = b \cos kt_0 - tbk \sin kt_0, z = ct_0^2 + 2tct_0;$

$$xa \cos kt_0 - yb \sin kt_0 + 2zct_0 = (a^2 - b^2) \sin kt_0 \cos kt_0 + \frac{2c^2t_0^3}{k}$$

5. $x = \sqrt{t^2 + t - 1}, y = \sqrt{1 - t^2}, z = t; \dfrac{\sqrt{5} - 1}{2} \leq t \leq 1$

9. $L = \sqrt{8} \int_1^3 \sqrt{t^2 + 8}\, dt = \sqrt{8} \left\{\frac{3}{2}(\sqrt{17} - 1) + 4 \ln \dfrac{3 + \sqrt{17}}{4}\right\}$

11. $8 + \ln 3$

Section 9

1. $(6, 26°34', 131°49'); (2\sqrt{5}, 120°, 26°34')$

3. $(6/5, 8/5, 6); (0, -10, 4)$

5. $r^2(1 + \sin 2\theta) = z - 5$; $r^2z^2 = 25$; $r^2(b^2 \cos^2 \theta + a^2 \sin^2 \theta) = a^2b^2$; $r^2 + z^2 - cz = r(a \cos \theta + b \sin \theta)$
7. $\rho = 2 \cos \phi$
9. a. $r^2 = 2z$ b. $\rho = 2 \csc \phi \cot \phi$
11. Cone, $4(x^2 + y^2) = z^2$
13. $x = \sin \theta \cos \theta,\ y = \sin^2 \theta,\ z = \cos \theta,\ 0 \le \theta \le \pi$
$x = -\sin \theta \cos \theta,\ y = -\sin^2 \theta,\ z = \cos \theta,\ \pi \le \theta < 2\pi$

Review

1. Sphere
3. Hyperbolic paraboloid (saddle-shaped)
5. Hyperbolic cylinder
7. Hyperboloid of one sheet
9. Two perpendicular planes
11. Cylinder
13. $s\sqrt{\frac{2}{3}}$
15. b. $2x - 6y + 13z + 14 = 0$
17. $(\frac{8}{5},0,\frac{4}{5})$
19. $15x + y - 7z + 2 = 0$
21. $\frac{1}{2}\sqrt{2237}$
23. a. $\dfrac{x-2}{2} = \dfrac{y}{2} = \dfrac{z-1}{1}$; $2x + 2y + z = 5$

b. $x = 2,\ \dfrac{y}{-3} = \dfrac{z - \pi^3/8}{\pi^2/4}$; $\dfrac{3\pi^2}{4} z - 3y = \dfrac{3\pi^5}{32}$

25. $\dfrac{\sqrt{2}}{2}(3 + \ln 2)$

CHAPTER 15

Section 1

1. Domain $\{(x,y) \mid x \ge 0, y \ge 0\}$, range $\{z \mid z \ge 0\}$
3. Domain $\{(x,y) \mid x \ne \pm y\}$, range $\{z \mid z \ne 0\}$
5. Domain $\{(x,y) \mid x^2 + y^2 < 1\}$, range $\{z \mid z > 0\}$
7. a. $g(f(x,y)) = \dfrac{x^2y^2}{(x+y)^2}$ b. $h(f(x,y)) = 1 - \dfrac{xy}{x+y}$

c. $f(g(t),h(t)) = \dfrac{t^2(1-t)}{t^2 + 1 - t}$ d. $\frac{1}{2}$

11. 2
15. All (x,y)
17. All (x,y)
19. Discontinuous for $x = -1$
21. Discontinuous at (0,0)
23. Consider limits along the lines $y = 0$ and $y = x$.

Section 2

1. $f_1 = y \cos xy$, $f_2 = x \cos xy$, $f_{12} = f_{21} = \cos xy - xy \sin xy$

3. $f_1 = \dfrac{2y}{(x+y)^2}, f_2 = -\dfrac{2x}{(x+y)^2}, f_{12} = f_{21} = \dfrac{2(x-y)}{(x+y)^3}$

5. $3 \ln 2 - 1$

7. $-(x+y)^{-2}, -(x+y)^{-2}$

9. 0, 0

11. $\dfrac{2}{(z+w)^3}$

13. Harmonic for $(x,y) \neq (0,0)$

15. Harmonic for all (x,y)

17. Harmonic for all (x,y)

Section 3

1. $\Delta f(P,\mathbf{v}) = (2ab - 3)h + (a^2 + 2b)k + bh^2 + 2ahk + h^2k + k^2$

3. $\Delta f(P,\mathbf{v}) = \dfrac{bh - ak}{b(b+k)}$

5. $\frac{1}{5}$

7. $\frac{1}{3}(r + s + t)$

9. $h + k + l$

11. Let $w = \sqrt{x^2 + y^2 + z^2}$, $P = (100,200,200)$, $\mathbf{v} = \langle 0,-1,1 \rangle$; $P + \mathbf{v} = (100,199,201)$.

$$w(P + \mathbf{v}) - w(P) \doteq \nabla w(P) \cdot \langle 0,-1,1 \rangle.$$

$$\nabla w(P) = \left\langle \frac{x}{w}, \frac{y}{w}, \frac{z}{w} \right\rangle_P = \langle \tfrac{1}{3}, \tfrac{2}{3}, \tfrac{2}{3} \rangle.$$

$$w(P + \mathbf{v}) - 300 \doteq \langle \tfrac{1}{3}, \tfrac{2}{3}, \tfrac{2}{3} \rangle \cdot \langle 0,-1,1 \rangle = 0.$$

13. Let $z = \sin x \cos y$, $P = \left(\dfrac{\pi}{4}, \dfrac{\pi}{6}\right)$, $\mathbf{v} = \left\langle -\dfrac{\pi}{180}, \dfrac{\pi}{180} \right\rangle$, $P + \mathbf{v} = \left(\dfrac{44\pi}{180}, \dfrac{31\pi}{180}\right)$.

$$z(P + \mathbf{v}) - z(P) \doteq \nabla z(P) \cdot \left\langle -\frac{\pi}{180}, \frac{\pi}{180} \right\rangle.$$

$$\nabla z(P) = \langle \cos x \cos y, -\sin x \sin y \rangle_P = \left\langle \frac{\sqrt{3}}{2\sqrt{2}}, -\frac{1}{2\sqrt{2}} \right\rangle.$$

$$z(P + \mathbf{v}) - \frac{\sqrt{3}}{2\sqrt{2}} \doteq -\frac{\pi}{180}\,\frac{1 + \sqrt{3}}{2\sqrt{2}}.$$

15. In general, if $V = F(x,y,z)$, $P = (l,w,h)$, $Q = (l + \Delta l, w + \Delta w, h + \Delta h)$,

$$V(Q) - V(P) = \Delta V \doteq \nabla V(P) \cdot \langle \Delta l, \Delta w, \Delta h \rangle$$

$$\doteq \{F_1\, \Delta l + F_2\, \Delta w + F_3\, \Delta h\}_{(x,y,z)=(l,w,h)}$$

Here, $V = xyz$. Thus

$$\frac{\Delta V}{V} = \frac{\Delta l}{l} + \frac{\Delta w}{w} + \frac{\Delta h}{h}$$

where $l = 10.2$, $w = 20.4$, $h = 9.9$, $V = 2059.992$, $\Delta l = \pm .05$, $\Delta w = \pm .05$, $\Delta h = \pm .05$.

$$\frac{\Delta V}{2060} \doteq \pm .05\left(\frac{1}{10.2} + \frac{1}{20.4} + \frac{1}{9.9}\right)$$

or

$$\Delta V \lesssim (.0125)(2060) \doteq 25$$

Since writing the volume as 2060.0 means the error is within $\pm .05$, we are not justified in the way we rounded off the number.

Section 4

1. $\dfrac{dw}{dt} = \dfrac{2(t-1)}{t^3} + \dfrac{2t}{(t+1)^3}$

3. $\dfrac{dw}{dt} = t^3(1 + 3\ln t)\cos t^3 \ln t + \sin t^3 \ln t$

5. $\dfrac{\partial z}{\partial u} = 2u, \dfrac{\partial z}{\partial v} = 0$

7. $\dfrac{\partial z}{\partial u} = \dfrac{2u}{\sqrt{1-(u^2-v^2)^2}}, \dfrac{\partial z}{\partial v} = \dfrac{-2v}{\sqrt{1-(u^2-v^2)^2}}$

9. $\dfrac{\partial z}{\partial u} = \dfrac{5v}{(u+2v)^2}\, e^{(2u-v)/(u+2v)}$

11. $\dfrac{\partial w}{\partial r} = \dfrac{3(r^2+s^2+t^2) - 2r(3r-s-2t)}{(r^2+s^2+t^2)^2}$

13. $\dfrac{\partial w}{\partial u} = \sqrt{1+v^2}, \dfrac{\partial w}{\partial v} = \dfrac{uv}{\sqrt{1+v^2}}$

15. By the chain rule,

$$z_u = z_x x_u + z_y y_u, \tag{1}$$

$$z_{uu} = z_x x_{uu} + z_y y_{uu} + x_u \frac{\partial z_x}{\partial u} + y_u \frac{\partial z_y}{\partial u}.$$

To obtain $\dfrac{\partial z_x}{\partial u}$, just replace z by z_x in (1), i.e.,

$$\frac{\partial z_x}{\partial u} = z_{xx} x_u + z_{yx} y_u.$$

Similarly for $\dfrac{\partial z_y}{\partial u}$.

17.
$$\begin{aligned} z_x &= 1 + f'(u)u_x, & u_x &= y, \\ z_y &= f'(u)u_y, & u_y &= x. \end{aligned}$$

$$\frac{z_x - 1}{z_y} = \frac{u_x}{u_y} = \frac{y}{x}.$$

19.
$$\begin{aligned} z &= f(x), & x &= u^2 + v^2 \\ z_u &= f'(x)x_u, & x_u &= 2u, \end{aligned}$$

$$z_v = f'(x)x_v, \qquad x_v = 2v,$$

$$\frac{z_u}{z_v} = \frac{x_u}{x_v} = \frac{u}{v}.$$

21. $F(x,y) = f(u) + g(v), \qquad u = y + ax, \qquad v = y - ax,$
$F_x = f'(u)a + g'(v)(-a), \qquad F_{xx} = f''(u)a^2 + g''(v)a^2.$
$F_y = f'(u) + g'(v), \qquad F_{yy} = f''(u) + g''(v).$
$F_{xx} = a^2 F_{yy}.$

23. $\left(\dfrac{\partial H}{\partial V}\right)_P = \left(\dfrac{\partial E}{\partial V}\right)_P + P = \dfrac{\partial g}{\partial V} + P$

25. $\dfrac{dz}{dx} = \dfrac{\partial f}{\partial x} + \dfrac{\partial f}{\partial y}\dfrac{dg}{dx},$

$$\frac{d^2z}{dx^2} = \frac{\partial^2 f}{\partial x^2} + 2\frac{\partial^2 f}{\partial x\,\partial y}\frac{dg}{dx} + \frac{\partial^2 f}{\partial y^2}\left(\frac{dg}{dx}\right)^2 + \frac{\partial f}{\partial y}\frac{d^2 g}{dx^2}$$

27. $\dfrac{\partial w}{\partial x} = \dfrac{\partial f}{\partial x} + \dfrac{\partial f}{\partial y}\dfrac{\partial g}{\partial x} + \dfrac{\partial f}{\partial z}\dfrac{\partial h}{\partial x},$

$$\frac{\partial w}{\partial u} = \frac{\partial f}{\partial y}\frac{\partial g}{\partial u}, \qquad \frac{\partial w}{\partial v} = \frac{\partial f}{\partial z}\frac{\partial h}{\partial v}$$

Section 5

1. $(8a + 2b - 6c)/5$

3. $\frac{11}{3}$

5. $4\sqrt{2}$

7. $\frac{3}{5}$

9. $\langle 4/\sqrt{185}, 13/\sqrt{185}\rangle$

11. $\langle 1/\sqrt{3}, 1/\sqrt{3}, 1/\sqrt{3}\rangle$

13. $\dfrac{\mathbf{u}}{|\mathbf{u}|}$, where $\mathbf{u} = \left\langle \dfrac{1}{4}, \dfrac{1}{8}, -\dfrac{1 + \ln 2}{16}\right\rangle$

15. **a.** $D_{\mathbf{v}}f = (2a + 2b)\cos\theta + (2a + 2b)\sin\theta$ where $\mathbf{v} = \langle\cos\theta, \sin\theta\rangle$.
Max $D_{\mathbf{v}}f = 2\sqrt{2}(a + b)$, min $D_{\mathbf{v}}f = -2\sqrt{2}(a + b)$.
b. $D_{\mathbf{v}}f = 3(\cos\theta - \sin\theta)$ where $\mathbf{v} = \langle\cos\theta, \sin\theta\rangle$.
Max $D_{\mathbf{v}}f - 3\sqrt{2}$, min $D_{\mathbf{v}}f = -3\sqrt{2}$.

17. Let $\mathbf{u} = \langle\cos\theta, \sin\theta\rangle$, $\mathbf{v} = \langle -\sin\theta, \cos\theta\rangle$.
Expand $[D_{\mathbf{u}}f(P)]^2 + [D_{\mathbf{v}}f(P)]^2$ and simplify.

Section 6, Part I

1. $8x - 4y + z + 8 = 0$

3. $x + 2y + 2z = 9$

5. $4x = 8y + z$

7. $x + 4y + 13 = 18z$

9. $x - y - 2z = 2$

11. Follows from 14.7 and 15.25.

13. $x = 1 + 2t,\ y = 2 + t,\ z = 2 - 2t$

15. $x = -2 + 2t,\ y = 1 + t,\ z = -1 - 3t$

Section 6, Part II

1. Equation of the tangent plane is $x/\sqrt{a} + y/\sqrt{b} + z/\sqrt{c} = \sqrt{k}$. Sum of the intercepts is $(\sqrt{a} + \sqrt{b} + \sqrt{c})\sqrt{k} = \sqrt{k}$.

Section 7

1. $\dfrac{dy}{dx} = -\dfrac{1 + y\cos xy}{2y + x\cos xy}$

3. $\dfrac{dy}{dx} = \dfrac{y}{x - 3y^2(x^2 + y^2)}$

5. $\dfrac{dy}{dx} = \dfrac{3x^2 - 4xy^2 - 3y}{4x^2y - 3y^2 + 3x}$

7. $(-3,6)$ max, $(3,-6)$ min

9. $\dfrac{\partial z}{\partial x} = -\dfrac{2xz + y^2}{x^2 + 2yz - 3z^2}$

11. $\dfrac{\partial z}{\partial y} = \dfrac{3y^2z + z^3}{3(x^3 + y^3) + (x + y)z^2}$

13. Differentiating both equations with respect to u:

$$x + ux_u + vy_u + 2x^2u^2 + 2xx_uu^2 + 2yy_uv^2 = 0,$$
$$2xu + x^2x_u + x^2 + 2xx_uu - y_uv^2 - 2yy_uv = 0.$$

Now solve these simultaneous linear equations for x_u and y_u. To obtain x_v and y_v, differentiate the original pair of equations with respect to v and solve.

15. $$f''(x) = -\frac{\left(\dfrac{\partial F}{\partial y}\right)^2 \dfrac{\partial^2 F}{\partial x^2} - 2\dfrac{\partial F}{\partial x}\dfrac{\partial F}{\partial y}\dfrac{\partial^2 F}{\partial x\,\partial y} + \left(\dfrac{\partial F}{\partial x}\right)^2 \dfrac{\partial^2 F}{\partial y^2}}{\left(\dfrac{\partial F}{\partial y}\right)^3}$$

Section 8

1. $f(0,0) = 0$, min

3. $F(2,0) = -16$, min

5. No extremum

7. $f(-\frac{1}{2},4) = -6$, max

9. $F(2^{-1/5},\pm 2^{3/10}) = 5\cdot 2^{-2/5}$, min

11. $(\frac{3}{7})\sqrt{14}$

13. Let the coordinates of one vertex of the parallelopiped be (x,y,z). Then

$$V = 8xyz,$$

or

$$\frac{V^2}{8a^2b^2c^2} = \frac{x^2}{a^2}\cdot\frac{y^2}{b^2}\cdot\frac{z^2}{c^2}.$$

Let

$$r = \frac{x^2}{a^2}, \qquad s = \frac{y^2}{b^2}, \qquad t = \frac{3^2}{c^2}.$$

We now have to maximize

$$\bar{V} = rst \qquad \text{subject to } r + s + t = 1.$$
$$\bar{V} = rs(1 - r - s).$$
$$\bar{V}_r = s(1 - r - s) - rs = 0 = \bar{V}_s = r(1 - r - s) - rs.$$

Thus, $r = s = t = \frac{1}{3}$ and $V = 8abc/3\sqrt{3}$.

15. We are to maximize $V = xyz$ subject to $xy + 2xz + 2yz = s$. Then

$$V_x = yz + xy\frac{\partial z}{\partial x} = 0 \quad \text{where} \quad y + 2z + 2(x + y)\frac{\partial z}{\partial x} = 0,$$

$$V_y = xz + xy\frac{\partial z}{\partial y} = 0 \quad \text{where} \quad x + 2z + 2(x + y)\frac{\partial z}{\partial y} = 0.$$

Thus

$$z = \frac{x(y + 2z)}{2(x + y)} = \frac{y(x + 2z)}{2(x + y)} \quad \text{and} \quad x = y = 2z.$$

These are the same subsidiary equations as in Example 4. Now substitute back in the surface equation.

19. $(0,0,\pm 1)$

Section 9

1. $2\int_0^{\pi} (1 - x\cos t)\sin t\, dt$ **3.** $3x^{7/2}$ **5.** 0

7. a. $\displaystyle\int \frac{1}{(t^2 + x^2)^2}\, dt = \frac{1}{2x^3}\tan^{-1}\frac{t}{x} + \frac{t}{2x^2(t^2 + x^2)} + C$

b. $\displaystyle\int \frac{1}{(x^2 - t^2)^{3/2}}\, dt = \frac{t}{x^2\sqrt{x^2 - t^2}} + C$

c. $\displaystyle\int \frac{1}{t(t^2 - x^2)^{3/2}}\, dt = -\frac{[\sec^{-1} t/x + x(t^2 - x^2)^{-1/2}]}{x^3} + C$

d. $\displaystyle\int t^x \ln t\, dt = \frac{t^{x+1}}{(x + 1)^2}\{(x + 1)\ln t - 1\} + C$

Review

5. 1.642

7. $\dfrac{\partial^2 z}{\partial r^2} = f_{xx}\cos^2\theta + 2f_{xy}\sin\theta\cos\theta + f_{yy}\sin^2\theta$

9. $\dfrac{\partial z}{\partial y} = \dfrac{2yz}{3x^2 + 3y^2 + z^2}$

11. $D_v f(P) = \frac{1}{2}$

13. $2(x - 1) + 6(y - 3) + z = 0$

15. $(x - 1) - 4(y - 2) + (z - 3) = 0$

17. $x = \sqrt{3} - 4t,\ y = \frac{1}{2} + 2\sqrt{3}t,\ z = \frac{3}{4} - 2\sqrt{3}t$

19. Minimum 0 for each point on the line $x = 0$

21. Maximum at each point above the curve $2x = \ln y$, saddle point $(1, e^2, -1)$

23. $3 \cdot 2^{1/3} \times 3 \cdot 2^{1/3} \times 9 \cdot 4^{-1/3}$

25. A cube

27. $\dfrac{\partial x}{\partial u} = \dfrac{y - 3u}{3x + 2y}, \dfrac{\partial y}{\partial u} = \dfrac{x + 2u}{3x + 2y},$ provided $3x + 2y \neq 0$.

CHAPTER 16

Section 1

1. $l_2 = 20, u_2 = 60, l_4 = 30, u_4 = 50, l_n = 40 - 40/n, 40$
3. $l_4 = 34, u_4 = 76, l_n = 8(19n^2 - 27n + 8)/n^2, \frac{152}{3}$
5. 32

Section 3

1. $\frac{1}{72}$ **3.** $\pi^2/4$

Section 5, Part I

1. $\frac{153}{2}$ **3.** $\pi^2(1 - e^2)/8e$ **5.** $\frac{11}{2}$
7. $-\frac{5}{24}$ **9.** $40 \tan^{-1} 2 - \frac{80}{3}$ **11.** $\frac{9}{2}$
13. $(\pi + 12)/6\pi$ **15.** $\frac{1}{12}$ **17.** 5
19. 22 **21.** $\frac{32}{5}$ **23.** 16π
25. 9 **27.** 6 **29.** $r^3/3$
31. 5π

33. $\int_0^r dx \int_x^r \sqrt{r^2 - y^2}\, dy; \int_{-3}^0 dy \int_{-y/3}^1 (3x + y)\, dx + \int_0^1 dy \int_0^{\sqrt{1-y^2}} (3x + y)\, dy;$

$\int_0^2 dy \int_0^{\sqrt{4-y^2}} (x^2 + 4y^2)\, dy;$

$\int_0^2 dx \int_{2-x}^2 (x^2 + 4y^2)\, dy + \int_2^6 dx \int_{(x-2)/2}^2 (x^2 + 4y^2\,)dy$

Section 5, Part II

1. $8 \int_0^a dx \int_0^{b\sqrt{1-x^2/a^2}} c\sqrt{1 - x^2/a^2 - y^2/b^2}\, dy$

3. $V = \int_0^a dx \int_0^{(a^{2/3}-x^{2/3})^{3/2}} (a^{2/3} - x^{2/3} - y^{2/3})^{3/2}\, dy$ (in first octant)

Section 6, Part I

1. $4\pi(27 - 5\sqrt{5})/3$ **3.** $8\pi/3$ **5.** $10\pi/3$

7. $\frac{\pi}{2} \ln \frac{b}{a}$ **9.** $\frac{\pi a^4}{2}$ **11.** $\frac{\pi}{4} \ln \frac{b}{a}$

Section 6, Part II

1.

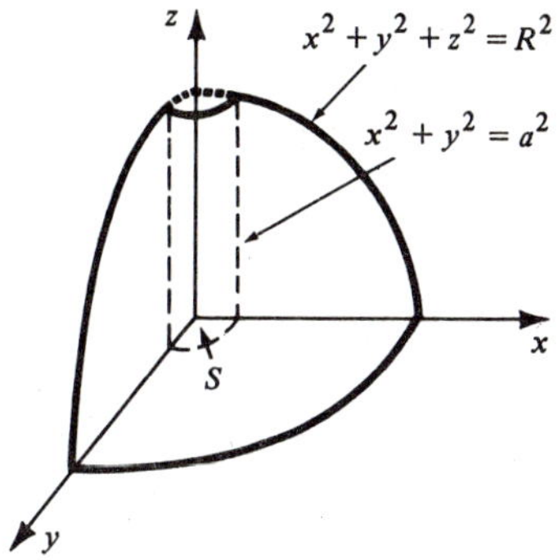

$$V = 8 \int_S \sqrt{R^2 - x^2 - y^2}$$

or

$$V = 2 \int_{x^2+y^2 \leq a^2} \sqrt{R^2 - x^2 - y^2}.$$

In polar coordinates,

$$V = 2 \int_0^{2\pi} d\theta \int_0^a r\sqrt{R^2 - r^2}\, dr = \frac{4\pi}{3} [R^3 - (R^2 - a^2)^{3/2}].$$

Section 7, Part I

1. $(\frac{12}{5}, \frac{3}{4})$
3. $(2b/3, h/2)$
5. $(8a(\pi - 2)/3\pi^2, 16a/3\pi^2)$
7. $([612 - 3\sqrt{2} \ln (3 + 2\sqrt{2})]/256, 0)$
9. $(16\sqrt{2}/35, 0)$
11. $h/2$
13. $(0, 8a/5\pi)$

Section 7, Part II

1. $M = \int_0^a dx \int_0^{(\sqrt{a}-\sqrt{x})^2} xy\, dy,$

$$M_y = \int_0^a dx \int_0^{(\sqrt{a}-\sqrt{x})^2} x^2 y\, dy,$$

$$\bar{x} = \frac{M_y}{M} = \frac{2a}{9}.$$

By symmetry, $\bar{x} = \bar{y}$.

Section 8, Part I

1. $|BC|/\sqrt{3}$
3. $|BD|/\sqrt{3}$
5. $\sqrt{|AB|^2 + |BC|^2}/\sqrt{12}$
7. $a/2$
9. $a\sqrt{5}/2$
11. $a\sqrt{15}/5$
13. $ch^3/12$ (h the altitude on side c)
15. $\sqrt{\pi}/4$

Section 8, Part II

1.

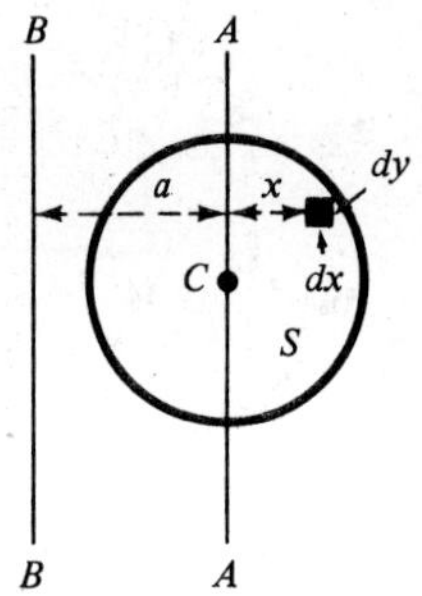

$$I_A = \int_S x^2 \rho, \qquad I_B = \int_S (x + a)^2 \rho,$$

where ρ = density and C = center of mass.

$$I_B = \int_S x^2 \rho + 2a \int_S x\rho + a^2 \int_S \rho$$

or

$$I_B = I_A + 0 + Ma^2.$$

($\int_S x\rho = 0$ since C lies on axis AA.)

Section 9, Part I

1. 27, 0
3. 48π, 0
5. $16\sqrt{2}/3$, $-288\sqrt{2}/35$
7. $(1 - \ln 2)/2$, $\frac{1}{12}$
9. $2a^3/3$, $a^6/24$

Section 9, Part II

1. $$V = \int_0^a dx \int_0^{b(1-\sqrt{x/a})^2} dy \int_0^{c(1-\sqrt{x/a}-\sqrt{x/b})^2} dz = \frac{abc}{90},$$

$$I = \int_0^a dx \int_0^{b(1-\sqrt{x/a})^2} dy \int_0^{c(1-\sqrt{x/a}-\sqrt{x/b})^2} xyz\, dz = \frac{(abc)^2}{277{,}200}.$$

(To simplify the integrations let $y/b = s^2$ and $x/a = t^2$.)

Section 10

1. $(a/4, b/4, c/4)$ 3. $(0, 0, \frac{1}{3})$ 5. $(\frac{15}{8}, 0, -\frac{3}{4})$

7. $([2\sqrt{3} + \ln(2 + \sqrt{3})]/[4\sqrt{3} - 4\ln(2 + \sqrt{3})],$
$0, [14\sqrt{3} - 17\ln(2 + \sqrt{3})]/[16\sqrt{3} - 16\ln(2 + \sqrt{3})])$

9. $(9\pi a/64, 3a/8, 9\pi a/64)$

11. $\int_{-a}^{a} dx \int_{-\sqrt{a^2-x^2}}^{\sqrt{a^2-x^2}} dy \int_{0}^{\sqrt{a^2-x^2-y^2}} \rho(x^2 + y^2)\, dz$

13. $\int_{-\sqrt{h}}^{\sqrt{h}} dx \int_{-\sqrt{h-x^2}}^{\sqrt{h-x^2}} dy \int_{x^2+y^2}^{h} \rho(x^2 + y^2)\, dz$

15. $\int_{2}^{4} dx \int_{-\sqrt{x^2-4}}^{\sqrt{x^2-4}} dx \int_{0}^{2-x/2} \rho(x^2 + y^2)\, dz$

17. $\int_0^a dx \int_0^b dy \int_0^c \rho(x^2 + y^2)\, dz = \rho abc \dfrac{a^2 + b^2}{3}$

Section 11

1. Cone has equation $\phi = \tan^{-1} a/h$; (a) $(3h/4, 0, 0)$; (b) $\pi a^4 h/10$; (c) $\pi a^2 h(3a^2 + 2h^2)/60$.

3. (a) $([3(b^4 - a^4)]/[8(b^3 - a^3)], 0, 0)$; (b) $4\pi(b^5 - a^5)/15$; (c) $4\pi(b^5 - a^5)/15$

5. (a) $3\pi a^4 b/32$; (b) $(2a/3, 0, 5a^2 b/18)$

7. $\dfrac{\pi a^5}{20}$ 9. $\frac{2}{3}\pi(\sqrt{2} - 1)(b^3 - a^3)$ 11. $\dfrac{\pi a^4}{4}$

13. $\frac{3}{4}\pi a^3$

Review, Part I

1. 3 3. 1 5. $\frac{16}{3}$

7. **a.** Frustrum of a right cylinder erected on the rectangle (0,0), (0,2), (1,2), (1,0) in the xy plane and cut by the plane $z = x + y$
   b. same as (a), the rectangle being (0,0), (0,1), (2,1), (2,0)
   c. frustrum of a right cylinder erected on the triangle (0,0), (2,0), (2,2) in the xy plane, and cut by the paraboloid $z = x^2 + y^2$
   d. frustrum of a right cylinder erected on the rectangle (0,0), (0,2), $(\sqrt{2},2)$, $(\sqrt{2},0)$ in the xy plane and cut by the plane $z = x + 2y$ and by the right cylinder $x = y^2$.

9. $\int_0^1 dx \int_{3x}^{4-x^2} f(x,y)\, dy$ 11. $\frac{1}{4}(e^4 - 1)$

13. $\dfrac{\pi}{4} - \dfrac{1}{2}$

15. $V = 4\int_0^1 dx \int_0^{\sqrt{1-x^2}} \frac{dy}{x^2+y^2}$; $\bar{x} = \bar{y} = 0$

$$\bar{z} = \frac{4}{V}\int_0^1 dx \int_0^{\sqrt{1-x^2}} dy \int_0^{1/(x^2+y^2)} z\,dz$$

17. $V = 4\int_{\sqrt{3\pi/2}}^{\sqrt{5\pi/2}} dx \int_{\sqrt{3\pi/2-x^2}}^{\sqrt{5\pi/2-x^2}} \cos(x^2+y^2)\,dy$; $\bar{x} = \bar{y} = 0$

$$\bar{z} = \frac{4}{V}\int_{\sqrt{3\pi/2}}^{\sqrt{5\pi/2}} dx \int_{\sqrt{3\pi/2-x^2}}^{\sqrt{5\pi/2-x^2}} dy \int_0^{\cos(x^2+y^2)} z\,dz$$

19. $I_z = 8\int_0^a dx \int_0^{b\sqrt{1-(x/a)^2}} dy \int_0^{c\sqrt{1-(x/a)^2-(y/b)^2}} \rho(x^2+y^2)\,dz$

Review, Part II

1. By symmetry, if the left-hand repeated integral $= I$, the right-hand integral $= -I$. Thus it suffices to show $I \neq 0$.

$$\int_0^1 \frac{(x^2-y^2)}{(x^2+y^2)^2}\,dy = 2x^2\int_0^1 \frac{1}{(x^2+y^2)^2}\,dy - \int_0^1 \frac{1}{x^2+y^2}\,dy = \frac{1}{x^2+1},$$

since

$$\int \frac{dy}{x^2+y^2} = \frac{1}{x}\arctan\frac{y}{x}, \qquad \int \frac{dy}{(x^2+y^2)^2} = \frac{1}{2x^3}\arctan\frac{y}{x} + \frac{y}{2x^2(x^2+y^2)}.$$

Then

$$I = \int_0^1 \frac{1}{x^2+1}\,dx = \frac{\pi}{4}.$$

CHAPTER 17

Section 1

1. $\frac{23}{44}$ **3.** 0 **5.** $\frac{1}{6}$ **7.** 0, 2π

9. $2\pi^2 + 4\pi b - \pi ab$, $a^2/2 + ab/3 + ac/4 + ab^2c/4 + a^2b^2c^3/5$

11. 6, $14\frac{2}{3}$.

Section 2, Part I

1. $f(x,y) = x\sqrt{1-y^2} + \sin^{-1} y + C$

3. $f(x,y) = x^4 + 3x^2y^2 + 7xy + y^3 + C$

5. $f(x,y) = e^{xy} + \tan^{-1}(x/y) + C$

7. 42 **9.** 1

Section 2, Part II

5. 1

Section 3

1. $-\frac{1}{44}$

3. $\frac{\pi}{2}$

5. $(b - a)$ times area enclosed by λ

Section 4, Part I

1. a. Circles through the origin

   b. $x = \dfrac{u}{u^2 + v^2}$, $y = \dfrac{v}{u^2 + v^2}$

3. $\{(u,v) \mid 0 \le v \le \sqrt{1 - u^2}, (u,v) \ne (0,0)\}$
5. $\rho^2 = \cos 2\phi$, where ρ and ϕ are polar coordinates in the uv plane
7. $(D_1u)(D_1v) = -(D_2u)(D_2v)$ at each point of the domain of F

Section 4, Part II

1. $D_1u = D_2v$ and $D_2u = -D_1v$ for every point in the domain of F

Section 5

1. $\dfrac{3\pi}{4}$

3. $\sqrt{2} - 1$

5. b. $\frac{14}{3}$

7. $\frac{1}{4}(e - e^{-1})$

9. b. $\frac{1}{6}a^2b^2c^2$

11. $\frac{28}{75}$

Review, Part I

1. $-\int_0^1 [t^2 + (t + 1)^2 + (t - 1)^2 2t]\, dt = -\frac{17}{6}$

3. π

5. $\dfrac{\pi}{4}$

Review, Part II

1. $\frac{3}{2}a^2$

3. $\dfrac{\pi}{4} - \dfrac{1}{2}$

CHAPTER 18

Section 1

9. $(2x^3y - x^4)\dfrac{dy}{dx} + 2x^3y - y^4 = 0$

11. $\dfrac{d^2y}{dx^2} - 4\dfrac{dy}{dx} + 5y = 0$

13. $y^2 = x^2 + C$

15. $y = Cx$

17. $y = Cx^2$

19. $y = x \ln |x| + C_1x + C_2$

Section 3

1. $xy' = y$
3. $y' = 2x$
5. $yy' = x$
7. $(xy' - yy')^2 + (y - x)^2 = (x + yy')^2$
9. $y = x^2 + C$
11. $xy = C$
13. $(x - 2)^2 + y^2 = C$
15. $r^3 + 3\cos\theta = 1 + 3/\sqrt{2}$
17. $y^2 = (x + 1)^2 + 8$
23. $x^3 + 3y = C'$
25. $2x^2 + y^2 = C'$
27. $2x^2 + y^2(2\ln y - 1) = C'$
29. $y = 0$ and $y = 2$ are the envelopes of the given family of circles.

Section 5

1. $x^2 + 2xy + 2y^2 = C$
3. $2y(1 + e^x) - x^2 = C$
5. $y^2 + 2x\cos y = C$
7. $x^2y + e^{xy} = C$
9. $y\tan x = C$
11. $y^2 - 2e^x\cos y = C$
13. $x^2 + y^4 = 2xy + 1$
15. a. Differentiating with respect to x:

$$M(x,y) + \int_{x_0}^{x} \frac{\partial M(s,y)}{\partial y}\frac{dy}{dx}\,ds + N(x_0,y)\frac{dy}{dx} = 0.$$

Since $\dfrac{\partial M(s,y)}{\partial y} = \dfrac{\partial N(s,y)}{\partial s}$ (exactness condition), we get

$$M(x,y) + \frac{dy}{dx}\int_{x_0}^{x} \frac{\partial N(s,y)}{\partial s}\,ds + N(x_0,y)\frac{dy}{dx} = 0$$

or

$$M(x,y) + \frac{dy}{dx}\{N(x,y) - N(x_0,y)\} + N(x_0,y)\frac{dy}{dx}$$

$$= 0 = M(x,y) + N(x,y)\frac{dy}{dx}.$$

The boundary-condition is obviously satisfied (just let $x = x_0$, $y = y_0$ to give $0 = 0$).

17. $y = Cx$
19. $e^x(x + y - 1) = C$
21. $\ln(x^2 + y^2) + 2\tan^{-1}\dfrac{x}{y} = C$
23. a. $\dfrac{x\,dy - y\,dx}{x^2 - y^2} = dx$ or $\frac{1}{2}d\ln\dfrac{x + y}{x - y} = dx$; $x + y = C(x - y)e^{2x}$

    b. $\dfrac{x\,dy - y\,dx}{x^2 + y^2} = x\,dx$ or $2\tan^{-1}\dfrac{y}{x} = x^2 + C$

Section 6

1. $x^2 = C(x - y)$
3. $x^2e^{-y/x} + y^2 = C$

5. $x^2(y - 2C) + C^2y = 0$

7. $y \ln \left|\sec \frac{x}{y} + \tan \frac{x}{y}\right| + Cy + 1 = 0$

9. $x^3 + 3xy^2 = C$

11. $\ln|y + 2x - 1| + (x - 2)/(y + 2x - 1) = C$

13. $\ln|x^2 + 3xy + y^2 + x - y - 1| - \frac{1}{\sqrt{5}} \ln \left|\frac{2y + 2 + (3 - \sqrt{5})(x - 1)}{2y + 2 + (3 + \sqrt{5})(x - 1)}\right| = C$

15. $(x - 1)\left[\left(\frac{x - 1}{y + 1}\right)^2 - 1\right]^2 \doteq C$

17. $y = C_1e^x + C_2$

19. $y + \sin x = C_1x + C_2$

21. $y = C_1e^x + e^{-x}/4C_1 + C_2$

Section 7

1. $y = 1 + Ce^{-x^2/2}$

3. $y = c/b + ke^{-bx}, b \neq 0$

5. $y = \sin x + C \cos x$

7. $-y = e^{-2x} + Ce^{3x}$

9. $y = (x^2 + 1)(x^2 + C)$

11. $y(1 + Ce^{x^2/2}) = 1$

13. $3y^2 + 2e^x = Ce^{4x}$

15. $xy^2 = \frac{y^6}{6} + C$

17. $\frac{1}{x} + e^{2y} = Ce^y$

19. $y^2 + 1 = Ce^{-x^2}$

21. $2s^2 = 2t - 1 + Ce^{-2t}$

Section 8

1. $r^2x = g \ln \cosh rt$

3. $y = 4P(x + P)$, Solution 2

5. $I(t) = (1 - Ce^{-4t})/20$

7. $LI(t) = e^{-Rt/L}\left\{\int E(t)e^{Rt/L}\, dt + C\right\}$

9. 140 lb

11. Approx. $\frac{500}{29}$ min. Maximum quantity of salt = 3000 lb

13. $(5 - x)/(10 - x) = \frac{1}{2}e^{-5kt}$

15. $(b - x)/(a - x) = a^{-1}be^{k(b-a)t}$

17. $(a - x)/(a - B_0) = e^{-k_1t}$; then $D_t(ye^{k_2t}) - k_2xe^{k_2t}$

19. $(3x^2 + 7y^2)x^5 = C'$

21. $x^2 + y^2 = C'x$

Section 9

1. $5 + 5i$

3. $5 - 4i$

5. $\frac{1}{2} - i$

7. $3 \pm \sqrt{2}i$

9. $z = \pm(3 + i)$

11. -1

13. $i, -\frac{1}{2} - \frac{\sqrt{3}}{2}i, \frac{1}{2} - \frac{\sqrt{3}}{2}i$

15. $\pm\sqrt{2}, \frac{1 \pm \sqrt{3}i}{\sqrt{2}}, \frac{-1 \pm \sqrt{3}i}{\sqrt{2}}$

17. $1, -\frac{1}{2} \pm \frac{\sqrt{3}}{2}i, -2, 1 \pm \sqrt{3}i$

Section 10

1. (a) $\frac{1}{2} + \frac{\sqrt{3}}{2}i$ (b) e

(c) $e(\sin 1 + i \cos 1)$ (d) $2(\cos 3 + i \sin 3)$

(e) $e^{\pi/5}(\cos (2\pi/5) - i \sin (2\pi/5))$

5. $z = \ln 2 + (2n + 1)\pi i$, for $i = 0, \pm 1, \pm 2, \ldots$

Section 11

3. $y = C_1e^{3x} + C_2e^{-3x}$ **5.** $y = e^{-x}[C_1e^{x\sqrt{2}} + C_2e^{-x\sqrt{2}}]$

7. $y = C_1e^{5x} + C_2e^{-2x}$

9. $y = e^{-x/2}[C_1 \cos x\sqrt{11}/2 + C_2 \sin x\sqrt{11}/2]$

11. $y = e^x[C_1 \cos x\sqrt{2} + C_2 \sin x\sqrt{2}]$

13. $y = e^{x/\sqrt{2}}[C_1 \cos x\sqrt{26}/2 + C_2 \sin x\sqrt{26}/2]$

15. $y = e^{x/3}[C_1 \cos x\sqrt{14}/3 + C_2 \sin x\sqrt{14}/3]$

17. $y = e^{x\sqrt{3}}[C_1 \cos \pi x + C_2 \sin \pi x]$

19. $y = 0$ **21.** $y = 2e^{2(x-1)} - e^{3(x-1)}$

23. Simple harmonic motion: $S = S_0 \cos wt$

25. $y = C_1e^x + C_2e^{-x} + C_3e^{-2x}$

27. $y = e^{x/\sqrt{2}}\left(C_1 \cos \frac{x}{\sqrt{2}} + C_2 \sin \frac{x}{\sqrt{2}}\right) + e^{-x/\sqrt{2}}\left(C_3 \cos \frac{x}{\sqrt{2}} + C_4 \sin \frac{x}{\sqrt{2}}\right)$

Section 12

1. $L(\sin x) = -\sin x - 3 \cos x$, $L(e^{3x}) = 0$

3. $2y = e^x[C_1 \cos x + C_2 \sin x] + (x + 1)^2$

5. $13y = e^{-x}[C_1e^{\sqrt{2x}} + C_2e^{-\sqrt{2x}}] - 8 \sin x - \cos x$

7. $y = C_1 \sin x + \{C_2 - \ln |\sec x + \tan x|\} \cos x$

9. $5y = C_1e^x + C_2e^{-x} - e^x[2 \cos x + \sin x]$

11. $4y = C_1 - [x^2 + x + C_2]e^{-2x}$

13. $y = C_1 \cos 2x + C_2 \sin 2x + (x/4) \sin 2x$

15. $2y = C_1e^{x\sqrt{2}} + C_2e^{-x\sqrt{2}} - 5x$

17. $2y = C_1 + C_2e^{-x} + e^x$

19. $3y = C_1 \sinh x(3 + \sqrt{13})/2 + C_2 \cosh x(3 + \sqrt{13})/2 - e^x$

21. $y = C_1 \cos x + C_2 \sin x + xe^x - 2e^x$

23. $I = A \sinh t(\sqrt{23} + 5)/10 + B \cosh t(\sqrt{23} - 5)/10 + (50 \sin t - 49 \cos t)/905$

25. $y = Ae^x + Be^{2x} + Ce^{3x}$

27. $y = Ae^{3x} + B \cos x + C \sin x$

Section 13

1. Use the same approach as in the proof of Theorem 18.37 or else note that

$$(D - a)(D^2 + aD + a^2)y = 0$$

where $a = \sqrt[3]{k}$. Thus,

$$y = C_1 e^{ax} + e^{-ax/2}\left(C_2 \cos \frac{ax\sqrt{3}}{2} + C_3 \sin \frac{ax\sqrt{3}}{2}\right)$$

which is analytic.

3. $y = C_0 \sum_{k=1}^{\infty} \frac{x^{2k}}{2^k k!} = C_0 e^{x^2/2}$. Infinite radius of convergence.

5. $$y = C_0\left[1 - \frac{x^3}{3!} + \frac{4^2x^6}{6!} - \frac{4^2 \cdot 7^2 x^9}{9!} + \cdots (-1)^r \frac{4^2 \cdot 7^2 \cdot 10^2 \cdots (3r-2)^2 x^{3r}}{(3r)!} + \cdots\right]$$

$$+ C_1\left[x - \frac{2^2x^4}{4!} + \frac{2^2 \cdot 5^2 x^7}{7!} - \cdots (-1)^r \frac{2^2 \cdot 5^2 \cdot 8^2 \cdots (3r-1)^2 x^{3r+1}}{(3r+1)!} + \cdots\right]$$

Infinite radius of convergence.

9. $y = C_1(1 - 3x^2) + C_2(x - \frac{1}{3}x^3)$

Review

1. $y = C\sqrt{1 + x^2}$

3. $x^2 - xy + 2kx + k^2 = 0$

5. $(x + 5y + a)^4 = C(x + 2y + 3)$

7. $x^2 + 4xy - 3y^2 = C$

9. $xy^3 = \frac{y^6}{6} + C$

11. $y^3(x^2 + Cx) = 1$

13. $x^3 + 3x^2y^2 + \frac{4}{3}y^3 = C$

15. $r \ln r\, ds + (s - r)\, dr = 0$

17. Separable; $y = ae^{x^3}$, $y = a\left(1 + \frac{x^3}{1!} + \frac{x^6}{2!} + \frac{x^9}{3!} + \cdots\right)$

19. $(D + 2x)y' = 2x$, linear in y'.

$$y' = 1 + C_1 e^{-x^2}, \qquad y = x + C_2 + C_1 \int e^{-x^2}\, dx,$$

$$y = x + C_2 + C_1\left(x - \frac{x^3}{3 \cdot 1!} + \frac{x^5}{5 \cdot 2!} - \frac{x^7}{7 \cdot 3!} + \cdots\right)$$

21. $y = Ae^{6x} + Be^{-x}$ **23.** $y = Ae^{6x} + Be^{-x} + C$

25. $y = A \sin 2x + B \cos 2x + (2x^2 - 1)/8$

27. $y = (A \sin x\sqrt{15}/2 + B \cos x\sqrt{15}/2)e^{x/2} + (\sin x - \cos x)/10$

29. $y = Ae^{x} + Be^{2x} + Ce^{-3x} + e^{-x}/12$

Index

A

B

C

D

E

F

Q

R

S